The Genetic Code

First position	Second position				Third position
	U	C	A	G	
U	Phe	Ser	Tyr	Cys	U
	Phe	Ser	Tyr	Cys	C
	Leu	Ser	Stop	Stop	A
	Leu	Ser	Stop	Trp	G
C	Leu	Pro	His	Arg	U
	Leu	Pro	His	Arg	C
	Leu	Pro	Gln	Arg	A
	Leu	Pro	Gln	Arg	G
A	Ile	Thr	Asn	Ser	U
	Ile	Thr	Asn	Ser	C
	Ile	Thr	Lys	Arg	A
	Met	Thr	Lys	Arg	G
G	Val	Ala	Asp	Gly	U
	Val	Ala	Asp	Gly	C
	Val	Ala	Glu	Gly	A
	Val	Ala	Glu	Gly	G

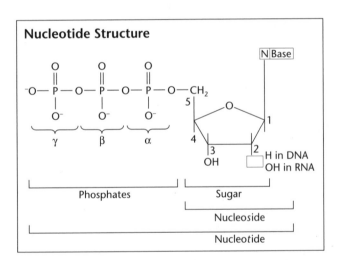

Names of Nucleic Acid Subunits

Base	Nucleoside	Nucleotide	Abbreviation RNA	DNA
Adenine	Adenosine	Adenosine triphosphate	ATP	dATP
Guanine	Guanosine	Guanosine triphosphate	GTP	dGTP
Cytosine	Cytidine	Cytidine triphosphate	CTP	dCTP
Thymine	Thymidine	Thymidine triphosphate		dTTP
Uracil	Uridine	Uridine triphosphate	UTP	

Molecular Genetics *of* Bacteria

Larry Snyder and Wendy Champness

Department of Microbiology
Michigan State University
East Lansing, Michigan

ASM Press
Washington, D.C.

Copyright © 1997 American Society for Microbiology
1325 Massachusetts Avenue, N.W.
Washington, DC 20005

Library of Congress Cataloging-in-Publication Data

Snyder, Larry.
    Molecular genetics of bacteria / Larry Snyder and Wendy Champness.
        p.   cm.
    Includes bibliographical references and index.
    ISBN 1-55581-102-7 (hardcover)
    1. Bacterial genetics. 2. Bacteriophages—Genetics. 3. Molecular genetics.
I. Champness, Wendy. II. Title.
QH434.S96 1997
572.8'293—dc21   96-46770
                                                                         CIP

10 9 8 7 6 5 4 3 2

All Rights Reserved
Printed in the United States of America

Cover and interior design: Susan Brown Schmidler
Cover illustration: Terese Winslow

Contents

v

Chapter 2

Introduction to Macromolecular Synthesis: Gene Expression   43

Chapter 10

DNA Repair and Mutagenesis 237

Evidence for DNA Repair 238

Specific Repair Pathways 239

General Repair Mechanisms 248

Summary of Repair Pathways in *E. coli* 260

Chapter 14
Genetic Analysis in Bacteria   363

Chapter 15
Recombinant DNA Techniques and Cloning Bacterial Genes   393

Boxes

Preface

In writing this textbook, a process that began about 5 years ago, we attempted to solve some basic problems involved in presenting bacterial molecular genetics. First of all, we felt that the material could not be fully grasped without some exposure to the experiments that yielded the knowledge. Therefore, throughout the book we describe actual experiments, choosing ones that are not too technical, are conceptually simple, or are illustrative of the type of experiments that illuminated the subject under discussion. Where possible, we present these experiments as the material is being discussed; otherwise, we present them at the end of the section. The experiments that we chose are not always the most recent ones because often only the older reports describe a broad, basic approach to understanding a phenomenon.

Another problem involves how to engage a reader's interest and give perspective on how the information being presented relates to developments in other areas without detracting from the essential material. Any basic book discussing the molecular genetics of bacteria is bound to appear "colicentric" simply because vastly more is understood about *Escherichia coli* than about any other bacterium; yet in many cases, similar phenomena occur in other bacteria and even in higher organisms, including humans. Presenting this material helps to hold the students' interest and emphasizes the interconnectedness of basic cell and molecular biology as applied to all organisms on Earth; but, in our experience, too many digressions only confuse students. To address this problem, we discuss information that is interesting and important but, in our opinion, less central to understanding bacterial molecular genetics in separate boxes, giving one or two references.

In organizing the book, our aim was to prepare a text that could be used for a one- or two-semester undergraduate course or as background reading for a beginning graduate course. Accordingly, the first 12 chapters are relatively descriptive, presenting necessary background material along with some experiments that give a feeling for the origins of the science. We used chapters 1 through 10 and parts of chapters 11 and 12 for our own one-semester undergraduate course this year. With the last four chapters, which focus on experimental methods and the interpretation of data, the book is also well suited to a two-semester course. In particular, chapters 13 through 16 give examples of molecular genetic analysis, a subject not covered in most textbooks. They also contain some more advanced material relevant to current applications of molecular genetics in biotechnology, which is of interest to many students. All of the chapters offer Suggested Readings, in which we list both older papers that describe some of the experiments we discuss and more recent papers or review articles that can be used for further research. Moreover, the boxes that appear in many of the chapters can serve as starting points for extra reading or special reports.

An extraordinary number of researchers have made major contributions to the field of bacterial molecular genetics. We could not reasonably expect students to learn even a small fraction of their names. If we put in some names, where should we stop? Therefore, we decided to use only those names that have become icons in the field because they are associated with certain experiments or concepts, such as the Watson and Crick structure of DNA and the Luria and Delbrück experiment, and to leave out all the others, even the names of those who made very major contributions. We redress these omissions somewhat in the Suggested Readings, where we give some of the original references that led to the developments under discussion.

For the past 3 years, we have used draft copies of this book in our undergraduate class in microbial genetics and as background reading in the corresponding graduate course. Thanks to feedback from our students, especially Nana Alford, the book has improved dramatically over the years. In addition, many other people have helped with the development of this textbook. One advantage of publishing through ASM has been the access to outside reviewers who found errors and made suggestions as to what should or should not be included in a textbook at this level. These reviewers remain anonymous, so we cannot thank them by name. We also welcome further comments from readers.

We especially thank our editor, Patrick Fitzgerald, formerly Director of ASM Press, who first approached us about writing a textbook and offered us plentiful advice and encouragement throughout the process. Special thanks also go to our developmental editor, Amanda Suver, who improved the writing throughout.

We are indebted to the professionals who performed the magic of transforming our manuscript and hand-drawn figures into a printed book. Susan Birch, Production Manager at ASM Press, oversaw the entire production process. Yvonne Strong copyedited the manuscript and illustrations. Susan Brown Schmidler created the book and cover design and supervised the work of Precision Graphics artists, who rendered our hand-drawn sketches into clear, attractive figures. Terese Winslow created the cover illustration. It was a pleasure to work with this team, and we are delighted with the result of their efforts.

Finally, we wrote parts of this book while on a short sabbatical leave, and we thank our colleagues at the University of Tel Aviv and the John Innes Institute in Norwich, England, for their hospitality.

Introduction

THE GOAL OF THIS TEXTBOOK is to introduce the student to the field of bacterial molecular genetics. Bacteria are relatively simple organisms, and some are quite easy to manipulate in the laboratory. For these reasons, many methods in molecular biology and recombinant DNA technology have been developed around bacteria, and these organisms often serve as model systems for understanding cellular functions and developmental processes in more complex organisms.

However, bacteria are not just important as laboratory tools; they are important and interesting in their own right. For instance, they play an essential role in the ecology of the Earth. They are the only organisms that can "fix" atmospheric nitrogen, that is, convert N_2 to ammonia, which can be used to make nitrogen-containing cellular constituents such as proteins and nucleic acids. Without bacteria, the natural nitrogen cycle would be broken. Bacteria are also central to the carbon cycle of the Earth because of their ability to degrade recalcitrant natural polymers such as cellulose and lignin. Bacteria and some types of fungi thus prevent the Earth from being buried in plant debris and other carbon-containing material. Toxic compounds including petroleum, many of the chlorinated hydrocarbons, and other products of the chemical industry can also be degraded by bacteria. For this reason, these organisms are essential in water purification and toxic waste cleanup. Moreover, bacteria produce most of the naturally occurring so-called greenhouse gases, such as methane and carbon dioxide, which are in turn degraded by other types of bacteria. This cycle helps maintain climate equilibrium. Bacteria have even had a profound effect on the geology of the Earth, being responsible for some of the major iron ore and other types of deposits in the Earth's crust.

Another unusual feature of bacteria is their ability to live in extremely inhospitable environments, many of which are devoid of life except for bacteria. These organisms are the only ones living in the Dead Sea, where the water's salt concentration is very high. Some types of bacteria live in hot springs at temperatures close to the boiling point of water, and others survive in atmospheres devoid of oxygen, such as eutrophic lakes and swamps.

Bacteria that live in inhospitable environments sometimes enable other organisms to survive in those environments through symbiotic relationships. For example, symbiotic bacteria make life possible for tubular worms next to hydrothermal vents on the ocean floor, where the atmosphere is hydrogen sulfide rather than oxygen. In this symbiosis, the bacteria fix carbon dioxide by using the reducing power of the hydrogen sulfide given off by the hydrothermal vents, thereby furnishing food in the form of high-energy carbon compounds for the worms. Symbiotic cyanobacteria allow fungi to live in the Arctic tundra in the form of lichens. The bacterial partners in the lichens fix atmospheric nitrogen and make carbon-containing molecules through photosynthesis to allow their fungal partners to grow on the tundra in the absence of nutrient-containing soil. Symbiotic nitrogen-fixing *Rhizobium* and *Azorhizobium* spp. in the nodules on the roots of legumes and some other types of higher plants allow plants to grow in nitrogen-deficient soils. Other types of symbiotic bacteria digest cellulose to allow cows and other ruminant animals to live on a diet of grass. Chemiluminescent bacteria even generate light for squid and other marine animals, allowing individuals to find each other in the darkness of the deep ocean.

Bacteria are also worth studying because of their role in disease. They cause many human, plant, and animal diseases, and new ones are continuously appearing. Knowledge gained from the molecular genetics of bacteria will help in the development of new ways to treat or otherwise control these diseases.

Bacteria and their phages (i.e., viruses that infect bacteria) are also the source of many useful substances such as many of the enzymes used in biotechnology and other industries. Moreover, bacteria make antibiotics and chemicals such as benzene and citric acid.

In spite of substantial progress, we have only begun to understand the bacterial world around us. Bacteria are the most physiologically diverse organisms on Earth, and the importance of bacteria to life on Earth and the potential uses to which bacteria can be put can only be guessed at. Thousands of different types of bacteria are known, and new insights into their cellular mechanisms and their applications constantly emerge from research

with bacteria. Moreover, it is estimated that less than 1% of the types of bacteria living in the soil and other environments have ever been isolated. Who knows what interesting and useful functions the undiscovered bacteria might have. Clearly, studies with bacteria will continue to be essential to our future efforts to understand, control, and benefit from the biological world around us, and bacterial molecular genetics will be an essential tool in these efforts. But before discussing this field, we must first briefly discuss the evolutionary relationship of the bacteria to other organisms.

The Biological Universe
The Eubacteria

According to a present view, all organisms on Earth belong to three major divisions: the eubacteria, the archaea (formerly archaebacteria), and the eukaryotes (Figure 1). Most of the familiar bacteria such as *Escherichia coli, Streptococcus pneumoniae,* and *Staphylococcus aureus* are **eubacteria.** These organisms can differ greatly in their physical appearance. Although most are single celled and rod shaped or spherical, some are multicellular and undergo complicated developmental cycles. The cyanobacteria (formerly called blue-green algae) are eubacteria, but they have chlorophyll and can be filamentous, which is why they were originally mistaken for algae. The antibiotic-producing actinomycetes, which include *Streptomyces* spp., are also eubacteria, but they form hyphae and stalks of spores, making them resemble fungae. Another eubacterial group, the *Caulobacter* spp., have both free-swimming and sessile forms that attach to surfaces through a holdfast structure. One of the most dramatic-appearing eubacteria of all is the genus *Myxococcus,* which can exist as free-living single-celled organisms but can also aggregate to form fruiting bodies much like slime molds. Eubacterial cells are usually much smaller than the cells of higher organisms, but recently a eubacterium was found that is 1 mm long, longer than even most eukaryotic cells. Because eubacteria come in so many shapes and sizes, they cannot be distinguished by their physical appearance but only by biochemical criteria and by the absence of organelles.

GRAM-NEGATIVE AND GRAM-POSITIVE EUBACTERIA

The eubacteria can be further divided into two major subgroups, the **gram-negative** and **gram-positive** eubacteria. This division is based on the response to a test called the Gram stain. Gram-negative eubacteria will be pink after this staining procedure, whereas gram-posi-

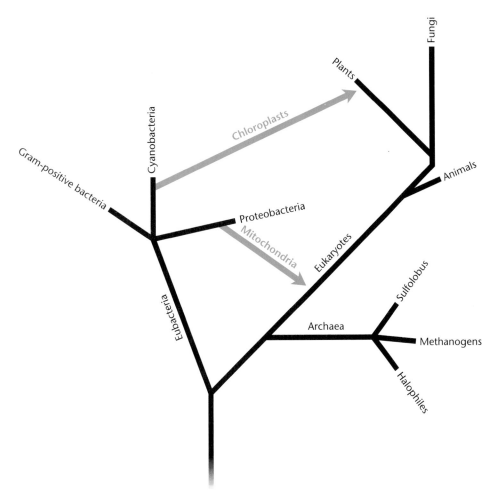

Figure 1 The evolutionary tree showing the points of divergence of eubacteria, archaea, and eukaryotes. The points of transfer of the eubacterial symbiotic organelles—mitochondria and chloroplasts—to eukaryotes are shown in blue.

tive bacteria will turn deep blue. The difference in staining reflects the fact that only gram-negative eubacteria have a well-developed outer membrane. However, the difference between these groups seems to be more fundamental than the possession of an outer membrane. Individual types of gram-negative bacteria are in general more closely related to other gram-negative bacteria than they are to gram-positive bacteria, suggesting that the eubacteria separated into these two groups long before modern bacterial species arose.

The Archaea

The **archaea** (formerly called archaebacteria) are single-celled organisms that resemble eubacteria but are very different biochemically. The archaea are mostly represented by extremophiles (or "extreme-condition-loving" organisms) that, as their name implies, live under extreme conditions where other types of organisms cannot survive, such as at the very high temperatures in sulfur springs, at high pressures on the ocean floor, and at

very high osmolality such as in the Dead Sea. Some of the archaea also perform relatively unusual biochemical functions such as making methane.

The separate classification of archaea from the true bacteria or eubacteria is fairly recent and is mostly based on the sequence of their ribosomal RNAs (rRNAs) and the structures of their RNA polymerase and lipids (see Olsen et al., Suggested Reading). In fact, recent evidence obtained by comparing the sequences of translation factors and membrane ATPases suggests that the archaea may be more closely related to eukaryotes than they are to eubacteria (see Figure 1 and Iwabe et al., Suggested Reading). The archaea themselves form a very diverse group of organisms and are sometimes divided into two different kingdoms.

Although substantial progress is being made, much less is known about the archaea than about the eubacteria. The examples in this book come mostly from the eubacteria, which will be referred to as just "bacteria" throughout.

The Eukaryotes

The **eukaryotes** are members of the third kingdom of organisms on Earth. They include organisms as seemingly diverse as plants, animals, and fungi. The name "eukaryotes" is derived from their nuclear membrane. They usually have a nucleus (the genus *Giardia* is a known exception), and the word "karyon" in Greek means "nut"—which is what the nucleus must have resembled to early cytologists. The eukaryotes can be unicellular like yeasts and protozoans and some types of algae, or they can be multicellular like plants and animals. In spite of their widely diverse appearances, lifestyles, and relative complexity, however, all eukaryotes are remarkably similar at the biochemical level, particularly in their pathways for macromolecular synthesis.

The Prokaryotes and the Eukaryotes

Organisms on Earth are also sometimes divided into prokaryotes and eukaryotes. This classification is based on whether the organism has a nucleus and some other organelles. In contrast to eukaryotes, both archaea and eubacteria lack a nuclear membrane, which caused them to be lumped together as **prokaryotes,** which means "before the nucleus." They were given this name because they were thought to be the most primitive organisms, existing before the development of "higher organisms," or eukaryotes that have a nucleus.

The presence or absence of a nuclear membrane greatly influences the mechanisms available to make proteins in the cell. Messenger RNA (mRNA) synthesis and translation can occur simultaneously in prokaryotes, since no nuclear membrane separates the ribosomes (which synthesize proteins) from the DNA. However, in most eukaryotes, the DNA is physically separated from the ribosomes. Therefore, mRNA made in the nucleus must be transported through the nuclear membrane before it can be translated into protein in the cytoplasm, and transcription and translation cannot occur simultaneously.

Besides lacking a nucleus, prokaryotic cells also lack many other cellular constituents common to eukaryotes, including mitochondria and chloroplasts, as well as such organelles as the Golgi apparatus and the endoplasmic reticulum. The absence of mitochondria, chloroplasts, and most organelles generally gives prokaryotic cells a much simpler appearance under the microscope.

MITOCHONDRIA AND CHLOROPLASTS

All eukaryotic cells contain mitochondria. In addition, plant cells and some unicellular eukaryotic cells contain chloroplasts. The mitochondria of eukaryotic cells are the sites of efficient ATP generation through respiration, and the chloroplasts are the sites of photosynthesis.

Recent evidence suggests that the mitochondria and chloroplasts of eukaryotes are descended from free-living eubacteria that formed a symbiosis with eukaryotes. In fact, these organelles resemble bacteria in many ways. For instance, they contain DNA that encodes ribosomes, transfer RNAs (tRNAs), translation factors, and other components of the transcription-translation apparatus. Even more striking, the mitochondrially encoded and chloroplast-encoded RNAs and proteins, as well as the membranes of the organelles, more closely resemble those of the eubacteria than they do those of eukaryotes.

Not only is there little doubt of the eubacterial origin of mitochondria and chloroplasts, but also it is possible to guess to which eubacterial families they are most closely related. Comparisons of the sequences of highly conserved organelle genes, such as those for the rRNAs, with those of eubacteria suggest that mitochondria are descended from the proteobacteria and that chloroplasts are descended from the cyanobacteria (Figure 1).

Mitochondria and chloroplasts may have come to be associated with early eukaryotic cells when these cells engulfed eubacteria to take advantage of their superior energy-generating systems or the ability to obtain energy from light through photosynthesis. The engulfed bacteria eventually lost many of their own genes and thereby their autonomy and became permanent symbionts of the eukaryotic cells. In fact, modern-day eukaryotes called dinoflagellates sometimes engulf cyanobacteria, allowing the dinoflagellates to photosynthesize.

What Is Genetics?

Genetics can be simply defined as the manipulation of DNA to study cellular and organismal functions. Since DNA encodes all of the information needed to make the cell and the complete organism, the effects of changing this molecule can give clues to the normal functions of the cell and organism.

Before the advent of methods for manipulating DNA in the test tube, the only genetic approaches available for studying cellular and organismal functions were those of **classical genetics.** In this type of analysis, mutants (i.e., individuals that differ from the normal, or wild-type, members of the species by a certain observable attribute, or phenotype) that are altered in the function being studied are isolated. The changes in the DNA, or mutations, responsible for the altered function are then localized in the chromosome by genetic crosses. The mutations are then grouped into genes by allelism tests to determine how many different genes are involved. The functions of the genes can then sometimes be deduced from the specific effects of the mutations on the organism. The ways in which mutations in genes involved in a biological system can alter the biological system provide clues to the normal functioning of the system.

Classical genetic analyses continue to contribute greatly to our understanding of developmental and cellular biology. A major advantage of the classical genetic approach is that mutants altered in a function can be isolated and characterized without any a priori understanding of the molecular basis of the function. Classical genetic analysis is also often the only way to determine how many gene products are involved in a function and, through suppressor analysis, to find other genes whose products may interact either physically or functionally with the products of these genes.

The development of **molecular genetic techniques** has greatly expanded the range of methods available for studying genes and their functions. These techniques include methods for isolating DNA and identifying the regions of DNA that encode particular functions, as well as methods for altering or mutating DNA in the test tube and then returning the mutated DNA to cells to determine the effect of the mutation on the organism.

The approach of first cloning a gene and then altering it in the test tube before reintroducing it into the cells to determine the effect of the alterations is sometimes called **reverse genetics** and is essentially the reverse of a classical genetic analysis. In classical genetics, a gene is known to exist only because a mutation in it has caused an observable change in the organism. With the molecular genetic approach, a gene can be isolated and mutated in the test tube without any knowledge of its function. Only after the mutated gene has been returned to the organism does its function become apparent.

Rather than one approach supplanting the other, molecular genetics and classical genetics can be used to answer different types of questions, and the two approaches often complement each other. In fact, the most remarkable insights into biological functions have sometimes come from a combination of classical and molecular genetic approaches.

Bacterial Genetics

In bacterial genetics, genetic techniques are used to study bacteria. Applying genetic analysis to bacteria is no different in principle from applying it to other organisms. However, the methods that are available differ greatly. Some types of bacteria are relatively easy to manipulate genetically. As a consequence, more is known about some bacteria than is known about any other organism on Earth. Some of the properties of bacteria that facilitate genetic experiments are listed below.

Bacteria Are Haploid

One of the major advantages of bacteria for genetic studies is that they are **haploid.** This means that they have only one copy or **allele** of each gene. This property makes it much easier to identify cells with a particular type of mutation.

In contrast, most higher organisms are diploid, with two alleles of each gene, one on each homologous chromosome. Most mutations are recessive, which means that they will not cause a phenotype in the presence of a normal copy of the gene. Therefore, in diploid organisms, most mutations will have no effect unless both copies of the gene in the two homologous chromosomes have the mutation. Backcrosses between different organisms with the mutation are usually required to produce offspring with the mutant phenotype, and even then only some of the progeny of the backcross will have the mutated gene in both homologous chromosomes. With a haploid organism like a bacterium, however, most mutations have an immediate effect, and there is no need for backcrosses.

Short Generation Times

Another advantage of some bacteria for genetic studies is that they have very short generation times. The **generation time** is the length of time the organism takes to reach maturity and give rise to offspring. If the generation time of an organism is too long, it can limit the number of possible experiments. Some strains of the bacterium *Escherichia coli* can reproduce themselves every 20 min under ideal conditions. With such rapid multiplication, cultures of the bacteria can be started in the morning and the progeny can be examined later in the day.

Asexual Reproduction

Another advantage of bacteria is that they multiply asexually, by cell division. Sexual reproduction, in which individuals of the same species must mate with each other to give rise to progeny, can complicate genetic experiments because the progeny are never identical to their parents. To achieve purebred lines of a sexually reproducing organism, a researcher must repeatedly cross the individuals with their relatives. However, if the organism multiplies asexually by cell division, all the progeny will be genetically identical to their parent and to each other. Genetically identical organisms are called **clones.** Some lower eukaryotes and some types of plants can also multiply asexually to form clones.

Colony Growth on Agar Plates

Genetic experiments often require that numerous individuals be screened for a particular property. Therefore, it helps if large numbers of individuals of the species being studied can be propagated in a small space.

With some types of bacteria, thousands, millions, or even billions of individuals can be screened on a single agar-containing petri plate. Once on an agar plate, these bacteria will divide over and over again, with all the progeny remaining together on the plate until a visible lump or **colony** has formed. Each colony will be composed of millions of bacteria, all derived from the original bacterium and hence all clones of the original bacterium.

Colony Purification

The ability of some types of bacteria to form colonies through the multiplication of individual bacteria on plates allows **colony purification** of bacterial strains and mutants. If a mixture of bacteria containing different mutants or strains is placed on an agar plate, individual mutant bacteria or strains in the population will each multiply to form colonies. However, these colonies may be too close together to be separable or may still contain a mixture of different strains of the bacterium. However, if the colonies are picked and the bacteria are diluted before replating, discrete colonies that result from the multiplication of individual bacteria may appear. No matter how crowded the bacteria were on the original plate, a pure strain of the bacterium can be isolated in one or a few steps of colony purification.

Serial Dilutions

To count the number of bacteria in a culture or to isolate a pure culture, it is often necessary to obtain discrete colonies of the bacteria. However, because bacteria are so small, a concentrated culture contains billions of bacteria per milliliter. If such a culture is plated directly on a petri plate, the bacteria will all grow together and discrete colonies will not form. **Serial dilutions** offer a practical method for diluting solutions of bacteria before plating to obtain a measurable number of discrete colonies. The principle is that if greater dilutions are repeated in succession, they can be multiplied to produce the total dilution. For example, if a solution is diluted in three steps by adding 1 ml of the solution to 99 ml of water, followed by adding 1 ml of this dilution to another 99 ml of water and finally by adding 1 ml of the second

dilution to another 99 ml of water, the final dilution is $10^{-2} \times 10^{-2} \times 10^{-2} = 10^{-6}$, or one in a million. To achieve the same dilution in a single step, 1 ml of the original solution would have to be added to 1,000 liters (about 250 gallons) of water. Obviously, it is more convenient to handle three solutions of 100 ml each than to handle a solution of 250 gallons, which weighs about 1,000 lb!

Selections

Probably the major advantage of bacterial genetics is the opportunity to do **selections,** by which very rare mutants and other types of strains can be isolated. To select a rare strain, billions of the bacteria are plated under conditions where only the desired strain, not the bulk of the bacteria, can grow. In general, these conditions are called the **selective conditions.** For example, a nutrient may be required by most of the bacteria but not by the strain being selected. Agar plates lacking the nutrient then present selective conditions for the strain, since only the strain being selected will multiply to form a colony in the absence of the nutrient. In another example, the desired strain may be able to multiply at a temperature that would kill most of the bacteria. Incubating agar plates at that temperature would provide the selective condition. After the strain has been selected, a colony of the strain can be picked and colony purified away from other contaminating bacteria under the same selective conditions.

The power of selections with bacterial populations is awesome. Using a properly designed selection, a single bacterium can be selected from among billions placed on an agar plate. If we could apply such selections to humans, we could find one individual in the entire human population of the Earth.

Storing Stocks of Bacterial Strains

Most types of organisms must be continuously propagated; otherwise they will age and die off. Propagating organisms requires continuous transfers and replenishing of the food supply, which can be very time-consuming. However, many types of bacteria can be stored in a dormant state and therefore do not need to be continuously propagated. The conditions used for storage depend upon the type of bacteria. Some bacteria sporulate and so can be stored as dormant spores. Others can be stored by being frozen in glycerol or being dried. Storing organisms in a dormant state is particularly convenient for genetic experiments, which often require the accumulation of large numbers of mutants and other strains. The strains will remain dormant until the cells are needed, at which time they can be revived.

Genetic Exchange

Genetic experiments with an organism usually require some form of exchange of DNA or genes between members of the species. Most types of organisms on Earth are known to have some means of genetic exchange, which presumably accelerates evolution and increases the adaptability of a species.

Exchange of DNA from one bacterium to another can occur in one of three ways. In **transformation**, DNA released from one cell enters another cell of the same species. In **conjugation**, plasmids, which are small autonomously replicating DNA molecules in bacterial cells, transfer DNA from one cell to another. Finally, in **transduction**, a bacterial virus accidentally picks up DNA from a cell it has infected and injects this DNA into another cell. The ability to exchange DNA between strains of a bacterium makes possible genetic crosses and complementation tests as well as the tests essential to genetic analysis.

Phage Genetics

Some of the most important discoveries in genetics have come from studies with the viruses that infect bacteria, called **bacteriophages**, or **phages** for short. Phages are not alive; instead, they are just genes wrapped in a protective coat of protein and/or membrane, as are all viruses. Because phages are not alive, they cannot multiply outside a bacterial cell. However, if a phage encounters a type of bacterial cell that is sensitive to phages, the phage will enter the cell and direct it to make more phage.

Phages are usually identified by the holes, or **plaques**, they form in layers of sensitive bacteria. In fact, the name "phage" (Greek for "eat") derives from these plaques, which look like eaten-out places. A plaque can form when a phage is mixed with large numbers of susceptible bacteria and the mixture is placed on an agar plate. As the bacteria multiply, one may be infected by the phage, which will multiply and eventually break open or **lyse** the bacterium, releasing more phage. As the surrounding bacteria are infected, the phage will spread, even as the bacteria multiply to form an opaque layer called a **bacterial lawn**. Wherever the original phage infected the first bacterium, the plaque will disrupt the lawn, forming a clear spot on the agar. Despite its appearance, this location will contain millions of the phage.

Phages offer many of the same advantages for genetics as bacteria. Thousands or even millions of phages can be put on a single plate. Also, like bacterial colonies, each plaque contains millions of genetically identical phage. By analogy to the colony purification of bacterial strains, individual phage mutants or strains can be isolated from other phages through **plaque purification.**

Phages Are Haploid

Phages are, in a sense, haploid, since they have only one copy of each gene. As with bacteria, this property makes isolation of phage mutants relatively easy, since all mutants will immediately exhibit their phenotypes without the need for backcrosses.

Selections with Phages

Selection of rare strains of a phage is possible; as with bacteria, it requires conditions under which only the desired phage strain can multiply to form a plaque. For phage, these selective conditions may be a bacterial host in which only the desired strain can multiply or a temperature at which only the phage strain being selected can multiply. Note that the bacterial host must be able to multiply under the same selective conditions; otherwise, a plaque cannot form.

As with bacteria, selections allow the isolation of very rare strains or mutants. If selective conditions can be found for the strain, millions of phages can be mixed with the bacterial host and only the desired strain will multiply to form a plaque. A pure strain can then be obtained by picking the phage from the plaque and plaque purifying the strain under the same selective conditions.

Crosses with Phages

Phage strains can be crossed very easily. The same cells are infected with different mutants or strains of the phage. The DNA of the two phages will then be in the same cell, where the molecules can interact genetically with each other, allowing genetic manipulations such as gene mapping and allelism tests.

A Brief History of Bacterial Molecular Genetics

Because of the ease with which they can be handled, bacteria and their phages have long been the organisms of choice for understanding basic cellular phenomena, and their contributions to this study are almost countless. The following chronological list should give a feeling for the breadth of these contributions and the central position that bacteria have occupied in the development of modern molecular genetics. Some original references are given at the end of the chapter under Suggested Reading.

Inheritance in Bacteria

In the early part of this century, biologists agreed that inheritance in higher organisms follows Darwinian principles. According to Darwin, changes in the hereditary properties of organisms occur randomly and are passed on to the progeny. In general, the changes that happen to be beneficial to the organism are more apt to be passed on to subsequent generations.

With the discovery of the molecular basis for heredity, Darwinian evolution now has a strong theoretical foundation. The properties of organisms are determined by the sequence of their DNA, and as the organisms multiply, changes in this sequence sometimes occur randomly and without regard to the organism's environment. However, if a random change in the DNA happens to be beneficial in the situation in which the organism finds itself, the organism will have an improved chance of surviving and reproducing.

As late as the 1940s, many bacteriologists believed that inheritance in bacteria was different from inheritance in other organisms. It was thought that rather than enduring random changes, bacteria could adapt to their environment by some sort of "directed" change and that the adapted organisms could then somehow pass on the change to their offspring. Such beliefs were encouraged by the observations of bacteria growing under selective conditions. For example, in the presence of an antibiotic, all the bacteria in the culture soon become resistant to the antibiotic. It seemed as though the resistant bacterial mutants appeared in response to the antibiotic.

One of the first convincing demonstrations that inheritance in bacteria follows Darwinian principles was made in 1943 by Salvador Luria and Max Delbrück (see Suggested Reading). Their work demonstrated that particular phenotypes, in their case resistance to a virus, occur randomly in a growing population, even in the absence of the virus. By the directed change or adaptive mutation hypothesis, the resistant mutants should have appeared only in the presence of the virus.

The demonstration that inheritance in bacteria follows the same principles as inheritance in higher organisms set the stage for the use of bacteria in studies of basic genetic principles common to all organisms. However, the directed-change hypothesis still persists in altered forms for some types of bacterial inheritance (see Box 3.2).

Transformation

As discussed at the beginning of the Introduction, most organisms exhibit some mechanism for exchanging genes. The first demonstration of genetic exchange in bacteria was made by Fred Griffith in 1928. He was studying two variants of pneumococci, now called *Streptococcus pneumoniae*. One variant formed smooth-appearing colonies on plates and was pathogenic in mice. The other variant formed rough-appearing colonies on plates and did not kill mice. Only live, and not dead, smooth-colony-forming bacteria could cause disease, since the disease requires that the bacteria multiply in the infected mice. However, when Griffith mixed dead smooth-colony formers with live rough-colony formers and injected the mixture into mice, the mice became sick and died. Moreover, he isolated live smooth-colony formers from the dead mice. Apparently, the dead smooth-colony formers were "transforming" some of the live rough-colony formers into the pathogenic, smooth-colony-forming type. The "transforming principle" given off by the dead smooth-colony formers was later shown to be DNA, since addition of purified DNA from the dead smooth-colony formers to the live rough-colony formers in a test tube transformed some of the rough type to the smooth type (see Avery et al., Suggested Reading). This method of exchange is called **transformation,** and this experiment provided the first direct evidence that genes are made of DNA. Later experiments by Alfred Hershey and Martha Chase in 1952 (see Suggested Reading) showed that phage DNA alone is sufficient to direct the synthesis of more phages.

Conjugation

In 1946, Joshua Lederberg and Edward Tatum (see Suggested Reading) discovered a different type of gene exchange in bacteria. When they mixed some strains of *Escherichia coli* with other strains, they observed the appearance of recombinant types that were unlike either parent. Unlike transformation, which requires only that DNA from one bacterium be added to the other bacterium, this means of gene exchange requires direct contact between two bacteria. It was later shown to be mediated by plasmids and is named **conjugation.**

Transduction

In 1953, Norton Zinder and Joshua Lederberg discovered yet a third mechanism of gene transfer between bacteria. They showed that a phage of *Salmonella typhimurium* could carry DNA from one bacterium to another. This means of gene exchange is named **transduction** and is now known to be quite widespread.

Recombination within Genes

At the same time, experiments with bacteria and phages were also contributing to the view that genes were linear arrays of nucleotides in the DNA. By the early

1950s, recombination had been well demonstrated in higher organisms, including fruit flies. However, recombination was believed to occur only between mutations in different genes and not between mutations in the same gene. This led to the idea that genes were like "beads on a string" and that recombination is possible between the "beads," or genes, but not within a gene. In 1955, Seymour Benzer disproved this hypothesis by using the power of phage genetics to show that recombination is possible within the rII genes of phage T4. He mapped numerous mutations in the rII genes, thereby demonstrating that genes are linear arrays of mutable sites in the DNA. Later experiments with other phage and bacterial genes showed that the sequence of nucleotides in the DNA directly determines the sequence of amino acids in the protein product of the gene.

Semiconservative DNA Replication

In 1953, James Watson and Francis Crick published their structure of DNA. One of the predictions of this model is that DNA replicates by a semiconservative mechanism, in which specific pairing occurs between the bases in the old and the new DNA strands. In 1958, Matthew Meselson and Frank Stahl used bacteria to confirm that DNA replicates by this semiconservative mechanism.

mRNA

The existence of mRNA was also first indicated by experiments with bacteria and phage. In 1961, Sydney Brenner, François Jacob, and Matthew Meselson used phage-infected bacteria to show that ribosomes are the site of protein synthesis and confirmed the existence of a "messenger" RNA that carries information from the DNA to the ribosome.

Genetic Code

Also in 1961, phages and bacteria were used by Francis Crick and his collaborators to show that the genetic code is unpunctuated, three lettered, and redundant. These researchers also showed that not all possible codons designated an amino acid and that some were nonsense. These experiments laid the groundwork for Marshall Nirenberg and his collaborators to decipher the genetic code, in which a specific three-nucleotide set encodes one of 20 amino acids. The code was later verified by the examination of specific amino acid changes due to mutations in the lysozyme gene of T4 phage.

The Operon Model

François Jacob and Jacques Monod published their operon model for the regulation of the lactose utiliza-

tion genes of E. coli in 1961 as well. They proposed that a repressor blocks RNA synthesis on the lac genes unless the inducer, lactose, is bound to the repressor. Their model has served to explain gene regulation in other systems, and the lac genes and regulatory system continue to be used in molecular genetic experiments, even in systems as far removed from bacteria as animal cells and viruses.

Enzymes for Molecular Biology

The early 1960s saw the start of the discovery of many interesting and useful bacterial and phage enzymes involved in DNA and RNA metabolism. In 1960, Arthur Kornberg demonstrated the synthesis of DNA in the test tube by an enzyme from E. coli. The next year, a number of groups independently demonstrated the synthesis of RNA in the test tube by RNA polymerases from bacteria. From this time on, other interesting and useful enzymes for molecular biology were isolated from bacteria and their phages, including polynucleotide kinase, DNA ligases, topoisomerases, and many phosphatases.

From these early observations, the knowledge and techniques of molecular genetics exploded. For example, in the early 1960s, techniques were developed for detecting the hybridization of RNA to DNA and DNA to DNA on nitrocellulose filters. These techniques were used to show that RNA is made on only one strand in specific regions of DNA, which later led to the discovery of promoters and other regulatory sequences. By the late 1960s, restriction endonucleases had been discovered in bacteria and shown to cut DNA in specific places (see Linn and Arber, Suggested Reading). By the early 1970s, these restriction endonucleases were being exploited to introduce foreign genes into E. coli (see Cohen et al., Suggested Reading), and by the late 1970s, the first human gene had been expressed in a bacterium. Also in the late 1970s, methods were developed to sequence DNA by using enzymes from phages and bacteria.

In 1988, a thermally stable DNA polymerase from a thermophilic bacterium was used to invent the technique called the polymerase chain reaction (PCR). This extremely sensitive technique allows the amplification of genes and other regions of DNA, facilitating their cloning and study.

These examples illustrate that bacteria and their phages have been central to the development of molecular genetics and recombinant DNA technology. Contrast the timing of these developments with the timing of comparable major developments in physics (early 1900s) and chemistry (1920s and 1930s), and you can see that molecular genetics is arguably the most recent major conceptual breakthrough in the history of science.

What's Ahead

This textbook emphasizes how molecular genetic approaches can be used to solve biological problems. As an educational experience, the methods used and the interpretation of experiments are at least as important as the conclusions drawn. Therefore, whenever possible, the experiments that led to the conclusions will be presented. The first two chapters of the textbook mostly review the concepts of macromolecular synthesis that are essential to understanding bacterial molecular genetics. The next six chapters are more descriptive accounts of the mechanisms of mutations and gene exchange in bacteria and phages, giving overviews of the experiments that contributed to this knowledge where feasible. Chapters 9 through 12 describe the current state of knowledge about the molecular basis of recombination, repair, mutagenesis, and gene regulation in bacteria, again presenting experimental evidence where feasible. Finally, chapters 13 through 16 give more detailed examples of how classical and molecular genetic analyses have been used to study gene structure and regulation in bacteria.

SUGGESTED READING

Avery, O. T., C. M. MacCleod, and M. McCarty. 1944. Studies on the chemical nature of the substance inducing transformation of pneumococcal types. I. Induction of transformation by a desoxyribonucleic acid fraction isolated from pneumococcus type III. *J. Exp. Med.* **79:**137.

Brenner, S., F. Jacob, and M. Meselson. 1961. An unstable intermediate carrying information from genes to ribosomes for protein synthesis. *Nature* (London) **190:**576.

Cairns, J., G. S. Stent, and J. D. Watson. 1966. *Phage and the Origins of Molecular Biology.* Cold Spring Harbor Laboratory Press, Cold Spring Harbor, N.Y.

Cohen, S. N., A. C. Y. Chang, H. W. Boyer, and R. B. Helling. 1973. Construction of biologically functional bacterial plasmids *in vitro. Proc. Natl. Acad. Sci. USA* **70:**3240–3244.

Crick, F. H. C., L. Barnett, S. Brenner, and R. J. Watts-Tobin. 1961. General nature of the genetic code for proteins. *Nature* (London) **192:**1227–1232.

Hershey, A. D., and M. Chase. 1952. Independent functions of viral protein and nucleic acids in growth of bacteriophage. *J. Gen. Physiol.* **36:**393.

Iwabe, N., K. Kuma, M. Hasegawa, S. Osawa, and T. Miyata. 1989. Evolutionary relationship of archaebacteria, eubacteria and eukaryotes inferred from phylogenetic trees of duplicated genes. *Proc. Natl. Acad. Sci. USA* **86:**9355–9359.

Jacob, F., and J. Monod. 1961. Genetic regulatory mechanisms in the synthesis of proteins. *J. Mol. Biol.* **3:**3183.

Lederberg, J., and E. L. Tatum. 1946. Gene recombination in *E. coli. Nature* (London) **158:**558.

Linn, S., and W. Arber. 1968. Host specificity of DNA produced by *Escherichia coli.* X. *In vitro* restriction of phage fd replicative form. *Proc. Natl. Acad. Sci. USA* **59:**1300–1306.

Luria, S., and M. Delbrück. 1943. Mutations of bacteria from virus sensitivity to virus resistance. *Genetics* **28:**491–511.

Meselson, M., and F. W. Stahl. 1958. The replication of DNA in *Escherichia coli. Proc. Natl. Acad. Sci. USA* **44:**671.

Olsen, G. J., C. R. Woese, and R. Overbeek. 1994. The winds of (evolutionary) change: breathing new life into microbiology. *J. Bacteriol.* **176:**1–6. (Minireview.)

Zinder, N. D., and J. Lederberg. 1952. Genetic exchange in *Salmonella. J. Bacteriol.* **64:**679–699.

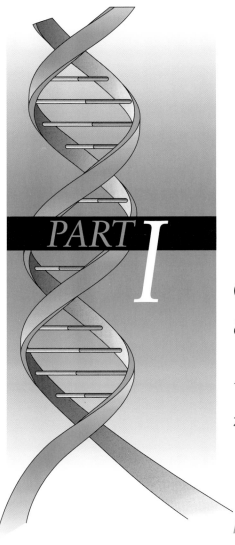

PART *I*

Genes: Replication and Expression

1 Introduction to Macromolecular Synthesis: Chromosome Structure and Replication

2 Introduction to Macromolecular Synthesis: Gene Expression

*T*HE MODERN SCIENCE OF GENETICS requires an understanding of the structure and synthesis of DNA, RNA, and proteins, called **macromolecules** because of their large size. Much has been learned in recent years about how macromolecules are made and how their structures influence their properties. Chapters 1 and 2 give the necessary background information about the structure and synthesis of macromolecules in the eubacteria, with occasional references to the eukaryotes and archaea in those instances where important differences are known to exist. There are many useful biochemistry and molecular biology textbooks which deal with the structure and synthesis of macromolecules in more detail; some of these are listed in the Suggested Reading sections at the ends of chapters 1 and 2.

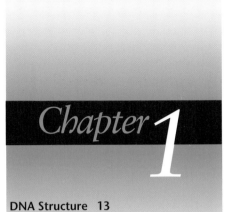

Chapter 1

Introduction to Macromolecular Synthesis: Chromosome Structure and Replication

DNA Structure

The science of molecular genetics began with the determination of the structure of DNA. Experiments with bacteria and phages (i.e., viruses that infect bacteria) in the late 1940s and early 1950s, as well as the presence of DNA in chromosomes of higher organisms, had implicated this macromolecule as the hereditary material (see Introduction). In the 1930s, biochemical studies of the base composition of DNA by Erwin Chargaff established that the amount of guanine always equals the amount of cytosine and the amount of adenine always equals the amount of thymine, independent of the total base composition of the DNA. In the early 1950s, X-ray diffraction studies by Rosalind Franklin and Maurice Wilkins showed that DNA is a double helix. Finally, in 1953, Francis Crick and James Watson put together the chemical and X-ray diffraction information in their famous model of the structure of DNA. This story is one of the most dramatic in the history of science and has been the subject of many historical treatments, some of which are listed at the end of this chapter.

Figure 1.1 illustrates the Watson and Crick structure of DNA, in which two strands wrap around each other to form a double helix. These strands can be extremely long, even in a simple bacterium, extending up to 1 mm—a thousand times longer than the bacterium itself. In a human cell, the strands that make up a single chromosome (which is one DNA molecule) are hundreds of millimeters, or many inches, long.

The Deoxyribonucleotides

If we think of DNA strands as chains, deoxyribonucleotides form the links. Figure 1.2 shows the basic structure of deoxyribonucleotides, called deoxynucleotides for short. Each is composed of a **base,** a **sugar,** and a **phos-**

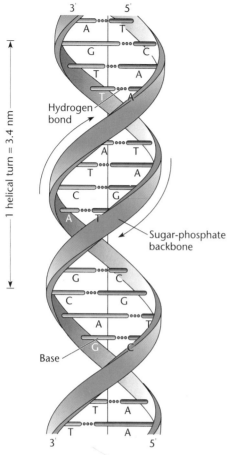

3′ 5′

1 helical turn = 3.4 nm

Hydrogen
bond

Sugar-phosphate
backbone

Base

3′ 5′

Figure 1.1 A schematic drawing of the Watson-Crick structure of DNA, showing the helical sugar-phosphate backbones of the two strands held together by hydrogen bonding between the bases.

phate group. DNA bases are **adenine (A), cytosine (C), thymine (T),** and **guanine (G),** which have either one or two rings, as shown in Figure 1.2. The bases with two rings (A and G) are the **purines,** and those with only one ring (T and C) are **pyrimidines.** A third pyrimidine, uracil (U), replaces thymine in RNA. All four DNA bases are attached to the five-carbon sugar **deoxyribose.** This sugar is identical to ribose, which is found in RNA, except that it does not have an oxygen attached to the second carbon, hence the name *deoxy*ribose. The carbons in the sugar of a nucleotide are numbered 1, 2, 3, and so on (Figure 1.2). The nucleotides also have one or more phosphate groups attached to the fifth carbon of the deoxyribose sugar, as shown. The carbon to which the phosphate group is attached is numbered and primed, i.e., 3′ or 5′.

The components of the deoxynucleotides have special names. A **deoxynucleoside** (rather than -*tide*) is a base attached to a sugar but lacking a phosphate. Without

phosphates, the four deoxynucleosides are called **deoxyadenosine, deoxycytidine, deoxythymidine,** and **deoxyguanosine.** As shown in Figure 1.2, the deoxynucleotides have one, two, or three phosphates attached to the sugar and are known as deoxynucleoside *mono*phosphates, *di*phosphates, or *tri*phosphates, respectively. The individual deoxynucleoside monophosphates, called deoxyguanosine monophosphate, etc., are often abbreviated dGMP, dAMP, dCMP, and dTMP, where the d stands for *deoxy*; the G, A, C, or T stands for the base; and the MP stands for *mono*phosphate. In turn, the *di*phosphates and *tri*phosphates are abbreviated dGDP, dADP, dCDP, dTDP, and dGTP, dATP, dCTP, dTTP, respectively. Collectively, the four deoxynucleoside triphosphates are often referred to as dNTPs.

The DNA Chain

Phosphodiester bonds join each deoxynucleotide link in the DNA chain. As shown in Figure 1.3, the phosphate attached to the last (5′) carbon of the deoxyribose sugar of one nucleotide is attached to the third (3′) carbon of the sugar of the next nucleotide, thus forming one strand of nucleotides connected 5′ to 3′, 5′ to 3′, etc.

The 5′ and 3′ Ends

Clearly, at the ends of DNA will be nucleotides that are linked to only one other nucleotide. At one end of the DNA chain, a nucleotide will have a phosphate attached to its 5′ carbon that does not connect it to another nucleotide. This end of the strand is called the **5′ end** or the **5′ phosphate end** (Figure 1.3B). On the other end, the last nucleotide lacks a phosphate at its 3′ carbon. Because it only has a hydroxyl group (the OH in Figure 1.3B), this end is called the **3′ end** or the **3′ hydroxyl end.**

Base Pairing

The sugar and phosphate groups of DNA form what is often called a **backbone** to support the bases, which jut out from the chain. This structure allows the bases to form hydrogen bonds with each other, thereby holding together two separate nucleotide chains (Figure 1.3B). This role of the four bases was first suggested by their ratios in DNA.

First, Erwin Chargaff found that no matter the source of the DNA, the concentration of guanine (G) always equals the concentration of cytosine (C) and the concentration of adenine (A) always equals the concentration of thymine (T). These ratios, named Chargaff's Rules, gave Watson and Crick one of the essential clues to the structure of DNA. They proposed that the two strands of the DNA are held together by specific hydrogen bonding between the bases in opposite strands, as

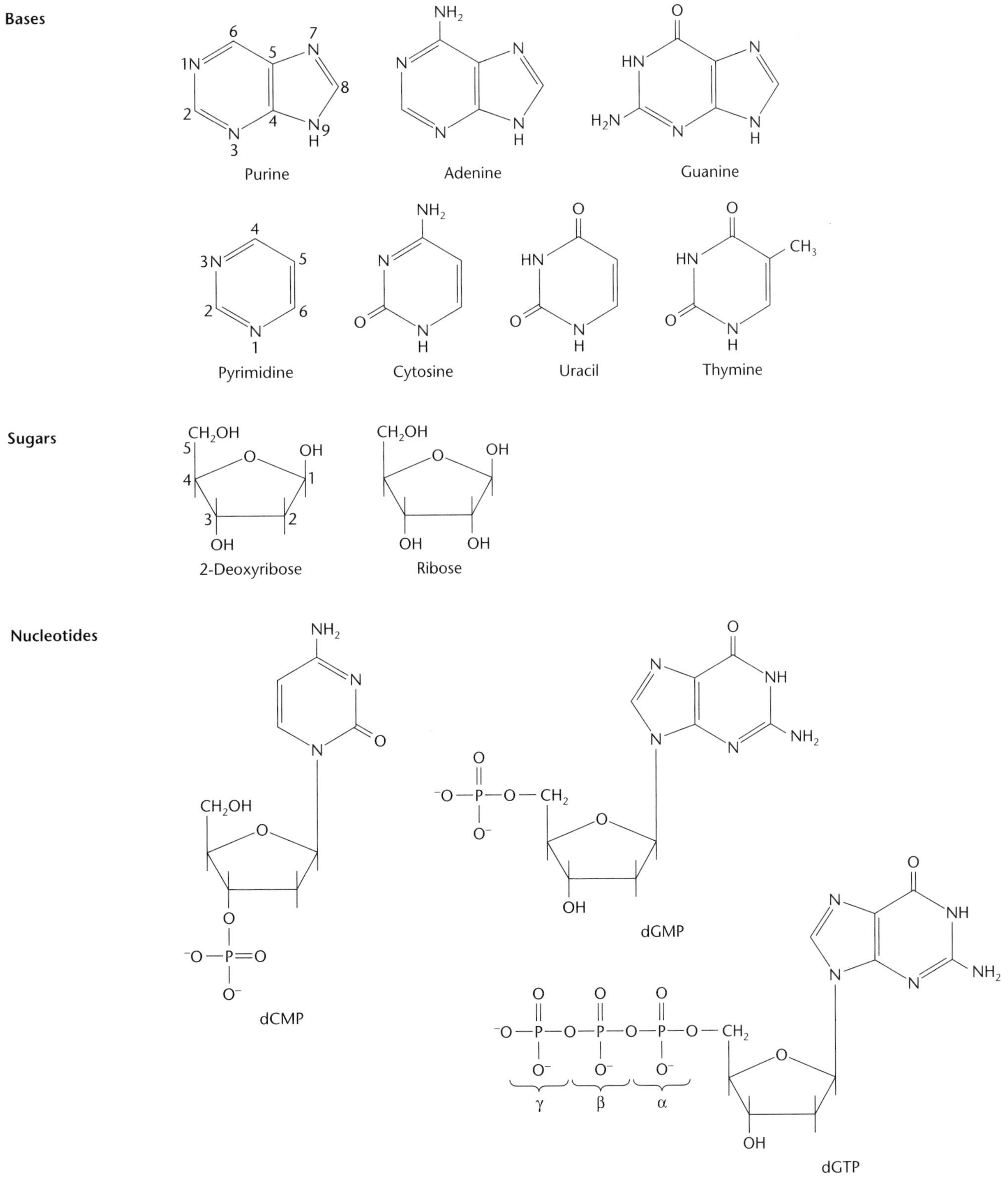

Bases

Sugars

Nucleotides

Figure 1.2 The chemical structures of deoxynucleotides, showing the bases and sugars and how they are assembled into a deoxynucleotide.

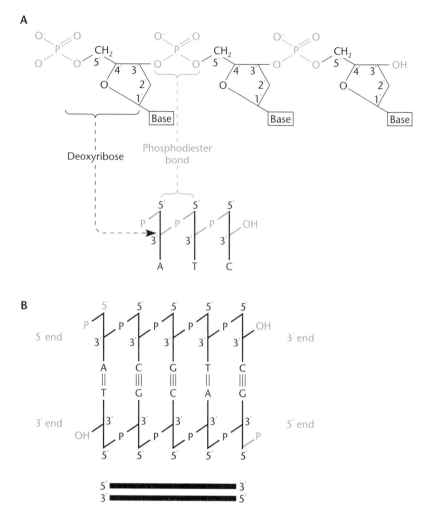

Figure 1.3 (A) Schematic drawing of a DNA chain showing the 3'-to-5' attachment of the phosphates to the sugars, forming phosphodiester bonds. (B) Two strands of DNA bind at the bases in an antiparallel arrangement of the phosphate-sugar backbones.

shown in Figure 1.4. Thus, the amounts of A and T and of C and G are always the same because A's will pair only with T's and G's will pair only with C's to hold the DNA strands together. Each A-and-T pair and each G-and-C pair in DNA is called a **complementary base pair,** and the sequences of two strands of DNA are said to be **complementary** if one strand always has a T where there is an A in the other strand and a G where there is a C in the other strand.

It did not escape the attention of Watson and Crick that the complementary base-pairing rules essentially explain heredity. If A pairs only with T and G pairs only with C, then each strand of DNA can replicate to make a complementary copy of itself, so that the two replicated DNAs will be exact copies of each other. Offspring containing these DNAs would have the same sequence of nucleotides in their DNAs as their parents and so would be exact copies of their parents.

Antiparallel Construction

As mentioned at the beginning of this section, the complete DNA molecule consists of two long chains wrapped around each other in a double helix (Figure 1.1). The double-stranded molecule can be thought of as being like a circular staircase, with the alternating phosphates and deoxyribose sugars forming the railings and the bases connected to each other forming the steps. But because of the base-pairing rules, the two chains run in opposite directions, with the phosphates on one strand attached 5' to 3', 5' to 3', etc., to the sugars and those on the other strand attached 3' to 5', 3' to 5', etc. This arrangement is called **antiparallel.** In addition to phosphodiester bonds running in opposite directions, the antiparallel construction causes the 5' phosphate end of one strand and the 3' hydroxyl end of the other to be on the same end of the double-stranded DNA molecule (Figure 1.3B).

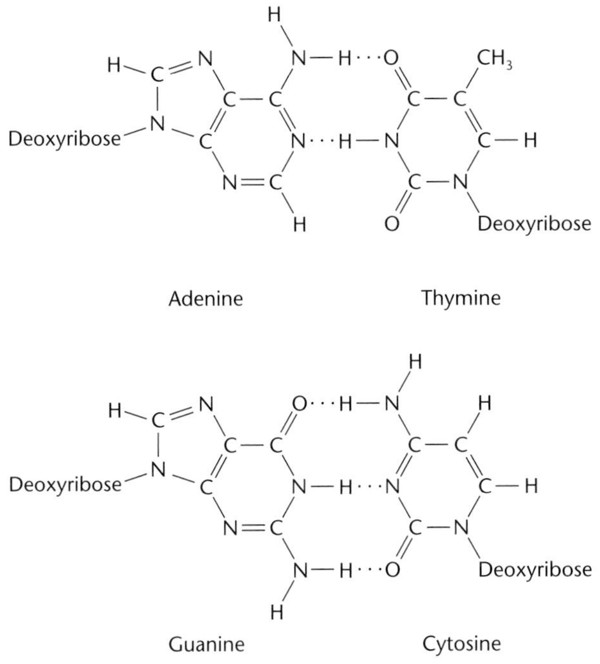

Figure 1.4 The two complementary base pairs found in DNA. Two hydrogen bonds form in adenine-thymine base pairs. Three hydrogen bonds form in guanine-cytosine base pairs.

The Major and Minor Grooves

Because the two strands of the DNA are wrapped around each other to form a double helix, the helix will have two grooves between the two strands (Figure 1.1). One of these grooves is wider than the other, so it is called the **major groove**. The other, narrower groove is called the **minor groove**. Most of the modifications to DNA that are discussed in later chapters occur in the major groove of the helix.

Mechanism of DNA Replication

The molecular details of DNA replication are probably similar in all organisms on Earth. The basic process of replication involves **polymerizing,** or linking, the nucleotides of DNA into long chains or strands, using the sequence on the other strand as a guide. Because the nucleotides must be made before they can be put together into DNA, the nucleotides are called the **precursors** of DNA synthesis.

Deoxyribonucleotide Precursor Synthesis

The precursors of DNA synthesis are the four deoxyribonucleoside triphosphates, dATP, dGTP, dCTP, and

dTTP. The triphosphates are synthesized from the corresponding ribose nucleoside diphosphates by the pathway shown in Figure 1.5. In the first step, the enzyme **ribonucleotide reductase** reduces (i.e., removes an oxygen from) the ribose sugar to produce the deoxyribose sugar by changing the hydroxyl group at the 2' position (the second carbon) of the sugar to a hydrogen. Then an enzyme known as a **kinase** adds a phosphate to the deoxynucleoside diphosphate to make the deoxynucleoside triphosphate precursor.

The deoxynucleoside triphosphate dTTP is synthesized by a somewhat different pathway from the other three. The first step is the same. Ribonucleotide reductase synthesizes the nucleotide dUDP (deoxyuridine diphosphate) from the ribose UDP. However, from then on, the pathway differs. Rather than adding a phosphate to make dUTP, the dUDP is converted to dUMP by a phosphatase that removes one of the phosphates. This molecule is then converted to dTMP by the enzyme thymidylate synthetase, using tetrahydrofolate to donate a methyl group. Kinases then add two phosphates to the dTMP to make the precursor dTTP.

Deoxynucleotide Polymerization

The complex process of DNA replication involves many enzymes and other cellular components. In the end, complementary copies of each of the extremely long strands of a double-stranded DNA must be made. In this section, we discuss the obstacles that DNA replication must overcome and how the various functions involved overcome these obstacles.

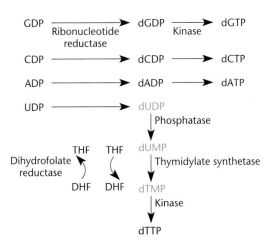

Figure 1.5 The pathways for synthesis of the deoxynucleotides from the ribonucleotides. Some of the enzymes referred to in the text are identified. THF, tetrahydrofolate; DHF, dihydrofolate.

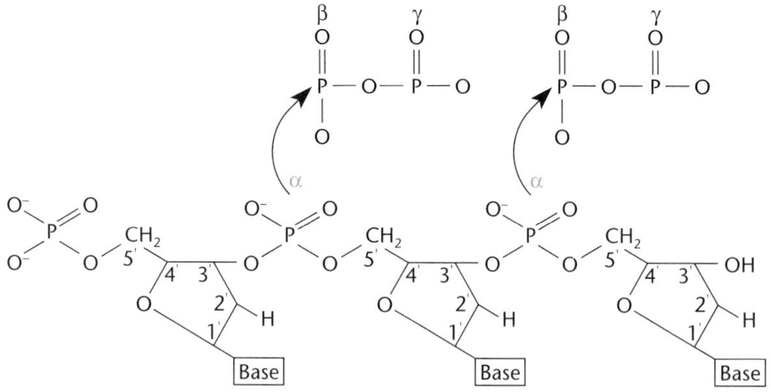

Figure 1.6 Polymerization of the deoxynucleotides during DNA synthesis. The β and γ phosphates of each deoxynucleoside triphosphate are cleaved off to give energy for the polymerization reaction.

DNA POLYMERASE

The properties of the DNA polymerases, the enzymes that actually join the deoxynucleotides together to make the long chains, are the best guides to an understanding of the replication of DNA. These enzymes make DNA by linking one deoxynucleotide to another to make a long chain of DNA. This process is called DNA **polymerization,** hence the name DNA *polymerases.*

Figure 1.6 shows the basic process of DNA polymerization by DNA polymerase. The DNA polymerase attaches the first phosphate (called α) of one deoxynucleoside triphosphate to the 3' carbon of the sugar of the next deoxynucleoside triphosphate, in the process releasing the last two phosphates (called the β and γ phosphates) of the first deoxynucleoside triphosphate to produce energy for the reaction. Then the α phosphate of another deoxynucleoside triphosphate is attached to the 3' carbon of this deoxynucleotide, and the process continues until a long chain is synthesized.

PRIMERS

DNA polymerases cannot start the synthesis of a new strand of DNA; they can only attach deoxynucleotides to a preexisting 3' OH group. The 3' OH group to which DNA polymerase adds a deoxynucleotide is called the **primer** (Figure 1.7).

RNA PRIMERS

The requirement of a primer for DNA polymerase creates an apparent dilemma in DNA replication. When a new strand of DNA is synthesized, there is no DNA upstream (i.e., on the 5' side) to act as a primer. The cell usually solves this problem by using RNA as the primer to initiate the synthesis of new strands. Unlike DNA polymerase, RNA polymerase does not require a primer to initiate the synthesis of new strands. The RNA primers that are used to initiate the synthesis of new

strands of DNA are either made by the RNA polymerase, which makes all the other RNAs including mRNA, tRNA, and rRNA, or by special enzymes called **primases.** During DNA replication, special enzymes recognize and remove the RNA primer.

TEMPLATE STRAND

DNA polymerases also need a **template strand.** The template strand directs which nucleotide will be inserted at each step of the polymerization reaction by complementary base-pairing, as shown in Figure 1.7. As mentioned in the base-pairing section, whenever there is an A in the template strand, the DNA polymerase inserts a T into the new strand. When there is a C in the template strand,

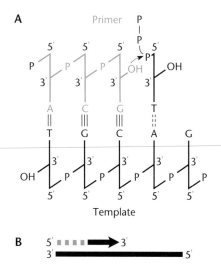

Figure 1.7 Functions of the primer and template in DNA replication. (A) The DNA polymerase adds deoxynucleotides to the 3' end of the primer by using the template strand to direct the selection of each base. (B) Simple illustration of 5' → 3' DNA synthesis. Dotted line, primer.

the DNA polymerase inserts a G, and so forth. The DNA polymerase moves along the template strand in the 3'-to-5' direction, linking deoxynucleotides in the new strand in the 5'-to-3' direction. When replication is finished, the product is a double-stranded DNA with two antiparallel strands, one of which is the old template strand and the other of which is the newly synthesized strand.

ACCESSORY PROTEINS

Some proteins travel with the DNA polymerase as part of a **DNA replication complex** as the polymerase moves along the template strand. These proteins are called **DNA polymerase accessory proteins.** Some form a clamp that helps keep the DNA polymerase from falling off the template strand; others are exonucleases that serve an **editing function** to correct mistakes the DNA polymerase may make (see the section on editing, below). Table 1.1 lists many of the DNA replication proteins.

Semiconservative Replication

The process of replication described above is called **semiconservative replication** because each time a DNA molecule replicates, the two old strands are *conserved*

TABLE 1.1	Proteins involved in *E. coli* DNA replication	
Protein	**Gene**	**Function**
DnaA	*dnaA*	Initiator protein; primosome (priming complex) formation
DnaB	*dnaB*	DNA helicase
DnaC	*dnaC*	Delivers DnaB to replication complex
SSB	*ssb*	Binding to single-stranded DNA
Primase	*dnaG*	RNA primer synthesis
DNA Pol I	*polA*	Primer removal; gap filling
DNA Pol III (holoenzyme)		
α	*dnaE*	Polymerization
ε	*dnaQ*	3' → 5' editing
θ	*holE*	Present in core (αεθ)
τ	*dnaX*	Dimerizes core
β	*dnaN*	Sliding clamp
γ	*dnaX*	Clamp loading
δ	*holA*	Clamp loading
δ'	*holB*	Clamp loading
χ	*holC*	Clamp loading
ψ	*holD*	Clamp loading
DNA ligase	*lig*	Sealing DNA nicks
DNA gyrase		Supercoiling
α	*gyrA*	Nick closing
β	*gyrB*	ATPase

to become part of two new DNA molecules. Each of the new molecules will consist of one old conserved strand and one newly synthesized strand, so that each new molecule is only partly conserved (or *semi*conserved). The next time the DNA molecule replicates, each strand will serve as a template, becoming the conserved strand in a new double-stranded molecule.

THE MESELSON-STAHL EXPERIMENT

The semiconservative mechanism was suggested by the structure of DNA. However, other mechanisms of DNA replication are also possible. Soon after Watson and Crick published their structure for DNA, Matthew Meselson and Frank Stahl performed an experiment that showed that DNA does replicate by the proposed semiconservative mechanism.

One prediction of the semiconservative replication hypothesis is that after replication, each DNA molecule should have one newly synthesized strand and one old strand from the original molecule. If it could be demonstrated that newly synthesized DNA had one new strand and one old strand, this prediction would have been fulfilled. Figure 1.8 illustrates the details of their experiment. First, they chose the heavy isotopes of nitrogen, carbon, and hydrogen, three of the types of atoms contained in DNA, as markers for new strands. DNA synthesized from these heavy isotopes is more dense than DNA synthesized from the normal atoms.

To incorporate the heavy isotopes into newly synthesized DNA, they grew the bacterium *Escherichia coli* on a medium containing the heavy isotopes instead of one containing the normal isotopes for about the time the cells took to divide. Then they extracted the DNA from the cells and analyzed its density by means of density gradient equilibrium centrifugation. As Figure 1.8 shows, the density of the DNA will be reflected by the position of the DNA in the gradient; heavier DNA, composed of atoms of the heavy isotopes, bands farther down. If the DNA replicates by a semiconservative mechanism, then after one cycle of replication in the presence of the heavy isotopes, one strand should be made from precursors with the heavy isotopes and the newly replicated molecule should then consist of one heavy strand and one light strand. Consequently, its density will be intermediate between those of light DNA, with no heavy isotopes, and heavy DNA, in which both new strands contain heavy isotopes. After a short time of replication, Meselson and Stahl did observe DNA of intermediate density, composed of one light strand and one heavy strand; thus, their results supported the semiconservative mechanism of replication.

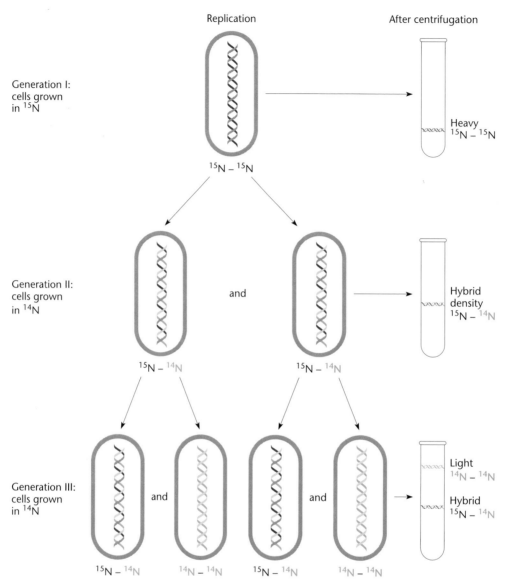

Figure 1.8   The Meselson-Stahl experiment. Synthesis of a "light" complementary strand gives a double-stranded DNA of hybrid light and heavy density. The light-light, heavy-light, and heavy-heavy DNAs can be separated by equilibrium density centrifugation.

Replication of Double-Stranded DNA

The replication of most long DNA molecules such as bacterial DNA begins at one point and moves in both directions from there. In the process, the two old strands of DNA are separated and used as templates to synthesize new strands. The structure where the two strands are separated and the new synthesis is occurring is referred to as the **replication fork.** The DNA polymerase enzyme cannot separate the two strands of a bacterial chromosome, replicate them, and separate the daughter DNAs by itself. Many other proteins are required for replication, as we discuss in this section.

HELICASES AND HELIX-DESTABILIZING PROTEINS

One task the DNA polymerase cannot perform in the replication of double-stranded DNA is the separation of the strands of DNA at the replication fork. The strands of DNA must be separated for the two strands to serve as templates. The bases of the DNA are inside of the double helix, where they are not available to pair with the incoming deoxynucleotides, to direct which nucleotide will be inserted at each step. The strands of DNA are separated by proteins called **DNA helicases.** Separating the strands of DNA requires energy, and ATP is cleaved by the helicases to provide this energy.

Once the strands of DNA have been separated, they also must be prevented from coming back together (or from annealing to themselves if they happen to be complementary over short regions). The separation of the strands is maintained by proteins called **helix-destabilizing proteins** or single-strand-binding proteins. These are proteins that bind preferentially to single-stranded DNA and prevent double-stranded helical DNA from re-forming prematurely.

OKAZAKI FRAGMENTS AND THE REPLICATION FORK

Another problem in replicating double-stranded DNA is created because the two strands are antiparallel. As already discussed, in one strand the phosphates connect the sugars 3' to 5' and in the other strand they connect the sugars 5' to 3'. However, DNA polymerase can move only in the 3'-to-5' direction on the template strand, synthesizing the new DNA in the 5'-to-3' direction. How can the replication fork move in one direction on a double-stranded DNA and make complementary copies of both strands at the same time? Because of the antiparallel structure, the DNA polymerase on one of the two strands would have to be moving in the wrong direction overall.

This problem is overcome by replicating the two strands differently (Figure 1.9). There are actually two types of DNA polymerases in E. coli that participate in replication, called DNA polymerase I (Pol I) and DNA polymerase III (Pol III) (Table 1.1). On one template strand, DNA polymerase III initiates synthesis from an RNA primer and moves along the template DNA in the 3'-to-5' direction. This newly synthesized strand is referred to as the **leading strand** (Figure 1.9). On the other strand, DNA polymerase III also moves in the 3'-to-5' direction but works in opposition to the movement of the replication fork as a whole. This second synthesized strand is known as the **lagging strand,** and to produce it, the DNA polymerase makes short pieces called **Okazaki fragments** (Figure 1.10). Synthesis of each Okazaki fragment requires a new RNA primer, about 10 to 12 nucleotides long, synthesized by DnaG primase. The RNA primer allows DNA polymerase III to synthesize DNA in the 5'-to-3' direction (Figures 1.9 and 1.10). As the replication fork progresses, DnaG primase will produce a new RNA primer about once every 2 kilobases, recognizing the sequence 3'-GTC-5' and beginning synthesis opposite the T. RNA primers are replaced by DNA, for which the upstream Okazaki fragment serves as the primer (Figure 1.9). DNA poly-

merase I, which has both a 5' **exonuclease** and a DNA polymerase activity, removes each RNA primer and replaces it with DNA. Figure 1.11 shows the concerted 5' exonuclease and DNA-polymerizing activities of DNA polymerase I. The 5' exonuclease can act on nicked duplex DNA or RNA-DNA hybrids. In E. coli, the polA gene encodes DNA polymerase I. The Okazaki fragments are then joined together by **DNA ligase** before the replication fork moves on, as shown in Figure 1.9.

What actually happens at the replication fork may be more complicated than suggested by this simple picture. For one thing, this picture ignores the overall topological restraints on the DNA that is replicating. The **topology** of a molecule refers to its position in space. Because the circular DNA is very long and its strands are wrapped around each other, pulling the two strands apart introduces stress into other regions of the DNA in the form of **supercoiling.** Unless the two strands of DNA were free to rotate around each other, supercoiling would cause the chromosome to look like a telephone cord wound up on itself. To relieve this stress, enzymes called **topoisomerases** undo the supercoiling ahead of the replication fork. We shall discuss DNA supercoiling and topoisomerases later in the chapter.

Other evidence suggests that the picture of the two strands of DNA replicating independently may be too simple. The two strands may actually replicate simultaneously, with the template for the lagging strand being looped out of the replicating complex before the synthesis of both strands is completed and the replication fork moves on, as shown in Figure 1.12. This has been called the "trombone" model of replication at the replication fork because of its superficial similarity to the musical instrument, with a periodic looping out of one of the two template strands.

THE GENES FOR REPLICATION PROTEINS

Most of the genes for replication proteins have been found by isolating mutants defective in DNA replication but not RNA or protein synthesis. Since a mutant cell that cannot replicate its DNA will die, any mutation (for definitions of mutants and mutations, see the section on replication errors, below, and chapter 3) that inactivates a gene whose product is required for DNA replication will kill the cell. Therefore for experimental purposes, only a type of mutant called a **temperature-sensitive mutant** can be usefully isolated with mutations in DNA replication genes. These are mutants in which the product of the gene is inactive at one temperature but active at another. The mutant cells can be propagated at the temperature at which the protein is active. Then, the effects of inactivating the protein can be

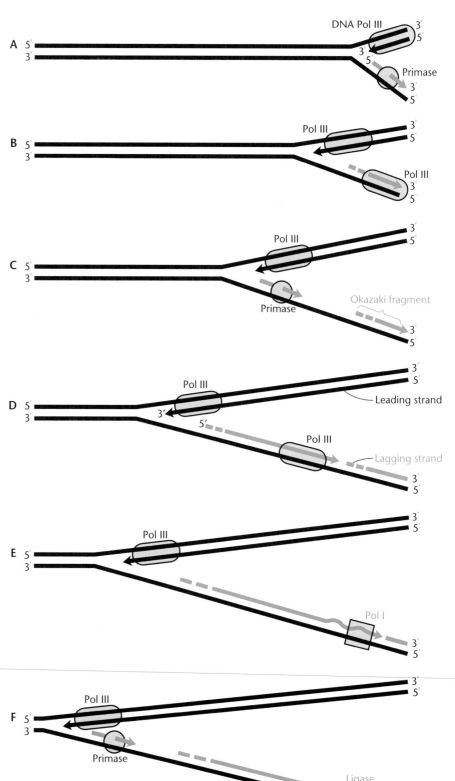

Figure 1.9 Discontinuous synthesis of one of the two strands of DNA during chromosome replication. The functions of the primase and ligase are shown. The short RNA primers in the lagging strand are removed after replication and resynthesized as DNA by using upstream DNA as primer.

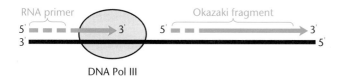

Figure 1.10   Synthesis of short Okazaki fragments from an RNA primer.

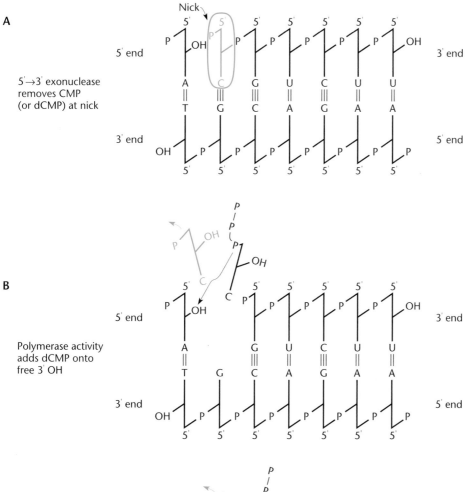

A

5'→3' exonuclease removes CMP (or dCMP) at nick

B

Polymerase activity adds dCMP onto free 3' OH

C

5'→3' exonuclease removes GMP (or dGMP) at nick

Nick translation

Figure 1.11   DNA polymerase I can remove the nucleotides of an RNA primer by using its "nick translation" activity. (A) A break, or nick, in the DNA strand occurs when the DNA polymerase III holoenzyme incorporates the last deoxynucleotide before it encounters a previously synthesized RNA primer. (B) In the example, the 5'-to-3' exonuclease activity of DNA polymerase I removes the CMP at the nick and its DNA polymerase activity incorporates a dCMP onto the free 3' hydroxyl. (C) This process continues, moving the nick in the 5'-to-3' direction.

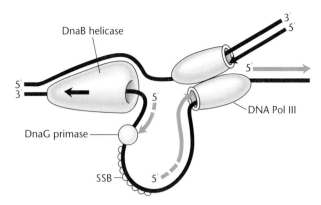

Figure 1.12 A "trombone" model for how both the leading and lagging strands of the DNA helix might be simultaneously replicated at the replication fork. SSB, single-strand-binding protein. Adapted with permission from T. A. Baker and S. H. Wickner, *Annu. Rev. Genet.* **26**:447–478, 1992.

tested by shifting to the other temperature. We shall discuss the molecular basis for temperature-sensitive mutants in more detail in chapter 3.

The immediate effect of a temperature shift on a mutant with a mutation in a DNA replication gene depends on whether the product of the gene is continuously required for replication at the replication forks or is involved only in the initiation of rounds of replication. For example, if the mutation is in a gene for DNA polymerase III or in the gene for the DnaG primase, replication will immediately cease. However, if the temperature-sensitive mutation is in a gene whose product is required only for initiation of DNA replication, for example, the gene for DnaA or DnaC (see section on initiation of chromosome replication, below), the replication rate for the population will slowly decline. Unless the cells have been somehow synchronized in their cell cycle, each cell will be at a different stage of replication, with some cells having just finished a round of replication and other cells having just begun a new round. Cells in which rounds of chromosome replication were under way at the time of the temperature shift will complete their replication but not start a new round. Therefore, the rate of replication will decrease until the rounds of replication in all the cells are completed.

Replication Errors

To maintain the stability of a species, replication of the DNA must be almost free of error. Otherwise, changes in the DNA sequence called **mutations** will occur, and

these changes will be passed on to subsequent generations. Depending on where these changes occur, they can severely alter the protein products of genes or other cellular functions. To avoid such instability, the cell has mechanisms that reduce the error rate.

As DNA replicates, the wrong base is sometimes inserted into the growing DNA chain. For example, Figure 1.13 shows the incorrect incorporation of a T opposite a G. A base pair in which the bases are paired wrongly is called a **mismatch.** Mismatches can occur when the bases take on forms called **tautomers,** which pair differently from the normal form of the base (see chapter 3). After the first replication in Figure 1.13, the mispaired T will usually be in its normal form and pair correctly with an A, causing a GC-to-AT change in the sequence of one of the two progeny DNAs and thus changing the base pair at that position on all subsequent copies of the mutated DNA molecule.

Editing

One way the cell reduces mistakes during replication is through **editing** functions. Sometimes these functions are performed by separate proteins, and sometimes they are part of the DNA polymerase itself! Editing proteins are aptly named because they go back over the newly replicated DNA looking for mistakes, recognizing and removing incorrectly inserted bases (Figure 1.14). If the last nucleotide inserted in the growing DNA chain creates a mismatch, the editing function will stop the replication until the offending nucleotide is removed. The replication then continues, inserting the correct nucleotide. Because the DNA chain grows in the 5'-to-3' direction, the last nucleotide added is at the 3' end. The enzyme activity that removes this nucleotide is therefore called a 3' **exonuclease.** The editing proteins probably recognize a mismatch because the mispairing (between A and G in the example) causes a minor distortion in the structure of the double-stranded helix of the DNA.

In many cellular organisms, some viruses, and some repair DNA polymerases, for example, DNA polymerase I, the 3' exonuclease editing activity is part of the DNA polymerase itself. In bacterial chromosome replication, the editing functions are accessory proteins encoded by separate genes whose products travel along the DNA with the DNA polymerase during replication. In *E. coli*, the 3' exonuclease editing function is encoded by the *dnaQ* gene (Table 1.1), and *dnaQ* mutants, also called *mutD* mutants (i.e., cells with a mutation in this gene that inactivates the 3' exonuclease function), show much higher rates of spontaneous mutagenesis than do the **wild-type,** or normally functioning, cells. Because of their high spontaneous mutation rates, *mutD* mutants

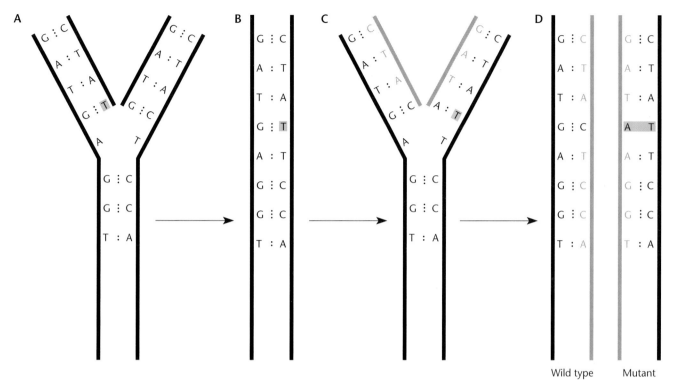

Figure 1.13 Mistakes in base pairing can lead to changes in the DNA sequence called mutations. A T mistakenly put opposite a G during replication (A) can lead to an AT base pair replacing a GC in the progeny DNA (B to D).

of *E. coli* are often used to introduce random mutations into plasmids and bacteriophages.

RNA PRIMERS AND EDITING

The importance of the editing functions in lowering the number of mistakes during replication may explain why DNA replication is primed by RNA rather than by DNA. When the replication of a DNA chain has just initiated, the helix may be too short for distortions in its structure to be easily recognized by the editing proteins. The mistakes may then go uncorrected. However, if the first nucleotides inserted in a growing chain are ribonucleotides rather than deoxynucleotides, an RNA primer will be synthesized rather than a DNA primer. The RNA primer can be removed and resynthesized as DNA by using preexisting upstream DNA as primer. Under these conditions, the editing functions will be active and mistakes will be avoided.

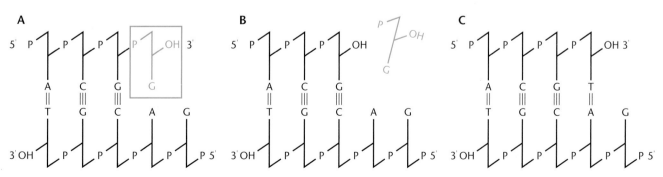

Figure 1.14 The editing function of DNA polymerase. (A) A G is mistakenly put opposite an A while the DNA is replicating. (B and C) The DNA polymerase stops while the G is removed and replaced by a T before the replication continues.

Methyl-Directed Mismatch Repair

Sometimes, the wrong base pair will be inserted into DNA in spite of the vigilance of the editing functions. However, the cell still has another chance to prevent a permanent mistake or mutation: the wrong base can be recognized by another repair system called the **mismatch repair system**. This system recognizes the mismatch and removes it as well as DNA in the same strand around the mismatch, leaving a gap in the DNA that is refilled by the action of DNA polymerase, which inserts the correct nucleotide.

The mismatch repair system is very effective at removing mismatches from DNA. However, by itself, it would not lower the rate of spontaneous mutagenesis unless it repaired the correct strand of DNA at the mismatch. In the example shown in Figure 1.13, a T was mistakenly incorporated opposite an G. If the mismatch repair system changes the T to the correct C in the *newly* replicated DNA, a GC base pair will be restored at this position and no change in the sequence or mutation will have occurred. However, if it repairs the G in the *old* DNA in the mismatch to an A, the mismatch will have been removed and replaced by a AT base pair with correct pairing, but a GC-to-AT change would have occurred in the DNA sequence at the site of the mismatch, creating a mutation. To prevent mutations, the mismatch repair system must have some

way of distinguishing the newly synthesized strand from the old strand, so that it can repair the correct strand.

The state of methylation of the DNA strands allows the mismatch repair system of *E. coli* to distinguish the new from the old strands after replication. In *E. coli*, the A's in the symmetric sequence GATC/CTAG are methylated at the 6' position of the larger of the two rings of the adenine base. These methyl groups are added to the bases by the enzyme **deoxyadenosine methylase** (Dam methylase), but this occurs only *after* the nucleotide has been incorporated into the DNA. Since DNA replicates by a semiconservative mechanism, the newly synthesized strand of the GATC/CTAG will remain temporarily unmethylated after replication of a region containing this sequence. The DNA at this site is said to be **hemimethylated** if the bases on only one strand are methylated. Figure 1.15 shows that a hemimethylated GATC/CTAG tells the mismatch repair system which strand is newly synthesized and should be repaired. For this reason, the repair system is called the **methyl-directed mismatch repair system.**

The mismatch repair system also repairs many types of chemical damage to DNA. which we discuss in more detail in chapter 10. Methylation of DNA in *E. coli* plays other roles as well, including protecting against cutting by restriction endonucleases (chapter 13) and

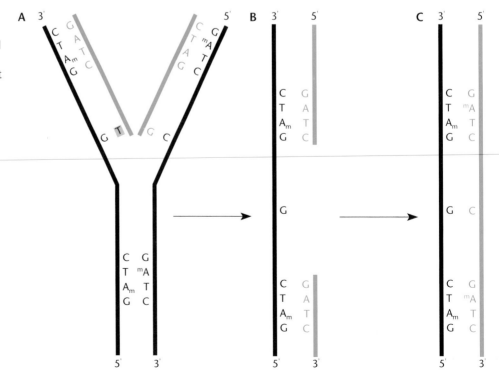

Figure 1.15 The methyl-directed mismatch repair system. (A) The newly replicated DNA contains a GT mismatch, and the GATC sequences are not methylated. (B) The repair system excises the mismatched T and adjacent bases of the unmethylated strand. (C) The gap has been filled in, and the GATC sequences have been methylated.

helping to time the initiation of chromosomal DNA replication (see below).

Role of Editing and Mismatch Repair in Maintaining Replication Fidelity

It is possible to estimate how much each of the repair systems lowers the mistake rate of replication. On the basis of normal mutation rates, the *E. coli* replication mechanism probably makes a mistake about once every 10^{10} times it incorporates a nucleotide. Since there are about 4.7×10^6 nucleotides in each strand of *E. coli* DNA, the cell makes $1/10^{10} \times 4.7 \times 10^6 = 4.7 \times 10^{-4}$ mistake every time it replicates a chromosome. In other words approximately one in every 2,000 progeny bacteria will have a mistake in its DNA, since the entire bacterial DNA must replicate once every time the cell divides. Mistake levels like these are apparently low enough to be tolerated, and some mistakes may even be desirable because they increase diversity in the population and speed up evolution.

In contrast, mutant bacteria (i.e., those that differ from normal, or wild-type, bacteria by a definable mutation; see chapter 3) that lack either the editing functions or the mismatch repair system have unacceptably high mistake rates. An *E. coli* bacterium that lacks the editing function will make a mistake on average once every 10^6 times a nucleotide is added during replication, or 10,000 times more frequently than the wild-type bacteria. This means that every such cell will suffer an average of about five mutations each time its DNA replicates. If the mismatch repair system is also inactivated, the mistake rate will be increased another 10-fold or more, for an average of 50 or more mistakes every time the DNA replicates and the cell divides. It would not be very long before these bacteria were severely compromised by mutations and bore little resemblance to their ancestors. By lowering the spontaneous mutation rate, DNA correction systems are important for maintaining replication fidelity and the stability of the species.

Replication of the Bacterial Chromosome and Cell Division

So far we have discussed the details of DNA replication, but we have not discussed how the bacterial DNA as a whole replicates, nor have we discussed how the replication process is coordinated with division of the bacterial cell. To simplify the discussion, we shall consider only bacteria that grow as individual cells and divide by binary fission to form two cells of equal size, even though this is far from the only type of multiplication observed among bacteria.

The replication of the bacterial DNA occurs during the cell division cycle. The **cell division cycle** is the time during which a cell is born, grows larger, and divides into two progeny cells. **Cell division** is the process by which the larger cell splits into the two new cells. The **division time,** or **generation time,** is the time that elapses from the point when a cell is born until it divides. This time is usually approximately the same for all the individuals in the population under certain growth conditions. Before cell division, the original cell is called the **mother cell** and the two progeny cells after division are the **daughter cells.**

Structure of the Bacterial Chromosome

The DNA molecule of a bacterium that carries most of its normal genes is commonly referred to as its **chromosome,** by analogy to the chromosomes of higher organisms. This name distinguishes the molecule from plasmid DNA, which can be almost as large as chromosomal DNA but usually carries genes that are not always required for growth of the bacterium (see chapter 4).

Most bacteria only have one chromosome; in other words, there is only one *unique* DNA molecule per cell that carries most of the normal genes. This does not mean that there is necessarily only one copy of the chromosomal DNA in bacterial cells. Bacterial cells whose chromosomes replicate but for some reason do not divide have more than one copy of the chromosomal DNA per cell. However, these DNAs are not unique, since they are derived from the same original molecule.

The structure of bacterial DNA differs significantly from that of the chromosomes of higher organisms. For example, DNA in the chromosomes of most bacteria is **circular** (for exceptions, see Box 1.1), with a circumference of approximately 1 mm. In contrast, eukaryotic chromosomes are linear with free ends. As we shall discuss later, the circularity of bacterial chromosomal DNA allows it to replicate in its entirety without using telomeres, as eukaryotic chromosomes do, or terminally redundant ends, as some phages do. Another difference between the DNA of bacteria and eukaryotes is that the DNA in eukaryotes is wrapped around proteins called histones to form nucleosomes. Bacteria contain histone-like proteins including HU, HN-S, Fis, and IHF around which DNA is often wrapped. However, in general, DNA is much less structured in bacteria than in eukaryotes.

Replication of the Bacterial Chromosome

The replication of the circular bacterial chromosome initiates at a unique site in the DNA called the **origin** of

Linear Chromosomes in Bacteria

Not all bacteria have circular chromosomes. Some, including *Borrelia burgdorferi*, the causative agent of Lyme disease, *Streptomyces lividans*, and *Rhodococcus fasciens*, can have linear chromosomes. It is not clear how these linear chromosomes replicate in their entirety. They do not seem to replicate like eukaryotic chromosomes, which are also linear. At their ends, eukaryotic chromosomes contain enzymatically synthesized repeated sequences called telomeres, which do not need complementary sequences to be synthesized from the template during replication. However, the ends of some linear bacterial chromosomes have inverted repeated sequences that are sometimes covalently joined to each other. These features may be why bacterial linear chromosomes replicate differently from eukaryotic ones. Interestingly, bacteria that can have linear chromosomes also often contain linear plasmids (see chapter 4).

Reference

Hinnebursch, J., and K. Tilly 1993. Linear plasmids and chromosomes in bacteria. *Mol. Microbiol.* **10:**917–922. (MicroReview.)

chromosomal replication, or *oriC*, and proceeds in both directions around the circle. On the *E. coli* chromosome, *oriC* is located at 84.3 min. At the positions where polymerases add the nucleotides, the double-stranded DNA splits and forms two new double-stranded DNAs. As mentioned earlier, the place in DNA at which replication is occurring is known as the replication fork. The two replication forks proceed around the circle until they meet and **terminate** chromosomal replication. As discussed below in the section on termination, many bacteria do not terminate replication at a unique site in the DNA but, rather, where the two replication forks meet. Each time the two replication forks proceed around the circle and meet, a **round of replication** has been completed and two new DNAs, called the **daughter DNAs**, are created.

Initiation of Chromosome Replication

Much has been learned about the molecular events occurring during the initiation of replication. Some of this information has a bearing on how the initiation of chromosome replication is regulated and serves as a model for the interaction of proteins and DNA.

Two types of functions are involved in the initiation of chromosome replication. One consists of the sites or sequences on DNA at which proteins act to initiate replication. These are called *cis*-acting sites. The prefix *cis* means "on this side of," and these sites act only on the same DNA. The proteins involved in initiation of replication are called *trans*-acting functions. The prefix *trans* means "on the other side of," and these functions can act on any DNA in the same cell, not just the DNA from which they were made. These concepts will be used again as the book progresses.

ORIGIN OF CHROMOSOMAL REPLICATION

One *cis*-acting site is the *oriC* site, at which replication initiates. The sequence of *oriC* is well defined and is similar in most bacteria. Figure 1.16 shows the structure of the origin of replication of *E. coli*. Less than 250 base pairs of DNA is required for initiation at this site. Within *oriC*, similar sequences of 9 bases, the so-called **DnaA boxes,** are repeated four times. In addition, regions 13 base pairs long with a higher than average AT base pair frequency are repeated three times. These two types of repeated sequences are thought to be very important for the initiation of chromosome replication.

INITIATION PROTEINS

Many *trans*-acting proteins are also required for the initiation of DNA replication, including the DnaA, DnaB, and DnaC proteins. DnaA is required only for initiation, but both DnaB and DnaC are also required for primer synthesis once DNA replication is under way. Many proteins used in other cellular functions are also involved, such as the primase (DnaG) and the normal RNA polymerase that makes most of the RNA in the cell.

Figure 1.17 outlines how DnaA, DnaB, DnaC, and other proteins participate in the initiation of chromosome replication. In the first step, 10 to 12 molecules of the DnaA protein bind to the DnaA boxes in the *oriC* region. This has the effect of wrapping the DNA around

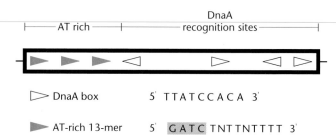

Figure 1.16 Structure of the *oriC* region of *E. coli*. Shown are the position of the AT-rich region and the position of some of the DnaA-binding sequences (DnaA boxes) referred to in the text.

A

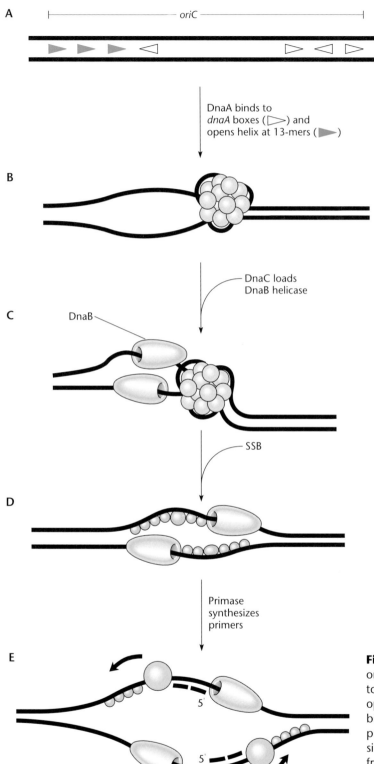

DnaA binds to
dnaA boxes (▷—) and
opens helix at 13-mers (▶—)

DnaC loads
DnaB helicase

SSB

Primase
synthesizes
primers

Figure 1.17   (A) Initiation of replication at the *E. coli* origin (*oriC*) region. (B) A number of DnaA proteins bind to the origin, wrapping the DNA around themselves and opening the helix. (C) DnaC helps the DnaB helicase to bind. (D and E) The DnaG primase synthesizes RNA primers, initiating replication in both directions. SSB, single-strand-binding protein. Adapted with permission from T. A. Baker and S. H. Wickner, *Annu. Rev. Genet.* **26:**447–478, 1992.

the aggregated DnaA proteins, as shown in the figure. The bending helps separate the strands of the DNA in the region of the bend as shown.

Once the strands are partially opened, the DnaB protein binds to the *oriC* region with the help of the DnaC protein. This binding is also aided by supercoiling at the origin (see the section on supercoiling, below) and by the helix-destabilizing protein, or single-strand-binding protein (SSB in Figure 1.17), which helps keep the helix from re-forming. The DnaB protein is a helicase that opens the strands further for priming and replication. The DnaC protein may then leave. This structure, with many copies of DnaA as well as DnaB and other proteins, is called a **primosome.** The primosome may help the DnaG primase or another RNA polymerase to synthesize an RNA primer to start replication.

RNA PRIMING OF INITIATION

Initiation of DNA replication requires RNA primers, but which RNA polymerase makes them is not completely clear. The RNA polymerase that synthesizes most of the RNA molecules, including mRNA, in the cell (see chapter 2) is needed to initiate rounds of replication. However, its role may be to help separate the strands of DNA in the *oriC* region by transcribing through this region, because the strands may have to be separated before the DnaA protein can bind. In this case, the RNA primers themselves may be synthesized by the DnaG protein, the same RNA polymerase that makes RNA primers for lagging-strand synthesis at the replication fork. Alternatively, the normal RNA polymerase may make the primer that initiates leading-strand synthesis, while the DnaG protein makes only

the primers that initiate lagging-strand synthesis. More experiments are needed to answer this question.

Termination of Chromosome Replication

After the replication of the chromosome initiates in the *oriC* region and proceeds around the circular chromosome in both directions, the two replication forks must meet somewhere on the other side of the chromosome and the two daughter chromosomes must separate. Do they meet and terminate replication at a certain well-defined site in the DNA, or do they terminate replication wherever they happen to meet? Also, are specific proteins required for termination of chromosome replication?

As with most cellular processes, the process of termination of chromosome replication is best understood in *E. coli.* In this bacterium, chromosome replication usually terminates in a certain region but not at a well-defined unique site. This termination region, called *ter,* contains sites called *ter* sequences, which are only 22 base pairs long. These sites act somewhat like the one-way gates in an automobile parking lot, allowing the replication forks to pass through in one direction but not in the other.

Figure 1.18 shows how the one-way nature of *ter* sequences causes replication to terminate in the *ter* region. In the illustration, two *ter* sites called *terA* and *terB* bracket the termination region. Replication forks can pass site *terA* in the clockwise direction but not the counterclockwise direction. The opposite is true for *terB.* Thus, the clockwise-moving replication fork can pass through *terA,* but if it gets to *terB* before it meets the counterclockwise-moving fork, it will stall because it cannot move clockwise through the *terB* site. Simi-

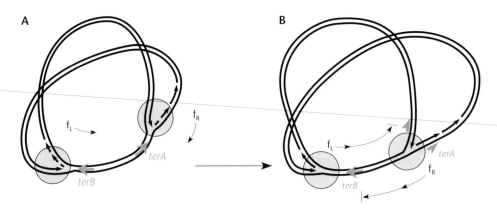

Figure 1.18 Termination of chromosome replication in *E. coli.* (A) The replication forks can traverse *terA* and *terB* in only one direction, opposite to that indicated by the arrows. (B) When they meet, between or at one of the two sites, chromosome replication terminates. f_L is the fork that initiated to the left and moved in the counterclockwise direction. f_R is the fork that initiated to the right and moved in the clockwise direction.

larly, the replication fork moving in the counterclockwise direction will stall at site *terA* and wait for the clockwise-moving replication fork. When the counterclockwise and clockwise replication forks meet, at *terA*, *terB*, or somewhere in between them, the two forks will terminate replication, releasing the two daughter DNAs.

In addition to the *ter* sequences, a protein called Tus is required for termination of chromosome replication. The Tus protein may be required to stall the replication fork at a wrong-way *ter* site.

It is intriguing that the *ter* region in *E. coli* contains very few essential genes of the bacterium. Instead, it often contains "junk DNA" such as cryptic prophages, which the cell does not normally need for growth. It is as though the *ter* sites were once together and termination occurred at a well-defined site in the DNA, but over time a lot of junk DNA was inserted between them. The situation seems to be similar in the gram-positive bacterium *Bacillus subtilis*, which also has a number of *ter* sites and a protein analogous to Tus (see Franks et al., Suggested Reading).

The significance of the termination regions for chromosome replication is unclear. The *ter* region of the *B. subtilis* chromosome can be deleted and the bacterium can still multiply. Apparently, the chromosome can terminate replication wherever the forks happen to meet, but the presence of the *ter* site normally favors termination in the *ter* region. There are probably subtle advantages to terminating chromosome replication in the *ter* region, but these advantages are not apparent in a laboratory situation.

Coordination of Cell Division with Replication of the Chromosome

Somehow the replication of the chromosome and the division of the cells must be closely coordinated, so that the cells divide only after chromosome replication is completed. Otherwise, each of the daughter cells would not get a complete copy of the DNA. How cell division and replication are coordinated is an interesting problem in bacterial molecular genetics but unfortunately is one for which there are still no clear answers, only a lot of relevant information.

SEPARATION OF THE TWO DAUGHTER DNAs AFTER REPLICATION
Once the DNA has replicated, the two daughter DNAs may still be intertwined or looped through one another like two links in a chain. If so, the two daughter DNAs must be separated, or **resolved** (by passing the double-stranded DNAs through each other), before the cell divides so that each progeny cell gets one of the daughter DNAs. This can be accomplished with the aid of topoisomerases (see below) or by recombination. Recombination is the process by which DNA molecules are broken and joined in new combinations (see chapter 9). By breaking the strands of one DNA and passing the other DNA through the breaks, the two DNA molecules can be resolved.

In *E. coli*, recombination to separate the chromosomes after replication occurs at a specific site named *dif* (see Blakely et al., Suggested Reading). Special proteins called **recombinases** act at the *dif* sequence on the two daughter chromosomes, promoting recombination between them. This special type of recombination will be discussed in detail in chapter 8.

PARTITIONING OF THE CHROMOSOME AFTER REPLICATION
After the two daughter DNAs separate, they must be faithfully segregated into the two daughter cells because each daughter cell must get one of the new copies of the chromosome to be viable. This process is called **partitioning**. The fact that so few daughter cells are born without a chromosome argues that there must be an efficient mechanism to achieve proper partitioning of the chromosome. However, this mechanism is not well understood. The bacterial membrane and/or cell wall may play a role (see the section on sequestration, below). If the newly replicating chromosomes were attached to the membrane or cell wall, as the cell grew they would be pulled apart into the daughter cells as the cell divided. In this model, the bacterial membrane or cell wall plays the role of the mitotic apparatus of eukaryotes. However, there is no direct evidence for this model. Partitioning is much better understood for plasmids and is discussed in more detail in chapter 4.

HOW IS CHROMOSOME REPLICATION COORDINATED WITH CELL DIVISION?
Two different hypotheses attempt to explain how chromosome replication is coordinated with cell division. In one, the cell waits a certain length of time after a round of chromosome replication has initiated before beginning cell division. Perhaps something generated during the initiation process sets in motion the mechanism of cell division. The other hypothesis is that the cell begins to divide as soon as a round of chromosome replication is completed. The second mechanism seems more satisfying because, if something goes wrong during the course of replication of the chromosome so that replication is delayed, the cell will not divide until the chromosome has completely replicated. That way, both daughter cells

can get a complete copy of the chromosome, as they must if multiplication is to be successful.

The putative signal announcing the end of a round of chromosome replication has never been identified. However, experiments were designed to determine the relationship between the time of chromosome replication and the cell cycle in *E. coli* (see Helmstetter and Cooper, Suggested Reading). The conclusions have never been challenged, so it is worth going over them in some detail.

These scientists recognized that if the DNA content of cells at different stages in the cell cycle could be measured, it would be possible to determine how far chromosome replication had proceeded at that time in the cell cycle. Since bacterial cells are too small to allow observation of DNA replication in a single cell, it was necessary to measure DNA replication in a large number of cells. However, cells growing in culture are all at different stages in their cell cycles. Therefore, to know how far replication had proceeded at a certain stage in the cell cycle, it was necessary to **synchronize** cells in the population so that all were the same age or point in their life cycle at the same time.

Helmstetter and Cooper accomplished this by using what they called a bacterial "baby machine." Their idea was to first label the DNA of a growing culture of bacterial cells by adding radioactively labeled nucleosides and then fix the bacterial cells on a membrane. When the cells on the filter divided, one of the two daughter cells would no longer be attached and would be released into the medium. All of the daughter cells released at a given time would be newborns and so would be the same age. This means that cells that divided to release the daughter cells at a given time would also be the same age and would have DNA in the same replication state. The amount of radioactivity in the released cells is then a measure of how much of the chromosome had replicated in cells of this age. This experiment was done under different growth conditions to show how the timing of replication and the timing of cell division are coordinated under different growth conditions.

Figure 1.19 shows the results of these experiments. For convenience, the following letters were assigned to each of the intervals during the cell cycle. The letter I denotes the time from when the last round of chromosome replication initiated until a new round begins. The letter C is the time it takes to replicate the entire chromosome, and the letter D is the time from when a round of chromosome replication is completed until cell division occurs. The top of the figure shows the relationship of I, C, and D when the cells are growing very slowly with a generation time of 70 min. Under these conditions, I is

70 min, C is 40 min, and D is 20 min. However, when the cells are growing in a richer medium and they are dividing more rapidly with a generation time of only 30 min, the pattern changes. The C and D intervals remain about the same, but the I interval is much shorter, only about 30 min.

Some conclusions could be drawn from these data. One conclusion is that the C and D intervals remain about the same independent of the growth rate. At 37°C, the time it takes the chromosome to replicate is always about 40 min, and it takes about 20 min from the time a round of replication terminates until the cell divides. However, the I interval gets shorter when the cells are growing faster and have shorter generation times. In fact, the I interval is approximately equal to the generation time—the time it takes a newborn cell to grow and divide. This makes sense because, as we shall discuss later, initiation of chromosome replication occurs every time the cells reach a certain size. They will reach this size once every generation time, independent of how fast they are growing.

Another point apparent from the data is that in cells growing rapidly with a short generation time, the I interval can be shorter than the C interval. If I is shorter than C, a new round of chromosomal DNA replication will begin before the old one is completed. This explains the higher DNA content of fast-growing cells than of slow-growing cells. It also explains the observation that genes closer to the origin of replication are present in more copies than are genes closer to the replication terminus.

Despite providing these important results, this elegant analysis does not allow us to tell whether division is coupled to initiation or termination of chromosomal DNA replication. The fact that the I interval always equals the generation time suggests that the events leading up to division are set in motion at the time a round of chromosome replication is initiated and are completed 60 min later independent of how fast the cells are growing. However, it is also possible that the act of termination of a round of chromosome replication sets in motion a cell division 20 min later. More experiments are needed to resolve these issues.

Coordination of the Cell Cycle with Replication

A new round of replication must be initiated each time the cell divides, or the amount of DNA in the cell would increase until the cells were stuffed full of it or decrease until no cell had a complete copy of the chromosome. Clearly, initiation of replication is exquisitely timed. In cells growing very rapidly, in which the next rounds of replication initiate before the last ones are completed, so that they contain a number of origins of replication, all

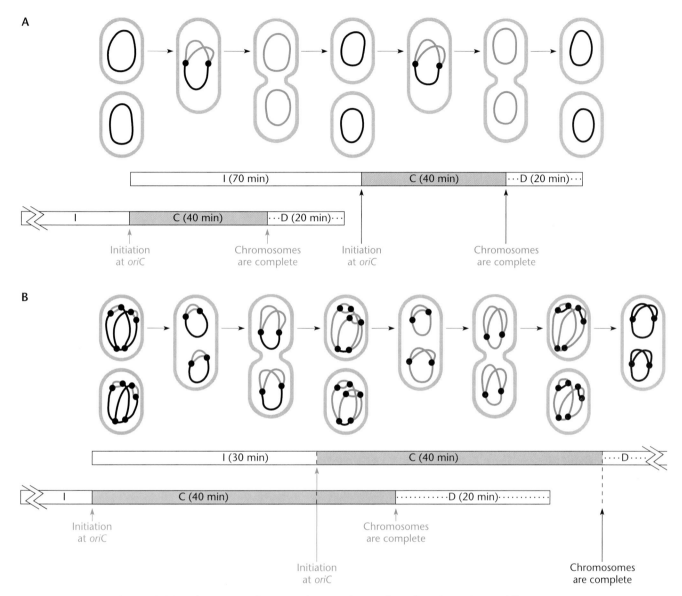

Figure 1.19   The timing of DNA replication during the cell cycle, with two different generation times. Only the time between initiations (I) changes. See the text for definitions of I, C, and D.

of the origins in the cell "fire" simultaneously, indicating tight control. In spite of its obvious interest, the regulation of initiation of chromosome replication is still not completely understood, even in *E. coli*. However, a few interesting facts point to how the coordination of cell division and the initiation of replication might be achieved.

REGULATION OF THE SYNTHESIS OF THE DnaA PROTEIN

Since the binding of the DnaA protein to the *oriC* region is the first step in initiating chromosome replication, it is tempting to speculate that a certain amount of the DnaA protein in the cell triggers the initiation of chromosome replication. According to this hypothesis, when the concentration of the DnaA protein in the cell reaches a certain level, the origins of replication will fire. Once replication begins, all the DnaA protein degrades. More DnaA protein is then synthesized until the critical concentration required to trigger a new round of replication is achieved.

Although this simple hypothesis is attractive, evidence indicates that the concentration of the DnaA protein could not be the only deciding factor in the initiation of chromosome replication. If it were, the amount of the DnaA protein per cell would be proportional to the growth rate. However, the concentration of DnaA is the same in fast- and slow-growing cells, yet fast-growing cells initiate rounds of replication more frequently. Therefore, if the concentration of the DnaA protein coordinates the initiation of chromosome replication with the cell cycle, it must do so in conjunction with other factors.

SEQUESTRATION

Dam methylase may play a role in regulating initiation by preventing the *oriC* region from starting a new replication round until the cell is ready. This action, known as **sequestration,** may be essential for the timing of chromosome replication, because immediately after initiation, conditions are optimal for new rounds of replication to begin: there is an excess of DnaA protein in the cell, and the cell is at the right stage of the cell cycle for initiation to occur. Making the *oriC* regions unavailable, or sequestered, helps delay new initiations until the proper time in the cell cycle.

We discussed the Dam methylase earlier in this chapter in connection with repair of mismatches created by replication errors. The Dam methylase methylates the two A's in the DNA sequence GATC/CTAG. The methylation occurs only after the DNA is synthesized, so that the A in a newly synthesized strand of this sequence will not be immediately methylated. Intriguingly, the sequence GATC/CTAG appears 11 times in only 245 base pairs in the *oriC* region of the chromosome, much more often than would be expected by chance alone. Furthermore, the promoter region of the *dnaA* gene, the region in which mRNA synthesis initiates for the DnaA protein, also has GATC/CTAG sequences, and no DnaA protein will be synthesized unless these sequences are fully methylated.

Figure 1.20 depicts a model of how the Dam methylase may sequester *oriC* regions after initiation. Immediately after an *oriC* region has been used to initiate replication, the GATC/CTAG sequences in the *oriC* region will be hemimethylated; only the A in the old strand of the sequence will be fully methylated. According to the model, the hemimethylated *oriC* region is sequestered by binding to the membrane, a process that renders it nonfunctional for the initiation of new rounds of replication and prevents it from being further methylated by the Dam methylase. In addition, a protein called SeqA (*seq*uestration protein A) may be involved in binding hemimethylated *oriC* regions to the membrane.

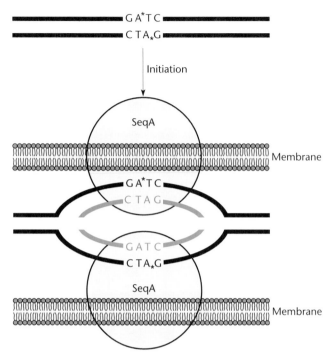

Figure 1.20 Model of the sequestration of the *oriC* region of the *E. coli* chromosome after initiation of chromosome replication. Before initiation, the *oriC* region is methylated in both strands. After initiation, only one of the two strands is methylated; hence, the region is hemimethylated. A protein called SeqA helps bind the hemimethylated *oriC* to the membrane, thereby sequestering it and preventing further initiation and methylation. The newly synthesized strand is shown in blue, and the methylated bases are starred. Only 1 of the 11 GATC/CTAG sequences in the *oriC* region is shown.

The role played by the binding of *oriC* regions to the membrane in regulating the initiation of rounds of chromosome replication is intriguing. There is longstanding speculation that the membrane of bacteria may play a role analogous to that of the mitotic spindles of eukaryotic cells in separating two daughter chromosomes after replication (see the section on partitioning, above).

There is some evidence to support the role of methylation in *oriC* sequestration and regulation of DnaA protein synthesis after initiation (see Crooke, Suggested Reading). For instance, the GATC/CTAG sequences in the *oriC* region and in the promoter region for the *dnaA* gene remain hemimethylated much longer after replication than GATC/CTAG sequences elsewhere in the chromosome. Also, increasing the amount of Dam methylase in the cell causes premature initiation of replication, as might be expected if higher than normal levels of Dam methylase fully methylated the

GATC/CTAG sequences, which would unsequester *oriC* sooner than the hemimethylated sequences would allow. Finally, hemimethylated *oriC* regions bind more readily to membranes than do fully methylated *oriC* regions. Many more details of the regulation of chromosome replication initiation remain to be uncovered.

Supercoiling

The Bacterial Nucleoid

As mentioned at the beginning of this chapter, the DNA of even a simple bacterium is approximately 1 mm long, while bacteria themselves measure only micrometers in length. Therefore, the DNA is about 1,000 times longer than the bacterium itself and must be condensed to fit in the cell, but it also must be folded in such a way that it is available for transcription, recombination, and other functions.

Figure 1.21 shows a picture of a thin section of an *E. coli* cell. The chromosome is not spread all over the cell but is condensed in only one part. Obviously, something holds the DNA so that it is not free to diffuse over the entire cell. As the chromosome replicates, this con-

densed mass becomes larger until it finally separates into two masses of DNA, just before cell division.

Condensed bacterial DNA isolated from bacteria is shown in Figure 1.22. This condensed structure is

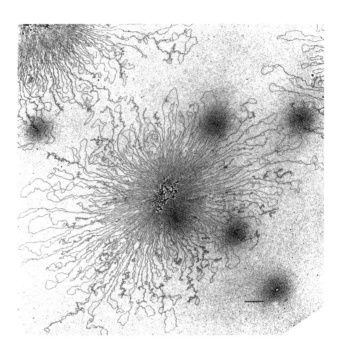

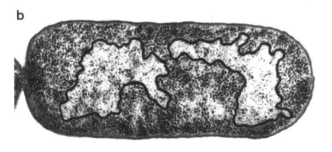

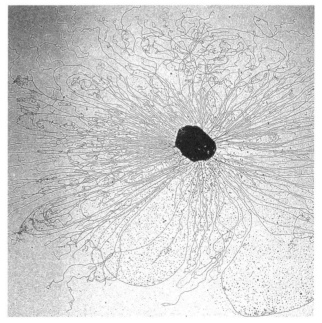

Figure 1.21   Thin section of *E. coli* showing condensed DNA. From F. C. Neidhardt, R. Curtiss III, J. L. Ingraham, E. C. C. Lin, K. B. Low, B. Magasanik, W. S. Reznikoff, M. Riley, M. Schaechter, and H. E. Umbarger (ed.), *Escherichia coli and Salmonella: Cellular and Molecular Biology*, ASM Press, Washington, D.C., 1996.

Figure 1.22   Electron micrographs of bacterial nucleoids. From F. C. Neidhardt, R. Curtiss III, J. L. Ingraham, E. C. C. Lin, K. B. Low, B. Magasanik, W. S. Reznikoff, M. Riley, M. Schaechter, and H. E. Umbarger (ed.), *Escherichia coli and Salmonella: Cellular and Molecular Biology*, ASM Press, Washington, D.C., 1996.

called the bacterial **nucleoid.** The nucleoid is composed of 30 to 50 loops of DNA emerging from a more condensed region, or **core.** Seeing this tangle of loops, it is difficult to imagine that the DNA in this complicated structure is actually one long, continuous circular molecule. Yet somehow the DNA strands are fastened periodically to the core region. Whether core attachment sites on the DNA are unique or random is not clear. However, repeated sequences called *rep* sequences have been implicated as sites at which the DNA might attach to the core of the nucleoid. These *rep* sequences are almost identical in all bacteria.

Supercoiling in the Nucleoid

One of the most noticeable features of the nucleoid is that most of the DNA loops are twisted up on themselves. This twisting is the result of **supercoiling** of the DNA.

Figure 1.23 illustrates supercoiling. In this example, the ends of a DNA molecule have been rotated in opposite directions and the DNA has become twisted up on itself to relieve the stress. The DNA will remain supercoiled as long as its ends are constrained and so cannot rotate, and a circular DNA has no free ends that can rotate. Therefore, a supercoiled circular DNA will remain supercoiled, but a linear DNA will immediately lose its supercoiling unless the ends are somehow constrained.

Even a circular DNA will lose its supercoiling if we cut one of the strands of the DNA, thereby allowing the strands to rotate around each other. The phosphodiester bond connecting the two deoxyribose sugars on the other strand will serve as a swivel and rotate, resulting in **relaxed** (i.e., not supercoiled) DNA. A DNA with a phosphodiester bond broken in one of the two strands is said to be **nicked.**

When nucleoids are prepared, some of the loops are usually relaxed, probably by nicks introduced during the extraction process. The fact that only some, and not all, of the loops of DNA in the nucleoid are relaxed tells us something about the structure of the nucleoid. A break or nick in a circular DNA should relax the whole DNA unless portions of the molecule are periodically attached to something that prevents rotation of the strands.

SUPERCOILING OF NATURAL DNAs

It is possible to estimate the supercoiling of natural DNAs. According to the Watson-Crick structure, the two strands are wrapped around each other about once every 10.5 base pairs to form the double helix. Therefore, in a DNA of 2,100 base pairs, the two strands should be wrapped around each other about 2,100/10.5 = 200 times. In a supercoiled DNA of this size, however,

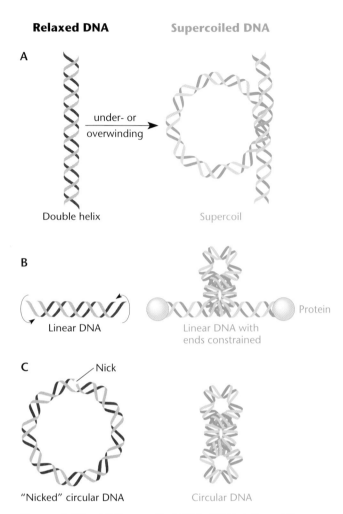

Figure 1.23 (A) Supercoiled DNA. (B) Twisting of the ends in opposite directions causes linear DNA to wrap up on itself. The supercoiling will be lost if the ends of the DNA are not somehow constrained. (C) A break, or nick, in one of the two strands of a circular DNA will relax the supercoils.

the two strands will be wrapped around each other either more or less than 200 times. If they are wrapped around each other more than once every 10.5 base pairs, the DNA is said to be **positively supercoiled;** if less than once every 10.5 base pairs, **negatively supercoiled.**

Most DNA in bacteria is negatively supercoiled, with an average of one negative supercoil for every 300 base pairs, although there are localized regions of higher or lower negative supercoiling. For example, in some regions, such as ahead of a transcribing RNA polymerase, the DNA may be positively supercoiled.

SUPERCOILING STRESS

Supercoiled DNA is under topological stress that can provide energy for some types of reactions. However, much of this stress is usually relieved. Circular chromosomes relieve stress due to negative supercoiling by twisting up on themselves, as shown in Figure 1.23; by wrapping around other objects such as proteins; or by separating their strands. Some reactions such as replication, recombination, and the activation of promoters require that the two strands of the DNA be separated; so, supercoil stress may enhance these reactions.

Topoisomerases

The supercoiling of DNA in the cell is modulated by enzymes called **topoisomerases** (see Wang, Suggested Reading) All organisms have these proteins, which manage to remove the supercoils from a circular DNA without permanently breaking either of the two strands. They perform this feat by binding to DNA, breaking one or both of the strands, and passing the DNA strands through the break before resealing it. As long as the enzyme holds the cut ends of the DNA so that they do not rotate, this process, known as **strand passage,** will either introduce or remove supercoils from DNA.

The topoisomerases are classified into two groups, type I and type II (Figure 1.24). These two types differ in how many strands are cut and how many strands pass through the cut. The type I topoisomerases cut one

strand and pass the other strand through the break before resealing the cut. The type II topoisomerases cut both strands and pass two other strands from somewhere else in the DNA through the break before resealing it. Hence, type I topoisomerases change DNA one supercoil at a time, whereas type II topoisomerases change DNA two supercoils at a time, as shown in Figure 1.24.

TYPE I TOPOISOMERASES
Bacteria have several type I topoisomerases. The major bacterial type I topoisomerase removes negative supercoils from DNA. In *E. coli* and *Salmonella typhimurium*, the gene *topA* encodes this type I topoisomerase. As expected, DNA isolated from *E. coli* with a *topA* mutation is more highly negatively supercoiled than normal.

TYPE II TOPOISOMERASES
Bacteria also have more than one type II topoisomerase. Because type II topoisomerases can break both strands and pass two other DNA strands through the break, they can either separate two linked circular DNA molecules or link them up. Linkage sometimes happens after replication or recombination. The major type II topoisomerase in bacteria is called **gyrase** instead of topoiso-

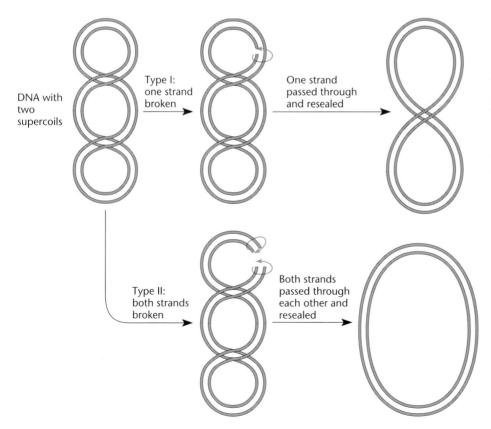

Figure 1.24 Action of the two types of topoisomerases. The type I topoisomerases break one strand of DNA and pass the other strand through the break, removing one supercoil at a time. The type II topoisomerases break both strands and pass another part of the same DNA through the breaks, introducing or removing two supercoils at a time.

DNA with two supercoils

Type I: one strand broken

One strand passed through and resealed

Type II: both strands broken

Both strands passed through each other and resealed

merase II because rather than removing negative supercoils like most type II topoisomerases, this enzyme adds them. Gyrase acts by first wrapping the DNA around itself and then cutting the two strands before passing another part of the DNA through the cuts, thereby introducing two negative supercoils. Adding negative supercoils increases the stress in the DNA and so requires energy; hence, gyrase needs ATP for this reaction.

The gyrase of *E. coli* is made up of four polypeptides, two of which are encoded by the *gyrA* gene, whereas the other two are encoded by *gyrB*. These genes were first identified by mutations that make the cell resistant to antibiotics that affect gyrase. The GyrA subunits seem to be responsible for breaking the DNA and holding it as the strands pass through the cuts. The GyrB subunits have the ATP site that furnishes the energy for the supercoiling.

OTHER FUNCTIONS OF BACTERIAL
TOPOISOMERASES
Gyrase is responsible for generating the supercoiled loops found in bacterial nucleoids. Gyrase also seems to play a more direct role in the initiation of chromosome replication and could help separate interlocked DNA molecules after replication. Hence, this enzyme is required for growth of the bacterium. Inactivating either the *gyrA* or *gyrB* gene or inactivating the gyrase enzyme with antibiotics will cause cell death.

The function of the major type I topoisomerase is less clear. As mentioned above, mutants that lack this enzyme have higher than normal levels of negative supercoiling; therefore, type I topoisomerase probably functions to relieve supercoiled stress in the DNA, particularly in front of replication forks and transcribing RNA polymerases. The enzyme may also modulate the supercoils in DNA introduced by the gyrase. As evidence, *topA* mutations alleviate the effects of mutations in *gyrA* and *gyrB* that reduce the activity of gyrase, an observation suggesting that the two types of topoisomerases cooperate to keep supercoiling from getting too high or two low.

In spite of its seemingly important role, however, the major type I topoisomerase is not absolutely required for growth, at least in *E. coli* and *S. typhimurium*. Mutations, including deletions, that inactivate the *topA* gene do not kill the cell, although they do make the bacteria grow more slowly. Presumably, the higher levels of negative supercoiling exhibited by *topA* mutants make the cells sick but do not kill them, at least in some types of bacteria.

Antibiotics That Affect Replication and DNA Structure

Antibiotics are substances that block the growth of cells. Many antibiotics are naturally synthesized chemical compounds made by soil microorganisms, especially actinomycetes, to help them compete with other soil microorganisms. Consequently, this group of compounds has a broad spectrum of activity and target specificity. Antibiotics have proven useful in understanding cellular functions as well as in disease therapy. Many stop the growth of bacteria by specifically blocking DNA replication or by changing the molecule's structure. Table 1.2 lists a few representative antibiotics that affect DNA, along with their targets in the cell and their sources. Because some parts of the replication machinery have remained relatively unchanged throughout evolution, many of these antibiotics work against essentially all types of bacteria. Some even work against eukaryotic cells and so are used as antifungal agents and in tumor chemotherapy.

Antibiotics That Block Precursor Synthesis

As discussed above, DNA is polymerized from the deoxynucleoside triphosphates. Any antibiotic that blocks the synthesis of these nucleotide precursors will block DNA replication.

INHIBITION OF DIHYDROFOLATE REDUCTASE
Some of the most important precursor synthesis blockers are antibiotics that inhibit the enzyme dihydrofolate

TABLE 1.2	Antibiotics that block replication	
Antibiotic	**Source**	**Target**
Trimethoprim	Chemically synthesized	Dihydrofolate reductase
Hydroxyurea	Chemically synthesized	Ribonucleotide reductase
5-Fluorodeoxyuridine	Chemically synthesized	Thymidylate synthetase
Nalidixic acid	Chemically synthesized	*gyrA* subunit of gyrase
Novobiocin	*Streptomyces sphaeroides*	*gyrB* subunit of gyrase
Mitomycin C	*Streptomyces caespitosus*	Cross-links DNA

reductase. One such compound, trimethoprim, works very effectively in bacteria, and the antitumor drug methotrexate (amethopterin) inhibits the dihydrofolate reductase of eukaryotes. Methotrexate is used as an antitumor agent.

These antibiotics block the synthesis of deoxythymidine monophosphate (dTMP). Inhibition of dihydrofolate reductase reduces the amount of tetrahydrofolate, which is needed in large amounts by the thymidylate synthetase enzyme. Thymidylate synthetase transfers a methyl group from tetrahydrofolate to the uracil ring of dUMP to make dTMP. Apparently, the thymidylate synthetase is the biggest user of tetrahydrofolate in the cell, because DNA replication is the only cellular function significantly affected by trimethoprim.

There is more than one mechanism by which cells can achieve trimethoprim resistance. They can have an altered dihydrofolate reductase to which trimethoprim cannot bind or can have more copies of the gene so that they make more enzyme. Some plasmids and transposons carry genes for resistance to trimethoprim. These genes encode dihydrofolate reductases that are much less sensitive to trimethoprim and so can act even in the presence of large concentrations of the antibiotic.

INHIBITION OF RIBONUCLEOTIDE REDUCTASE
The antibiotic hydroxyurea inhibits the enzyme ribonucleotide reductase, which is required for the synthesis of all four precursors of DNA synthesis (Figure 1.5). The ribonucleotide reductase catalyzes the synthesis of the deoxynucleoside diphosphates dCDP, dGDP, dADP, and dUDP from the ribonucleoside diphosphates, an essential step in deoxynucleoside triphosphate synthesis. Mutants resistant to hydroxyurea have an altered ribonucleotide reductase.

COMPETITION WITH dUMP
5-Fluorodeoxyuridine and the related 5-fluorouracil have monophosphate forms resembling dUMP, the substrate for the thymidylate synthetase. By competing with the natural substrate for this enzyme, they inhibit the synthesis of deoxythymidine monophosphate. Mutants resistant to these compounds have an altered thymidylate synthetase. These are useful antibiotics for the treatment of fungal as well as bacterial infections.

Antibiotics That Block Polymerization of Nucleotides

The polymerization of deoxynucleotide precursors into DNA would also seem to be a tempting target for antibiotics. However, there seem to be surprisingly few antibiotics that directly block this process. Most antibiotics that block polymerization do so indirectly, by binding to DNA or by mimicking the deoxynucleotides and causing chain termination, rather than by inhibiting the DNA polymerase itself.

DEOXYNUCLEOTIDE PRECURSOR MIMICS
Dideoxynucleotides are similar to the normal deoxynucleotide precursors except that they lack a hydroxyl group on the 3' carbon of the deoxynucleotide. Consequently, they can be incorporated into DNA, but then replication stops because they cannot link up with the next deoxynucleotide. These compounds are not useful antibacterial agents, probably because they are not phosphorylated well in bacterial cells. However, this property of prematurely terminating replication has made them useful for DNA sequencing by the so-called "dideoxy" method (see chapter 15).

CROSS-LINKING
Mitomycin C blocks DNA synthesis by cross-linking the guanine bases in DNA to each other. Sometimes the cross-linked bases are in opposing strands. If the two strands are attached to each other, they cannot be separated during replication. Even one cross-link in DNA that is not repaired will prevent replication of the chromosome. This antibiotic is also a useful antitumor drug, probably for the same reason.

Antibiotics That Affect DNA Structure

ACRIDINE DYES
The acridine dyes include proflavin, ethidium, and chloroquine. These compounds insert between the bases of DNA and thereby cause frameshift mutations and inhibit DNA synthesis, particularly in the kinetoplasts of trypanosomes and the mitochondria of eukaryotic cells.

Their ability to insert themselves between the bases in DNA has made acridine dyes very useful in genetics and molecular biology. We discuss some of these applications in later chapters. Members of this large family of antibiotics have long been used as antimalarial drugs.

THYMIDINE MIMIC
5-Bromodeoxyuridine (BUdR) is similar to thymidine and is efficiently phosphorylated and incorporated in its place. However, BUdR incorporated into DNA often mispairs and increases replication errors. DNA containing BUdR is also more sensitive to some wavelengths of ultraviolet (UV) light (which makes BUdR useful in enrichment schemes for isolating mutants; see chapter 14). Moreover, DNA containing BUdR has a different density from DNA containing exclusively thymidine (another feature of BUdR that is useful in experiments).

Antibiotics That Affect Gyrase

The gyrase in bacteria is a target for many different antibiotics. These antibiotics will kill the bacterial cell because gyrase is required for bacterial growth. Because this enzyme is similar among all bacteria, these antibiotics have a broad spectrum of activity and will kill many types of bacteria.

GyrA INHIBITION

Nalidixic acid specifically binds to the GyrA subunit, which is involved in cutting the DNA and in strand passage. This activity makes nalidixic acid and its many derivatives, including oxolinic acid and chloromycetin, very useful antibiotics. Another antibiotic that binds to the GyrA subunit, ciprofloxacin, is used for treating gonorrhea.

The mechanism of killing by these antibiotics is not completely understood. They are known to cause degradation of the DNA and, after mild-detergent treatment, can cause the DNA to become covalently linked to gyrase, presumably trapping it in an intermediate state in the process of strand passage. Bacteria resistant to nalidixic acid have an altered *gyrA* gene.

GyrB INHIBITION

Novobiocin and its more potent relative coumermycin bind to the GyrB subunit, which is involved in ATP binding. These antibiotics do not resemble ATP, but by binding to the gyrase, they somehow prevent ATP cleavage, perhaps by changing the conformation of the enzyme. Mutants resistant to novobiocin have an altered *gyrB* gene.

SUMMARY

1. DNA consists of two strands wrapped around each other in a double helix. Each strand consists of a chain of nucleotides held together by phosphates joining their deoxyribose sugars. Because the phosphate joins the third carbon of one sugar to the fifth carbon of the next sugar, the DNA strands have a directionality, or polarity, and have distinct 5' phosphate and 3' hydroxyl ends. The two strands of DNA are antiparallel, so that the 5' end of one is on the same end as the 3' end of the other.

2. DNA is synthesized from the precursor deoxynucleoside triphosphates by DNA polymerase. The first phosphate of each nucleotide is attached to the 3' hydroxyl of the next deoxynucleotide, giving off the terminal two phosphates to provide energy for the reaction.

3. DNA polymerases require both a primer and a template strand. The pairing of the bases between the incoming deoxynucleotide and the base on the template strand dictates which deoxynucleotide will be added at each step, with A always pairing with T and G always pairing with C. The DNA polymerase synthesizes DNA in the 5'-to-3' direction, moving in the 3'-to-5' direction on the template.

4. DNA polymerases cannot put down the first deoxynucleotide, so RNA is usually used to prime the synthesis of a new strand. Afterward, the RNA primer is removed and replaced by DNA using upstream DNA as a primer. The use of RNA primers helps reduce errors by allowing editing.

5. DNA polymerase does not synthesize DNA by itself but needs other proteins to help it replicate DNA. These other proteins are helicases that separate the strands of the DNA, ligases to join two DNA pieces together, primases to synthesize RNA primers, and other accessory proteins to keep it on the DNA and reduce errors.

6. Both strands of double-stranded DNA are usually replicated from the same end, so that the overall direction of DNA replication is from 5' to 3' on one strand and from 3' to 5' on the other strand. Because DNA polymerase can polymerize only in the 5'-to-3' direction, it must replicate one strand in short pieces and ligate these afterward to form a continuous strand.

7. The DNA in a bacterium that carries most of the genes is called the bacterial chromosome. The chromosome of most bacteria is a long, circular molecule that replicates in both directions from a unique origin of replication *oriC*. Replication of the chromosomes initiates each time the cells reach a certain size. For fast-growing cells, new rounds of replication initiate before old ones are completed. This accounts for the fact that fast-growing cells have a higher DNA content than slower-growing cells.

8. Chromosome replication terminates and the two daughter DNAs separate when the two replication forks meet on the other side of the circle. Termination usually seems to occur within a restricted region on the chromosome.

9. The chromosomal DNA of bacteria is usually one long, continuous circular molecule about 1,000 times as long as the cell itself. This long DNA is condensed in a small part of the cell called the nucleoid. In this structure, the DNA loops out of a central condensed core region. Some of these loops of DNA are negatively supercoiled. In *E. coli*, most DNAs have one supercoil about every 300 bases.

10. The enzymes that modulate DNA supercoiling in the cell are called topoisomerases. There are two types of topoisomerases in cells. Type I topoisomerases add or remove supercoils one at a time by breaking only one strand and

(continued)

SUMMARY (continued)

passing the other strand through the break. Type II topoisomerases remove supercoils two at a time by breaking both strands and passing another region of the DNA through the break. The enzyme responsible for adding the negative supercoils to DNA in bacteria is a type II topoisomerase called gyrase.

11. Some antibiotics block DNA replication or affect the structure of DNA. The most useful of these inhibit the synthesis of deoxynucleotides or inhibit the gyrase enzyme. Antibiotics that inhibit deoxynucleotide synthesis include inhibitors of dihydrofolate reductase (trimethoprim, methotrexate) and inhibitors of ribonucleotide reductase (hydroxyurea). Inhibitors of gyrase include novobiocin and nalidixic acid. Acridine dyes also affect the structure of DNA by intercalating between the bases and are used as antimalarial drugs as well as in molecular biology.

QUESTIONS FOR THOUGHT

1. Some viruses avoid the problem of lagging-strand synthesis by replicating the individual strands of the DNA in the 3'-to-5' direction simultaneously from both ends so that eventually the entire molecule will be replicated. Why don't bacterial chromosomes replicate this way?

2. Why are DNA molecules so long? Wouldn't it be easier to have lots of shorter pieces of DNA? What problems would this present for the cell?

3. Why do cells have DNA as their hereditary material instead of RNA?

4. What effect would shifting a temperature-sensitive mutant with a mutation in the *dnaA* gene for initiator protein DnaA have on the rate of DNA synthesis? Would the rate drop linearly or exponentially? Would the slope of the curve be affected by the growth rate of the cells at the time of the shift? Explain.

5. The gyrase inhibitor novobiocin inhibits the growth of almost all types of bacteria. What would you predict about the gyrase of the bacterium, *Streptomyces sphaeroides*, that makes this antibiotic? How would you test your hypothesis?

6. How do you think chromosome replication and cell division are coordinated in bacteria like *E. coli*? How would you go about testing your hypothesis?

7. Why does the terminus of chromosome replication seldom carry essential genes?

PROBLEMS

1. You are synthesizing DNA on the template 5'ACCTTACCGTAATCC3' from an upstream primer. You add three of the deoxynucleotides but leave out the fourth deoxynucleotide, deoxycytosine triphosphate, from the reaction. What DNA would you make? Draw a picture.

2. You are synthesizing DNA from the same template and with the same upstream primer, but instead of just deoxythymidine triphosphate you add an equal mixture of the inhibitor of replication, dideoxythymidine triphosphate, and the normal nucleotide deoxythymidine triphosphate in addition to the other three deoxynucleoside triphosphates. What DNAs would you make? Draw a picture.

3. You are growing *E. coli* with a generation time of only 25 min. How long will the I periods, C periods, and D periods be? Draw a picture showing when the various events occur during the cell cycle.

4. You are growing *E. coli* with a generation time of 90 min. Now how long will the I, C, and D periods be? Draw a picture.

5. You are measuring the supercoiling of the nucleoid from a *topA* mutant of *E. coli* which lacks the major type I topoisomerase and comparing it with the normal *E. coli* (without the *topA* mutation). Would you expect there to be more or fewer negative supercoils in the nucleoid of the mutant? Why?

SUGGESTED READING

Blakely, G., G. May, R. McCulloch, L. K. Arciszewska, M. Burke, S. T. Lovett, and D. J. Sherratt. 1993. Two related recombinases are required for site-specific recombination at *dif* and *cer* in *E. coli*. *Cell* 75:351–361.

Crooke, E. 1995. Regulation of chromosome replication in *E. coli*: sequestration and beyond. *Cell* 82:877–880. (Minireview.)

Franks, A. H., A. A. Griffiths, and R. G. Wake. 1995. Identification and characterization of new DNA replication terminators in *Bacillus subtilis*. *Mol. Microbiol*. 17:13–23.

Helmstetter, C. E., and S. Cooper. 1968. DNA synthesis during the division cycle of rapidly growing *Escherichia coli* B/r. *J. Mol. Biol*. 31:507–518.

Hill, T. M. 1996. Features of the chromosomal terminus region, p. 1602–1614. *In* F. C. Neidhardt, R. Curtiss III, J. L. Ingraham, E. C. C. Lin, K. B. Low, B. Magasanik, W. S. Reznikoff, M. Riley, M. Schaechter, and H. E. Umbarger (ed.), *Escherichia coli and Salmonella: Cellular and Molecular Biology*, 2nd ed. ASM Press, Washington, D.C.

Hill, T. M., J. M. Henson, and P. L. Kuempel. 1987. The terminus region of the *Escherichia coli* chromosome contains two separate loci that exhibit polar inhibition of replication. *Proc. Natl. Acad. Sci. USA* **84:**1754–1758.

Kornberg, A., and T. Baker. 1992. *DNA Replication*, 2nd ed. W. H. Freeman and Co., New York.

Langer, U., S. Richter, A. Roth, C. Weigel, and W. Messer. 1996. A comprehensive set of DnaA-box mutations in the replication origin, *oriC*, of *Escherichia coli. Mol. Microbiol.* **21:**301–311.

Linn, S. 1996. The DNases, topoisomerases, and helicases of *Escherichia coli* and *Salmonella*, p. 764–772. *In* F. C. Neidhardt, R. Curtiss III, J. L. Ingraham, E. C. C. Lin, K. B. Low, B. Magasanik, W. S. Reznikoff, M. Riley, M. Schaechter, and H. E. Umbarger (ed.), *Escherichia coli and Salmonella: Cellular and Molecular Biology*, 2nd ed. ASM Press, Washington, D.C.

Marians, K. 1996. Replication fork propagation, p. 749–763. *In* F. C. Neidhardt, R. Curtiss III, J. L. Ingraham, E. C. C. Lin, K. B. Low, B. Magasanik, W. S. Reznikoff, M. Riley, M. Schaechter, and H. E. Umbarger (ed.), *Escherichia coli and Salmonella: Cellular and Molecular Biology*, 2nd ed. ASM Press, Washington, D.C.

Marinus, M. G. 1996. Methylation of DNA, p. 782–791. *In* F. C. Neidhardt, R. Curtiss III, J. L. Ingraham, E. C. C. Lin, K. B. Low, B. Magasanik, W. S. Reznikoff, M. Riley, M. Schaechter, and H. E. Umbarger (ed.), *Escherichia coli and Salmonella: Cellular and Molecular Biology*, 2nd ed. ASM Press, Washington, D.C.

Messer, W. and C. Weigel. 1996. Initiation of chromosome replication, p. 1579–1601. *In* F. C. Neidhardt, R. Curtiss III, J. L. Ingraham, E. C. C. Lin, K. B. Low, B. Magasanik, W. S. Reznikoff, M. Riley, M. Schaechter, and H. E. Umbarger (ed.), *Escherichia coli and Salmonella: Cellular and Molecular Biology*, 2nd ed. ASM Press, Washington, D.C.

Nordstrom, K., and S. J. Austin. 1993. Cell-cycle-specific initiation of replication. *Mol. Microbiol.* **10:**457–463. (MicroReview.)

Olby, R. 1974. *The Path to the Double Helix*. Macmillan Press, London.

Wang, J. C. 1985. DNA topoisomerases. *Annu. Rev. Biochem.* **54:**665–698.

Watson, J. D. 1968. *The Double Helix*. Atheneum, New York.

Zyskind, J. W., and D. W. Smith. 1993. DNA replication, the bacterial cell cycle and cell growth. *Cell* **69:**5–8. (Minireview.)

Chapter 2

Introduction to Macromolecular Synthesis: Gene Expression

U NCOVERING THE MECHANISM of protein synthesis, and therefore of gene expression, was one of the most dramatic accomplishments in the history of science. The process of protein synthesis is sometimes called the **Central Dogma** of molecular biology, which states that information in DNA is copied into RNA to be translated into protein. We now know of many exceptions to the Central Dogma. For example, information does not always flow from DNA to RNA but sometimes in the reverse direction, from RNA to DNA. The information in RNA is often changed after it has been copied from the DNA. Moreover, the information in DNA may be translated differently depending on where it is in a gene. Despite these exceptions, however, the basic principles of the Central Dogma remain sound.

This chapter outlines the process of protein synthesis and gene expression. The discussion is meant to be only a broad overview but one that is sufficient to understand the chapters that follow. For more detailed treatments, consult any modern biochemistry textbook.

Overview

DNA carries the information for the synthesis of RNA and proteins in regions called **genes**. The first step in expressing a gene is to **transcribe**, or copy, an RNA from one strand in that region. The word *transcription* is descriptive because the RNA is copied in the same language as DNA, a language written in a sequence of nucleotides. If the gene carries information for a protein, this RNA transcript is called **messenger RNA (mRNA)**. An mRNA is a messenger because it carries the gene's message to a **ribosome**. Once on the ribosome, the information in the mRNA can be **translated** into the protein. *Translation* is another descriptive word because one language—the sequence of nucleotides in DNA and RNA—is interpreted into a different language—a sequence of amino acids in a protein. The mRNA is translated as it moves along the ribosome, three nucleotides at a time. Each three-nucleotide sequence, called a **codon,** carries information for a specific amino acid. The assignment of each of the possible codons to amino acids is called the genetic code and is shown in Table 2.1 (also see below).

The actual translation from the language of nucleotide to amino acid sequences is performed by small RNAs called transfer RNAs (tRNAs) and enzymes called aminoacyl-tRNA synthetases. The enzymes attach specific amino acids to each tRNA. Then a tRNA specifically pairs with a codon in the mRNA as it moves through the ribosome, and the amino acid it carries is inserted into the growing protein. The tRNA pairs with the codon in the mRNA through a complementary three-nucleotide sequence called the **anticodon.** The base-pairing rules for codons and anticodons are basically the same as the base-pairing rules for DNA replication, except that RNA has uracil (U) rather than thymine (T) and the pairing between the last of the three bases in the codon and the anticodon is less stringent.

This basic outline of gene expression leaves many important questions unanswered. How does mRNA synthesis begin and end at the correct places and on the correct strand in the DNA? Similarly, how does transla-

tion start and stop at the correct places on the mRNA? What actually happens to the tRNA and ribosomes during translation? The answers to these questions and many others are important for the interpretation of genetic experiments, so we will discuss the structure of RNA and proteins and the processes by which they are synthesized in much more detail.

The Structure and Function of RNA

In this section, we review the basic components of RNA and how it is synthesized. We also review how structure varies among different types of cellular RNAs and the role each type plays in cellular processes.

Types of RNA

There are many different types of RNA in cells. Some of these, including mRNA, ribosomal RNA (rRNA), and tRNA, are involved in protein synthesis. Each of these types of RNAs has special properties that we discuss later in the section. Others are involved in regulation and replication.

RNA Precursors

RNA is similar to DNA in that it is composed of a chain of nucleotides. However, RNA nucleotides contain the sugar ribose instead of deoxyribose. These five-carbon sugars differ in the second carbon, which is attached to a hydroxyl group in ribose rather than the hydrogen

TABLE 2.1	The genetic code				
First position (5' end)	**Second position**				**Third position (3' end)**
	U	**C**	**A**	**G**	
U	Phe	Ser	Tyr	Cys	U
	Phe	Ser	Tyr	Cys	C
	Leu	Ser	Stop	Stop	A
	Leu	Ser	Stop	Trp	G
C	Leu	Pro	His	Arg	U
	Leu	Pro	His	Arg	C
	Leu	Pro	Gin	Arg	A
	Leu	Pro	Gin	Arg	G
A	Ile	Thr	Asn	Ser	U
	Ile	Thr	Asn	Ser	C
	Ile	Thr	Lys	Arg	A
	Met	Thr	Lys	Arg	G
G	Val	Ala	Asp	Gly	U
	Val	Ala	Asp	Gly	C
	Val	Ala	Glu	Gly	A
	Val	Ala	Glu	Gly	G

found in deoxyribose (see chapter 1). Figure 2.1A shows the structure of a **ribonucleoside triphosphate,** so named because of the different sugar.

The only other difference between RNA and DNA chains when they are first synthesized is in the bases. Three of the bases—adenine, guanine, and cytosine—are the same, but RNA has uracil instead of the thymine found in DNA (Figure 2.1B). The RNA bases can also be modified more extensively, as we discuss later.

Figure 2.1C shows the basic structure of an RNA polynucleotide chain. As in DNA, RNA nucleotides are held together by phosphates that join the 5′ carbon of one ribose sugar to the 3′ carbon of the next. This arrangement ensures that, as with DNA chains, the two ends of an RNA polynucleotide chain will be different from each other, with the 5′ end terminating in a phosphate and the 3′ end terminating in a hydroxyl. When it is first synthesized, the 5′ end of an RNA chain has three phosphates attached to it, although two of the phosphates are usually removed soon after.

According to convention, the sequence of bases in RNA is given from the 5′ to the 3′ end, which is actually the direction in which the phosphates are attached 3′ to 5′, 3′ to 5′, etc. Also, by convention, regions in RNA that are closer to the 5′ end in a given sequence are referred to as **upstream** and regions that are 3′ as **downstream** because RNA is both made and translated in the 5′-to-3′ direction.

RNA Structure

Except for the sequence of bases and minor differences in the pitch of the helix, little distinguishes one DNA molecule from another. However, RNA chains generally have more structural properties than DNA, tending to be folded into complex structures, and can be extensively modified.

PRIMARY STRUCTURE

All RNA is created equal. No matter what their function, all RNA transcripts are made the same way, from a DNA template. Only the sequences of their nucleotides and their lengths are different. The sequence of nucleotides in RNA is the **primary structure** of the RNA. In some cases, the primary structure of an RNA is changed after it is transcribed from the DNA (see the section on processing and modification, below).

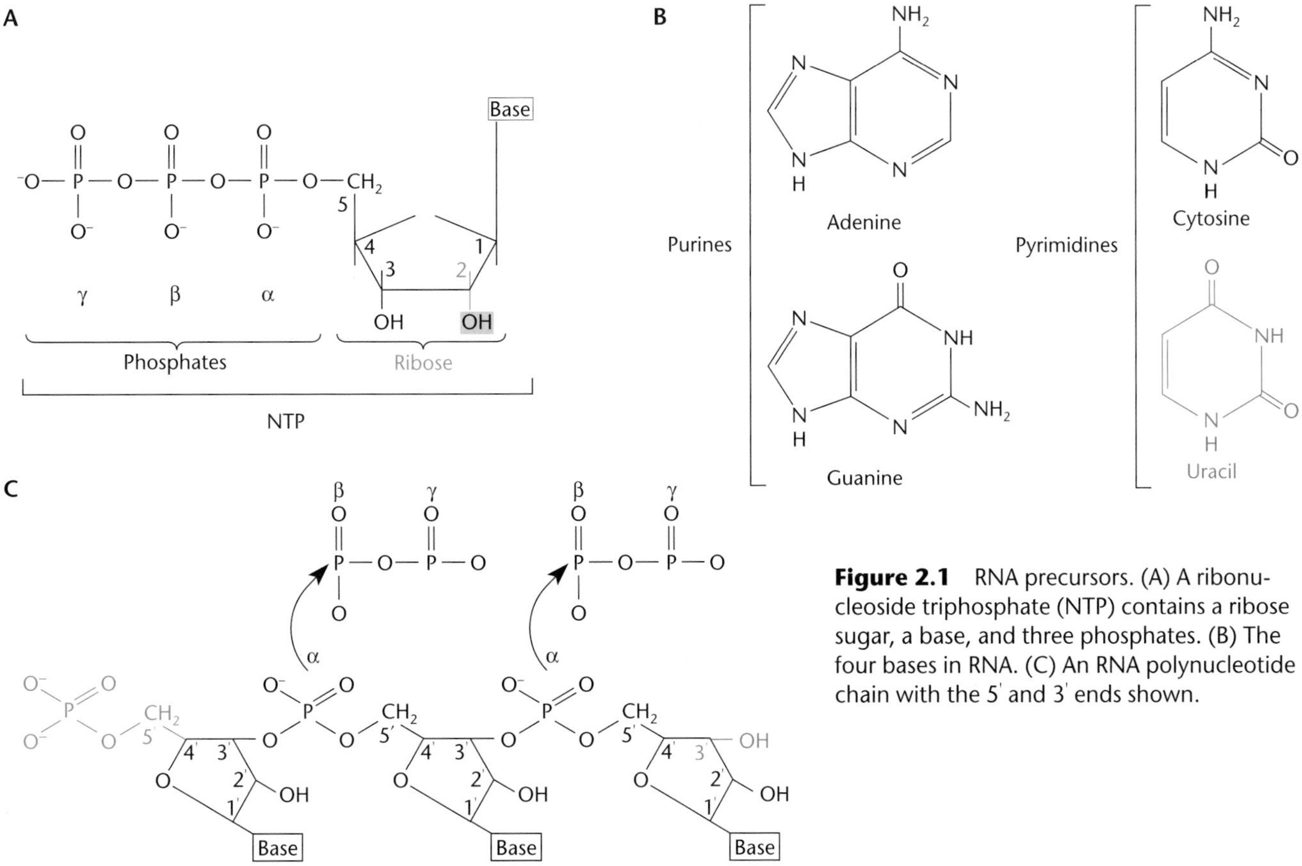

Figure 2.1 RNA precursors. (A) A ribonucleoside triphosphate (NTP) contains a ribose sugar, a base, and three phosphates. (B) The four bases in RNA. (C) An RNA polynucleotide chain with the 5′ and 3′ ends shown.

BOX 2.1

RNA World

In evolutionary biology, there is the long-standing argument about which came first, the chicken or the egg? In more modern molecular terms, this question has been rephrased as which came first: RNA, DNA, or protein? Were the most primitive self-replicating molecules, which presumably gave rise to living organisms, DNA molecules, RNA molecules, or protein molecules? Strong arguments can be made that RNA came first, that it was the primordial molecule. First, both RNA and DNA are more likely candidates than proteins for the primordial molecule, since they carry the information to duplicate themselves through complementary base pairing, even though neither can actually duplicate itself without the help of enzymes such as DNA and RNA polymerases. Of the two nucleic acids, the arguments seem to favor the RNA because these molecules can also be enzymes, or "ribozymes," and catalyze chemical reactions. The first RNAs that were shown to be enzymes were self-splicing introns and the RNA component of RNase P, which is involved in processing rRNAs and tRNAs. Furthermore, as part of the ribosomes, RNAs are essential components of the translational machinery involved in making proteins. Most convincing, however, is the recent demonstration that the 23S rRNA (28S rRNA in eukaryotes) is actually the enzyme peptidyltransferase, which makes peptide bonds.

Hence, it seems most likely that when life began, the 23S RNA began to make proteins to help it with the task of replicating itself, and DNA was made only later to store the information needed to make those proteins. This argument, carried to its extreme, suggests that life on Earth, as we know it, exists so that 23S rRNA can duplicate itself!

References

Noller, H. F. 1993. Peptidyltransferase: protein, ribonucleoprotein, or RNA? *J. Bacteriol.* **175**:5297–5300. (Minireview.)

Weiner, A. M., and N. Maizels. 1987. tRNA-like structures tag the 3' ends of genomic RNA molecules for replication: implications for the origins of protein synthesis. *Proc. Natl. Acad. Sci. USA* **84**:7383–7387.

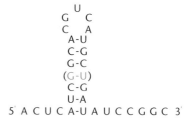

Figure 2.2 Secondary structure in an RNA. The RNA folds back on itself to form a hairpin loop. The presence of a G:U pair (in parentheses) does not disrupt the structure.

Figure 2.2 shows an example of RNA secondary structure, in which the sequence 5'AUCGGCA3' has paired with the complementary sequence 5'UGCU-GAU3' somewhere else in the molecule. As in double-stranded DNA, the paired strands of RNA are antiparallel; i.e., pairing occurs only when the two sequences are complementary when read in opposite directions (5' to 3' and 3' to 5'). However, the pairing rules for double-stranded RNA are slightly different from the pairing rules for DNA. In RNA, guanine can pair with uracil as well as with cytosine. Because these GU pairs do not share hydrogen bonds, they do not contribute to the stability of the double-stranded RNA. Thus, the GU pair shown in Figure 2.2 does not disrupt the pairing of the others, although it does not help hold the structure together.

Each base pair that forms in the RNA makes the secondary structure of the RNA more stable. Consequently, the RNA will generally fold so that the greatest number of continuous base pairs can form. The stability of a structure can be predicted by adding up the energy of all its hydrogen bonds that contribute to the structure. By eye, it is very difficult to predict what regions of a long RNA will pair to give the most stable structure. However, computer software is available that, given the sequence of bases (primary structure) of the RNA, will predict the most stable secondary structure.

TERTIARY STRUCTURE

Double-stranded regions of RNAs created by base pairing are stiffer than single-stranded regions. As a result, an RNA that has secondary structure will have a more rigid shape than one without double strands. Also, the intermingled paired regions cause the RNA to fold back on itself extensively. Together, these effects give many RNAs a well-defined three-dimensional shape, called its **tertiary structure.** Proteins or other cellular constituents recognize RNA forms by their tertiary structure, which also gives ribozymes their enzymatic activity (see Box 2.1).

SECONDARY STRUCTURE

Unlike DNA, RNA is usually single stranded. However, pairing between the bases in some regions of the molecule may cause it to fold up on itself to form a double-stranded region. Such double-stranded regions are called the **secondary structure** of the RNA. All RNAs, including mRNAs, probably have extensive secondary structure.

RNA Processing and Modification

The alterations to the RNA molecule caused by secondary and tertiary structure are examples of **noncovalent changes**, because only hydrogen bonds, not chemical (**covalent**) bonds, are formed or broken. However, once the RNA is synthesized, covalent changes can occur during **RNA processing** and **RNA modification**.

RNA processing involves forming or breaking phosphate bonds in the RNA after it is made. For example, the terminal phosphates at the 5' end may be removed, or the RNA may be cut into smaller pieces and even religated into new combinations, requiring the breaking and making of many phosphate bonds.

RNA modification, by contrast, involves altering the bases or sugars of RNA. Examples include methylation of the bases or sugars of rRNA and enzymatic alteration of the bases of tRNA. In eukaryotes, "caps" of inverted methylated nucleotides are added to the 5' ends of some types of mRNA. In bacteria, mRNAs are not capped, and only the stable rRNAs and tRNAs are extensively modified.

PROCESSING AND MODIFICATION OF tRNA

The tRNAs are probably the most highly processed and modified RNAs in cells. Figure 2.3 shows a "mature" tRNA that was originally cut out of a much longer molecule that may also have included the rRNAs. Then, some of the bases were modified by specific enzymes, creating altered bases such as pseudouracil and thiouracil. Finally, an enzyme called CCA transferase added the sequence CCA to the 3' end. Clearly, much had to be done to this molecule after it was synthesized so that it could become a functional, mature tRNA.

Transcription

Transcription is the synthesis of RNA on a DNA template. The process of transcription is probably fairly similar in all organisms, but it is best understood in bacteria.

Bacterial RNA Polymerase

The transcription of DNA into RNA is the work of **RNA polymerase**. In bacteria, the same RNA polymerase makes all the cellular RNAs, including rRNA, tRNA, and mRNA. Only the primer RNAs of Okazaki fragments are made by a different RNA polymerase. In contrast, eukaryotes have three nuclear RNA polymerases, as well as a mitochondrial RNA polymerase, which make their cellular RNAs.

Figure 2.4 shows a schematic structure of a typical bacterial RNA polymerase, which has five subunits and

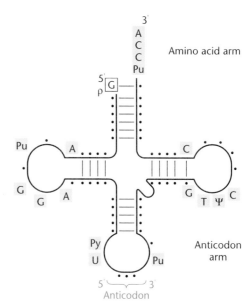

Figure 2.3   The structure of mature tRNAs, showing secondary structure and modifications. Shown is the base pairing that holds the molecule together and some of the standard modifications. Ψ is the modified base pseudouracil. tRNAs also contain thiouracil (T) and methylated bases such as thymine among other modifications. Boxed letters represent the bases found in the same places in all tRNAs. If a pyrimidine or purine is usually at a position, it is designated Py or Pu, respectively. The CCA end, where an amino acid is attached, is boxed.

a molecular weight of more than 400,000, making it one of the largest bacterial enzymes. As shown in Figure 2.4, the five subunits of the RNA polymerase holoenzyme are two identical α subunits, two very large subunits called β and β', and the σ factor. The α, β, and β' subunits are permanent parts of the RNA polymerase, but the σ factor is required only for initiation and cycles off the enzyme after initiation of transcription. Without the σ factor, the RNA polymerase is called the core enzyme. Eukaryotic and archaeal RNA polymerases have many more subunits and are much more complex.

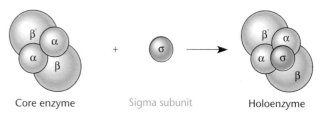

Figure 2.4   The composition of a typical bacterial RNA polymerase. The core enzyme contains two α subunits, a β subunit, and a β' subunit. The fifth subunit, the σ factor, cycles off after initiation of RNA synthesis.

A

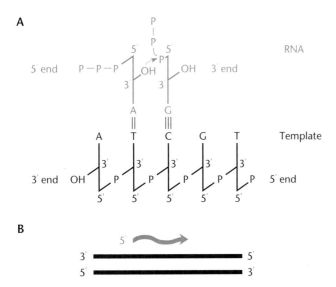

B

Figure 2.5 RNA transcription. (A) The polymerization reaction in which the newly synthesized RNA pairs with the template strand of DNA during transcription. (B) RNA polymerase synthesizes RNA in the 5'-to-3' direction, moving 3' to 5' on the template. RNA is shown as a wavy line, and both strands of DNA are shown as straight lines.

Transcription Initiation

Much like DNA polymerase (see chapter 1), the RNA polymerase makes a complementary copy of a DNA template, building a chain of RNA by attaching the 5' phosphate of a ribonucleotide to the 3' hydroxyl of the one preceding it (Figure 2.5). However, in contrast to DNA polymerase, RNA polymerases do not need a preexisting primer to initiate the synthesis of a new chain of RNA.

PROMOTERS

RNA transcripts are copied from only selected regions of the DNA, rather than from the whole molecule, so the RNA polymerase must start making an RNA chain from a double-stranded DNA only at certain sites. These DNA regions are called **promoters,** and the RNA polymerase will recognize a particular T or C in the promoter region as a **start point** of transcription. Thus, usually the first base in the chain is an A or a G laid down opposite to a T or C, respectively (Figure 2.5).

The RNA polymerase recognizes different types of promoters on the basis of which type of σ factor is attached. The most common promoters are those recognized by the RNA polymerase with σ⁷⁰. The σ factors are named for their size, and this one has a molecular weight of 70,000 (or a molecular mass of 70 kDa).

Even promoters of the same type are not identical to each other, but they do share certain sequences known as **consensus sequences** by which they can be distinguished. Figure 2.6 shows the consensus sequence of the σ⁷⁰ promoter in *E. coli*. The promoter sequence has two important regions: a short AT-rich region about 10 base pairs upstream of the transcription start site, known as the **–10 sequence,** and a region about 35 base pairs upstream of the start site, called the **–35 sequence.** RNA polymerase must bind to both sequences to start transcription.

The Polymerization Reaction

To begin transcription, the RNA polymerase binds to the promoter sequence and separates the strands of the DNA, exposing the bases. Unlike DNA polymerase, which requires helicases to separate the strands, the RNA polymerase can complete this step by itself. Then a ribonucleoside triphosphate complementary to the nucleotide at the transcription start site bonds to the template. The second ribonucleoside triphosphate comes in, pairing with the next complementary base in the DNA template. RNA polymerase catalyzes the reaction in which the α phosphate of the second nucleotide joins with the 3' hydroxyl of the first nucleotide. Then the

Figure 2.6 (A) Typical structure of a σ⁷⁰ bacterial promoter with –35 and –10 regions. (B) The consensus sequences of such promoters. (C) The relationship of these sequences with respect to the start site of transcription on the two strands. RNA synthesis typically starts with an A or a G, and no primer is required. N, any nucleotide.

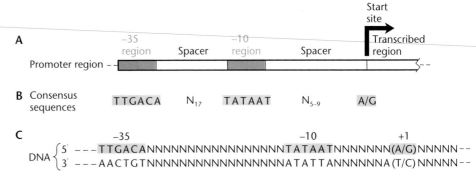

A

DNA { 5′ A T T A C G A C C T A C G C A T 3′   Coding strand
       3′ T A A T G C T G G A T G C G T A 5′   Template strand
RNA   5′ A U U A C G A C C U A C G C A U 3′

B

DNA { 5′ ————————————————— 3′
       3′ ————————————————— 5′
RNA   5′ ～～～～～～～～～→

Figure 2.7   The coding strand has the same sequence as the mRNA. The template strand is the strand to which the mRNA is complementary if read in the 3′-to-5′ direction.

third nucleotide comes in and bonds to the second, and so forth.

The template strand of DNA that is copied is called the **transcribed strand**. The other strand, which has the same sequence as the RNA copy, is called the **coding strand**. As shown in Figure 2.7, the sequence of a gene is usually written as the sequence of the coding strand.

As the RNA chain begins to grow, the RNA polymerase holoenzyme releases its σ factor, and the four-subunit enzyme continues moving along the DNA strand in the 3′-to-5′ direction, polymerizing RNA in the 5′-to-3′ direction. Inside an opening approximately 18 bases long, called a **transcription bubble,** the elongating RNA and the complementary strand of DNA pair with each other to form a DNA-RNA hybrid, which has a double-helix structure similar to that of a double-stranded DNA molecule. At the end of the polymerization process, an antiparallel complementary copy, or **transcript,** has been made on one strand of the DNA in a particular region. Figure 2.8 shows an overview of the transcription process.

Transcription Termination

Once the RNA polymerase has initiated transcription at a promoter, it will continue along the DNA, polymerizing ribonucleotides, until it encounters a transcription termination site in the DNA. These sites are not necessarily at the end of individual genes. In bacteria, more than one gene is often transcribed into a single RNA, so that a transcription termination site will not occur until the end of the cluster of genes being transcribed. Even if only a single gene is being transcribed, the transcription termination site may occur far downstream of the gene.

Bacterial DNA has two basic types of transcription termination sites: **factor independent** and **factor dependent**. As their names imply, these types are distinguished by whether they work with just RNA polymerase and DNA alone or need other factors before they can terminate transcription.

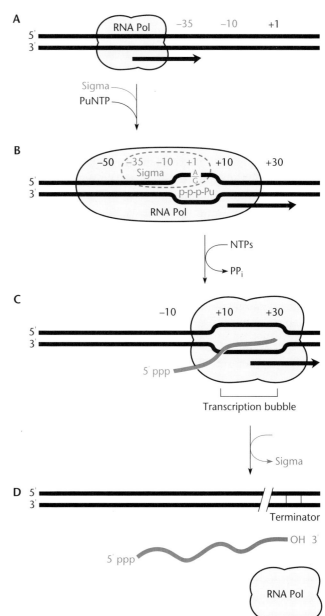

Figure 2.8   Transcription begins at a promoter and ends at a transcription terminator. (A and B) RNA polymerase (Pol) moves along the DNA until it recognizes a promoter. Transcription begins when the strands of DNA are opened at the promoter. (C) As the RNA polymerase moves along the DNA, polymerizing ribonucleotides, it forms a transcription bubble containing an RNA-DNA double-stranded hybrid, which helps hold the RNA polymerase on the DNA. The sigma factor is released. (D) The RNA polymerase encounters a transcription terminator and comes off the DNA, releasing the newly synthesized RNA.

FACTOR-INDEPENDENT TERMINATION

The factor-independent transcription terminators are easy to recognize because they share similar properties. As shown in Figure 2.9A, a typical factor-independent terminator sequence consists of two sequences. The first is an inverted repeat, which is one sequence followed by another that is similar when read in the opposite direction on the other strand, as though the sequence were turned around, or "inverted." The inverted repeat is followed by a short string of A's. Transcription then terminates somewhere in the string of A's in the DNA, leaving a string of U's at the end of the RNA.

Figure 2.9 also shows how a factor-independent transcription terminator might work. When an inverted repeat is transcribed into RNA, the RNA can form a secondary structure called a **hairpin** (Figure 2.9B). According to this model, the RNA hairpin formed as a result of transcription of the inverted repeat can destroy the RNA-DNA hybrid in the transcription bubble that helps hold the RNA polymerase on the DNA. The RNA-RNA hairpin may be more stable than the DNA-RNA hybrid in the transcription bubble and therefore may pull the RNA from the RNA-DNA hybrid. Then, when this destabilized RNA polymerase hits the AT-rich region, the AU base pairs that form are less stable than the AT base pairs in the original DNA, so the RNA polymerase and RNA spontaneously fall off the DNA, terminating transcription.

FACTOR-DEPENDENT TERMINATION

The factor-dependent transcription terminators have very little sequence in common with each other and so are not readily apparent. *E. coli*, in which factor-dependent termination is best understood, has three transcription termination factors, Rho (ρ), Tau (τ), and NusA. Since τ and NusA are not as specific and their role is less well understood, we shall concentrate on the ρ factor.

Any model for how the ρ factor terminates transcription at ρ-dependent termination sites has to incorporate the following facts about ρ-dependent termination. First, ρ usually causes the termination of RNA synthesis only if the RNA is not being translated. In bacteria, which lack a nuclear membrane, transcription and translation can occur at the same time, so this is an important criterion (see Introduction). Second, ρ is an RNA-dependent ATPase, so it will cleave ATP to get energy, but this will occur only if RNA is present. Finally, ρ is also an RNA-DNA helicase. It is similar to the DNA helicases that separate the strands of DNA during replication, but it will unwind only a double helix with RNA in one strand and DNA in the other.

Figure 2.10 illustrates a current model for how ρ terminates transcription: First, ρ binds to certain sequences in the RNA before the RNA polymerase reaches a ρ-dependent termination site. These sequences, called *rut* for *rho utilization* sites, are not well characterized but have a lot of C's and not much secondary structure. If that portion of the RNA is being translated, the ρ protein cannot bind to the *rut* sequences because the ribosomes get in the way. However, if a ribosome is not translating, the ρ protein can bind to the RNA; it then moves along it in the 5'-to-3' direction, chasing the RNA polymerase. This movement requires the energy that ρ obtains from cleaving ATP. If ρ catches up with the RNA polymerase because the enzyme pauses, or stalls, at the termination site, the RNA-DNA helicase activity of ρ will unwind

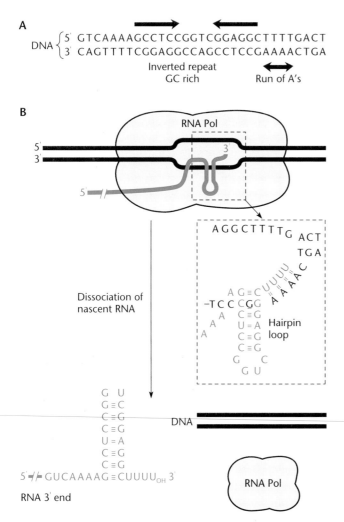

Figure 2.9 Transcription termination at a factor-independent termination site (A). (B) The RNA hairpin loop forms because the inverted repeat in the RNA is more stable than the DNA-RNA hybrid in the transcription bubble. Loss of the transcription bubble destabilizes the polymerase on the DNA, causing it to come off at a string of A's in the template.

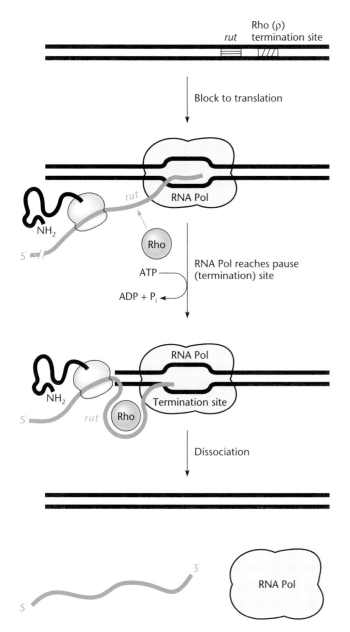

Figure 2.10   A model for factor-dependent transcription termination at a ρ termination site. The ρ-factor attaches to the mRNA at a *rut* site if the mRNA is not being translated and then chases the RNA polymerase along the mRNA until it catches up when the RNA polymerase pauses at a ρ-termination site. The ρ factor then dissociates the RNA-DNA hybrid in the transcription bubble, causing the RNA polymerase to come off the DNA.

the RNA-DNA hybrid, causing the RNA polymerase to be released from the DNA and transcription to terminate. The coupling of transcription termination to translation blockage ensures that after translation terminates at the end of an RNA transcript, transcription

of the gene will stop as soon as the next ρ-dependent transcription signal is encountered. This type of ρ-dependent termination occurs at the end of transcribed regions and also accounts for polarity (see the section on polarity below).

According to these models, the basic difference between a factor-independent termination site and a ρ-dependent one is the dissociation of the RNA-DNA hybrid in the transcription bubble. The factor-independent site has a symmetric sequence that results in an RNA-RNA loop that is more stable than the alternative RNA-DNA hybrid and so prevents it from forming, whereas at the ρ-dependent site, the ρ protein dissociates the RNA-DNA hybrid.

rRNAs and tRNAs and Their Synthesis

Transcription of the genes for all the RNAs of the cell is basically the same. However, rRNAs and tRNAs have special roles in protein synthesis, so their fate after transcription differs from that of mRNAs.

The ribosomes are some of the largest organelles in bacterial cells and are composed of both proteins and RNA. Bacterial ribosomes contain three types of rRNA: 16S, 23S, and 5S. These names derive from the sedimentation rates of these molecules, that is, their rate of movement toward the bottom of the tube in an ultracentrifuge. In general, the higher the S value, the larger the RNA.

In addition to their structural role in the ribosome, the rRNAs may play a more direct role in translation. The 23S rRNA is the peptidyltransferase enzyme, which joins amino acids into protein on the ribosome, making it a ribozyme (see Box 2.1). The 16S RNA is directly involved in both initiation and termination of translation.

The rRNAs and tRNAs are the most common RNAs in cells for two reasons. In a rapidly growing bacterial cell, about half of the total RNA synthesis is devoted to making these RNAs. Also, the rRNAs and tRNAs are far more stable than mRNA. They are not usually degraded until a long time after they are synthesized. With this combination of high synthesis rate and high stability, the rRNAs and tRNAs together amount to more than 95% of the total RNA in a bacterial cell.

Not only do the rRNAs physically associate in the ribosome, but also they are synthesized together as long precursor RNAs containing all three species, or forms, of rRNAs. The precursors often contain one or more tRNAs as well (Figure 2.11). After the precursor RNA is synthesized, the individual rRNAs and tRNAs are cut from it. At some point during the processing, the RNAs are modified to make the mature rRNAs and tRNAs.

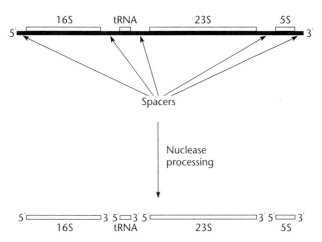

Figure 2.11 The precursor of rRNA. The long molecule contains the 16S, 23S, and 5S rRNAs, as well as one or more tRNAs. Nucleases cut the individual RNAs out of the long precursor after it is synthesized.

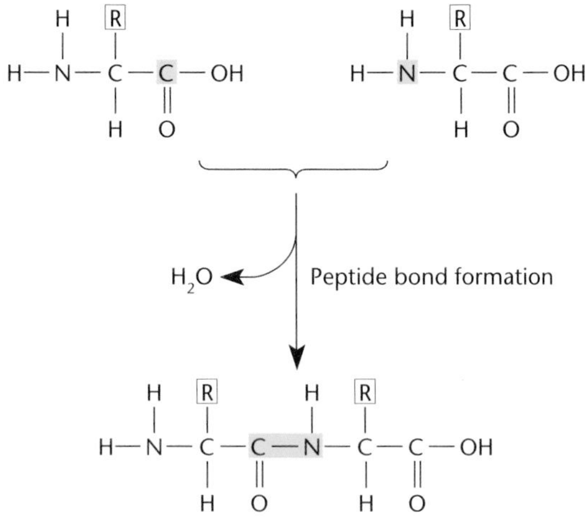

Figure 2.12 Two amino acids joined by a peptide bond. The bond connects the amino group on one amino acid to the carboxylic acid (carboxy) group on the one preceding it. R is the side group of the amino acid that differs in each type of amino acid.

Ribosomes are the site of protein synthesis; therefore, cells can increase their growth rate by increasing their number of ribosomes, because the faster a cell makes proteins, the faster it grows. In many bacteria, the coding sequences for the rRNAs are repeated in 7 to 10 different places around the genome. Duplication of these genes leads to higher rates of rRNA synthesis in these bacteria. However, although the precursor RNAs encoded by these different regions have identical rRNAs, they each have different tRNAs and so-called spacer regions.

Translation

The proteins do most of the work of the cell. Most of the enzymes that make and degrade energy sources and make cell constituents are proteins. Also, proteins make up much of the structure of the cell. Because of these diverse roles, cells have many more types of proteins than other types of constituents. Even in a relatively simple bacterium, there are thousands of different types of proteins, and most of the DNA sequences in bacteria are dedicated to genes encoding proteins.

Protein Structure

Unlike DNA and RNA, which consist of a chain of nucleotides held together by phosphodiester bonds between the sugars and phosphates, proteins consist of chains of 20 different amino acids held together by **peptide bonds**. Figure 2.12 shows an example of a peptide bond between two amino acids. The peptide bond is formed by joining the amino group (NH_2) of one amino

acid to the carboxyl group (COOH) of another. These amino acids will in turn be joined to other amino acids by the same type of bond, making a chain. A short chain of amino acids is called an **oligopeptide,** and a long one is called a **polypeptide.**

Like RNA and DNA, polypeptide chains have direction. However, in polypeptides, the direction is defined by their carboxyl and amino groups. One end of the chain, the **amino terminus** or **N terminus,** will have an unattached amino group. The amino acid at this end is called the **N-terminal amino acid.** On the other end of the polypeptide, the unattached carboxyl group is called the **carboxy terminus** or **C terminus,** and the amino acid is called the **C-terminal amino acid.** As we shall see, proteins are synthesized from the N terminus to the C terminus.

Protein structure terminology is the same as that for RNA structures. Proteins have primary, secondary, and tertiary structures, as well as quatenary structures. All of these are shown in Figure 2.13.

PRIMARY STRUCTURE
Primary structure refers to the sequence of amino acids and the length of a polypeptide. Because polypeptides are made up of 20 amino acids instead of just four nucleotides, as in RNA, many more primary structures are possible for polypeptides than for RNA chains.

SECONDARY STRUCTURE
Also like RNA, polypeptides can have a **secondary structure,** in which parts of the chain are held together

Amino acids
Primary structure

Secondary structures

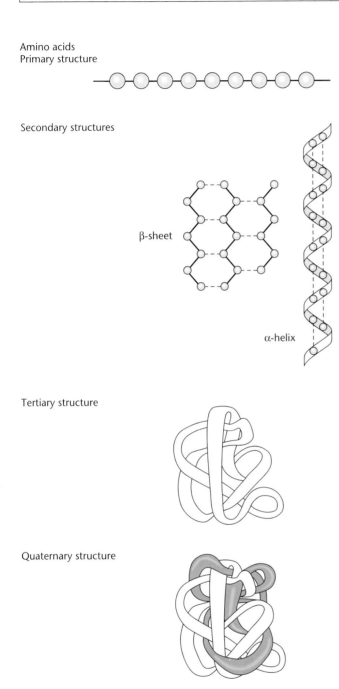

β-sheet

α-helix

Tertiary structure

Quaternary structure

Figure 2.13   Primary, secondary, tertiary, and quaternary structures of proteins.

by hydrogen bonds. However, because many more types of pairings are possible between amino acids than between nucleotides, the secondary structure of a polypeptide is more difficult to predict. The two basic forms of secondary structures in polypeptides are α-helices, where a short region of the polypeptide chain forms a helix owing to the pairing of each amino acid with the one before and the one after it, and β-sheets,

in which stretches of amino acids pair with other stretches to form loops or sheetlike structures (Figure 2.13). Computer software programs are available to help predict which secondary structures of a polypeptide are possible on the basis of its primary structure. However, none of these programs are entirely reliable, and a technique called X-ray crystallography is the only way to be certain of the secondary structure of a polypeptide.

TERTIARY STRUCTURE

Polypeptides usually also have a well-defined **tertiary structure**, in which they fold up on themselves, with hydrophobic amino acids such as leucine and isoleucine, which are not very soluble in water, on the inside and charged amino acids such as glutamate and lysine, which are more water soluble, on the outside.

QUATERNARY STRUCTURE

Proteins made up of more than one polypeptide chain also have **quaternary structure**. Such proteins are called **multimeric proteins**. When the polypeptides are the same, the protein is a **homomultimer**. When they are different, the protein is a **heteromultimer**. Other names reflect the number of polypeptides composing the protein. For example, the term **homodimer** describes a protein composed of two identical polypeptides whereas **heterodimer** describes a protein made of two different chains encoded by different genes. The names trimer, tetramer, and so on refer to increasing numbers of polypeptides.

The polypeptide chains in a protein are usually held together by hydrogen bonds. However, the polypeptides in a protein are sometimes held together by covalent **disulfide bonds** between cysteines in the two polypeptides. Usually only proteins that are active outside the cell, which must be particularly stable, are held together by disulfide bonds.

Even though the secondary, tertiary, and quaternary structures are usually specific to a protein, determining the structure of even simple proteins is laborious and generally requires analysis of X-ray diffraction patterns of the crystallized protein.

Ribosome Structure

As mentioned above in the section on rRNA, the ribosome is one of the largest and most complicated organelles in a bacterial cell, consisting of both RNA and protein components. However, to understand translation, we need to know more of the structural details of the ribosome itself.

RIBOSOMAL SUBUNITS

Figure 2.14 shows the components of a ribosome. The complete ribosome, called the 70S ribosome, consists of two subunits, the 30S subunit, which contains 16S rRNA, and the 50S subunit, which contains both 23S and 5S rRNA. Each subunit also contains **ribosomal proteins;** the 30S subunit contains 21 different proteins, while the 50S subunit contains 31 different proteins. Like the different terms for rRNA, the names of ribosomal subunits are derived from their sedimentation rates, and the S values are not additive. Indeed, the 30S and 50S subunits normally exist separately in the cell; only when they are translating an mRNA do they come together to form the complete 70S ribosome.

The two ribosomal subunits play very different roles in translation. To initiate translation, the 30S subunit binds to the mRNA. Then the 50S ribosome binds to the 30S subunit, which is already on the mRNA, to make the 70S ribosome. The complete molecule then moves along the mRNA, allowing tRNA anticodons to

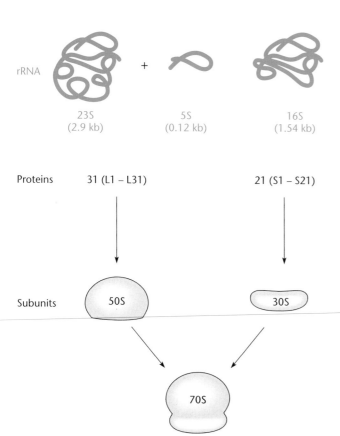

Figure 2.14 The composition of a ribosome containing one copy each of the 16S, 23S, and 5S rRNAs as well as many proteins. The proteins of the large 50S subunit are designated L1 to L31. The proteins of the small 30S subunit are designated S1 to S21.

pair with the mRNA codons and translate the nucleic acid chain into a polypeptide. After the polypeptide chain is completed, the ribosome separates again into the 30S and 50S subunits. We discuss the role of the subunits in more detail below in the section on initiation of translation.

Details of Protein Synthesis

In this section, we return to the process of translation in more detail. First, we discuss reading frames. Then we discuss translation elongation, or what happens as the 70S ribosome moves along the mRNA, translating its nucleotides into amino acids. Finally, we discuss how translation is initiated and terminated.

READING FRAME

As discussed in the overview of translation, each three-nucleotide sequence, or codon, in the mRNA encodes a specific amino acid, and the assignment of the codons is known as the genetic code. Because there are three nucleotides in each codon, an mRNA can be translated in three different frames in each region. Usually, initiation of translation at a initiator codon establishes the **reading frame of translation**. Once translation has begun, the ribosome moves three nucleotides at a time through the coding part of the mRNA. If the translation is occurring in the proper frame for protein synthesis, we say the translation is in the **zero frame** for that protein. If the translation is occurring in the wrong reading frame, it can be displaced either back by one nucleotide in each codon (the −1 frame) or forward by one nucleotide (+1 frame). Sometimes, translational frameshifts will occur that change the reading frame even after translation has initiated (see Box 2.2).

TRANSLATION ELONGATION

Before translation can begin, a specific amino acid is attached to each tRNA by its **cognate aminoacyl-tRNA synthetase** (Figure 2.15). Each of these enzymes specifically recognizes only one type of tRNA, hence the name *cognate*. How each cognate tRNA-synthetase recognizes its own tRNA varies, but the anticodon (i.e., the three tRNA nucleotides that base pair with the complementary mRNA sequence [see the overview section]) is not the only determinant. Often, if the anticodon changes in a given tRNA, the cognate synthetase will still attach the amino acid for the original tRNA, and that amino acid will be inserted for a different codon in the mRNA. This is the reason for nonsense suppression, which is discussed later in chapter 3. Finally, the tRNA with its amino acid becomes bound to a protein called **translation elongation factor Tu (EF-Tu)**.

BOX 2.2

Exceptions to the Code

Normally, translation begins at an initiation codon and then proceeds three bases at a time through the mRNA in the 5'-to-3' direction, inserting the amino acid at a codon according to the genetic code. However, there are exceptions to this general rule. Although the code is mostly universal, in some situations a codon can mean something else. We gave the example of initiation codons that encode different amino acids when internal to a gene (see the text). Also, some organelles and primitive microorganisms use a different code. Mitochondria are notorious for using different code words for some amino acids and for termination. Mitochondria from different species sometimes even use different codons from each other! Also, some protozoans use the nonsense codons UAA and UAG for glutamine. In these organisms, UGA is the only nonsense codon. In bacteria, the only known exceptions to the Universal Code involve the codon UGA. Usually, UGA is a nonsense codon that causes termination of translation at the end of an ORF or coding sequence. However, it encodes the amino acid tryptophan in bacteria of the genus *Mycoplasma*, which are responsible for some plant and animal diseases.

Some exceptions to the code occur only at specific sites in the mRNA. For example, UGA can encode the rare amino acid selenocysteine in some contexts. This amino acid exists at one or a very few positions in certain bacterial and eukaryotic proteins. It has its own unique aminoacyl-synthetase, translation elongation factor Tu (EF-TU), and tRNA, to which the amino acid serine is added and converted into selenocysteine. Then the selenocysteine is inserted for the codon UGA, but only at a very few, unique positions in proteins. But how does the tRNA insert selenocysteine at the few correct UGA codons instead of at the numerous other UGAs, which usually signify the end of a polypeptide? The answer seems to be that the specific selenocystyl EF-Tu recognizes the sequence around the selenocystyl codon and that only if the sequence around the UGA codon is right will this EF-Tu allow its tRNA to enter the ribosome. It is a mystery why the cell goes to so much trouble to insert selenocysteine in a specific site in only a very few proteins. In some instances where selenocysteine was replaced by cysteine, the mutated protein still functioned. Nevertheless, this amino acid has persisted throughout evolution, existing in a bacterial as well as human proteins.

Other examples of specific deviations from the code are high-level frameshifting and readthrough of nonsense codons. In high-level frameshifting, the ribosome can back up one base or go forward one base before continuing translation. High-level frameshifting usually occurs where there are two cognate codons next to each other in the RNA, for example, in the sequence *UU*U*UC*, where both UUU and UUC are phenylalanine codons that are presumably recognized by the same tRNA through wobble. Then the ribosome with the tRNA bound can slip back one codon before it continues translating, creating a frameshift. Sometimes both the normal protein and the frameshifted protein, which will have a different carboxy end, can function in the cell. Frameshifting can also allow the readthrough of nonsense codons to make "polyproteins," as occurs in many retroviruses such as the AIDs virus. Moreover, high-level frameshifting can play a regulatory role, for example, in the regulation of the RF2 gene in *E. coli*.

In the most extreme cases of frameshifting, the ribosome can skip large sequences in the mRNA. This is known to occur in gene *60* of bacteriophage T4 and the *trpR* gene of *E. coli*. Somehow, the ribosome quits translating the mRNA at a certain codon and "hops" to the same codon further along. Presumably, the secondary and tertiary structures of the mRNA between the two codons cause the ribosome to hop. In the case of gene *60* of T4, the hopping occurs almost 100% of the time and the protein that results is the normal product of the gene. In the *E. coli trpR* gene, the hopping is less efficient and the physiological significance of the hopped form is unknown.

High-level readthrough of nonsense codons can also give rise to more than one protein from the same ORF. Instead of stopping at a particular nonsense codon, the ribosome often continues making a longer protein. Examples are the synthesis of the head proteins in the RNA phage Qβ and the synthesis of Gag and Pol proteins in some retroviruses. Many plant viruses also make readthrough proteins. Again, it seems to be the sequence around the nonsense codon that destines it for high-level readthrough. However, it is important to emphasize that these are all exceptions and normally the codons on an mRNA are translated one after the other from the translational initiation region.

References

Atkins, J. F., R. B. Weiss, and R. F. Gesteland. 1990. Ribosomal gymnastics—degree of difficulty 9.5, style 10.0. *Cell* **62**:413–423.

Benhar, I., and H. Engelberg-Kulka. 1993. Frameshifting in the expression of the *E. coli trpR* gene occurs by the bypassing of a segment of its coding sequence. *Cell* **72**:121–130.

Bock, A., K. Forchhammer, J. Heider, W. Leinfelder, G. Sawers, B. Veprek, and F. Zinoni. 1991. Selenocysteine: the 21st amino acid. *Mol. Microbiol.* **5**:515–520. (Microreview.)

Parker, J. 1989. Errors and alternatives in reading the universal code. *Microbol. Rev.* **53**:273–298.

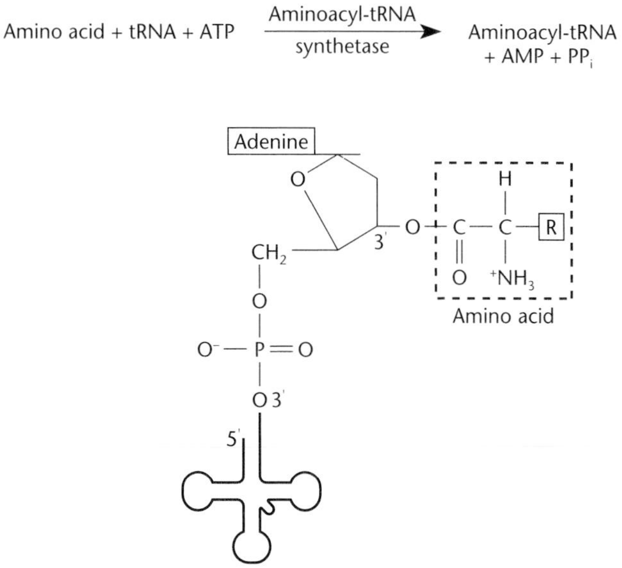

Amino acid + tRNA + ATP $\xrightarrow[\text{synthetase}]{\text{Aminoacyl-tRNA}}$ Aminoacyl-tRNA + AMP + PP$_i$

Figure 2.15 Aminoacylation of a tRNA by its cognate aminoacyl-tRNA synthetase.

During translation, the ribosome moves three nucleotides at a time along the mRNA in the 5'-to-3' direction, allowing tRNAs to pair with the larger molecule. Which tRNA can enter the ribosome depends on the sequence of the mRNA codon occupying the ribosome at that time. At a particular place in one of its loops, the tRNA must have three nucleotides that are complementary to the mRNA bases, so that the tRNA and mRNA can pair with each other. As mentioned above, this tRNA sequence is called the anticodon (Figure 2.16). To pair, the two RNA sequences must be complementary when read in opposite directions. In other words, the 3' to 5' sequence of the anticodon must be complementary to the 5' to 3' sequence of the codon.

If the anticodon is complementary to the mRNA codon passing through the ribosome, the complex of tRNA plus EF-Tu can enter, binding to a site called the

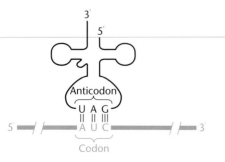

Figure 2.16 Complementary pairing between a tRNA anticodon and an mRNA codon.

A site (for acceptor site) on the ribosome (Figure 2.17). The pairing of only three bases is enough to direct the right tRNA to the A site on the ribosome; in fact, sometimes the pairing of only two bases is sufficient to direct the anticodon-codon interaction (see the section on wobble, below). However, the hydrogen bonding in base pairs is not strong enough to stably hold the tRNA at this site. Apparently, the presence of the bound EF-Tu and the structure of the ribosome at the A site help to stabilize the binding.

After the tRNA has bound, the EF-Tu is released, and another enzyme, **translation elongation factor G** (**EF-G**), moves, or **translocates**, the tRNA to the P site (for peptidyl site) to make room for another at the A. Now two tRNAs are bound to the ribosome, as schematized in Figure 2.17, and the tRNA at the P site has the growing polypeptide attached to it. The enzyme peptidyl-transferase (which is actually the 23S rRNA [see Box 2.1]) catalyzes the formation of a peptide bond between the amino acid attached to the A-site tRNA and the one at the P site. The P-site tRNA releases the polypeptide chain, which is now attached to the tRNA at the A site (Figure 2.17). Then EF-G translocates the tRNA and polypeptide (now longer by one amino acid) to the P site, freeing the A site for the binding of the next tRNA. (We discuss the P site in the section on translation initiation, below.) The ribosome moves in the 3' direction on the mRNA until the next three nucleotides are in the A position, and the process is repeated.

The translation of even a single codon in an mRNA requires a lot of energy. First, ATP must be cleaved for an aminoacyl-tRNA synthetase to attach an amino acid to a tRNA (Figure 2.15). Also, EF-Tu requires that one or possibly two GTPs be cleaved to GDP before it can be released from the ribosome after the tRNA is bound (Figure 2.17). Yet another GTP must be cleaved to GDP for the EF-G to move the tRNA with the attached polypeptide to the P site (Figure 2.17C). In all, the energy of three or possibly four nucleoside triphosphates is required for each step of translation.

The Genetic Code

As mentioned in the overview, the **genetic code** determines which amino acid will be inserted into a protein for each three-nucleotide set, or **codon,** in the mRNA. More precisely, the genetic code is the assignment of each possible combination of three nucleotides to one of the 20 amino acids. The code is universal (with a few minor exceptions, see Box 2.2, Exceptions to the Code), meaning it's the same in all organisms, from bacteria to humans. The assignment of each codon to its amino acid appears in Table 2.1.

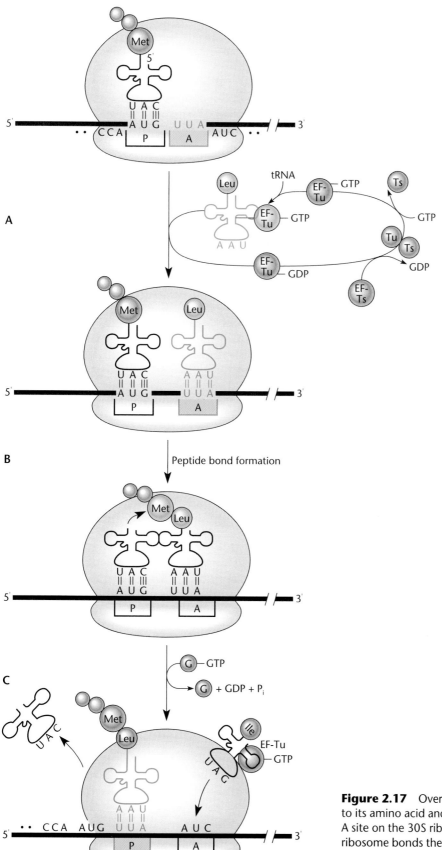

Figure 2.17 Overview of translation. (A) The tRNA bound to its amino acid and complexed with EF-Tu comes into the A site on the 30S ribosome. (B) Petidyltransferase on the 50S ribosome bonds the next amino acid to the growing polypeptide. (C) The tRNA is moved to the P site by EF-G, making room at the A site for another tRNA.

REDUNDANCY

In the genetic code, more than one codon often encodes the same amino acid. This feature of the code is called **redundancy**. There are 4 × 4 × 4 = 64 possible codons that can be made of four different nucleotides taken three at a time. Thus, without redundancy, there would be far too many codons for only 20 amino acids.

WOBBLE

Codons that encode the same amino acid often differ only by their third base (Table 2.1). For example, the two codons for the amino acid lysine are AAA and AAG. Often, this allows the same tRNA to pair with all the codons for the same amino acid. As mentioned in the overview, each tRNA has a set of three nucleotides called the **anticodon** opposite the attached amino acid (Figure 2.3); the anticodon is the sequence that pairs with mRNA. One tRNA can pair with multiple mRNA codons because of **wobble,** or less stringent binding, of the third nucleotide base of the codon to the first base of the tRNA anticodon, particularly when the first base of the anticodon is modified.

The binding at the third position is not totally random, however, and certain rules apply (Figure 2.18). For example, a G in the first position of the anticodon might pair with either a C or a U in the third position of the codon but not with an A or a G, explaining why UAU and UAC, but not UAA or UAG, are codons for tyrosine. Looking at the figure, we might predict the existence of a tyrosine tRNA with the anticodon GUA. The rules for wobble are difficult to predict, however, because the bases in tRNA are sometimes modified, and a modified base in the first position of an anticodon can have altered pairing properties.

NONSENSE CODONS

Not all codons stipulate an amino acid; of the 64 possible, only 61 nucleotide combinations actually encode an amino acid. The other three, UAA, UAG, and UGA, are **nonsense codons** in most organisms (for the exceptions see Box 2.2). The nonsense codons are usually used to terminate translation at the end of genes (see the section on termination of translation below).

AMBIGUITY

In general, each codon specifies a single amino acid, but some can specify a different amino acid depending upon where they are in the mRNA. For example, the codons AUG and GUG encode formylmethionine if they are at the beginning of the coding region but they encode methionine or valine, respectively, if they are internal to the coding region. The codons CUG, UUG, or even AUU

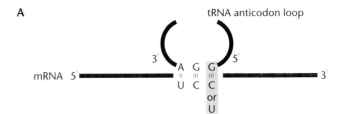

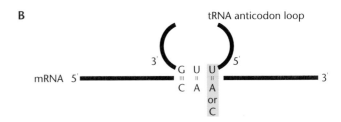

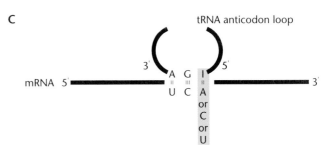

Figure 2.18 Wobble pairing between the anticodon on the tRNA and the codon in the mRNA. Many pairing interactions are possible in the third position of the codon. Alternative pairings for the anticodon base are shown: guanine (A), uracil (B), and inosine (a purine base found only in tRNAs) (C).

also sometimes encode formylmethionine if they are at the beginning of a coding sequence.

UGA is another exception. This codon is usually used for termination but encodes the amino acid selenocysteine in a few positions in genes (see Box 2.2) and encodes tryptophan in some types of bacteria.

CODON USAGE

Just because more than one codon can encode an amino acid does not mean that all the codons will be used equally in all organisms. The same amino acid may be preferentially encoded by a different codon in different organisms. This codon preference may reflect higher concentrations of certain tRNAs or may be related to the base composition of the DNA of the organism. While mammals and other higher eukaryotes have an average GC content of about 50% (so that there are about as many AT base pairs in the DNA as there are GC base pairs), some bacteria and their viruses have very high or very low GC contents. How the GC content can influence codon preference is illustrated by some members of the genera *Pseudomonas* and *Streptomyces*. These organisms have GC contents of almost 75%. To maintain such high GC contents, the protein-synthesizing mechanisms of these bacteria use the codon that has the most G's and C's for each amino acid.

Translation Initiation

The process of initiating the synthesis of a new polypeptide chain is very different from the process of translation once it is under way. For example, the 30S, but not the 50S, ribosomal subunit works with other factors unique to initiation. Initiation of translation in bacteria is somewhat different from initiation in eukaryotic organisms, and we shall point out some of these differences as we go along.

TRANSLATIONAL INITIATION REGIONS

In the chain of thousands of nucleotides that make up an mRNA, the ribosome must bind and initiate translation at the correct site. If the ribosome starts working at the wrong initiation codon, the protein will have the wrong N-terminal amino acid or the mRNA will be translated out of frame and all of the amino acids will be wrong. Hence, mRNA have sequences called **translational initiation regions** (**TIRs**) that flag the correct first codon for the ribosome. In spite of extensive research, it is still not possible to predict with 100% accuracy whether a sequence is a TIR. However, some general features of TIRs are known.

Initiation Codon

All TIRs have an **initiation codon,** which codes for a definable amino acid. The three bases in these codons are usually AUG or GUG but in rare cases are CUG, UUG, or other sequences. There is even one known case in *E. coli* of a gene with the initiation codon AUU.

The initiation codon does not have to be the first sequence in the mRNA chain. In fact, the RNA copy of the gene's promotor may be some distance from the TIR and the initiation codon; this region is called the 5' **untranslated region,** or **leader sequence.**

Regardless of which amino acid these sequences call for in the genetic code (Table 2.1), if they are serving as initiation codons, they encode methionine (actually formylmethionine [see below]) as the N-terminal amino acid. After translation, this methionine is usually cut off (see the section on removal of the methionine, below). Notice that for the initiation codons, there seems to be "wobble in reverse," with the first position being the one that can wobble instead of the third position. The significance of this is unknown.

Shine-Dalgarno Sequence

Given that the initiation codons code for amino acids other than methionine when internal to a coding region, the presence of one codon is clearly not enough to define a translational inititation region. These sequences may also occur out of frame, in which case they would not read as an amino acid. They could even appear in an mRNA sequence that is not translated at all. Obviously, other regions around these three bases must help define them as a place to begin translation.

Many bacterial genes have 5 to 10 nucleotides on the 5' side (upstream) of the initiation codon that define a ribosome binding site. These sequences, named the **Shine-Dalgarno sequence** after the two scientists who first noticed them, are complementary to short sequences within certain regions of the 16S RNA. Figure 2.19 shows an example of a ribosome-binding site with a characteristic Shine-Dalgarno sequence. In all likelihood, by pairing with their complementary sequences on the 16S rRNA, Shine-Dalgarno sequences define TIRs by properly aligning the mRNA on the ribosome. However, these sequences are not always so easy to identify because they can be very short and do not have a distinct sequence, being complementary to different regions of the 16S rRNA. Moreover, not all bacterial genes have Shine-Dalgarno sequences. The inititation codon sometimes resides at the extreme 5' end of the mRNA, leaving no room for these markers. In such cases, the sequence that interacts with the 16S rRNA of the ribosome may be downstream of the initiation codon.

Because of this lack of universality, often the only way to be certain that translation is initiated at a particular initiation codon is to sequence the N terminus of the polypeptide to see if the N-terminal amino acids correspond to the codons immediately adjacent to the putative initiation codon.

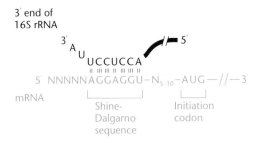

Figure 2.19 Structure of a typical bacterial ribosome-binding site showing the pairing between the Shine-Dalgarno sequence in the mRNA and a short sequence close to the 3' end of the 16S rRNA. The initiator codon, typically AUG or GUG, is 5 to 10 bases downstream of the Shine-Dalgarno sequence. N designates any base.

THE STEPS OF INITIATION

The steps in the initiation process are outlined in Figures 2.20 and 2.21. First the 30S subunit binds to the ribosome-binding site of the mRNA. Then a complex of an initation factor and fMet-tRNA$_f^{Met}$ binds to the P site on the 30S ribosomal subunit. Next the 50S subunit binds to form the 70S preinitiation complex. The

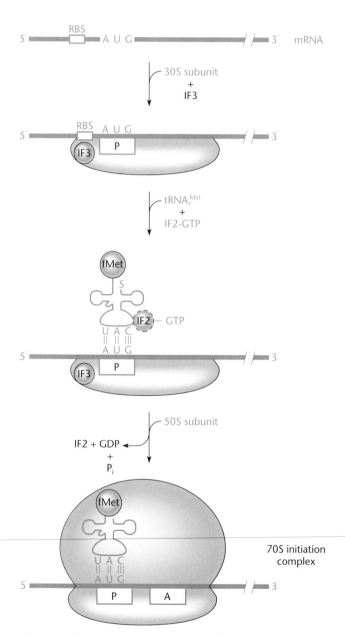

Figure 2.20   Initiation of translation. The 30S subunit and the mRNA associate. IF2 brings fMet-tRNA$_f^{Met}$ into the P site of the ribosome. The 50S subunit then binds. Another aminoacylated-tRNA can enter the A site. RBS, ribosome-binding site.

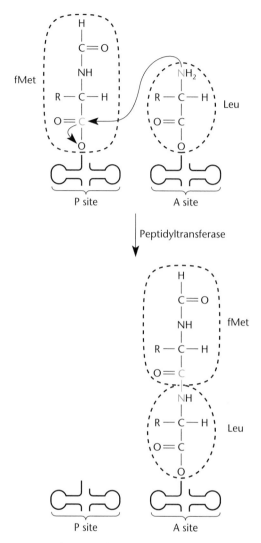

Figure 2.21   The peptidyltransferase reaction.

aminoacylated-tRNA dictated by the codon immediately following the initiation codon then binds to the A site (Figure 2.17). The peptidyltransferase on the 50S subunit forms a bond between the formylmethionine on the first tRNA and whatever amino acid is on the second tRNA (Figure 2.21). Finally, elongation factor G (EF-G) moves the second tRNA to the P site along with the attached 2-amino-acid peptide (as discussed above), and the process of translation is under way.

The Initiator tRNA

Translation is initiated by a special tRNA called the **formylmethionyl-tRNA$_f^{Met}$**, or just **fMet-tRNA$_f^{Met}$**, to distinguish it from the normal methionyl-tRNA called tRNA^{Met} (Figure 2.22). A specific aminoacyl-tRNA synthetase puts methionine on a tRNA, and a **trans-**

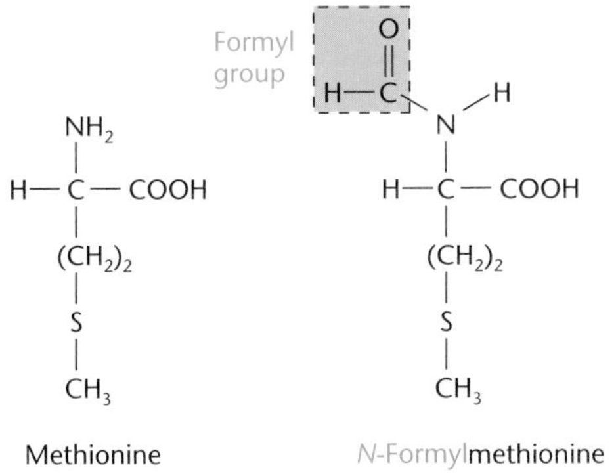

Figure 2.22 Comparison between methionine (Met) and N-formylmethionine (fMet).

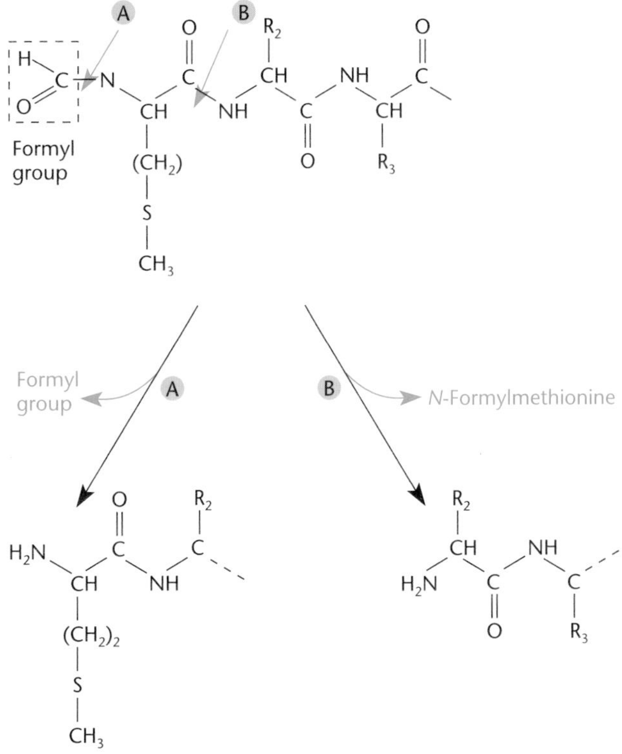

Figure 2.23 Removal of the N-terminal formyl group by peptide deformylase (A) and the N-terminal methionine by methionine aminopeptidase (B).

formylase adds a formyl group to the amino group of the methionine. This tRNA with its formylmethionine can then bind to the ribosome in response to the initiation codon in a TIR.

Because of its special structure, fMet-tRNA$_f^{Met}$ does not bind to the ribosome the same way as other tRNAs. For instance, fMet-tRNA$_f^{Met}$ does not enter the ribosome bound to the normal EF-Tu but, rather, is bound to its own special EF-Tu called **initiation factor 2** (**IF2**). Also, fMet-tRNA$_f^{Met}$ does not bind to the A site of the ribosome like other incoming tRNAs but, rather, binds to the P site, which is unoccupied because translation has not yet begun. It may be that the presence of the formyl group on the methionine, which (as shown in Figure 2.22) looks something like a peptide bond, directs tRNA$_f^{Met}$ to the P site on the ribosome.

Initiation Factors for Translation

Three translation initiation factors—IF1, IF2, and IF3—are required for initiation of translation in bacteria. As mentioned, IF2 seems to play a special EF-Tu-like role in that it binds the fMet-tRNA$_f^{Met}$ and brings it into the ribosome. The other initiation factors may help keep the 30S and 50S subunits disassociated during initiation and allow them to reassociate after initiation has occurred.

Removal of the Formyl Group and the N-Terminal Methionine

Normally, polypeptides do not have a formyl group attached to their N terminus. In fact, they usually do not even have methionine as their N-terminal amino acid. The formyl group is removed from the polypeptide after it is synthesized by a special enzyme called **peptide deformylase** (Figure 2.23). The N-terminal methionine is also usually removed by an enzyme called **methionine aminopeptidase**.

TRANSLATION INITIATION IN ARCHAEA AND EUKARYOTES

Translation initiation in the archaea is similar to that in the eubacteria. Like bacteria, archaea use well-defined ribosome-binding sites and formylmethionine for initiation of translation. In contrast, eukaryotes do not seem to use special ribosome-binding sites but may merely use the first AUG from the 5' end of the mRNA as the initiation codon. Also, although eukaryotes have a special methionine tRNA that responds to the first AUG codon, the methionine attached to the eukaryotic initiator tRNA is never formylated. The first methionine is, however, usually removed by an aminopeptidase after the protein is synthesized. Eukaryotes also seem to need many more initiation factors than do bacteria, although the exact role of most of these initiation factors is unknown. The mechanism of translation initiation in the archaea operates more like its counterpart in eubacteria, while transcription initiation in the archaea seems more like that in eukaryotes. Thus, in some ways the archaea seem more like eukaryotes and in others more like eubacteria (see Box 2.3).

Molecular Phylogeny

The advent of DNA sequencing has made possible the sequencing of genes from a number of different organisms. A comparison of these sequences has given rise to the science of molecular phylogeny, which is based on the relatedness of sequences from different organisms rather than just on similarities in physiological and morphological characteristics. The idea is that the more recently two types of organisms separated during evolution, the more similar will be the sequences of their genes. Two properties of the translational machinery have made these components particularly useful for such studies. First, some components of the translational apparatus are very highly conserved. Second, particularly in bacteria, these components are probably not often exchanged horizontally during evolution. In other words, these are not genes that are likely to be carried by plasmids or transposons.

The conservation of components of the translation apparatus is so high that evolutionary trees can be made that include eukaryotes and archaebacteria. Such trees are usually not too different from what has been obtained from physiological and other comparisons, but there are sometimes surprises. For example a 1-mm-long organism found in sea clams around thermal vents in the sea floor was recently shown to be a bacterium on the basis of its 16S rRNA sequence. Also, the sequence of the translation elongation factors led to the suggestion that the archaebacteria are more closely related to eukaryotes than they are to other bacteria, prompting the change of their name to archaea.

References

Ibawe, N., K.-I. Kuma, M. Hasegawa, S. Osawa, and T. Migata. 1989. Evolutionary relationships of archaebacteria, eubacteria and eukaryotes inferred from phylogenetic trees of duplicated genes. *Proc. Natl. Acad. Sci. USA* **86**:9355–9359.

Woese, C. R. 1987. Bacterial evolution. *Microbiol. Rev.* **51**:221–271.

Translation Termination

Once initiated, translation proceeds along the mRNA, one codon at a time, until the ribosome encounters the codon UAA, UAG, or UGA. These codons do not encode an amino acid, so they have no corresponding tRNA. When a ribosome comes to a nonsense codon, translation stops. Similar to the positioning of translation initiators, the nonsense codon that terminates translation may not be at the end of the mRNA molecule. The region between this last codon and the 3' end of the mRNA is called the **3' untranslated region**.

RELEASE FACTORS

In addition to a codon for which there is no tRNA, termination of translation requires **release factors**. These proteins recognize the specific nonsense codons and promote the release of the ribosome from the mRNA.

In *E. coli*, there are three translational release factors, RF1, RF2, and RF3, which respond to specific nonsense codons with termination. RF1 responds to UAA and UAG, whereas RF2 responds to UAA and UGA. RF3 seems to be a more general release factor that helps the other two.

Figure 2.24 outlines the process of translation termination. After translation stops at the nonsense codon, the release factors free the polypeptide chain. Termination is more efficient when it occurs in the proper

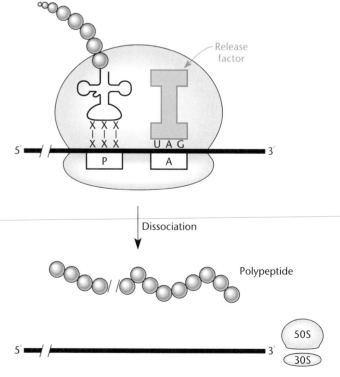

Figure 2.24 Termination of translation at a nonsense codon. A specific release factor interacts with the ribosome stalled at the nonsense codon, causing dissociation of the ribosome from the mRNA.

context, that is, when a nonsense codon is surrounded by certain sequences.

RELEASE OF THE POLYPEPTIDE

How the last tRNA is removed from the polypeptide after termination remains a mystery. An enzyme called **peptidyl tRNA hydrolase** can remove tRNA from a polypeptide, but this enzyme is not associated with the ribosome, where normal release is thought to occur. The peptidyl tRNA hydrolase is also not essential for growth, as you might expect of a protein playing such an important role in translation. Instead, tRNA removal might normally be caused by EF-G in combination with another ribosomal protein called **ribosome release factor**.

Polycistronic mRNA

In bacteria and archaea, the same mRNA can encode more than one polypeptide. Such mRNAs, called **polycistronic mRNAs**, must have more than one TIR to allow simultaneous translation of more than one sequence of the mRNA.

The name "polycistronic" is derived from *cistron*, which is the genetic definition of the coding region for each polypeptide, and *poly*, which means many. Figure 2.25 shows a typical polycistronic mRNA, in which the coding sequence for one polypeptide is followed by the coding sequence for another. The space between two coding regions can be very short, and the coding sequences may even overlap. For example, the coding region for one polypeptide may end with the nonsense

codon UAA, but the last A may be the first nucleotide of the initiator codon AUG for the next coding region. Even if the two coding regions overlap, the two polypeptides on an mRNA can be translated independently by different ribosomes.

Polycistronic mRNAs do not exist in eukaryotes, in which TIRs are much less well defined and translation usually initiates at the AUG codon closest to the 5′ end of the RNA. In eukaryotes, the synthesis of more than one polypeptide from the same mRNA usually results from differential splicing of the mRNA or to high-level frameshifting during the translation of one of the coding sequences (see Box 2.2 and the section on reading frames, above). Polycistronic RNA leads to phenomena unique to bacteria, i.e., polarity and translational coupling, which are described in the following sections.

POLARITY

When the coding sequences for two genes are transcribed onto the same polycistronic mRNA, a mutation that blocks the *translation* of one gene's RNA codons, for example, one that results in a nonsense codon, can prevent the *transcription* of the next gene on the mRNA. This phenomenon, called **polarity** (Figure 2.26), occurs if there is at least one ρ-dependent termination site between the point where translation was blocked in the first gene and the TIR of the second. If translation in the first gene has been blocked by a mutation, the ρ factor will bind to the mRNA and stop transcription of the second gene (see the section on transcription termination, above).

TRANSLATIONAL COUPLING

Two or more polypeptides encoded by the same polycistronic mRNA can also cause **translational coupling**. Two genes are **translationally coupled** if translation of the upstream gene is required for translation of the gene immediately downstream.

Figure 2.27 shows an example of how two genes could be translationally coupled. The TIR including the AUG initiation codon of the second gene is inside a hairpin on the mRNA, and so it cannot be recognized by a ribosome. However, a ribosome arriving at the UGA stop codon for the first gene can open up this secondary structure, allowing another ribosome to bind and initiate translation on the second gene. Thus, translation of the second gene depends on the translation of the first.

Superficially, polarity and translational coupling have similar effects; in both cases, blocking the translation of one polypeptide affects the synthesis of another polypeptide encoded on the same mRNA. However, as

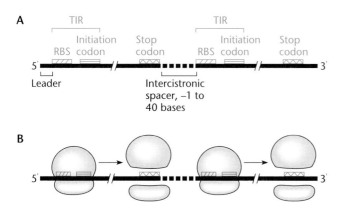

Figure 2.25 Structure of a polycistronic mRNA. The coding sequence for each polypeptide is between the TIR and the stop codon. The region 5′ of the first initiation codon is called the leader sequence, and the untranslated region between a stop codon for one gene and the next translational initiation region is known as the intercistronic spacer. RBS, ribosome-binding site.

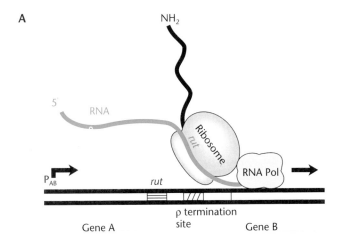

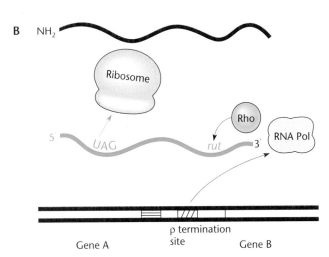

Figure 2.26 Polarity in transcription of a polycistronic mRNA. (A) RNA polymerase transcribes a polycistronic mRNA while ribosomes translate it. (B) If translation is blocked in the transcript of gene A, the ρ factor can cause transcription termination before the RNA polymerase reaches gene B.

we have seen, the molecular bases of the two phenomena are completely different.

Protein Folding

Translating the mRNA into a polypeptide chain is only the first step in making an active protein. To be an active protein, the polypeptide must fold into its final conformation. This is the most stable state of the protein and is determined by the primary structure of its polypeptides. In principle, every protein would eventually fold into its most stable structure. However, without the help of other factors, folding might take too long for the protein to be useful.

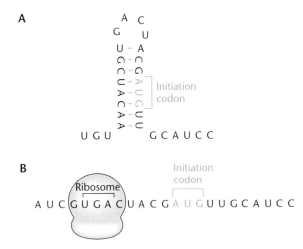

Figure 2.27 Model for translational coupling in a polycistronic mRNA. The secondary structure of the RNA blocks translation of the second polypeptide (A) unless it is disrupted by a ribosome translating the first coding sequence (B).

Chaperones

As polypeptides are made, proteins known as **chaperones** prevent them from folding prematurely. In addition, chaperones help fold proteins that are partially denatured after cells are briefly heated (see the section on heat shock in chapter 12) and aid in the transport of membrane proteins (see the section on membrane proteins, below). Chaperones are among the most evolutionarily conserved cellular components, being very similar in human and bacterial cells.

Protein Disulfide Isomerases

As mentioned above in the section on protein structure, the amino acid cysteine often plays a unique role because of covalent linkages called disulfide bonds that can form between the sulfur atoms of cysteines in different parts of the protein and help hold proteins together. Disulfide linkages are found only in periplasmic and extracellular proteins, which often face extreme conditions and need the stability provided by disulfide linkages.

Protein disulfide isomerases are the enzymes that catalyze the formation of disulfide bonds—a complicated endeavor. These enzymes probably exist only in the periplasm, where they form disulfide linkages as proteins pass out of the cell. Despite their necessity in some proteins, disulfide bonds should not be formed prematurely on cysteines that should not be linked in the final protein. It is not clear how many disulfide isomerases *E. coli* contains or how such enzymes distinguish between cysteines that should be linked and those that should not, but this distinction is probably determined by the final conformation of the protein.

Membrane Proteins

Some proteins are destined to enter or pass through the cellular membrane after they are synthesized. Many of these proteins have a short sequence of mostly hydrophobic amino acids at their N terminus, called the **signal sequence**, that helps them enter the membrane. This sequence is removed as the protein enters the membrane.

As already mentioned, chaperones help membrane proteins reach their final destination in the membrane. Although chaperones perform other tasks in the cell (see the section on chaperones, above), they are so named because they bind to membrane proteins being synthesized and "escort" them to the cell membrane. They do not actually guide the newly synthesized proteins to their destination, but they keep them from folding prematurely before they have entered the membrane. If a membrane protein folds prematurely, the hydrophobic signal sequence could get buried in the interior of the

protein, where it would not be available to help transport the protein into or through the membrane. Even if the signal sequence is not buried, premature folding may still interfere with insertion of the membrane protein or its eventual secretion outside the cell.

Useful Concepts

We have introduced a lot of detail in this chapter, so it is worth going over the most important concepts and words, which will be used throughout the book and so should be very familar. Some concepts addressed here are presented for the first time, but the following paragraphs offer a brief review.

Figure 2.28 illustrates some of these concepts. Figure 2.28 shows a typical gene with a promoter and terminator. The mRNA is transcribed from the gene beginning at the promoter and ending at the transcription terminator. An RNA polymerase synthesizes mRNA in the 5′-to-3′ direction, moving 3′ to 5′ on the transcribed strand of DNA. The opposite strand of DNA has the same sequence as the RNA and so is called the coding strand. Sequences 5′ of a particular site on the transcribed

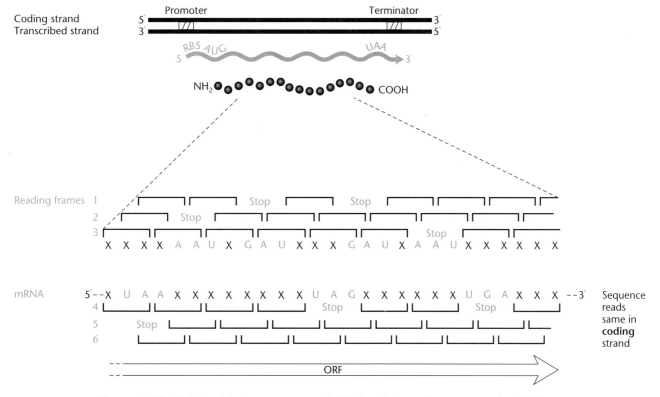

Figure 2.28 Relationship between a gene in DNA and the coding sequence in mRNA. There are a total of six different sequences in the two strands of DNA that may contain ORFs, but generally only one ORF encodes a polypeptide in each region.

strand of DNA are upstream of the site, while sequences 3′ are downstream.

Figure 2.29 shows the terms for the regions around a promoter. The position on the DNA at which transcription begins is +1, and nucleotides downstream of this region have positive numbers (e.g., +10) while those upstream have negative numbers (e.g., −10).

Because mRNAs are both made and translated in the 5′-to-3′ direction, an mRNA can be translated while it is still being made. Such coupling of transcription and translation is not possible in eukaryotes, in which the mRNA must pass out of the nucleus into the cytoplasm before it can be translated.

It is important to distinguish promoters from translational initiation regions (TIRs) and to distinguish transcription termination from translation termination. The promoter may be some distance from the first TIR on the mRNA; therefore, a long sequence at the 5′ end of the mRNA may not be translated. An untranslated region on the 5′ end of an mRNA is called the 5′ untranslated region. Similarly, a nonsense codon is *not* a transcription terminator, only a translation terminator. The transcription terminator may not be at the end of a gene and may be some distance from the nonsense codon for the last gene. The distance from the last nonsense codon for a gene to the 3′ end of the mRNA is the 3′ untranslated region. Polycistronic mRNAs contain more than one coding sequence for polypeptides. These mRNAs have a separate TIR and nonsense codon for each gene.

Open Reading Frame

Figure 2.28 also shows an **open reading frame (ORF)**. This sequence is not necessarily a gene but just a string of potential codons for amino acids in DNA that is not broken by nonsense codons. Computer software is available that shows where all the ORFs are in a sequence, and most DNA sequences will have many ORFs on both strands, although most of these will be short. The region shown in the figure contains many ORFs, but only the longest is likely to actually encode a polypeptide.

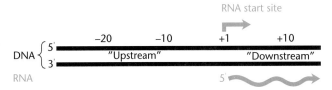

Figure 2.29 Some concepts in molecular genetics. Upstream in DNA refers to the direction of the 5′ end of the coding strand in that region. Downstream is in the 3′ direction on the coding strand.

It is sometimes difficult to determine which, if any, of the ORFs in a DNA sequence actually encode(s) a polypeptide. The presence of even a long ORF does not necessarily indicate that the sequence encodes a protein, and fairly long ORFs will often occur by chance. If an ORF does encode a polypeptide, it will begin with a TIR, but, as we have discussed above, TIRs are sometimes difficult to identify. If the DNA of the organism has a high GC or AT content, it is sometimes possible to tell if an ORF is likely to encode a protein just from the choice of the third base in the codon for each amino acid (see the section on codon usage, above). However, more information is usually required to establish which, if any, of the ORFs in a sequence encode(s) a protein.

One way to determine if an ORF actually encodes a protein is to ask which polypeptides are made from the DNA in an in vitro coupled transcription-translation system. In these systems, extracts are made of cells, typically of *E. coli*. The DNA is removed from the extracts, but the RNA polymerase, ribosomes, and other components of the translation apparatus remain. DNA with the ORFs under investigation and radioactive amino acids are added to these extracts, and radioactive proteins are synthesized from the added DNA. If the size of one of the radioactive polypeptides synthesized corresponds to the size of an ORF on the DNA, this ORF probably encodes a protein.

TRANSCRIPTIONAL AND TRANSLATIONAL FUSIONS

Probably the most convenient way to determine which of the possible ORFs on the two strands of DNA in a given region are translated into proteins is to make **transcriptional** and **translational fusions** to the ORFs. These methods make use of **reporter genes** such as *lacZ* (β-galactosidase), *lux* (luciferase), *cat* (chloramphenicol acetyltransferase), or other genes whose products are easy to detect.

An ORF can be translated only if it is transcribed into RNA. Transcriptional fusions can be used to determine whether this has occurred. To make a transcriptional fusion, a reporter gene with the sequence for a TIR but no promoter of its own is fused immediately downstream of the promoter. If the gene is transcribed into mRNA, the reporter gene will also be transcribed and its product will be detectable in the cell.

Translational fusions are a more direct way to determine if an ORF is translated into a polypeptide, but they are more difficult to construct than are transcriptional fusions. To make a translational fusion, a reporter gene without the sequence for a TIR is fused downstream of the ORF so that translation initiated at a potential TIR for the ORF will continue into the reporter gene coding region, thus making a fusion protein. The reporter gene product must retain its activity even when fused to the potential polypeptide encoded by the ORF; otherwise it will not be detectable. Many reporter genes have been chosen because their products remain active when fused to other polypeptides. We shall return to gene fusion methods in more detail in later chapters.

Antibiotics That Block Transcription and Translation

As in chapter 1, we devote the remainder of this chapter to a discussion of antibiotics because these compounds not only are useful therapeutic agents but also allow mutants to be isolated for genetic studies. Studies on how antibiotics affect transcription and translation have greatly contributed to our understanding of these processes.

Antibiotic Inhibitors of Transcription

Some of the components of the transcription apparatus are the targets of many antibiotics used in treatment of bacterial infections and in tumor therapy. Some of these antibiotics are made by soil bacteria and fungi and some have been synthesized chemically. Table 2.2 lists examples, along with their sources and their targets.

INHIBITORS OF RIBONUCLEOSIDE TRIPHOSPHATE SYNTHESIS
Some antibiotics that inhibit transcription do so by inhibiting the synthesis of the ribonucleoside triphosphates. An example is azaserine, which inhibits purine biosynthesis. Antibiotics that block the synthesis of the ribonucleotides are usually not specific to transcription, since the ribonucleotides have many other uses in the cell.

INHIBITORS OF RNA SYNTHESIS INITIATION
Some species of soil actinomycetes secrete rifampin and other streptovarcin-type antibiotics. These antibiotics bind to the β subunit of RNA polymerase and specifically block the initiation of RNA synthesis. This property of blocking only initiation of transcription has

made these antibiotics very useful in the study of transcription. For example, they have been used to analyze the steps in initiation of RNA synthesis, to study the stability of RNA and proteins in the cell, and to study the role of RNA polymerase in the initiation of chromosome replication. They bind to the RNA polymerases of essentially all types of bacteria but not to those of eukaryotes, so they are not toxic to humans and are also useful therapeutic agents. Many derivatives have been made from them.

In rifampin- and streptovarcin-resistant mutants, one or more amino acids are changed in the β subunit of RNA polymerase because of a mutation in the sequence of the gene. Thus, rifampin can no longer bind to the altered β subunit. Although such resistant mutants are fairly common and have limited the usefulness of these antibiotics, they are still part of standard antibiotic therapy for tuberculosis and some other diseases.

INHIBITORS OF RNA ELONGATION
Streptolydigin also binds to the β subunit of RNA polymerases of bacteria but can block RNA synthesis after it is under way. It has a weaker affinity for RNA polymerase than rifampin so only blocks transcription when added at higher concentrations.

BLOCKS ON THE DNA TEMPLATE
Actinomycin D and bleomycin block transcription by binding to or nicking the DNA, respectively. While many such drugs are useful for studying transcription in bacteria, they are not very useful in antibacterial therapy because they are very toxic and not specific to bacteria. They are, however, used in antitumor therapy.

Antibiotic Inhibitors of Translation

The translation apparatus is the most common target of antibiotics made by soil microorganisms. Some of these antibiotics are very useful in combating bacterial and fungal diseases and in cancer chemotherapy. They have also proved useful in studies aimed at understanding the translation process itself. Commonly used antibiotics directed against translation are listed in Table 2.3, which also lists their target and source. We discuss some of these antibiotics in more detail below. Where bacter-

TABLE 2.2	Antibiotics that block RNA synthesis	
Antibiotic	Source	Target or action
Streptolydigin	*Streptomyces lydicus*	β subunit of RNA polymerase
Actinomycin D	*Streptomyces antibioticus*	Binds DNA
Rifampin	*Nocardia mediterranei*	β subunit of RNA polymerase
Bleomycin	*Streptomyces verticulus*	Cuts DNA

TABLE 2.3	Antibiotics that block translation	
Antibiotic	**Source**	**Target**
Puromycin	*Streptomyces alboniger*	A site of ribosome
Kanamycin	*Streptomyces kanamyceticus*	16S rRNA
Neomycin	*Streptomyces fradiae*	16S rRNA
Streptomycin	*Streptomyces griseus*	30S ribosome
Thiostrepton	*Streptomyces azureus*	23S rRNA
Gentamicin	*Micromonospora purpurea*	16S rRNA
Tetracycline	*Streptomyces rimosus*	A site of ribosome
Chloramphenicol	*Streptomyces venezuelae*	Peptidyltransferase
Erythromycin	*Saccharopolyspora erythraea*	23S rRNA
Fusidic acid	*Fusidium coccinuem*	Translation elongation factor G
Kirromycin	*Streptomyces collinus*	Translation elongation factor Tu

ial, fungal, and tumor cells can develop resistance to the action of these antibiotics, some of the resistance mechanisms are also discussed.

tRNA MIMICRY
Puromycin mimics a tRNA with an amino acid attached. It enters the ribosome as an aminoacylated-tRNA, and the peptidyltransferase attaches it to the growing polypeptide. However, it does not translocate properly from the A site to the P site and is released from the ribosome, terminating translation. Studies with puromycin have contributed greatly to our understanding of translation. The model of the A and P sites in the ribosome and the concept that the 50S ribosome contains the enzyme for peptidyl bond formation, which was recently shown to be the 23S RNA itself, came from studies with this antibiotic. Puromycin is not a very useful antibiotic for treating bacterial diseases, however, because it also inhibits translation in eurkaryotes, making it toxic in humans and animals.

INHIBITORS OF PEPTIDE BOND FORMATION
Chloramphenicol inhibits translation by binding to ribosomes and preventing the formation of peptide bonds. It is effective at low concentrations and therefore has been one of the most useful antibiotics for studying cellular functions. For example, it has been used to determine the time in the cell cycle at which proteins required for cell division and for initiation of chromosomal replication are synthesized. It is also not very toxic because it is fairly specific for the translation apparatus of bacteria and does not inhibit the translation of chromosomal genes in humans or animals; hence, it is quite useful. Whatever toxicity it does have may result from inhibition of the translational apparatus of mitochondria, which is similar to the translation apparatus of bacteria (see Introduction). Like many an-

tibiotics that block translation, chloramphenicol is a bacteriostatic agent in that it stops the growth of bacteria without actually killing them.

Many changes are required to make bacteria resistant to chloramphenicol. Many ribosomal proteins must be changed so that the antibiotic can no longer bind to the ribosome. As with numerous antibiotics, bacteria sometimes make enzymes that inactivate chloramphenicol. The genes for these enzymes are often carried on plasmids and transposons, interchangeable DNA elements that will be discussed in subsequent chapters. The best-characterized chloramphenicol resistance gene is the *cat* (or Cam[r]) gene of the transposon Tn9, which acetylates chloramphenicol, thereby inactivating it. This gene is extensively used as a reporter gene to study gene expression in both bacteria and eukaryotes.

Erythromycin is a member of a large group called the macrolide antibiotics. These antibiotics inhibit translation by blocking either the peptidyltransferase reaction or the translocation step, although different assays give somewhat different results with the different macrolides. These antibiotics are also bacteriostatic.

Mutants resistant to macrolides have an altered ribosomal protein L4. Some plasmids and transposons confer resistance to the macrolide antibiotics by methylating the 23S rRNA.

INHIBITORS OF BINDING OF tRNA TO THE A SITE
Tetracycline inhibits translation by inhibiting the binding of the tRNA to the A site of the ribosome. This is another useful antibiotic, although it is somewhat toxic to humans because it also inhibits the eukaryotic translation apparatus.

In some types of bacteria, ribosomal mutations confer low levels of resistance to tetracycline by changing protein S10 of the ribosome. Transposons and plasmids can carry genes that confer resistance to tetracycline. One of these, *tetM*, carried by the conjugative transposon Tn*916* and its relatives, confers resistance by methylating certain bases in the 16S rRNA. This gene is ubiquitous; related genes occur in both gram-positive and gram-negative bacteria. The product of other tetracycline resistance genes, such as the *tet* (or Tet[r]) genes carried by transposon Tn*10* and plasmid pSC101 of *E. coli,* are membrane proteins that confer resistance by pumping tetracycline out of the cell. The *tet* (or Tet[r]) genes of Tn*10* and pSC101 are specific for *E. coli* and do not confer tetracycline resistance in many other types of bacteria. Nevertheless, these tetracycline resistance genes are extensively used as reporter genes and as markers for genetic analysis in *E. coli.* The *tet* gene from pSC101 has been introduced into many plasmid cloning vectors (see chapter 4).

INHIBITORS OF TRANSLOCATION

Fusidic acid has been very useful in studies of the function of ribosomes because it specifically inhibits the translation elongation factor G, EF-G (called EF-2 in eukaryotes), probably by preventing its dissociation from the ribosome after GTP cleavage. This is the protein required for translocation. In *E. coli,* mutations that confer resistance to fusidic acid are in the *fus* gene, which encodes EF-G.

Kanamycin and its close relatives, neomycin and gentamicin, are members of a larger group of antibiotics, the aminoglycoside antibiotics, which also includes streptomycin. These antibiotics have a very broad spectrum of action, inhibiting translation in plants and animal cells as well as in bacteria. Their ability to block translation in plants and animals has made them very useful in technology. They are discussed further in chapter 16.

These antibiotics probably inhibit translation by a very different mechanism, although they also cause misreading of mRNA by the ribosome. Their mechanism of action is somewhat obscure, but they seem to affect some aspect of translocation. The high level of translation errors they cause could be lethal.

Bacterial mutants resistant to these antibiotics are quite rare, and multiple mutations are required to confer high levels of resistance. The fact that resistant mutants are rare has contributed to the usefulness of kanamycin and its relatives in biotechnology.

Some genes confer resistance to these antibiotics by phosphorylating or acetylating them, thereby inactivating them. For example, the *neo* (or Kan[r]) gene for kanamycin and neomycin resistance, from transposon Tn*5,* phosphorylates these antibiotics. This gene has been very important in genetics and biotechnology because it expresses kanamycin resistance in almost all gram-negative bacteria and even makes plant and animal cells resistant to kanamycin (more accurately, a derivative, G418), provided that it is transcribed and translated in the plant or animal cells!

BINDING TO 23S RNA

Thiostrepton has been useful for studying ribosome function because it binds to 23S rRNA in the region of the ribosome involved in the peptidyltransferase reaction. Thiostrepton is specific to gram-positive bacteria; it does not enter gram-negative bacterial cells.

Thiostrepton-resistant mutants are missing the ribosomal protein L11 from the 50S ribsomal subunit. Ribosomal protein L11 seems not to be required for protein synthesis but plays a role in guanosine tetraphosphate (ppGpp) synthesis (see chapter 12). Plasmids and transposon genes can confer thiostrepton resistance by methylating certain ribose sugars of the 23S rRNA in certain positions. Eukaryotes may be insensitive to this antibiotic because the ribose sugars of the analogous eukaryotic 28S rRNAs are normally extensively methylated.

BINDING TO THE 30S RIBOSOME

Streptomycin is one of the most useful antibiotics because it is specific for bacteria and works at low concentrations. It is used for treating bacterial diseases of humans and animals as well as of plants.

Of all the antibiotics, the mechanism of action of streptomycin has been studied most extensively. Streptomycin binds very tightly to the 30S ribosome, probably directly to the 16S RNA, and distorts the A site, thereby preventing the correct positioning of tRNA[fMet] for initiation. Other effects are evident at lower concentrations, such as an increased misreading of mRNA, causing the cell to accumulate defective proteins. However, cell killing probably results from the alteration of the A site.

The molecular basis for streptomycin resistance in bacteria has also been well characterized. *E. coli* strains that are streptomycin-resistant mutants have an altered 30S ribosomal protein, S12, which is the product of the *rpsL* (*strA*) gene. Somehow, the altered S12 protein prevents streptomycin binding to the 16S RNA. Mutations that change the *rpsL* gene can make cells almost totally resistant to streptomycin.

Some plasmids and transposons can also carry genes that confer resistance to streptomycin by encoding enzyme products that specifically inactivate the antibiotic. Some of these enzymes adenylate (transfer an AMP to) the antibiotic, while others phosphorylate it.

SUMMARY

1. RNA is a polymer made up of a chain of ribonucleotides. The bases of the nucleotides—adenine, cytosine, uracil, and guanine—are attached to the five-carbon sugar ribose. Phosphate bonds connect the sugars to make the RNA chain, attaching the third (3') carbon of one sugar to the fifth (5') carbon of the next sugar. The 5' end of the RNA is the nucleotide that has a free phosphate attached to the 5' carbon of its sugar. The 3' end has a free hydroxyl group at the 3' carbon, with no phosphate attached. RNA is both made and translated from the 5' end to the 3' end.

2. After they are synthesized, RNAs can undergo extensive processing and modification. Processing is when phosphate bonds are broken or new phosphate bonds are formed. Modification is when the bases or the sugars of the RNA are chemically altered, for example, by methylation. The rRNAs and tRNAs, but not the mRNAs, of bacteria are extensively modified.

3. The primary structure of an RNA is its sequence of nucleotides. The secondary structure is formed by hydrogen bonding between bases in the same RNA to give localized double-stranded regions. The tertiary structure is the three-dimensional shape of the RNA due to the stiffness of the double-stranded regions of secondary structure. All RNAs including mRNA, rRNA, and tRNA probably have secondary and tertiary structure.

4. The enzyme responsible for making RNA is called RNA polymerase. One of the largest enzymes in the cell, the bacterial RNA polymerase, has four subunits plus another detachable subunit, the σ factor, which comes off after the initiation of transcription.

5. Transcription begins at well-defined sites on DNA called promoters. What type of promoter is used depends upon the type of σ factor bound to the RNA polymerase.

6. Transcription stops at sequences in the DNA called transcription terminators, which can be either factor dependent or factor independent. The factor-independent terminators have a string of A's that follows a symmetric sequence. This sequence allows the RNA transcribed from that region to fold back on itself to form a loop or hairpin, which causes the RNA molecule to fall off the DNA template. The factor-dependent terminators do not have such a well-defined sequence. The ρ protein is the best-characterized termination factor in *E. coli*.

7. Most of the RNA in the cell falls into three groups: messenger (mRNA), ribosomal (rRNA), and transfer (tRNA). Of these, mRNA is very unstable, existing for only a few minutes before being degraded. rRNAs in bacteria are further divided into three types: 16S, 23S, and 5S. Both rRNA and tRNA are very stable and account for about 95% of the total RNA. Other RNAs include the primers for DNA replication and small RNAs involved in regulation or RNA processing.

8. Ribosomes, the site of protein synthesis, are made up of two subunits, the 30S subunit and the 50S subunit, as well as many proteins. The 16S rRNA is in the 30S subunit, while the 23S and 5S rRNAs are in the 50S subunit.

9. Polypeptides are chains of the 20 amino acids, which are held together by peptide bonds between the amino group of one amino acid and the carboxy group of another. The amino terminus or N terminus of the polypeptide has the amino acid with an unattached amino group. The carboxy terminus or C terminus of a polypeptide has the amino acid with a free carboxy group.

10. Translation is the synthesis of polypeptides from mRNA. During translation, the mRNA moves in the 5'-to-3' direction along the ribosome three nucleotides at a time. Three reading frames are possible depending on how the ribosome is positioned at each triplet.

11. The genetic code is the assignment of each possible three-nucleotide codon sequence in mRNA to 1 of 20 amino acids. The code is redundant, with more than one codon sometimes encoding the same amino acid. Because of wobble, the first position of the tRNA anticodon does not have to be complementary to the third position of the antiparallel codon sequence.

12. Initiation of translation occurs at translation initiation regions (TIRs) on the mRNA that consist of an initiation codon, usually AUG or GUG, and often a Shine-Dalgarno sequence, a short sequence that is complementary to part of the 16S rRNA and precedes the initation codon.

13. The first tRNA to enter the ribosome is a special methionyl-tRNA called fMet-tRNA$_f^{Met}$, which carries the amino acid formylmethionine. After the polypeptide has been synthesized, the formyl group and often the first methionine are removed.

14. Translation termination occurs when one of the terminator or nonsense codons UAA, UAG, or UGA is encountered as the ribosome moves down the mRNA. Proteins called ribosome release factors (RFs) are also required for release of the polypeptide.

15. The primary structure of a polypeptide is the sequence of amino acids in the polypeptide. Proteins can be made up of more than one polypeptide chain, which can be the same as or different from each other. The secondary structure results from hydrogen bonding of the amino acids to form α-helical regions and β-sheets. Tertiary structure refers to how the chains fold up on themselves, and quatenary structure refers to one or more different polypeptide chains folding

(continued)

SUMMARY (continued)

up on each other. Proteins can also be held together by disulfide linkages between cysteines in the protein.

16. An open reading frame (ORF) is a string of amino acid codons in DNA unbroken by a nonsense codon. In vitro transcription-translation systems or transcriptional and translation fusions are often required to prove that an ORF in DNA actually encodes a protein.

17. The strand of DNA from which the mRNA is made is the transcribed strand. The opposite strand, which has the same sequence as the mRNA, is the coding strand.

18. A sequence 5′ on the coding strand of DNA is said to be upstream, whereas a sequence 3′ is downstream.

19. The TIR sequence of a gene does not necessarily occur at the beginning of the mRNA. The 5′ end of the mRNA is called the 5′ untranslated region. Similarly, the sequence downstream of the nonsense codon is the 3′ untranslated region.

20. Because mRNA is both transcribed and translated in the 5′-to-3′ direction, it can be translated as it is transcribed in bacteria, which have no nuclear membrane.

21. Bacteria often make polycistronic mRNAs with more than one polypeptide coding sequence on an mRNA. This makes possible polarity of transcription and translation coupling, phenomena unique to bacteria.

22. Many naturally occurring antibiotics attack components of the transcription and translation apparatuses. Some of the more useful are rifampin, streptomycin, tetracycline, thiostrepton, chloramphenicol, and kanamycin. Besides treating bacterial infections and being used in tumor chemotherapy, such antibiotics have also helped us understand the workings of the transcription and translation apparatuses. In addition, the genes that confer resistance to these antibiotics have served as selectable genetic markers and reporter genes in molecular genetics studies of both bacteria and eukaryotes.

QUESTIONS FOR THOUGHT

1. Which do you think came first, DNA, RNA, or protein? Why?

2. Why is the genetic code universal?

3. Why do you suppose prokaryotes have polycistronic messages but eukaryotes do not?

4. Why do you suppose mitochondrial genes differ in their genetic code from chromosomal genes?

5. Why is selenocysteine inserted into proteins of almost all organisms but only into a few sites in a few proteins in these organisms?

6. Why do so many antibiotics inhibit the translation process as opposed to, say, amino acid biosynthesis?

PROBLEMS

1. What is the longest open reading frame (ORF) in the mRNA sequence 5′AGCUAACUGAUGUGAUGUCAACGUCCUACUCUAGCGUAGUCUAAAG3′? Remember to look in all three frames.

2. Where do you think translation is most likely to start in the mRNA sequence 5′UAAGUGAAAGAUGUGAAUGAAGUAGCCACCAAAGUCACUAAUGCUUCCAACA3′? Why?

3. Which of the following is more likely to be a factor-independent transcription termination site? Note that in each case, only the transcribed strand of the DNA is shown in the 3′-to-5′ direction.

a. 3′AACGACTAGTACGACATACTAGTCGTTGGCAAAAAAAAATGCA5′

b. 3′ACTAGCCTAAGCATCTTGCATCAGGCACAGAAAAAAAAAATCGCA5′

SUGGESTED READING

Alberts, B., D. Bray, J. Lewis, M. Raff, K. Roberts, and J. D. Watson. 1994. *Molecular Biology of the Cell*, 3rd ed. Garland Publishing Inc., New York.

Bardwell, J. C. A., and J. Beckwith. 1993. The bonds that tie: catalyzed disulfide bond formation. *Cell* **74**:769–771. (Minireview.)

Brenner, S., F. Jacob, and M. Meselson. 1961. An unstable intermediate carrying information from genes to ribosomes for protein synthesis. *Nature* (London) **190**:576–581.

Draper, D. 1996. Translation initiation, p. 902–908. *In* F. C. Neidhart, R. Curtiss III, J. L. Ingraham, E. C. C. Lin, K. B. Low, B. Magasanik, W. S. Reznikoff, M. Riley, M. Schaechter, and H. E. Umbarger (ed.), *Escherichia coli and Salmonella: Cellular and Molecular Biology*, 2nd ed. ASM Press, Washington, D.C.

Helmann, J. D., and M. J. Chamberlin. 1988. Structure and function of bacterial sigma factors. *Annu. Rev. Biochem.* **57**:839–872.

Hershey, J.W. B. 1987. Protein synthesis, p. 613–647. In F. C. Niedhardt, J. L. Ingraham, K. B. Low, B. Magasanik, M. Schaechter, and H. E. Umbarger (ed.), *Escherichia coli and Salmonella typhimurium: Cellular and Molecular Biology*. American Society for Microbiology, Washington, D.C.

Mathews, C. K., and K. E. van Holde. 1996. *Biochemistry*, 2nd ed. Benjamin-Cummings, Redwood City, Calif.

Nirenberg, M.W., and J. H. Matthei. 1961. The dependence of cell-free protein synthesis in *E. coli* upon naturally occurring or synthetic polyribonucleotides. *Proc. Natl. Acad. Sci. USA* **47**:1588–1602.

Richardson, J. P., and J. Greenblatt. 1996. Control of RNA chain elongation and termination, p. 822–848. *In* F. C. Neidhardt, R. Curtiss III, J. L. Ingraham, E. C. C. Lin, K. B. Low, B. Magasanik, W. S. Reznikoff, M. Riley, M. Schaechter, and H. E. Umbarger (ed.), *Escherichia coli and Salmonella: Cellular and Molecular Biology*, 2nd ed. ASM Press, Washington, D.C.

Stevens, A. 1960. Incorporation of the adenine ribonucleotide into RNA by cell fractions from *E. coli* B. *Biochem. Biophys. Res. Commun.* **3**:92–96.

Stryer, L. 1995. *Biochemistry*, 4th ed. W. H. Freeman and Co., New York.

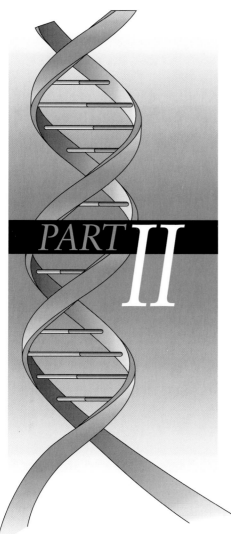

PART *II*

Genes and Genetic Elements

ALL OF THE FEATURES OF A LIVING ORGANISM are determined either directly or indirectly by the products of the genes of the organism. The geneticist manipulates the genes of an organism to study how the organism normally performs its various functions and perhaps to alter these functions in potentially useful ways.

Not all genes are a normal part of the DNA of the organism. Some genes are carried on DNA elements. A DNA element is a DNA sequence which is not common to all of the members of a species of organism and can sometimes be exchanged from one organism to another. Plasmids, transposons, and bacteriophages are DNA elements found in bacteria. These DNA elements can exist either as an integral part of the normal chromosomal DNA or autonomously in the cell as a self-replicating entity. While these DNA elements can be considered parasites and depend upon the host cell for their replication and maintenance, they often carry genes that change the properties of the bacterium and sometimes confer advantages on the bacterium. Bacteria are also unusual among organisms on Earth in that they depend on their DNA elements to exchange genetic information from one bacterium to another. Detailed knowledge about bacterial DNA elements has been exploited to study gene structure and function in both bacteria and eukaryotic cells.

In part II, we show how mutations, which are changes in the DNA sequence, manifest themselves in bacteria. We also discuss the properties of the different types of mutations and how they can be distingished. We give the properties of the different types of DNA elements that exist in bacteria and how these DNA elements affect the bacteria that carry them and how they can transfer themselves as well as normal bacterial genes between members of the population. These chapters give the necessary background for later chapters where we give more specific examples to illustrate how this knowledge can be used to study many of the functions of bacteria.

Chapter 3

Mutations in Bacteria

A S MENTIONED IN THE INTRODUCTORY CHAPTER, the relative ease with which bacteria can be handled genetically has made them very useful model systems for understanding many life processes, and much of the information on basic macromolecular synthesis that we discussed in the first two chapters came from genetic experiments with bacteria. In this chapter, we introduce the genetic concepts and definitions that will be used in later chapters.

Definitions

In genetics, as in any field of knowledge, we need definitions. However, words do not mean much when taken out of context, and so here we define only the most basic terms. We will define other important terms, as we go along; these appear in boldface.

Terms Used in Genetics

These words are common to all types of genetic experiments, whether with prokaryotic or eukaryotic systems.

MUTANT
The word **mutant** refers to an organism that is the direct offspring of a normal member of the species (the **wild type**) but is different. Organisms of the same species isolated from nature that have different properties are usually not called mutants but, rather, variants or **strains**, because even if one of the strains has recently arisen from the other in nature, we have no way of knowing which one is the mutant and which is the wild type.

PHENOTYPE

The phenotypes of an organism are all the observable properties of that organism. Usually in genetics, the term **phenotype** means **mutant phenotype**, or the characteristics of the mutant organism that differ from those of the wild type. The corresponding normal property is sometimes referred to as the **wild-type phenotype**.

GENOTYPE

The **genotype** of an organism refers to the actual sequence of its DNA. If two organisms have the same genotype, they are genetically identical to each other. Identical twins have almost the same genotype.

MUTATION

A **mutation** refers to any heritable change in the DNA sequence. Practically every imaginable type of change is possible, and all changes are called mutations. However, the word "heritable" must be emphasized. Changes or damage that are repaired, so that the original sequence is restored, are not inherited. Hence, only a permanent change in the sequence of deoxynucleotides constitutes a mutation.

ALLELE

Different forms of the same gene are called **alleles**. For example, if one form of a gene has a mutation and the other has the wild-type sequence, the two forms of the gene are different alleles of the same gene. In this case, one gene is a mutant allele and the other gene is the wild-type allele. Diploid organisms can have two different alleles of the same gene, one on each homologous chromosome. The term "allele" can also refer to genes with the same or similar sequences that appear at the same chromosomal location in closely related species. However, similar gene sequences occurring in different chromosomal locations are not alleles; rather, they are **copies** of the gene.

USE OF GENETIC DEFINITIONS

The following example illustrates the use of the definitions in the previous section. For an explanation of the methods, see the introductory chapter.

A culture of *Pseudomonas fluorescens* normally grows as bright green colonies on agar plates. However, suppose that one of the colonies is colorless. It probably arose through multiplication of a **mutant** organism. The **mutant phenotype** is "colorless colony," and the corresponding **wild-type phenotype** is "green colony." The mutant bacterium that formed the colorless colony probably had a **mutation** in a gene for an enzyme required to make the green pigment. Perhaps the mutation consists of a base pair change in the gene, causing the insertion of a wrong amino acid in the polypeptide, and the resulting enzyme cannot function. Thus, the mutant and wild-type bacteria have different **alleles** of this gene, and we can refer to the gene in the colorless colony-forming bacteria as the **mutant allele** and the gene in the green-colony-forming bacteria as the **wild-type allele**.

In the example above, we only know that a mutation has occurred because of the lack of color. However, recall that any heritable change in the DNA sequence is a mutation, and so mutations can occur without changing the organism's phenotype. Many such changes, called **silent mutations**, have been found by sequencing the DNA directly.

Genetic Names

There are some commonly accepted rules for naming mutants, phenotypes, and mutations in bacteria, although different publications sometimes use different notations. We use the terms recommended by the American Society for Microbiology.

NAMING MUTANT ORGANISMS

The mutant organism can be given any name as long as the designation does not refer specifically to the phenotype or the gene thought to have been mutated. This rule helps to avoid confusion if the gene with the mutation is introduced into another strain or if other mutations occur or are transferred into the original organism. Quite often, someone who has isolated a mutant names it after himself or herself, giving it his or her initials and a number (e.g., *Escherichia coli* AB2497). This notation is not intended to gratify the person's ego but to inform others where they can obtain the mutant strain and get advice about its properties. If another mutation alters the mutant strain, this new strain is usually given another name, such as *E. coli* AB2498.

NAMING GENES

Bacterial genes are designated by three lowercase italicized letters that usually refer to the function of the gene's product, when this is known. For example, the name *his* refers to a gene whose product is an enzyme required to synthesize the amino acid histidine. Sometimes more than one gene encodes a product with the same function, or an enzymatic pathway requires more than one different polypeptide. In these cases, a capital letter designating each individual gene follows the three lowercase letters. For example, the *hisA* and *hisB* genes both encode polypeptides required to synthesize histidine. A mutation that inactivates either gene will make the cell unable to synthesize histidine.

NAMING MUTATIONS

Hundreds of different types of mutations can occur in a single gene, and so all alleles of a particular gene have a specific allele number. For example, *hisA4* refers to the *hisA* gene with mutation number 4, and the *hisA* gene with mutation number 4 is referred to as the *hisA4* allele.

If a mutation is known to inactivate the product of a gene, a superscript minus sign, or simply the word "mutation," may be added to the gene or allele name. For example, *hisA⁻* or a *hisA* mutation inactivates the product of the *hisA* gene. Alternatively, the designation *hisA⁺* refers specifically to the wild-type form of the *hisA* gene, which encodes a functional gene product.

Different nomenclatural rules apply if a mutation is a deletion or insertion. We shall defer a discussion of these rules until we discuss these types of mutations (see below).

NAMING PHENOTYPES

Phenotypes are also denoted by three-letter names, but the letters are not italicized and the first letter is capitalized. As with genotypes, superscripts are often used to distinguish mutant from wild-type phenotypes. For example, His⁻ describes the phenotype of an organism with a mutated *his* gene that will not grow without histidine in its environment. The corresponding wild-type organisms will grow without histidine, so they are phenotypically His⁺. Another example, Rifʳ, describes resistance to the antibiotic rifampin, which blocks RNA synthesis (see chapter 2). A mutation in the *rpoB* gene, which encodes a subunit of the RNA polymerase, makes the cell resistant to this antibiotic. The corresponding wild-type phenotype is rifampin sensitivity, or Rifˢ.

Useful Phenotypes in Bacterial Genetics

What phenotypes are useful for genetic experiments depends on the organism being studied. For bacterial genetics, the properties of the colonies formed on agar plates are the most useful phenotypes (see the introductory chapter).

The visual appearance of colonies sometimes provides useful mutant phenotypes, such as the colorless colony discussed earlier. Colonies formed by mutant bacteria might also be smaller than normal or smooth instead of wrinkled. The mutant bacterium may not multiply to form a colony at all under some conditions, or, conversely, it may multiply when the wild type cannot.

Many mutant phenotypes have been used to study cellular processes such as DNA recombination and repair, mutagenesis, and development. The following sections describe a few of the more commonly used

phenotypes. In later chapters, we shall discuss many more types of mutants and demonstrate how mutations can be used to study life processes.

Auxotrophic Mutants

Some of the most useful bacterial mutants are **auxotrophic mutants**, or auxotrophs. Of the two types of these mutants, one cannot multiply without a particular growth supplement that is not required by the original, wild-type isolate. For example a His⁻ auxotrophic mutant cannot grow unless the medium is supplemented with the amino acid histidine, while the wild type could grow without added histidine. Similarly, a Bio⁻ auxotrophic mutant cannot grow without the vitamin biotin, which is not needed by the wild type.

The other type of auxotrophic mutant cannot use a particular substance for growth that can be used by the wild type. For example, the wild-type bacteria may be able to use the sugar maltose as a sole carbon and energy source, but a Mal⁻ auxotrophic mutant must be given another carbon and energy source.

Even though these two types of auxotrophs seem opposite, their molecular basis is similar. In both types, a mutation has altered a gene encoding an enzyme of a metabolic pathway, thereby inactivating the enzyme. The only difference is that in the first case, the inactivated enzyme was in a **biosynthetic** pathway, which is required to synthesize a substance, while in the second case, the inactivated enzyme was in a **catabolic** pathway, which is required to degrade a substance to use it as a carbon and energy source.

ISOLATING AUXOTROPHIC MUTANTS

Figure 3.1 shows a simple method for isolating mutants auxotrophic for histidine and biotin. In this experiment, eight colonies from plate 1, which contains all the nutrients the bacteria need, including histidine and biotin, were picked up with a loop and transferred onto two other plates. These plates are the same as plate 1 except that plate 2 lacks biotin but has histidine and plate 3 lacks histidine but has biotin. The bacteria from most of the colonies can multiply on all three types of plates. However, the bacteria in colony 2 grow only on plate 2; they are mutants that require added histidine, that is, they are His⁻. However, these mutants do not require biotin, and so are Bio⁺. Similarly, the bacteria in colony 6 are Bio⁻ but His⁺, since they can grow on plate 3 but not on plate 2. Under real conditions, mutants that require histidine or biotin would not be this frequent, and thousands of colonies would have to be tested to find one mutant that required histidine, biotin, or indeed any growth supplement not required by the wild type.

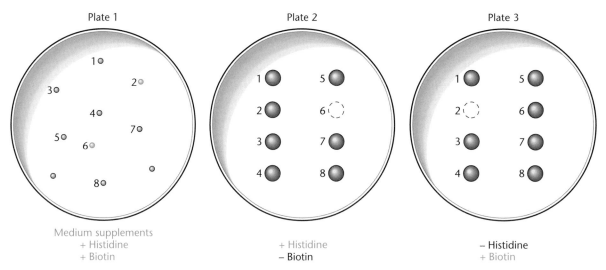

Figure 3.1 Detection of auxotrophic mutants. Colonies were scraped with a loop from plate 1 and transferred onto plates 2 and 3. Colony 6 contains bacteria that cannot multiply without biotin and so are Bio⁻. The bacteria in colony 2 are His⁻.

In principle, it should be possible to find auxotrophic mutants unable to synthesize any compound required for growth or unable to use any carbon and energy source. However, auxotrophic mutants must be supplied with the compound they cannot synthesize, and these compounds must enter the cell. Yet many bacteria cannot take in some compounds that have a high electrical charge, such as nucleotides, so that some types of auxotrophs are very difficult to isolate.

Conditional Lethal Mutants

Auxotrophic mutants can be isolated because they have mutations in genes whose products are required under only certain conditions. The cells can be grown under conditions where the product of the mutated gene is not required and tested under conditions where it is required. However, many gene products of the cell are essential for growth no matter what conditions the bacteria find themselves in. The genes that encode such functions are called **essential genes**. Examples of essential genes include those for RNA polymerase, ribosomal proteins, DNA ligase, and helicases. Cells with mutations that inactivate essential genes cannot be isolated unless the mutations inactivate the gene under only some conditions. Hence, any mutants that are isolated will have **conditional lethal mutations**, because these DNA changes are lethal only under some conditions.

TEMPERATURE-SENSITIVE MUTANTS

The most generally useful conditional lethal mutations in bacteria are mutations that make the mutant **temperature sensitive** for growth. Usually, such mutations

change an amino acid of a protein so that the protein no longer functions at higher temperatures but still functions at lower temperatures. The higher temperatures are called the **nonpermissive temperatures** for the mutant, whereas the temperatures at which the protein still functions are the **permissive temperatures** for the mutant.

Mutations can affect the temperature stability of proteins in various ways. Often, an amino acid required for the protein's stability at the nonpermissive temperature is changed, causing the protein to unfold, or denature, partially or completely. The protein could then remain in the inactive state or be destroyed by cellular proteases that remove abnormal proteins. If the protein remains, it sometimes spontaneously renatures (refolds) when the temperature is lowered; then growth can resume immediately. With other mutations, the protein is irreversibly denatured and must be resynthesized before growth can resume.

The temperature ranges used to isolate temperature-sensitive mutations depend on the organism. Bacteria can be considered poikilothermic, or cold-blooded, organisms, with a cell temperature that varies with the outside temperature. Therefore, their proteins are designed to function over a wide range of temperatures. However, different species of bacteria differ greatly in their preferred temperature range. For example, a "mesophilic" bacterium such as *E. coli* may grow well in a range of temperatures from 20 to 42°C. In contrast, a "thermophilic" bacterium such as *Bacillus stearothermophilus* may grow well only between 42 and 60°C. For *E. coli*, a temperature-sensitive mutation may make a protein nonfunctional at 42°C, the nonpermissive temperature, but not at 33°C, the permissive temperature. For *B. stearothermophilus*, the temperature-sensitive mutation may make a protein inactive at the nonpermissive temperature of 55°C but leave it functional at the permissive temperature of 47°C.

Isolating Temperature-Sensitive Mutants

In principle, temperature-sensitive mutants are as easy to isolate as auxotrophic mutants. If a mutation that makes the cell temperature sensitive occurs in a gene whose protein product is required for growth, the cells will stop multiplying at the nonpermissive temperature. To isolate such mutants, the bacteria are incubated on a plate at the permissive temperature until colonies appear and then the colonies are transferred to a plate incubated at the nonpermissive temperature. Bacteria that can form colonies at the permissive temperature but not at the nonpermissive temperature are temperature-sensitive mutants. However, temperature-sensitive mutants are usually much rarer than auxotrophic mutants. Many changes in a protein will inactivate it, but very few will make a protein functional at one temperature and nonfunctional at another. We shall discuss the frequency of occurrence of different types of mutations later in the chapter.

COLD-SENSITIVE MUTANTS

Cells with proteins that fail to function at lower temperatures are called specifically **cold-sensitive mutants**. Mutations that make a bacterium cold sensitive for growth are often in genes whose products must form a larger complex such as the ribosome. The increased movement at the higher temperature may allow the mutated protein, despite its altered shape, to enter the complex, but it will not be able to do so at lower temperatures.

NONSENSE MUTATIONS

Mutations that change a codon in a gene to one of the three nonsense codons—UAA, UAG, or UGA—can also be conditional lethal mutations. A nonsense mutation will cause translation to stop within the gene unless the cell has a "nonsense suppressor" tRNA, as we shall explain later in the chapter. Because nonsense mutations are more generally useful in viral genetics than bacterial genetics, we discuss them in more detail in chapter 13.

Resistant Mutants

Among the most useful types of bacterial mutants to isolate are resistant mutants. If a substance kills or inhibits the growth of a bacterium, mutants resistant to the substance can often be isolated merely by plating the bacteria in the presence of the substance.

The numerous mechanisms of resistance depend on the basis for toxicity and on the options available to prevent the toxicity (examples are given in Table 3.1). For example, the mutation may destroy a cell surface receptor to which the toxic substance must bind to enter the cell. If the substance cannot enter the cell, the mutant will not be killed by it. Alternatively, a mutation might change the "target" affected by the substance inside the cell. For example, an antibiotic might normally bind to a ribosomal protein and affect protein translation. However, if the antibiotic cannot bind to a mutant (but still functional) protein, it cannot kill the cell. In some cases, the substance added to the cells is not toxic until one of the cell's own enzymes changes it. A mutation inactivating the enzyme that converts the nontoxic substance into the toxic one could make the cell resistant to that substance.

Inheritance in Bacteria

Salvador Luria and Max Delbrück were among the first people to attempt to study inheritance quantitatively in bacteria. They published a now-classic paper in the journal *Genetics* in 1943. This paper is still very much worth reading and is listed in the Suggested Reading section at the end of the chapter. As discussed in the introductory chapter, the experiments and reasoning of

TABLE 3.1	Some resistance mutations	
Substance	**Toxicity**	**Resistance mutation**
Bacteriophage T1	Infects and kills	Inactivates *tonB* outer membrane protein; phage cannot absorb
Streptomycin	Binds to ribosomes; inhibits translation	Changes ribosomal protein S12 so that it no longer binds
Chlorate	Converted to chlorite, which is toxic	Inactivates nitrate reductase, which converts chlorate to chlorite
High concentrations of valine, no isoleucine	Feedback inhibits acetolactate synthetase; starves for isoleucine	Activates a valine-insensitive acetolactate synthetase

Luria and Delbrück helped debunk what was then a popular misconception among bacteriologists. At the time of the Luria and Delbrück studies, it was generally believed that bacteria were different from other organisms on Earth in their inheritance. It was generally accepted that heredity in higher organisms followed "neo-Darwinian" principles. According to Darwin, random mutations occur, and if one happened to confer a desirable phenotype, organisms with this mutation would be selected by the environment and become the predominant members of the population. Undesirable as well as desirable mutations would continuously occur, but only the desirable mutations would be passed on to future generations.

However, many bacteriologists believed that heredity in bacteria followed different principles. They thought that bacteria, rather than changing as the result of random mutations, somehow "adapt" to the environment by a process of directed change, after which the adapted organism would pass the adaptation on to its offspring. This process is called Lamarckian Inheritance, and belief in it was encouraged by the observation that all the bacteria in a culture exposed to a toxic substance seem to become resistant to that substance in response to it (Figure 3.2).

The Luria and Delbrück Experiment

The Luria and Delbrück experiment was designed to test two hypotheses for how mutants arise in bacterial cultures: the **random-mutation** hypothesis and the **directed-change** hypothesis. The random-mutation hypothesis predicts that the mutants appear prior to the addition of the selective agent, whereas the directed-change hypothesis predicts that mutants appear only in response to a selective agent.

Figure 3.3 illustrates that mutations that occur early in the culture will have a disproportionate effect on the number of mutants in the culture. In culture 1, only one

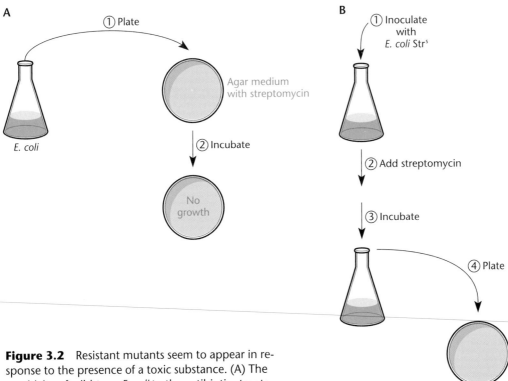

Figure 3.2 Resistant mutants seem to appear in response to the presence of a toxic substance. (A) The sensitivity of wild-type *E. coli* to the antibiotic streptomycin. Plating an *E. coli* culture on streptomycin-containing agar medium results in a lack of colony growth after incubation. (B) The emergence of streptomycin-resistant *E. coli* mutants. A flask containing wild-type *E. coli* is incubated in the presence of streptomycin. When the contents of the flask are transferred to an agar medium also containing streptomycin, streptomycin-resistant mutants form colonies.

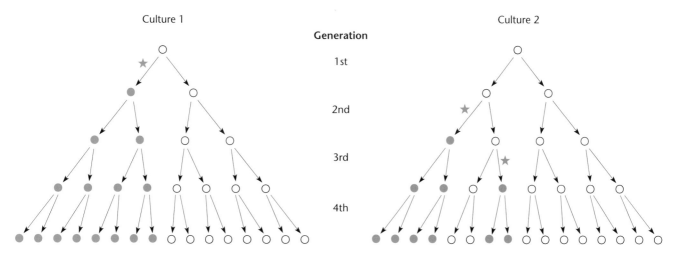

Figure 3.3 Earlier mutations give rise to more mutant progeny in a growing culture. Only one mutation occurred in culture 1, but it gave rise to eight mutant progeny because it occurred in the first generation. In culture 2, two mutations occurred, one in the second generation and one in the third. However, because these mutations occurred later, they gave rise to only six mutant progeny. The mutant cells are shaded.

mutation occurred, but this mutation gave rise to eight resistant mutants because it occurred early. In culture 2, two mutations arose, but they gave rise to only six resistant mutants because they occurred later.

To determine if mutants appear before or after addition of a selective agent, one can grow several cultures in the absence of the selective agent, add the agent to all the cultures at the same time, and then measure the fraction of bacteria resistant to the selective agent in each culture. If the random-mutation hypothesis is correct, the number of mutant colonies will vary among all the cultures, depending on when the mutations occurred, as was illustrated in Figure 3.3. In contrast, if the directed-change hypothesis is correct, each bacterium will have the same chance of becoming a mutant, but only after the selective agent is added, and so the same percentage of the bacteria should become resistant in all the cultures. Therefore, a result in which the number of mutants per culture varies greatly will favor the random-mutation hypothesis but a result in which the number of mutants in a series of cultures is about the same will favor the directed-change hypothesis.

In their experiments, Luria and Delbrück used *E. coli* as the bacterium and bacteriophage T1 as the selective agent. As shown in Table 3.1 and Figure 3.4, phage T1 will kill wild-type *E. coli*, but a mutation in the gene for an outer membrane protein called TonB can make these cells resistant to killing by the phage. If bacteria are spread on an agar plate with the phage, only those resistant to the phage will multiply to form a colony. All the others will be killed. The number of colonies on the

plate is therefore a measure of the number of bacteria resistant to the bacteriophage in the culture.

Figure 3.5 shows the two experiments that Luria and Delbrück performed, which seem superficially similar but gave very different results. In experiment 1, they

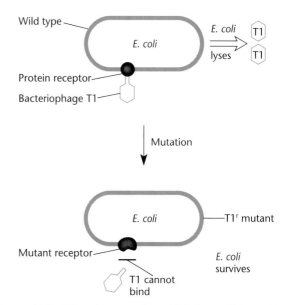

Figure 3.4 When bacteriophage T1 infects wild-type *E. coli*, it binds to a receptor in the outer membrane, protein TonB (Table 3.1). After phage replication, the *E. coli* cell is lysed and new phage are released. A mutation in the *tonB* gene results in an altered (mutant) receptor to which T1 can no longer bind or eliminates the receptor, and so the cells survive.

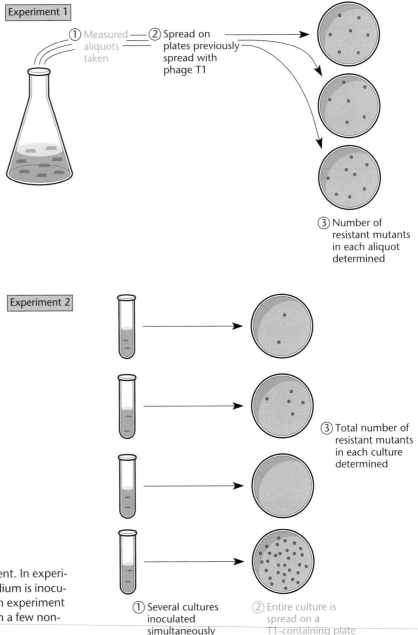

Figure 3.5 The Luria and Delbrück experiment. In experiment 1, a single flask containing standard medium is inoculated with bacteria and incubated overnight. In experiment 2, a number of smaller cultures are started with a few nonmutant bacteria. See the text for details.

started one culture of bacteria. After incubating it, they took out small aliquots and plated them with and without phage T1 to measure the number of resistant mutants as well as the total number of bacteria in the culture. They then calculated the fraction of resistant mutants. In experiment 2, they started a large number of relatively smaller cultures. After incubating these cultures, they measured the number of resistant mutants and the total number of bacteria in each culture.

Table 3.2 displays some representative results. In experiment 1, the number of resistant mutants in each

aliquot is almost the same, subject only to sampling errors and statistical fluctuations. However, in experiment 2, a very large variation in the number of resistant bacteria per culture was found. Some cultures had no resistant mutants, while some had many. One culture even had 107 resistant mutants! Luria and Delbrück referred to this and the other mutant-rich cultures as "jackpot" cultures. Apparently, these are cultures in which a mutation to resistance occurred very early. Hence, these results fulfill the predictions of the random-mutation hypothesis. In contrast, the directed-change hypothesis

TABLE 3.2	The Luria and Delbrück experiment		
Experiment 1		**Experiment 2**	
Aliquot no.	No. of resistant bacteria	Culture no.	No. of resistant bacteria
1	14	1	1
2	15	2	0
3	13	3	3
4	21	4	0
5	15	5	0
6	14	6	5
7	26	7	0
8	16	8	5
9	20	9	0
10	13	10	6
		11	107
		12	0
		13	0
		14	0
		15	1
		16	0
		17	0
		18	64
		19	0
		20	35

predicts that the results of the two experiments should be the same, and certainly no jackpot cultures should appear in the second experiment. Box 3.1 presents these predictions in statistical terms.

The Newcombe Experiment

The analysis of Luria and Delbrück was fairly sophisticated mathematically and so was not generally understood. Some people still held to the belief that bacteria were somehow different from other organisms in their inheritance. Consequently, Howard Newcombe in 1951 devised an experiment that was conceptually simpler and so convinced many skeptics. In his experiment, he also used *E. coli* mutants resistant to phage T1.

The principle behind Newcombe's experiment is that if the random-mutation hypothesis is correct, mutants should be **clonal**; that is, one mutant bacterium will give rise to more, even in the absence of the selective agent (see the introductory chapter). However, if the directed-change hypothesis is correct, mutants should not be clonal, because they will not be multiplying to form a colony before the selective agent is added. Instead, all the mutants will first appear at the time the selective agent is added.

To detect clones of resistant bacteria, Newcombe analyzed cultures grown on agar plates. According to the random-mutation hypothesis, the number of colonies

BOX 3.1

Statistical Analysis of the Number of Mutants per Culture

A simple statistical analysis shows us that the number of mutants in experiment 2 of Luria and Delbrück does not follow a normal distribution. If the number of mutants per culture follows a normal distribution, the variance would be approximately equal to the mean:

$$\text{Variance} = \sum_{i=1}^{n} \frac{(M_i - \overline{M})^2}{n-1} = \text{mean} = \overline{M} = \sum_{i=1}^{n} \frac{M_i}{n}$$

where M_i is the number of mutants in each culture and n is the number of cultures.

In experiment 1 of Luria and Delbrück, the variance was 18.23 and the mean was 16.7, so they are approximately equal and vary owing to statistical fluctuations and pipetting errors. In experiment 2, however, the variance was 752.38 and the mean was 11.35—very different values. Therefore, the number of mutants per culture does *not* follow a normal distribution, and the result is not consistent with the directed-change hypothesis; however, it is consistent with the random-mutation or neo-Darwinian hypothesis.

due to resistant mutants on an agar plate will vary depending on whether the colonies are left alone or disturbed by having been spread out on the plate. When the colonies on a plate are not disturbed, all the descendants of a particular bacterium will remain together in the same colony. However, if the colonies on a plate are disturbed, each resistant bacterium should give rise to a separate colony of resistant bacteria. Consequently, a spread plate will have many more resistant colonies than an unspread plate. However, the directed-change hypothesis predicts that the mutants need not arise from each other (i.e., are not clonal), so that the number of resistant colonies on the undisturbed and disturbed plates should be about the same, because the resistant bacteria will have appeared only at the time the phage were added.

How Newcombe did his experiment is illustrated in Figure 3.6, and some of his actual data are given in Table 3.3. He first spread the same number of bacteria (an average of 5.1×10^4) on plates and incubated them. (He actually used many more plates than the six pictured in Figure 3.6, but this number serves as an illustration.) After 5 h, he removed three of the plates and sprayed one with the virus without disturbing the bacteria. This is the unsp (*unsp*read) plate in Table 3.3. He treated the second plate in the same way, except that he spread the bacteria around before spraying them with the virus. This is the sp (*sp*read) plate in Table 3.3. He washed the bacteria off the third plate, diluted them, and plated them without the virus to determine the total number of bacteria on the plates at this time. After a 6-h incubation, he took the remaining three plates out of the incubator and subjected them to the same treatment. He incubated all the plates overnight, and the next day he counted the colonies produced by phage-resistant mutants. The data in Table 3.3 show that the spread plates have many more resistant colonies than the corresponding unspread plates, a result supporting the random-mutation hypothesis. The difference became greater the longer the plates were incubated. After 5 h, only a few resistant colonies had appeared, and there was not much difference between the unspread and spread plates (8

Figure 3.6 The Newcombe experiment. See the text for details.

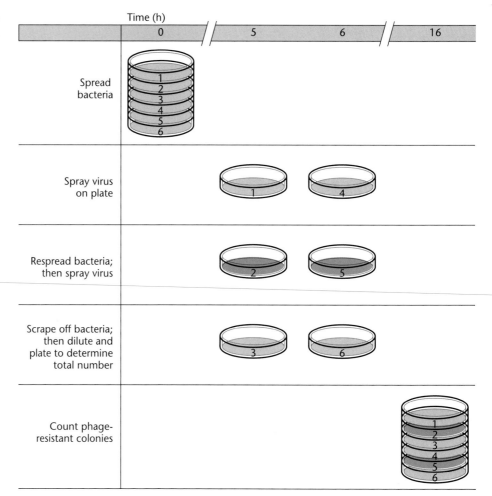

TABLE 3.3	The Newcombe experiment		No. of resistant colonies[a]	
Incubation time (h)	No. of bacteria plated	Ending no. of bacteria	unsp	sp
5	5.1×10^4	2.6×10^8 (plate 3)	8 (plate 1)	13 (plate 2)
6	5.1×10^4	2.8×10^9 (plate 6)	49 (plate 4)	3,719 (plate 5)

[a]unsp, unspread; sp, spread.

and 13 colonies, respectively), because the mutants had not had much time to multiply, and so each colony contained few resistant bacteria. However, after 6 h, the numbers of unspread and spread colonies are very different (49 and 3,719, respectively). Note also that the number of resistant bacteria, as measured by the resistant colonies on the spread plates, increased faster than the total population. At 5 h, there were 13 resistant bacteria in 2.6×10^8 total bacteria, so the fraction of resistant bacteria was $13/(2.6 \times 10^8)$, or 1 mutant for every 2×10^7 bacteria. By 6 h, there were 3,719 resistant mutants in 2.8×10^9 total bacteria, so that the fraction was $3,719/(2.8 \times 10^9)$, or 1 mutant for every 7×10^5 bacteria. Therefore, the fraction of resistant bacteria rose about 30-fold in just 1 h. In other words, the number of resistant mutants increased about 30 times faster than the total number of bacteria during this hour. This fulfills a prediction of the random-mutation hypothesis, as we explain below in the section on mutation rates.

The Lederbergs' Experiment

The experiments that really buried the directed-change hypothesis, at least as the sole explanation for some types of resistant mutants, were the replica-plating experiments of the Lederbergs (see Lederberg and Lederberg, Suggested Reading). They spread millions of bacteria on a plate without an antibiotic and allowed the bacteria to form a lawn during overnight incubation. This plate was then replicated (see chapter 14) onto another plate containing the antibiotic. After incubating the antibiotic-containing plate, the Lederbergs could determine where antibiotic-resistant mutants had arisen on the original plate by aligning the two plates and marking the regions on the first plate where antibiotic-resistant mutants had grown on the second. They cut these regions out of the original plate, diluted the bacteria, and repeated the experiment. This time, there were many more resistant mutants than previously. Eventually, by repeating this process, they obtained a pure culture of bacteria, all of which were resistant to the antibiotic even though they had never been exposed to it! Therefore, the bacteria must have acquired the resistance independently of exposure to the antibiotic and passed the resistance on to their offspring.

The experiments of Luria and Delbrück, Newcombe, and the Lederbergs prove that at least some types of mutants of *E. coli* arise through Darwinian inheritance. Recently, however, researchers have proposed that some mutations may arise in response to the environment. Others argue that the data suggesting the existence of directed mutations can be explained in terms of the random-mutation hypothesis (see Box 3.2). At present, there are no satisfying mechanisms for how directed mutations could occur.

Mutation Rates

As defined above, a mutation is any heritable change in the DNA sequence of an organism, and we usually know a mutation has occurred because of a phenotypic change in the organism. The **mutation rate** can be loosely defined as the chance of mutation to a particular phenotype. Mutation rates can differ because mutations to some phenotypes occur much more often than mutations to other phenotypes. When many possible different mutations in the DNA can give rise to a particular phenotype, the chance that a mutation to that phenotype will occur is relatively high. However, if only a very few types of mutations can cause a particular phenotype, the mutation rate for that phenotype will be relatively low.

For example, the spontaneous mutation rate for the His⁻ phenotype is hundreds of thousands of times higher than the mutation rate for Str^r. Approximately 11 gene products are required for histidine biosynthesis, and each has hundreds of amino acids, many of which are essential for activity. Changing any of these amino acids will inactivate the enzyme. By contrast, streptomycin resistance results from a change in one of very few amino acids in a single ribosomal protein, S12, so that the mutation rate for streptomycin resistance is very low. Hence, a mutation to Str^r will occur spontaneously in about 1 in 10^{10} to 10^{11} cells, whereas a mutation to His⁻ will occur in about 1 in 10^6 to 10^7 cells.

Generally, if the mutation rate for a phenotype is high, the phenotype probably results from inactivation of the product of a gene or genes. If the mutation rate is low, the phenotype probably is due to a subtle change in

BOX 3.2

The Directed- or Adaptive-Mutation Controversy

One of the basic tenets of neo-Darwinian inheritance is that mutations occur randomly and then are selected by the environment. Mutations that happen to confer a selective advantage on the organism are preferentially passed on to future generations. Some of the details of the theory—for example, the role of random genetic drift in this process—are under debate, but the basic theory seems unassailable. The mutations themselves are random, and the mutations that are favorable are not necessarily more frequent than those that are deleterious or neutral. In other words, mutations are random and not "adaptive" or "directed" based on the needs of the cell.

Experiments by scientists such as Luria and Delbrück supported this conclusion and extended this principle to bacteria, at least for the phenotype of resistance to phage T1. As we have seen, most of the mutations to phage T1 resistance occur prior to the selective pressure, for example, addition of the phage, and those few bacteria that happen to be resistant to the phage as the result of the prior mutation can then multiply, so that their descendents become the predominant bacteria in the population. Neo-Darwinian inheritance makes perfect sense in terms of our understanding of how mutations occur and the mechanism of macromolecular synthesis. Recently, however, the conclusion that mutations in bacteria are always random has been challenged. The initial observation made by John Cairns and his collaborators was that mutations that restored the ability to use the sugar lactose were more frequent under conditions where lactose was the only energy source available. This observation was subsequently extended by the Cairns group and others to include other types of mutations.

These observations have in turn been challenged on the grounds that the state of individual bacteria on plates is difficult to determine. For example, the state of starvation itself may be mutagenic and induce many more random mutations, some of which happen to be advantageous. Also, it has been proposed that a mispairing created during replication could allow an mRNA to be synthesized on one strand of the DNA that encodes a functional protein, even though the other strand still has the original sequence. The synthesis of some functional protein could have an immediate salubrious effect, increasing the energy of the cell and causing the replication fork to move on more quickly, thereby increasing the chances of the mismatch not being repaired and therefore fixed as a mutation. None of these explanations disprove the basic tenets of the neo-Darwinian theory of random mutations, however. In the absence of any persuasive models for how adaptive mutations might occur, the burden of proof must lie with the adaptive-mutation hypothesis. In any case, this controversy has stimulated much experimentation on mechanisms of mutagenesis in bacteria and has forced people to think more clearly about the population genetics of bacteria.

References

Drake, J. W. 1991. Spontaneous mutation. *Annu. Rev. Genet.* **25:** 125–146.

Foster, P. L. 1993. Adaptive mutation: the uses of adversity. *Annu. Rev. Microbiol.* **47:**467–504.

Galitski, T., and J. R. Roth. 1996. A search for a general phenomenon of adaptive mutability. *Genetics* **143:**645–659.

Lenski, R. E., and J. E. Mittler. 1993. The directed mutation controversy and neo-Darwinism. *Science* **259:**188–194.

the properties of a gene product. An extremely high mutation rate to a particular phenotype may indicate not a mutation but rather the loss of a plasmid or prophage or the occurrence of some programmed recombination event such as inversion of an invertible sequence. We shall discuss plasmids and prophages and other gene rearrangements in subsequent chapters.

Calculating Mutation Rates

To calculate mutation rates, we shall define them as the chance of a mutation each time a cell grows and divides. This is a reasonable definition because, as discussed in chapter 1, DNA replicates once each time the cell di-

vides, and most mutations occur during this process. The number of times a cell grows and divides in a culture is called the number of **cell generations**. Therefore, the mutation rate is the number of mutations to a particular phenotype that have occurred in a growing culture divided by the total number of cell generations that have occurred in the culture during the same time.

DETERMINING THE NUMBER OF CELL GENERATIONS

The total number of cell generations that have occurred in an exponentially growing culture is easy to calculate and is simply the the total number of cells in the culture minus the number of cells in the starting innoculum. To

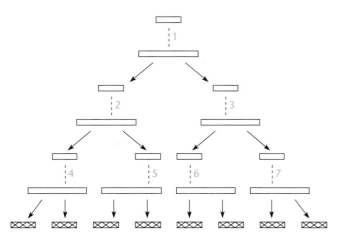

Figure 3.7 The number of cell generations (7) equals the total number of cells in an exponentially growing culture (8) minus the number at the beginning (1).

understand this, look at Figure 3.7. In this illustration, a culture that was started from one cell multiplies to form eight cells in seven cell divisions, or cell generations. This number equals the final number of cells (8) minus the number of cells at the beginning (1). In general, the number of cell generations that have occurred in the culture equals $N_2 - N_1$ if N_2 equals the number of cells at time 2 and N_1 equals the number of cells at time 1.

Therefore, from the definition, the mutation rate (a) is given by

$$a = \frac{m_2 - m_1}{N_2 - N_1}$$

where m_2 and m_1 are the number of mutations in the culture at time 2 and time 1, respectively.

Usually, a culture is started with a few cells and ends with many, so we can often ignore the initial cells and just call the number of cell generations N, where N is the total number of cells in the culture. Then the mutation rate equation can be simplified to

$$a = m/N$$

where m is the number of mutations that have occurred in the culture and N is the number of bacteria. This equation assumes that there were no or an insignificant number of mutants in the culture when it was started, which is likely if the culture was started with only a few cells.

DETERMINING THE NUMBER OF MUTATIONS THAT HAVE OCCURRED IN A CULTURE

From the equations above, it looks as though it might be easy to calculate the mutation rate. The total number of mutations must simply be divided by the total num-

ber of cells. The problem comes in determining the number of mutational events that have occurred in a culture, because mutant cells, not mutational events, are usually what are detected. Moreover, recall from Figure 3.3 that one mutant cell resulting from a single mutational event can give rise to many mutant cells, depending upon when the mutation occurred during the growth of the culture. Therefore, one cannot determine the number of mutations in a culture merely by counting the number of mutant cells. However, in some cases the number of mutant cells can form the basis of a calculation of the number of mutational events and, by extension, the mutation rate.

Using the Data of Luria and Delbrück To Calculate the Mutation Rate

Luria and Delbrück used the data in Table 3.2 to calculate the mutation rate to T1 phage resistance. They assumed that even though the number of mutants per culture does not follow a normal distribution, the number of mutations per culture should, because each cell has the same chance of acquiring a mutation to T1 phage resistance each time it grows and divides. For convenience, the Poisson distribution can be used to approximate the normal distribution. According to the Poisson distribution, if P_i is the probability of having i mutations in a culture, then

$$P_i = \frac{m^i e^{-m}}{i!}$$

where m is the average number of mutations per culture, the number they wanted to know. Therefore, if they knew how many cultures had a certain number of mutations, they could calculate the average number of mutations per culture.

By this method and with the data from experiment 2 in Table 3.2, Luria and Delbrück calculated the average number of mutations per culture. The data give the number of T1-phage-resistant mutants per culture but do not indicate how many of the cultures had one, two, three, or more mutations. For example, cultures with one mutant probably had one mutation, but others, even the one with 107 mutants, might also have had only one mutation. However, how many cultures had zero mutations is clear—those with zero mutants, or 11 of 20. Therefore, the probability of having zero mutations equals 11/20. Applying the formula for the Poisson distribution, the probability of having zero mutations is given by

$$\frac{11}{20} = \frac{m^0 e^{-m}}{0!} = \frac{1 \cdot e^{-m}}{1} = e^{-m}$$

and $m = -\ln 11/20 = 0.59$. Therefore, in this experiment, an average of 0.59 mutation occurred in each culture. From the equation for mutation rate,

$$a = m/N = 0.59/5.6 \times 10^8 = 1.06 \times 10^{-9}$$

Therefore, there is 1.06×10^{-9} mutation per cell generation if there were a total of 5.6×10^8 total bacteria per culture. In other words, a mutation for phage T1 resistance will occur about once every 10^9, or every billion, times a cell divides.

Calculating the Mutation Rate from Newcombe's Data

Newcombe's data can be used more directly to calculate the number of mutations per culture (refer to Figure 3.6). On the unspread plate, each mutation will give rise to only one resistant colony. Therefore, the number of resistant colonies on his unspread plates equals the number of mutational events that have occurred at the time of incubation.

According to Newcombe's data in Table 3.3, from 5 to 6 h there were $49 - 8 = 41$ new resistant colonies on the unspread plates. Therefore, 41 mutations to phage resistance must have occurred during that time interval. During this time, the total number of bacteria went from 2.6×10^8 to 2.8×10^9 based on the total number of bacteria on the plates (see Table 3.3). From the equation for mutation rate,

$$a = \frac{m_2 - m_1}{N_2 - N_1} = \frac{49 - 8}{2.8 \times 10^9 - 2.6 \times 10^8}$$

the mutation rate to T1 resistance is $41/(2.54 \times 10^9) = 1.6 \times 10^{-8}$ mutation per cell generation. In other words, Newcombe's data indicate that a mutation to resistance to phage T1 will occur a little more than once every hundred million times a cell divides. Notice that Newcombe's data give a mutation rate about 10 times higher than that derived from the data of Luria and Delbrück. This discrepancy can be explained by the phenotypic lag, as we explain later.

Using the Increase in the Fraction of Mutants To Measure Mutation Rates

As mentioned, Newcombe's data fulfilled one prediction of the random-mutation hypothesis, i.e., that the number of mutants should increase faster than the total population. In other words, the fraction of mutants in the population should increase as the population grows.

It seems surprising that the total number of mutants increases faster than the total population until one thinks about where mutants come from. If the multiplication of old mutants were the only source of mutants,

the fraction of mutants would remain constant, or even drop, if the mutants do not multiply as rapidly as the normal type (which is often the case). However, new mutations occur constantly, and their progeny are also multiplying. Therefore, new mutations are continuously adding to the total number of mutants.

This fact can also be used to measure mutation rates. The higher the mutation rate, the faster the proportion of mutants will increase (Figure 3.8). In fact, if we plot the fraction of mutants (M/N) against time (in doubling times), as in Figure 3.8, the slope of this curve will be the mutation rate. In theory, this fact could be used to calculate mutation rates. In practice, however, mutation rates are usually low and the bacteria we can conveniently work with are few, so each new mutation makes too large a contribution to the number of mutants and we do not get a straight line. To make this method practicable, we would have to work with trillions of bacteria in a large chemostat.

The fact that other mutations are causing some mutants to become wild type again (in a process known as reversion [see later sections]) will affect the results shown in Figure 3.8. However, reversion of mutants becomes significant only when the number of mutants is very large and the number of mutants times the mutation rate back to the wild type (called the reversion rate) begins to approximate the forward mutation rate times the number of nonmutant bacteria. At earlier stages of culture growth, the vast majority of bacteria will be nonmutant. Also, for reasons that we discuss later in the chapter, reversion rates are generally much lower than forward mutation rates. Therefore, the reversion of mutants to the wild type can generally be ignored.

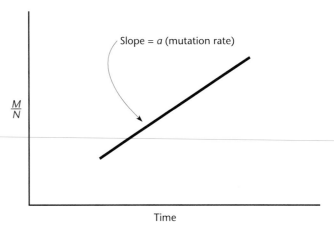

Figure 3.8 The fraction of mutants increases as a culture multiplies, and the slope is the mutation rate. *M* is the number of mutants, *N* is the total number of cells, and time is the total time elapsed divided by the time it takes the culture to double in mass (i.e., the doubling time).

PHENOTYPIC LAG

Some of the difficulty in accurately determining mutation rates results from **phenotypic lag**. Most phenotypes are not immediately evident after a mutation but appear some time later. The length of the lag depends on the molecular basis for the phenotype.

Mutations to phage T1 resistance would be expected to show a phenotypic lag. Recall that resistance to phage T1 derives from the alteration or loss of the protein product of a gene, *tonB*. This is an outer membrane protein to which the phage binds to start the infection. The mutant bacteria survive because they lack the wild-type protein in their outer membrane, so they cannot be infected by the virus. However, when the mutation first occurs, the mutant bacteria still have wild-type TonB in their outer membrane and so are T1 sensitive. Only after a few generations is all the wild-type TonB diluted out, so that the progeny cells can no longer absorb virus and are resistant.

It may seem that phenotypic lag would not be a serious problem in measuring mutation rates, because bacteria go through so many generations in a culture. However, in an exponentially growing culture, one-half of all mutations occur in the last generation time. Therefore, when the bacteria are plated with the selective agent, in our example phage T1, more than half of the mutations will not be counted because they have not yet been expressed and the bacteria are still sensitive. Obviously, ignoring more than half of all mutations will introduce a significant error in the mutation rate.

Some methods for measuring mutation rates are also influenced by differences in the growth rate of mutants relative to the original or wild type. Quite often, mutants grow measurably more slowly than the wild type even under nonselective conditions. Note that the two methods that have been described for determining mutation rates will not be affected by such differences.

PRACTICAL IMPLICATIONS
OF POPULATION GENETICS

The fact that the proportion of mutants increases as the culture grows presents both opportunities and problems in genetics. This fact can be advantageous in the isolation of a rare mutant such as one resistant to streptomycin. If we grow a culture from a few bacteria and plate 10^9 bacteria on agar containing streptomycin, we might not find any resistant mutants, since they occur at a frequency of only about 1 in 10^{11} cell generations. However, if we add a large number of bacteria to fresh broth, grow the broth culture to saturation, and then repeat this process a few times, the fraction of streptomycin-resistant mutants will increase. Then when we plate 10^9 bacteria, we may find many streptomycin-resistant mutants.

On the other hand, if we allow a culture to go through enough generations, it will become a veritable "zoo" of different kinds of mutants—virus resistant, antibiotic resistant, auxotrophic, and so on. To deal with this problem, most researchers store cultures under nongrowth conditions (e.g., as spores or lyophilized cells or in a freezer) that still maintain cell viability. An alternative is to periodically colony purify bacteria in the culture to continuously isolate the progeny of a single cell (see the introductory chapter). The progeny of a single cell are not likely to be mutated in a way that could confound our experiments.

Summary

Two very important points emerge from this discussion of mutations and mutation rates. First, measuring mutation rates is not as simple as one might think. The mutation rate is not simply the number of mutants with a particular phenotype divided by the total number of organisms in the culture. To calculate the mutation rate, we must use special methods to measure the number of mutations or must apply statistical methods to the data. Second, mutants of all kinds accumulate in cultures as we grow them. Consequently, it is best to store bacteria without growing them or to periodically isolate a single cell before mutants have had a chance to become a significant proportion of the total population.

Types of Mutations

As defined above, any heritable change in the sequence of nucleotides in DNA is a mutation. A single base pair may be changed, deleted, or inserted; a large number of base pairs may be deleted or inserted; or a large region of the DNA may be duplicated or inverted. Regardless of how many base pairs are affected, a mutation is considered to be a **single mutation** if only one error in replication, recombination, or repair has altered the DNA sequence.

As discussed earlier in this chapter, to be considered a mutation, the change in the DNA sequence must be permanent. Damage to DNA, by itself, is not a mutation, but a mutation can occur when the cell attempts to repair damage or replicate over it and a strand of DNA is synthesized that is not completely complementary to the original sequence. The wrong sequence will then be faithfully replicated through subsequent generations and thus be a mutation.

Lethal changes in the DNA sequence (as also mentioned earlier) do occur but cannot be scored as mutations since the cells do not survive. Ordinarily, to be scored as a mutation, the change must be heritable and so cannot be lethal. Breakage of DNA in bacteria is not usually a mutation because linear broken DNA will not generally be capable of replication and so a cell with broken DNA will not be able to multiply. For example, deletion of a gene required for growth is usually lethal because bacteria are haploid and usually have only one gene of each type.

The properties and causes of the different types of mutations are probably not very different in all organisms, but they are more easily studied with bacteria. A geneticist can often make an educated guess about what type of mutation is causing a mutant phenotype merely by observing some of its properties.

One property that distinguishes mutations is whether or not they are **leaky**. The term "leaky" means something very specific in genetics. It means that in spite of the mutation, the gene product still retains some activity.

Another property of mutations is whether or not they **revert**. If the sequence has been changed to a different sequence, it can often be changed back to the original sequence by a subsequent mutation. The organism in which a mutation has reverted is called a **revertant**, and the **reversion rate** is the rate at which the mutated sequence in DNA returns to the original wild type sequence.

Usually, the reversion rate is much lower than the mutation rate that gave rise to the mutant phenotype. As an illustration, consider the previously discussed example, histidine auxotrophy (His⁻). Any mutation that

inactivates any of the approximately 11 genes whose products are required to make histidine will cause a His⁻ phenotype. Since thousands of changes can result in this phenotype, the mutation rate for His⁻ is relatively quite high. However, once a *his* mutation has occurred, the mutation can revert only through a change in the mutated sequence that restores the original sequence. Everything else being equal, the reversion rate to His⁺ revertants would be expected to be thousands of times lower than the forward mutation rate to His⁻.

Some types of revertants are very easy to detect. For example, His⁺ revertants can be obtained by plating large numbers of His⁻ mutants on a plate with all the growth requirements except histidine. Most of the bacteria cannot multiply to form a colony. However, any His⁺ revertants in the population will multiply to form a colony. The appearance of His⁺ colonies when large numbers of a His⁻ mutant are plated would be evidence that the *his* mutation can revert.

Base Pair Changes

A **base pair change** is when one base pair in DNA, for example a GC pair, is changed into another base pair, for example an AT pair. Base pair changes can be classified as **transitions** or **transversions** (Figure 3.9). In a transition, the purine (A and G) in a base pair is replaced by the other purine and the pyrimidine (C and T) is replaced by the other pyrimidine. Thus, an AT pair would

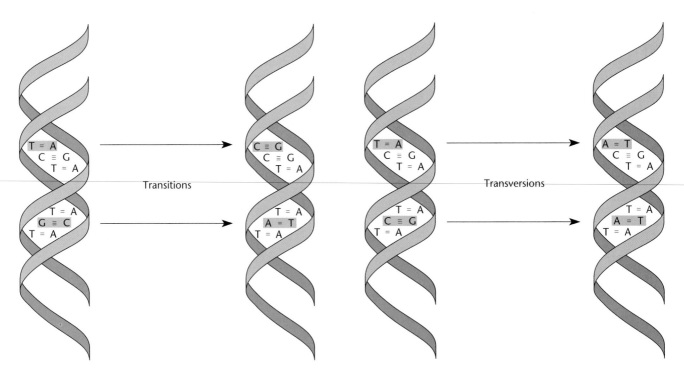

Figure 3.9 Transitions versus transversions. The mutations are shown in blue.

become a GC pair or a CG would become a TA. In a transversion, by contrast, the purines change into pyrimidines and vice versa. For example, a GC could become a TA, or a CG could become an AT.

BASE PAIR CHANGES RESULTING FROM MISPAIRING

Base pair changes can be the result of mistakes in replication, recombination, or repair. Figure 3.10A shows an example of mispairing during replication. In this example, a T instead of the usual C is mistakenly placed opposite a G as the DNA replicates. In the next replication, this T will usually pair correctly with an A, causing a GC-to-AT transition in one of the two daughter DNAs. Mistakes in pairing may occur because the bases are sometimes in a different form called the enol form, which causes them to pair differently (Figure 3.10B).

Mispairing between a purine and a pyrimidine will cause a transition, whereas mispairing between two purines or two pyrimidines will cause a transversion. Because a pyrimidine in the enol form still pairs with a purine and a purine in the enol form still pairs with a pyrimidine, mispairing during replication usually leads to transition mutations. Furthermore, all four bases can undergo the shift to the enol form, and either the base in the DNA template or the incoming base can be in the enol form and cause mispairing. Thus, the thymine in the enol form pictured in Figure 3.10B might be in the template, in which case the transition would be AT to GC, or it could be the incoming base, resulting in a GC-to-AT transition.

Mistakes during replication leading to mutations are not random, however, and some sites are much more prone to base pair changes than are others. Mutation-

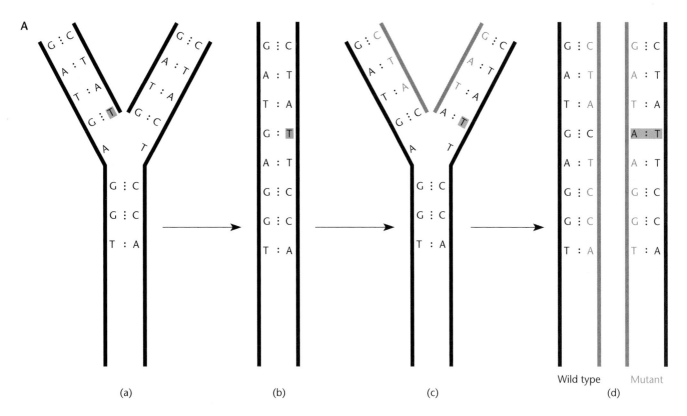

(a)   (b)   (c)   (d)

Wild type   Mutant

Figure 3.10 (A) A mispairing during replication can lead to a base pair change in the DNA (shown in blue). The blue line shows the second generation of newly replicated DNA. The mispaired base is boxed. (B) Pairing between guanine and an alternate enol form of thymine can cause G-T mismatches.

Guanine   Thymine (enol form)

prone sites are called **hot spots**; we discuss these in greater detail in chapter 13 under genetic analysis.

Mispairing occurs fairly often during replication, and it is of obvious advantage for the cell to reduce the number of base pair change mutations that occur during replication. In chapter 1, we discussed some of the mechanisms that cells use to reduce these base pair changes.

DEAMINATION OF BASES IN DNA

Deamination, or the removal of an animo group, can also cause base pair changes. This is particularly true of cytosine, which becomes uracil when deaminated, since the only difference between cytosine and uracil is the amino group at the 6 position of the cytosine ring (see chapters 1 and 2 for structures). However, uracil pairs with adenine instead of guanine. Therefore, unless the uracils due to deamination of cytosines are removed from DNA, they will cause CG-to-TA transitions the next time the DNA replicates.

Because of the special problems caused by deamination of cytosine, cells have evolved a special mechanism for removing uracil from DNA whenever it appears (Figure 3.11). An enzyme called uracil-*N*-glycosylase, the product of the *ung* gene in *E. coli*, recognizes the uracil as unusual in DNA and removes the uracil base. The DNA strand in the region where the uracil was removed is then degraded and resynthesized, and the correct cytosine is inserted opposite the guanine. As expected, *ung* mutants of *E. coli* show high rates of spontaneous mutagenesis, and most of the mutations are GC-to-AT transitions.

OXIDATION OF BASES

Reactive forms of oxygen such as peroxides and oxygen free radicals are given off as by-products of oxidative metabolism, and these forms can react with and alter the bases in DNA. A common example is the altered guanine base, 8-oxoG, which sometimes mistakenly pairs with adenine instead of cytosine, causing GC-to-TA transversion mutations. We discuss repair systems specific to damage such as deamination and oxidation in more detail in chapter 10.

CONSEQUENCES OF BASE PAIR CHANGES

Whether or not a base pair change causes a detectable phenotype depends, of course, on where the mutation occurs and what the actual change is. Even a change in an open reading frame (ORF) that encodes a polypeptide may not result in an altered protein. If the mutated base is the third in a codon, the amino acid inserted into the protein may not be different because of degeneracy of the code (see the section on the genetic code in chap-

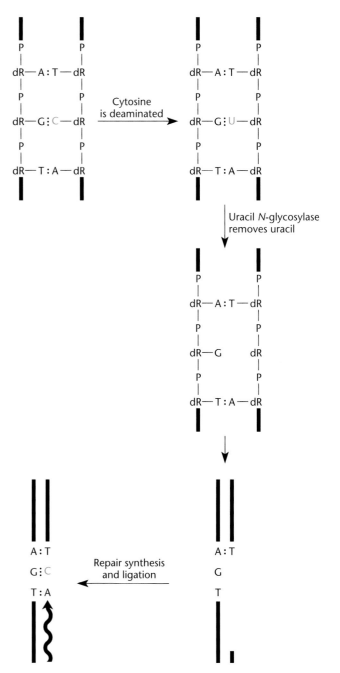

Figure 3.11 Removal of deaminated cytosine (uracil) from DNA by uracil-*N*-glycosylase. The uracil base is cleaved off, and the DNA strand is degraded and resynthesized with cytosine.

ter 2). Mutations in the coding region of a gene that do not change the amino acid sequence of the polypeptide product are called **silent mutations**.

The change may also occur in a region that does not encode a polypeptide but, rather, is a regulatory sequence such as an operator or promoter. Alternatively, the mutation may occur in a region that has no detectable function. We shall first discuss mutations that change the coding region of a polypeptide.

MISSENSE MUTATIONS

Most base pair changes in bacterial DNA will cause one amino acid in a polypeptide to be replaced by another. These mutations are called **missense mutations** (Figure 3.12). Not all of them inactivate proteins. If the original and new amino acids have similar properties, the change may have little or no effect on the activity of a protein. For example, a missense mutation changing an acidic amino acid, such as glutamate, into another acidic amino acid, such as aspartate, may have less effect on the functioning of the protein than does a mutation that substitutes a basic amino acid, such as arginine, for an acidic one.

The consequences also depend on which amino acid is changed. Certain amino acids in any given protein sequence will be more essential to activity than others, and a change at one position can have much more effect than a change elsewhere. Investigators often use this fact to determine which amino acids are essential for activity in different proteins. Such methods, called site-specific mutagenesis, will be discussed in chapter 15.

NONSENSE MUTATIONS

Instead of changing a codon into one coding for a different amino acid, base pair changes sometimes produce one of the nonsense codons, UAA, UAG, or UGA. These changes are called **nonsense mutations**.

While nonsense mutations are base pair changes and have the same causes as other base pair changes, the consequences are very different. Because the nonsense codons are normally used to signify the end of a gene, these codons are normally recognized by release factors (see chapter 2), which cause release of the translating ribosome and the polypeptide chain. Therefore, if a mutation to one of the nonsense codons occurs in an ORF for a protein, the protein translation will terminate prematurely at the site of the nonsense codon and the shortened or truncated polypeptide will be released from the ribosome (Figure 3.13). For this reason, nonsense mutations are sometimes called chain-terminating mutations. These mutations almost always inactivate the protein product of the gene in which they occur. If, however, they occur in a noncoding region of the DNA or in a region that encodes an RNA rather than a protein, such as a gene for a tRNA, they are indistinguishable from other base pair changes.

The three nonsense codons—and their corresponding mutations—are sometimes referred to by color designations: **amber** for UAG, **ochre** for UAA, and **umber** or **opal** for UGA. These names have nothing to do with the effects of the mutation. Rather, when nonsense mutations were first discovered at Cal Tech, their molecular basis was unknown. The investigators thought that descriptive names might be confusing later on if their interpretations were wrong, so they followed the lead of

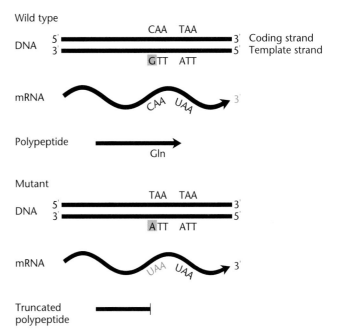

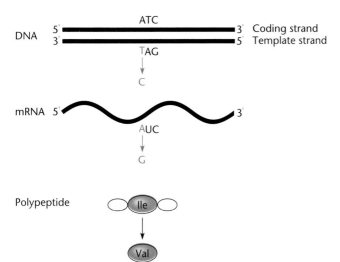

Figure 3.12   A mutation that changes T to C in the DNA template strand will result in an A-to-G change in the mRNA. The mutant codon GUC will be translated as valine.

Figure 3.13   Nonsense mutation. Changing codon CAA, encoding glutamine (Gln), to UAA, a nonsense codon, causes truncation of the polypeptide gene product.

physicists with their "quarks" and "barns." The first nonsense mutations to be discovered, to UAG, were called amber mutations. Following suit, UAA and UGA mutations were also named after colors—ochre and umber, respectively.

PROPERTIES OF BASE PAIR CHANGE MUTATIONS

Base pair changes are often leaky. A substituted amino acid may not work nearly as well as the original at that position in the chain, but the protein can retain some activity. Even nonsense mutations are usually somewhat leaky because sometimes an amino acid will be inserted for a nonsense codon, albeit at a low frequency. In wild-type *E. coli*, UGA tends to be most leaky, followed by UAG; the nonsense codon UAA tends to be the least leaky.

Base pair mutations also revert. If the base pair has been changed to a different base pair, it can also be changed back to the original base pair by a subsequent mutation. Moreover, base pair changes are a type of point mutation, as we shall discuss in chapters 13 and 14.

Frameshift Mutations

A high percentage of all spontaneous mutations are **frameshift mutations** (Figure 3.14). This type of mutation occurs when a base pair or a few base pairs are removed from or added to the DNA, causing a shift in the reading frame if they occur in the sequence of an ORF. Because the code is three lettered, any addition or subtraction that is not a multiple of 3 will cause a frameshift in the translation of the remainder of the gene. For example, adding or subtracting 1, 2, or 4 base pairs will cause a frameshift, but adding or subtracting 3 or 6 base pairs will not. Mutations that remove or add base pairs are usually called frameshift mutations even if they do not occur in an ORF and do

not actually cause a frameshift in the translation of a polypeptide.

CAUSES OF FRAMESHIFT MUTATIONS

Spontaneous frameshift mutations often occur where there is a short repeated sequence that can slip. As an example, Figure 3.15 shows a string of AT base pairs in the DNA. Since any one of the A's in one strand can pair with any T in the other strand, the two strands could slip with respect to each other, as in the illustration. Slippage during replication could leave one T unpaired, and an AT base pair would be left out on the other strand when it replicates. Alternatively, the slippage could occur before the base was added, and an extra AT base pair could appear in one strand as shown.

PROPERTIES OF FRAMESHIFT MUTATIONS

Frameshift mutations are usually not leaky and almost always inactivate the protein, because every amino acid in the protein past the point of the mutation will be wrong. The protein will also most often be truncated, because a nonsense codon will usually be encountered while the gene is being translated in the wrong frame. Because, in general, 3 of the 64 codons are the nonsense codons, one of these should be encountered by chance about every 20 codons when the region is being translated in the wrong frame.

Another property of frameshift mutations is that they revert. If a base pair has been subtracted, one can be added to restore the correct reading frame and vice versa. More often, frameshift mutations do not revert but are suppressed by the addition or subtraction of a base pair close to the site of the original mutation that restores the original reading frame. We shall discuss this means of frameshift suppression later in this chapter. Finally, some frameshift mutations are a type of point mutation, as we discuss in chapters 13 and 14.

Some types of pathogenic bacteria apparently take advantage of the frequency and high reversion rate of frameshift mutations to avoid host immune systems. In such bacteria, genes required for the synthesis of cell surface components that are recognized by host immune systems often have repeated sequences. Consequently, these genes can be turned off and on by frameshift mutations and subsequent reversion. We discuss how one such frameshift mutation aids in the synthesis of virulence gene products by *Bordetella pertussis*, the causative agent of whooping cough, in chapter 12. Frameshift mutations are also used to reversibly inactivate genes of *Haemophilus influenzae* and *Neisseria gonorrhoeae*, which cause spinal meningitis and gonorrhea, repetively (for a review of this subject see Moxon et al., Suggested Reading).

Wild type

```
          Gln        Ser        Arg
   5'---⌐C  A  A⌐⌐U  C  C⌐⌐C  G  G⌐ ────────→  etc.
```

Mutant

```
   5'---⌐C  A  A⌐A  U  C⌐⌐C  C  G⌐ ──────────→
        ⌐Gln⌐    ⌐Ile⌐    ⌐Pro⌐     All amino acids
                                     downstream
                                     are changed
```

Figure 3.14 Frameshift mutation. The wild-type mRNA is translated glutamine (Gln)-serine (Ser)-arginine (Arg)- etc. Addition of an A (boxed) would shift the reading frame, so that the codons would be translated glutamine (Gln)-isoleucine (Ile)-proline (Pro)-, etc., with all downstream amino acids being changed.

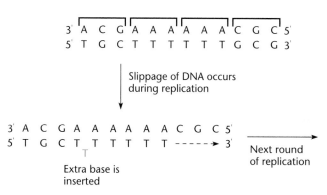

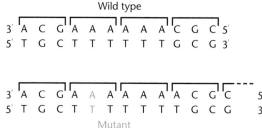

Figure 3.15 Slippage of DNA at a repeated sequence can cause a frameshift mutation.

Deletion Mutations

Deletion mutations may be very long, removing thousands of base pairs and possibly many genes. Often, the only limitation on these mutations in bacterial DNA is that they cannot delete any essential genes, since haploid bacteria generally only have one copy of each gene. Some deletions in bacteria can be quite long, however, since bacterial genomes often possess long stretches of genes that can be deleted without losing cell viability. Deletion mutations often constitute a high percentage of spontaneous mutations; for example, in *E. coli*, almost 5% of all spontaneous mutations are deletions.

CAUSES OF DELETIONS

Deletions can be caused by recombination between different regions of the DNA. Recombination usually occurs between the same regions in two DNAs because the two DNAs will be identical in the same region. However, recombination can sometimes mistakenly occur between two different regions if they are similar enough in sequence. The strands of two DNA molecules are broken and rejoined in new combinations, hence the name **recombination**. This process is discussed in detail in chapter 9.

As shown in Figure 3.16, deletions will result from recombination between two sequences that are **direct repeats**, that is, two sequences that are similar or identical when read in the 5'-to-3' direction on the same strand of DNA. Bacterial DNA contains several types of repeats, the longest of which include insertion sequence (IS) elements and the rRNA genes, which are often repeated in many places in the DNA. We shall discuss IS elements and other types of repeated elements in more detail in subsequent chapters.

In Figure 3.16, mistaken pairing occurs between direct repeats in the two daughter DNA molecules. The two regions are then broken and rejoined, removing the sequence between the two direct repeats as shown. Al-

ternatively, the two direct repeats on the same DNA molecule could pair, "looping out" the intervening sequences. Breaking and rejoining the DNA would remove the looped-out sequences as shown. For purposes

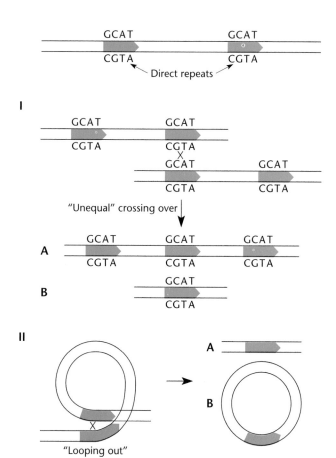

Figure 3.16 Recombination between directly repeated sequences can cause deletion mutations. The recombination can occur between repeated sequences on different DNAs (unequal crossing over) (I) or between repeated sequences on the same molecule, resulting in looping out (II).

of illustration, the directly repeated sequences shown in the figure are much shorter than would normally be required for recombination. Usually, direct repeats that promote mistaken recombination are hundreds or even thousands of base pairs long.

PROPERTIES OF DELETION MUTATIONS

Deletions have very distinctive properties. They are usually not leaky; deleting part or all of a gene will usually totally inactivate the gene product. Mutations that inactivate more than one gene simultaneously are most often deletions. Moreover, deletion mutations sometimes fuse one gene to another, sometimes putting one gene under the control of another.

The most distinctive property of long deletion mutations is that they never revert. Every other type of mutation will revert at some frequency, but for a deletion to revert, the missing sequence would somehow have to be found and reinserted. Deletions also behave differently from point mutations in genetic crosses, as discussed in chapters 13 and 14.

NAMING DELETION MUTATIONS

Deletion mutations are named differently from other mutations. The Greek letter Δ, for *del*etion, is written in front of the gene designation and allele number, e.g., Δ*his8*. Often, deletions remove more than one gene, and so if known, the deleted regions are shown, followed by a number to indicate the particular deletion. For example, Δ(*lac-proAB*)195 is deletion number 195 extending through the *lac* and *proAB* genes on the *E. coli* chromosome. Often a deletion removes one or more known genes but extends into a region of unknown genes, so that the endpoints of the deletion are not known. In this case, the deletion is often named after the known gene. For example, the Δ*his8* deletion may delete the entire *his* operon but also extend an unknown distance into neighboring genes.

Inversion Mutations

Sometimes a DNA sequence is not removed, as in a deletion, but, rather, is flipped over, or **inverted**. After such an **inversion**, all the genes in the inverted region face in the opposite orientation.

CAUSES OF INVERSIONS

Inversions are caused in the same way as deletions, by recombination between repeats. However, recombination between inverted sequences rather than directly repeated sequences produces inversions. Inverted repeats read almost the same in the 5'-to-3' direction on opposite strands (see chapter 2). Also unlike deletions, the re-

combination that produces inversions must occur between two regions on the same DNA (Figure 3.17).

PROPERTIES OF INVERSION MUTATIONS

Unlike deletions, inversion mutations can generally revert. Recombination between the inverted repeats that caused the mutation will "reinvert" the affected sequence, recreating the original order. However, an inversion might occur between very short inverted repeats or repeats that are not exactly the same. Then the recombination event would have to occur between the exact bases involved in the first recombination to restore the correct sequence. Such a recombination could be a very rare event, and reversions of such a mutation would be very rare.

Inversion mutations often cause no phenotype. If the inversion involves a longer sequence, including many genes, generally the only affected regions will be those in the **inversion junctions**, where the recombination occurred. Most of the genes in the inverted region will still be intact, although they will be present in the reverse order. Consequently, even very long inversion mutations often cause no obvious phenotypes. Like deletions, inversion mutations sometimes fuse one gene to another gene. This property provides a mechanism for detecting

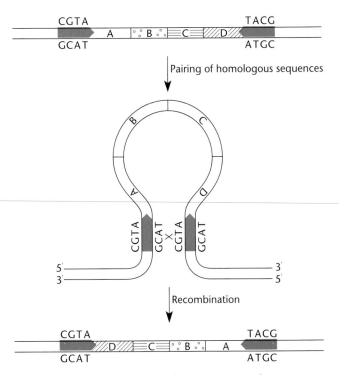

Figure 3.17 Recombination between inverted repeats can cause inversion mutations. The order of genes within the inversion is reversed after the recombination.

them. The occurrence of inversions in evolution is discussed in Box 3.3.

NAMING INVERSIONS

A mutation known to be an inversion is given the letters IN followed by the genes in which the inversion junctions occur, provided that these are known, followed by the number of the mutation. For example, IN(*purB-trpA*)*3* is inversion number 3 in which the inverted region extends from somewhere within the gene *purB* to somewhere within the gene *trpA*.

Tandem Duplication Mutations

In a duplication mutation, a sequence is copied from one region of the DNA to another. The most common, a **tandem duplication**, consists of a sequence immediately followed by its duplicate. Tandem duplications occur frequently and can be very long.

CAUSES OF TANDEM DUPLICATIONS

Like deletions, tandem duplications can result from recombination between directly repeated sequences in DNA. In fact, as shown in Figure 3.18, they are probably

---

BOX 3.3

Inversions and the Genetic Map

Even a single large inversion mutation will cause a dramatic change in the genetic map, or order of genes in the DNA, of an organism. The order of all the genes will be reversed between the sites of the recombination that led to the inversion. We would also expect inversions to be fairly frequent because repeated sequences often exist in inverted orientation with respect to each other. In spite of this, inversions seem to have occurred very infrequently in evolution. As evidence, consider the genetic maps of *S. typhimurium* and *E. coli.* These bacteria presumably diverged billions of generations ago. Nevertheless, the maps are very similar except for one short inverted sequence between about 25 and 27 min on the *E. coli* map. At present, we can only speculate on why the genetic maps are so highly conserved. Perhaps organisms with this gene order have some selective advantage, or perhaps other sequences in the DNA cannot be inverted without disadvantaging the organism.

Termination of chromosome replication after sites like *terA* and *terB* (see chapter 1) may help explain why so few large inversions seem to have occurred in the evolution of bacteria. There may be sequences that resemble *terA* and *terB* distributed around the chromosome, but because they are on the wrong strand, they do not cause termination. However, an inversion mutation would reverse their orientation, so that if the *terA* site were preceded by a *terB*-like site, the DNA between the two sites would not be replicated. This situation would be lethal. However, there are probably other explanations for the rarity of large inversions.

Reference
Mahan, M. J., and J. R. Roth. 1991. Ability of a bacterial chromosome to invert is dictated by included material rather than flanking sequences. *Genetics* **129**:1021–1032.

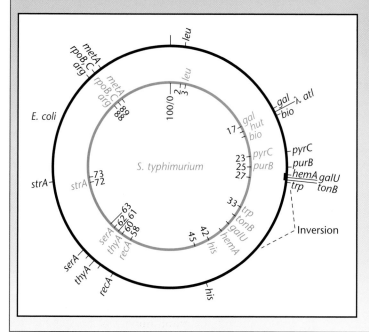

Genetic maps of *S. typhimurium* and *E. coli*, showing a high degree of conservation. The region from *hemA* to the 40-min position is inverted in *E. coli* relative to that in *S. typhimurium*.

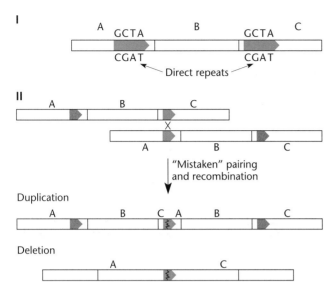

Figure 3.18 Formation of tandem duplication mutations by recombination between directly repeated sequences on different DNA molecules. A deletion, the reciprocal recombinant, is created at the same time as the duplication. The symbol ⚡ designates the duplication junction.

often created at the same time as a deletion. Pairing between two directly repeated sequences in different DNAs, followed by recombination, can give rise to a tandem duplication and a deletion as the two products.

PROPERTIES OF TANDEM DUPLICATION MUTATIONS

Although the mechanism by which tandem duplications arise is probably similar to the mechanism that creates deletions and inversions, except that the recombination that creates a tandem duplication occurs between direct repeats that must be on different DNAs, the properties of tandem duplications are very different. Tandem duplication mutations that occur within a single gene will usually inactivate the gene and not be leaky. However, if the duplicated region is long enough to include one or more genes, no genes will be inactivated, including those in which the recombination occurred—the **duplication junctions**. This conclusion may seem surprising, but consider the example shown in Figure 3.18. Direct repeats in genes A and C on different DNAs pair with each other. The repeats in the two DNAs are then broken and rejoined to each other, creating a duplication in one DNA and a deletion in the other. Only part of gene A exists in the duplicate, but an entire gene A exists upstream. Conversely, only part of gene C exists upstream, but the entire gene exists in the duplicate. There are now two copies of gene B, both of which are unaltered.

Therefore, intact genes A, B, and C still exist after the duplication, and there would be no indication that a mutation had even occurred unless there happened to be a phenotype associated with the presence of two copies of gene B.

Like deletions and inversions, duplications can sometimes fuse two genes to put expression of one gene under the control of a different gene. In the example in Figure 3.18, part of gene A has been fused to gene C, which might put gene A under the control of gene C.

The most characteristic property of tandem duplications is that they are very unstable and revert at a high frequency. Even though the recombinations that lead to a duplication are usually rare, recombination anywhere within the duplicated segments can delete them, restoring the original sequence. The instability of tandem duplications is often the salient feature that allows their identification.

ROLE OF TANDEM DUPLICATION MUTATIONS IN EVOLUTION

Tandem duplication mutations may play an important role in evolution. Ordinarily, a gene cannot change without loss of its original function, and if the lost function was a necessary one, the organism will not survive. However, when a duplication has occurred, there will be two copies of the genes in the duplicated region, and now one of these is free to evolve to a different function. This mechanism would allow organisms to acquire more genes and become more complex. However, how tandem duplications could persist long enough for some of the duplicated genes to evolve is not clear.

Insertion Mutations

Insertion mutations are caused by the insertion of a large piece of DNA being into a region, usually by transposons "hopping" into the DNA. **Transposons** are DNA elements that can promote their own movement from one place in the DNA to another. In doing so, they create insertion mutations. Although these elements are usually thousands of base pairs long, sometimes only part of a transposon moves, or hops, producing a shorter insertion. Indeed, the movement of relatively short transposons, known as **insertion elements**, produces the majority of insertion mutations. These elements, which are only about 1,000 base pairs long, carry no easily identifiable genes. Most bacteria carry several insertion elements in their chromosome.

PROPERTIES OF INSERTION MUTATIONS
Insertion of DNA into a gene almost always inactivates the gene; therefore, insertion mutations are usually not

leaky. Transposons also contain many transcription termination sites, and so their insertion results in polarity (see chapter 2), which prevents the transcription of genes normally copied onto the same mRNA as the gene with the insertion. Finally, insertion mutations seldom revert, because the inserted DNA must be precisely removed, with no DNA sequences remaining. These last two unusual properties of insertion mutations led to their discovery.

SELECTING INSERTION MUTATIONS

A significant percentage of all spontaneous mutations are insertion mutations, but their phenotypes are difficult to distinguish from those of other types of mutations. However, transposons can serve as useful tools in genetics experiments, because many carry a selectable gene, such as one for antibiotic resistance. The insertion of such a transposon into a cell's DNA will then make the cell antibiotic resistant and easy to isolate. Moreover, transposon insertions are relatively easy to map, both genetically and physically. The methods of transposon mutagenesis are central to bacterial molecular genetics and biotechnology and so will be discussed in some detail in later chapters.

NAMING INSERTION MUTATIONS

An insertion mutation in a particular gene is represented by the gene name, two colons, and the name of the insertion. For example, galK::Tn5 denotes the insertion of the transposon Tn5 into the galK gene. If more than one Tn5 insertion exists in galK, the mutations can be numbered to distinguish them (e.g., galK35::Tn5). When insertion mutations are constructed for use in genetic experiments, they are denoted with the capital Greek letter Ω (omega) followed by the name of the insertion. For example, pBR322Ω::kan is a kanamycin resistance gene inserted into plasmid pBR322.

Reversion versus Suppression

Reversion mutations are often detected through the restoration of a mutated function. As discussed above, a reversion actually restores the original sequence of a gene. However, sometimes the function that was lost because of the original mutation can be restored by a second mutation elsewhere in the DNA. Whenever one mutation in the DNA relieves the effect of another mutation, that mutation has been **suppressed** and the second mutation is called a **suppressor mutation**. The following sections present some of the mechanisms of suppression.

Intragenic Suppressors

Suppressor mutations in the same gene as the original mutation are called **intragenic suppressors**. These mutations can restore the activity of a mutant protein by many means. For example, the original mutation may have made an unacceptable amino acid change that inactivated the protein, but changing another amino acid somewhere else in the polypeptide could restore the protein's activity. This form of suppression is not uncommon and is often interpreted to indicate an interaction between the two amino acids in the protein.

The suppression of one frameshift mutation by another frameshift mutation in the same gene is another example of intragenic suppression. If the original frameshift resulted from the removal of a base pair, the addition of another base pair close by could return translation to the correct frame. The second frameshift can restore the activity of the protein product provided that ribosomes, while translating out of frame, do not encounter any nonsense codons or insert any amino acids that alter the activity of the protein.

Intergenic Suppressors

Intergenic (or **extragenic**) suppressors do not occur in the same gene as the original mutation. For example, in some cases, if the gene for a step in a biochemical pathway is mutated, a toxic intermediate in that pathway can accumulate, causing cell death. However, a suppressing mutation in another gene of the pathway may prevent the accumulation, allowing the cell, now known as a **double mutant**, to survive.

The suppression of galE mutations by galK mutations provides an illustration of such an intergenic suppressor. Many types of cells use the sugar galactose by first converting it to glucose in the pathway shown in Figure 3.19. In the first step, galactose is phosphorylated by the product of the galK gene, a galactose kinase. The second step is the transfer of galactose 1-phosphate to uridine diphosphoglucose (UDPglucose) by the product of the galT gene, a transferase. The glucose produced is used as a carbon and energy source. The third step is the isomerization of the galactose on UDPgalactose to UDPglucose by the product of the galE gene, an isomerase. The newly synthesized UDPglucose can then cycle back into the pathway to convert more galactose to glucose.

Cells with galE mutations are galactose sensitive (galactosemic), and their growth will be inhibited by galactose in the medium, because the absence of the GalE epimerase will permit the accumulation of both phosphorylated galactose and UDPgalactose, which are

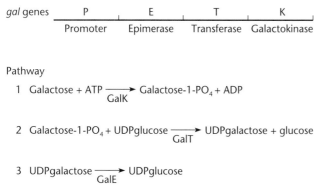

gal genes | P | E | T | K |

Promoter | Epimerase | Transferase | Galactokinase

Pathway

1 Galactose + ATP $\xrightarrow{\text{GalK}}$ Galactose-1-PO$_4$ + ADP

2 Galactose-1-PO$_4$ + UDPglucose $\xrightarrow{\text{GalT}}$ UDPgalactose + glucose

3 UDPgalactose $\xrightarrow{\text{GalE}}$ UDPglucose

Figure 3.19 The pathway to galactose utilization in *E. coli* and most other organisms. *galK* mutations suppress *galE* mutations because they prevent the accumulation of the toxic intermediate galactose 1-phosphate.

toxic to cells in high concentrations. Consequently, if we plate large numbers of a *galE* mutant strain on plates containing galactose, the cells will be inhibited and only a few mutants will multiply to make colonies. Most of these mutants will not have undergone reversion of the *galE* mutation but will be double mutants with the original *galE* mutation and a suppressing *galK* mutation. The *galK* mutation blocks the first step of the pathway, so that no toxic intermediates accumulate. Revertants with reversion of the original *galE* mutation could also grow. However, we would expect *galE* reversion mutations to be much rarer than *galK* suppression mutations because many changes will inactivate the *galK* gene but only one base pair change will cause the *galE* mutation to revert.

Also, the *galE*+ revertants can be distinguished from *galE galK* double mutants because *galE*+ revertants will be Gal+ and will grow on galactose as the sole carbon and energy source. In contrast, the *galE galK* double mutants are still Gal−, and so another carbon source must be provided in the medium.

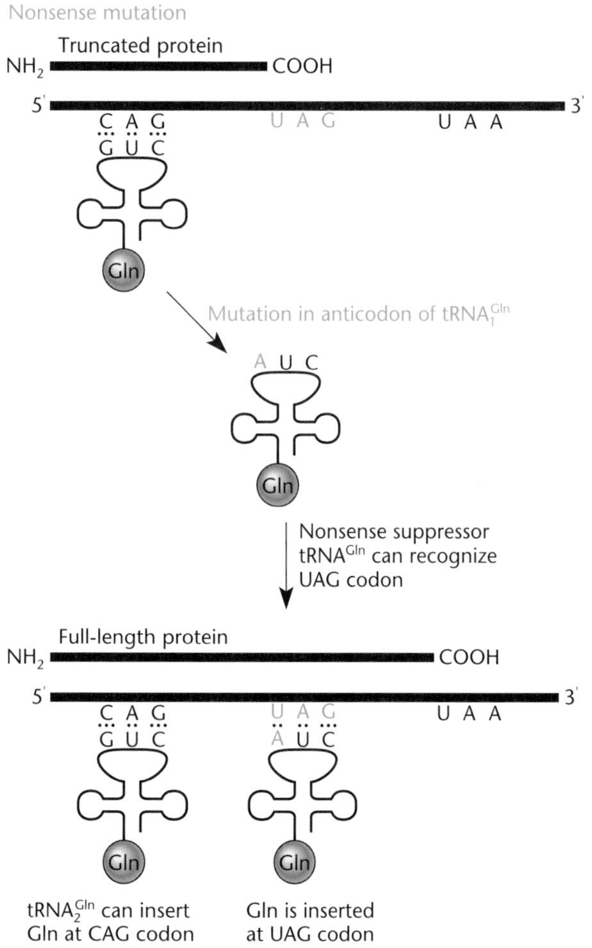

Nonsense Suppressors

Nonsense suppressors are another type of intergenic suppressor. A **nonsense suppressor** is usually a mutation in a tRNA gene that changes the anticodon of the tRNA product of the gene, so that it now recognizes a nonsense codon. In Figure 3.20, for example, the gene for a tRNA with the anticodon 3′GUC5′ (so that it normally recognizes the glutamine codon 5′CAG3′) mutates, causing the anticodon to become 3′AUC5′. This altered anticodon can pair with the nonsense codon UAG instead of CAG. However, the anticodon mutation does not significantly change the tertiary shape of the tRNA,

Figure 3.20 Formation of a nonsense suppressor tRNA. A nonsense mutation causes truncation of a protein gene product. One of two genes that encode glutamine tRNAs that can recognize the codon CAG is mutated so that the tRNA product now recognizes the nonsense codon UAG. In the suppressor strain with the mutated tRNA, the translational machinery will now sometimes insert glutamine for the nonsense codon UAG. The tRNA product of the second tRNA gene can still recognize the codon CAG.

which means that the cognate aminoacyl tRNA synthetase will still load it with glutamine. Therefore, this mutated tRNA will bind with the amber codon UAG, allowing insertion of glutamine into the growing polypeptide instead of translation termination. This can lead to synthesis of the active polypeptide and suppression of the amber mutation.

The mutated tRNA is called a **nonsense suppressor tRNA**, and nonsense suppressors themselves are referred to as amber suppressors, ochre suppressors, or umber suppressors depending on whether they suppress

TABLE 3.4	Some *E. coli* nonsense suppressor tRNAs		
Suppressor name	tRNA	Anticodon change	Suppressor type
supE	tRNA^{Gln}	CU<u>G</u>-CUA	Amber
supF	tRNA^{Tyr}	<u>G</u>UA-CUA	Amber
supB	tRNA^{Gln}	UU<u>G</u>-UUA	Ochre/amber
supL	tRNA^{Lys}	UU<u>U</u>-UUA	Ochre/amber

UAG, UAA, or UGA mutations, respectively. Table 3.4 lists several *E. coli* nonsense suppressor tRNAs.

Nonsense suppressors can also be classified as **allele-specific suppressors** because they suppress only one type of allele of a gene, that is, one with a particular type of nonsense mutation. In contrast, the *galK* suppressing mutations discussed above suppress any *galE* mutation and so are not allele specific.

SUPPRESSED POLYPEPTIDES ARE NOT NORMAL

The polypeptide synthesized as the result of a nonsense suppressor is not always fully active. Usually, the amino acid inserted at the site of a nonsense mutation is not the same amino acid that was coded for by the original gene. This changed amino acid will sometimes cause the polypeptide to be almost inactive or temperature sensitive.

TYPES OF NONSENSE SUPPRESSORS

Not all tRNA genes can be mutated to form a nonsense suppressor. Generally, if there is only one type of tRNA to respond to a particular codon, the gene for that tRNA cannot be mutated to make a suppressor tRNA. The original codon to which the tRNA responded would be "orphaned," and no tRNA would respond to it in an mRNA. In the example, two different tRNAs encoded by different genes recognize the codon CAG, one of which continues to recognize CAG after the other has been mutated to recognize UAG.

Wobble (see chapter 2) offers the only exception to the rule that a tRNA can be mutated to a nonsense suppressor only if there is another tRNA to respond to the original codon. Because of wobble, the same tRNA can sometimes respond both to its original codon and to one of the nonsense codons. For example, in a particular organism, there may be only one tRNA that recognizes the codon for tryptophan, 5'UGG3'. If the anticodon, 3'ACC5', is mutated to ACU, by wobble the tRNA might be able to recognize both the tryptophan codon UGG and the nonsense codon UGA, so that the suppressor strain could be viable. Wobble also allows the same suppressor tRNA to recognize more than one

nonsense codon. In *E. coli,* all naturally occurring ochre suppressors also suppress amber mutations. From the wobble rules (see chapter 2), we know that a suppressor tRNA with the anticodon AUU could recognize both the UAG and UAA nonsense codons in mRNA (Table 3.4) Note that Table 3.4 anticodons are written 5'-3'.

EFFICIENCY OF SUPPRESSION

Nonsense suppression is never complete, because the nonsense codons are also recognized by release factors, which free the polypeptide from the ribosome (see chapter 2). Therefore, translation of the complete protein depends on the outcome of a race between the release factors and the suppressor tRNA. If the tRNA can base pair with the nonsense codon before the release factors terminate translation at that point, translation will continue. Sequences around the nonsense codon influence the outcome of this race and determine the efficiency of suppression of nonsense mutations at particular sites.

NONSENSE SUPPRESSOR STRAINS ARE USUALLY SICK

It would seem that nonsense suppressors would tend to translate through the proper nonsense codons at the ends of genes, resulting in proteins longer than normal. However, because nonsense suppressors are never 100% efficient, some of the correct proteins will always be synthesized. Moreover, since the efficiency of suppression depends upon the sequence of nucleotides in the gene around the nonsense codon, the nonsense codons at the ends of genes presumably have a "context" that favors termination rather than suppression. Also, often more than one type of nonsense codon lies in frame at the end of genes, presumably to avoid suppression by any particular tRNA suppressor.

Nevertheless, cells do pay a price for nonsense suppression. Cells with nonsense suppressors usually grow more slowly. Only lower organisms such as bacteria, fungi, and roundworms seem to tolerate nonsense suppressors, which are known to be lethal in higher organisms, including fruit flies and humans.

SUMMARY

1. A mutation is any heritable change in the sequence of DNA of an organism. The organism with a mutation is called a mutant, and that organism's mutant phenotype includes all of the characteristics of the mutant organism that are different from the wild-type, or normal, organism.

2. The mutation rate is the chance of occurrence of a mutation to a particular phenotype each time a cell divides. The mutation rate offers clues to the molecular basis of the phenotype. Quantitative determination of mutation rates is often difficult.

3. One important conclusion from an analysis of population genetics is that the fraction of mutants increases as the population grows. This causes practical problems in genetics, which can be partially overcome by storing organisms without growth or by periodically isolating a single or very few organisms.

4. Base pair changes revert and are often leaky. Frameshift mutations also revert but are seldom leaky.

5. Deletions, inversions, and tandem duplication mutations can be caused by recombination between different regions in the DNA. Deletions never revert and are seldom leaky. Inversion mutations usually revert and often do not inactivate genes except at the inversion junction. Tandem duplications often have no observable phenotypes, and they revert with a very high frequency.

6. If a secondary mutation returns the DNA to its original sequence, we say that the mutation has reverted. If a mutation somewhere else in the DNA restores function, we say that the mutation was suppressed. Examples of suppressors are secondary mutations which restore the reading frame in frameshift mutations and nonsense suppressor mutations which alter the tRNA product of a tRNA gene so that it now recognizes a nonsense codon.

QUESTIONS FOR THOUGHT

1. A single inversion mutation will greatly alter a genetic map, or the order of genes in the DNA. Then why are the genetic maps of *Salmonella typhimurium* and *E. coli* so similar?

2. Do you suppose that duplication mutations play a role in evolution? If so, why are they not always destroyed as quickly as they form?

3. Why do you suppose that nonsense suppressors are possible in lower organisms but not higher organisms?

4. Can you propose a mechanism by which "directed" mutations might occur?

PROBLEMS

1. Which phenotype would you expect to have the higher mutation rate, rifampin resistance or Arg⁻ (arginine auxotrophy)? Rifampin inhibits transcription by binding to RNA polymerase, and Rif^r mutations change the RNA polymerase so that it no longer binds rifampin.

2. Luria and Delbrück grew 100 cultures of 1 ml each to 2×10^9 bacteria per ml. They then measured the number of bacteria resistant to T1 phage in each culture: 20 cultures had no resistant bacteria, 35 had one resistant mutant, 20 had two resistant mutants, and 25 had three or more resistant mutants. Calculate the mutation rate to T1 resistance by using the Poisson distribution.

3. Newcombe spread an equal number of bacteria on each of four plates. After 4 h of incubation, he sprayed plate 1 with T1 and put it back in the incubator. At the same time, he washed the bacteria off of plate 2, diluted them 10^7-fold, and replated them to determine the total number of bacteria. After a further 2 h of incubation, he sprayed plate 3 and washed the bacteria off of plate 4 and diluted them 10^8-fold before replating them. The next morning, he counted the colonies on each plate:

	Plate 1	Plate 2	Plate 3	Plate 4
No. of colonies	10	20	120	22

Calculate the mutation rate to T1 resistance.

4. You have isolated Arg⁻ auxotrophs of *Klebsiella pneumoniae*.

a. If you plate one mutant with a mutation, *arg-1*, on plates without arginine, they multiply to make tiny colonies. If you plate 10^8 mutants, you get some large, rapidly growing colonies. What kind of mutation is *arg-1* likely to be?

b. If you plate another mutant with a different mutation, *arg-2*, you get no growth on plates without arginine. Even if you plate large numbers of mutant bacteria ($>10^8$), you get no colonies. What type of mutation is *arg-2* likely to be?

c. What are some other possible explanations for (a) and (b)?

5. Design an experiment to show that *dam* mutations of *E. coli* are mutagenic, i.e., show higher than normal rates of spontaneous mutations. Assume that you can get a Dam⁻ mutant in the mail, and do not have to isolate one in the laboratory.

SUGGESTED READING

Lederberg, J., and E. M. Lederberg. 1952. Replica plating and the indirect selection of bacterial mutants. *J. Bacteriol.* **63:**399.

Luria, S., and M. Delbrück. 1943. Mutations of bacteria from virus sensitivity to virus resistance. *Genetics* **28:**491–511.

Moxon, E. R., P. B. Rainey, M. A. Nowak, and R. E. Lenski. 1994. Adaptive evolution of highly mutable loci in pathogenic bacteria. *Curr. Biol.* **4:**24–33.

Newcombe, H. 1949. Origin of bacterial variants. *Nature* (London) **164:**150–151.

Chapter 4

Plasmids

What Is a Plasmid?

In addition to the chromosome, bacterial cells often contain **plasmids**. These DNA molecules are found in essentially all types of bacteria and, as we discuss below, play a significant role in bacterial adaptation and evolution. They also serve as important tools in studies of molecular biology. We address such uses later in the chapter.

Plasmids, which vary widely in size from a few thousand to hundreds of thousands of base pairs (a size comparable to that of the bacterial chromosome), are most often circular molecules of double-stranded DNA. However, some bacteria have linear plasmids (see Box 4.1), and plasmids from gram-positive bacteria can accumulate single-stranded DNA owing to aberrant rolling-circle replication (discussed below). The number of copies also varies among plasmids, and bacterial cells can harbor more than one type. Thus, a cell can harbor, for example, two different types of plasmids, with hundreds of copies of one plasmid type and only one copy of the other type.

Like chromosomes, plasmids encode proteins and RNA molecules and replicate as the cell grows, and the replicated copies are usually distributed into each daughter cell when the cell divides. However, unlike chromosomes, plasmids generally do not encode functions essential to bacterial growth. They instead provide gene products that can benefit the bacterium under certain circumstances (see Table 4.1; Box 4.2). In this chapter we'll discuss the molecular basis for these and other features of plasmids.

Naming Plasmids

Before methods for physical detection of plasmids became available, plasmids made their presence known by conferring phenotypes on the cells har-

BOX 4.1

Linear Plasmids

Not all plasmids are circular. Linear plasmids have been found in many bacteria, including members of the genus *Streptomyces,* which are soil bacteria responsible for making antibiotics, and *Borrelia burgdorferi,* the causative agent of Lyme disease. The *B. burgdorferi* linear plasmids solve the "primer problem" by having telomeric sequences on their ends much like some eukaryotic viruses. The *Streptomyces* plasmids have a protein at their ends that may serve as a primer (see Box 7.2). The primer problem occurs because DNA polymerases require a primer and so cannot replicate the extreme 3' ends of a linear DNA; it is discussed in more detail in chapter 7 in the section on replication of linear phage DNAs. The figure depicts the ends of a 16-kb plasmid from *B. burgdorferi.* Such plasmids carry genes for the major surface proteins of the bacteria, breaking the rule that essential genes are never found on plasmids. It is also possible that the chromosomes from these bacteria are linear and the plasmids are directly derived from the chromosome, perhaps through the deletion of large chromosomal segments.

Not only are these plasmids linear, but also their ends are covalently joined, as shown in the figure. If these plasmids are denatured by strand separation, they will form one large, single-stranded, circular molecule. The ends have the same sequence, but they are inverted. Somehow these features allow the plasmids to replicate all the way through without losing DNA from the ends of each replication. If the DNA polymerase could somehow replicate around the end, then this DNA could serve as a primer for replication of the other strand.

Another interesting feature of the ends of these linear plasmids is their similarity to the ends of the linear DNAs of poxviruses and African swine fever virus, as shown in the figure. Whether this similarity is significant in terms of the origins of these bacteria and viruses remains to be seen.

Reference
Hinnebusch, J., and A. G. Barbour. 1991. Linear plasmids of *Borrelia burgdorferi* have a telomeric structure and sequence similar to those of a eukaryotic virus. *J. Bacteriol.* 173:7233–7239.

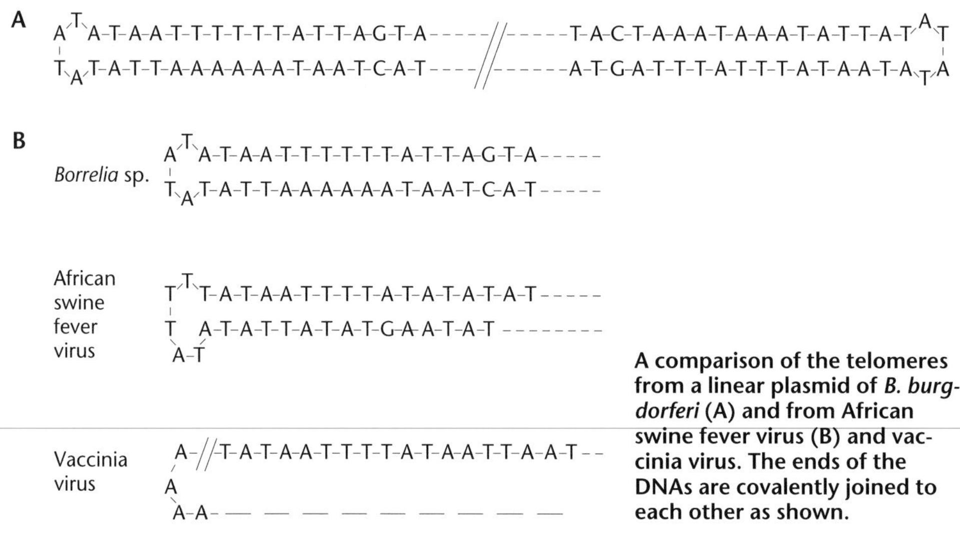

A comparison of the telomeres from a linear plasmid of *B. burgdorferi* (A) and from African swine fever virus (B) and vaccinia virus. The ends of the DNAs are covalently joined to each other as shown.

boring them. Consequently, many plasmids were named after the genes they carry. For example, R-factor plasmids contain genes for resistance to several antibiotics (hence the name R for *resistance*). These were the first plasmids discovered, when *Shigella* and *Escherichia coli* strains resistant to a number of antibiotics were isolated from the fecal flora of patients in Japan in the late

1950s. The ColE1 plasmid carries a gene for the protein colicin E1, a bacteriocin that kills bacteria that do not carry this same plasmid. The Tol plasmid contains genes for the degradation of toluene, and the Ti plasmid of *Agrobacterium tumefaciens* carries genes for *tumor ini*tiation in plants. This system of nomenclature has led to some confusion, because plasmids carry various genes

TABLE 4.1	Some naturally occurring plasmids and the traits they carry	
Plasmid	**Trait**	**Original source**
ColE1	Bacteriocin which kills *E. coli*	*E. coli*
Tol	Degradation of toluene and benzoic acid	*Pseudomonas putida*
Ti	Tumor initiation in plants	*Agrobacterium tumefaciens*
pJP4	2,4-D (dichlorophenoxyacetic acid) degradation	*Alcaligenes eutrophus*
pSym	Nodulation on roots of legume plants	*Rhizobium meliloti*
SCP1	Antibiotic methylenomycin biosynthesis	*Streptomyces coelicolor*
RK2	Resistance to ampicillin, tetracycline, and kanamycin	*Klebsiella aerogenes*

besides the ones for which they were originally named. Also, we have altered many of these plasmids beyond recognition to make plasmid cloning vectors (see below) and for other purposes.

To avoid further confusion, the naming of plasmids is now standarized. Plasmids are given number and letter names much like bacterial strains. A small "p," for *p*lasmid, precedes capital letters that describe the plasmid or sometimes give the initials of the person or persons who isolated or constructed it. These letters are often followed by numbers to identify the particular construct. When the plasmid is further altered, a different number is assigned to indicate the change. For example, the plasmid pBR322 was constructed by *Bolivar* and *Rodriguez* and is derivative number 322 of the plasmids they constructed. pBR325 is pBR322 with a

BOX 4.2

Plasmids and Bacterial Pathogenesis

Plasmids often carry virulence genes required for bacterial pathogenicity. For example, to be pathogenic, strains of the genus *Shigella* must carry a large plasmid longer than 200 kb that contains genes for cell invasion and cell adhesion. Interestingly, the genes for regulating the virulence genes on the plasmids are in the chromosome rather than in the plasmid. The genes for the Shiga toxin are also in the chromosome. The Shiga toxin is related structurally to the diphtheria toxin but has an *N*-glycosylase activity that removes a specific adenine base in the 28S rRNA, thus blocking translation. This distribution of virulence genes between the plasmid and the chromosome in strains of *Shigella* demonstrates the close relationship between plasmids and their hosts.

Most strains of *Salmonella* also require a large plasmid for their pathogenicity. However, the precise virulence genes that are carried on the plasmid are not known, and *Salmonella typhi* does not require a plasmid to cause typhoid fever.

One of the clearest examples of plasmid virulence genes is in the genus *Yersinia*. The three species of *Yersinia*, *Y. enterocolitica*, *Y. pseudotuberculosis*, and *Y. pestis*, all cause disease, ranging in severity from mild enteritis in the case of *Y. enterocolitica* and *Y. pseudotuberculosis* to bubonic plague in the case of *Y. pestis*. To be pathogenic, all three species must harbor closely related plasmids that are about 70 kb long.

These plasmids encode outer membrane proteins called Yops, which may help the bacteria avoid host phagocytosis. The Yops are synthesized only under conditions of limiting calcium ions and high concentrations of sodium ions—conditions that may mimic the environment inside eukaryotic cells. The "plague bacillus," *Y. pestis*, must also carry two other smaller plasmids to cause disease. One of these encodes a toxin and an antiphagocytic protein similar to Yops, and the other encodes a protease that increases invasiveness. *Y. pestis* also has an iron-scavenging system that allows it to extract iron directly from hemin. The ion-scavenging system is encoded by a large, normally nonessential, chromosomal element that might be an integrated plasmid or prophage. Why so many genes required for pathogenesis are carried on plasmids and other DNA elements is a subject of speculation.

References
Brubaker, R. R. 1991. Factors promoting acute and chronic disease by the yersiniae. *Clin. Microbiol. Rev.* 4:309–324.

Gulig, P. A., H. Danbara, D. G. Guiney, A. J. Lax, F. Norel, and M. Rhen. 1993. Molecular analysis of *spv* virulence genes of the *Salmonella* virulence plasmids. *Mol. Microbiol.* 7:825–830.

Tobe, T., C. Sasakawa, N. Okada, Y. Honma, and M. Yoshikawa. 1992. *vacB*, a novel chromosomal gene required for expression of virulence genes on the large plasmid of *Shigella flexneri*. *J. Bacteriol.* 174:6359–6367.

chloramphenicol resistance gene inserted. The new number 325 distinguishes this plasmid from pBR322.

Functions Encoded by Plasmids

Depending on their size, plasmids can encode a few or hundreds of different proteins. However, as mentioned above, plasmids rarely encode gene products that are always essential for growth, such as RNA polymerase, ribosomal subunits, or enzymes of the tricarboxylic acid cycle. Instead, plasmid genes usually give bacteria a selective advantage under only some conditions.

Table 4.1 lists a few naturally occurring plasmids and some traits they encode, as well as the host in which they were originally found. Gene products encoded by plasmids include enzymes for the utilization of unusual carbon sources such as toluene, resistance to substances such as heavy metals and antibiotics, synthesis of antibiotics, and synthesis of toxins and proteins that allow the successful infection of higher organisms. We can use the fact that plasmids generally carry only nonessential genes to distinguish them from the chromosome, particularly when the plasmid is almost as large as the chromosome.

If plasmid genes, such as those for antibiotic resistance and toxin synthesis, were part of the chromosome, all bacteria of the species, not just the ones with the plasmid, would have the benefits of those genes. Consequently, all the members of that species would be more competitive in environments where these traits were desirable. So why are plasmid genes not simply part of the chromosome? Maybe having some genes on plasmids makes the host species able to survive in more environments without the burden of a larger chromosome. Bacteria must be able to multiply very quickly under some conditions to obtain a selective advantage, and smaller bacterial chromosomes can replicate faster than larger ones. Plasmids encoding different traits can then be distributed among different members of the population, where they do not burden any single bacterium too heavily. However, if the environment abruptly changes so that the genes carried on one of the plasmids become essential, the bacteria that carry the plasmid will suddenly have a selective advantage and will survive, thereby ensuring survival of the species. In this way, plasmids allow bacteria to occupy a larger variety of ecological niches and contribute to the evolutionary success of not only the bacterial species but also the plasmids found in that species.

Plasmid Structure

Most plasmids are circular with no free ends. All of the nucleotides in each strand are joined to another nu-

cleotide on each side by covalent bonds to form continuous strands that are wrapped around each other. Such DNAs are said to be **covalently closed circular (CCC)**. This structure prevents the strands from separating, and there are no ends to rotate, so that the plasmid can be supercoiled. As discussed in chapter 1, in a DNA that is supercoiled, the two strands are wrapped around each other more or less often than once in about 10.5 base pairs, as predicted from the Watson-Crick double-helical structure of DNA. If they are wrapped around each other more often than once every 10.5 base pairs, the DNA is positively supercoiled; if they are wrapped around each other less often, the DNA is negatively supercoiled. Like the chromosome, covalently closed circular plasmid DNAs are usually negatively supercoiled (see chapter 1). Because DNA is stiff, the negative supercoiling introduces stress, and this stress is partially relieved by the plasmid wrapping up on itself, as illustrated in Figure 4.1. In the cell, the DNA wraps around proteins, which relieves some of the stress. The remaining stress facilitates some reactions involving the plasmid, such as separation of the two DNA strands for replication or transcription.

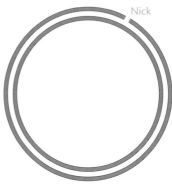

Relaxed, no supercoiling

Supercoiled, covalently
closed circular DNA

Figure 4.1 Supercoiling of a covalently closed circular plasmid. A break in one strand relaxes the DNA, eliminating the supercoiling.

Properties of Plasmids

Replication

To exist free of the chromosome, plasmids must have the ability to replicate independently. DNA molecules that can replicate autonomously in the cell are called **replicons**. The concept of a replicon was first proposed in a paper by Jacob et al. in 1963 (see Suggested Reading). Plasmids, phage DNA, and the chromosomes are all replicons, at least in the same type of cells.

To be a replicon in a particular type of cell, a DNA molecule must have at least one origin of replication, or *ori* site, where replication begins (see chapter 1). In addition, the cell must contain the proteins that enable replication to initiate at this site. Plasmids encode only a few of the proteins required for their own replication. In fact, many encode only one of the proteins needed for initiation at the *ori* site. All of the other required proteins, including DNA polymerases, ligases, primases, helicases, and so on, are borrowed from the host.

Each type of plasmid replicates by one of two general mechanisms (Figure 4.2), which is determined along with other properties by the genes surrounding

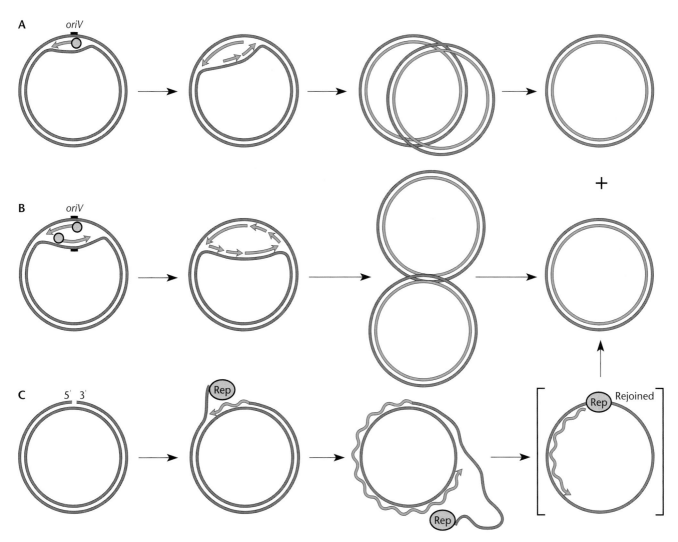

Figure 4.2   Some common schemes of plasmid replication. The origin region is designated *oriV*. (A) Unidirectional replication. Replication terminates when the replication fork gets back to the origin. (B) Bidirectional replication. Replication terminates when the replication forks meet somewhere on the DNA molecule opposite the origin. (C) Rolling-circle replication. Replication from the 3' end at a nick displaces one of the two strands. The ends of the displaced strand are rejoined, and the complement of the displaced strand is made from RNA primers synthesized at unique sites.

the molecule's *ori* (see the section on the *ori* region, below). The plasmid replication origin is named *oriV* (*ori v*ector). Most of the evidence for the mechanisms described below came from observations of replicating plasmid DNA with the electron microscope.

THETA REPLICATION

Some plasmids begin replication by opening the two strands of DNA at the *ori* region, creating a structure that looks like the Greek letter θ—hence the name **theta replication** (Figure 4.2). In this process, an RNA primer begins replication, which can proceed in one or both directions around the plasmid. In the first case, a single replication fork moves around the molecule until it returns to the origin and then the two daughter DNAs separate. In the other case (bidirectional replication), two replication forks move out from the *ori* region, one in either direction, and replication is complete (and the two daughter DNAs separate) when the two forks meet somewhere on the other side of the molecule.

The theta mechanism is the most common form of DNA replication. It is used not only by most plasmids, including ColE1, RK2, F, and P1, but also by the chromosome (see chapter 1).

ROLLING-CIRCLE REPLICATION

In the other type of replication, one strand is nicked at the *ori* region and the 3' OH end at the nick serves as a primer for replication. The displaced strand can then serve as a template to make another double-stranded DNA molecule. The term **rolling-circle replication** describes how the template strand can be imagined to roll as it replicates (Figure 4.2).

To begin, rolling-circle replication requires a nick in one strand of the double-stranded replicative form (RF) at the so-called plus origin. Then replication proceeds around the circle, displacing the opposite, or minus, strand. On the displaced strand, replication initiates at specific sites, "minus-strand origins," to make another double-stranded replicative form. If these minus-strand origins do not function well in a particular host, single minus strands will accumulate. In fact, the so-called singled-stranded plasmids found in some gram-positive bacteria have large amounts of single-stranded DNA, apparently as a result of defective rolling-circle replication.

Plasmids that replicate by a rolling-circle mechanism include pUB110 and pC194, isolated from *Staphylococcus aureus,* and pIJ101, isolated from *Streptomyces lividans.* Some types of phage also replicate by this mechanism (see chapter 7).

Functions of the *ori* Region

In most plasmids, the genes for proteins required for replication are located very close to the *ori* sequences at which they act. Thus, only a very small region surrounding the plasmid *ori* site is required for replication. As a consequence, the plasmid will still replicate if most of its DNA is removed, provided the *ori* region remains and the plasmid DNA is still circular. Smaller plasmids are easier to use as cloning vectors, as discussed later in the chapter.

In addition, the genes in the *ori* region often determine many other properties of the plasmid. Therefore, any DNA molecule with the *ori* region of a particular plasmid will have other characteristics of that plasmid. The following sections describe the major plasmid properties determined by the *ori* region.

HOST RANGE

The **host range** of a plasmid includes all the types of bacteria in which the plasmid can replicate, and the host range is usually determined by the *ori* region. Some plasmids, such as those with *ori* regions of the ColE1 plasmid type, including pBR322 and pET and pUC, have **narrow host ranges**. These plasmids will replicate only in *E. coli* and some other closely related bacteria such as *Salmonella* and *Klebsiella* species. In contrast, plasmids with a **broad host range** include the RK2 and RSF1010 plasmids, as well as the rolling-circle plasmids from gram-postive bacteria mentioned above. The host range of these plasmids is truly remarkable. Plasmids with the *ori* region of RK2 will replicate in most types of gram-negative bacteria, and RSF1010-derived plasmids will even replicate in some tyes of gram-positive bacteria. Many of the plasmids isolated from gram-positive bacteria also have quite broad host ranges. For example, pUB110, which was first isolated from the gram-positive *S. aureus,* will replicate in many other gram-positive bacteria, including *Bacillus subtilis.* However, most plasmids isolated from gram-negative bacteria will not replicate in gram-positive bacteria and vice versa, which reflects the evolutionary divergence of these groups (see the introductory chapter).

It is perhaps surprising that the same plasmid can replicate in bacteria which are so distantly related to each other. Broad-host-range plasmids must encode all of their own proteins required for initiation of replication, and so they do not have to depend on the host cell for any of these functions. They also must be able to express these genes in many types of bacteria. Apparently, the promoters and ribosome initiation sites for the replication genes of broad-host-range plasmids have evolved so that they will be recognized in a wide variety of bacteria.

REGULATION OF COPY NUMBER

Another characteristic of plasmids that is determined mostly by their *ori* region is their **copy number**, or the average number of a particular plasmid per cell. More precisely, we can define the copy number as the number of copies of the plasmid in a newborn cell immediately after cell division. All plasmids must regulate their replication; otherwise they would fill up the cell and become too great a burden for the host, or their replication would not keep up with the cell replication and they would be progressively lost during cell division. Some plasmids, such as pIJ101 of *Streptomyces coelicolor*, replicate enough to populate the cell with hundreds of copies. However, others, such as the F plasmid of *E. coli*, replicate only once or a few times during the cell cycle. Table 4.2 lists the copy numbers of these and other plasmids.

The regulation mechanisms used by plasmids with higher copy numbers often differ greatly from that used by plasmids with lower copy numbers. Plasmids that have high copy numbers, such as ColE1 plasmids, need only have a mechanism that inhibits the initiation of plasmid replication when the number of plasmids in the cell reaches a certain level. Consequently, these molecules are called **relaxed plasmids**. By contrast, low-copy-number plasmids such as F must replicate only once or very few times during each cell cycle and so must have a tighter mechanism for regulating their replication. Hence, these are called **stringent plasmids**. Much more is understood about the regulation of replication of relaxed plasmids than about the regulation of replication of stringent plasmids. The following paragraphs describe some of the better-known examples.

ColE1-Derived Plasmids: Regulation by an RNA

Figure 4.3 shows the genetic map of the original ColE1 plasmid. This molecule has been put to use in numerous molecular biology studies, and many vectors have been derived from the closely related pMB1, including pBR322, the pUC plasmids, and the pET series of plasmids discussed in chapter 7. Although the genetic maps

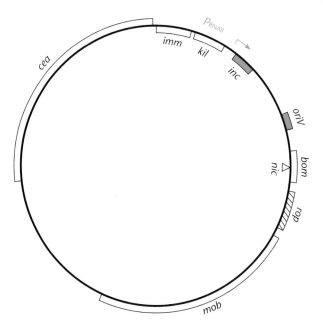

Figure 4.3 Genetic map of plasmid ColE1. The plasmid is 6,646 bp long. On the molecule, *oriV* is the origin of replication; p_{RNAII} is the promoter for the primer RNA II, *inc* encodes RNA I, *rop* encodes a protein that helps regulate copy number, *bom* is a site that is nicked at *nic*, *cea* encodes the colicin ColE1, and *mob* encodes functions required for mobilization (discussed in chapter 5). Redrawn from F. C. Neidhardt, J. L. Ingraham, K. B. Low, B. Magasanik, M. Schaechter, and H. E. Umbarger, *Escherichia coli and Salmonella typhimurium: Cellular and Molecular Biology*, p. 1620, American Society for Microbiology, Washington, D.C., 1987.

TABLE 4.2	Copy numbers of some plasmids
Plasmid	**Approximate copy number**
F	1
P1 prophage	1
RK2	4–7 (in *E. coli*)
pBR322	16
pUC18	~30–50
pIJ101	40–300

of these engineered plasmids, known as cloning vectors (discussed in more detail below), have changed beyond recognition, they all have the *ori* region of the original ColE1-like plasmid and hence they share many of its properties, including the mechanism of replication regulation.

As shown in Figure 4.4, replication is regulated mostly through the effects of a small plasmid-encoded RNA called RNA I. This small RNA inhibits plasmid replication by interfering with the processing of another RNA called RNA II, which forms the primer for plasmid DNA replication. RNA II forms an RNA-DNA hybrid at the replication origin. RNA II is cleaved by the RNA endonuclease RNase H, releasing a 3' hydroxyl group that serves as the primer for replication catalyzed by DNA polymerase I. Unless RNA II is processed properly, it will not function as a primer and replication will not ensue.

RNA I and RNA II are complementary to each other, since they are made in the same region of the DNA but from opposite strands, as shown. Because they are complementary, the two RNAs can base pair to form a dou-

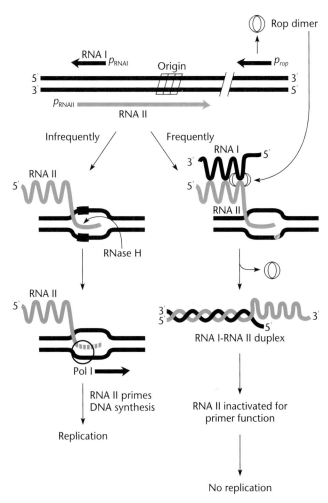

Figure 4.4 Regulation of replication of ColE1-derived plasmids. RNA II must be processed by RNase H before it can prime replication. "Origin" indicates the transition point between the RNA primer and DNA. Most of the time, RNA I binds to RNA II and inhibits the processing, thereby regulating the copy number. p_{RNAI} and p_{RNAII} are the promoters for RNA I and RNA II transcription, respectively. RNA II is shown in color. The Rop protein dimer enhances the initial pairing of RNA I and RNA II.

ble-stranded RNA helix. The formation of this double-stranded RNA interferes with the processing of RNA II, and replication will not ensue.

This mechanism provides an explanation for how the copy number of ColE1 plasmids is maintained. Since RNA I is synthesized from the plasmid, more RNA I will be made when the concentration of the plasmid is high. A high concentration of RNA I interferes with the processing of most of the RNA II, and replication is inhibited. The inhibition of replication is almost complete when the concentration of the plasmid reaches about 16 copies per cell, the copy number of the ColE1 plasmid.

In addition to RNA I, a plasmid-encoded protein called Rop helps maintain the copy number. This protein, which forms a dimer, enhances the pairing between RNA I and RNA II, so that processing of the primer can be inhibited even at relatively low concentrations of RNA I.

This model for the regulation of ColE1 plasmid replication has received support from genetic experiments indicating that the regulation of plasmid copy number requires complementarity between RNA I and RNA II. Furthermore, these experiments indicate that only a short region of RNA I must be complementary to RNA II and that changing the sequence of this region changes the incompatability group of the plasmids, a subject we discuss below. These experiments are discussed in some detail in chapter 16 on molecular genetic analysis (see also Lacatena and Cesareni, Suggested Reading, chapter 16).

The R1 Plasmid: Regulation by RNA and Protein

The regulation of replication of the IncFII family of plasmids, represented by the plasmid R1, is also quite well understood. As for the ColE1 plasmid, an RNA regulates the replication of the R1 plasmid, but only indirectly.

Figure 4.5 illustrates the regulation of R1 plasmid replication. The plasmid encodes a protein called RepA, which is required for the initiation of replication. The *repA* gene can be transcribed from two promoters. One of these promoters, called p_{copB} in Figure 4.5, transcribes both the *repA* and *copB* genes, making an mRNA that can be translated into the proteins RepA and CopB. The second promoter, p_{repA}, is in the *copB* gene and so makes an RNA that can encode only the RepA protein. Because the p_{repA} promoter is repressed by the CopB protein, it is turned on only immediately after the plasmid enters a cell and before any CopB protein is made. The short burst of synthesis of RepA from p_{repA} after the plasmid enters a cell causes the plasmid to replicate until it attains its copy number. Then the p_{repA} promoter is repressed by CopB protein, and the *repA* gene can be transcribed only from the p_{copB} promoter.

Once the plasmid has attained its copy number, the regulation of synthesis of RepA, and therefore the replication of the plasmid, is regulated by an RNA called CopA in Figure 4.5. The *copA* gene is transcribed from its own promoter and the RNA product affects the stability of the mRNA made from the p_{copB} promoter. Because the CopA RNA is made from the same region encoding the translation initiation region for the *repA* gene, but from the other strand of the DNA, the two mRNAs will be complementary and can pair to make double-stranded RNA. Then a ribonucle-

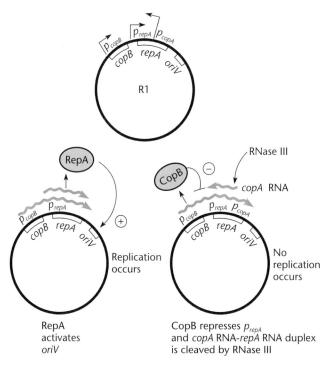

Figure 4.5 Regulation of replication of the IncFII plasmid R1. The *copA* RNA regulates copy number by binding to the RNA for the RepA protein, creating a substrate for RNase III. The action of RNase III leads to degradation of the *repA* mRNA. RepA is required for initiation of plasmid replication.

ase called RNase III, a chromosomally encoded enzyme that cleaves only double-stranded RNA, will cleave the CopA-RepA mRNA structure. The cleavage will in turn cause the mRNA for RepA to be degraded, and no more RepA protein will be made.

The destabilization of the mRNA for the RepA protein by CopA RNA ensures that the replication of the R1 plasmid and other IncFII plasmids will be regulated by the concentration of CopA RNA, which in turn will depend on the concentration of the plasmid. The higher the concentration of the plasmid, the more CopA RNA will be made and the less RepA protein will be synthesized, maintaining the concentration of plasmid around the plasmid copy number.

The Iteron Plasmids: Regulation by a Protein
Most commonly studied plasmids, including pSC101, F, R6K, P1, and the RK2- and RP4-related plasmids, use a protein to regulate their replication. The *ori* region of these plasmids contains several repeats of a certain set of DNA bases called an **iteron sequence**, from which they derive their name. These iteron sequences are typically 17 to 22 bp long and exist in about three to seven copies in the *ori* region. In addition, there are usually

additional copies of these repeated sequences a short distance away.

One of the simplest of the iteron plasmids is pSC101. For our purposes, the essential features of this plasmid's *ori* region (Figure 4.6) are the gene *repA*, which encodes the RepA protein required for initiation of replication, and three repeated iteron sequences, R1, R2, and R3, through which RepA regulates the copy number. The RepA protein is the only plasmid-encoded protein required for the replication of the pSC101 plasmid and many other iteron plasmids. It serves as a positive activator of replication, much like the RepA protein of the R1 plasmid. The host chromosome encodes the other proteins that bind to this region to allow initiation of replication, which include DnaA, DnaB, DnaC, and DnaG (see chapter 1).

Iteron plasmid replication is regulated by two superimposed mechanisms. First, the RepA protein represses its own synthesis by binding to its own promoter region and blocking transcription of its own gene. Therefore, the higher the concentration of plasmid, the more RepA protein will be made and the more it will repress its own synthesis. Thus, the concentration of RepA protein is maintained within narrow limits and the initiation of replication is strictly regulated. This type of regulation, known as **transcriptional autoregulation**, is also discussed in later chapters.

In the second mechanism, the iteron sequences of different plasmids sometimes associate, joining the plasmids. The hypothesis for this interaction, illustrated in Figure 4.7, is called the **coupling** or **handcuffing model**. The model is the result of experimental evidence indicating that the replication of iteron plasmids is controlled not solely through the amount of

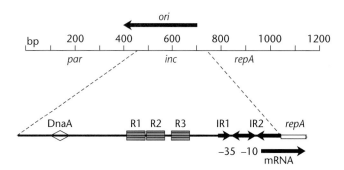

Figure 4.6 The *ori* region of pSC101. R1, R2, and R3 are the three iteron sequences (CAAAGGTCTAGCAGCAGAATT-TACAGA for R3) to which RepA binds to handcuff two plasmids. RepA autoregulates its own synthesis by binding to the inverted repeats IR1 and IR2. The location of the partitioning site *par* (see the section on Partioning) and the binding sites for the host protein DnaA are also shown.

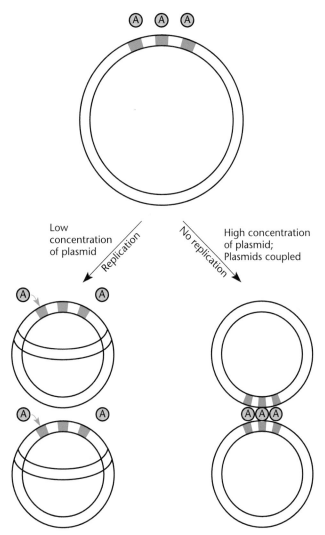

Figure 4.7 The "handcuffing" or "coupling" model for regulation of iteron plasmids. At low concentrations of plasmids, the RepA protein only binds to one plasmid at a time, initiating replication. At high plasmid concentrations, the RepA protein binds to two plasmids simultaneously, handcuffing them and inhibiting replication.

RepA protein but also by the concentration of plasmid or, more precisely, the concentration of iteron sequences (see McEachern et al., Suggested Reading). If the piece of DNA containing several iteron sequences is introduced on another plasmid into a cell containing an iteron plasmid, the copy number of the iteron plasmid will decrease.

The most important feature of the handcuffing model is that the RepA protein can link two plasmids by binding to their iteron sequences and thereby preventing them from initiating replication. By this mechanism, the replication of iteron plasmids will depend both on the

concentration of RepA protein and on the concentration of the plasmids themselves. At low concentrations of RepA protein or plasmids, the plasmids will not be coupled and replication will proceed. At higher concentrations of RepA and plasmid, the plasmids will be coupled and replication will be inhibited. We shall discuss the molecular evidence for the handcuffing model of regulation in more detail in chapter 16 on molecular genetic analysis.

Mechanisms To Prevent Curing of Plasmids

Cells that have lost a plasmid during cell division are said to be **cured** of the plasmid. Several mechanisms prevent curing, including plasmid addiction sytems (see Box 4.3, Plasmid Addiction), site-specific recombinases that destroy multimers, and partitioning systems.

RESOLUTION OF MULTIMERIC PLASMIDS
The possibility of losing a plasmid during cell division is increased if the plasmids form multimers during replication. A multimer consists of individual copies of the plasmid molecules linked to each other. The multimers probably occur as a result of defects in replication termination or by recombination of monomers. The formation of multimers effectively lowers the effective copy number because each multimer will segregate into the daughter cells as a single plasmid. Therefore, multimers greatly increase the chance of a plasmid being lost during cell division.

To avoid this problem, many plasmids have site-specific recombination systems that destroy multimers. These systems promote recombination between specific sites on the plasmid if the same site occurs more than once in the molecule, as it would in a multimer. This recombination has the effect of resolving multimers into separate molecules.

The best known example of this mechanism is the *cer*-Xer site-specific recombination system of the ColE1 plasmid (see Guhathakurta et al., Suggested Reading). The *cer* site is analogous to the *dif* site (discussed in chapter 1), and the same Xer proteins act on this site as act on the *dif* site to resolve the bacterial chromosomes after replication and allow them to separate into the daughter cells. The Xer proteins are site-specific recombinases because they act only on specific sites, the *dif* site in the chromosome and the *cer* site in ColE1 plasmids. We discuss site-specific recombinases in more detail in chapter 8.

PARTITIONING
Plasmids also avoid being lost from dividing cells by carrying **partitioning** systems, which ensure at least one

---

BOX 4.3

Plasmid Addiction

Even with their partitioning functions, plasmids are often lost from multiplying cells. In a kind of ironic revenge, some plasmids encode proteins that kill cells cured of their type. Such functions have been described for the F plasmid, the R1 plasmid, and the P1 prophage, which replicates as a plasmid. These plasmids encode a toxic protein that will kill the cell if it is expressed. The toxic protein kills the cell by various mechanisms, depending on the source. For example, the toxic protein of the F plasmid, Ccd, kills the cells by altering DNA gyrase, so it causes double-stranded breaks in the DNA. The killer protein, Hok, of the plasmid R1 destroys the cellular membrane potential, causing loss of cellular energy. The mechanism of killing by the P1-encoded killer protein, Doc, is not known, but it does not seem to work by either of the above mechanisms.

Why does the toxic protein kill cells only immediately after they have been cured of the plasmid? When cells contain the plasmid, other proteins or RNAs of the addiction system act as an antidote and either prevent the synthesis of the toxic protein or bind to it and prevent its activity. These other gene products are much more unstable than the toxic protein, however, so that they will be more rapidly inactivated. If a cell is cured of the plasmid, neither the toxic protein nor the antidote will be synthesized, but since the antidote is more unstable, it will not be long before the cured cells have only the toxic protein and are killed. These systems make the cell addicted to the plasmid once it has been acquired, and they prevent cured cells from accumulating.

References

Gerdes, K., P. B. Rasmussen, and S. Molin. 1986. Unique type of plasmid maintenance function. Postsegregational killing of plasmid-free cells. *Proc. Natl. Acad. Sci. USA* **83**:3116–3120.

Jaffe, A., T. Ogura, and S. Hiraga. 1985. Effects of the *ccd* function of the F plasmid on bacterial growth. *J. Bacteriol.* **163**:841–849.

Lehnherr, H., E. Maguin, S. Jafri, and M. B. Yarmolinsky. 1993. Plasmid addiction genes of bacteriophage P1: *doc*, which causes cell death on curing of prophage, and *phd*, which prevents host death when prophage is retained. *J. Mol. Biol.* **233**:414–428.

---

copy of the plasmid segregates into each daughter cell during cell division. The functions involved in these systems are called *par* functions.

Using combinatorial probability, we can calculate how often cells will be cured of a plasmid if the plasmid has no *par* system. In the simple example shown in Figure 4.8, the copy number of the plasmid is 4. Immediately after cell division, a cell contains four copies, and immediately before the cell divides, it contains eight copies. If the plasmids are equally divided into the two daughter cells, each will get four plasmids. However, the plasmids usually will not be equally distributed between the two daughter cells, and one daughter cell will get more than the other. In fact, with a certain probability, one cell will get all of the plasmids and the other cell will be cured. Since each plasmid can go into either one cell or the other, the probability that one daughter cell will be cured of the plasmid is the same as the probability of tossing eight heads (or tails) of a coin in a row. Thus, the probability that the first plasmid will go into one cell is 1/2, and the probability that the first two plasmids will go into the same cell is 1/2 times 1/2 = 1/4, and so on. The probability that all eight will go into one cell is therefore $(1/2)^8$ or 1/256. Since it is irrelevant which of the two cells is cured, the frequency of curing will be twice this value, so $2(1/2)^8$ or 1/128 of the cells will be cured each time the cells divide. In general, for a plasmid with a copy number n, the frequency of curing is $2(1/2)^{2n}$, since the number of plasmids at the time of division is twice the copy number. Also, as the cells divide once every generation time, the frequency of cured cells in the population will be roughly equal to the number of generation times that have elapsed times this number. This is the frequency of curing if the sorting of the plasmids into daughter cells is completely random. Therefore, if the fraction of the cells that are cured of the plasmid is less than $2(1/2)^{2n}$ times the number of generation times that have elapsed, the plasmid must have some sort of partitioning function.

This calculation indicates that few cells would be cured of a high-copy-number plasmid each generation, even without a partitioning mechanism. However, a significant fraction of cells would be cured of a low-copy-number plasmid each generation. In fact, with a plasmid with a copy number of only 1, such as F or P1, $2(1/2)^2$ or 1/2 of the cells would be cured each generation. Since cells are seldom cured of even low-copy-number plasmids, some mechanism must ensure that plasmids, especially those with low copy numbers, will be partitioned faithfully into the daughter cells each time the cell divides.

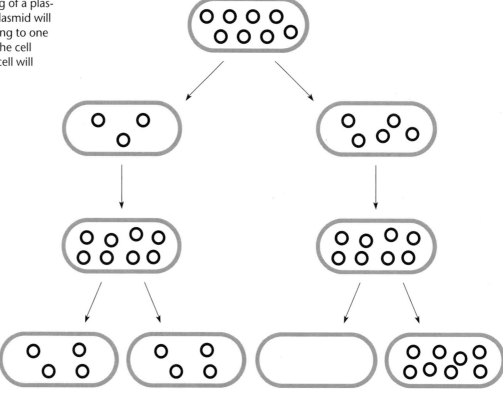

Figure 4.8 Random curing of a plasmid with no *par* site. Each plasmid will have an equal chance of going to one of the daughter cells when the cell divides, and the occasional cell will inherit no plasmids.

The *par* Functions

Some plasmids, including F, P1, and R1, have short regions that are known to enhance proper partitioning. If such a region is removed from the plasmid, the frequency of curing will be much higher, close to that predicted above. The partitioning system of the P1 plasmid has been studied most extensively. The P1 plasmid *par* region consists of a *cis*-acting *par* sequence and the genes for two proteins, ParA and ParB, one of which binds to the *par* site.

Mechanism of Partitioning

Figure 4.9 shows two general ideas for how *par* sites might ensure proper partitioning of a plasmid (see Austin and Nordstrom, Suggested Reading). According to these models, the bacterial membrane functions like the mitotic spindle in eukaryotic organisms, pulling plasmids apart prior to cell division. According to both models, the *par* site is the region of the plasmid that binds to the cellular membrane, and as the membrane grows during cell division, the two copies of the plasmid are pulled apart into the two daughter cells. The two models differ in explaining how binding to the membrane ensures that at least one copy of the plasmid enters each daughter cell at the time of division.

According to the model shown in Figure 4.9A, the *par* sites bind to putative sites on the bacterial membrane. In the model, each type of plasmid has its own unique site on the membrane. As the cell grows, these sites will duplicate and bind only one copy of the plasmid. Because these sites on the membrane move apart as the cell grows and divides, one copy of the plasmid will be distributed into each of the daughter cells. Even if there are more copies of the plasmid and these are randomly distributed during division, the cell will not be cured provided that at least one copy of the plasmid is bound to the membrane of each daughter cell. That single copy can then multiply many times before the next division.

The problem with this model is the required number of unique sites on the membrane—one for each type of plasmid. It seems unlikely that there would be as many unique sites on the membrane as there are different types of plasmids with partitioning sites.

Model A remains tenable yet avoids the requirement for unique membrane sites if we propose that the sites to which the plasmids bind are on the chromosome rather than the membrane. Because of the complexity of sequence in the chromosome, an almost infinite number of unique sites is possible, and these sites will double each time the chromosome replicates. The only unique site on the membrane need be the site to which the chro-

A

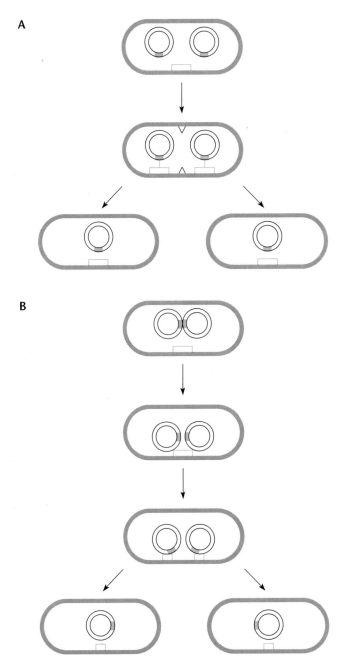

B

Figure 4.9 Two models for the function of *par* sites. In model A, the plasmid binds to a unique site on the bacterial membrane and the two copies of the plasmid are pulled apart as the site on the membrane divides. In model B, the two copies of the plasmid bind to each other before they bind to a site on the membrane. One then associates with each site when the site on the membrane divides.

mosome binds, and the plasmid will partition along with the chromosome.

Model B is similar but avoids the unique-site problem altogether. In this model, two copies of the plasmid

must pair with each other at their *par* sites before the compound *par* site can bind to the membrane. Then the two plasmids separate as the site on the membrane is pulled apart when the cell divides. One copy of the plasmid will partition properly into each of the daughter cells provided that two plasmids will not pair with each other unless they have the same *par* site and that only paired *par* sites will bind to the membrane.

High-copy-number plasmids often have sites that increase proper partitioning of the plasmid, but these regions do not have the same structure and may not work by the same mechanism. For example, the *par* region of pSC101 increases supercoiling of this plasmid. How increased supercoiling aids in proper partitioning is not clear, but perhaps it enhances replication of plasmids immediately after division so that they are less apt to be cured in the next division.

Incompatibility

Many bacteria, as they are isolated from nature, contain more than one type of plasmid. These plasmid types stably coexist in the bacterial cell and remain there even after many cell generations. In fact, bacterial cells containing multiple types of plasmids will not be cured of each plasmid any more frequently than if the other plasmids were not there.

However, not all types of plasmids can stably coexist in the cells of a bacterial culture. Some types will interfere with each other's replication or partitioning so that if two such plasmids are introduced into the same cell, one or the other will be lost at a higher than normal rate when the cell divides. This phenomenon is called **plasmid incompatibility**, and two plasmids that cannot stably coexist are members of the same **incompatibility**, or **Inc, group**. If two plasmids can stably coexist, they belong to different Inc groups. There may be hundreds of different Inc groups, and plasmids are usually classified by their group. For example, RP4 and RK2 are both IncP (incompatibility group P) plasmids. In contrast, RSF1010 is an IncQ plasmid, so can be stably maintained with either RP4 or RK2 because it belongs to a different Inc group.

Plasmids of the same Inc group can be incompatible because they share the same replication control mechanism and/or because they share the same partitioning (*par*) functions. The following sections address each of these reasons.

INCOMPATIBILITY DUE
TO REPLICATION CONTROL

Two plasmids that share the same mechanism of replication control will be incompatible. The replication control system will not recognize the two as different, and so either may be randomly selected for replication.

Figure 4.10A and B contrast distribution at cell division of plasmids of the same Inc group with plasmids of different Inc groups. Figure 4.10A shows a cell contain-

ing two types of plasmids that belong to different Inc groups and use different replication control systems. In the illustration, both plasmids exist in equal numbers before cell division, but after division, the two daughter cells are not likely to get the same number of each plas-

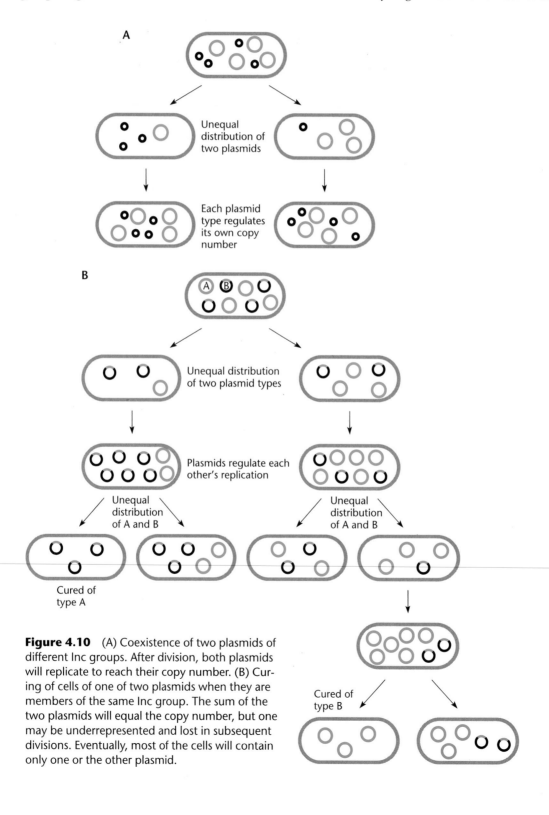

Figure 4.10 (A) Coexistence of two plasmids of different Inc groups. After division, both plasmids will replicate to reach their copy number. (B) Curing of cells of one of two plasmids when they are members of the same Inc group. The sum of the two plasmids will equal the copy number, but one may be underrepresented and lost in subsequent divisions. Eventually, most of the cells will contain only one or the other plasmid.

mid. However, in the new cells, each plasmid will replicate to reach its copy number, so that at the time of the next division, both cells will again have the same numbers of the plasmids. This process will be repeated each generation, so very few cells will be cured of either plasmid.

Now consider Figure 4.10B, in which the cell has two plasmids in the same Inc group, both of which use the same replication control system. As in the first example, both plasmids originally exist in equal numbers, and when the cell divides, chances still are that the two daughter cells will not receive the same number of the two plasmids. Note that in the original cell, the copy number of each plasmid is only half its normal number; both plasmids contribute to the total copy number since they both have the same *ori* region and inhibit each other's replication. After cell division, the two plasmids will replicate until the *total* number of plasmids in each cell equals the copy number. The underrepresented plasmid (recall that the daughters may not receive the same number of plasmids) will not necessarily replicate more than the other plasmid, so that the imbalance of plasmid number might remain or become even worse. At the next cell division, the underrepresented plasmid has less chance of being distributed to both daughter cells since there are fewer copies of it. Consequently, in subsequent cell divisions, the daughter cells are much more likely to be cured of one of the two plasmid types.

Incompatibility due to copy number control is probably more detrimental to low-copy-number plasmids than to high-copy-number plasmids. If the copy number is only 1, then only one of the two plasmids can replicate, and each time the cell divides, a daughter will be cured of one of the two types of plasmids.

INCOMPATIBILITY DUE TO PARTITIONING

Two plasmids can also be incompatible if they share the same *par* function. When coexisting plasmids share the same *par* function, one or the other will always be distributed into the daughter cells during division. However, sometimes one daughter cell will receive one type of plasmid and the other cell will get the other plasmid type, producing cells cured of one or the other plasmid.

Plasmid Genetics

Methods for Detecting and Isolating Plasmids

The many methods developed for detecting plasmids generally take advantage of the small size and/or the circularity of these molecules. Cloning manuals usually give detailed protocols for these methods (see chapter

15, Suggested Reading), but we review them briefly in this section.

ALKALI LYSIS METHOD

The alkali lysis method takes advantage of the relatively small size of plasmids to separate them from chromosomal DNA. First, treatment of the cell lysate with alkali results in separation, or denaturation, of the DNA strands. Then increasing the salt concentration causes the chromosomal DNA, which is very large and bound to proteins, to precipitate.

The relatively small plasmids do not precipitate readily at high salt concentrations and remain in the supernatant. Because plasmids are covalently closed, their DNA strands cannot completely separate if the DNA is denatured. Their intertwining also allows the two plasmid DNA strands to quickly find their complementary sequences and renature to form double-stranded DNA once the denaturing conditions are removed. The chromosome, in contrast, will not quickly renature. While circular, the chromosome is so large that it is usually broken at early stages of purification procedures, so that the strands separate in the presence of a denaturant. Also, because the chromosome is so large, the complementary sequences take hours to renature once they have been denatured.

CsCl-EtBr PURIFICATION

Other methods for purifying plasmids are based on the fact that covalently closed circular chromsomes bind less ethidium bromide (EtBr) than do linear or nicked circular DNAs (Figure 4.11). EtBr intercalates (inserts itself) between DNA bases, pushing the bases apart and rotating the two strands of the DNA around each other. The two strands of the covalently closed circular plasmid are not free to rotate, and the binding of EtBr therefore increases the stress on the DNA until no more EtBr can bind. In contrast, because the strands of the chromosomal DNA have usually been broken during the first steps of the purification procedure, they are free to rotate and chromosomal DNA can bind much higher concentrations of EtBr.

EtBr bound to DNA makes it *less* dense in salt solutions made with heavy atoms such as cesium chloride (CsCl). As a consequence, if the DNA is mixed with a solution of CsCl and EtBr and centrifuged to establish a gradient of CsCl concentration, the covalently closed circular plasmid DNAs will band lower, at a position where the solution is more dense (see Figure 4.12).

ELECTROPHORESIS

The methods discussed above work well with plasmids that have many copies per cell and are not too large.

Figure 4.11 Less EtBr can bind to a covalently closed circular DNA than to a linear or nicked circular DNA. Also, progressively higher EtBr concentrations shift DNA supercoiling from negative to positive. At about 2 μg of EtBr per ml, most DNAs are completely relaxed. The arrow indicates the free rotation of linear DNA.

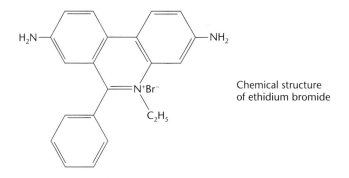

Chemical structure of ethidium bromide

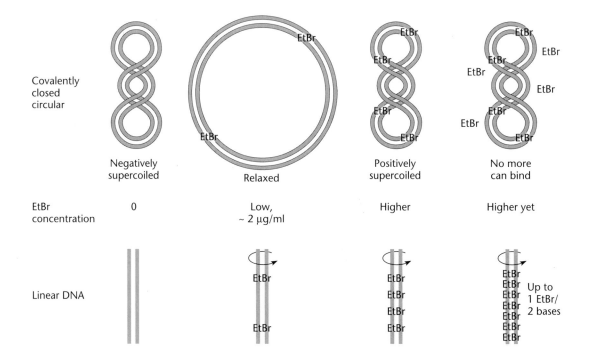

Covalently closed circular	Negatively supercoiled	Relaxed	Positively supercoiled	No more can bind
EtBr concentration	0	Low, ~ 2 μg/ml	Higher	Higher yet
Linear DNA				Up to 1 EtBr/ 2 bases

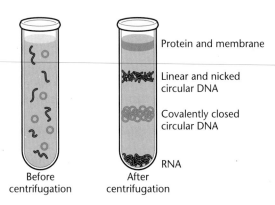

Protein and membrane

Linear and nicked circular DNA

Covalently closed circular DNA

RNA

Before centrifugation   After centrifugation

Figure 4.12 Separation of covalently closed circular plasmid DNA from linear and nicked circular DNAs on EtBr-CsCl gradients. After centrifugation, the plasmid DNA will band below the other DNAs because it has a higher buoyant density as a result of less binding of EtBr.

However, large, low-copy-number plasmids are much more difficult to detect, and other methods must be used. Large plasmids are much more susceptible to breakage than are smaller plasmids. Consequently, the cell-lysing procedures must be gentle. Also, like chromosomes, large plasmids will precipitate at high salt concentrations and will not renature spontaneously after the strands have separated.

Most methods for detecting large plasmids involve separating them from the chromosome directly by electrophoresis on agarose gels. In the example shown in Figure 4.13, the cells were broken open directly on the agarose gel to avoid breaking the plasmid DNA. Then application of an electric field caused the DNA to migrate, or run, along the gel. After electrophoresis, the DNA bands were stained with EtBr, and UV light was shone on the gel to reveal the position of the DNA. EtBr

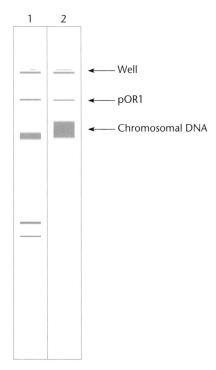

Figure 4.13 Agarose gel electrophoresis of DNA from cells containing a large plasmid. Lane 1, total DNA from *Stenotrophomonas maltophilia* ORO2; lane 2, total DNA from *E. coli* HB101 transformed with pOR1. (Courtesy of John Caguiat and Julius Jackson.)

will fluoresce if it absorbs UV light when bound to DNA. The plasmid, because of its unique size, makes a band on the gel that is distinct from that due to chromosomal DNA.

PULSED-FIELD GEL ELECTROPHORESIS

Standard agarose gels cannot be used to separate larger DNAs. Note that in Figure 4.13 the chromosomal DNAs form one band, even though these molecules have been broken into many pieces of different lengths. Also, although the plasmid is smaller, it runs more slowly (travels a shorter distance) than the chromosome on the gel. The reason for this anomalous behavior is that as long, linear molecules move through the gel, they tend to orient themselves in the direction they are going. They will then move faster and all at the same rate independent of their length. If a piece of DNA is shorter, or circular like the plasmid, it will not orient itself as readily and will travel a shorter distance.

Methods such as pulsed-field gel electrophoresis (PFGE) have been devised to allow the separation of long pieces of DNA based on size. These methods depend on periodic changes in the direction of the electric field. The molecules will attempt to reorient themselves each time the field shifts, and the longer molecules will reorient themselves more slowly than the shorter ones and so will move more slowly. Such methods have allowed the separation of DNA molecules hundreds of thousands of base pairs long and the detection of very large plasmids. We discuss them in more detail in chapter 16.

Determining the Incompatibility Group

To classify a plasmid by its incompatibility group, we must determine if it can coexist with other plasmids of known incompatibility groups. In other words, we must measure how frequently cells are cured of the plasmid when it is introduced into cells carrying another plasmid of a known incompatibility group. But we can know that cells have been cured of a plasmid only when it encodes a selectable trait, such as resistance to an antibiotic. Then the cells will become sensitive to the antibiotic if the plasmid is lost.

The experiment shown in Figure 4.14 is designed to measure the curing rate of a plasmid that contains the Cam^r gene, which makes cells resistant to the antibiotic chloramphenicol. To measure the frequency of plasmid curing, we grow the cells containing the plasmid in medium containing all the growth supplements and *no* chloramphenicol. At different times, we take a sample of the cells, dilute it, and plate the dilutions on agar containing the same growth supplements but again *no* chloramphenicol. After incubation of the plates, we pick the individual colonies with a loop or toothpick and transfer them to the same type of plate *with* chloramphenicol. To streamline this process, we might replicate the plate onto another plate containing chloramphenicol (see chapter 14). If we do not observe any growth for a colony, the bacteria in that colony must all have been sensitive to the antibiotic and hence the original bacterium that had multiplied to form the colony must have been cured of the plasmid. The percentage of colonies that contain no resistant bacteria is the percentage of bacteria that were cured of the plasmid at the time of plating.

To apply this test to determine if two plasmids are members of the same Inc group, both plasmids must contain different selectable genes, for example, genes encoding resistance to different antibiotics. Then one plasmid is introduced into cells containing the other plasmid. Resistance to both antibiotics is selected for. Then cells containing both plasmids are incubated without either antibiotic and finally grown on antibiotic-containing plates, as above. The only difference is that the colonies are transferred onto two plates, each

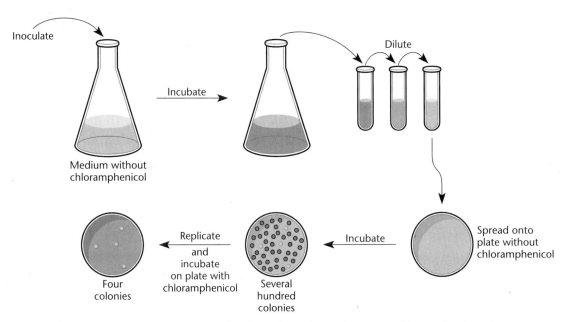

Figure 4.14 Measuring curing of a plasmid carrying resistance to chloramphenicol. See the text for details.

containing one or the other antibiotic. If the percentage of cells cured of one or the other plasmid is no higher than the percentage cured of either plasmid when it was alone, the plasmids are members of different Inc groups. We continue to apply this test until we find a known plasmid, if any, which is a member of the same Inc group as our unknown plasmid.

Maintaining Plasmids Belonging to the Same Incompatibility Group

We can also use selectable markers such as antibiotic resistance to maintain a population of cells in which most retain two high-copy-number relaxed plasmids, even if the two plasmids belong to the same Inc group. Two relaxed plasmids that are members of the same Inc group will coexist for a limited number of generations during growth of a culture, although one or the other will be cured with a higher than normal frequency (see above). Yet if the plasmids carry genes for resistance to different antibiotics, we can reduce the curing rates of both. For example, consider a cell in which one plasmid contains a gene for tetracycline resistance and the other, which lacks that gene, has a gene for chloramphenicol resistance. By growing the cells in media containing both antibiotics, we can maintain both plasmids. Loss of one plasmid or the other will occur at a high frequency during growth of the culture, but a cell that has lost one of the plasmids will have a limited life span, and it, or its immediate progeny, will die. Therefore, most of the surviving cells at any time will have both plasmids.

Determining the Host Range

The actual host range of most plasmids is unknown because it is sometimes difficult to determine if a plasmid can replicate in other hosts. First, we must have a way of introducing the plasmid into other bacteria. Transformation systems (see chapter 6) have been developed for some but not all types of bacteria, and these can be used to introduce plasmids. Plasmids that are self-transmissible or mobilizable (see chapter 5) can sometimes be introduced into other types of bacteria by conjugation, a process in which DNA is transferred from one cell to another.

Even if we can introduce the plasmid into other types of bacteria, we still must be able to select cells that have received the plasmid. The plasmid must carry a selectable gene that will be expressed in the other bacterium, but most genes will not be expressed in bacteria distantly related to those in which it was originally found. Fortunately, there are some exceptions. For example, the kanamycin resistance gene, first found in the Tn5 transposon, will be expressed in most gram-negative bacteria, making them resistant to the antibiotic kanamycin.

Most plasmids as they are isolated from nature do not carry a convenient selectable marker, and we must introduce one. We can either clone a marker gene into the plasmid or introduce a transposon carrying a selectable marker into the plasmid by methods we shall discuss in chapter 16.

If all goes well and we have a way to introduce the plasmid into other bacteria and the plasmid carries a marker that is likely to be expressed in other bacteria, we can see if the plasmid can replicate in bacteria other than its original host. Clearly, this could be a laborious process, since the mechanisms for introducing DNA into different types of bacteria differ and there are many barriers to plasmid transfer between species. Therefore, the host range of plasmids can be extrapolated from only a few examples but cannot be precisely determined.

Finding the Plasmid *ori* Region

As discussed above, usually only a very small region of a plasmid, the *ori* region, is responsible for most of its properties, including replication, copy number control, and partitioning. How do we find this region and determine its functions? In general, the classical genetic method of isolating mutants is not very applicable to studying the origins of plasmid replication. We cannot study plasmid mutations that prevent the function of the *ori* region, since if the origin does not function properly, the plasmid will not replicate and will be lost.

Recombinant DNA techniques are particularly suited for locating and studying these regions. As shown in Figure 4.15, the plasmid is cut into several pieces (arrows in the figure) and the pieces are ligated (joined) to another piece of DNA that has a selectable marker, such as resistance to ampicillin (Amp^r). For the experiment to work, the second piece of DNA cannot have a functional origin of replication. The ligated mixture is then used to transform bacteria, and the antibiotic-resistant transformants are selected by plating the mixture on agar plates containing the antibiotic. The only DNA molecules able to replicate and also confer antibiotic resistance on the cells will be hybrids with both the *ori* region of the plasmid and the piece of DNA with the antibiotic resistance gene. Therefore, only cells harboring these hybrid molecules will grow on the antibiotic-containing medium. We can determine which plasmid fragment these have by using methods described in chapter 15, and this will tell us approximately where the *ori* region is located on the plasmid.

We can further localize the *ori* region by cutting the fragment known to contain it into even smaller pieces and repeating the process above until the smallest piece from the plasmid that can function as an origin has been identified.

This same type of analysis also can be used to identify the partitioning region and the region responsible for incompatibility. To locate the *par* sequences, smaller and smaller pieces of DNA that retain origin function are tested to see if they also confer the ability to partition properly. If the plasmid containing the smaller origin region cures at a higher frequency, it has lost its *par*

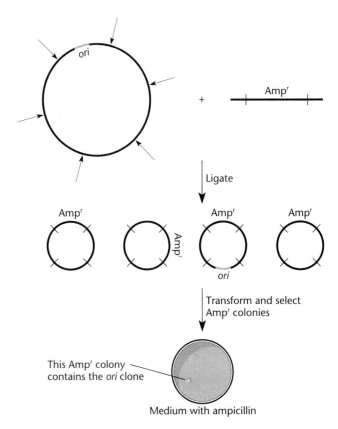

Figure 4.15 Finding the origin of replication in a plasmid. Random pieces of the plasmid are ligated to a piece of DNA containing a selectable gene but no origin of replication and introduced into cells. Cells that can form a colony on the selective plates contain the selectable gene ligated to the piece of DNA containing the origin.

region. Once we identify a plasmid that lacks a *par* region, we can clone pieces of the same type of plasmid back into it to find a piece that restores proper partitioning. In fact, the *par* regions of plasmids were discovered by this means.

Studying the Requirements for Plasmid Replication

To understand the function of plasmid origins of replication, we need to know which sequences and gene products encoded by the *ori* region are responsible for origin function. As discussed above, we cannot isolate mutant plasmids that cannot replicate, because we cannot propagate them. Furthermore, different methods are used depending on whether we are studying *trans*-acting proteins or RNA required for replication or *cis*-acting DNA sequences on which these functions act to initiate replication (see chapter 1 for the definition of *trans*- and *cis*-acting functions).

In a typical analysis of an *ori* region of a plasmid, we sequence the *ori* region and find an ORF we suspect encodes a protein required for replication. To determine if this protein is indeed required, we can provide the protein from another plasmid. First, we clone the ORF into a plasmid-cloning vector (see the next section) of a different incompatibility group, so that the protein is synthesized from the other plasmid. Then we disrupt the ORF on the original plasmid, perhaps by deleting part of it or by interrupting the coding sequence by cloning a small piece of DNA into it. If the protein encoded by the ORF is required for replication, then the plasmid with the disrupted ORF will transform only cells that contain the other plasmid expressing the ORF protein. If the protein is expressed from an inducible promoter, we could regulate the amount of the protein in the cell and see how this affects plasmid copy number.

Studies of *cis*-acting DNA sequences in the *ori* region that are required for replication call for alternative techniques. These sequences will function only on the plasmid that has them: therefore, they cannot be provided by another plasmid. However, inactivating these sequences will always prevent replication of the plasmid. To circumvent this problem, we could use a plasmid that has two origins of replication, the one we wish to study and another taken from a different plasmid active only under some conditions, for example, at low temperatures. As long as a functioning *ori* region remains in the molecule, we can change the sequences of the origin under study and then examine which changes inactivate this origin. In our example, this would be at high temperatures, when the other origin cannot function, so that replication of the plasmid would depend on the origin we have mutated.

Plasmid Cloning Vectors

A cloning vector is an autonomously replicating DNA into which other DNAs can be inserted. Any DNA inserted into the cloning vector will then replicate passively with the vector, so that many copies of the original piece of DNA can be obtained. Cloning vectors can be made from essentially any DNA that can replicate autonomously in cells.

Plasmids offer many advantages as cloning vectors. They generally do not kill the cell and are relatively easy to purify to obtain the cloned DNA. They can also be made relatively small, because they need very few plasmid-encoded functions for their replication. As a result, plasmids are particularly popular tools in some types of cloning applications. In fact, in one of the first DNA cloning experiments, a frog gene was cloned into

TABLE 4.3	Replication origins of several *E. coli* plasmid vectors	
Plasmid	*ori*	**Copy number**
pBR322	pMB1	15–20
pUC vectors	pMB1 mutant	100s
pET vectors	pMB1 mutant	100s
pBluescript	pMB1 mutant	100s
pACYC184	p15A	10–12
pSC101	pSC101	5

plasmid pSC101. The replication origins and copy numbers for several *E. coli* plasmid vectors are given in Table 4.3.

Desirable Features of Plasmid Cloning Vectors

Most naturally occurring plasmids are not convenient cloning vectors. However, plasmids can often be "engineered" to make a cloning vector. To be useful as a cloning vector, a plasmid should have at least some of the following properties.

1. It should be small, so that it can be easily isolated and introduced into various bacteria.
2. It should have a relatively high copy number, so that it can be easily purified in sufficient quantities.
3. It should carry an easily selectable trait, such as a gene conferring resistance to an antibiotic, which can be used to select cells that contain the plasmid.
4. It should have one or a few sites for restriction endonucleases, which cut DNA and allow the insertion of foreign DNAs. Also, these sites should ideally occur in selectable genes to facilitate the detection of plasmids that have foreign DNA inserts by a process called insertional inactivation.

Many plasmid cloning vectors have other special properties that aid in particular experiments. For example, some contain the sequences recognized by phage packaging systems (*pac* or *cos* sites), so that they can be packaged into phage heads (see chapter 7). Expression vectors can be used to make foreign proteins in bacteria. Mobilizable plasmids have mobilization (*mob*) sites and so can be transferred by conjugation to other cells (see chapter 5). Some broad-host-range vectors have *ori* regions that allow them to replicate in many types of bacteria or even in organisms from different kingdoms. Shuttle vectors contain more than one type of replication origin and so can replicate in unrelated organisms. We shall deal with these and some other types of speciality plasmid cloning vectors in more detail in later chapters.

CLONING VECTOR pBR322

Figure 4.16 shows a map of pBR322, which embodies many of the desirable traits of a cloning vector. This plasmid is fairly small (only 4,360 bp) and has a relatively high copy number (~16 copies per cell), making it easy to isolate. The vector was constructed by removing all but the essential *ori* region from pMB1, a ColE1-like plasmid, and adding two resistance genes for the antibiotics tetracycline and ampicillin, which were taken from plasmid pSC101 and transposon Tn*3*, respectively. The plasmid also has several unique sites for restriction endonucleases, including *Bam*HI, *Eco*RI, and *Pst*I. These enzymes cut DNA at specific sites, allowing a piece of foreign DNA to be inserted into the plasmid and then studied.

Ordinarily in cloning experiments, most of the cloning vectors will not have picked up another piece of DNA and will remain unchanged. Some way is needed to detect cells that have received the vector with a piece of DNA inserted. Insertional inactivation offers a simple genetic test for determining whether a cell contains a plasmid with a foreign DNA insert. In this technique, a gene is inactivated by having a piece of DNA inserted into it. To illustrate, Figure 4.17 shows a piece of foreign DNA inserted into the *Bam*HI site in the tetracycline resistance (Tet^r) gene in pBR322. The piece of foreign DNA in the *Bam*HI site will disrupt the Tet^r gene and cause the plasmid to lose the ability to confer tetracycline resistance on a bacterium that carries it. The plasmid will still confer ampicillin resistance, however, since the Amp^r gene remains intact. Therefore, cells containing a plasmid with a foreign DNA insert will be ampicillin resistant but tetracycline sensitive,

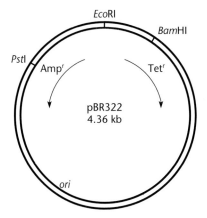

Figure 4.16 The plasmid cloning vector pBR322. Unique restriction sites are highlighted. Amp^r, ampicillin resistance; Tet^r, tetracycline resistance.

which is easy to test on agar plates containing one or the other antibiotic.

Broad-Host-Range Cloning Vectors

Many of the common *E. coli* cloning vectors such as pBR322, the pUC plasmids, and the pET plasmids have been constructed with the pMB1 *ori* region and thus are very narrow in their host range. They will replicate only in *E. coli* and a few of its close relatives. However, some cloning applications require a plasmid cloning vector that will replicate in other gram-negative bacteria, and so cloning vectors have been derived from the broad-host-range plasmids RSF1010 and RP4, which will replicate in most gram-negative bacteria. In addition to the broad-host-range *ori* region, these cloning vectors sometimes contain a *mob* site called an *oriT* site, which can allow them to be transferred into other bacteria (see chapter 5). This latter trait is very useful, because other ways of introducing DNA have not been developed for many types of bacteria.

SHUTTLE VECTORS

Sometimes, a plasmid cloning vector from one organism must be transferred into another organism. If the two organisms are not related, the same plasmid *ori* region is not likely to function in both organisms. Such situations require the use of **shuttle vectors,** so named because they can be used to "shuttle" genes between the two organisms. A shuttle vector has two origins of replication, one that functions in each organism. Shuttle vectors also must contain selectable genes that can be expressed in both organisms.

In most cases, one of the organisms in which the shuttle vector can replicate is *E. coli*. The genetic tests can be performed in the other organism, but the plasmid can be purified and otherwise manipulated by the refined methods developed for *E. coli*.

Some shuttle vectors can replicate in gram-positive bacteria and *E. coli*, whereas others can be used in lower or even higher eukaryotes. For example, plasmid YEpl3 (Figure 4.18) has the replication origin of the 2μm circle, a plasmid found in the yeast *Saccharomyces cerevisiae*, so it will replicate in *S. cerevisiae*. It also has the pBR322 *ori* region and so will replicate in *E. coli*. In addition, the plasmid contains the yeast gene *LEU2* as well as Amp^r, which confers ampicillin resistance in *E. coli*. Similar shuttle vectors that can replicate in mammalian cells and *E. coli* have been constructed. Some of these plasmids have the replication origin of the animal virus simian virus 40 and the ColE1 origin of replication.

Figure 4.17 Insertional inactivation of the tetracycline resistance (Tet^r) gene of pBR322 by insertion of a foreign DNA into the *Bam*HI site. The Tet^r gene will be disrupted, but the plasmid will still confer ampicillin resistance because the Amp^r gene will still be active.

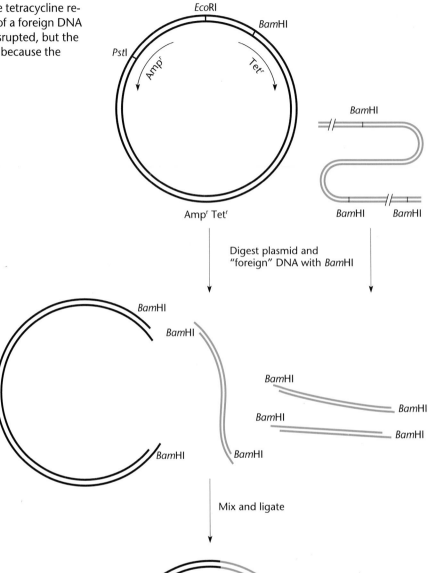

Figure 4.18 The shuttle plasmid YEp13. The plasmid contains origins of replication that will function in *S. cerevisiae* and *E. coli*. It also contains genes that can be selected in *S. cerevisiae* and *E. coli*.

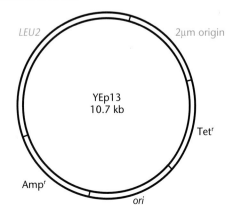

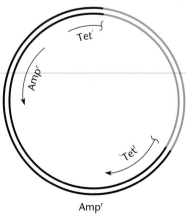

SUMMARY

1. Plasmids are DNA molecules that exist free of the chromosome in the cell. Most plasmids are circular, but some are linear. The sizes of plasmids range from a few thousand base pairs to almost the length of the chromosome itself.

2. Plasmids usually carry genes for proteins that are necessary or beneficial to the host under some situations but are not essential under all conditions. By carrying nonessential genes on plasmids, bacteria are able to keep their chromosome small but still respond quickly to changes in the environment.

3. Plasmids replicate from a unique origin of replication, or *oriV* region. Many of the characteristics of a given plasmid derive from this *ori* region. These include the mechanism for replication, copy number control, partitioning, and incompatibility. If other genes are added or deleted from the plasmid, it will retain most of its original characteristics, provided that the *ori* region remains.

4. The copy number of a plasmid is the number of copies of the plasmid per cell immediately after cell division. Different types of plasmids use different mechanisms to regulate their initiation of replication and therefore their copy number.

5. Some plasmids have a special partitioning mechanism to ensure that each daughter cell will get one copy of the plasmid as the cells divide.

6. If two plasmids cannot stably coexist in the cells of a culture, they are said to be incompatible or members of the same Inc group. They can be incompatible if they share the same copy number control system or the same partitioning function.

7. The host range for replication of a plasmid is all the different organisms in which the plasmid can replicate. Some plasmids are very broad in their host range and can replicate in a wide variety of bacteria. Others are very narrow in their host range and can replicate in only very closely related bacteria.

8. Many plasmids have been engineered for use as cloning vectors. They make particularly desirable cloning vectors for some applications because they do not kill the host, can be small, and are easy to isolate.

QUESTIONS FOR THOUGHT

1. Why are genes whose products are required for normal growth not carried on plasmids? List some genes which you would not expect to find on a plasmid and some genes you might expect to find on a plasmid.

2. Why do you suppose some plasmids are broad host range for replication? Why are all plasmids not broad host range?

3. How do you imagine a partitioning system for a single-copy plasmid like F could work? How might a copy number control mechanism work?

4. How would you find the genes required for replication of the plasmid if they are not all closely linked to the *ori* site?

5. How would you determine which of the replication genes of the host *E. coli* (e.g., *dnaA* and *dnaC*) are required for replication of a plasmid you have discovered?

PROBLEMS

1. The IncQ plasmid RSF1010 carries resistance to the antibiotics streptomycin and sulfonamide. Suppose you have isolated a plasmid that carries resistance to kanamycin. Outline how you would determine whether your new plasmid is an IncQ plasmid.

2. A plasmid has a copy number of 6. What fraction of the cells will be cured of the plasmid each time the cells divide if the plasmid has no partitioning mechanism?

3. You wish to clone pieces of human DNA cut with the restriction endonuclease *Bam*HI into the *Bam*HI site of pBR322. You cut both human DNA and pBR322 with *Bam*HI and ligate them. You transform the ligation mix into *E. coli*, selecting for Amp^r. Outline how you would determine which of the transformants probably contain a plasmid with a human DNA insert.

SUGGESTED READING

Abeles, A., L. D. Reaves, B. Youngren-Grimes, and S. J. Austin. 1995. Control of P1 plasmid replication by iterons. *Mol. Microbiol.* 18:903–912.

Austin, S., and K. Nordstrom. 1990. Partition-mediated incompatibility of bacterial plasmids. *Cell* 60:351–354. (Minireview)

Bagdasarian, M., R. Lurz, B. Ruckert, F. C. H. Franklin, M. M. Bagdasarian, J. Frey, and K. N. Timmis. 1981. Specific purpose plasmid cloning vectors. II. Broad host, high copy number, RSF1010-derived vectors, and a host vector system for cloning in *Pseudomonas. Gene* 16:237–247 del Solar, G., J. C. Alonso, M. Espinosa, and R. Diaz-Orejas. 1996. Broad-host-range plasmid replication: an open question. *Mol. Microbiol.* (MicroReview) **21**:661–666.

Firshein, W., and P. Kim. 1997. Plasmid replication and partition in *Escherichia coli*: is the membrane the key? *Mol. Microbiol.* **23**:1–10.

Freifelder, D., A. Folkmanis, and I. Kirschner. 1971. Studies on *Escherichia coli* sex factors: evidence that covalent circles exist within cells and the general problem of isolation of covalent circles. *J. Bacteriol.* **105**:722–727.

Guhathakurta, A., I. Viney, and D. Summers. 1996. Accessory proteins impose site selectivity during ColE1 dimer resolution. *Mol. Microbiol.* **20**:613–620.

Helinski, D. R., A. E. Toukdarian and R. P. Novick. 1996. Replication control and other stable maintenance mechanisms of plasmids, p. 2295–2324, In F. C. Neidhardt, R. C. Curtiss III, J. L. Ingraham, E. C. C. Lin, K. B. Low, B. Magasanik, W. S. Reznikoff, M. Riley, M. Schaechter, and H. E. Umbarger (ed.), *Escherichia coli and Salmonella: Cellular and Molecular Biology*, 2nd ed. ASM Press, Washington, D.C.

Jacob, F., S. Brenner, and F. Cuzin. 1963. On the regulation of DNA replication in bacteria. *Cold Spring Harbor Symp. Quant. Biol.* **28**:329–348.

McEachern, M. J., M. A. Bott, P. A. Tooker, and D. R. Helinski. 1989. Negative control of plasmid R6K replication: possible role of intermolecular coupling of replication origins. *Proc. Natl. Acad. Sci. USA* **86**:7942–7946.

Meacock, P. A., and S. N. Cohen. 1980. Partitioning of bacterial plasmids during cell division: a *cis*-acting locus that accomplishes stable plasmid maintenance. *Cell* **20**:529–542.

Miller, C. A., S. L. Beaucage, and S. N. Cohen. 1990. Role of DNA superhelicity in partitioning of the pSC101 plasmid. *Cell* **62**:127–133.

Novick, R. P., and F. C. Hoppensteadt. 1978. On plasmid incompatibility. *Plasmid* **1**:421–434.

Radloff, R., W. Bauer, and J. Vinograd. 1967. A dye-buoyant-density method for the detection and isolation of closed circular duplex DNA: the closed circular DNA in HeLa cells. *Proc. Natl. Acad. Sci. USA* **57**:1514–1521.

Studier, F. W., and B. A. Moffatt. 1986. Use of T7 RNA polymerase to direct selective high-level expression of cloned genes. *J. Mol. Biol.* **189**:113–130.

Conjugation

A REMARKABLE FEATURE OF MANY PLASMIDS is the ability to transfer themselves and other DNA elements from one cell to another in a process called **conjugation**. Joshua Lederberg and Edward Tatum first observed this process in 1947, when they found that mixing some strains of *Escherichia coli* with others resulted in strains that were genetically unlike either of the originals. As discussed later in this chapter, Lederberg and Tatum suspected that bacteria of the two strains exchanged DNA—that is, two parental strains mated to produce progeny unlike themselves but with characteristics of both parents. At that time, however, plasmids were unknown, and it was not until later that the basis for the mating was understood.

Overview

During conjugation, the two strands of a plasmid separate in a process resembling rolling-circle replication (see the section on mechanism of DNA transfer during conjugation in gram-negative bacteria, below), and one strand moves from the bacterium originally containing the plasmid—the **donor**—into a **recipient** bacterium. Then the two single strands serve as templates for the replication of complete double-stranded DNA molecules in both the donor cell and the recipient cell. A recipient cell that has received DNA as a result of conjugation is called a **transconjugant**.

Most naturally occurring plasmids are either **self-transmissible** or **mobilizable**, a fact that suggests that plasmid conjugation is advantageous for plasmids and their hosts. Self-transmissible plasmids encode all the functions they need to move among cells, and sometimes they also aid in the transfer of chromosomal DNA and mobilizable plasmids. The last encode

129

some but not all of the proteins required for transfer and consequently need the help of self-transmissible plasmids to move.

Any bacterium harboring a self-transmissible plasmid is a potential donor, because it can transfer DNA to other bacteria. In gram-negative bacteria, such cells produce a structure, called a **sex pilus**, that facilitates conjugation (discussed in a later section). Bacteria that lack self-transmissible plasmids are potential recipients, and conjugating bacteria are known as **parents**. Donors are sometimes referred to as **male**.

Self-transmissible plasmids probably exist in all types of bacteria, but those which have been studied most extensively are from the gram-negative genera *Escherichia* and *Pseudomonas* and the gram-positive genera *Enterococcus, Streptococcus, Bacillus, Staphylococcus,* and *Streptomyces*. The best-known transfer systems are those of plasmids isolated from *Escherichia* and *Pseudomonas* species, and so we focus our attention on these gram-negative systems and do not address conjugation in gram-positive bacteria until the end of the chapter.

Classification of Self-Transmissible Plasmids

Bacterial plasmids have many different types of transfer systems, which are encoded by the plasmid *tra* genes (see the section on transfer genes, below). But as discussed in chapter 4, plasmids are usually classified by their incompatibility (Inc) group. Accordingly, the F-type plasmids use a transfer system known as the Tra system of IncF plasmids, and the RP4 plasmid uses the Tra system of IncP plasmids.

Despite this nomenclatural link, transfer systems bear little relation to the replication and partitioning functions of a plasmid, the characteristics that determine its Inc group. In fact, the genes for these functions and the transfer genes are located in different regions of the plasmid, and there is no a priori reason for any correlation between them. Nevertheless, there is a high degree of correlation between the type of transfer system and Inc group. There may be a good reason for this. Some products of plasmid transfer genes inhibit the entry of plasmids with the same Tra functions (see below). If the *tra* genes did not correlate with the Inc group, a plasmid would sometimes transfer into a cell that already had a plasmid of the same Inc group, and one of the two plasmids would subsequently be lost.

Interspecies Transfer of Plasmids

Many plasmids have transfer systems that enable them to transfer DNA between unrelated species. These are known as **promiscuous plasmids** and include the IncW

plasmids, represented by R388; the IncP plasmids, represented by RP4; and the IncN plasmids, represented by pKM101 (Figure 5.1). The IncP plasmids can transfer themselves or mobilize other plasmids from *E. coli* into essentially any gram-negative bacterium. Recent studies showed that plasmids of this group transfer at a low frequency into cyanobacteria, gram-positive bacteria such as *Streptomyces* species, and even plant cells! The F plasmid, which was not known to be particularly promiscuous, can transfer itself from *E. coli* into yeast cells (see Box 5.1).

Transfer of DNA by promiscuous plasmids probably plays an important role in evolution. Such transfer could explain why genes with related functions are often very similar to each other regardless of the organism that harbors them. These genes could have been transferred by promiscuous plasmids fairly recently in evolution, which would account for their similarity relative to the other genes of the two organisms.

The interspecies transfer of plasmids also has important consequences for the use of antibiotics in treating human and animal diseases. Many of the most promiscuous plasmids, including those of the IncP group, such as RP4, and the IncW plasmid R388 were isolated in hospital settings. These large plasmids (commonly called R-plasmids because they carry genes for antibiotic resistance) presumably have become prevalent in recent years in response to the indiscriminate use of antibiotics in medicine and agriculture. The source of the resistance genes may be soil bacteria, such as actinomycetes, that are the producers of antibiotics. In chapter 8, we discuss how transposons might have helped assemble these promiscuous R-plasmids.

Whatever their source, the emergence of R-plasmids indicates why antibiotics should be used only when they are absolutely necessary. In humans or animals treated indiscriminately with antibiotics, bacteria that carry R-plasmids will be selected from the normal flora. R-plasmids can be quickly transferred into an invading pathogenic bacterium, making it antibiotic resistant. Consequently the infection will be difficult to treat.

Mechanism of DNA Transfer during Conjugation in Gram-Negative Bacteria

As mentioned at the beginning of the chapter, whole plasmids do not actually move during conjugation. Only one strand of the plasmid is transferred, so that after transfer, both the donor and recipient contain the plasmid sequence. A more detailed mechanism of DNA transfer appears in Figure 5.2. After contact has been made between a donor cell and a recipient cell, a break

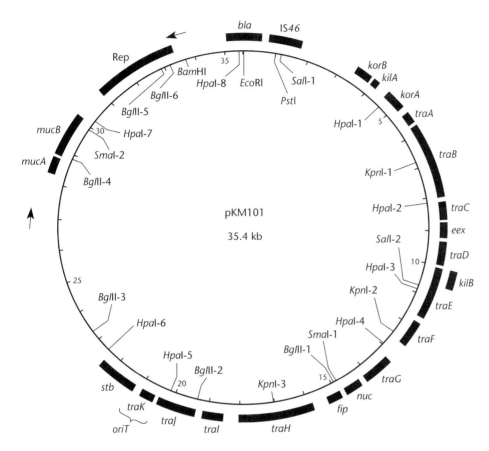

Figure 5.1 Physical and genetic map of the self-transmissible plasmid pKM101. *oriT* is the origin of plasmid transfer. The *tra* genes encode the many tranfer functions discussed in this chapter. The *mucA* and *mucB* genes enhance UV mutagenesis by coding for analogs of the *E. coli umuD* and *umuC* genes (see chapter 10). From S. Winans and G. Walker, *J. Bacteriol.* **161**:402–410, 1985.

is made in one strand of the plasmid DNA at the *oriT* (origin of transfer; see below) site by a specific DNA endonuclease encoded by one of the plasmid *tra* genes. This *oriT* site is different from the origin of replication—called *oriV* (see chapter 4)—from which the plasmid normally replicates. A specific helicase, which is also encoded by a plasmid *tra* gene, displaces the broken strand, and other *tra* gene products transfer it to the recipient cell. The pathway for DNA transfer remains unclear. Once the entire strand is inside the recipient cell, the two ends are rejoined to make a circular molecule. Yet another *tra* gene product presumably holds the two ends of the cut strand together at the *oriT* site so that they can be rejoined in the recipient cell.

Usually, a new complementary strand of DNA is synthesized in the donor as the old strand is being displaced by the helicase. The synthesis of this new strand is presumably primed by the broken 3' end of the DNA at the *oriT* sequence; hence, this type of replication is analogous to rolling-circle replication (see chapter 4). However, during plasmid transfer, the helicase instead of the replication fork separates the two DNA strands. Separation of the strands requires energy, and the helicase consumes copious amounts of ATP in this process.

Replication of complementary strands in the donor and the recipient cells does not seem to be essential for transfer, which can give rise to single-stranded plasmid molecules in either cell. The presence of single-stranded plasmids in both the donor and the recipient after transfer furnished early evidence for the model we have just described.

Transfer (*tra*) Genes

Conjugation is a complicated process that requires the products of many genes. As mentioned earlier, the genes required for transfer are called the **tra** genes. The products of the *tra* genes are *trans* acting and can act on another plasmid in the same cell. The map of pKM101 (Figure 5.1) shows that this self-transmissible plasmid contains at least 11 of these genes, as well as genes such as *eex* (entry exclusion) that are not required for transfer but play related roles, such as preventing the entry of other plasmids with the same Tra functions. In addition to the *tra* genes, a *cis*-acting site called *oriT* is required for transfer. With so many genes and sites involved, a large portion of a self-transmissible plasmid is taken up with transfer-related functions, and by necessity, these plasmids are quite large.

BOX 5.1

Gene Exchange between Kingdoms

Not only can some plasmids transfer themselves into other types of bacteria, but also they can sometimes transfer themselves into eukaryotes, that is, into organisms of a different kingdom.

Agrobacterium tumefaciens and Crown Gall Tumors in Plants

The first known example of transfer of bacterial plasmids into eukaryotes occurs in the plant disease crown gall, in which a tumor appears on the plant, usually where the roots join the stem (the crown). Crown gall disease is caused by *Agrobacterium tumefaciens*, and the interaction between this bacterium and the plant was the first known example of gene exchange between bacteria and a eukaryote. Virulent strains of *A. tumefaciens* contain a plasmid called the Ti plasmid, for tumor initiation. This plasmid has *tra* genes as well as *vir* genes involved in tumor formation. The presence of the plant cells is recognized by the products of the *vir* genes, which apparently sense phenolic compounds given off by the plant and induce the transcription of the *tra* genes. When the bacterium makes contact with the plant cell, these *tra* genes mobilize a small part of the plasmid, called the T DNA, into the plant cell, where it integrates into the plant DNA. In the plasmid, the T DNA is bracketed by two 25-bp sequences that, in a sense, act as *oriT* sites, allowing the T DNA to be mobilized into the plant cells by Tra functions on the Ti plasmid.

Once the T DNA integrates into the plant cell DNA, the plant cells begin to proliferate. The genes of T DNA are expressed in the plant because they have promoters and translation initiation regions (TIRs) recognized by the transcription functions of the plant. Some of these genes encode plant hormones that cause the multiplication of plant cells and therefore the production of tumors. Other genes encode enzymes that catalyze the synthesis of unusual compounds—opines—which are composed of a carbohydrate such as pyruvate bonded to an amino acid such as arginine. Because these compounds are unusual, most organisms cannot degrade them. However, *A. tumefaciens* has genes on its Ti plasmid that allow it to catabolize the particular opine compound it induces the plant cell to make, furnishing the bacterium with a carbon, nitrogen, and energy source. In this way, the Ti plasmid allows *A. tumefaciens* to make an ecological niche for itself in which it can compete effectively with other bacteria, most of which cannot use that particular opine.

The ability of the *A. tumefaciens* Ti plasmid to transfer part of itself into plant cells has become the basis for the genetic engineering of plants. Any gene introduced into the T DNA of the Ti plasmid will be transferred into the plant chromosome. Thus plants could be made resistant to certain diseases or insects. However, for this technology to work, methods must be available to regenerate the whole plant from plant cells, and such methods have not yet been developed for all crop plants. Also, the genes on the T DNA for inducing tumors must be inactivated so that a normal plant can be cultured.

Transfer of Broad-Host-Range Plasmids into Eukaryotes

The Ti plasmid is obviously designed to transfer part of itself into plant cells. The surprising result of recent studies is that other bacterial plasmids can also transfer themselves or mobilize other plasmids into eukaryotic cells. One striking example is the mobilization of other plasmids into plant cells by the Ti plasmid. As mentioned, the sequences bracketing the T DNA in the Ti plasmid can be thought of as *oriT* sites, and the Tra functions of the Ti plasmid can be thought of as mobilizing the T DNA into plant cells. Plasmid RSF1010, and plasmids derived from it, can also be mobilized into plant cells by the Ti plasmid, provided that they contain the correct *mob* sequence.

Not only do plasmids transfer into plants, but they also transfer into lower fungi. This observation is very surprising because the cell surfaces of bacteria and eukaryotes are very different. So are the surfaces of plant cells, but in the case of the Ti plasmid, we can assume that the transfer functions have evolved to recognize plant cells. However, there is no apparent reason why bacterial plasmids should have evolved to transfer into lower fungi. Whatever the reason, the transfer of genes between eukaryotes and bacteria may play an important role in evolution.

References

Buchanan-Wollasten, U., J. E. Passiatore, and F. Cannon. 1987. The *mob* and *oriT* mobilization functions of a bacterial plasmid promote its transfer to plants. *Nature* (London) **328**:172–175.

Farrand, S. K. 1993. Conjugal transfer of *Agrobacterium* plasmids, p. 255–291. *In* D. B. Clewell (ed.), *Bacterial Conjugation*. Plenum Press, New York.

Heinemann, J. A., and G. F. Sprague, Jr. 1989. Bacterial conjugative plasmids mobilize DNA transfer between bacteria and yeast. *Nature* (London) **340**:205–209.

Horsch, R. B., J. E. Fry, N. L. Hoffmann, D. Eichholtz, S. G. Rogers, and R. T. Fraley. 1985. A simple and general method for transferring genes into plants. *Science* **227**:1229–1231.

Zambryski, P. 1988. Basic processes underlying *Agrobacterium*-mediated DNA transfer to plant cells. *Annu. Rev. Genet.* **22**:1–30.

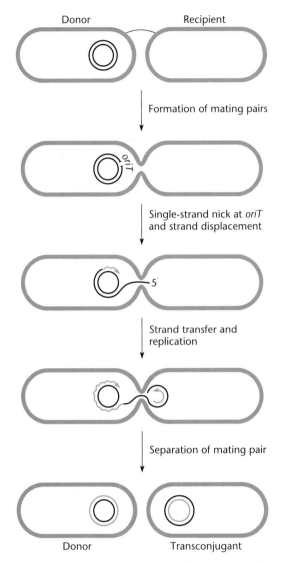

Donor   Recipient

Formation of mating pairs

oriT

Single-strand nick at *oriT*
and strand displacement

5'

Strand transfer and
replication

Separation of mating pair

Donor   Transconjugant

Figure 5.2 Mechanism of DNA transfer during conjugation. The donor cell produces a pilus, which is encoded by the plasmid and contacts a potential recipient cell that does not contain the plasmid. Retraction of the pilus brings the cells into close contact, and a pore forms in the adjoining cell membranes. Formation of the mating pair signals the plasmid to begin transfer from a single-stranded nick at *oriT*. The nick is made by plasmid-encoded *tra* functions. The 5' end of a single strand of the plasmid is transferred to the recipient through the pore. During transfer, the plasmid in the donor is replicated, its DNA synthesis being primed by the 3' OH of the *oriT* nick. Replication of the single strand in the recipient proceeds by a different mechanism with RNA primers. Both cells now contain double-stranded plasmids, and the mating pair separates.

THE SEX PILUS

The products of most of the *tra* genes are devoted to making the sex pilus. This appendage sticks out from the cell surface and may serve to bring two cells together for mating. Only one protein, **pilin**, the product of one of the *tra* genes, makes up the sex pilus. The 10 or so other *tra* gene products of pKM101 (Figure 5.1) are involved in transporting pilin through the cell membrane and assembling the pilus on the cell surface or in DNA transfer itself (see next section).

The structure of the sex pilus depends on the type of self-transmissible plasmid. For example, pKM101 makes a long, rigid pilus; the F-type plasmids make a long, flexible pilus; and plasmids such as RP4 make a short, rigid pilus. The specific structure is also the target of some phages, which use the sex pili as adsorption sites (see chapter 7). Susceptibility to a given phage often serves as an indicator of the plasmid type carried by a cell.

THE DNA TRANSFER GENES

As we mentioned at the beginning of the section on mechanism, some *tra* products are not involved in pilus assembly but are involved more directly in DNA transfer. Some are endonucleases that act directly on the *oriT* sequence to promote transfer. Others are helicases that unwind the DNA or primases that make RNA to prime the synthesis of complementary DNA (cDNA) strands during transfer.

The *oriT* Sequence

The *oriT* site is not only the site at which plasmid transfer initiates but also the site at which the DNA ends rejoin to recyclize the plasmid after transfer. Plasmid transfer initiates specifically at the *oriT* site because the specific endonuclease encoded by one of the *tra* genes will cut DNA only at this sequence. Also, presumably, the plasmid-encoded helicase will enter DNA only at this sequence to separate the strands. Moreover, after transfer, the two ends of the DNA are probably held together at the *oriT* sequence so that they can be religated. Therefore, to be transferred, the plasmid must have this specific *oriT* sequence. In fact, a self-transmissible plasmid will mobilize any DNA that contains its *oriT* sequence, as we discuss below.

The essential features of *oriT* sequences are currently being investigated. The *oriT* sequence of the F plasmid is known to be less than 300 bp long and contains inverted repeated sequences and a region rich in AT base pairs. The importance of these sequences for *oriT* function is under investigation.

Function of Plasmid Primases in Transfer

Many self-transmissible plasmids encode their own primases. The advantage for a transmissible plasmid to make a primase may not be immediately clear. The synthesis of double-stranded DNA after transfer requires only that RNA primers be made in the recipient cell, since the complementary strand synthesized in the donor cell is probably primed by the 3' end of the DNA created by the nick at the *oriT* sequence. However, even though it is used only in the recipient cell, the primase must be made in the donor cell, since it cannot be made in the recipient cell. This is because double-stranded DNA is needed for transcription into mRNA to make proteins (including primase), but only a single strand of DNA is transferred. Therefore, if the primase is needed to make double-stranded DNA in the recipient, it would be needed to make itself, an obvious impossibility. The answer is that plasmids transfer the primase protein along with the DNA, so that it is available to prime the synthesis of the complementary strand immediately upon entry. In fact, reasoning such as this inspired the first experiments to show that proteins as well as DNA are transferred during conjugation (see below).

Another question is why plasmids encode their own primase rather than just using the primase of the host. A clue comes from the fact that mutations in the *tra* genes encoding the primases for some plasmids can prevent conjugation with some hosts but not others. Presumably, the hosts into which the mutant plasmids cannot transfer are those whose primases will not function on the plasmid to prime replication. Yet if the plasmid is transferred into a cell whose own primase is similar to the plasmid primase, a complementary strand of the plasmid could be synthesized and conjugation could result in a functioning plasmid. By synthesizing their own primases, plasmids increase the range of hosts into which they can be transferred.

Efficiency of Transfer

One of the striking features of many transfer systems is their efficiency. Under optimal conditions, some plasmids can transfer themselves into other cells in almost 100% of cell contacts. This high efficiency has been exploited in the development of methods for transferring cloned genes between bacteria and in transposon mutagenesis, both of which require highly efficient transfer of DNA. We shall discuss such methods in subsequent chapters.

REGULATION OF THE *tra* GENES

Many naturally occurring plasmids transfer with a high efficiency for only a short time after they are introduced into cells and then transfer only sporadically thereafter. The rest of the time, the *tra* genes are repressed, and without the synthesis of pilin and other *tra* gene functions, the pilus is lost. For unknown reasons, the repression is relieved occasionally in some of the cells, allowing this small percentage of cells to transfer their plasmid at a given time.

Plasmids may normally repress their *tra* genes to prevent infection by some types of phages. The pilus serves as the adsorption site for some phages. If all the cells in a population had a pilus all the time, such a phage could multiply quickly, infecting and killing all the bacteria carrying the plasmid.

This property of only periodically expressing their *tra* genes probably does not prevent the plasmids from spreading quickly through a population of bacteria that does not contain them. When a plasmid-containing population of cells encounters a population that does not contain the plasmid, the plasmid *tra* genes in one of the plasmid-containing cells will eventually be expressed and the plasmid will transfer to another cell. Then when the plasmid first enters a new cell, efficient expression of the *tra* genes leads to a cascade of plasmid transfer from one cell to another. As a result, the plasmid will eventually occupy most of the cells in the population.

AN EXAMPLE: REGULATION OF *tra* GENES IN IncF PLASMIDS

Regulation of the *tra* genes of IncF plasmids has been studied more extensively than that of other types. Transfer of these plasmids depends on TraJ, a transcriptional activator. A **transcriptional activator** is a protein required for initiation of RNA synthesis at a particular promoter (see chapter 11). If the *traJ* gene were always transcribed, the other *tra* gene products would always be made and the cell would always have a pilus. However, the transcription of the *traJ* gene is normally repressed by the concerted action of the products of two plasmid genes, *finP* and *finO*, which encode an RNA and a protein, respectively. The FinP RNA is an antisense RNA that is transcribed constitutively from a promoter within and in opposite orientation to the *traJ* gene. Complementary pairing of the FinP RNA and the *traJ* transcript prevents translation of TraJ. The FinO protein stabilizes the FinP antisense RNA. When the plasmid first enters a cell, neither FinP RNA nor FinO protein is present, and so the *traJ* gene product is transcribed and the other *tra* gene products are made. Consequently, a pilus appears on the cell, and the plasmid can be transferred. After the plasmid has become established in the double-stranded state, however, the FinP RNA and FinO protein will be synthesized and will repress the *traJ* gene, so that the plasmid can no longer transfer. Why the *tra* genes are later expressed intermittently is not understood.

The F plasmid was the first transmissible plasmid discovered (Figure 5.3; Table 5.1), and its discovery may have resulted from a happy coincidence involving its *finO* gene. Because of an insertion mutation in this gene (IS*3* in Figure 5.3), the F plasmid is itself a mutant that always expresses the *tra* genes. Consequently, a sex pilus almost always extends from the surface of cells harboring the F plasmid, and the F plasmid can always transfer, provided that recipient cells are available, increasing the efficiency of transfer and facilitating the discovery of conjugation. Mutations that increase the efficiency of plasmid transfer, thereby increasing their usefulness in gene cloning and other applications, have been isolated in other commonly used transfer systems.

TABLE 5.1	Some F-plasmid genes and sites
Symbol	**Function**
ccdAB	Inhibition of host cell division
incBCE	Incompatibility
oriT	Site of initiation of conjugal DNA transfer
oriV	Origin of bidirectional replication
parABCL	Partitioning
traABCEFGHKLQUVW	Pilus biosynthesis/assembly
traGN	Mating-pair stabilization
traI	*oriT*-specific nickase; DNA helicase
traY	Accessory for nickase
traJ, finOP	Regulation of transfer
traST	Entry exclusion

Mobilizable Plasmids

Mobilizable plasmids lack the 10 or so genes required for pilus synthesis; therefore, to transfer, they must use the pilus of a self-transmissible plasmid in the same cell. The reliance on other plasmids allows mobilizable plasmids to be much smaller.

To be mobilizable, a plasmid need contain only the *oriT* site of a self-transmissible plasmid, because the *tra* gene products of the self-transmissible plasmid will act on the same *oriT* site in any other plasmid and mobilize it. Plasmids that include only the *oriT* region of a self-transmissible plasmid do not occur naturally but have been constructed in the laboratory.

Naturally occurring mobilizable plasmids also have genes, called *mob* genes, for the transfer of DNA and a unique *oriT* sequence. These genes are located in a region called **mob** and are analogous to the *tra* genes involved in the transfer of self-transmissible plasmids. Their products increase the range of self-transmissible plasmids by which their plasmid can be mobilized.

MOLECULAR MECHANISM OF MOBILIZATION
The mechanism by which one plasmid mobilizes another plasmid is similar to self-transfer (Figure 5.4). Like self-transfer, mobilization is initiated by contact with a potential recipient cell, which results in a break in one strand at the *oriT* sequence, and the transfer proceeds as with self-transmissible plasmids: One strand is displaced by a helicase and transferred to the recipient cell, where it is recyclized. Concomitantly, the complementary strands are synthesized in both the donor and recipient.

Usually, when one plasmid mobilizes another, the self-transmissible plasmid transfers as well, so that the recipient cell will end up with both plasmids.

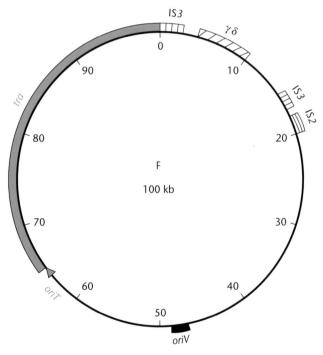

Figure 5.3 Partial genetic and physical map of the 100-kbp self-transmissible plasmid F. The regions IS*3* and IS*2* are insertion sequences; γδ is also known as transposon Tn*1000*. *oriV* is the origin of replication; *oriT* is the origin of conjugal transfer; the *tra* region encodes numerous *tra* functions. Redrawn from N. Willetts and R. Skurray, Structure and function of the F factor and mechanism of conjugation, p. 1110–1133, *in* F. C. Neidhardt, J. L. Ingraham, K. B. Low, B. Magasanik, M. Schaechter, and H. E. Umbarger (ed.), *Escherichia coli and Salmonella typhimurium: Cellular and Molecular Biology,* American Society for Microbiology, Washington, D.C., 1987.

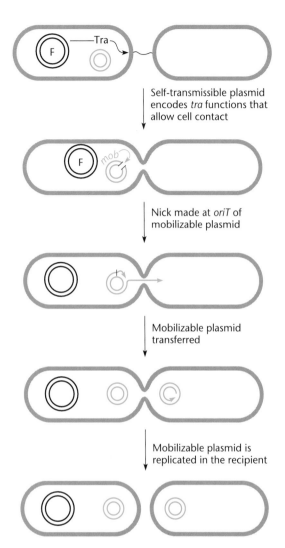

Figure 5.4 Mechanism of plasmid mobilization. The donor cell carries two plasmids, a self-transmissible plasmid, F, which encodes the *tra* functions that promote cell contact and plasmid transfer, and a mobilizable plasmid (blue). The *mob* functions encoded by the mobilizable plasmid make a single-stranded nick at *oriT* in the *mob* region. Transfer and replication of the mobilizable plasmid then occur. The self-transmissible plasmid may also transfer.

Self-transmissible plasmid encodes *tra* functions that allow cell contact

Nick made at *oriT* of mobilizable plasmid

Mobilizable plasmid transferred

Mobilizable plasmid is replicated in the recipient

THE *mob* REGION

Like the various *tra* genes, *mob* genes presumably include one that encodes a specific endonuclease that cleaves at the *oriT* site and another that encodes a helicase that displaces one strand. Also as in self-transmissible plasmids, the *oriT* site of mobilizable plasmids is usually far removed from *oriV*, the normal origin of replication. However, the structure of some mobilizable plasmids deviates from this rule; for instance, the

oriT site of RSF1010 is close to *oriV*. In this plasmid, some of the replication functions may play a dual role in mobilization.

IMPORTANCE OF PLASMID MOBILIZATION IN CLONING TECHNOLOGY AND BIOTECHNOLOGY

Mobilizable plasmids are very useful in biotechnology. Because they can be much smaller than self-transmissible plasmids, they are generally more convenient as cloning vectors. Genes cloned into a mobilizable plasmid can be efficiently mobilized into other strains of bacteria by the Tra functions of a promiscuous self-transmissible plasmid. For biotechnological applications, the self-transmissible plasmid used to mobilize the cloning vector will usually contain a mutation in its own *oriT* sequence that prevents it from also being transferred.

However, mobilizable plasmids still present complications. To meet regulatory requirements or for other reasons, genetic engineers often have to prove that a plasmid containing recombinant genes that confer desirable properties on one bacterium will not be mobilized into another, unknown, bacterium, where the genes might be harmful. But how do we really know that a plasmid does not contain an *oriT* site that will be recognized by some set of Tra functions? Unfortunately, negative evidence of mobilization by all the known self-transmissible plasmids does not mean that a given plasmid cannot be mobilized by *some* plasmid.

Chromosome Transfer by Plasmids

Usually during conjugation, only a plasmid is transferred to another cell. However, plasmids sometimes transfer the chromosomal DNA of their bacterial host, a fact which has been put to good use in bacterial genetics. Without the transfer of genes, bacterial genetics is not possible, and conjugation is one of only three ways in which chromosomal and plasmid genes can be exchanged among bacteria (transduction and transformation are the others).

Formation of Hfr Strains

Sometimes plasmids integrate into chromosomes, and when such plasmids attempt to transfer, they take the chromosome with them. Bacteria that can transfer their chromosome because of an integrated plasmid are called **Hfr strains**, where Hfr stands for *high-frequency recombination*. As we discuss, the name derives from the fact that many recombinants can appear when such a strain is mixed with another strain of the same bacterium.

The integration of plasmids into the chromosome can occur by several different mechanisms, including recombination between sequences on the plasmid and sequences on the chromosome. For normal recombination to occur, the two DNAs must share a sequence (see chapter 9). Most plasmid sequences are unique to the plasmid, but sometimes the plasmid and the chromosome share an **insertion element** (**IS element**; see chapter 3). These small transposons often exist in several copies in the chromosome and may appear in plasmids (see chapter 8); recombination between these common sequences can result in integration of the plasmid.

Figure 5.5 shows how recombination between the IS2 element in the F plasmid and an IS2 in the chromosome of *E. coli* can lead to integration of the F plasmid. Once integrated, the F plasmid will be bracketed by two copies of the IS2 element. This bacterium is now an Hfr strain. The *E. coli* chromosome contains 20 sites for IS-mediated Hfr formation.

Transposition can also lead to the integration of plasmids into the chromosome. Plasmids often carry transposons, and integration of a plasmid through transposition can occur even if there are no sequences common to both the plasmid and the chromosome. We defer a discussion of transposons and transposition until chapter 8.

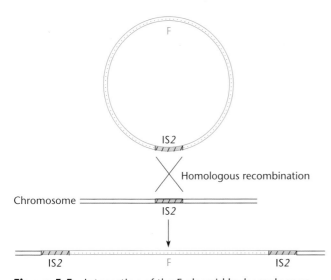

Figure 5.5 Integration of the F plasmid by homologous recombination between IS2 elements in the plasmid and in the chromosome, forming an Hfr cell. Integration can also occur through recombination of the IS3 or γδ sequences on the F plasmid (see Figure 5.3).

Transfer of Chromosomal DNA by Integrated Plasmids

We mentioned at the start of the chapter that self-transmissible plasmids were first detected in 1947 by Joshua Lederberg and Edward Tatum. These scientists observed recombinant types after mixing some strains of *E. coli* with other strains. **Recombinant types** differ from the two original, or parental, strains and in this case resulted from the transfer of chromsomal DNA from one strain to another by an F plasmid integrated into the chromosome. In retrospect, it was fortuitous that the strains used by Lederberg and Tatum included some with an integrated F plasmid. Such strains are not common, and, as mentioned, the F plasmid is a mutant that is always ready to transfer, so that recombinant types are more likely to be produced. Also, recall that in 1947, plasmids had yet to be discovered.

Figure 5.6 illustrates the process by which chromosomal DNA is transferred in an Hfr strain. The integrated plasmid will express its *tra* genes, and a pilus will be synthesized. Upon contact with a recipient cell, the DNA will be nicked at the *oriT* site in the integrated plasmid, and one strand will be displaced into the recipient cell. Now, however, after transfer of the *oriT* sequence and the portion of the plasmid on one side of the nick, chromosomal DNA will also be transferred into the new cell. If the transfer continues long enough (approximately 100 minutes at 37°C), the entire bacterial chromosome will eventually be transferred, ending with the remaining plasmid *oriT* sequences. However, transfer of the entire chromosome (and thus the whole integrated plasmid) is rare, perhaps because the union between the cells is frequently broken, or because the DNA is often broken during conjugation.

FORMATION OF RECOMBINANT TYPES

Because the entire chromosome is seldom transferred, the entire plasmid is usually not transferred, and the DNA will not be able to recyclize after it enters the recipient cell. Hence, the transferred DNA will be lost unless it recombines with the chromosome in the recipient cell. In fact, it would not be possible to tell that an Hfr transfer had occurred unless the transferred chromosomal DNA recombines with the recipient cell chromosome. If this happens and if the donor and recipient bacteria are different in their genotypes, recombinant types might arise that can be identified because they are different from both the Hfr donor bacteria and the recipient bacteria.

In the example shown in Figure 5.7, the donor Hfr strain has an *arg* mutation and therefore will not form

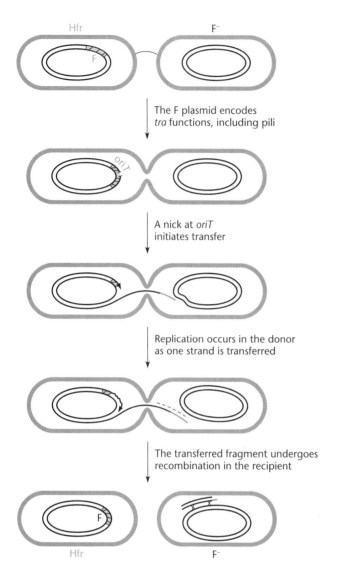

Figure 5.6 Transfer of chromosomal DNA by an integrated plasmid. Formation of mating pairs, nicking of the F *oriT* sequence, and transfer of the 5′ end of a single strand of F DNA proceed as in transfer of the F plasmid. Transfer of the covalently linked chromosomal DNA will also occur as long as the mating pair is stable. Complete chromosome transfer rarely occurs, and so the recipient cell remains F⁻, even after mating. Replication in the donor usually accompanies DNA transfer. Some replication of the transferred single strand may also occur. Once in the recipient cell, the transferred DNA may recombine with homologous sequences in the recipient chromosome.

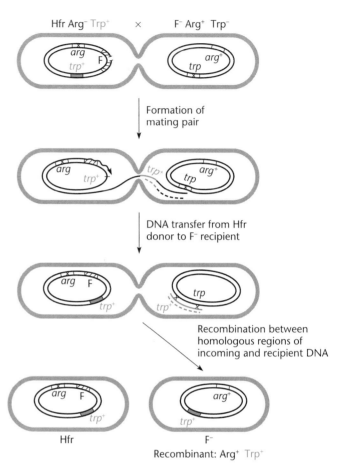

Figure 5.7 Formation of recombinant types after DNA transfer by an Hfr strain. If the *trp* region is transferred from the donor to the recipient, it can recombine to replace the homologous region in the recipient, giving rise to a Trp⁺ Arg⁺ recombinant.

colonies on agar plates containing all the growth supplements except arginine, while the recipient strain has a *trp* mutation and therefore will not form colonies on agar plates lacking tryptophan. When the two are mixed, the Hfr strain can transfer DNA into the recipient cell, and sometimes this DNA will replace the recipient DNA at the *trp* allele, replacing it with the wild-type *trp*⁺ allele, as shown. The recipient bacteria will then require neither arginine nor tryptophan and will multiply to form colonies on minimal plates containing no growth supplements. These Trp⁺ Arg⁺ bacteria are recombinant types because they are genetically unlike either parent. This is basically the experiment that allowed Lederberg and Tatum to discover conjugation. They mixed different strains of *E. coli* that required different growth supplements and showed that some mixtures gave rise to recombinant types with neither growth requirement.

As mentioned above, the donor chromosome is almost never transferred in its entirety to the recipient cell during an Hfr mating. Because of premature interruption of mating, genes that are transferred first are transferred at a much higher frequency than genes on the opposite side of the nick. Consequently, the genes exhibit a **gradient of transfer** from one side of the integrated plasmid around the bacterial genome. This gradient of transfer can be used for genetic mapping in bacteria and will be discussed in more detail in chapter 14.

Chromosome Mobilization

The Tra functions of self-transmissible plasmids can also mobilize the chromosome, provided that the chromosome has the *oriT* sequence of the plasmid. Chromosome transfer will begin at the *oriT* sequence. This has allowed the mapping of genes by gradient of transfer in many genera of bacteria. We discuss such methods in more detail in chapter 16.

Prime Factors

Chromosomal genes can also be transferred when they are incorporated into plasmids. When such a plasmid, called a **prime factor**, transfers itself, it of course takes the chromosomal genes with it. Prime factors are usually designated by the name of the plasmid followed by a prime symbol, for example, F' factor. An R-plasmid such as RK2 carrying bacterial chromosomal DNA is an R' factor.

CREATION OF PRIME FACTORS

Like Hfr strains, prime factors can be created through either transposition or homologous recombination. An illustration of the latter process appears in Figure 5.8. A prime factor forms from the chromosome of an Hfr strain, in which the plasmid is bracketed by two copies of an IS element (Fig. 5.5 and 5.8). The repeated IS sequences make the Hfr chromosome somewhat unstable, and the plasmid sometimes excises by looping out as a result of recombination between the flanking IS elements. However, sometimes the recombination occurs not between the IS elements immediately flanking the plasmid but between other DNA sequences repeated elsewhere in the chromosome, as shown in Figure 5.8. This excision will create a larger plasmid—the prime factor—that carries the chromosomal DNA that lay between the recombining DNA sequences. A prime factor can form from recombination between any repeated sequences, including identical IS elements or genes for rRNA, which often exist in more than one copy in bacteria.

Note that a deletion forms in the chromosome when the prime factor loops out. Some of the genes deleted from the chromosome may have been essential for the growth of the bacterium. Nevertheless, the cells will not die, because the prime factor still contains the essential

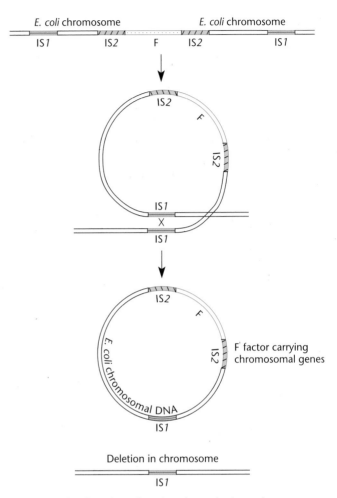

Figure 5.8 Creation of a prime factor by homologous recombination. Recombination may occur between homologous sequences, such as IS sequences, in the chromosomal DNA outside the F factor. The F factor will then contain chromosomal sequences, and the chromosome will carry a deletion.

genes, which should be passed on to daughter cells when the plasmid replicates. However, cells that lose the prime factor will die.

Prime factors can be very large, almost as large as the chromosome itself. In general, the larger a prime factor is, the less stable it is. Maintaining large prime factors in the laboratory requires selection procedures designed so that cells will die if they lose some or all of the prime factor. However, most prime factors are small enough to be transferred in their entirety. Because prime factors contain an entire self-transmissible plasmid, a cell receiving a prime factor becomes a donor and can transfer this DNA into other bacteria. Moreover, because prime factors are replicons with their plasmid origin of replication, they can replicate in any new bacterium that falls within the plasmid host range (see chapter 4). These properties can be used to select cells containing prime factors, as we discuss in chapter 14.

When a prime factor is transferred into a recipient cell, the cell becomes diploid for the chromosomal genes carried on the prime factor. Such cells often are referred to as **partial diploids** or **merodiploids** and are useful for complementation tests because the prime factor and the chromosome will carry different alleles of some of the same genes. Prime factors have been very important in the genetic analysis of bacteria, and in chapter 14 we discuss how Jacob and Monod used them to analyze the regulation of the lactose operon. The use of prime factors in bacterial genetics has been largely supplanted by cloning techniques, but prime factors continue to have special applications.

ROLE OF PRIME FACTORS IN EVOLUTION
Prime factors formed with promiscuous, broad-host-range plasmids probably play an important role in bacterial evolution. Once chromosomal genes are on a broad-host-range plasmid, they can be transferred into distantly related bacteria, where they will be maintained as part of the broad-host-range plasmid replicon. They may then become integrated into the chromosome through recombination or transposition. The similarities between some types of genes, even in distantly related bacteria, suggest that certain genes have been exchanged fairly recently in evolution, and prime factors may have been one of the mechanisms.

Genetics of Plasmid Transmission

So far in this chapter, we have given a mostly descriptive account of plasmid transmission without discussing the genetic experiments that led to this knowledge. Because of the relatively small size of plasmids, physical methods such as restriction endonuclease mapping, transposon mutagenesis, and recombinant DNA techniques are very useful for analyses of plasmid structure. Leaving a detailed treatment of these methods to later chapters, here we give a general idea of the types of genetic experiments that can be used to study plasmid transmission.

Determining Whether a Plasmid Is Self-Transmissible

Assume that we have isolated from nature a bacterial strain that contains a large plasmid. We want to determine whether the plasmid is self-transmissible. The process by which we do so is simple in principle. Donor cells containing the plasmid are mixed with potential re-

cipient cells, which are tested to determine if any have received the plasmid.

Some plasmids transfer more efficiently when the mating cells are trapped on a filter, so we might try a **filter mating**, in which the potential donor and recipient cells are mixed and collected on a filter. The filter is then placed on the surface of an agar plate containing rich medium. After incubation, the bacteria are washed off the filter and plated under selective conditions.

SELECTABLE GENES
Clearly, our experimental design must include some way to distinguish cells that carry the plasmid from those that do not. In chapter 4, we discussed the use of selectable genes in plasmid genetics, and these principles also apply here. Recall that selectable genes might confer resistance to an antibiotic or give the cell the ability to use an unusual carbon source.

If the new plasmid contains no convenient selectable gene, we may have to introduce one into it. This can be done by transposon mutagenesis or cloning, as we discuss in subsequent chapters.

CHOICE OF RECIPIENT
Choosing the proper recipient bacterium to mix with our donor may present a major problem. The plasmid may not transfer into or replicate in a distantly related host, or there may be other barriers to transfer such as different restriction modification systems (see chapter 16).

To avoid such problems, we should choose a recipient as closely related to the donor as possible. However, it is not always easy to determine which laboratory strain is most closely related to a natural isolate. Accordingly, the safest recipient cell is the same bacterium cured of the plasmid. However, as discussed in chapter 4, curing cells of plasmids is not always straightforward.

COUNTERSELECTING THE DONOR
Not only must we distinguish cells that carry the plasmid from those that do not, but also we must differentiate between the original donors and the transconjugant strains. By defining conditions under which the donor cannot multiply, we can separate donors from recipients; in other words, we must **counterselect** the donor. One way to do so is to introduce a different selectable gene into the recipient, for example, by isolating a rifampin-resistant (Rif^r) or streptomycin-resistant (Str^r) mutant. Most bacteria are naturally sensitive to these antibiotics, and resistant mutants can easily be obtained by plating large numbers of the recipient bacteria on

plates containing either compound and waiting for colonies to form (see chapter 3). The mutations that make the colony-forming bacteria resistant to rifampin or streptomycin are usually found in a chromosomal gene for either RNA polymerase or a ribosomal protein, respectively.

SELECTING TRANSCONJUGANTS

With our antibiotic-resistant recipient bacteria in hand, we are finally ready to determine if our plasmid is self-transmissible. The procedure is illustrated in Figure 5.9. We mix the donor bacteria, whose plasmid also carries a resistance gene, with the recipient bacteria, either on filters or in liquid, depending upon the plasmid. After incubating the mixture for a sufficient time to allow transfer, we spread the donor, the recipient, and the mating mixture onto plates that will allow growth of only the recipient containing the plasmid. For example, as shown in Figure 5.9, if the plasmid carries a kanamycin resistance gene and the recipients are Rif^r, the plates should contain both kanamycin and rifampin. The appearance of many more colonies on

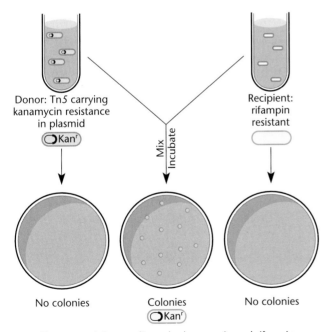

Donor: Tn5 carrying kanamycin resistance in plasmid
◯Kan^r

Mix Incubate

Recipient: rifampin resistant

No colonies

Colonies
◯Kan^r

No colonies

Plates containing medium plus kanamycin and rifampin

Figure 5.9 Determining if a plasmid is self-transmissible. Donor cells carrying the plasmid with a selectable marker, in this case kanamycin resistance, are mixed with closely related recipient cells with a rifampin resistance mutation in the chromosome. After being incubated, the cells are plated on medium containing kanamycin and rifampin. Colonies may arise if the plasmid is self-transmissible.

mixed-culture plates than on plates containing either the donor or the recipient alone is evidence that our plasmid is self-transmissible. For added assurance, we should show that the recipient bacteria now do contain the plasmid and that the plasmid can transfer itself out of them into other bacteria. We should also show that contact between the cells is required for the transfer and that transfer is not due to spontaneous lysis and transformation or to transduction, as we discuss in later chapters.

Determining Whether a Plasmid Is Promiscuous

Once we have determined that a plasmid is self-transmissible, we may wish to find whether it is promiscuous, i.e., whether it can transfer into other species. Failure to observe transconjugants when cells containing the plasmid are mixed with an unrelated host does not necessarily mean that the plasmid cannot transfer into that host. In the new host, the selectable gene in the plasmid might not be expressed or the plasmid might not replicate in the potential recipient bacterium.

To circumvent these problems with a particular host, we could introduce a broad-host-range replication origin into the plasmid or construct a shuttle vector with one origin of replication that allows it to replicate in the unrelated host and another that permits replication in the original donor (see chapter 4). A selectable gene that is known to be expressed in the unrelated host must also be introduced into the plasmid. This general approach has been used to show that the Tra systems of IncP plasmids can transfer a plasmid into gram-positive bacteria, including *Streptomyces* species, and that the F-plasmid Tra functions can transfer a plasmid into yeast cells (see Box 5.1).

Determining Whether a Plasmid Is Mobilizable

Our plasmid may not be self-transmissible, but it might still be mobilizable. The process of determining whether a plasmid can be mobilized is similar to investigating whether it can transfer itself, but it presents additional problems. The plasmid to be tested must carry a selectable marker, and a suitable recipient must be identified. However, we must also introduce a self-transmissible plasmid capable of mobilizing the plasmid into the donor cell. The potential donor cells containing both the larger, self-transmissible plasmid and the smaller plasmid under study are mixed with the recipient cells. After a sufficient time for transfer, the mating mixture is plated under conditions where only the recipient cells that have received the smaller plasmid will grow. The appearance of colonies under the selective conditions would indicate that the smaller plasmid can be mobilized by the self-transmissible plasmid.

However, the appearance of transconjugants containing the smaller plasmid does not confirm that the plasmid is mobilizable. The smaller plasmid may not have been truly mobilized but, rather, may have fused somehow to the larger plasmid. In this case, it would be transferred as part of the self-transmissible plasmid rather than being moved as a unique entity. This is a less likely possibility if some cells that received the smaller plasmid did not receive the larger, self-transmissible plasmid as well.

If a plasmid is not mobilized by one self-transmissible plasmid, it may still be mobilized by others. As mentioned, earlier, plasmids with a particular *oriT* site will be mobilized efficiently by the Tra functions of some, but not all, self-transmissible plasmids.

Triparental Matings

Mobilization of a plasmid into a recipient cell is often part of a cloning procedure or other experiment. Mobilizable plasmids have an advantage over self-transmissible plasmids in their smaller size. Nevertheless, difficulties will be encountered before plasmids can be mobilized. The self-transmissible plasmid and the plasmid to be mobilized may be members of the same Inc group and so will not stably coexist in the same cell. Also, the self-transmissible plasmid may express its *tra* genes only for a short time after entering a recipient cell.

Triparental matings help overcome these barriers to efficient plasmid mobilization. Figure 5.10 illustrates the general method, in which three bacterial strains participate in the mating mixture. One strain contains the self-transmissible plasmid, the second contains the plasmid to be mobilized, and the third is the eventual recipient. After the cells are mixed, some of the self-transmissible plasmids in the first strain will transfer into the strain carrying the plasmid to be mobilized. The smaller plasmid will then immediately be mobilized into the final recipient strain, where we want it for our cloning application. Even if the two plasmids are members of the same Inc group, they will coexist long enough for the mobilization to occur.

Mapping Plasmid Functions

Once we have identified the properties of an unknown plasmid, we will probably want to map its *oriT* region and its *tra* (or *mob*) genes. First, we examine how to go about locating the *oriT* site, and then we look at *tra* or *mob* genes and the products they encode.

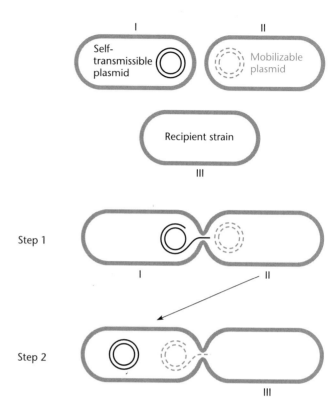

Figure 5.10 Triparental matings. In step 1, a self-transmissible plasmid from parent I transfers into parent II. In step 2, the self-transmissible plasmid transfers the mobilizable plasmid into parent III. This method will work even if the self-transmissible plasmid and the mobilizable plasmid are members of the same Inc group (see the text) and if the self-transmissible plasmid cannot replicate in parent II.

IDENTIFYING THE *oriT* SEQUENCES OF A PLASMID

As discussed, the *oriT* region of the plasmid is where some of the *tra* gene products initiate transfer. We can localize *tra* genes on the plasmid by mutating them (see the next section); however, this does not work for *oriT*, because mutations that inactivate this sequence will always prevent plasmid transfer, even if the Tra functions can be provided from another source.

Figure 5.11 shows a procedure for locating the *oriT* sequence of a plasmid that is based on the fact that any plasmid carrying the *oriT* sequence will be mobilized efficiently by the Tra functions of the same self-transmissible plasmid. In this procedure, fragments of the plasmid are randomly joined, or ligated, into a plasmid cloning vector of a different Inc group. By itself, this plasmid cloning vector should not be mobilizable and should carry a selectable gene different from any selectable genes on the self-transmissible plasmid. The ligated DNA mixture is then used to transform cells containing

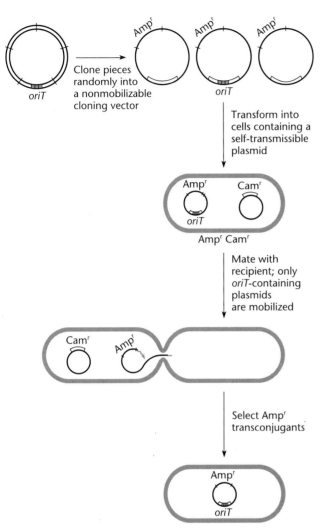

Figure 5.11 Identifying the *oriT* site on a plasmid. Pieces of the plasmid are cloned randomly into a nonmobilizable cloning vector. The mixture is transformed into cells containing the self-transmissible plasmid and mixed with a proper recipient. Pieces of DNA that allow the cloning vector to be mobilized contain the *oriT* site of the plasmid.

the self-transmissible plasmid, selecting the gene on the plasmid cloning vector. The transformants are then pooled and mated with a recipient, again selecting for the gene on the cloning vector. Only the vector plasmids that contain a fragment with the *oriT* region will have been mobilized into the recipient cells. We can prepare the plasmid vector containing the fragment from the transconjugants and retest it to be certain that it is mobilized efficiently. The fragment containing the *oriT* region can then be broken into yet smaller pieces and the process repeated to identify the smallest piece that has *oriT* activity. We further discuss methods for making and cloning DNA fragments in chapter 16.

IDENTIFYING THE *tra* GENES OF A PLASMID

Having established that a plasmid is self-transmissible, the next step is to determine how many *tra* genes are on the plasmid and where they are located and maybe get some idea of their function. Any mutation in the plasmid that inactivates a *tra* gene will make the plasmid unable to transfer (Tra⁻); to map these genes, we must first mutagenize the plasmid to obtain a large number of transferless (Tra⁻) mutants. These mutations are then assigned to genes by means of complementation tests (see chapter 16), and the genes are mapped with restriction endonucleases. Experiments involving this type of method showed that some self-transmissible plasmids of gram-negative bacteria carry 20 or more *tra* genes. Furthermore, the *tra* genes were often clustered in one or a few regions of the plasmid. We present more detailed coverage of methods for mutagenizing and mapping plasmids in chapter 16.

Identifying the *tra* genes and locating them on the plasmid are fairly straightforward, if somewhat laborious, tasks, but determining the exact function of the product of each of gene is much more difficult. In fact, only for the F plasmid are the functions of very many *tra* gene products known (Table 5.1).

Despite this difficulty, the *tra* gene products can be divided into two broad groups: those required for pilus synthesis and those more directly involved in DNA transfer.

Identifying the *tra* Genes for Pilus Formation

The *tra* genes whose products are involved in pilus synthesis can be easily identified if a phage is known that uses the pilus as its adsorption site. Then *tra* mutations that prevent pilus synthesis will make the cells resistant to the phage. We merely need to determine if the phage can multiply in bacteria carrying the plasmid with the *tra* mutation.

Isolating *tra* mutations that prevent pilus synthesis in the F plasmid is particularly easy because the F plasmid always expresses a pilus. Many phages, such as the single-stranded DNA phage M13, use the pilus of the F plasmid as their adsorption site. Therefore, these phages will form plaques on plates containing *E. coli* only if the *E. coli* cells harbor a pilus-expressing F plasmid. A mutation in a *tra* gene whose product is required for pilus synthesis will prevent the phages from forming plaques on the F-plasmid-containing bacteria. Therefore, to find which of the *tra* genes of the F plasmid participate in pilus synthesis, we merely screen a collection of *tra* mutations to see which ones prevent plaque formation by M13 or other F-plasmid-specific phages. Since each type of pilus is the target of a specific phage, other types of phage must be used to identify *tra* genes involved in pilus formation in other plasmids.

Unlike F, most self-transmissible plasmids do not always express a pilus; therefore, most cells containing these plasmids will not be sensitive to the phage at any one time. Finding the *tra* genes involved in pilus synthesis in these plasmids can be more difficult. The phage will not form plaques on bacteria containing these plasmids, because bacteria that are not expressing the pilus at any one time will grow in the plaques, obscuring them. In such cases, the only way to tell that some of the cells are expressing a pilus is to measure the phage titer after infecting the cells. The phage will adsorb to the bacteria that are expressing a pilus and will multiply, increasing the titer of the phage. Mutations in those *tra* genes required for pilus expression will prevent the increase in phage titer after infection of the mutant plasmid-containing cells.

Determining the exact role that each gene product plays in pilus expression involves even greater difficulties. The one exception may be pilin, the protein that makes up the pilus itself. If we can purify the pili, we can make antibodies to the pilin protein and use these antibodies to see which of the *tra* genes encode(s) this antigen.

Identifying *tra* Genes More Directly Involved in DNA Transfer

Most of the *tra* genes that do not play a role in pilus expression take part in DNA transfer. These gene products include the endonuclease that cleaves the plasmid at the *oriT* site, the helicase that specifically separates the DNA strands in the *oriT* region, and a primase. Often, the only way to determine the function of these *tra* genes is to use in vitro assays for the enzymes. For example, a nicking activity that cuts the plasmid might be deleted in extracts, and the mutant extracts can be tested for this activity.

Experiments To Demonstrate Transfer of Primase Protein

As mentioned earlier, the primase of self-transmissible plasmids is made in the donor cell but transferred to the recipient to prime the synthesis of the complementary strand of plasmid DNA after transfer. Transfer of primase during conjugation was first demonstrated in a genetic experiment with the self-transmissible plasmid Col1b-P9 (see Chatfield and Wilkins, Suggested Reading). A temperature-sensitive mutation in the *dnaG* gene encoding the *E. coli* primase will block DNA replication at high temperature, but this block will not occur in mutant cells containing the Col1b-P9 plasmid. Apparently, the primase of Col1b-P9 can substitute for the *E. coli*

primase in the initiation of chromosome replication. In this experiment, the investigators found that the plasmid could also allow DNA replication after conjugation into a recipient cell with the *dnaG* mutation, even if protein synthesis was blocked in the recipient. This result indicated that the primase was being transferred along with the DNA into the recipient.

Another experiment showed that the primase protein is required for replication only in the recipient cell. As mentioned, mutant plasmids that cannot make a primase can transfer into some types of bacteria but not others, presumably because the primase protein of some types of bacteria can be used to make the complementary strand of the plasmid after DNA transfer. In this experiment, the Col1b-P9 plasmid had an amber mutation in the *tra* gene that encodes the primase, but if the donor bacterium had an amber suppressor, the amber mutation in the Col1b-P9 primase gene did not prevent transfer. However, the amber mutation did prevent transfer if the recipient bacterium but not the donor had an amber suppressor, so that the capacity to synthesize the primase protein only in the recipient did not permit transfer.

Transfer Systems of Gram-Positive Bacteria

Self-transmissible plasmids have also been found in many types of gram-positive bacteria, including species of *Bacillus*, *Streptococcus*, *Staphylococcus*, and *Streptomyces*. However, much less is known about the transfer systems of these bacteria. The transfer regions of gram-positive plasmids are often much smaller than those of gram-negative bacteria. For example, the transfer region of the *Streptomyces nigrifaciens* plasmid pSN22 may have less than 7,000 bp of DNA and may need only about five genes for transfer. The *tra* regions can be smaller because gram-postive bacteria may not require a sex pilus to exchange plasmids, since they lack the elaborate outer membrane of gram-negative bacteria.

Many plasmids of gram-positive bacteria are also promiscuous. For example, plasmid pAMβ1, originally isolated from *Enterococcus faecalis,* can transfer into *Staphylococcus aureus* and many other gram-positive bacteria.

Plasmid-Attracting Pheromones

Some strains of *E. faecalis* excrete pheromone-like compounds that can stimulate mating with donor cells (see Dunny et al., Suggested Reading). These pheromones are small peptides, each of which stimulates mating with cells containing a particular plasmid. The pheromone-like peptides act by specifically stimulating the expression of *tra* genes in the plasmids of neighboring bacteria, thereby inducing aggregation and mating. Cells that already contain the plasmid no longer excrete the specific peptide, but they continue to excrete other peptides that will stimulate mating with cells containing other plasmid types.

The advantages of a pheromone system for stimulating the expression of the *tra* genes and transfer of the plasmid seem obvious. The plasmid need not express its *tra* genes unless an appropriate recipient, a bacterium of the same species that does not already possess the plasmid, is nearby. This not only conserves energy for the host but also prevents the expression of Tra proteins, some of which are surface proteins that may serve as attachment sites for phages. The existence of this system emphasizes the importance of plasmid transfer to bacteria, since *E. faecalis* seems to actively encourage the uptake of at least some types of plasmids.

Other Types of Transmissible Elements

Plasmids are not the only DNA elements in gram-positive bacteria that are capable of transferring themselves. Some transposons, called **conjugative transposons**, also encode Tra functions to promote their own transfer.

An example, Tn916 from *E. faecalis,* is not a replicon capable of autonomous replication, but it can transfer itself from the chromosome of one bacterium to the chromosome of another bacterium without transferring chromosomal genes. Presumably, it transiently excises from one DNA, transfers itself into another cell, and then transposes into the DNA of the recipient bacterium upon entry.

The Tn916 conjugative transposon and its relatives are known to be promiscuous and will transfer into many types of gram-positive bacteria and even into some gram-negative bacteria. The antibiotic resistance gene they carry, *tetM*, has also been found in many types of gram-positive and gram-negative bacteria. It is tempting to speculate that conjugative transposons such as Tn916 are responsible for the widespread dissemination of the *tetM* gene. We shall discuss the transposition of conjugative transposons in more detail in chapter 8.

Other elements similar to conjugative transposons have been found in the genus *Bacteroides*. These elements will transfer not only themselves but also other small DNA elements in the chromosome (see Salyers et al., Suggested Reading).

SUMMARY

1. Self-transmissible plasmids can transfer themselves to other bacterial cells. Mobilizable plasmids cannot transfer themselves but can be transferred by certain self-transmissible plasmids.

2. Self-transmissible plasmids have several *trans*-acting *tra* genes that encode products required for transfer, as well as a *cis*-acting *oriT* site at which plasmid transfer initiates.

3. Most of the *tra* gene products are required to synthesize a sex pilus, a protrusion on the cell surface that promotes contact with other cells that do not contain the plasmid. Other *tra* gene products are more directly involved in DNA transfer.

4. Mobilizable plasmids have a *mob* region that encodes a few *mob* gene products plus an *oriT*. The *mob* region of mobilizable plasmids encodes functions directly involved in DNA transfer but not functions involved in sex pilus synthesis.

5. The process by which self-transmissible plasmids promote their own transfer or the transfer of other plasmids into cells is called conjugation. The original cell that contained the plasmid is the donor, and the cell into which the plasmid has been transferred is the recipient. Recipient cells into which the plasmid has been transferred are called transconjugants.

6. Most plasmids express their *tra* genes only transiently after the plasmid enters a recipient cell. Presumably, this

helps them avoid infection by male-specific phages that use the sex pilus as their absorption site.

7. After contact is made with a potential recipient cell, DNA transfer is initiated by cutting one of the two strands of the plasmid DNA at the *oriT* sequence. Then the broken strand is displaced by a helicase and transferred into the recipient cell, where it is recycled. Normally, the complementary strands are synthesized at the same time, so after transfer, both the recipient cell and the donor cell have a double-stranded plasmid.

8. Under some conditions, self-transmissible plasmids can transfer chromosomal genes from one cell to another. Either the plasmid will have inserted into the bacterial chromosome, or chromosomal DNA will have been integrated into the plasmid. The transfer of chromosomal DNA is usually indicated by the appearance of recombinant types unlike either parent.

9. Bacterial strains with a self-transmissible plasmid inserted into their chromosome are called Hfr strains. Plasmids that have incorporated chromosomal DNA are called prime factors.

10. Plasmids that have the ability to transfer themselves between different species are said to be promiscuous. Some

(continued)

SUMMARY (continued)

promiscuous plasmids can transfer themselves or mobilize other plasmids into distantly related strains of bacteria and even into eukaryotic cells. The promiscuous transfer of plasmids among bacterial species has important implications in medicine and biotechnology and has also presumably played an important role in bacterial evolution.

11. Self-transmissible plasmids also exist in gram-positive bacteria. However, these plasmids do not encode a sex pilus. Some gram-positive bacteria have the interesting

property of excreting small pheromone-like compounds that stimulate mating with certain plasmids, presumably by inducing their *tra* genes. The existence of such systems emphasizes the importance of plasmid exchange to bacteria.

12. Some transposons of gram-positive bacteria are also self-transmissible. These so-called conjugative transposons can transfer themselves into other cells even though they are not replicons.

QUESTIONS FOR THOUGHT

1. Why do you suppose the Tra functions are usually the same for all the members of the same Inc group?

2. Why are the *tra* genes whose products are directly involved in DNA transfer usually adjacent to the *oriT* site?

3. Why do you suppose plasmids with a certain *mob* site are mobilized by only certain types of self-transmissible plasmids?

4. Why do self-transmissible plasmids usually encode their own primase function?

5. What do you think is different about the cell surfaces of gram-positive and gram-negative bacteria that causes only the self-transmissble plasmids of gram-negative bacteria to encode a pilus?

6. Why do so many types of phages use the sex pilus of plasmids as their adsorption site?

7. Why are so many plasmids either self-transmissible or mobilizable? Promiscuous?

PROBLEMS

1. After mixing two strains of a bacterial species, you observe some recombinant types that are unlike either parent. These recombinant types seem to be the result of conjugation, because they appear only if the cells are in contact with each other. How would you determine which is the donor strain and which is the recipient? Whether the transfer is due to an Hfr strain or to a prime factor?

2. How would you determine which of the *tra* genes of a self-transmissible plasmid encodes the pilin protein? The site-specific DNA endonuclease that cuts at *oriT*? The helicase?

3. How would you show that only one strand of the plasmid DNA enters the recipient cell during plasmid transfer?

4. You have discovered that tetracycline resistance can be tranferred from one strain of a bacterial species to another. How would you determine whether the teracycline resistance gene being transferred is on a self-transmissible plasmid or on a conjugative transposon?

SUGGESTED READING

Bagdasarian, M., A. Bailone, J. F. Angulo, P. Scholz, M. Bagdasarian, and R. Devoret. 1992. PsiB, an anti SOS protein, is transiently expressed by the F sex factor during its transmission to an *Escherichia coli* recipient. *Mol. Microbiol.* 6:885–894

Bagdasarian, M., R. Lurz, B. Rukert, F. C. H. Franklin, M. M. Bagdasarian, J. Frey, and K. N. Timmis. 1981. Specific-purpose plasmid cloning vectors. II. Broad host range, high copy number, RSF1010-derived vectors, and a host-vector system for gene cloning in *Pseudomonas*. *Gene* 16:237–247.

Chatfield, L. K., and B. M. Wilkins. 1984. Conjugative transfer of IncI plasmid DNA primase. *Mol. Gen. Genet.* 197:461–466.

Derbyshire, K. M., G. Hatfull, and N. Willets. 1987. Mobilization of the nonconjugative plasmid RSF1010: a genetic

and DNA sequence analysis of the mobilization region. *Mol. Gen. Genet.* 206:161–168.

Dunny, G. M., B. A. B. Leonard, and P. J. Hedberg. 1995. Pheromone-inducible conjugation in *Enterococcus faecalis*: interbacterial and host-parasite chemical communication. *J. Bacteriol.* 177:871–876.

Firth, N., K. Ippen-Ihler, and R. A. Skurray. 1996. Structure and function of the factor and mechanism of conjugation, p. 2377–2401. *In* F. C. Neidhardt, R. C. Curtiss III, J. L. Ingraham, E. C. C. Lin, K. B. Low, B. Magasanik, W. S. Reznikoff, M. Riley, M. Schaechter, and H. E. Umbarger (ed.), *Escherichia coli and Salmonella: Cellular and Molecular Biology*, 2nd ed. ASM Press, Washington, D.C.

Haase, J., R. Lurz, A. M. Grahn, D. H. Bamford, and E. Lanka. 1995. Bacterial conjugation mediated by plasmid

RP4: RSF1010 mobilization, donor-specific phage propagation, and pilus production require the same *tra2* core components of a proposed DNA transport process. *J. Bacteriol.* **177:**4779–4791.

Lederberg, J., and E. L. Tatum. 1946. Gene recombination in *E. coli. Nature* (London) **158:**558.

Low, K. B. 1968. Formation of merodiploids in matings with a class of Rec⁻ recipient strains of *E. coli* K12. *Proc. Natl. Acad. Sci. USA* **60:**160–167.

Read, T. D., A. T. Thomas, and B. M. Wilkins. 1992. Evasion of Type I and Type II DNA restriction systems by IncII plasmid ColIb-P9 during transfer by bacterial conjugation. *Mol. Microbiol.* **14:**1933–1941.

Salyers, A. A., N. B. Shoemaker, A. M. Stevens and L.-Y. Li. 1995. Conjugative transposons: an unusual and diverse set of integrated gene transfer elements. *Microbiol. Rev.* **59:**579–590.

Watanabe, T., and T. Fukasawa. 1961. Episome-mediated transfer of drug resistance in *Enterobacteriaceae.* 1. Transfer of resistance factors by conjugation. *J. Bacteriol.* **81:**669–678.

Willetts, N., and C. Crowther. 1981. Mobilization of the nonconjugative IncQ plasmid RSF1010. *Genet. Res. Camb.* **37:**311–316.

Zhang, S., and R. M. Meyer. 1995. Localized denaturation of *oriT* DNA within relaxosomes of the broad host range plasmid R1162. *Mol. Microbiol.* **17:**727–735.

Transformation

D NA CAN BE EXCHANGED among bacteria in three ways: conjugation, transduction, and transformation. Chapter 5 covered conjugation, in which a plasmid or other self-transmissible DNA element transfers itself and sometimes other DNA into another bacterial cell. In transduction, a subject of chapter 7, a phage carries DNA from one bacterium to another. In this chapter, we discuss **transformation**, in which cells take up free DNA directly from their environment.

Transformation is one of the cornerstones of molecular genetics because it is often the best way to reintroduce experimentally altered DNA into cells. Since transformation was first discovered in bacteria, ways have been devised to transform many types of animal and plant cells as well.

Discussions of transformation use terms similar to those used in discussions of conjugation. DNA is derived from a **donor bacterium** and taken up by a **recipient bacterium**, which is then called a **transformant**.

Natural Transformation

Most types of bacteria will not take up DNA efficiently unless they have been exposed to special chemical or electrical treatments to make them more permeable. However, **naturally transformable** bacteria can take up DNA from their environment without special treatment. The state of bacteria in which they can take up DNA is called **natural competence**. Naturally competent transformable bacteria are found in several genera and include both gram-negative and gram-positive bacteria such as *Bacillus subtilis*, a soil bacterium; *Haemophilus influenzae*, the causitive agent of spinal meningitis; *Neisseria gonorrhoeae*, which causes gonorrhea; *Streptococcus*

149

pneumoniae, which causes throat infections; and species of cyanobacteria from the genus *Synechococcus.*

Discovery of Transformation

Transformation was the first mechanism of gene exchange in bacteria to be discovered. In 1928, Fred Griffith found that one form of the pathogenic pneumococci (now called *Streptococcus pneumoniae*) could be mysteriously "transformed" into another form. Griffith's experiments were based on the fact that *S. pneumoniae* makes two different types of colonies, one pathogenic and the other nonpathogenic. Because they excrete a polysaccharide capsule, the pathogenic strains make colonies that appear smooth on agar plates. Apparently because the capsule allows them to survive in a vertebrate host, these bacteria can infect and kill mice. However, rough-colony-forming mutants that cannot make the capsule and are nonpathogenic in mice sometimes arise from the smooth-colony formers.

In his experiment, Griffith mixed dead *S. pneumoniae* cells that made smooth colonies with live nonpathogenic cells that made rough colonies and injected the mixture into mice (Figure 6.1). Mice given injections of only the rough-colony-forming bacteria survived, but mice that received a mixture of dead smooth-colony formers and live rough-colony formers died. Furthermore, Griffith isolated live smooth-colony-forming bacteria from the blood of the dead mice. Concluding that the dead pathogenic bacteria gave off a "transforming principle" that changed the live nonpathogenic rough-colony-forming bacteria into the pathogenic smooth-colony form, he speculated that this transforming principle was the poly-

saccharide itself. Later, other researchers obtained transformation of rough-colony formers into smooth-colony formers by mixing the rough forms with extracts of the smooth forms in a test tube. Then, about 16 years after Griffith did his experiments with mice, Oswald Avery and his collaborators used the in vitro system to purify the "transforming principle" and showed that it is DNA (see Avery et al., Suggested Readings). The work of Avery and colleagues helped show that DNA, not proteins or other factors in the cell, is the hereditary material (see the introductory chapter).

Competence

Most naturally transformable bacteria can take up DNA only late in their growth cycle, usually just before they reach the stationary phase. Bacteria that are capable of taking up DNA are said to be **competent.**

As with most cell functions, the first step in understanding the development of competence is to identify the genes whose products are required for competence and understand how these genes are regulated. In this regard, much more is known about *B. subtilis* than about most other types of naturally transformable bacteria, and so the discussion here centers on this species.

REGULATION OF COMPETENCE IN *B. SUBTILIS*
The regulation of competence in *B. subtilis* is achieved through a **two-component regulatory system** analogous to those used to regulate many other systems in bacteria (see Box 12.3). First, information that the cell is running out of nutrients and the population is reaching a high density is registered by ComP, a **sensor protein** in the membrane (Figure 6.2). The high cell density causes this

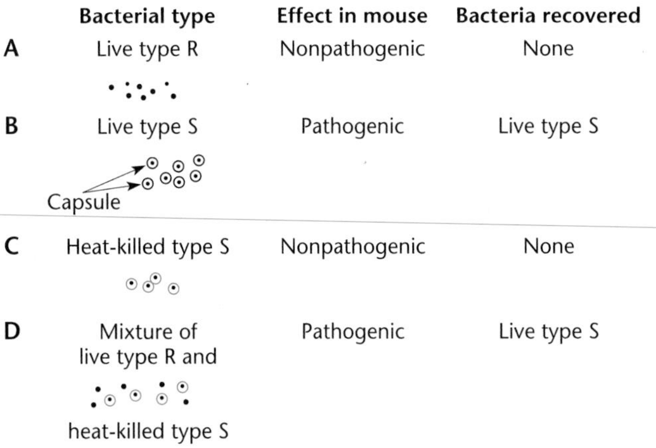

	Bacterial type	Effect in mouse	Bacteria recovered
A	Live type R	Nonpathogenic	None
B	Live type S	Pathogenic	Live type S
C	Heat-killed type S	Nonpathogenic	None
D	Mixture of live type R and heat-killed type S	Pathogenic	Live type S

Figure 6.1 The Griffith experiment. Heat-killed pathogenic encapsulated bacteria can convert nonpathogenic noncapsulated bacteria to the pathogenic capsulated form. Type R indicates rough-colony formers and type S indicates smooth-colony formers.

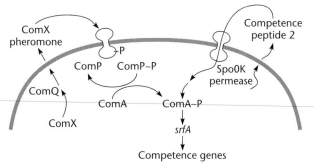

Figure 6.2 Regulation of competence development in *Bacillus subtilis.* The ComP protein in the membrane senses high cell density and is phosphorylated. The phosphate is then transferred to ComA, which allows transcription of competence genes. The ComP protein phosphorylation involves two competence pheromone peptides, one of which cooperates with an oligopeptide permease, the product of the *spo0K* gene.

sensor-kinase protein to phosphorylate itself. The phosphate is then transferred from ComP to ComA, a **response regulator protein**. In the phosphorylated state, the ComA protein is a transcriptional activator (see chapter 11) for several genes, including the operon *srfA*, which is required for competence. At least one of the genes in this operon affects the expression of other genes required for competence.

COMPETENCE PHEROMONES
High cell density is required for competence of *B. subtilis* because of small peptides called **competence pheromones** that are excreted by the bacteria as they multiply (see Solomon et al., Suggested Reading). Cells become competent only in the presence of high concentrations of these peptides, which are reached only when the concentration of cells giving them off is high. The requirement for competence pheromones ensures that cells will be able to take up DNA only when there are other *B. subtilis* cells nearby that are giving off DNA to be taken up.

One competence pheromone peptide is cut out of a longer polypeptide, the product of the *comX* gene. Another gene, *comQ*, which is immediately upstream of *comX*, is also required for synthesis of the competence pheromone, and its product may be the protease enzyme that cuts the competence pheromone out of the longer polypeptide. Once the peptide has been cut out of the longer molecule, it can trigger the phosphorylation of ComP, although the mechanism remains unknown.

B. subtilis cells produce a second competence peptide that also signals a critical density in the population, probably with the help of a product of the *spo0K* gene (see the next section), an oligopeptide permease.

RELATIONSHIP BETWEEN SPORULATION AND COMPETENCE
At about the same time *B. subtilis* develops competence for transformation at high cell density, it also begins to sporulate. Sporulation, a developmental process common to many bacteria, allows a bacterium to enter a dormant state and survive adverse conditions such as starvation, irradiation, and heat. During sporulation, the bacterial chromosome is packaged into a resistant spore, where it remains viable until conditions improve and the spore can germinate into an actively growing bacterium.

Some of the regulatory genes required for sporulation are also required for the development of competence. The *spo0K* gene is an example. This gene was first discovered because of its role in sporulation. A *spo0K* mutant is blocked in the first stage, the "0" stage, of sporulation. The *K* means that it was the 11th gene (as K is the 11th letter in the alphabet) involved in sporulation to be discovered in that collection. The regulation of sporulation seems to be even more complex than the regulation of competence, because cells will not sporulate unless they are also deprived of certain nutrients.

CELL PERMEABILITY
Although the regulation of the development of sporulation is fairly well understood, much less is known about the gene products that actually make the cell permeable to DNA. Some of these proteins presumably bind the DNA, and others must form the cell membrane structure through which the DNA passes. There is evidence that polyhydroxybutyrate, previously thought to be only a storage polymer, may form such channels in the membrane.

Uptake of DNA during Natural Transformation

Experiments directed toward an understanding of DNA uptake during natural transformation have sought to answer three obvious questions. (i) How efficient is DNA uptake? (ii) Can only DNA of the same species enter a given cell? (iii) Are both of the complementary DNA (cDNA) strands taken up and incorporated into the cellular DNA?

EFFICIENCY OF DNA UPTAKE
The efficiency of uptake is fairly easy to measure biochemically For example, Figure 6.3 shows an experiment based on the fact that transport of free DNA into the cell

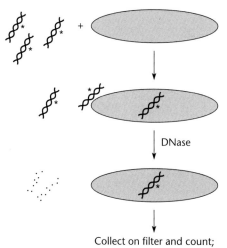

Collect on filter and count;
degraded DNA will pass through filter

Figure 6.3 Determining the efficiency of DNA uptake during transformation. DNA in the cell is insensitive to DNase. Degraded DNA will pass through a filter. The asterisk refers to radioactively labeled DNA.

will make those strands insensitive to DNases, which cannot enter the cell, because competent cells permit only DNA, not proteins, to enter. In the experiment, labeled DNA is mixed with competent cells and the mixture is then treated with DNase. Any DNA that is not degraded and survives intact must have been taken up by the cells. Such DNA can be distinguished from degraded DNA because intact DNA can be precipitated with acid and collected on a filter whereas degraded DNA will not precipitate and will pass through the filter. Therefore, if the radioactivity on the filter is counted and compared with the total radioactivity of the DNA that was added to the cells, the percentage of DNA that is taken up, or the efficiency of DNA uptake, can be calculated. Experiments such as these have shown that some competent bacteria take up DNA very efficiently.

SPECIFICITY OF DNA UPTAKE
The second question, i.e., whether DNA from only the same species is taken up, is also fairly easy to answer. By using the same assay of resistance to DNases, it has been determined that some types of bacteria will take up DNA from only their own species whereas others can take up DNA from any source. The first group includes *Neisseria gonorrhoeae* and *Haemophilus influenzae*.

A specific **uptake sequence** is required by bacteria that take up DNA only from their own species. Figure 6.4 shows the sequences for *H. influenzae* and *N. gonorrhoeae*. Uptake sequences are short, only about 10 to 12 bp long, but are long enough that they almost never occur by chance in other DNAs. In contrast, bacteria such as *B. subtilis* seem to take up any DNA. Possible reasons why some bacteria should preferentially take up DNA from their own species while others take up any DNA are subjects of speculation and will be discussed later.

Mechanism of DNA Uptake during Transformation

Although the genetic requirements for transformation are best known for *B. subtilis*, the more efficient uptake of DNA by some other naturally transformable bacteria has allowed biochemical experiments on the uptake of DNA by these species. The general pathway

was first worked out for *Streptococcus pneumoniae* but is probably similar in most other naturally transformable bacteria.

TRANSFORMATION IN *S. PNEUMONIAE*
Figure 6.5 shows a general scheme for DNA uptake during the transformation of *S. pneumoniae*. In the first step, double-stranded DNA released by lysis of the donor bacteria is bound to specific receptors on the cell surface of the recipient bacterium. The bound DNA is then broken into smaller pieces by endonucleases; one of the two complementary strands is degraded by an exonuclease; and the remaining strand is transported into the cell. The transforming DNA integrates into the cellular DNA in a homologous region by means of strand

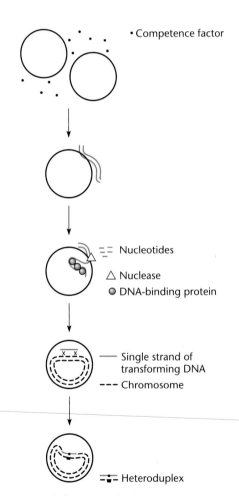

Figure 6.5 Transformation in *Streptococcus pneumoniae*. Competence factors accumulate as the cells reach a high density. Double-stranded DNA binds to the cell, and one strand is degraded. The remaining single strand replaces the strand of the same sequence in the chromosome, creating a "heteroduplex" in which one strand comes from the donor and one comes from the recipient.

Haemophilus influenzae	5′	AAGTGCGGTCA 3′
Neisseria gonorrhoeae	5′	GCCGTCTCAA 3′

Figure 6.4 The uptake sequences on DNA for some types of bacteria. Only DNA with these sequences will be taken up by the bacteria indicated. Only one strand of the DNA is shown.

displacement, a mechanism in which the new strand invades the double helix and displaces an old strand with the same sequence. The old strand is then degraded. If the donor DNA and recipient DNA sequences differ slightly in this region, recombinant types can appear.

Evidence for this model comes from several different experiments, some of which are discussed below. Also, the gene for a membrane-bound DNase that may be involved in degrading one of the two strands of the incoming DNA has been found in *S. pneumoniae* (see Puyet et al., Suggested Reading).

TRANSFORMASOMES

The basic scheme described above probably differs among different types of naturally competent bacteria. For example, *Haemophilus influenzae* may first take up double-stranded DNA in subcellular compartments called **transformasomes** (Figure 6.6). The new DNA may not become single stranded until it enters the cytoplasm. However, the basic process of all natural transformation is the same. Only one strand of the DNA enters the interior of the cell and integrates with the cellular DNA to produce recombinant types.

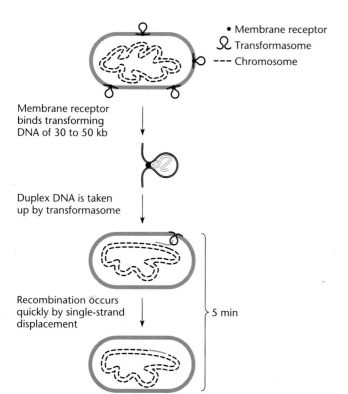

Membrane receptor binds transforming DNA of 30 to 50 kb

• Membrane receptor
Ω Transformasome
--- Chromosome

Duplex DNA is taken up by transformasome

Recombination occurs quickly by single-strand displacement

} 5 min

Figure 6.6 Transformation in *Haemophilus influenzae*. Double-stranded DNA is first taken up in transformasomes. One strand is degraded, and the other strand invades the chromosome, displacing one chromosome strand.

Genetic Evidence for Single-Strand Uptake

Genetic experiments taking advantage of the molecular requirements for transformation can be used to study the molecular basis for transformation; in other words, tranformation can be used to study itself. Evidence that DNA has transformed cells is usually based on the appearance of recombinant types after transformation. A recombinant type can form only if the donor and recipient bacteria differ in their genotypes and if the incoming DNA from the donor bacterium changes the genetic composition of the recipient bacterium. The chromosome of a recombinant type will have the DNA sequence of the donor bacterium in the region of the transforming DNA.

Experiments have shown that only double-stranded DNA can bind to the specific receptors on the cell surface, so that only double-stranded—not single-stranded—DNA can transform cells and yield recombinant types. However, we can also conclude from these experiments that the cells actually take up only *single*-stranded DNA, because the DNA enters an "eclipse" phase in which it cannot transform. For example, in the experiment shown in Figure 6.7, an Arg$^-$ mutant requiring arginine for growth is used as the recipient strain, and the corresponding Arg$^+$ prototroph is the source of donor DNA. At various times after the donor DNA has been mixed with the recipient cells, the recipients are treated with DNase, which cannot enter cells but will destroy any DNA remaining in the medium. The surviving DNA in the recipient cells is then extracted and used for retransformation of more auxotrophic recipients, and Arg$^+$ transformants are selected on agar plates without the growth supplement arginine. Any Arg$^+$ transformants must have been due to double-stranded donor DNA in the recipient cells.

Whether or not transformants were observed depends on the time the DNA was extracted from the cells. When the DNA is extracted at time 1, while it is still outside the cells and accessible to the DNase, no Arg$^+$ transformants are observed because the Arg$^+$ donor DNA is all destroyed by the DNase. At time 2, some of the DNA is now inside the cells, where it cannot be degraded by the DNase, but this DNA is single stranded. It has not yet recombined with the chromosome, and so Arg$^+$ transformants are still not observed in step 4. Only at time 3, when some of the DNA has recombined with the chromosomal DNA and so is again double stranded, will Arg$^+$ transformants appear in step 4. Thus, the transforming DNA enters the eclipse period for a short time after it is added to competent cells, as expected if it enters the cell in a single-stranded state.

Step 1
Mix Arg⁺ DNA
and recipient
cells

Time 1

Time 2

Time 3

Step 2
Treat mixture
with DNase at
various times

DNA is
extracellular
and so is
degraded

Intracellular DNA
is single stranded
and so cannot bind
to recipient

Intracellular DNA has
recombined with
chromosome and so
is double stranded

Step 3
Extract DNA
and mix with
Arg⁻ recipients

Double-stranded
DNA can transform
recipient

Step 4
Select Arg⁺
transformants

No Arg⁺
transformants

No Arg⁺
transformants

Arg⁺ transformant

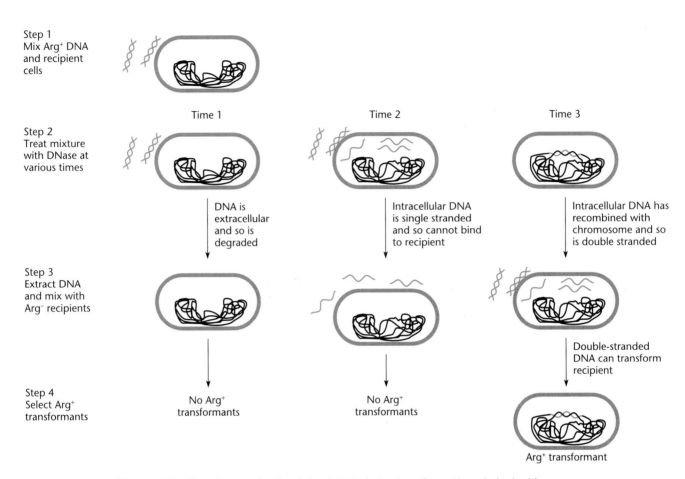

Figure 6.7   Genetic assay for the state of DNA during transformation. Only double-stranded DNA will bind to the cell to initiate transformation. The appearance of transformants in step 4 indicates that the transforming DNA was double stranded at the time of DNase treatment.

Plasmid Transformation and Transfection of Naturally Competent Bacteria

Chromosomal DNA can efficiently transform any bacterial cells from the same species that are naturally competent. However, neither plasmids nor phage DNAs can be efficiently introduced into naturally competent cells, because they must recycle to replicate. Natural transformation requires breakage of the double-stranded DNA and degradation of one of the two strands so that a linear single strand can enter the cell. Plasmid and phage DNAs are usually double stranded and must be double stranded to replicate autonomously. However, pieces of single-stranded plasmid or phage DNAs cannot recycle or make the complementary strand if there are no repeated or complementary sequences at their ends.

Plasmid or phage DNAs often do not have repeated DNA at their ends, and so transformation of naturally competent bacteria with plasmid or phage DNA usually occurs only with DNAs that are **dimerized**. A dimerized DNA is one in which two copies of the molecule are linked head to tail as illustrated in Figure 6.8. If a dimerized plasmid or phage DNA is cut only once, it will still have complementary sequences at its ends that can recombine to recyclize the plasmid, as illustrated in the figure. Such dimers form naturally while plasmid or phage DNA is replicating, so that most preparations of plasmid or phage DNAs contain some dimers. The fact that only dimerized plasmid or phage DNAs can transform naturally competent bacteria supports the model of transformation described earlier.

The Role of Natural Transformation

The fact that so many gene products play a direct role in competence indicates that the ability to take up DNA from the environment is advantageous. Below, we discuss three possible advantages and the arguments for and against them.

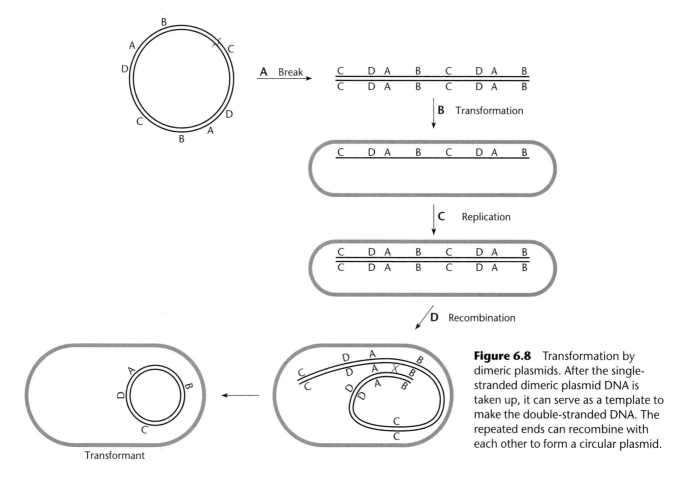

Figure 6.8 Transformation by dimeric plasmids. After the single-stranded dimeric plasmid DNA is taken up, it can serve as a template to make the double-stranded DNA. The repeated ends can recombine with each other to form a circular plasmid.

NUTRITION

Organisms may take up DNA for use as a carbon and nitrogen source (see Redfield, Suggested Reading). The major argument against this hypothesis is that taking up whole DNA strands for degradation inside the cell may be more difficult than degrading the DNA outside the cell and then taking up the nucleotides. In fact, some of the same bacteria excrete DNases, which degrade DNA so that it can be taken up more easily. Also, the fact that some bacteria take up only DNA of their own species seems to argue against this general explanation, since DNA from other organisms should offer the same nutritional benefits. Moreover, the fact that competence develops only in a minority of the population, at least in *B. subtilis*, argues against the nutrition hypothesis, since all the bacteria in the population would presumably need the nutrients.

These arguments are attractive but do not disprove the nutrition hypothesis. The bacteria may consume DNA of only their own species because of the danger inherent in taking up foreign DNAs, which might contain prophages, transposons, or other elements that could become parasites of the organism. Furthermore, consumption of DNA from the same species may be a normal part of colony development; cell death and cannibalism are thought to be part of some prokaryotic developmental processes. These processes would require that only some of the cells in the population become DNA consumers. The special circumstances of laboratory cultures may also explain the fact that only a minority of cells become competent. In some natural environments, all of the bacteria may become competent at the same time in response to starvation.

REPAIR

Cells may take up DNA from other cells to repair damage to their own DNA (see Mongold, Suggested Reading). Figure 6.9 illustrates this hypothesis, in which a population of cells is exposed to UV irradiation. The radiation damages the DNA, causing pyrimidine dimers and other lesions (see chapter 10). DNA leaks out of some of the dead cells and enters other bacteria. Because the damage to the DNA will not have occurred at exactly the same places, undamaged incoming DNA sequences can replace the damaged regions in the recipient, allowing at least some of the bacteria to survive.

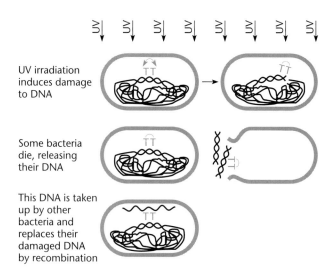

UV irradiation induces damage to DNA

Some bacteria die, releasing their DNA

This DNA is taken up by other bacteria and replaces their damaged DNA by recombination

Figure 6.9 Repair of DNA damage by transforming DNA. A region containing a thymine dimer (TT) induced by UV irradation is replaced by the same, but undamaged, sequence from the DNA of a neighbor killed by the radiation. This mechanism could allow survival of the species.

This scenario explains why some bacteria will take up DNA of only the same species, since, in general, this is the only DNA that can recombine and thereby participate in the repair.

If natural transformation helps in DNA repair, we might expect that repair genes will be induced in response to developing competence and that competence will develop in response to UV irradiation or other types of DNA damage. In fact, in some bacteria, including *B. subtilis* and *S. pneumoniae*, the *recA* gene required for recombination repair is induced in response to the development of competence (see Haijema et al., Pierce et al., and Raymon-Denise and Guillen, Suggested Reading). However, in other bacteria, such as *H. influenzae*, the *recA* gene is not induced in response to competence. There is also no evidence that competence genes are induced in response to DNA damage. Nevertheless, the need for DNA repair is an attractive explanation for why at least some types of bacteria develop competence.

RECOMBINATION
The possibility that transformation allows recombination between individual members of the species is also an attactive hypothesis that is difficult to prove. According to this hypothesis, transformation serves the same function that sex serves in higher organisms: it allows the assembly of new combinations of genes and thereby increases diversity and speeds up evolution. Bacteria do not have an obligatory sexual cycle; therefore, without some means of genetic exchange, any genetic changes

that a bacterium accumulates during its lifetime will not be exchanged with other members of the species.

The gene exchange function of transformation is supported by the fact that cells of some naturally transformable bacteria leak DNA as they grow. It is hard to imagine what function this leakage could perform unless the leaked DNA is to be taken up by other bacteria.

In several *Neisseria* species, including *N. gonorrhoeae*, transformation may enhance antigenic variability, allowing the organism to avoid the host immune system (see Box 6.1). In mixed laboratory cultures, transformation does contribute substantially to the antigenic diversity in this species. However, under natural conditions, it is debatable whether most of this antigenic diversity results from recombination between DNAs brought together by transformation or simply from recombination between sequences within the chromosomal DNA of the bacterium.

We still do not know why some types of bacteria are naturally transformable. In fact, transformation may serve different purposes in different organisms. Perhaps transformation is used for DNA repair in soil bacteria such as *B. subtilis*, but it is used to increase genetic variability in obligate parasites such as *N. gonorrhoeae*. Moreover, most types of bacteria may be naturally transformable at low levels, which would help explain phenomena such as mutations that appear to be directed (see Box 3.2).

Whatever its purpose for individual bacterial species, natural transformation has many uses in molecular genetics. It has been used to map genetic markers in chromosomes and to reintroduce DNA into cells after the DNA has been manipulated in the test tube.

Artificially Induced Competence

Most types of bacteria are not naturally transformable, at least not at easily detectable levels. Left to their own devices, these bacteria will not take up DNA from the environment. However, even these bacteria can sometimes be made competent by certain chemical treatments, or DNA can be forced into them by a strong electric field in a process called electroporation.

Calcium Ion Induction

Treatment with calcium ions (see Cohen et al., Suggested Reading) can make some bacteria competent, including *Escherichia coli* and *Salmonella* spp. as well as some species of *Pseudomonas*, although the reason is not understood.

Chemically induced transformation is usually inefficient, and only a small percentage of the cells are ever

BOX 6.1

Antigenic Variation in *Neisseria gonorrhoeae*

Many types of pathogenic microorganisms avoid the host immune system by changing the antigens on their cell surface. Well-studied examples include trypanosomes, which cause sleeping sickness, and *Neisseria gonorrhoeae*, which causes a sexually transmitted disease.

The pili of *N. gonorrhoeae* are involved in attaching the bacteria to the host epithelial cells. These pili can undergo spontaneous alterations that can change the specificity of binding and confound the host immune system. *N. gonorrhoeae* appears to be capable of making millions of different pili.

The mechanism of pilin variation in *N. gonorrhoeae* is understood in some detail. The major protein subunit of the pilus is encoded by the *pilE* gene. In addition, silent copies of *pilE*, called *pilS*, lack promoters or have various parts deleted. These silent copies share some conserved sequences with each other and with *pilE* but differ in the so-called variable regions. Pilin protein is usually not expressed from these silent copies. However, recombination between a *pilS* gene and *pilE* can change the *pilE* gene and result in a somewhat different pilin protein. This recombination is a type of gene conversion because reciprocal recombinants are not formed (see chapter 9).

Because *N. gonorrhoeae* is naturally transformable, not only could recombination occur between a *pilS* gene and the *pilE* gene in the same organism, but also transformation could allow even more variation through the exchange of *pilS* genes with other strains. Experiments indicate that pilin variation is affected by the presence of DNase. Also experiments with marked *pil* genes indicate that transformation can result in the exchange of *pil* genes between bacteria. These experiments suggest but do not prove that transformation plays an important role in pilin variation during infection.

References

Gibbs, C. P., B.-Y. Reimann, E. Schultz, A. Kaufmann, R. Haas, and T. F. Meyer. 1989. Reassortment of pilin genes in *Neisseria gonorrhoeae* occurs by two disinct mechanisms. *Nature* (London) **338:**651–652.

Meyer, T. F., C. P. Gibbs, and R. Haas. 1990. Variation and control of protein expression in *Neisseria. Annu. Rev. Microbiol.* **44:**451–477.

Seifert, H. S., R. S. Ajioka, C. Marchal, P. F. Sparling, and M. So. 1988. DNA transformation leads to pilin antigenic variation in *Neisseria gonorrhoeae. Nature* (London) **336:**392–395.

transformed. Accordingly, the cells must be plated under conditions selective for the transformed cells. Therefore, the DNA used for the transformation should contain a selectable gene such as one encoding resistance to an antibiotic.

TRANSFORMATION BY PLASMIDS

In contrast to naturally competent cells, cells made permeable to DNA by calcium ion treatment will take up both single-stranded and double-stranded DNA. Therefore, both linear and double-stranded circular plasmid DNAs can be efficiently introduced into chemically treated cells. This fact has made calcium ion-induced competence very useful for cloning and other applications that require the introduction of plasmid and phage DNAs into cells.

TRANSFECTION

In addition to plasmid DNAs, viral DNAs or genomic RNAs can often be introduced into cells by transformation, thereby initiating a viral infection. This process is called **transfection** rather than transformation, although the principle is the same. To detect transfection, the potentially transfected cells are usually mixed with indicator bacteria and plated (see the introductory chapter). If the transfection is successful, a plaque will form where the transfected cells had produced phage, which then infected the indicator bacteria.

Some viral infections cannot be initiated merely by transfection with the viral DNA. These viruses cannot transfect cells, because in a natural infection, proteins in the viral head are normally injected along with the DNA, and these proteins are required to initiate the infection. For example, the *E. coli* phage N4 carries a phage-specific RNA polymerase in its head that is injected with the DNA and used to transcribe the early genes. Transfection with the purified phage DNA will not initiate an infection, because the early genes will not be transcribed without this RNA polymerase. Another example of a phage in which the infection cannot be initiated by the nucleic acid alone is the phage φ6 (see Box 7.1). This phage has RNA instead of DNA in the phage head and must inject an RNA replicase to initiate the infection; therefore, the cells cannot be infected by the RNA alone. Such examples of phages that inject required proteins are rather rare, and for most phages, the infection can be initiated by transfection.

As an aside, many animal viruses do inject proteins required for multiplication, and these proteins cannot be made after injection of the naked DNA or RNA. For example, a retrovirus such as human immunodeficiency virus (HIV), which causes AIDS, injects a reverse transcriptase required to make a DNA copy of the incoming RNA before it can be transcribed to make viral proteins. Therefore, human cells cannot be transfected with HIV RNA alone.

TRANSFORMATION OF CELLS WITH CHROMOSOMAL GENES

Transformation with linear DNA is one method used to replace endogenous genes with genes altered in vitro. However, most types of bacteria made competent by calcium ion treatment are transformed poorly by chromosomal DNA because the linear pieces of double-stranded DNA entering the cell are degraded by an enzyme called the RecBCD nuclease. This nuclease degrades DNA from the ends; therefore, it does not degrade circular plasmid and phage DNAs. However, inactivating the RecBCD nuclease by a mutation would preclude recombination between the incoming DNA and the chromosome, because the enzyme is required

for normal recombination in *E. coli* and other bacteria (see chapter 9).

Nevertheless, methods have been devised to transform competent *E. coli* with linear DNA. One way is to use a mutant *E. coli* lacking the D subunit of the RecBCD nuclease. These *recD* mutants are still capable of recombination, but because they lack the nuclease activity that degrades linear DNA, they can be transformed with linear double-stranded DNAs. We discuss such procedures further in chapter 16.

Electroporation

Another way in which DNA can be introduced into bacterial cells is by **electroporation**. In the electroporation process, the bacteria are mixed with DNA and briefly exposed to a strong electric field. The brief electric shock seems to open the cells up, and the DNA moves into them much as it moves along a gel during electrophoresis. Electroporation works with most types of cells, including most bacteria, unlike the methods mentioned above, which are very specific for certain species. Also, electroporation can be used to introduce linear chromosomal and circular plasmid DNAs into cells. However, electroporation requires specialized equipment.

SUMMARY

1. In transformation, DNA is taken up directly by cells. Transformation was the first form of genetic exchange to be discovered in bacteria, and the demonstration that DNA is the transforming principle was the first direct evidence that DNA is the hereditary material. The bacteria from which the DNA was taken are called the donors, and the bacteria to which the DNA has been added are called the recipients. Bacteria that have taken up DNA are called transformants.

2. Bacteria that are capable of taking up DNA are said to be competent.

3. Some types of bacteria can naturally take up DNA during part of their life cycle. Most types of naturally transformable bacteria usually become competent only during part of their cellular growth, when they have reached high densities and are entering the stationary phase. A number of genes whose products are required for competence have been identified.

4. The fate of the DNA during natural transformation is fairly well understood. The double-stranded DNA first binds to the cell surface and then is broken into smaller pieces by endonucleases. Then one strand of the DNA is degraded by an exonuclease. The single-stranded pieces of DNA then invade the chromosome in homologous regions,

displacing one strand of the chromosome at these sites. Through repair or subsequent replication, the sequence of the incoming DNA may replace the original chromosomal sequence in these regions.

5. Naturally competent cells can be transformed with linear chromosomal DNA but usually not with circular plasmid or circular phage DNAs. Transformation by plasmid DNA usually occurs with dimers or higher multimers of the plasmid that can recyclize by recombination between the repeated sequences at the ends.

6. Some types of bacteria, including *Haemophilus influenzae* and *Neisseria gonorrhoeae*, will take up DNA of only the same species. Their DNA contains short uptake sequences that are required for uptake of DNA into the cells. Other types of bacteria, including *Bacillus subtilis* and *Streptococcus pneumoniae*, seem to be capable of taking up any DNA.

7. There are three possible roles for natural competence: a nutritional function allowing competent cells to use DNA as a carbon, energy, and nitrogen source; a repair function in which cells use DNA from neighboring bacteria to repair damage to their own chromosomes, thus ensuring survival

(continued)

SUMMARY (continued)

of the species; and a recombination function in which bacteria exchange genetic material among members of their species, increasing diversity and accelerating evolution.

8. Some types of bacteria that do not show natural competence can nevertheless be transformed after some types of chemical treatment or by electroporation. The standard method for making *E. coli* permeable to DNA involves treatment with calcium ions. Cells made competent by calcium treatment can be transformed with plasmid and phage DNAs, making this method one of the cornerstones of molecular genetics.

9. If the cell is transformed with viral DNA to initiate an infection, the process is called transfection.

QUESTIONS FOR THOUGHT

1. Why do you think some types of bacteria are capable of developing competence? What is the real function of competence?

2. How would you determine if the competence genes of *Bacillus subtilis* are turned on by UV irradiation and other types of DNA damage?

3. How would you determine whether antigenic variation in *Neisseria gonorrhoeae* is due to transformation between bacteria or to recombination within the same bacterium?

PROBLEMS

1. How would you determine if a type of bacterium you have isolated is naturally competent? Outline the steps you would use.

2. Outline how you would isolate mutants of your bacterium that are defective in transformation. Distinguish those that are defective in recombination from those that are defective in the uptake of DNA.

3. How would you determine if a naturally transformable bacterium can take up DNA of only its own species or can take up any DNA?

4. How would you determine if a piece of DNA contains the uptake sequence for that species?

5. How would you determine if the DNA of a phage can be used to transfect *E. coli*?

SUGGESTED READING

Avery, O. T., C. M. MacLeod, and M. McCarty. 1944. Studies on the chemical nature of the substance inducing transformation of pneumococcal types. Induction of transformation by a desoxyribonucleic acid fraction isolated from pneumococcus type III. *J. Exp. Med.* 79:137–159.

Cohen, S. N., A. C. Y. Chang, and L. Hsu. 1972. Nonchromosomal antibiotic resistance in bacteria: genetic transformation of *Escherichia coli* by R-factor DNA. *Proc. Natl. Acad. Sci.* 69:2110

Dubnau, D. 1991. Genetic competence in *Bacillus subtilis*. *Microbiol. Rev.* 55:395–424.

Haijema, B. J., D. van Sinderen, K. Winterling, J. Kooistra, G. Venema, and L. W. Hamoen. 1996. Regulated expression of the *dinR* and *recA* genes during competence development and SOS induction in *Bacillus subtilis*. *Mol. Microbiol.* 22:75–86.

Hanahan, D., and F. R. Bloom. 1996. Mechanisms of DNA transformation, p. 2449–2459. *In* F. C. Neidhardt, R. Curtiss III, J. L. Ingraham, E. C. C. Lin, K. B. Low, B. Magasanik, W. S. Reznikoff, M. Riley, M. Schaechter, and H. E. Umbarger (ed.) *Escherichia coli and Salmonella: Cellular and Molecular Biology*, 2nd ed. ASM Press, Washington, D.C.

Mongold, J. A. 1992. DNA repair and the evolution of competence in *Haemophilus influenzae*. *Genetics* 132:893–898.

Pierce, B. J., A. M. Naughton, E. A. Campbell, and H. R. Masure. 1995. The *rec* locus, a competence-induced operon in *Streptococcus pneumoniae*. *J. Bacteriol.* 177:86–93.

Puyet, A., B. Greenberg, and S. A. Lacks. 1990. Genetic and structural characterization of *endA*, a membrane-bound nuclease required for transformation of *Streptococcus pneumoniae*. *J. Mol. Biol.* 213:727–738.

Raymond-Denise, A., and N. Guillen. 1992. Expression of the *Bacillus subtilis dinR* and *recA* genes after DNA damage and during competence. *J. Bacteriol.* 174:3171–3176.

Redfield, R. J. 1993. Genes for breakfast: the have your cake and eat it too of bacterial transformation. *J. Hered.* 84:400–404.

Solomon, J. M., R. Magnuson, A. Srivastavan, and A. D. Grossman. 1995. Convergent sensing pathways mediate response to two extracellular competence factors in *Bacillus subtilis*. *Genes Dev.* 9:547–558.

Bacteriophages

P ROBABLY ALL ORGANISMS on earth are parasitized by viruses, and bacteria are no exception. For purely historical reasons, viruses that infect bacteria are usually not called viruses but are called **bacteriophages** (**phages** for short), even though they have lifestyles similar to those of plant and animal viruses. As mentioned in the introduction, the name *phage* derives from the Greek verb "to eat," and it describes the eaten-out places, or **plaques**, that are formed on bacterial lawns. The plural of phage is phage, but we add an s (phages) when we are discussing more than one type of phage.

Like all viruses, phages are so small that they can be seen only under the electron microscope. As shown in Figure 7.1A, phages are often spectacular in appearance, with **capsids**, or isocahedral heads, and elaborate tail structures that make them resemble interplanetary landing modules. The tail structures allow them to penetrate bacterial membranes and cell walls to inject their DNA into the cell. Animal and plant viruses have much simpler shapes because they do not need such elaborate tail structures. They are either engulfed by the cell, in the case of animal viruses, or enter through wounds, in the case of plant viruses.

Phages differ greatly in their complexity. Smaller phages, such as MS2, usually have no tail, and their heads may consist of as few as two different types of proteins. The heads and tails of some of the larger phages such as T4 have up to 20 different proteins, each of which can exist in as few as 1 copy to as many as 1,000 copies depending on the structural role they play in the phage particle. Phages also infect specific bacterial hosts, and different phages have very different host ranges, as discussed below.

Like all viruses, phages are not live organisms but merely a nucleic acid—either DNA or RNA depending on the type of phage—wrapped in a

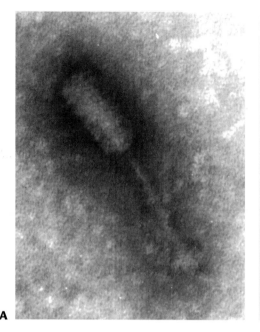

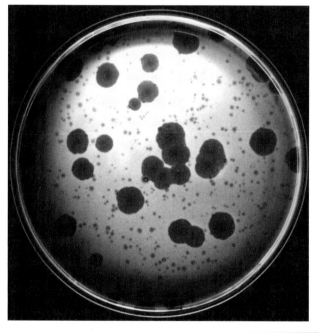

Figure 7.1 Electron micrographs and plaques of some bacteriophages. (A) Left panel, a phage of *Enterococcus* sp. (Electron micrograph by Sally Burns, Michigan State University.) Right panel, electron micrograph of phages T4 and lambda (left and right, respectively). (Provided by Arthur Zachary and Lindsay Black.) (B) Plaques of *E. coli* phages M13 (smaller plaques) and T3, a relative of phage T7 (larger plaques). (Photograph by Kurt Stepnitz, Michigan State University.)

protein and/or membrane coat for protection. This nucleic acid carries genes that direct the synthesis of more phage. In phages, either type of nucleic acid carried in the head is called the **phage genome**. These molecules can be very long, because the genome must be long enough to have at least one copy of each of the phage genes. The length of the DNA or RNA genome therefore reflects the size and complexity of the phage. For instance, the small phage MS2 has only four genes and a rather small RNA genome, whereas phage T4 has more

than 200 genes and a DNA genome that is almost 10 μm long. Long genomes, which can be as much as 1,000 times longer than the head, must be very tightly packed into the head of the phage.

Because phage are so small, they are usually only detected by the "holes" or plaques (Figure 7.1B) they form on **lawns** of susceptible host bacteria (see Introduction). Each type of phage will make plaques on only certain host bacteria, which define its **host range**. Mutations in the phage DNA can alter the host range of a phage or the conditions under which the phage can form a plaque, which is usually how mutations are detected. In this chapter, we discuss what is known about how some representative phages multiply and some of the genetic experiments that have contributed to this knowledge. First, however, we review some general features of phage development.

Bacteriophage Lytic Development Cycle

Because phages, like all viruses, are essentially genes wrapped in a protein or membrane coat, they cannot multiply without benefit of a host cell. The virus injects its genes into a cell, and the cell furnishes some or all of the means to express those genes and make more viruses.

Figure 7.2 illustrates the multiplication process for a typical large DNA phage. To start the infection, a phage adsorbs to an actively growing bacterial cell by binding to a specific receptor on the cell surface. In the next step, the phage injects its entire DNA into the cell, where transcription of RNA, usually by the host RNA polymerase, begins almost immediately. However, not all the genes of a phage are transcribed into mRNA when the

DNA first enters the cell. Only some of the genes of the phage have promoters that mimic those of the host cell DNA and so are recognized by the host RNA polymerase. Those transcribed soon after infection are called the **early genes** of the phage and encode mostly enzymes involved in DNA synthesis such as DNA polymerase, primase, DNA ligase, and helicase. With the help of these enzymes, the phage DNA begins to replicate and many copies accumulate in the cell.

Next, mRNA is transcribed from the rest of the phage genes, the **late genes**, which may or may not be intermingled with the early genes in the phage DNA, depending on the phage. These genes have promoters that are unlike those of the host cell and so are not recognized by the host RNA polymerase alone. Most of these genes encode proteins involved in assembly of the head and tail. After the phage particle is completed, the DNA is taken up by the heads. Finally, the cells break open, or **lyse**, and the new phage are released to infect another sensitive cell. This whole process, known as the **lytic cycle**, takes less than 1 h for many phages, and hundreds of progeny phage can be produced from a single infecting phage.

Actual phage development is usually much more complex than this basic process, proceeding through

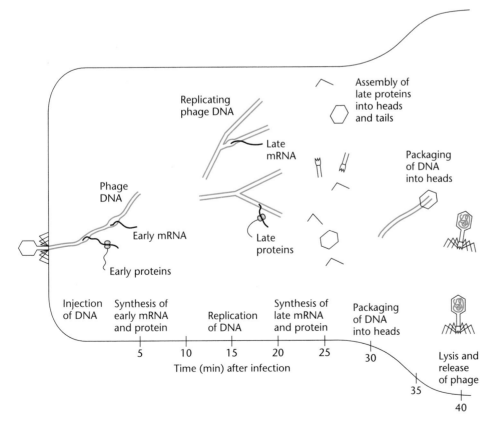

Figure 7.2 A typical bacteriophage multiplication cycle. After the phage injects its DNA, the early genes, most of which encode products involved in DNA replication, are transcribed and translated. Then DNA replication begins, and the late genes are transcribed and translated to form the head and tail of the phage. The DNA is packaged into the heads; the tails are attached; and the cells lyse, releasing the phage to infect other cells.

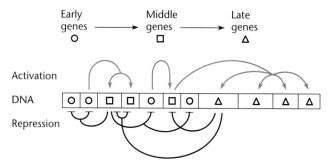

Figure 7.3 Transcriptional regulation during development of a typical large DNA phage. The arrows indicate activation of gene expression; the bars indicate repression of gene expression.

several intermediate stages in which the expression of different genes is regulated by specific mechanisms. Most of the regulation is achieved by having genes be transcribed into mRNA only at certain times; this type of regulation is called **transcriptional regulation** (see chapter 11). However, some genes undergo **posttranscriptional regulation**, which occurs after the mRNA has been made. For example, regulation may operate at the level of whether the mRNAs are translated; this is known as **translational regulation**. Other types of posttranscriptional regulation involve the stability of certain RNAs that quickly degrade unless they are synthesized at the right stage of development.

Figure 7.3 shows the basic process of phage gene transcriptional regulation, in which one or more of the gene products synthesized during each stage of development turns on the transcription of the genes in the next stage of development. The gene products synthesized during each stage can also be responsible for turning off the transcription of genes expressed in the preceding stage. Genes whose products are responsible for regulating the transcription of other genes are called **regulatory genes**, and this type of regulation is called a **regulatory cascade** because each step triggers the next step and stops the preceding step. By having such a cascade of gene expression, all the information for the step-by-step development of the phage can be preprogrammed into the DNA of the phage.

Regulatory genes can usually be easily identified by mutations. Mutations in most genes affect only the product of the mutated gene. However, mutations in regulatory genes can affect the expression of many other genes. This fact has been used to identify the regulatory genes of many phages. Below, we discuss some of these genes and their functions, selecting our examples either because they are the basis for cloning technologies or because of the impact they have had on our understanding of regulatory mechanisms in general. All the phages we present in this chapter contain DNA; however, Box 7.1 briefly describes the properties of some RNA phages.

Phage T7: a Phage-Encoded RNA Polymerase

Compared with some of the larger phages, phage T7 has a relatively simple program of gene expression after infection, with only two major classes of genes, the early and late genes. The phage has about 50 genes, many of which are shown on the genome map in Figure 7.4. After infection, expression of the T7 genes proceeds from left to right, with the genes on the extreme left of the genetic map, including gene *1.3*, expressed first. These are the early genes. The genes to the right of *1.3*—the middle DNA metabolism and late phage assembly genes—are transcribed after a few minutes' delay.

Nonsense and temperature-sensitive mutations were used to identify which of the early-gene products is responsible for turning on the late genes. Under nonpermissive conditions, in which the mutated genes were inoperable, amber and temperature-sensitive mutations in gene *1* prevented transcription of the late genes, and so gene *1* was a candidate for the regulatory gene. Later work showed that the product of gene *1* is an RNA polymerase that recognizes the promoters used to transcribe the late genes. The sequence of these promoters differs greatly from those recognized by bacterial RNA polymerases, so that these phage promotors will be recognized only by this T7-specific RNA polymerase.

The specificity of the RNA polymerases of T7 and other related phages for their own promoters has been capitalized upon to make expression vectors (see Studier et al., Suggested Reading, and chapter 15). Because the phage RNA polymerases are very active, these expres-

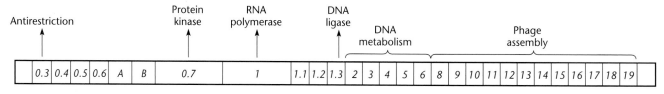

Figure 7.4 Genetic map of phage T7.

BOX 7.1

RNA Phages

The capsids (i.e., heads) of many viruses contain RNA instead of DNA. Some of these viruses, the so-called retroviruses, use enzymes called reverse transcriptases to transcribe the RNA into DNA, and these enzymes, because they are essentially DNA polymerases, need primers. In contrast, other RNA viruses, such as the influenza viruses, which cause flu, and the reoviruses, which cause colds, replicate their RNA by using RNA replicases and need no DNA intermediate. Because these RNA replicases have no need for primers, the genomes of RNA viruses can be linear without repeated ends. As we might expect, RNA viruses seem to have higher spontaneous mutation rates during replication, probably because their RNA replicases have no editing functions.

Some phages also have RNA as their genome. Examples include Qβ, MS2, R17, f2, and φ6. The *E. coli* RNA phages Qβ, MS2, R17, and f2 are similar to each other. All have a single-stranded RNA genome that encodes only four proteins: a replicase, two head proteins, and a lysin. Immediately after the RNA enters the cell, it serves as an mRNA and is translated into the replicase. This enzyme replicates the RNA, first by making complementary minus strands and then by using these as a template to synthesize more plus strands. The phage genomic RNA must serve as an mRNA to synthesize the replicase, because no such enzyme exists in *E. coli*. Interestingly, the phage Qβ replicase has four subunits, only one of which is encoded by the phage. The other three are components of the host translational machinery: two of the elongation factors for translation, EF-Tu and EF-Ts, and a ribosomal protein, S1. Because the genomes of these RNA phages also function as an mRNA, they have

served as a convenient source of a single species of mRNA in studies of translation.

Another RNA phage, φ6, was isolated from the bean pathogen *Pseudomonas syringae* subsp. *phaseolicola*. The RNA genome of this phage is double stranded and exists in three segments in the phage capsid, much like the reoviruses of mammals. Also like animal viruses, this phage is surrounded by membrane material derived from the host cell, i.e., an envelope, and the phage enters its host cells in much the same way that animal viruses enter their hosts. However, unlike most animal viruses, φ6 is released by lysis.

The replication, transcription, and translation of the double-stranded RNA of a virus such as φ6 present special problems. Not only must the phage replicate its double-stranded RNA, but it must also transcribe it into single-stranded mRNA since double-stranded RNA cannot be translated. The uninfected host cell will contain neither of the enzymes required for these functions, which therefore must be virus encoded and packaged into the phage head so that they enter the cell with the RNA. Otherwise, neither the transcriptase nor the replicase could be made. Another interesting question with this phage is how three separate RNAs are encapsidated in the phage head. This question has not yet been answered and is common to all viruses with segmented genomes.

References

Blumenthal, T., and G. G. Carmichael. 1979. RNA replication: function and structure of Qβ replicase. *Annu. Rev. Biochem.* **48:**525–548.

Ewen, M. E., and H. R. Revel. 1990. RNA-protein complexes responsible for replication and transcription of the double-stranded RNA bacteriophage φ6. *Virology* **178:**509–519.

sion vectors can be used to synthesize large amounts of the protein products of cloned genes in *Escherichia coli*. These vectors are also used to make specific RNA in vitro, which can be used as hybridization probes or as substrates for RNA-processing enzymes.

Phage T4: a New Sigma Factor and Replication-Coupled Transcription

Phage T4 is much larger than T7, with over 200 genes (the T4 genome map is shown in Figure 7.5), and the regulation of its gene expression is predictably more complex.

Figure 7.6 shows the time course of T4 protein synthesis after infection. Each band in the figure is the polypeptide product of a single phage T4 gene, and

some of the gene products are identified by the gene that encodes them (for details on how the bands were obtained, see the legend to Figure 7.6). For example, p37 is the *p*roduct of gene *37* (see the T4 map in Figure 7.5). Because of the way the polypeptides were labeled, the time at which a band first appears is the time at which that gene begins to be expressed, and the time a band disappears is the time that gene is shut off. Clearly, some genes of T4 are expressed immediately after infection. These are called the immediate-early genes. Other genes, called the delayed-early and middle genes, are expressed only a few minutes after infection. Later, expression begins of the true-late genes, so called to distinguish them from some of the delayed-early and middle genes that continue to be expressed throughout

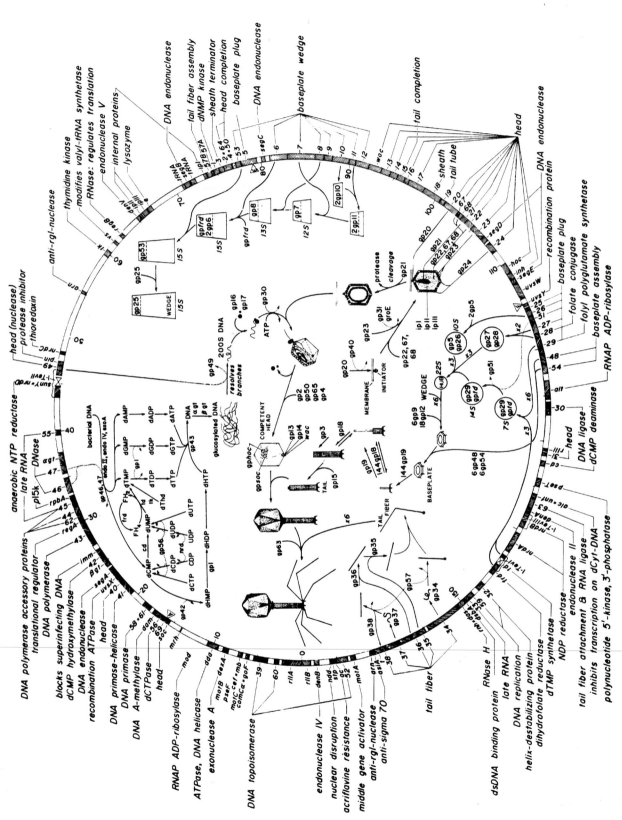

Figure 7.5 Genomic map of phage T4. Reprinted with permission from J. D. Karam (ed.), *Molecular Biology of Bacteriophage T4*, ASM Press, Washington, D.C., 1994.

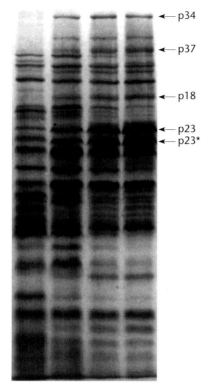

← p34

← p37

← p18

← p23
← p23*

Figure 7.6 Polyacrylamide gel electrophoresis of proteins synthesized during the development of phage T4. The proteins were labeled by adding amino acids containing the ^{14}C radioisotope of carbon at various times after the phage were added to the bacteria. Because phage T4 stops host protein synthesis after infection, radioactive amino acids will be incorporated only into phage proteins, and therefore only the phage proteins will become radioactive. Moreover, only those phage proteins made when the radioactive amino acids were added will be labeled. Hence, we can tell which phage proteins are made at any given time. To separate the proteins, the cells were broken open with the detergent sodium dodecyl sulfate, which also separates the polypeptides that are part of the multimeric proteins. Electrophoresis caused the polypeptides to migrate on an acrylamide gel, forming bands in columns. The smaller polypeptides move faster on these gels, so that the polypeptides are arranged by size from the smallest at the bottom to the largest at the top. The gel was subsequently dried on a piece of filter paper and then laid next to a photographic film, which was exposed to the high-energy light waves given off from the radioactive disintegrations. Each band represents the polypeptide product of a single T4 gene.

Lane 1 from the left, proteins synthesized between 5 and 10 min after infection; lane 2, 10 to 15 min; lane 3, 15 to 20 min; lane 4, 30 to 35 min.

infection. Overall, the regulation of protein synthesis during T4 phage development is very complex, as might be expected from such a large virus.

The assigments of the polypeptide products to genes were made by using amber mutations in the genes. If nonsuppressor cells (i.e., cells that lack a nonsense suppressor; see chapter 3) are infected by a phage with an amber mutation in a gene, the band corresponding to the product of that gene will be missing. Translation of the gene will stop at the amber mutation in a nonsuppressor host, leading to the synthesis of a shorter polypeptide, which can sometimes be detected elsewhere on the gel. Sometimes two or more bands are missing as the result of a single amber mutation, such as the two bands missing as a result of an amber mutation in gene *23*, identified as p23 and p23* in Figure 7.6. Several factors could account for the absence of multiple bands as a result of a single mutation. In this case, the polypeptide product of gene *23*, which makes up the phage head, is cleaved after it is synthesized. Normally, approximately 1,000 copies of this polypeptide are used to build every phage head. The head is first assembled with the p23 polypeptide, and then part of the N terminus of p23 is cut off to form the shorter polypeptide p23* as the head matures into its final form before DNA is encapsidated (i.e., put inside the head). Thus, by disrupting the synthesis of p23, the mutation also prevents the appearance of p23*. The T4 gene products are often referred to as gp23, etc., for *gene product of 23*.

Experiments like those described in the legend to Figure 7.6 were also used to identify the regulatory genes of phage T4. Mutations in these genes prevent the expression of many other genes and so cause the disppearance of many bands from the gel. Mutations in a gene named *mot* prevent the appearance of the middle gene products. Mutations in genes *33* and *55*, as well as mutations in many of the genes whose products are required for T4 DNA replication, prevent the appearance of the true-late gene products. Therefore, *mot*, *33*, and *55*, as well as some genes whose products are required for DNA replication, were predicted to be regulatory genes.

Many genetic and biochemical experiments have been directed toward understanding how these T4 regulatory gene products turn on the synthesis of other proteins. Here, we discuss only the function of genes whose products are required to turn on the true-late genes. In phage T4, as in T7, the true-late genes are transcribed from promoters that are different from those of its host (see Christensen and Young, Suggested Reading). As shown in Figure 7.7, the sequence TATAAATA, rather than the characteristic –35 and –10 sequences of a bacterial σ^{70} promoter (see chapter 2), resides in the –10 region of the T4 promoters. Because of this difference, the

−10   +1

TGAGAGTATAAATACTCCTGATACTGA

Figure 7.7 Sequence of a T4 late promoter. The underlined −10 sequence is recognized only by an *E. coli* RNA polymerase with the T4 gene *55*-encoded sigma factor attached.

host RNA polymerase will not normally recognize the T4 promoters. However, the product of the regulatory gene *55* is an alternate sigma factor that binds to the host RNA polymerase, changing its specificity so that it recognizes only the promoters for the T4 true-late genes (see Kassavetis and Geiduschek, Suggested Reading).

Phage T4 and its close relatives are not the only phages to use alternate sigma factors to activate the transcription of their late genes. For example, the *Bacillus subtilis* phage SPO1 also uses this regulatory mechanism. The SPO1 late promoters are very unlike the normal bacterial promoters, but they are also quite unlike the T4 late promoters. In fact, host RNA polymerases can be adapted to recognize a wide variety of promoter sequences merely through the attachment of an alternate sigma factor. This general strategy is also used during many bacterial developmental processes, as we discuss in later chapters.

REPLICATION-COUPLED TRANSCRIPTION
In addition to the alternate sigma factor encoded by gene *55*, many other gene products are required to turn on the transcription of the late genes. Most, notably those of genes *44*, *62*, and *45*, are required for replication of the phage DNA. This observation has led to the conclusion that T4 DNA replication is also required for the expression of the late genes and that the complex of the host RNA polymerase and the gene *55* protein will initiate RNA synthesis efficiently only if the T4 DNA is replicating. Coupling the transcription of the true-late genes to the replication of the phage DNA makes sense from a mechanistic standpoint. Many of the true-late genes encode parts of the phage particle. These proteins are not needed until phage DNA is available to be packaged inside them.

Figure 7.8 shows features of a model recently proposed to explain the requirement of T4 DNA replication for the activation of true-late gene transcription and related observations (see Herendeen et al., Suggested Reading). According to the model, the promoters for the true-late genes are activated when the gp33 protein, which is bound to RNA polymerase, makes contact with the gp45 protein, which is normally part of the replication apparatus. The gp45 protein is a DNA poly-

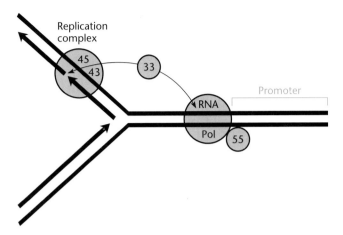

Figure 7.8 Model for replication activation of a T4 late-gene promoter in which the promoter is not activated until a replication fork approaches. The proteins are identified by the genes that encode them. The replication complex includes the DNA polymerase gp43 and the accessory protein gp45. gp33 binds both to the clamp formed by the gp45 on the DNA at the replication fork and to the RNA polymerase at the promoter, thereby activating transcription.

merase accessory protein that acts like the β protein in *E. coli* and wraps around the DNA to form a "sliding clamp," which moves with the DNA polymerase and helps prevent it from falling off the DNA during replication (see chapter 1). The gp44 and gp62 proteins help clamp the gp45 protein onto the DNA. According to the model for how the true-late promoters are activated, the gp45 protein sliding clamp can load on the DNA either at a nick or because it is associated with the replication apparatus. It can then slide along the DNA in either direction. If it makes contact with a gp33 protein bound to the RNA polymerase at a true-late promoter, it will allow the RNA polymerase to initiate transcription from the true-late promoter. This explains why the gp33 and gp45 proteins are required for optimal true-late-gene transcription, as well as gp44 and gp62, since the latter are required for gp45 protein to load on the DNA to form the sliding clamp.

Many other types of viruses, including the herpesviruses and adenoviruses of eukaryotes, seem to couple the transcription of some of their genes to the replication of their DNA. The replication-coupled transcription in T4 may share features in common with the regulatory systems of these other viruses.

Phage λ: Transcription Antitermination

Phage λ is fairly large, with a genome intermediate in size between those of T7 and T4 (Figure 7.9). Some gene products and sites encoded by λ are listed in Tables

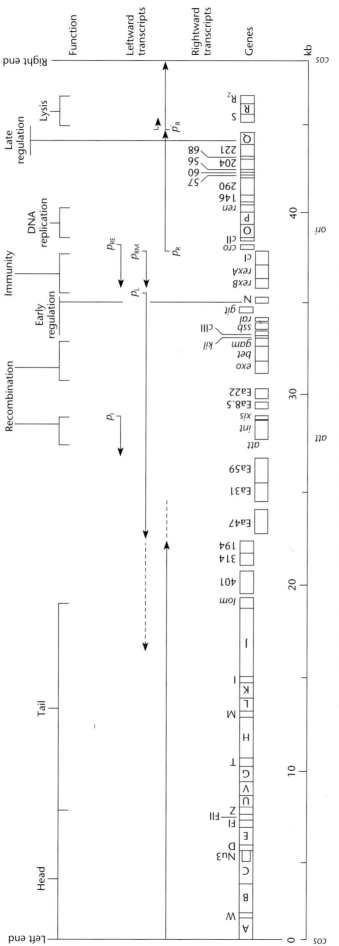

Figure 7.9 Genetic map of phage λ. Reprinted with permission from D. L. Daniels, J. L. Schroeder, W. Szybalski, F. Sanger, and F. R. Blattner, A molecular map of coliphage lambda, p. 473, *in* R. W. Hendrix, J. W. Roberts, F. W. Stahl, and R. A. Weisberg (ed.), *Lambda II*, Cold Spring Harbor Laboratory Press, Cold Spring Harbor, N.Y.

TABLE 7.1   Some λ gene products and their function

Gene product	Function
N	Antitermination protein acting at t_L^1, t_R^1, and t_R^2
O, P	Initiation of λ DNA replication
Q	Antitermination protein acting at t'_R
CI	Repressor; protein inhibitor of transcription from p_L and p_R
CII	Activator of transcription of cI and *int*
CIII	Stabilizer of CII
Cro	Protein inhibitor of CI synthesis
Gam	Protein required for rolling-circle replication
Red	Protein involved in λ recombination
Int	Integrase; protein required for site-specific recombination with chromosome
Xis	Excisionase; protein forms complex with Int and functions in excision of prophage

7.1 and 7.2, respectively. Phage λ goes through three major stages during development. The first λ genes to be expressed after infection are *N* and *cro*. Most of the genes expressed next play a role in replication and recombination. Finally, the late genes of the phage are expressed, encoding the head and tail proteins of the phage particle and enzymes involved in cell lysis.

The mechanisms used by phage λ and its relatives to regulate transcription differ from those used by T7 and

TABLE 7.2   Some sites involved in phage λ transcription and replication[a]

Sites	Function
p_L	Left promoter
p_R, p_R'	Right promoters
o_L	Operator for leftward transcription; binding sites for CI and Cro repressors
o_R	Operator for rightward transcription; binding sites for CI and Cro repressors
t_L^1, t_L^2	Termination sites of leftward transcription
t_R^1, t_R^2, t_R'	Termination sites of rightward transcription
nutL	N-utilization site for leftward transcribing RNA Pol (i.e., the site at which N binds to RNA Pol)
nutR	N-utilization site for rightward transcribing RNA Pol
p_{RE}	Promoter for repressor establishment; activated by CII
p_{RM}	Promoter for repressor maintenance; activated by CI
p_I	Promoter for *int* transcription; activated by CII
PoP'	Attachment site (attλ)
cos	Cohesive ends of λ genome (12-bp single-stranded ends in linear genome anneal to form circular genome after infection)

[a]In λ, essential genes have single-letter names while nonessential genes have more conventional three-letter names.

T4. One mechanism, discussed later in the chapter, uses a repressor to prevent transcription while λ remains in a lysogenic, or quiescent, state. Another mechanism, which was first discovered in phage λ, is called **antitermination** (Figure 7.10). Rather than regulate transcription through initiation at new promoters, as discussed for T7 and T4, antitermination in λ allows transcription to continue through termination sites once it is under way. When the λ DNA first enters the cell, transcription begins normally and terminates at transcription termination sites after two short RNAs are synthesized (Figure 7.10A). One of these RNAs encodes the Cro protein, which is an inhibitor of repressor synthesis, as discussed in the section on lysogeny (see below). However, the other encodes the N protein, the antitermination factor that permits the RNA polymerase to continue along the DNA, as shown in Figure 7.10B to D.

THE N PROTEIN

The molecular mechanism of antitermination by N protein has been the subject of many genetic and biochemical experiments. Figure 7.10C and D outlines the current picture for how N protein antiterminates, showing only rightward transcription. Initially, transcription initiated at the rightward p_R promoter terminates at the transcription terminator designated t_R^1. One of the sequences transcribed into RNA is *nutR* (for N utilization rightward). When the N protein appears, it will bind to the RNA polymerase after the *nutR* region on DNA has been transcribed (Figure 7.10D). The RNA polymerase with N protein riding on it does not stop for any termination site, transcribing past the t_R^1 terminator and any other terminator it encounters.

Similar events are occurring during leftward transcription. When the *nut* site on the other side, called *nutL* (for N utilization leftward), is transcribed, the N protein binds to the RNA polymerase and prevents any further termination, allowing transcription to continue into other genes, including *gam* and *red*.

It is not clear why N binds to the RNA polymerase only after the *nut* sequences are transcribed. Perhaps the N protein binds first to the *nut* sequence on RNA and only then to RNA polymerase, as shown in Figure 7.10D. We discuss the genetic experiments that led to the detection of the *nut* sequences in chapter 13.

In addition to the N gene product of the phage, N-mediated antitermination requires many host proteins called Nus proteins, for N utilization host (see Friedman et al., Suggested Reading). Nus proteins are involved in similar regulation in the uninfected bacterium.

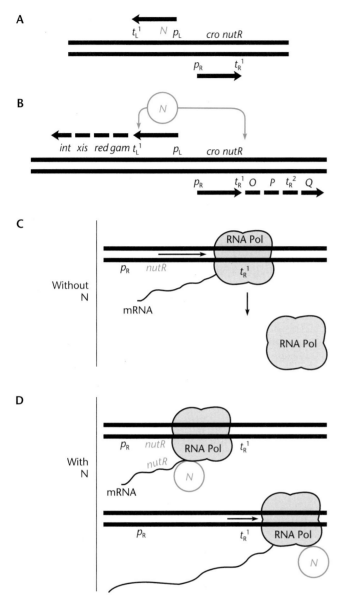

Figure 7.10 Antitermination of transcription in phage λ. (A) Before the N protein is synthesized, transcription starts at promoters p_L and p_R and stops at transcription terminators t_L^1 and t_R^1. (B) The N protein causes transcription to continue past t_L^1 and t_R^1 into *gam-red-xis-int* and *O-P-Q*, respectively. (C and D) Mechanism of antitermination by N, showing rightward transcription only. (C) In the absence of N, transcription initiated at p_R terminates at the terminator t_R^1. (D) If N has been made, it will bind to RNA polymerase as the polymerase transcribes the *nutR* site, possibly because N undergoes a conformational change when it binds to the *nut* sequence in the RNA. This change is required before N can bind to the RNA polymerase. With N bound, the RNA polymerase does not stop transcription at t_R^1. The sites and gene products shown here are defined in Tables 7.1 and 7.2.

THE Q PROTEIN

The *Q* gene product of λ is also an antiterminator, responsible for allowing transcription of the late genes. In the absence of Q protein, a very short RNA will be synthesized from the promoter p_R'. By the time the Q protein is made, the λ DNA ends have joined to form a circle linking the late head and tail genes to the lysis genes of the phage (see the section on phage λ replication below). The Q protein will antiterminate transcription and thereby allow the transcription initiated at p_R' to continue into the lysis genes and around the circle into the head and tail protein genes of the phage. As N must interact with the *nut* sequences, the Q protein must interact with a short sequence different from the terminator—the *qut* (Q utilization) site—to antiterminate transcription. It is not fully understood how the Q protein functions, but it could not work by the model shown above for N because the *qut* sequence is not entirely transcribed into RNA. The *qut* site bears little resemblance to *nut* sites and other known antitermination sequences.

ANTITERMINATION IN OTHER SYSTEMS

Like many regulatory systems, antitermination was first discovered in phage, but it is used in many other systems. This mechanism operates not only in related phages such as P22 but also for many bacterial genes. For example, antitermination regulates transcription of the rRNA genes of all bacteria, the *bgl* operon of *E. coli*, and the aminoacyl-tRNA synthetase genes of *B. subtilis* (see Box 11.3, Often, where regulation by antitermination of transcription occurs, sequences similar to λ *nut* sequences also are present (Figure 7.11). Antitermination may also be used to regulate eukaryotic genes. For example, the *myc* oncogene of mammals and the transcription of human immunodeficiency virus, which causes AIDS, may also be regulated through antitermination.

Phage DNA Replication

Unlike the replication of chromosomal or plasmid DNAs, which must be coordinated with cell division, phage DNA replication is governed by only one purpose: to make the greatest number of copies of the phage genome in the shortest possible time. Phage replication can be truly impressive. The one or a few phage DNA molecules that initially enter the cell can replicate to make hundreds or even thousands of copies to be packaged into phage heads in as little as 10 or 20 min. This unchecked replication often makes phage DNA replication easier to study than replication in other systems.

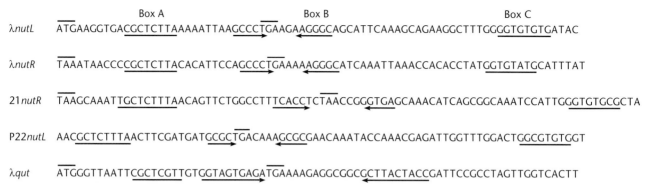

Figure 7.11 The sequences of some antitermination (*nut*) sites in bacteriophages and in bacterial rRNA genes. The sequences consist of similar box A, box B, and sometimes box C sequences. Adapted from D. I. Friedman and M. Gottesman, *in* R. W. Hendrix, J. W. Roberts, F. W. Stahl, and R. A. Weisberg (ed.), *Lambda II*, p. 29, Cold Spring Harbor Laboratory Press, Cold Spring Harbor, N.Y.

Nevertheless, phage replication has many of the same requirements as cellular replication in living organisms, and phage DNA replication has served as a model system to understand DNA replication in bacteria and other organisms.

The structures of many phage genomes present special problems for replication. Some phage genomes, for example, that of phage M13, are single-stranded circles. The DNA of most larger phages is not circular like that of M13 but, rather, is linear. As discussed in chapter 1, linearity presents problems for replication of the extreme 5′ ends of DNA, because DNA polymerases require a primer. Even if the primers for the 5′ ends are synthesized as RNA, once the RNA primer is removed there is no DNA upstream to serve as a primer for the primer's replacement as DNA. Because of this priming problem, a linear DNA would get smaller each time it replicated until essential genes were lost. Eukaryotic chromosomes have telomeres at their ends, but phages with linear genomes solve this primer problem in different ways. One is to use protein primers (see Box 7.2). Other phages have repeated sequences at the ends of their genomic DNA, as we discuss below. Phages use a surprising variety of mechanisms to solve their replication problems; some of these are described in this section.

Phage M13: Single-Stranded Circular DNA

The genome of some small phages consists of circular single-stranded DNA. The small *E. coli* phages that fit into this category can be separated into two groups. The representative phage of one group, φX174, has a spherical capsid with spikes sticking out, like the ball portion of the medieval weapon known as a morning star. In the other group, represented by M13 and f1, a single layer of protein covers the extended DNA molecule, making the phage filamentous in appearance.

M13 and other **filamentous phages** are **male-specific** phages because they specifically adsorb to the sex pilus encoded by certain plasmids and so will infect only "male" strains of bacteria (see chapter 5). Unlike most other phages, filamentous phages do not inject their DNA. Instead, the entire phage is ingested by the cell, and the protein coat is removed from the DNA as the phage passes through the inner cytoplasmic membrane of the bacterium. After the phage DNA has replicated, it is again coated with protein as it leaks back out through the cytoplasmic membrane. These phage do not lyse infected cells and leak out only slowly. Consequently, cells

---

BOX 7.2

Protein Priming

Some viruses, including the adenoviruses and the *Bacillus subtilis* phage φ29, have solved the primer problem by using proteins, rather than RNA, to prime their DNA replication. In the virus head, a protein is covalently attached to the 5′ end of the virus DNA. After infection, the DNA grows from this protein, with the first nucleotide attached to a specific serine on the protein. Thus, the virus DNA does not need to form circles or concatemers through pair bonding of repeated sequences at its ends.

Reference

Escarmis, D., D. Guirao, and M. Salas. 1989. Replication of recombinant φ29 DNA molecules in *Bacillus subtilis* protoplasts. *Virology* **169:**152–160.

infected with M13 or other filamentous phages are "chronically" rather than "acutely" infected. Nevertheless, the filamentous phages form visible plaques, because chronically infected cells grow more slowly than uninfected cells.

Figure 7.12 shows an overview of the replication of a typical single-stranded phage DNA. Immediately after the single-stranded phage DNA (usually called the **plus**, or **+, strand**) enters the cell, synthesis of the complementary strand (called the **minus**, or **–, strand**) begins, creating a double-stranded molecule called the **replicative form** (**RF**). The synthesis of the complementary minus strand does not require any phage-encoded gene products. In fact, because all of the phage mRNA is synthesized from the minus strand, none of its mRNA, and therefore none of its proteins, can be made without the minus strand. It must depend entirely on host enzymes. An RNA primer made at a specific site on the plus strand DNA by the host DnaG protein or RNA polymerase is required for minus-strand synthesis but is later removed by the exonuclease associated with the host DNA polymerase I and resynthesized as DNA by using upstream DNA as primer.

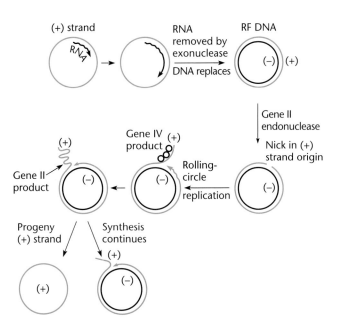

Figure 7.12 Replication of the circular single-stranded DNA phage M13. First, an RNA primer is used to synthesize the complementary minus strand (in black) to form double-stranded RF DNA. The product of gene II, an endonuclease, nicks the plus strand of the RF. Then more RFs are synthesized via rolling-circle replication. Later, the gene IV product binds to the plus strands as they are synthesized, preventing them from being used as templates for more RF synthesis and helping package them into phage heads.

ROLLING-CIRCLE REPLICATION

After the RF has formed, it replicates a few times to produce more double-stranded RF molecules by **rolling-circle replication**, which for phages is analogous to one form of plasmid replication and to the replication of DNA during conjugation (see chapters 4 and 5, respectively).

Figure 7.12 also illustrates rolling-circle replication of M13 RF DNA, in which a break, or nick, is made at a particular place in the plus strand and the free 3' end serves as a primer for DNA polymerase to make a new plus strand. As the new plus strand is synthesized, the old plus strand is peeled off and used as a template to synthesize another minus strand to form another RF. After a few RF molecules have accumulated, the plus strands displaced by rolling-circle replication are no longer used as templates. Instead, the phage coat proteins bind to the DNA at unique **packaging**, or *pac*, **sites** and coat it. Finally, these single-stranded DNA phage particles leak out of the cell.

GENETIC REQUIREMENTS FOR M13 DNA REPLICATION

The DNA of phage M13 encodes only two gene products required for the synthesis of more phage DNA plus strands, the products of genes II and IV. Gene II encodes the endonuclease that makes the nick in the plus strand required to initiate rolling-circle replication. The product of gene IV is a single-stranded DNA-binding protein that binds to and sequesters the single plus strands late in infection, so that these strands can be packaged into phage heads and will not be used as templates to synthesize more minus strands. All the other proteins required for phage DNA synthesis, including the DNA polymerase, DNA ligase, and helicases, are contributed by the host, albeit involuntarily. Apparently, the phage DNA mimics certain sequences of host DNA so that the host enzymes replicate it. As we discuss later, some of the larger phages do not depend so heavily on host functions, encoding many of their own replication functions.

M13 CLONING VECTORS

Because M13 and related phages have only one DNA strand, these phages provide a convenient vehicle for cloned DNA that we might want to sequence, use as a probe, or use in other applications. Also, because filamentous phages such as M13 have no fixed length and the phage particle will be as long as its DNA, foreign DNA of variable lengths can be cloned into the phage DNA, producing a molecule longer than normal without disrupting the phage's functionality of the phage.

Figure 7.13 shows the M13 cloning vector M13-mp18 (see Yanisch-Perron et al., Suggested Reading).

Like pUC plasmid vectors, the mp series of M13 phage vectors contain the α-fragment-coding portion of the *E. coli lacZ* gene, into which has been introduced some convenient restriction sites (shown as the polylinker cloning site in Figure 7.13, with the multiple restriction sites shown at the bottom of the figure). Phage with a foreign DNA insert in one of these sites can be identified easily by insertional inactivation (see chapter 4), because they will make colorless instead of blue plaques on plates containing X-Gal (5-bromo-4-chloro-3-in-dolyl-β-D-galactopyranoside).

To use a single-stranded DNA phage vector, the double-stranded RF must be isolated from infected cells, since most restriction endonucleases and DNA ligase require double-stranded DNA. A piece of foreign DNA is cloned into the RF by using restriction endonucleases, and the recombinant DNA is used to **transfect** competent bacterial cells. The term "transfection" refers to the artificial initiation of a viral infection by viral DNA (see chapter 6). When the RF containing the clone replicates

to form single-stranded progeny DNA, a single strand of the cloned DNA will be packaged into the phage head. The phage plaques obtained when these phage are plated are a convenient source of one strand of the cloned DNA.

Some plasmid cloning vectors have also been engineered to contain the *pac* site of a single-stranded DNA phage. If cells containing such a plasmid are infected with the phage, the plasmid will be packaged into the phage head in a single-stranded form.

Phage λ: Linear DNA That Replicates as a Circle

The λ DNA is linear in the phage head but cyclizes, that is, forms a circular molecule, after it enters the cell through pairing between its cohesive ends, or *cos* sites. These sites are single stranded and complementary to each other for 12 bases and so can join by complementary base pairing. Once the cohesive ends are paired, DNA ligase can join the two ends to form covalently

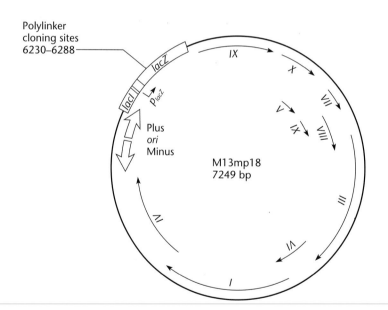

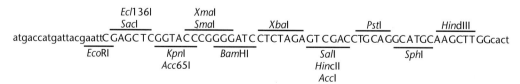

Figure 7.13 Map of the M13mp18 cloning vector. The positions of the genes of M13 and the polycloning site containing multiple restriction sites are shown below the map. Cloning into one of these sites will inactivate the portion of the *lacZ* gene on the cloning vector, a process called insertional inactivation. The cloning vector also contains the *lacI* gene, whose product represses transcription from the p_{LAC} promoter (see chapter 11).

closed circular λ DNA molecules. Circular DNA molecules can replicate because there will always be DNA upstream to serve as a primer (see chapter 1).

CIRCLE-TO-CIRCLE OR θ REPLICATION OF λ DNA

Once circular λ DNA molecules have formed in the cell, they can replicate by a mechanism similar to the θ replication described for the chromosome in chapter 1 and for plasmids in chapter 4. Replication initiates at the *ori* site in gene O (Figure 7.9) and proceeds in both directions, with both leading- and lagging-strand synthesis in the replication fork (Figure 7.14). When the two replication forks meet somewhere on the other side of the circle, the two daughter molecules separate.

ROLLING-CIRCLE REPLICATION OF λ DNA

After a few circular λ DNA molecules have accumulated in the cell by θ replication, the rolling-circle type of replication ensues. The initiation of λ rolling-circle replication is similar to that of M13 in that one strand of the circular DNA is cut and the free 3' end serves as a primer to initiate the synthesis of a new strand of DNA that displaces the old strand. DNA complementary to the displaced strand is also synthesized to make a new double-stranded DNA. The λ process differs in that the displaced individual single-stranded molecules are not released when replication around the circle is completed. Rather, the circle keeps rolling, giving rise to long tandem repeats of individual λ DNA molecules called con-

catemers (Figure 7.14). The concatemers are like the picture obtained by rolling an engraved ring, dipped in ink, across a piece of paper. The pattern on the ring will be repeated over and over again on the paper.

In the final step, the long concatemers are cut at the *cos* sites into λ-genome length pieces to be packaged into phage heads. Phage λ will package only DNA from concatemers, and at least two λ genomes must be linked end to end in a concatemer, because the packaging system in the λ head recognizes one *cos* site on the concatemeric DNA and takes up DNA until it arrives at the next *cos* site, which it cleaves to complete the packaging.

GENETIC REQUIREMENTS FOR λ DNA REPLICATION

The products of only two λ genes, O and P, are required for λ DNA replication. As Figure 7.14 illustrates, both of these proteins are required for priming DNA replication at the *ori* site. The O protein is thought to bend the DNA at this site by binding to repeated sequences, similar to the mechanism by which DnaA protein initiates chromosome replication (see chapter 1), and the P protein binds to the O protein and to components of the host replication machinery, thus commandeering them for λ DNA replication.

RNA synthesis must also occur in the *ori* region for λ replication to initiate. It normally initiates at the p_R promoter and may be required to separate the DNA strands at the origin or may serve as a primer for rightward replication.

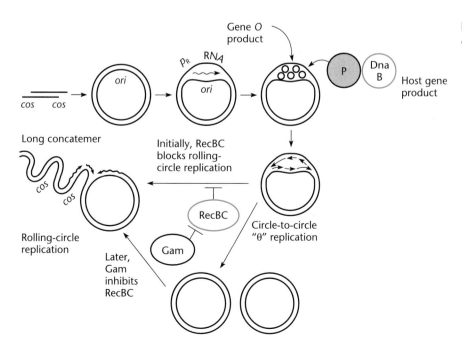

Figure 7.14 Overview of replication of phage λ. See the text for details.

A third λ protein, the product of the *gam* gene, is required for the shift to the rolling-circle type of replication, albeit indirectly. The RecBCD nuclease (an enzyme that facilitates recombination) of *E. coli* somehow inhibits the switch to rolling-circle replication, but Gam inhibits the RecBCD nuclease. Therefore, a *gam* mutant of λ is restricted to the θ mode of replication, and concatemers can form from individual circular λ DNA molecules only by recombination. Because λ requires concatemers for packaging, a *gam* mutant of λ will not multiply in the absence of a functional recombination system, either its own or the RecBCD pathway of its host. This fact led to the detection of *chi* sequences, which are important in RecBCD recombination and will be discussed further in chapter 9.

PHAGE λ CLONING VECTORS
The many cloning vectors derived from phage λ offer numerous advantages. The phage multiply to a high copy number, allowing for the synthesis of large amounts of DNA and protein. Also, it matters less if the cloned gene encodes a toxic protein than with plasmid cloning vectors. The toxic protein will not be synthesized until the phage infects the cell, and the infected cell is destined to die anyway. It is also relatively easy to store libraries in the relatively stable λ phage head.

COSMIDS
As mentioned, λ packages DNA into its head by recognizing *cos* sites in concatemeric DNA; therefore, any DNA containing a *cos* sequence will be packaged into phage heads. In particular, plasmids containing *cos* sites can be packaged into λ phage heads. Such plasmid cloning vectors, called **cosmids**, also offer many advantages for genetic engineering, including **in vitro packaging**. In this procedure, plasmid DNA is mixed with extracts of λ-infected cells containing heads and tails of the phage. The DNA will be taken up by the heads, and because λ particles will self-assemble in the test tube, the tails will be attached to the heads to make infectious λ particles, which can then be introduced into cells. Any λ cloning vector will serve in this method, and infection is a more efficient way of introducing DNA into bacteria than is transfection or transformation.

Another major advantage of cosmids is that the size of the cloned DNA is limited by the size of the phage head. If the piece of DNA cloned into a cosmid is too large, the *cos* sites will be too far apart and the DNA will be too long to fit into a phage head. However, if the cloned DNA is too small, the *cos* sites will be too close to each other and the phage heads will have too little DNA and be unstable. Therefore, the use of cosmids en-

sures that the pieces of DNA cloned into a vector will all be approximately the same size, which is sometimes important for making libraries (see chapter 15).

LYSOGENIC VECTORS
Phage λ sometimes infects cells and then becomes quiescent (see the section on lysogeny, below), and some λ cloning vectors retain this ability. If a gene is cloned into λ and the phage is then allowed to lysogenize a host, the cloned gene will exist in only one copy in the chromosome. This can be useful for complementation studies, which are often complicated if a gene is present in too many copies (see chapter 14).

Phage T7: Linear DNA That Never Cyclizes

Phage T7 has linear DNA in its head, but this DNA does not cyclize after infection. Therefore, T7 must use another mechanism to solve its primer problem. It also forms concatemers, but by pairing between the complementary ends of its DNA.

As shown in Figure 7.15, the replication begins at a unique *ori* site and proceeds toward both ends of the molecule, leaving the extreme 3' ends single stranded. Because T7 has the same sequence at both ends, these single strands are complementary to each other and so can pair, forming a concatemer with the genomes linked end to end. Consequently, the information missing as a result of incomplete replication of the 3' ends is provided by the complete information at the 5' end of the other daughter DNA molecule. Individual molecules are then cut out of the concatemers at the unique *pac* sites at the ends of the T7 DNA and packaged into phage heads.

GENETIC REQUIREMENTS FOR T7 DNA REPLICATION
In contrast to λ, which encodes only two of its own replication proteins and otherwise depends on the host replication machinery, T7 encodes many of its own replication functions, including DNA polymerase, DNA ligase, DNA helicase, and primase (Figure 7.4). The phage T7 RNA polymerase is also required to synthesize the initial primer for phage T7 DNA synthesis. In addition to these proteins, the phage encodes a DNA endonuclease and exonuclease that degrade host DNA to mononucleotides, thereby providing a source of deoxynucleotides for phage DNA replication. Analogous host enzymes can substitute for some of these T7-encoded gene products, so they are not absolutely required for T7 DNA replication. For example, the T7 DNA ligase is not required because the host ligase can act in its stead. T7 DNA replication is a remarkably simple

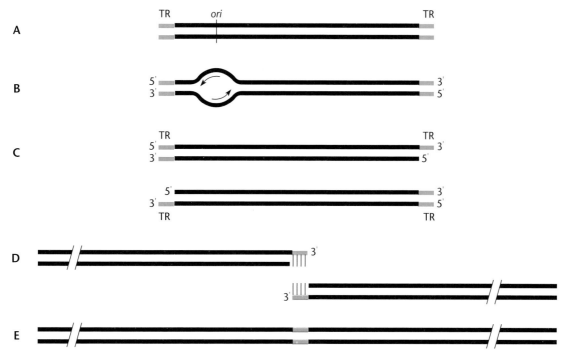

A

B

C

D

E

Figure 7.15 Replication of phage T7 DNA. Replication is initiated bidirectionally at the origin (*ori*). The replicated DNAs pair at their terminally repeated ends (TR) to give long concatemers.

process that requires fewer gene products overall than the replication of bacterial chromosomes and of many other large DNA phages.

Phage T4: Another Linear DNA That Never Cyclizes

Phage T4 also has linear DNA in its head that never cyclizes. It replicates much like T7, except that it forms concatemers by recombination between repeated sequences at its ends rather than by pairing between complementary single-stranded ends. However, T4 and T7 differ greatly in how the DNA replicates and is packaged.

REPLICATION FROM DEFINED ORIGINS

The replication of T4 DNA also proceeds through two stages (Figure 7.16). In the first stage, T4 replication initiates at unique origins much like T7 and λ replication. However, befitting its larger size, T4 has a number of such origins scattered around the T4 DNA. Initiation requires RNA primers synthesized by the host RNA polymerase as well as a T4-encoded DNA topoisomerase (see chapter 1), perhaps to unwind the DNA. Daughter molecules synthesized in this way recombine at their repeated ends to form large concatemers so that no sequences will be lost at the ends.

REPLICATION FROM RECOMBINATION INTERMEDIATES

Later in T4 infection, replication no longer initiates from defined sites (see Mosig et al., Suggested Reading). Instead, recombination intermediates furnish the free 3′ ends of DNA required to prime DNA synthesis, as

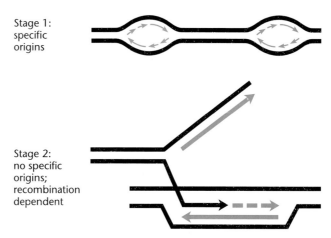

Stage 1: specific origins

Stage 2: no specific origins; recombination dependent

Figure 7.16 Replication of phage T4 DNA. In stage 1, replication initiates at specific origins, using RNA primers. In stage 2, recombinational intermediates furnish the primers for initiation.

shown in Figure 7.16. It is not clear what triggers the switch to the second stage of replication; perhaps it is the appearance of T4 recombination proteins. These proteins are discussed further in chapter 9.

GENETIC REQUIREMENTS OF T4 DNA
REPLICATION
Befitting its large size, as many as 30 T4 gene products participate in replication (Figure 7.5), including DNA polymerase, DNA ligase, primase, helicase, a single-stranded-DNA-binding protein, and other accessory proteins, including the aforementioned accessory proteins p45, p44, and p62, which may help to "clamp" the DNA polymerase onto the DNA so that it does not fall off during replication. The replication apparatus of phage T4 is remarkably like that of eukaryotes. Even the T4 DNA topoisomerase (see above) is more similar to the eukaryotic enzyme than to bacterial topoisomerases.

Because T4 replication proceeds through two different stages, mutations in genes whose products participate in the different stages have differing effects on T4 DNA replication. For example, amber mutations in the DNA topoisomerase genes, *39*, *52*, and *60*, or in the primase genes, *41* and *61,* all of which are required for the early stage of replication, will delay replication until it can be initiated from recombination intermediates. In contrast, amber mutations in such genes as *46*, *47*, and *49*, which encode the recombination functions of T4, will not disrupt the synthesis of DNA in stage 1, but replication will be arrested at stage 2 and concatemers will not accumulate.

HEADFUL PACKAGING
Like λ and T7 DNA, T4 DNA is packaged into phage heads from long concatemers. However, T4 does not have unique *pac* or *cos* sites where the long molecules can be cut. Instead, the concatemeric DNA enters each T4 head until the head is filled with DNA. Then the DNA is broken off and another "headful" of DNA is packaged (Figure 7.17). Because the length of T4 DNA

packaged into phage heads is not determined by the distance between *cos* or *pac* sites but, rather, by the amount of DNA a head will hold, this packaging mechanism is called **headful packaging**.

As mentioned above, formation of concatemers after infection requires recombination between the two ends of the phage DNA, which must be identical. This is accomplished by having the length of DNA that the heads will hold be somewhat longer than that needed to encode all the genes, as shown in Figure 7.17. In the illustration, the phage has genes A to Z, which are repeated over and over again in the concatemer. When the DNA is packaged into the head, the cut is made after the repetition of genes ABC, so that this DNA will have the genes ABC repeated at each of its ends. The next DNA will have genes DEF repeated at its ends, and so on. In other words, the DNA molecules in the T4 phage heads are cyclic permutations of each other. Recombination between these cyclically permuted DNA molecules gives rise to a circular linkage map, even though the DNA molecules themselves are never circular. Factors that influence genetic linkage maps are discussed in detail in chapter 13.

Generalized Transduction

Bacteriophages not only infect and kill cells but also sometimes transfer bacterial DNA from one cell to another in a process called **transduction**. There are two types of transduction in bacteria: **generalized transduction**, in which essentially any region of the bacterial DNA can be transferred from one bacterium to another, and **specialized transduction**, in which only certain genes close to the attachment site of a lysogenic phage in the chromosome can be transferred. These two types of transduction have fundamentally different mechanisms. In this section, we discuss only generalized transduction; coverage of specialized transduction is deferred until it can be addressed in the context of lysogeny.

Figure 7.18 illustrates the process of generalized transduction. While phages are packaging their own DNA, they sometimes mistakenly package the DNA of the bacterial host instead. These phages are still capable of infecting other cells, but progeny phage will not be produced. If the bacterial DNA is a piece of the bacterial chromosome, it may recombine with the host chromosome if they have sequences in common. If the injected DNA is a plasmid, it may replicate after it enters the cell and thus be maintained. If the incoming DNA contains a transposon, the transposon may hop, or insert itself, into a host plasmid or chromosome (see chapter 8).

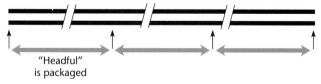

ABCDEFGHI XYZABCDEFGHI XYZABCDEFGHI XYZABCDEFGHI

"Headful"
is packaged

Figure 7.17 Headful packaging by phage T4. Vertical arrows indicate where cuts in concatemeric DNA are made. The letters represent fictional genes. Each packaged genome (horizontal arrows) is a cyclic permutation of the others.

rare, and transduced DNA must survive in the recipient cell to form a stable transductant. Since each of these steps has a limited probability of success, transductants can be detected only by powerful selection techniques, which are described, along with methods for analyzing the results of transduction experiments, in chapter 14.

What Makes a Transducing Phage?

Not all phages can transduce. The phage must not degrade the host DNA completely after infection, or no host DNA will be available to be packaged into phage heads. The packaging sites, or *pac* sites, of the phage must not be so specific that such sequences will not occur in host DNA. Some examples of transducing phages help illustrate these properties (see Table 7.3).

Phage P1, which infects gram-negative bacteria, is a good transducer because it has less *pac* site specificity than most phages and packages DNA via a headful mechanism; therefore, it will efficiently package host DNA. About 1 in 10^6 phage P1 particles will transduce a particular marker. It also has a very broad host range for adsorption and can transduce DNA from *E. coli* into a wide variety of other gram-negative bacteria including members of the genera *Klebsiella* and *Myxococcus*.

The *Salmonella typhimurium* phage P22 is also a very good transducer and, in fact, was the first transducing phage to be discovered (see Zinder and Lederberg, Suggested Reading). Like P1, P22 has *pac* sites that are not too specific and packages DNA by a headful mechanism. From a single *pac*-like site, about 10

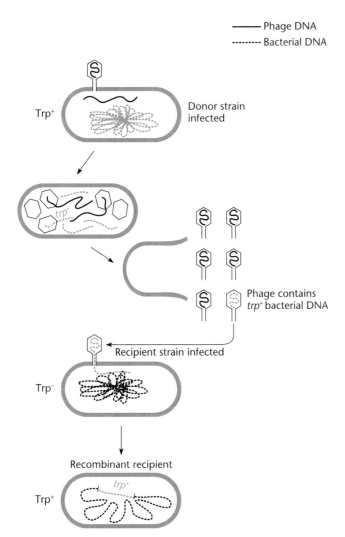

——— Phage DNA
-------- Bacterial DNA

Figure 7.18 An example of generalized transduction. A phage infects a Trp⁺ bacterium, and in the course of packaging DNA into heads, the phage mistakenly packages some bacterial DNA containing the *trp* region instead of its own DNA into a head. In the next infection, this transducing phage injects the Trp⁺ bacterial DNA instead of phage DNA into the Trp⁻ bacterium. If the incoming DNA recombines with the chromosome, a Trp⁺ recombinant transductant may arise. Only one strand of the DNA is shown.

The nomenclature of transduction is much like that of transformation and conjugation. Phages capable of transduction are called **transducing phages**. The original bacterial strain in which the transducing phage had multiplied is called the **donor strain**. The infected bacterial strain is called the **recipient strain**. Cells that have received DNA from another bacterium by transduction are called **transductants**.

Transduction occurs very rarely for a number of reasons. First, mistaken packaging of host DNA is itself

TABLE 7.3	Characteristics of generalized transducing phages P1 and P22	
Characteristic	**Phage**	
	P1	**P22**
Length (kb) of DNA packaged	100	44
Length (%) of chromosome transduced	2	1
Packaging mechanism	Sequential headful	Sequential headful
Specificity of markers transduced	Almost none	Some markers transduced at low frequency
Packaging of host DNA	Packaged from ends	Packaged from *pac*-like sequences
% of transducing particles in lysate	1	2
% of transduced DNA recombined into chromosome	1–2	1–2

headfuls of DNA can be packaged before the mechanism requires another *pac* site. Because of even this limited *pac* site specificity, however, some regions of *Salmonella* DNA will be transduced by P22 at a much higher frequency than others.

Other phages are not normally transducing phages but can be converted into them by special treatments. For example, T4 normally degrades the host DNA after infection but works extremely well as a transducing phage if its genes for the degradation of host DNA have been inactivated. Because phage T4 does not have *pac* sites, it packages any DNA including the host DNA with high efficiency.

In contrast, phage λ does not work well for generalized transduction, because it normally packages DNA between two *cos* sites rather than by a headful mechanism. It will very infrequently pick up host DNA by mistake, but then it does not cut the DNA properly when the head is filled unless another *cos*-like sequence lies the right distance along the DNA. Thus, potential transducing particles usually have DNA hanging out of them that must be removed with DNase before the tails can be added. Even with these and other manipulations, λ works poorly as a generalized transducer.

Transducing phages have been isolated for a wide variety of bacteria and have greatly aided genetic analysis of these bacteria. Transduction is particularly useful for moving alleles into different strains of bacteria and making isogenic strains that differ only in a small region of their chromosomes. However, if no transducing phage is known for a particular strain of bacterium, finding one can be very time-consuming. Therefore, generalized transduction has been largely replaced by conjugation, transposon mutagenesis, and DNA cloning, particularly in bacteria that are not well characterized. We discuss such methods in chapter 16.

The Role of Transduction in Bacterial Evolution

Phages may play an important role in evolution by promoting the horizontal transfer of genes between individual members of a species, as well as between distantly related bacteria. The DNA in phage heads is usually more stable than naked DNA and so may persist longer in the environment. Also, many phages have a broad host range for adsorption. Although incoming DNA from one species will not recombine with the chromosome of a different species if they share no sequences, stable transduction of genes between distantly related bacteria becomes possible when the transduced DNA is a broad-host-range plasmid that can replicate in the recipient strain or contains a transposon that can hop into the DNA of the recipient cell.

Lysogeny

In addition to the lytic cycle discussed in the preceding sections, some phages undergo a **lysogenic cycle**. In this cycle, the phages do not multiply but, instead, their DNA integrates into the host chromosome (often in an essential gene; see Box 7.3) or maintains itself as a plasmid, replicating only once or a few times each time the bacterial cell replicates.

A phage that is capable of entering this quiescent state is a lysogenic phage. The phage DNA in the quiescent state in the cell is called a **prophage**, and a bacterium harboring a prophage is a **lysogen**. Thus, a bacterium harboring the prophage P2 would be a P2 lysogen. In general, lysogens appear to be normal cells, and often the only evidence that they harbor prophages is that they are immune to **superinfection**, or infection by another phage of the same type. They may also synthesize toxins or other prophage-encoded gene products that affect the host only under certain circumstances. This process is called **lysogenic conversion** (see the section on lysogenic phages and bacterial pathogenesis, below).

Bacterial genomes often carry many prophages, many of which can be defective. A **defective prophage** has lost some of its genes and so can no longer become a phage. Nevertheless, it can synthesize gene products that affect its hosts in some circumstances. We mentioned in chapter 1 how the terminus region of chromosome replication in *E. coli* is the site of integration of many prophages, and some of these are defective.

In a lysogen, the prophage acts like any good parasite and does not place too great a burden upon its host. As mentioned, often the only indication that the host cell carries a prophage is the immunity of the cell to superinfection by the same phage. This state can continue almost indefinitely until the prophage is induced to make more phage, often after the host cell has suffered potentially lethal damage to its chromosomal DNA or has been infected by a phage of the same type. Then, like a rat leaving a sinking ship, the phage DNA will replicate to produce more phage. The released phage can then lysogenize other bacteria.

Although lysogeny was suspected as early as the 1920s, the first convincing demonstration that bacterial cells could carry phage in a quiescent state was not made until 1950 (see Lwoff, Suggested Reading). In this experiment, apparently uninfected bacteria began to produce phage after UV irradiation.

Phage λ Lysogeny

Phage λ is the classical example of a phage that can form lysogens. In the lysogenic state, very few λ genes are expressed, and essentially the only evidence that the

BOX 7.3

Phages That Integrate into Essential Genes of the Host

Until fairly recently, it seemed axiomatic that lysogenic phage could integrate only into sites on the bacterial DNA between genes; otherwise, they would inactivate bacterial genes and be deleterious to their host. The archetypal phage λ, which integrates into a nonessential region between the *gal* and *bio* operons of *E. coli*, supported this contention. However, we now know that some phages integrate directly into genes, often into genes for tRNAs. Examples include the *Salmonella* phage P22, which integrates into a threonine tRNA gene; the *E. coli* phage P4, which also integrates into a leucine tRNA gene; the *Haemophilus influenzae* phage HPc1, which integrates into a leucine tRNA gene; and the virus-like element SSV1 of the archebacterium *Sulfolobus* sp., which integrates into an arginine tRNA gene. It is not known why so many phages use tRNA genes as their attachment sites. Perhaps it is because tRNA genes are relatively highly conserved in evolution. A phage could lysogenize a different species of bacterium if the sequence of its attachment site were highly conserved, and thus it could be found in the new chromosome. Another possible explanation is that phages seem to prefer sequences with twofold rotational symmetry for their attachment sites. The sequences of tRNA genes have such symmetry, since the tRNA products of the genes can form hairpin loops, and in fact, most phage seem to integrate into the region of the tRNA gene that encodes the anticodon loop.

Not all phages that integrate into genes use tRNA genes for their attachment sites, however. For example, the phage,

φ21, a close relative of λ, and e14, a defective prophage, both integrate into the isocitrate dehydrogenase (*icd*) gene of *E. coli*. The product of the *icd* gene is an enzyme of the tricarboxylic acid cycle and is required for optimal utilization of most energy sources, as well as for the production of precursors for some biosynthetic reactions. Inactivation of the *icd* gene would cause the cell to grow poorly on most carbon sources.

How can phages integrate into genes and not inactivate the gene, thereby compromising the host? The answer is that they duplicate part of the gene in their *attP* site. The 3' end of the gene is repeated in the phage *attP* site with very few changes, so that when the phage integrates, the normal 3' end of the gene will be replaced by the very similar phage-encoded sequence. It is an interesting question in evolution how the bacterial sequence could have arisen in the phage. Perhaps the phage first arose as a specialized transducing particle, which then adapted to using the substituted bacterial genes as their normal attachment site in the chromosome.

References
Hill, C. W., J. A. Gray, and H. Brody. 1989. Use of the isocitrate dehydrogenase structural gene for attachment of e14 in *Escherichia coli* K-12. *J. Bacteriol.* **171**:4083–4084.
Reiter, W. D., P. Palm, and S. Yeats. 1989. Transfer RNA genes frequently serve as integration sites for prokaryotic genetic elements. *Nucleic Acids Res.* **17**:1907–1914.

cell harbors a prophage is that the lysogenic cells are immune to superinfection by more λ. The growth of the immune lysogens in the plaque is what gives the λ plaque its characteristic fried-egg appearance, with the lysogens forming the "yolk" in the middle of the plaque (see Figure 7.19).

Some phage λ mutants form plaques that are clear because they do not contain immune lysogens. These phage have mutations in the *cI*, *cII*, or *cIII* gene, where the "c" stands for *clear* plaque. These mutations prevent the formation of lysogens. Understanding the regulation of the λ lysogenic pathway and the function of the *cI*, *cII*, and *cIII* gene products in forming lysogens required the concerted effort of many people. Their findings illustrate the complexity and subtlety of biological regulatory pathways and serve as a model for other systems (see Ptashne, Suggested Reading).

THE *cII* GENE PRODUCT
After λ infects a cell, whether the phage enters the lytic cycle and makes more phage or forms a lysogen is determined by the outcome of a competition between the product of the *cII* gene, which acts to form lysogens, and the products of genes in the lytic cycle that replicate the DNA and make more phage particles. Most of the time, the lytic cycle wins, the λ DNA replicates, and more phage are produced. However, about 1% of the time, depending on environmental factors such as the richness of the medium, the *cII* gene product wins the race and a lysogen is formed.

The CII protein promotes lysogeny by activating the RNA polymerase to begin transcribing at two promoters, which are otherwise inactive (Figure 7.20). Proteins that enable RNA polymerase to begin transcription at certain promoters are called **transcriptional activators**

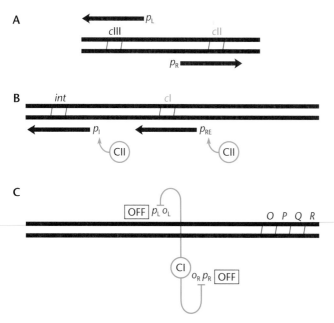

Figure 7.19 Phage λ plaques with a typical fried-egg appearance. (Photograph by Kurt Stepnitz.)

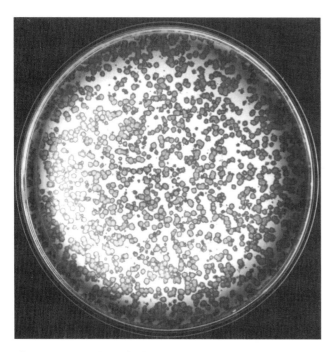

Figure 7.20 Formation of lysogens after λ infection. (A) the *cII* and *cIII* genes are transcribed from promoters p_R and p_L, respectively. (B) CII activates transcription from promoters p_{RE} and p_I, leading to the synthesis of CI repressor and the integrase Int, respectively. (C) The repressor shuts off transcription from p_L and p_R by binding to o_R and o_L. Finally, the Int protein integrates the λ DNA into the chromosome (see Fig. 7.21).

(see chapter 11). One of the promoters activated by CII is p_{RE}, which allows transcription of the *cI* gene. The product of this gene, the **CI repressor**, prevents transcription from the promoters p_R and p_L, which service many of the remaining genes of λ. We discuss the CI repressor in more detail below. The other promoter activated by the CII protein, p_I, allows transcription of the integrase (*int*) gene. The Int enzyme integrates the λ DNA into the bacterial DNA to form the lysogen.

The role of the *cIII* gene product in lysogeny is less direct. CIII inhibits a cellular protease that degrades CII. Therefore, in the absence of CIII, the CII protein will be rapidly degraded and no lysogens will form.

PHAGE λ INTEGRATION

As discussed above, the λ DNA forms a circle immediately after infection by pairing between the *cos* sequences at its ends. The Int protein can then promote the integration of the circular λ DNA into the chromosome, as illustrated in Figure 7.21. Int is an integrase that specifically promotes recombination between the attachment sequence (called *attP* for *att*achment *p*hage) on the phage DNA and a site on the bacterial DNA (called *attB* for *att*achment *b*acteria) that lies between the galactose (*gal*) and biotin (*bio*) operons in the chromosome of *E. coli*. Because the Int-promoted recombination does not occur at the ends of λ DNA but, rather, at the internal *attP* site, the prophage map is different from the map of DNA found in the phage head. In the phage head, the λ DNA has the *A* gene at one end and the *R* gene at the other end (see the λ map in Figure 7.9). In contrast, in the prophage, the *int* gene is on one end and the *J* gene is on the other (Figure 7.21). The relative order of the genes on the maps is still the same, but the prophage and phage maps are cyclic permutations of each other.

The recombination promoted by Int is called **site-specific recombination** because it occurs between specific sites, one on the the bacterial DNA and another on the phage DNA. This site-specific recombination is not normal homologous recombination but, rather, a type of nonhomologous recombination because the sequences of the phage and bacterial *att* sites are mostly dissimilar. They have a common core sequence, O, of only 15 bp—GCTTT(TTTATAC)TAA—flanked by two dissimilar sequences, B and B', in *attB*, and P and P' in *attP* (Figure 7.21, inset). The recombination always occurs within the bracketed 7-bp sequence. Because the region of homology is so short, this recombination would not occur without the Int protein, which recognizes both *attP* and *attB*. We discuss other examples of site-specific recombination in chapter 8.

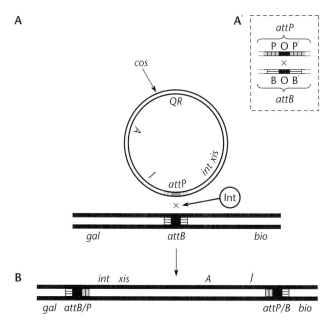

Figure 7.21 Integration of λ DNA into the chromosome of *E. coli*. (A) The Int protein promotes recombination between the *attP* sequence in the λ DNA and the *attB* sequence in the chromosome. The inset (A') shows the region in more detail, with sequences POP' and BOB'. The common core sequence of the two sites is shown in black. (B) The gene order in the prophage. The *cos* site is where the λ DNA is cut for packaging and recircularization after infection. The location of the *int*, *xis*, *A*, and *J* genes in the prophage is shown (refer to λ map in Figure 7.9). The *E. coli gal* and *bio* operons are on either side of the prophage DNA in the chromosome.

MAINTENANCE OF LYSOGENY

After the lysogen has formed, the *cI* repressor gene is one of the few λ genes to be transcribed. The CI repressor binds to two regions, called **operators.** These operators, *o*R and *o*L, are close to promoters *p*R and *p*L, respectively, and prevent transcription of most of the other genes of λ. Operators are discussed further in chapter 11. In the prophage state, the *cI* gene is transcribed from the *p*RM promoter (for *r*epression *m*aintenance), which is immediately upstream of the *cI* gene, rather than from the *p*RE promoter used immediately after infection. The *p*RM promoter is not used immediately after infection because its activation requires the CI repressor. We discuss the regulation of CI synthesis in more detail below.

REGULATION OF REPRESSOR SYNTHESIS IN THE LYSOGENIC STATE

The CI repressor is the major protein required to maintain the lysogenic state; therefore, its synthesis must be regulated even after a lysogen has formed. If the amount

of repressor drops below a certain level, the transcription of the lytic genes will begin and the prophage will be induced to produce phage. However, if the amount of repressor increases beyond optimal levels, cellular energy is wasted making excess repressor. The mechanism of regulation of repressor synthesis in lysogenic cells is well understood and has served as a model for gene regulation in other systems (Figure 7.22).

To function, the CI repressor protein must be a homodimer (see chapter 2) composed of two identical polypeptides encoded by the *cI* gene. At very low concentrations of CI polypeptide, the dimers will not form and the repressor will not be active.

Each CI polypeptide consists of two parts, or **domains.** One of these domains promotes the formation of a dimer by binding to another polypeptide. The other domain on each polypeptide binds to an operator sequence on the DNA. To illustrate this structure, the CI polypeptides are traditionally drawn as dumbbells, with the ends of each dumbbell indicating the two domains, and the complete dimer repressor protein is presented as two dumbbells bound to each other (Figure 7.22).

The repressor regulates its own synthesis by binding to the operator sequences and by being both a repressor and an activator of transcription. The important operator for regulating repressor synthesis is *o*R, to the right of the repressor gene, and so only this operator is shown,

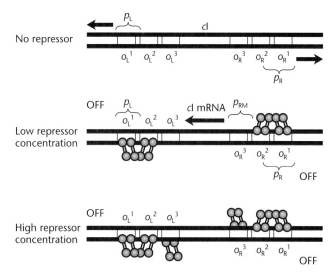

Figure 7.22 Regulation of repressor synthesis in the lysogenic state. The dumbbell shape represents the two domains of the repressor. The dimeric repressor, shown as two dumbbells, binds cooperatively to o_R^1 and o_R^2, (and o_L^1 and o_L^2), repressing transcription from p_R (and p_L) and activating transcription from p_{RM}. At higher repressor concentrations, it also binds to o_R^3 (and o_L^3), preventing transcription from p_{RM}.

although both operators have the same structure. The operator o_R can be divided into three repressor-binding sites, o_R^1, o_R^2, and o_R^3. If the concentration of repressor is low, only the o_R^1 site will be occupied, which is sufficient to repress transcription from promoter p_R, which overlaps the operator site (Figure 7.22). However, as the repressor concentration increases, o_R^2 will also be occupied. Only repressor bound at o_R^2 can activate transcription from the promoter p_{RM}, which is why p_{RM} is used to transcribe the repressor gene in the lysogen only when there is some repressor in the cell. With very high concentrations of repressor, o_R^3 will also be occupied, and transcription from p_{RM} will be blocked, blocking synthesis of more repressor, although there is less direct experimental evidence for this part of the model. By synthesizing more repressor when there is less in the cell and less when there is more, the cell maintains the levels of repressor within narrow limits.

Part of the regulation is based on the fact that repressor can bind to o_R^2 only if o_R^1 is already occupied by repressor and can bind to o_R^3 only if o_R^2 is already occupied. Apparently, the repressor protein bound at one site makes contact with another repressor molecule, allowing it to bind at the adjacent site. This is called **cooperative binding,** because protein bound at one site cooperates in the binding of a protein to an adjacent site. Cooperative binding is used in many regulatory systems, some of which we discuss in later chapters.

IMMUNITY TO SUPERINFECTION

The CI repressor in the cell of a lysogen will prevent not only the transcription of the other prophage genes by binding to operators o_L and o_R but also the transcription of the genes of any other λ phage infecting the lysogenic cell by binding to the operators of that phage. Thus, bacteria lysogenic for λ are immune to λ superinfection. However, λ lysogens can still be infected by any relative of λ phage that has different operator sequences. The λ CI repressor could not bind to such sequences. Any two phages that differ in their operator sequences are said to be **heteroimmune.**

THE INDUCTION OF λ

Phage λ will remain in the prophage state until the host cell DNA is damaged. Figure 7.23 outlines the process of induction of λ. When the cell attempts to repair the damage to its DNA, short pieces of single-stranded DNA accumulate and bind to the RecA protein of the host. The RecA protein with single-stranded DNA attached then binds to the λ CI repressor, causing it to cleave itself. The DNA-binding domain is separated from the domain involved in dimer formation. Without the dimerization do-

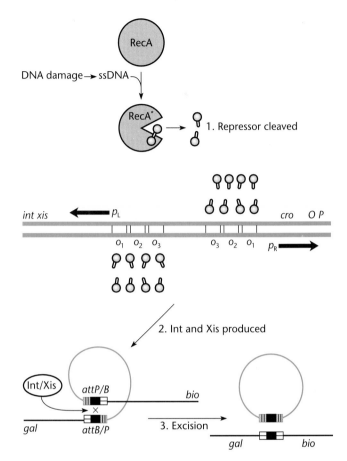

Figure 7.23 Induction of λ. Damage to the DNA results in activation of the RecA protein, which promotes the autocleavage of the CI repressor protein, separating the dimerization domain of the protein from the DNA-binding domain, so that the repressor can no longer form dimers and bind to DNA. Transcription of *int-xis* and *O* and *P* ensues, and the phage DNA (in blue) excises from the chromosome and replicates. ssDNA, single-stranded DNA.

main, the DNA-binding domains can no longer bind tightly to the operators. As the repressors drop off the operators, transcription will initiate from the promoters p_R and p_L, and the lytic cycle will begin.

The Cro Protein

Early during induction, more repressor could be made to interfere with later stages in lytic development or even reestablish lysogeny. However, the *cro* gene product, which is one of the first λ proteins to be made after induction, prevents the synthesis of more repressor. Cro does this by binding to the operator sequences, although in reverse order. It binds first to the o_R^3 site and then to the o_R^2 site, thereby preventing the CI repressor from binding to the o_R^2 site and activating its own synthesis from the p_{RM} promoter. Cro also binds to o_R^3, thereby preventing CI repressor binding to o_L. Thus, p_L will no longer be repressed.

Excision

Once the repressor is out of the way, transcription from p_L and p_R can begin in earnest. Among the genes transcribed from p_L are *int* and *xis*. Unlike after infection, when only Int is synthesized, after induction both the integrase (Int) and excisase (Xis) proteins are synthesized (see below). Excision requires both of these proteins, because the recombination occurs between different sequences from those used for integration. Integration requires recombination between the *attP* sequence on the phage and the *attB* sequence in the chromosome (see the section on phage λ integration above); however, excision requires recombination between hybrid *attP-attB* sequences that exist at the junctions between the prophage DNA and the chromosomal DNA. These sequences are different from either *attB* or *attP*. Therefore, Int alone is not capable of excising the prophage, and by making just Int after infection, the phage ensures that a newly integrated prophage will not be excised. The prophage will be excised only after induction, when both Int and Xis are made. Box 7.4 describes the molecular basis for the timing of Int and Xis synthesis.

While the Int and Xis proteins are excising λ DNA from the chromosome, the *O* and *P* genes are being transcribed from p_R. These proteins will promote the replication of the excised λ DNA. Therefore, a few minutes after the cellular DNA is damaged, the phage DNA is replicating, repressor levels are dropping, and the phage is irreversibly committed to lytic development. In about 1 h, depending on the medium and the temperature, the cell will lyse, spilling about 100 phage into the medium.

COMPETITION BETWEEN THE LYTIC AND LYSOGENIC CYCLES

As mentioned above, some cells infected by λ follow the lytic pathway, while others become lysogens. Figure 7.24 illustrates the competition for entry into the lysogenic cycle versus the lytic cycle. After infection, when there is no CI repressor in the cell, the *N* and *cro* genes are transcribed. As discussed earlier in this chapter, the *N* gene product acts as an antiterminator and allows the transcription of many genes, including *cII* and *cIII*, as well as the genes encoding the replication proteins O and P.

The phage will enter the lytic or the lysogenic cycle according to the outcome of a race between the CII activator protein and the Cro protein, which may be determined by chance or the host cell's metabolic state. If the CII protein wins, it will activate the synthesis of the CI repressor from the p_{RE} promoter and the integrase from the p_I promoter. The CI repressor will bind to the operators o_L and o_R and repress the synthesis of more Cro, as well as O and P, the DNA will integrate, and the lysogen will form. However, if the Cro protein wins, it will prevent the synthesis of more CI repressor. Then, without more CI repressor, some transcription will occur from genes *O* and *P* and replication of the λ DNA will begin. Eventually, there will be too much DNA for the repressor to bind to all of it, and transcription of *O* and *P* will increase further, followed by yet more DNA replication. The Q protein, which is also an antiterminator, will be synthesized next, allowing transcription of the head, tail, and lysis genes. Phage will then be assembled, the cells will lyse, and newly minted phage will spill out into the medium.

Specialized Transduction

Lysogenic phages are capable of another type of transduction called specialized transduction. Two properties distinguish specialized from generalized transduction. In generalized transduction, essentially any gene of the donor bacterium can be transduced into the recipient bacterium. However, in specialized transduction, only bacterial genes close to the attachment site of the prophage can be transduced. Also, the specialized transducing phage carries *both* bacterial genes and phage genes instead of only bacterial genes, like a generalized transducing phage.

Figure 7.25 illustrates how specialized transduction occurs in phage λ. In a λ lysogen of *E. coli*, the λ prophage is integrated close to and between the *gal* and *bio* genes in the chromosome. The *gal* gene products degrade galactose for use as a carbon and energy source, and the *bio* gene products make the vitamin biotin.

Specialized transduction can occur when a phage picks up neighboring bacterial genes during induction of the prophage. As shown in Figure 7.25, a specialized transducing phage carrying the *gal* genes, called λd*gal*, forms as the result of a mistake during the excision recombination. As mentioned in the preceding section, when the phage DNA is excised from the bacterial DNA, recombination occurs between the hybrid *attP-attB* sites at the junction between the prophage and host DNA. However, recombination sometimes occurs by mistake between the prophage DNA and a neighboring site in the bacterial DNA. The DNA later packaged into the head will include some bacterial sequences, as shown. Such transducing phage are very rare because the erroneous recombination that gives rise to them is extremely infrequent, occurring at one-millionth the frequency of normal excision. Furthermore, the recombi-

BOX 7.4

Retroregulation

The term **retroregulation** means that the expression of a gene is regulated by sequences downstream of it rather than upstream, such as at the promoter. How phage λ ensures that only Int will be made after infection but that both Int and Xis will be made after prophage induction is an example of retroregulation. Initially, after infection, both the *int* and *xis* genes are transcribed from the promoter p_L. However, the *int* and *xis* coding parts of this RNA do not survive long enough to be translated, because the RNA contains an RNase III cleavage site downstream of the *int* and *xis* coding sequences. The RNA is cleaved at this site and degraded past the *int* and *xis* coding sequences by a 3′ exonuclease, probably RNase II. This regulation was named retroregulation because mutations downstream of the *int* gene changed the RNase III cleavage site, preventing enzyme recognition, and thus stabilized the RNA and allowed both *int* and *xis* to be translated from the transcript initiated at p_L.

As discussed in the text, the *int* gene will also be transcribed from the promoter p_I, which is actually located in the *xis* gene (see Figure 7.9). This resulting RNA does not contain a *nut* site, and so the N protein cannot bind to the RNA

polymerase and allow it to proceed past termination signals as far as the coding sequence for the RNase III cleavage site; therefore, the RNA will be stable. Moreover, the *int* RNA contains all of the *int* sequence but only part of the *xis* sequence, so that only Int can be made from this RNA.

Immediately after induction, however, both Int and Xis can be made from the RNA produced from the p_L promoter, because *xis-int* RNA produced from this promoter will now be stable. As shown in the figure, during integration of the phage, the coding region for the RNase III cleavage site has been separated from the *xis-int* coding region, since this region is on the other side of the *attP* site, which is split during integration of the phage DNA. Therefore, the long RNA transcript initiated at p_L will no longer contain the RNase III cleavage site at its 3′ end and so will be stable. Both Int and Xis can be translated from this RNA after induction. This is an example of posttranscriptional regulation, because it occurs after the RNA synthesis has occurred on the gene (see chapter 11). It is interesting how the phage has taken advantage of the reorganization of its genes during integration to regulate the expression of its genes.

References

Guarneros, G., C. Montanez, T. Hernandez, and D. Court. 1982. Posttranscriptional control of bacteriophage λ *int* gene expression from a site distal to the gene. *Proc. Natl. Acad. Sci. USA* **79**:238–242.

Schmeissner, U., K. M. McKenny, M. Rosenberg, and D. Court. 1984. Removal of a terminator structure by RNA processing regulates *int* gene expression. *J. Mol. Biol.* **176**:39–53.

A After λ infection, *int* expressed from p_I

B After induction, *int* and *xis* expressed from p_L

(A) After infection, the *xis* and *int* genes cannot be expressed from the p_L promoter. Because of *N*, transcription from p_L will continue past *int* into an RNase III cleavage site. The RNA will be cleaved and digested back into *xis* and *int*, removing them from the RNA. Xis also cannot be expressed from p_I because the p_I promoter is in the *xis* gene. The RNA from p_I is stable, however, because this transcript does not contain a *nut* site and so will not continue through *int* to the RNase III cleavage site. The blue region indicates the location of the coding information for the RNase III site, but RNase III cleaves only the mRNA transcript. (B) In the prophage, however, the sequence encoding the RNase III cleavage site is separated from the *xis-int* coding sequence, so that the RNA made from p_L is stable.

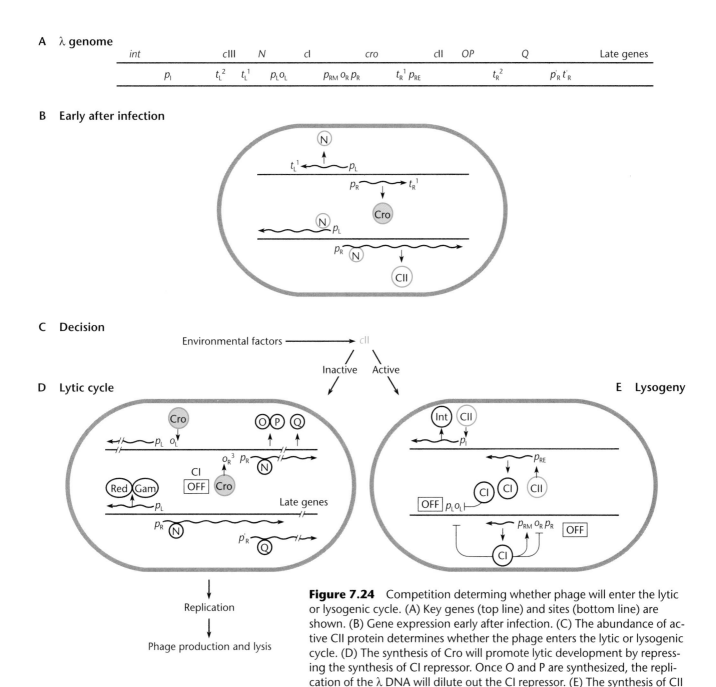

Figure 7.24 Competition determining whether phage will enter the lytic or lysogenic cycle. (A) Key genes (top line) and sites (bottom line) are shown. (B) Gene expression early after infection. (C) The abundance of active CII protein determines whether the phage enters the lytic or lysogenic cycle. (D) The synthesis of Cro will promote lytic development by repressing the synthesis of CI repressor. Once O and P are synthesized, the replication of the λ DNA will dilute out the CI repressor. (E) The synthesis of CII will promote lysogeny.

nation must, by chance, occur between two sites that are approximately a λ genome length apart, or the DNA would not fit into a phage head.

Because of the rarity of these transducing phage, powerful selection techniques are required to detect them. To select λ phage carrying *gal* genes of the host, induced phage are used to infect Gal⁻ recipient bacteria, and Gal⁺ transductants are selected on plates with galactose as the sole carbon source. In the rare Gal⁺ transductants, a λ phage carrying *gal* genes may have integrated into the chromosome, providing the *gal* gene product that the mutant lacks. If such a Gal⁺ lysogen is colony purified and the prophage is induced from it, all of the resultant phage progeny will carry the *gal* genes. Such a lysogenic strain produces an **HFT lysate** (for *h*igh-*f*requency *t*ransduction) because it produces phage that can transduce bacterial genes at a very high frequency.

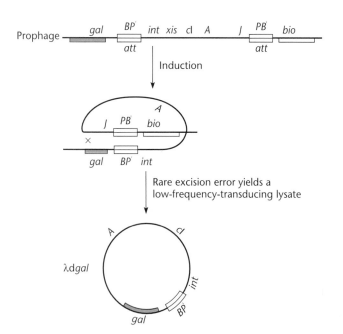

Figure 7.25 Formation of a λdgal transducing particle. A rare mistake in recombination between a site in the prophage DNA (in this case between A and J) and a bacterial site to the left of the prophage results in excision of a DNA particle in which some bacterial DNA has replaced phage DNA.

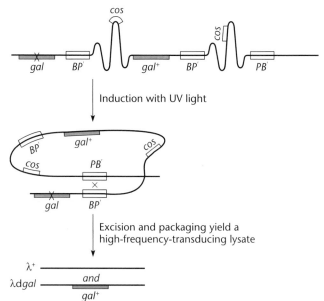

Figure 7.26 Induction of the λdgal phage from a dilysogen containing both λdgal and a wild-type λ in tandem. Recombination between the hybrid PB′ and BP′ sites at the ends excises both phages. The wild-type phage helps the λdgal phage to form phage particles. See the text for details.

Because the λ-phage head can hold DNA of only a certain length, the transducing particles will of necessity have lost some phage genes to make room for the bacterial genes. Which phage genes are lost depends on where the mistaken recombination occurs. If the recombination occurs to the left of the prophage, some of the head and tail genes will be replaced by *gal* genes of the host, as in Figure 7.25. However, if the mistake in recombination occurs to the right, the *int* and *xis* genes will be replaced by the *bio* genes of the host.

Clearly, the properties of the transducing particle will be determined by the lost phage genes. For example, the λdgal phage shown in Figure 7.25 lack essential head and tail genes and so cannot multiply without a wild-type λ **helper phage** to provide the missing head and tail proteins (Figure 7.26). These phage are therefore called λdgal, where the "d" stands for *defective*.

Transducing phage in which the *int* and *xis* genes have been replaced by *bio* will be able to multiply, since the genes on this side of *attP* are not required for multiplication. However, they cannot form a lysogen without the help of a wild-type phage. Because they can multiply and form plaques, *bio* transducing phages are called λp*bio*, in which the "p" stands for *plaque* forming.

Specialized transducing phage particles played a major role in the development of microbial molecular genetics, including the first isolation of genes and the discovery of IS elements in bacteria. Although their general use has been largely supplanted by recombinant DNA techniques, they continue to have special applications.

Other Lysogenic Phages

Phage λ was the first lysogenic phage to be extensively studied and thus serves as the archetypal lysogenic phage. However, many other types of lysogenic phages are known. Many use somewhat different strategies to achieve and maintain the prophage state. We briefly describe some of them here.

PHAGE P2

Phage P2 is another lysogenic phage of *E. coli*. Unlike λ, which almost always integrates into the same site in the *E. coli* chromosome, this phage can integrate into many sites in the bacterial DNA, although it uses some sites more than others. Like λ, P2 requires one gene product to integrate and two gene products to excise. P2 prophage is much more difficult to induce than λ, however. It is not inducible by UV light, and even temperature-sensitive repressor mutations cannot efficiently induce it.

PHAGE P4

Even viruses can have parasites! Phage P4 depends on phage P2 for its lytic development (see Kahn et al., Suggested Reading). Thus, it is a representative of a group called **satellite viruses**, which need other viruses to multiply. Phage P4 does not encode its own head and tail proteins but, rather, uses those of P2. Thus, P4 can multiply only in a cell that is lysogenic for P2 or that has been simultaneously infected with a P2 phage. When P4 multiplies in bacteria lysogenic for P2, it induces transcription of the head and tail genes of the P2 prophage, which are normally not transcribed in the P2 lysogen. How phage P4 bypasses the normal regulation of the P2 prophage is the object of current studies.

Because it wears the protein coat of P2, phage P4 particle looks similar to P2. Only the head of P4 is shorter, to accommodate the shorter DNA. While the DNAs of P2 and P4 have otherwise very different sequences, the *cos* sites at the ends of the DNA are the same, so that the head proteins of P2 can package either DNA.

Phage P4 can also form a lysogen; when it does so, it usually integrates into a unique site on the chromosome. It can also replicate autonomously as a circle in the prophage state, as does P1 (see below). Because of this ability to maintain itself as a circle, phage P4 has been engineered for use as a cloning vector.

Phages P2 and P4, as well as their many relatives, have a very broad host range and infect many members of the *Enterobacteriaceae* including *Salmonella* and *Klebsiella* spp. as well as some *Pseudomonas* spp. They are also related to phage P1, although their lifestyles and strategies for lytic development and lysogeny are very different.

PHAGE P1: A PLASMID PROPHAGE

Not all prophages integrate into the chromosome of the host to form a lysogen. Some, represented by P1, form a prophage that replicates autonomously as a plasmid. Because P1 can exist both as a phage and as a plasmid, it is very useful in studies of plasmid replication and partitioning during cell division, and some of the information on these subjects presented in chapter 4 was obtained with the P1 prophage. To maintain itself as a plasmid, the prophage must replicate autonomously once each time the chromosome replicates and distribute one copy faithfully to each daughter cell upon cell division. However, when it is induced to make a phage, it will make hundreds of copies of itself and be packaged into phage heads.

Use of Lysogenic Phages as Cloning Vectors

Lysogenic phages offer some distinct advantages as cloning vectors. Because they can multiply as a phage, obtaining large amounts of the cloned DNA is relatively easy. Since they can also integrate into the DNA of the host, a cloned gene will exist in only two copies, one in the prophage and one at the normal site, which is important in complementation studies with bacteria, as we discuss in chapter 14.

Cloning into lysogenic phage can also facilitate genetic mapping. In such experiments, a lysogenic phage that lacks its *att* site but has an easily selectable marker (such as resistance to an antibiotic) introduced into it is used. The gene of interest is cloned into the phage DNA, which is introduced into the cell either by in vitro packaging and infection or by transfection. Lysogens will form by recombination between the cloned gene in the phage and its counterpart in the chromosome (Figure 7.27). The phage will be integrated at this site in the chromosome rather than at the *attB* site because the phage *attP* site is deleted. When the antibiotic resistance gene on the phage DNA is mapped genetically by methods discussed in chapter 14, the original location in the chromosome of the gene will be known. This may be the preferred way to map a cloned gene for which no convenient phenotype is available, since antibiotic resistance markers are relatively easy to map genetically.

Lysogenic Phages and Bacterial Pathogenesis

In a surprising number of instances, prophages carry genes for virulence factors or toxins required by the pathogenic bacteria they lysogenize. For example, the

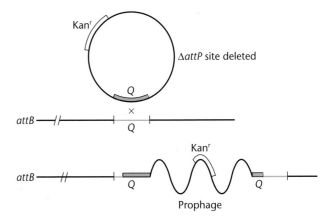

Figure 7.27 Use of a phage lacking its *attP* site to mark a cloned chromosomal gene for mapping. In the example, a phage lysate containing a DNA clone of the bacterial gene Q is used to infect the bacterium. Because the phage has its *attP* site deleted, it can only integrate by recombination between the Q gene in the phage and in the chromosome. The location of the Q gene in the chromosome can then be ascertained by mapping the kanamycin resistance (Kan^r) gene on the prophage.

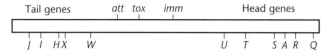

Figure 7.28 Partial genetic map of the *Corynebacterium diphtheriae* prophage β. *tox* is the toxin gene. Based on the data of R. K. Holmes and L. Barksdale, Genetic analysis of *tox+* and *tox-* bacteriophages of *Corynebacterium diphtheriae*, *J. Virol.* 3:586–598, 1969.

bacteria that cause diphtheria, scarlet fever, botulism, tetanus, and cholera all require lysogenic phages for virulence. This may also be true of bacteria that cause some types of toxic shock syndrome. Even λ phage carries genes that confer on its *E. coli* host serum resistance and the ability to survive in macrophages. As mentioned earlier, the process by which a prophage converts a non-pathogenic bacterium to a pathogen is an example of lysogenic conversion.

DIPHTHERIA

Diphtheria is the classic example of a disease caused in part by the product of a gene carried by a lysogenic phage. Pathogenic strains of *Cornyebacterium diphtheriae* differ from nonpathogenic strains in that they are lysogenic for phage β (see Freeman, Suggested Reading). The β prophage carries a gene, *tox*, for the diphtheria toxin. Figure 7.28 shows a partial genetic map of the β prophage, indicating the position of the *tox* gene in the prophage DNA (see Costa et al., Suggested Reading). The diphtheria toxin is an enzyme that kills eukaryotic cells by ADP-ribosylating (attaching adenosine diphosphate to) the EF-2 translation factor, thereby inactivating it and blocking translation in the cell. The *tox* gene of the β prophage is transcribed only when *C. diphtheriae* infects its eukaryotic host or under conditions that mimic this environment. We discuss the mechanism of action of the toxin and the regulation of the *tox* gene in chapter 12.

CHOLERA

The most recently discovered example of toxin genes carried by a lysogenic phage involves the bacterium that causes cholera, *Vibrio cholerae* (see Waldor and Mekalanos, Suggested Reading). In this case, the toxin genes are carried on a single-stranded filamentous phage related to M13. This phage can lysogenize *V. cholerae* either by integrating into the chromosome or, like phage P1, by replicating as an autonomous plasmid. The cholera toxin genes on the prophage are regulated by a chromosomal gene, *toxR*, that also regulates the synthesis of the pili that serve as the receptor sites for the phage. These pili are also important virulence determinants because they enable the bacteria to adhere to the intestinal mucosa. We also discuss the regulation of cholera toxin genes in chapter 12.

BOTULISM AND TETANUS

Other striking examples of diseases caused by toxins encoded by lysogenic phages are botulism and (probably) tetanus. These diseases are caused by different toxins, but both are encoded by lysogenic phages of the gram-positive genus *Clostridium*. Recent evidence indicates that these two toxins work by a common mechanism. They both cleave the same neuronal protein, synaptobrevin, in exactly the same position in the amino acid sequence, although the routes of entry of the toxins into the host and the symptoms they cause are very different (see Schiavo et al., Suggested Reading).

Why are toxin genes and other virulence factors often encoded by prophages instead of being normal genes in the chromosome? The argument is similar to that used to explain why genes are carried on plasmids (see chapter 4). Having toxin and other virulence genes on a movable DNA element like a phage may allow the bacterium to adapt to being pathogenic without normally having to carry extra genes. Furthermore, the nonpathogenic bacterium without the prophage can colonize the host without causing disease and will become pathogenic only if it is infected by the phage.

SUMMARY

1. Viruses that infect bacteria are called bacteriophages, or phages for short.

2. The productive developmental cycle of a phage is called the lytic cycle. The larger DNA phages undergo a complex program of gene expression during development.

3. The products of phage regulatory genes regulate the expression of other phage genes during development. One or more regulatory genes in each stage of development turn on the genes in the stage to follow and turn off the genes in the preceding stage, a regulatory cascade. In this way, all the information for the stepwise development of the phage can be preprogrammed into the phage DNA.

(continued)

SUMMARY (continued)

4. Phage T7 encodes an RNA polymerase that specifically recognizes the promoters for the late genes of the phage. Because of its specificity and because transcription from the T7 promoters is so strong, this system has been used to make cloning vectors for expressing large amounts of protein from cloned genes.

5. Phage T4 uses the host RNA polymerase but encodes a phage-specific sigma (σ) factor that binds to the host RNA polymerase and changes its specificity so that it can transcribe only the late genes of the phage. In addition, T4 couples the transcription of its late genes to the replication of the phage DNA, so that the late genes will not be transcribed until the phage DNA begins to replicate.

6. Phage λ regulates its transcription through antitermination. The N protein binds to the RNA polymerase as it transcribes short regions on λ DNA called the *nut* (N utilization) sites. Once N protein is bound to the RNA polymerase, it stops for no termination signals.

7. All phages encode at least some of the proteins required to replicate their nucleic acids. They borrow others from their hosts.

8. The requirement of DNA polymerases for a primer prevents the replication of the extreme 5' ends of linear DNAs. Phages solve this replication primer problem in different ways. Some replicate as circular DNA. Others form long concatemers by rolling-circle replication, by linking single DNA molecules end to end by recombination, or by pairing through complementary ends.

9. Transduction occurs when a phage accidentally packages bacterial DNA into a head and carries it from one host to another. In generalized transduction, essentially any region of bacterial DNA can be carried.

10. Some types of phage are also capable of a lysogenic cycle. In the lysogenic cycle, the phage DNA persists in the cell as a prophage, and the cell harboring a prophage is called a lysogen.

11. Whether or not λ enters the lysogenic state depends upon the outcome of a race between the *c*II gene product and the products of the lytic genes of the phage. The product of the *c*II gene is a transcriptional activator that activates the transcription of the *c*I and *int* genes after infection. The *c*I gene product is the repressor that blocks transcription of most of the genes of λ in the prophage state, and the *int* gene product is an enzyme that is required to integrate λ DNA into the bacterial chromosome.

12. The CI repressor protein of λ is a homodimer made up of two identical polypeptides encoded by the *c*I gene. The repressor blocks transcription by binding to operators o_R and o_L on both sides of the *c*I gene, preventing the utilization of two promoters, p_R and p_L, which are responsible for the transcription of genes to the right and to the left of the *c*I gene, respectively.

13. The synthesis of repressor is regulated in the lysogenic state through its binding to three repressor binding sites within o_R. These sites are named o_R^1, o_R^2, and o_R^3. The regulation of repressor synthesis has served as a model for regulation in other systems.

14. Damage to the DNA of its host can cause the λ prophage to be induced and produce phage. The activated RecA protein of the host causes the λ repressor protein to cleave itself so that it can no longer form dimers and be active. The process of excision of λ DNA is esentially the reverse of integration, except that excision requires both the *int* and *xis* gene products of λ.

15. Very rarely, when λ DNA excises, it picks up neighboring bacterial DNA and becomes a transducing particle. This type of transduction is called specialized transduction, because only bacterial genes close to the insertion site of the prophage can be transduced.

16. Lysogenic phages are often useful as cloning vectors. A bacterial gene cloned into a prophage will exist in only two copies, one in its normal site in the chromosome and another in the prophage. This can be important in complementation tests or in other applications. If the prophage is induced, the cloned DNA can be recovered in large amounts.

17. Lysogenic phages often carry genes for bacterial toxins. Examples include the toxins that cause diphtheria, botulism, scarlet fever, cholera, and, possibly, toxic shock syndrome.

QUESTIONS FOR THOUGHT

1. Why do you suppose phages regulate their gene expression during development so that genes whose products are involved in DNA replication are transcribed before genes whose products become part of the phage particle? What do you suppose would happen if they did not do this?

2. Why do you suppose some phages encode their own RNA polymerase while others change the host RNA polymerase?

3. Why do you suppose the λ prophage uses different promoters to transcribe the *c*I repressor gene immediately after infection and in the lysogenic state?

4. Why is only one protein, Int, required to integrate the phage DNA into the chromosome while two proteins, Int and Xis, are required to excise it? Why not just make one different Int-like protein that excises the prophage?

5. Why do some types of phage duplicate host functions and others not?

6. Can you propose an alternative explanation for why toxins and other proteins that make bacteria pathogenic are often encoded by phages?

7. Why do you suppose some types of prophage can be induced only if another phage of the same type infects the lysogenic cell containing them? What purpose does this serve?

PROBLEMS

1. Phage components (e.g., heads and tails) can often be seen in lysates by electron microscopy, even before they are assembled into phage particles. In studying a newly discovered phage of *Pseudomonas putida*, you have observed that amber mutations in gene *C* of the phage prevent the appearance of phage tails in a nonsuppressor host. Similarly, mutations in gene *T* prevent the appearance of heads. However, mutations in gene *M* prevent the appearance of either heads or tails. Which gene, *C*, *T*, or *M*, is most likely to be a regulatory gene? Why?

2. Would you expect to be able to isolate amber mutations in the *ori* sequence of a phage? Why or why not?

3. Lambda (λ) *vir* mutations cause clear plaques because they change the operator sequences so that they no longer bind repressor. How would you determine if a clear mutant you have isolated has a *vir* mutation instead of a mutation in any one of the three genes *c*I, *c*II, or *c*III?

4. Lambda (λ) integrates into the bacterial chromosome in the region between the galactose utilization (*gal*) genes and the biotin biosynthetic (*bio*) genes on the other side. Outline how you would isolate an HFT strain carrying the *bio* operon of *E. coli*. Would you expect your transducing phage to form plaques? Why or why not? A *vir* mutation changes the operator sequences. Would you expect λ with *vir* mutations in the o_R^1 and o_L^1 sites to form plaques on λ lysogens? Why or why not?

5. The DNA of a λ specialized transducing particle usually integrates next to a preexisting prophage. Would you expect just Int or both Int and Xis to be required for this integration? Why?

6. You have isolated a phage from a strain of *Staphylococcus aureus* known to cause food poisoning owing to production of a toxin. Outline how you would go about determining if the toxin is encoded by a prophage. Assume that you can detect the toxin by its ability to kill human cells in culture.

SUGGESTED READING

Campbell, A. M. 1996. Bacteriophages, p. 2325–2338. *In* F. C. Neidhardt, R. Curtiss III, J. L. Ingraham, E. C. C. Lin, K. B. Low, B. Magasanik, W. S. Reznikoff, M. Riley, M. Schaechter, and H. E. Umbarger (ed.), *Escherichia coli and Salmonella: Cellular and Molecular Biology*, 2nd ed. ASM Press, Washington, D.C.

Christensen, A. C., and E. T. Young. 1982. T4 late transcripts are initiated near a conserved DNA sequence. *Nature* (London) 299:369–371.

Costa, J. J., J. L. Michel, R. Rappuoli, and J. R. Murphy. 1981. Restriction maps of corynebacteriophages βc and βvir and physical location of the diphtheria *tox* operon. *J. Bacteriol.* 148:124–130.

Echols, H., and H. Murialdo. 1978. Genetic map of bacteriophage lambda. *Microbiol. Rev.* 42:577–591.

Freeman, V. J. 1951. Studies on the virulence of bacteriophage-infected strains of *Corneybacterium diphtheriae*. *J. Bacteriol.* 61:675–688.

Friedman, D. I., M. F. Baumann, and L. S. Baron. 1976. Cooperative effects of bacterial mutations affecting λ N gene expression. *Virology* 73:119–127.

Hendrix, R. W., J. W. Roberts, F. W. Stahl, and R. A. Weisberg (ed.). 1983. *Lambda II*. Cold Spring Harbor Laboratory Press, Cold Spring Harbor, N.Y.

Herendeen, D. R., K. P. Williams, G. A. Kassavetis, and E. P. Geidushek. 1990. An RNA polymerase binding protein that is required for communication between an enhancer and a promoter. *Science* 248:573–578.

Johnson, A. D., A. R. Poteete, G. Lauer, R. T. Sauer, G. K. Akers, and M. Ptashne. 1980. λ repressor and cro⁻ components of an efficient molecular switch. *Nature* (London) 294:217–223.

Johnson, L. P., M. A. Tomai, and P. M. Schlievert. 1986. Bacteriophage involvement in group A streptococcal pyrogenic exotoxin A production. *J. Bacteriol.* 166:623–627.

Kahn, M. L., R. Ziermann, G. Deho, D. W. Ow, M. G. Sunshine, and R. Calendar. 1991. Bacteriophage P2 and P4. *Methods Enzymol.* 204:264–280.

Kassavetis, G. A., and E. P. Geiduschek. 1984. Defining a bacteriophage T4 late promoter: bacteriophage T4 gene 55 protein suffices for directing late promoter recognition. *Proc. Natl. Acad. Sci. USA* 81:5101–5105.

Lwoff, A. 1953. Lysogeny. *Bacteriol. Rev.* 17:269–337.

Masters, M. 1996. Generalized transduction, p. 2421–2441. *In* F. C. Neidhardt, R. Curtiss III, J. L. Ingraham, E. C. C. Lin, K. B. Low, B. Magasanik, W. S. Reznikoff, M. Riley, M. Schaechter, and H. E. Umbarger (ed.), *Escherichia coli and Salmonella: Cellular and Molecular Biology*, 2nd ed. ASM Press, Washington, D.C.

Mosig, G., A. Luder, A. Ernst, and N. Canan. 1991. Bypass of a primase requirement for bacteriophage T4 DNA replication *in vivo* by a recombination enzyme, endonuclease VII. *New Biol.* 3:1195–1205.

Ptashne, M. 1992. *A Genetic Switch,* 2nd ed. Blackwell Scientific Publications, Cambridge, Mass.

Salstrom, J. S., and W. Szybalski. 1978. Coliphage λ *nutL:* a unique class of mutations defective in the site of N product utilization for antitermination of leftward transcription. *J. Mol. Biol.* **124:**195–222.

Schiavo, G., F. Benfenati, B. Poulain, O. Rossetto, P. Polverino de Laureto, B. R. dasGupta, and C. Montecucco. 1992. Tetanus and botulinum-B neurotoxins block neurotransmitter release by proteolytic cleavage of synaptobrevin. *Nature* (London) **359:**832–835.

Skalka, A. 1977. DNA replication-Bacteriophage lambda. *Curr. Top. Microbiol. Immunol.* **78:**201–211.

Studier, F. W. 1969. The genetics and physiology of bacteriophage T7. *Virology* **39:**562–574.

Studier, F. W., A. H. Rosenberg, J. J. Dunn, and J. W. Dubendoff. 1990. Use of T7 RNA polymerase to direct expression of cloned genes. *Methods Enzymol.* **185:**60–89.

Waldor, M. K., and J. J. Mekalanos. 1996. Lysogenic conversion by a filamentous phage encoding cholera toxin. *Science* **272:**1910–1914.

Yanisch-Perron, C., J. Vieira, and J. Messing. 1985. Improved M13 phage cloning vectors and host strains: nucleotide sequences of the M13mp18 and pUC19 vectors. *Gene* **33:**103–119.

Zinder, N. D., and J. Lederberg. 1952. Genetic exchange in *Salmonella. J. Bacteriol.* **64:**679–699.

Chapter 8

Transposition and Nonhomologous Recombination

R ECOMBINATION IS THE BREAKING and rejoining of DNA in new combinations. In homologous recombination, which accounts for most recombination in the cell, the breaking and rejoining occurs only between regions of two DNA molecules that have similar or identical sequences. Homologous recombination requires that the two DNAs pair through complementary base pairing (see chapter 9). However, other types of recombination, known as **nonhomologous recombination**, sometimes occur in cells. They do not require that the two DNAs have the same sequence of nucleotides. Rather, nonhomologous recombination depends on enzymes that recognize specific regions in DNA, which may or may not have sequences in common, and promote recombination between those regions. This chapter addresses some examples of nonhomologous recombination in bacteria and the mechanisms involved, including transposition by transposons and site-specific recombination that occurs during the integration and excision of prophages and other DNA elements, the inversion of invertible sequences, and resolution of cointegrates by resolvases.

Transposition

Transposons are DNA elements that can hop, or **transpose**, from one place in DNA to another. Transposable DNA elements were first discovered in corn by Barbara McClintock in the early 1950s and about 20 years later in bacteria by others. Transposons are now known to exist in all organisms on Earth, including humans. The movement by a transposon is called **transposition**, and the enzymes that promote transposition are called **transposases**. The transposon itself usually encodes its own transposases, so that it carries with it the ability to hop each time it moves. For this reason, transposons have been called "jumping genes."

Not all DNA elements that can move are true transposons. For example, "homing" DNA elements, which include some types of moveable RNA and protein introns, move by means of endonucleases that make a specific double-stranded break in the DNA at a given site. Then, through homologous recombination aimed at repairing the double-stranded break, the DNA element is inserted at that site. No specific transposases are required for the movement of homing DNA, and these DNA elements can move only into the same sequence in other DNA molecules that lack them. We discuss homing endonucleases in more detail in chapter 9 (see Box 9.1).

True transposons should also be distinguished from **retrotransposons**, so named because they behave like RNA retroviruses with a DNA intermediate. An RNA copy of a region is made and then copied into DNA by a reverse transcriptase. The DNA intermediate then integrates elsewhere by various mechanisms that may or may not be analogous to transposition. Although only a few examples of retrotransposons are known in bacteria, such elements are well known in fungi.

Although transposons probably exist in all organisms on Earth, they are best understood in bacteria, where they obviously play an important role in evolution. Transposons found in different bacterial genera may be more closely related to each other than are the bacteria in which they are found. This suggests that transposons move among different genera of bacteria with some regularity. As mentioned in previous chapters, transposons may enter other genera of bacteria during transfer of promiscuous plasmids or via transducing phage. Some transposons are themselves conjugative or can be induced to form phage (see Boxes 8.1 and 8.2). Transposons may offer a way of introducing genes from one bacterium into the chromosome of another bacterium to which it has little DNA sequence homology.

Overview of Transposition

The net result of transposition is that the transposon appears at a place in DNA different from where it was originally. Many transposons are essentially cut out of one DNA and inserted into another (Figure 8.1), whereas other transposons are copied and then inserted elsewhere. Regardless of the type of transposon, however, the DNA from which the transposon originated is called the **donor DNA** and the DNA into which it hops is called the **target DNA**.

In all transposition events, the transposase enzyme cuts the donor DNA at the ends of the transposon and then inserts the transposon into the target DNA. However, the details of the mechanism can vary. Some types of transposons may exist free of other DNA during the

---

BOX 8.1

Phage Mu: a Transposon Masquerading as a Phage

Phage Mu is a lysogenic phage that can integrate into the bacterial DNA after infection. It integrates randomly into the chromosome by what is essentially a transposition event. In the virus head, the Mu DNA is bracketed by short sequences of bacterial DNA from the previous infection. After infection, the Mu DNA can essentially transpose from this DNA into the chromosome. Therefore, like most transposons, Mu never exists free of other flanking DNA.

After induction, Mu does not excise from the bacterial DNA but, rather, transposes by replicative transposition, forming cointegrates. These cointegrates are apparently not resolved, however, and the bacterial DNA becomes scrambled and riddled with Mu genomes. At later times, the Mu DNA is packaged into virus heads much like any other phage DNA; enzymes cut specific *pac* sequences at the ends of the Mu DNAs inserted around the chromosome. However, rather than cut the Mu DNA precisely from the bacterial DNA, the cuts are made in the bacterial DNA outside of the Mu sequence. Consequently, the Mu DNA in the virus heads is bracketed by bacterial DNA, and each Mu DNA is bracketed by different bacterial DNA sequences derived from where it was inserted in the bacterial chromosome. Because it transposes randomly into the chromosome, Mu is useful for transposon mutagenesis and for creating random gene fusions. Besides its usefulness in bacterial genetics and its general interest as a phage, Mu is an effective tool for biochemical studies of transposition because it transposes so frequently. We shall discuss uses of Mu in making random gene fusions and in in vivo cloning in later chapters.

References
Ljungquist, E., and A. I. Bukhari. 1977. State of prophage Mu upon induction. *Proc. Natl. Acad. Sci. USA* **74:**3143–3147.

Toussaint, A., and A. Resibois. 1983. Phage Mu: transposition as a life style, p. 105–158. *In* J. A. Shapiro (ed.), *Mobile Genetic Elements.* Academic Press, Inc., New York.

BOX 8.2

Conjugative Transposons

The conjugative transposons are transposons combined with transfer functions, such as those of a self-transmissible plasmid (see the text and chapter 5). The first known conjugative transposon, Tn*916*, was isolated from *Enterococcus* (formerly *Streptococcus*) *faecalis* (Senghas et al.). It carries tetracycline resistance and is very large, presumably because it must encode transfer (Tra) as well as transposition functions. It probably excises itself from the DNA of the donor cell and transfers the excised DNA into the recipient cell, where it immediately transposes into recipient cell DNA. It can also "hitchhike" in on another conjugative plasmid and then immediately transpose into other DNA in the recipient cell.

Tn*916* transposes by the cut-and-paste, or conservative, mechanism. However, unlike other transposons known to transpose by this mechanism, Tn*916* seems to precisely excise itself from the donor DNA, leaving the original sequence intact. Moreover, unlike other known transposons, Tn*916* does not duplicate a short sequence in the target DNA when it moves. The transposase of this transposon is related to the phage λ integrase family. Target selection is not random, but there is no requirement for strict homology as there is for λ integration.

The conjugative transposons are remarkably broad in their host range and have been found to transfer into and transpose in a wide variety of bacteria. They usually carry a gene, *tetM*, that confers resistance to tetracycline. This gene has been found widely in gram-positive as well as gram-negative bacteria, and it is reasonable to suppose that the conjugative transposons have played a role in its dispersal.

References

Salyers, A. A., N. B. Shoemaker, A. M. Stevens, and L.-Y. Li. 1995. Conjugative transposons: an unusual and diverse set of integrated gene transfer elements. *Microbiol. Rev.* **59:**579–590.

Senghas, E., J. M. Jones, M. Yamamoto, C. Gawron-Burke, and D. B. Clewell. 1988. Genetic organization of the bacterial conjugative transposon, Tn*916*. *J. Bacteriol.* **170:**245–249.

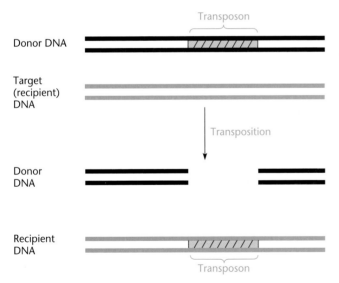

Figure 8.1 Overview of transposition. See the text for details.

act of transposition, but many transposons, before and after they hop, remain contiguous with other DNA molecules. Later in this chapter, we discuss more detailed models for transposition.

Transposition must be tightly regulated and occur only rarely; otherwise the cellular DNA would become riddled with the transposon, which would have many deleterious effects. Transposons have evolved elaborate mechanisms to regulate their transposition so that they hop very infrequently and do not often kill the host cell. The frequency of transposition varies from about once in every 10^3 cell divisions to about once in every 10^8 cell divisions, depending upon the type of transposon. Thus, the chance of a transposon hopping into a gene and inactivating it is not much higher than the chance that a gene will be inactivated by other types of mutations (see chapter 3).

Structure of Bacterial Transposons

There are many different types of bacterial transposons. Some of the smaller ones are about 1,000 bp long and carry only the genes for the transposases that promote their movement in DNA and the genes that regulate this movement. Larger transposons may also contain one or more other genes, such as those for resistance to an antibiotic.

One distinguishing feature of bacterial transposons is that all those identified so far contain **inverted repeats** at their ends (Figure 8.2). As discussed in chapters 2 and 3, two regions of DNA are inverted repeats if the sequence of nucleotides on one strand in one region, when read in the 5'-to-3' direction, is the same or almost the same as the 5'-to-3' sequence of the opposite strand in the other region.

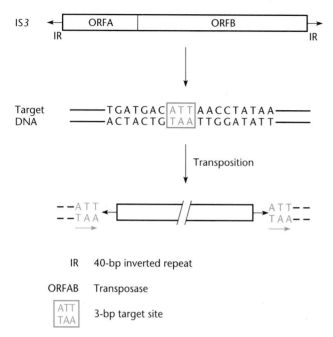

IR   40-bp inverted repeat

ORFAB   Transposase

ATT
TAA   3-bp target site

Figure 8.2 Structure of the insertion sequence element IS3. The inverted repeats are shown as arrows, and the 3-bp target sequence that will be duplicated after transposition is boxed.

Another common feature is the presence of short **direct repeats** of the target DNA that bracket the transposon (Figure 8.2). Direct repeats have the same or almost the same 5'-to-3' sequence of nucleotides on the same strand. As shown in Figure 8.2, the target DNA originally contains only one copy of the sequence at the place where the transposons insert. During the insertion of the transposon, this sequence is duplicated. Most transposons can insert into many places in DNA and so have little or no target specificity. Thus, the duplicated sequence will vary with the sequence at the site in the target DNA into which the transposon inserted. However, even though the duplicated sequences differ, the number of duplicated base pairs is characteristic of each transposon. Some duplicate as few as 3 bp and others as many as 9 bp. The molecular models for transposition discussed later in the chapter offer an explanation for the duplicated sequences.

Types of Bacterial Transposons

Each type of bacterium carries its own unique transposons, although many transposons are related across species as though they were recently exchanged. In this section, we describe some of the common types of transposons.

INSERTION SEQUENCE ELEMENTS

The smallest bacterial transposons are called **insertion sequence elements**, or IS elements for short. These transposons are usually only about 750 to 2,000 bp long and encode little more than the transposase enzymes that promote their transposition.

Insertion sequence elements carry no selectable genes, and they were discovered only because they inactivate a gene if they happen to hop into it. The first IS elements were detected as a type of *gal* mutation that was unlike any other known mutations. This type of mutation resembled deletion mutations in that it was nonleaky; however, unlike deletion mutations, it could revert, albeit at a lower frequency than base pair changes or frameshifts. Such anomalous *gal* mutations were also very polar and could prevent the transcription of downstream genes (see chapter 2). Later work showed that these mutations resulted from insertion of about 1,000 bp of DNA into a *gal* gene. Moreover, they were due to insertion of not just any piece of DNA but one of very few sequences.

Originally, four different IS elements were found in *Escherichia coli*: IS1, IS2, IS3, and IS4. Most strains of *E. coli* K-12 contain approximately six copies of IS1, seven copies of IS2, and fewer copies of the others. Almost all bacteria carry IS elements, with each species harboring its own characteristic IS elements, although sometimes related IS elements can be found in different bacteria. Plasmids also often carry IS elements, which are important in the formation of Hfr strains (see chapter 5).

Figure 8.2 shows the structure of the IS element IS3. In addition to the inverted repeats at its ends, it consists of two open reading frames (ORFAB). Translational frameshifting assembles the transposase from ORFA and ORFB. The smaller ORFA encodes a protein that regulates transcription of the transposase gene. The target site sequence that is duplicated in the target DNA upon insertion of IS3 is 3 bp long. As mentioned above, the length of such direct repeats is characteristic of each type of transposon.

Although the original IS elements were discovered only because they had hopped into a gene, causing a detectable phenotype, IS elements are now more often discovered during hybridization experiments with cloned regions of bacterial DNA as probes. Sometimes a cloned piece of bacterial DNA will hybridize to several sequences around the bacterial genome, suggesting that the sequence exists as more than one copy. When the cloned DNA is sequenced, it is often found to contain sequences characteristic of an IS element: inverted repeats at the ends flanked by short direct repeats.

COMPOSITE TRANSPOSONS

Sometimes two IS elements of the same type form a larger transposon, called a **composite transposon**, by bracketing other genes. Figure 8.3 shows the structures of three composite transposons, Tn5, Tn9, and Tn10. Tn5 consists of genes for kanamycin resistance (Kan^r) and streptomycin resistance (Str^r) bracketed by copies of an IS element called IS50. Tn9 has two copies of IS1 bracketing a chloramphenicol resistance gene (Cam^r). In Tn10, two copies of IS10 flank a gene for tetracycline resistance (Tet^r). Some composite transposons, such as Tn9, have the bracketing IS elements in the same orientation, whereas others, including Tn5 and Tn10, have them in opposite orientations.

Outside-End Transposition

Figure 8.4 illustrates transposition of a composite transposon. Each IS element can transpose independently as long as the transposase acts on both of its ends. However, because all the ends of the IS elements in a composite transposon are the same, a transposase encoded by one of the IS elements can recognize the ends of either. When such a transposase acts on the inverted repeats at the farthest ends of a composite transposon, the two IS elements will transpose as a unit, bringing along the genes between them. These two inverted repeats are called the "outside ends" of the two IS elements because they are the *farthest* from each other.

The two IS elements that form composite transposons are often not completely autonomous, because of mutation in the transposase gene of one of the ele-

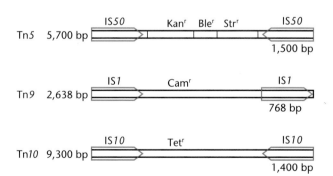

Figure 8.3 Structures of some composite transposons. The commonly used genes for kanamycin resistance, Kan^r, and the gene for chloramphenicol resistance, Cam^r, come from Tn5 and Tn9, respectively. The active transposase gene is in one of the two IS elements. Note that the IS elements can be in either the same or opposite orientation (arrows). Str^r, gene encoding streptomycin resistance; Tet^r, gene encoding tetracycline resistance; Ble^r, gene encoding bleomycin resistance.

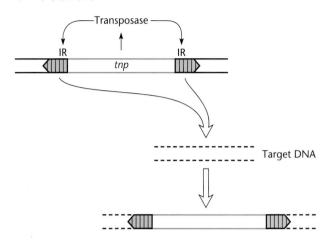

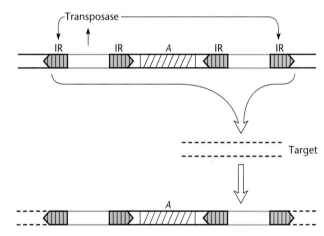

Figure 8.4 Two IS elements can transpose any DNA between them. (A) Action of the transposase at the ends of an isolated IS element causes it to transpose. (B) Two IS elements of the same type are close to each other in the DNA. Action of the transposase on their outside ends causes them to transpose together, carrying along the DNA between them. *A* denotes the DNA between the IS elements (hatched bar), and arrows indicate the inverted repeated (IR) sequences at the ends of the IS elements. Dashed lines represent the target DNA.

ments. Thus, only one of the IS elements encodes an active transposase. However, this transposase can act on the outside ends to promote transposition of the composite transposon.

Inside-End Transposition

The transposase encoded by one IS element in a composite transposon can also act on the "inside ends" of both IS elements, that is, the two ends that are closest to each other. Inside-end transposition presumably happens as often as outside-end transposition but has very different consequences.

One possible outcome of inside-end transposition is the creation of a new composite transposon, which was first demonstrated with Tn10 (see Foster et al., Suggested Reading). In these experiments, Tn10 was inserted into a small plasmid with an origin of replication (ori) and an ampicillin resistance gene (Amp^r). The plasmid served as the donor DNA. As shown in Figure 8.5, transposition with the outside ends of the IS10 element would move Tn10, with the tetracycline resistance gene, Tet^r, to another DNA. However, transposition from the inside ends would create a new composite transposon carrying the Amp^r gene and the plasmid origin of replication (ori) to another DNA. If this new composite transposon hops into a target DNA that does not have a functional origin of replication, it may confer on that DNA the ability to replicate. In the experiment, a λ phage with amber mutations in its replication genes *O* and *P* was used to infect amber-suppressing cells containing the donor plasmid with the transposon. The progeny phage were then plated on a non-amber-suppressor host. In this host, the phage could not replicate from the λ origin of replication because of the amber mutations in their replication genes. However, any phage into which the new composite transposon had hopped would be able to replicate by using the plasmid origin of replication. As expected, those few phages that formed plaques did contain the new composite transposon. The inside ends of the IS10 elements of Tn10 must have been used for its transposition.

Deletions and inversions can also be caused by inside-end transposition of a composite transposon to a nearby target on the *same* DNA (Figure 8.6). The neighboring sequences between the orginal site of insertion of the transposon and the site into which it is trying to transpose will be either deleted or inverted. Whether a deletion or inversion is created depends on how the inside ends of the IS elements in the transposon are attached to the target DNA. If the inside ends cross over each other before they attach, the neighboring sequences will be inverted. If they do not cross over each other, the neighboring sequences will be deleted. As

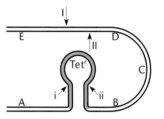
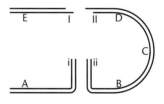
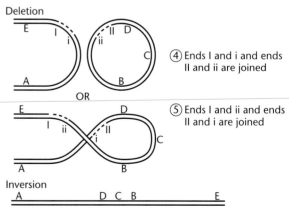

Figure 8.6 Rearrangements of DNA caused by composite transposons. Attempts to transpose by the inside ends of a composite transposon to a neighboring target sequence can cause either a deletion or an inversion of the intervening sequences, depending upon how the ends are attached. For an explanation of steps 1 and 2, see Figure 8.11.

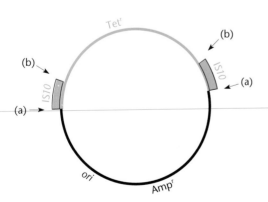

Figure 8.5 The outside or inside ends of the IS elements in a composite transposon can be used for transposition. Outside-end (a) transposition will transpose Tn10 (blue), including the gene encoding tetracycline resistance (Tet^r), whereas inside-end (b) transposition will transpose the plasmid, including the origin of replication and the gene for ampicillin resistance (Amp^r).

shown in Figure 8.6, the DNA between the two IS elements in the composite transposon will also be deleted, independent of what happens to the neighboring DNA. Therefore, these rearrangements are usually accompanied by the loss of any resistance gene on the composite transposon, which is how they are usually selected. For example, methods have been developed to select tetracycline-sensitive derivatives of *E. coli* harboring the Tn*10* transposon. Most of these tetracycline-sensitive derivatives will have deletions or inversions of DNA next to the site of insertion of the Tn*10* element. Presumably, inside-end transposition is responsible for most of the often-observed instability of DNA caused by composite transposons.

Assembly of Plasmids by IS Elements

Any time two IS elements of the same type happen to hop close to each other on the same DNA, a composite transposon is born. These transposons have not yet evolved a defined structure such as the named transposons (e.g., Tn*10*) described above. Nevertheless, the two IS elements can transpose any DNA between them. In this way, "cassettes" of genes bracketed by IS elements can be moved from one DNA molecule to another.

Many plasmids seem to have been assembled from such cassettes. Figure 8.7 shows a naturally occurring plasmid carrying genes for resistance to many antibi-

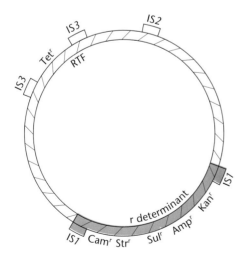

Figure 8.7 R-factors, or plasmids containing many resistance genes, may have been assembled by IS elements. The tetracycline resistance (Tetʳ) gene is bracketed by IS3 elements, and the region containing the other resistance genes (the r determinant, blue) is bracketed by IS1 elements. Reprinted with permission from H. Saedler, *in* A. I. Bukhari, J. A. Shapiro, and S. Adhya (ed)., *DNA Insertion Elements, Plasmids and Episomes*, p. 66, Cold Spring Harbor Laboratory Press, Cold Spring Harbor, N.Y., 1977.

otics. Such plasmids are historically called R-factors, because they confer resistance to so many different antibiotics (see chapter 4). Notice that many of the resistance genes on the plasmid are bracketed by IS elements. In partcular, IS3 flanks the tetracycline resistance gene and IS1 brackets the genes for resistance to many other antibiotics. Apparently, the plasmid was assembled in nature by resistance genes hopping onto the plasmid from some other DNA via the bracketing IS elements. In principle, any two transposons of the same type can move other DNA lying between them by a similar mechanism, but because IS elements are the most common transposons and often exist in more than one copy per cell, they probably play the major role in the assembly of plasmids.

NONCOMPOSITE TRANSPOSONS

Composite transposons are not the only ones to carry resistance genes. Such genes can also be an integral part of transposons known as **noncomposite transposons** (Figure 8.8). They are bracketed by short inverted repeats, but the resistance gene is part of the minimum transposable unit. These transposons may have evolved from composite transposons, but if that is so, little remains of the autonomous IS elements.

Noncomposite transposons seem to belong to a number of families in which the members are related to each other by sequence and structure. One well-studied family is the Tn*3* family. Interestingly, different members of transposon families, notably the Tn*21* family, often carry different resistance genes, even though they are almost identical otherwise. Apparently, the resistance genes can integrate into the transposon by means of specific integrase enzymes, much like lysogenic phages can integrate into the chromosome of bacteria (see section on integrases of transposon integrons, below).

Assays of Transposition

To study transposition, we must have assays for it. As mentioned above, insertion elements were discovered because they create mutations when they hop into a gene. However, this is usually not a convenient way to assay transposition, because transposition is infrequent and it is laborious to distinguish insertion mutations from the myriad other mutations that can occur. If the transposon carries a resistance gene, the job of assaying transposition is easier. But how do we know if a transposon has hopped in the cell? The cells are resistant to the antibiotic no matter where the transposon is inserted in the cellular DNA, and so hopping from one place to another makes no difference in the level of resistance of the cell. Obviously, detecting transposition requires special methods.

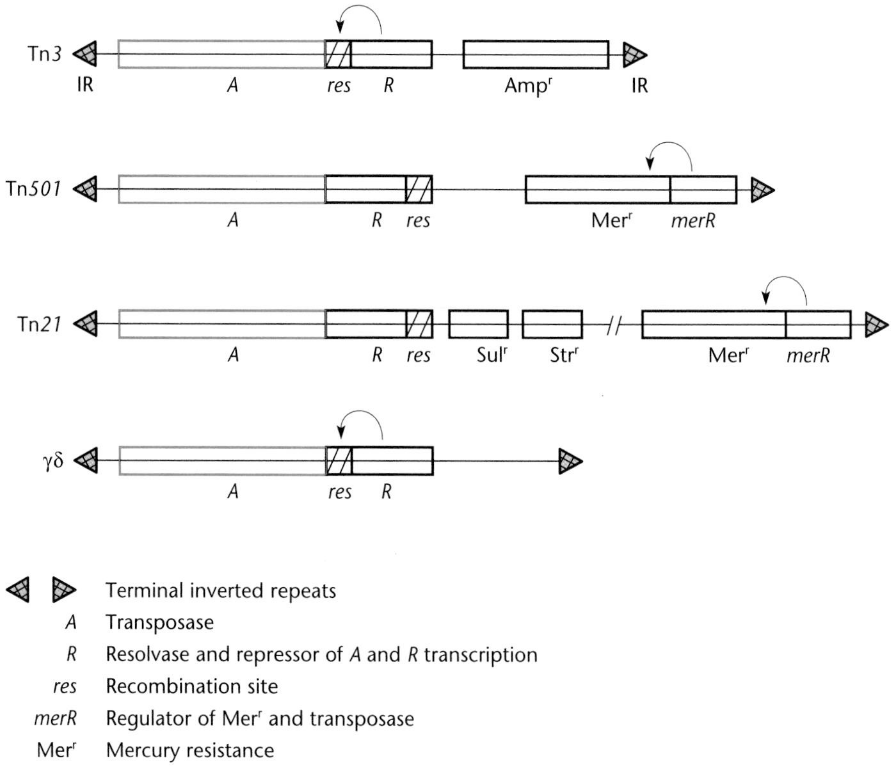

Figure 8.8 Some examples of noncomposite transposons. The open reading frames encoding the proteins are boxed. The terminal inverted repeats ends are shown as hatched arrows. *A* is the transposase; *R* is the resolvase and repressor of *A* transcription; *res* is the site at which resolvase acts; Mer^r^ is the mercury resistance region; and *merR* is the regulator of mercury resistance gene transcription. Tn*3* was originally found on the broad-host-range plasmid pR1drd19, Tn*501* was found on the *Pseudomonas* plasmid pUS1, Tn*21* was found on the *Shigella flexneri* plasmid R100, and γδ was found on the chromosome of *E. coli* and on the F plasmid.

◁ ▷	Terminal inverted repeats
A	Transposase
R	Resolvase and repressor of *A* and *R* transcription
res	Recombination site
merR	Regulator of Mer^r^ and transposase
Mer^r^	Mercury resistance

SUICIDE VECTORS

One way to assay transposition is with **suicide vectors.** Any DNA, including plasmid or phage DNA, that cannot replicate (i.e., is not a replicon) in a particular host can be used as a suicide vector. These DNAs are called suicide vectors because, by entering cells in which they cannot replicate, they essentially kill themselves. To assay transposition with a suicide vector, we use one to introduce a transposon carrying an antibiotic resistance gene into an appropriate host. How the suicide vector itself is introduced into the cells will depend upon its source. If it is a phage, the cells could be infected with that phage. If it is a plasmid, it could be introduced into the cells through conjugation.

Once in the cell, the suicide vector will remain unreplicated and will eventually be lost. The only way the transposon can survive and confer antibiotic resistance on the cells is by hopping to another DNA molecule that is capable of autonomous replication in those cells, for example a plasmid or the chromosome. Therefore, when the cells under study are plated on antibiotic-containing agar and incubated, the appearance of colonies, as a result of the multiplication of antibiotic-resistant bacteria, is evidence for transposition. These cells have been mutagenized by the transposon, since the transposon has hopped into a cellular DNA molecule—either

the chromosome or a plasmid. This is called **transposon mutagenesis.**

Phage Suicide Vectors

Some derivatives of the phage λ are designed to be used as suicide vectors. These phage have been rendered incapable of replication in nonsuppressing hosts by the presence of nonsense codons in their replication genes *O* and *P* (see chapter 7). They have also been rendered incapable of integrating into the host DNA by deletion of their attachment region, *attP*. Such a λ phage can be propagated on an *E. coli* strain carrying a nonsense suppressor. However, in a nonsuppressor *E. coli*, it cannot replicate or integrate. Because of the narrow host range of λ, these suicide vectors can normally only be used in strains of *E. coli* K-12.

Plasmid Suicide Vectors

Plasmid cloning vectors can also be used as suicide vectors. The plasmid containing the transposon with a gene for antibiotic resistance can be propagated in a host in which it can replicate and is then introduced into a cell in which it cannot replicate. In principle, any plasmid with a conditional-lethal mutation, nonsense or temperature sensitive, in a gene required for plasmid replication can be used as a suicide vector. The plasmid could be propagated in the permissive host or under permissive conditions and then introduced into a nonpermissive host or into the same host under nonpermissive conditions, depending upon the type of mutation. Alternatively, a narrow host range plasmid could be used. It can be propagated in a host in which it can replicate and introduced into a different species in which it cannot. We discuss such methods of transposon mutagenesis in chapter 16.

Plasmid suicide vectors can be introduced into the cells by transformation. However, conjugation is usually much more efficient, and so it is better to introduce plasmids by conjugation. If the plasmid contains a *mob* region, it can be mobilized into the recipient cell by using the Tra functions of a self-transmissible plasmid (see chapter 5).

THE MATING-OUT ASSAY FOR TRANSPOSITION

Transposition can also be assayed by using the "mating-out" assay. In this assay, a transposon in a nontransferable plasmid will not be transferred into other cells unless it hops into a plasmid that is transferable. Figure 8.9 shows a specific example of a mating-out assay with *E. coli*. In the example shown, the transposon Tn*10* carrying tetracycline resistance has been inserted into a small plasmid that is neither self-transmissible nor mobilizable (see chapter 5). This small plasmid is used to transform

cells containing F, a larger, self-transmissible plasmid. While the cells are growing, the transposon may hop from the smaller plasmid into the F plasmid in a few of the cells. Later, when these cells are mixed with streptomycin-resistant recipient cells, any F plasmid into which the transposon hopped will carry the transposon when it transfers to a new cell, thus conferring tetracycline resistance on that cell. Transposition can be detected by plating the mating mixture on agar containing tetracycline and counterselecting the donor with streptomycin.

The appearance of antibiotic-resistant transconjugants in a mating-out assay is not absolute proof of transposition. Some transconjugants could become antibiotic resistant by means other than transposition of the transposon into the larger, self-transmissible plasmid. The smaller plasmid containing the transposon could have been somehow mobilized by the larger plasmid, or the smaller plasmid could have been fused to the larger plasmid by recombination or by cointegrate formation (see below). A few representative transconjugants should be tested, for example, by using restriction digests (see chapter 15) to verify that they contain only the larger plasmid with the transposon inserted.

Molecular Models for Transposition

All bacterial transposons seem to transpose by one of two mechanisms. Some transposons, represented by Tn*3* and phage Mu, use a **replicative** mechanism of transposition, in which the entire transposon replicates during transposition, resulting in two copies of the transposon. Other transposons, represented by many of the IS elements and composite transposons such as Tn*10*, Tn*5*, and Tn*9*, transpose by a **cut-and-paste**, or conservative, mechanism, in which the transposon is removed from one place and inserted into another.

REPLICATIVE TRANSPOSITION

The first detailed model to be developed for transposition attempted to explain all of what was known about Tn*3* transposition and the transposition of other transposons such as phage Mu (see Shapiro, Suggested Reading). This model continues to be generally accepted for those types of transposons. The model incorporates the following observations, some of which we have already mentioned.

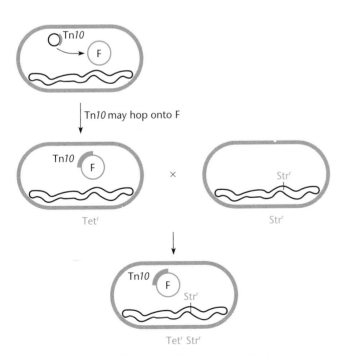

Figure 8.9 Example of a mating-out assay for transposition. See text for details.

1. Whenever a transposon such as Tn*3* hops into a site, a short sequence of the target DNA is duplicated. The number of bases duplicated is characteristic of each transposon. (For Tn*3*, 5 bp is duplicated; for IS*1*, 9 bp is duplicated.)

2. The formation of a **cointegrate**, in which the donor and target DNAs have become fused and encode

two copies of the transposon, is an intermediate step in the transposition process.

3. Once the cointegrate has formed, it can be resolved into separate donor and target DNA molecules either by the host recombination functions or by a transposon-encoded **resolvase** that promotes recombination at internal *res* sequences.

4. The donor DNA and target DNA molecules both have a copy of the transposon after resolution of the cointegrate. Therefore, the transposon does not actually move but duplicates itself, and a new copy appears somewhere else; hence the name "replicative transposition."

5. Neither transposition nor, for some transposons, resolution of cointegrates requires the normal recombination enzymes or extensive homology between the transposon and the target DNA.

Figure 8.10 shows the molecular details of the model for replicative transposition. In the first step, the transposase makes single-stranded breaks at each junction between the transposon and the donor DNA and a double-stranded break in the target DNA. The break in the target DNA is staggered so that the nicks in the two strands are separated by the same number of base pairs as will be duplicated in the target DNA during insertion of the transposon, as explained below. The cutting leaves two 5' ends and two 3' ends in the target DNA and a 5' end and a 3' end at each junction between the transposon and the donor DNA. The 5' ends in the target DNA are then joined (ligated) to the 3' ends of the transposon (Figure 8.10). Replication then proceeds in both directions over the transposon, with the free 3' ends of the target DNA serving as primers. After replication over the transposon, the 3' ends of the newly synthesized strands are ligated to the remaining free 5' ends of the donor DNA to form the cointegrate. The last step, **resolution** of the cointegrate, results from recombination between the two *res* sites in the cointegrate promoted by the resolvase of the transposon (see section on resolvases, below). Resolution of the cointegrate gives rise to two copies of the transposon, one at the former, or donor site, and a new one at the target site.

This model explains why cointegrates are obligate intermediates in replicative transposition. After replication has proceeded over the transposon in both directions, the donor DNA and the target DNA will be fused to each other, separated by copies of the transposon, as shown.

This model also explains why, after transposition, a short target DNA sequence of defined length has been duplicated at each end of the transposon. Because it makes a staggered break in the target DNA, the trans-

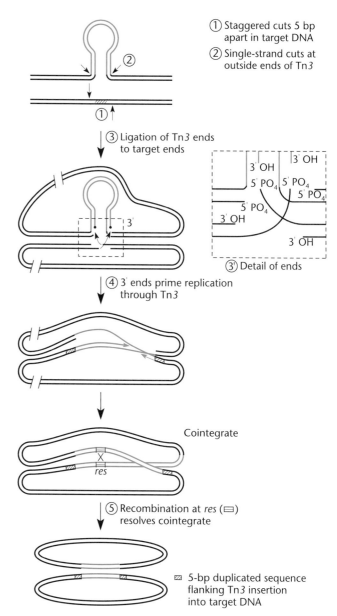

① Staggered cuts 5 bp apart in target DNA

② Single-strand cuts at outside ends of Tn3

③ Ligation of Tn3 ends to target ends

③' Detail of ends

④ 3' ends prime replication through Tn3

Cointegrate

⑤ Recombination at *res* (▱) resolves cointegrate

▱ 5-bp duplicated sequence flanking Tn3 insertion into target DNA

Figure 8.10 Replicative transposition of Tn3 (blue) and the formation and resolution of cointegrates. At sites labeled 1 and 2, breaks are made in the target DNA and at the ends of the transposon, respectively. (3) The 3' OH ends of the transposon (dots) are ligated to 5' PO₄ ends of the target DNA. The inset (3') shows details of the ends. (4) The free 3' ends of target DNA prime replication in both directions over the transposon to form the cointegrate. (5) The cointegrate is resolved by recombination promoted by the resolvase TnpR at the *res* sites.

posase causes a short region of the target DNA to be duplicated when replication proceeds from this staggered break over the transposon. The number of base pairs of target DNA duplicated at the ends of the transposon will be the same as the number of base pairs between the nicks in the two strands in the staggered break and will be characteristic of the transposase enzyme for each type of transposon.

Finally, this model explains why replicative transposition is independent of most host functions including DNA ligase and the normal recombination functions such as RecA (see chapter 9). The transposase cuts the target and donor DNAs and promotes ligation of the ends. Also, the normal recombination system is not needed to resolve the cointegrate into the original replicons, because the resolvase specifically promotes recombination between the *res* elements in the cointegrates. Although cointegrates can also be resolved by homologous recombination anywhere within the repeated copies of the transposon, the resolvase greatly increases the rate of resolution by actively promoting recombination between the *res* sequences.

This model for replicative transposition was based mostly on the results of a genetic analysis of Tn3 transposition. We discuss these results in more detail in chapter 16 (see Gill et al., Suggested Reading in chapter 16).

CUT-AND-PASTE TRANSPOSITION

Some bacterial transposons, including many of the IS elements and composite transposons, do not transpose by a replicative mechanism but, rather, by the simpler cut-and-paste mechanism illustrated in Figure 8.11. In this mechanism, the transposase makes double-stranded breaks at the ends of the transposon, cutting it out of the donor DNA, and then pastes it into the target DNA at the site of a staggered break. When the single-stranded gaps created by the nicks in the target DNA are filled in, a short sequence will be duplicated. For most types of transposons, removal of the transposon from the donor DNA probably leaves breaks in the donor DNA, which is consequently degraded, as shown in Figure 8.11.

Genetic and biochemical experiments have contributed the evidence for cut-and-paste transposition of some transposons. Some of the strongest evidence supporting this mechanism has come from recently developed in vitro transposition systems. In the next few sections, we describe in detail some of the ways in which cut-and-paste transposition differs from replicative transposition and the evidence for each of these features.

No Cointegrate Intermediate

In cut-and-paste transposition, cointegrates do not form as a necessary intermediate, as they do in the replicative mechanism. This conclusion is supported by indirect ev-

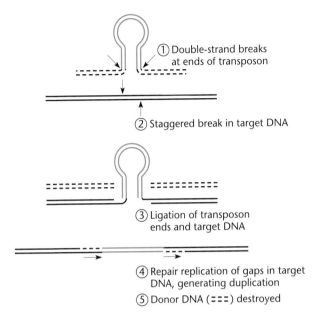

Figure 8.11   Cut-and-paste transposition. (1) Double-strand breaks are made at the ends of the transposon (arrows). (2) Staggered breaks are made in the target DNA (arrows). (3) The free 3' ends of the transposon are ligated to the 5' ends of the target DNA. The dashed lines represent the donor DNA, which will be degraded. (4) DNA polymerase fills in the gaps of the target DNA, producing the short duplication of target DNA at the ends of the transposon. (5) Donor DNA is destroyed.

idence suggesting that IS elements and composite transposons do not form these intermediates. For example, there are no mutants of the transposons Tn10 and Tn5 that accumulate cointegrates as there are for Tn3. Moreover, even if cointegrates are formed artificially by recombinant DNA techniques, there is no evidence that the cointegrates can be resolved except by the normal host recombination system. Therefore, the cut-and-paste transposons do not seem to encode their own resolvases, which they would be likely to do if cointegrates were a normal intermediate in their transposition process.

Both Strands of the Transposon Transpose

The primary difference between replicative and cut-and-paste transposition is that in the latter, both strands of the transposon move to the target DNA. The results of genetic experiments with Tn10 (outlined in Figures 8.12 and 8.13) support this conclusion (see Bender and Kleckner, Suggested Reading). The first step in these experiments was to introduce different versions of transposon Tn10 into a λ suicide vector. Both of the two Tn10 derivatives contained a copy of the *lacZ* gene as

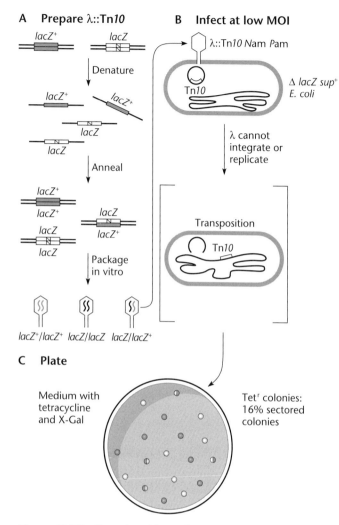

A Prepare λ::Tn*10*

B Infect at low MOI

λ cannot integrate or replicate

Δ *lacZ sup⁺ E. coli*

Transposition

C Plate

Medium with tetracycline and X-Gal

Tetʳ colonies: 16% sectored colonies

Figure 8.12 Genetic evidence for nonreplicative transposition by Tn*10*. (A) Preparation of λ::Tn*10 lacZ/lacZ⁺* heteroduplex DNA. (B) The λ::Tn*10* infects a nonpermissive *E. coli* host. Because the λ contains *Nam* and *Pam* mutations, it cannot integrate or replicate. If the transposon replicates during transposition, the bacteria in the Tetʳ colonies will get only one or the other strand of DNA in the heteroduplex, and the colonies will be all blue or all colorless. If both strands are transferred, some colonies will be sectored, part blue and part white. See the text for further details. MOI, multiplicity of infection.

well as the Tetʳ gene Tn*10* usually carries. However, one of the Tn*10* derivatives carried three missense mutations in the *lacZ* gene to inactivate it. The strands of the two λ DNAs were separated and reannealed to make heteroduplex DNA, in which each of the strands came from a λ phage carrying a different derivative of Tn*10*. Consequently, the heteroduplex DNAs had one strand with a good copy of *lacZ* and another strand with the mutated copy of *lacZ*.

In the next step, the heteroduplex DNA was packaged into λ heads in vitro (see chapter 7) and used to infect Lac⁻ cells. Because this λ was a suicide vector, the only cells that became Tetʳ were ones in which the Tn*10* derivatives had hopped into the chromosome. If the transposition had occurred by a replicative mechanism, the Tetʳ colonies would have contained either Lac⁺ or Lac⁻ bacteria (Figure 8.13), since the information in only one of the two strands could have been transferred. If, however, the transposition had occurred by a nonreplicative cut-and-paste mechanism, the cells in at least some Tetʳ colonies due to transposition would have received both the Lac⁺ and the Lac⁻ strands of the heteroduplex DNA. When these heteroduplex DNAs replicated, they would give rise to both Lac⁺ and Lac⁻ bacteria in the same colony, making "sectored" blue-and-white colonies on X-Gal plates (see chapter 14), as shown in Figure 8.12. In the experiment, sectored blue-and-white colonies were observed, supporting the conclusion that both strands were transferred.

Transposon Leaves the Donor DNA

A major difference between replicative and cut-and-paste transposition is the number of copies of the transposon after transposition. The replicative mechanism leaves the cell with two copies, and the cut-and-paste mechanism leaves it with one. It might seem easy to determine whether a copy of the transposon still exists in the donor DNA after transposition; however, it is usually difficult. For example, if the donor DNA containing the transposon is a multicopy plasmid and the transposition occurs by a cut-and-paste mechanism, only one copy of the plasmid will lose its transposon, leaving many plasmids intact with the transposon. Even if the transposition occurs from a DNA that normally exists in a single-copy, such as the chromosome or a single copy plasmid, a copy of the transposon remaining in the donor DNA could be attributed to replication of the donor DNA prior to transposition.

For much the same reasons, it is difficult to determine whether the donor DNA is resealed after the transposon is cut out of it during cut-and-paste transposition. It might seem that the donor DNA must be resealed after the transposon hops; otherwise, transposition would leave a double-stranded break in the donor DNA, which would be a lethal event in many cases. However, cells usually carry more than one copy of a plasmid, and even the chromosome is usually in a partial state of replication, so that many regions exist in more than one copy per cell. If a transposon were to leave an unreplicated portion of the chromosome by means of cut-and-paste transposition, the cell would die. However, transposition events are infrequent, and

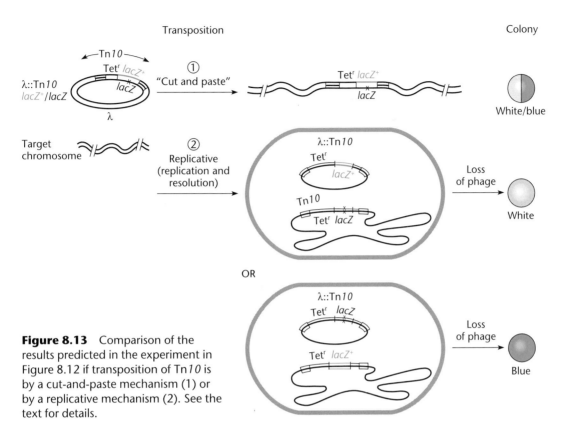

Figure 8.13 Comparison of the results predicted in the experiment in Figure 8.12 if transposition of Tn*10* is by a cut-and-paste mechanism (1) or by a replicative mechanism (2). See the text for details.

the few that are lethal will go undetected in a large population of bacteria.

The best evidence that the donor DNA is not resealed or, at least, is not resealed correctly after transposition by most cut-and-paste transposons comes from reversion studies. For a transposon insertion mutation to revert, the transposon must be completely removed from the DNA in a process called **precise excision**. Not a trace of the transposon can remain, including the duplication of the short target sequence, or the gene would probably remain disrupted and nonfunctional.

If the transposon were precisely excised and the donor DNA were resealed every time a transposon hopped by a cut-and-paste mechanism, then transposon insertion mutations would revert every time the transposon hopped. However, reversion of insertion mutations occurs at a much lower rate than does transposition itself. Moreover, mutations in the transposon itself that inactivate the transposase and render the transposon incapable of transposition do not further lower the reversion frequency, as might be expected if the few revertants that are seen resulted from another transposition event. Presumably, the rare precise excisions that cause transposon insertion mutations to revert are due to homologous recombination between the short duplicated target sequences bracketing the transposon and are unrelated to transposition itself.

Therefore, for most cut-and-paste transposons, the donor DNA is apparently left broken after the transposon is cut out of it, as shown in Figure 8.11. Only during the transposition of conjugative transposons such as Tn*916* is the donor DNA known to be correctly resealed after transposition. These transposons do not duplicate target DNA sequences when they insert (see Box 8.2). After excision, these transposons form transient circular molecules. The absence of duplicated target sequences facilitates precise excision of these transposons, since cutting out the transposon at its ends will not leave the short duplicated region.

RELATIONSHIP BETWEEN REPLICATIVE AND CUT-AND-PASTE TRANSPOSITION

Even though replicative transposition and cut-and-paste transposition seem different, they are actually mechanistically related. As shown in Figures 8.10 and 8.11, the major difference is in the number of strand cuts that the transposase enzyme makes in the junction between the transposon and the donor DNA. A cut-and-paste transposase makes cuts in both strands in the junction, whereas a replicative transposase cuts only one strand at the junction. Otherwise, the two mechanisms are similar. In both, the cut 5′ ends of the target DNA are joined to the free 3′ ends of the transposon. Then in both mechanisms, the free 3′ ends of the target DNA are used as primers for replication that proceeds until a free 5' end in the donor DNA, to which the newly replicated DNA can be joined, is reached. The difference comes from how much DNA must be replicated before this 5' end is reached. In cut-and-paste transposition, replication proceeds only a short distance owing to the staggered break in the target DNA before the growing 3' end reaches the free 5' end at the end of the transposon and the two ends join. As mentioned, this replication gives rise to the short duplications of the target DNA but does not duplicate the transposon itself. In replicative transposition, in contrast, the replication must proceed over the region of the staggered break and then over the entire transposon before the free 5' end is reached on the other side of the transposon. This process duplicates the target DNA as well as the transposon and gives rise to a cointegrate.

A dramatic confirmation of the similarity between the cut-and-paste and replicative mechanisms of transposition came with the demonstration that a cut-and-paste transposon can be converted into a replicative transposon by a single amino acid change in the transposase (see May and Craig, Suggested Reading). Transposon Tn7 normally transposes by a cut-and-paste mechanism, and different regions of the transposase make the cuts in the opposite strands of DNA at the ends of the transposon. If the region that makes one of the two cuts is inactivated by a mutation, the transposase will cut only one of the two strands, like a replicative transposase. The Tn7 transposon with such an altered transposase then transposes by the replicative mechanism, forming a cointegrate. Apparently, the transposase need only make the appropriate cuts and joinings, and the replication apparatus of the cell does the rest. It is somewhat surprising that the same transposase enzyme can support both types of transposition, since the transposon presumably needs different cellular replication machineries for each. Replication of tens of thousands of base pairs during replicative transposition would be a much more involved process than replication of a few base pairs during cut-and-paste transposition.

TARGET SITE SPECIFICITY

No transposable elements insert completely randomly into target DNA. For most, specific target sequences seem less important than local properties of the DNA such as supercoiling, methylation, or transcription. Tn5 is a transposon that is close to random in choice of insertion sites, although some hot spots can be observed. Tn10 transposes nonrandomly, with most insertions occurring in one or a few sites in a given gene. Tn7 is especially limited in transposition, inserting preferentially into one chromosomal site.

EFFECTS ON GENES ADJACENT TO THE INSERTION SITE

Most insertion element and transposon insertions cause polar effects if they insert into a gene transcribed as a polycistronic mRNA. The inserted element contains transcriptional stop signals and may also contain long stretches of sequence that is transcribed but not translated. The latter may cause Rho-dependent transcriptional termination.

Some insertions may enhance expression of a gene adjacent to the insertion site. This expression can result from transcription that originates within the transposon. For example, both Tn5 and Tn10 contain outward-facing promoters near their termini.

REGULATION OF TRANSPOSITION

Transposition of most transposons occurs rarely, as discussed above, because transposons self-regulate their transposition frequency. The regulatory mechanisms used by various transposons differ greatly. For some, transposition occurs primarily just after a replication fork has passed through the element. Newly replicated DNA is hemimethylated (see chapter 1), and hemimethylated DNA has been shown not only to activate the transposase promoter of Tn10 but also to increase the activity of the transposon ends. Most transposons employ mechanisms to modulate the level of transposase transcription and/or translation (see Craig, Suggested Reading).

Site-Specific Recombination

Another type of nonhomologous recombination, **site-specific recombination,** occurs only between specific sequences or sites on DNA. It is promoted by enzymes called **site-specific recombinases,** which recognize two specific sites in DNA and promote recombination between them. Even though the two sites generally have short sequences in common, the regions of homology are usually too short for normal homologous recombination to occur efficiently. Therefore, efficient recombination between the two sites requires the presence of a specific recombinase enzyme. In this section, we discuss some examples of site-specific recombination in bacteria and phages.

Developmentally Regulated Excision of Intervening DNA

Genes sometimes contain sequences that are not present in the final RNA or protein product of the gene. The extra or intervening sequences in genes are often removed from the RNA or the protein product by splicing after the RNA or protein is synthesized (see Box

9.1). However, sometimes the intervening sequences are cut out of the DNA iself, before the gene is expressed. One function of site-specific recombinases is to remove intervening DNA sequences from genes. Note that this type of gene rearrangement is possible only in cell lines that do not need to reproduce themselves, that is, those that are terminally differentiated. Otherwise, the intervening DNA sequences would be lost to subsequent generations.

The classic example of DNA rearrangements during terminal differentiation is in the vertebrate immune cells that make antibodies. In this case, site-specific recombination during the differentiation of the immune cells removes different lengths of sequences from the few germ line genes that encode antibodies, thereby creating hundreds of thousands of new genes encoding antibodies of different specificities. The antibody-encoding genes in the germ line cells (eggs and sperm) remain intact, however, so that no DNA sequences are lost to future generations and the genes can undergo similar rearrangements in subsequent generations.

Most bacteria multiply by cell division, and none of the cells undergo terminal differentiation. Therefore, irreversible DNA rearrangements are generally not possible in bacteria. Nevertheless, bacteria do manifest a few examples of terminal differentiation.

Some types of bacteria sporulate by forming endospores, a process during which the cell divides into a mother cell and a spore. Only the spore need contain the full complement of the bacterial DNA since the mother cell is required only to make the spore and will eventually lyse. Consequently, the mother cell terminally differentiates and therefore can undergo irreversible DNA rearrangements needed for the sporulation process. In *Bacillus subtilis* sporulation, an intervening sequence of 42 kbp is cut out of the gene for the sigma factor σ^K, which is required for transcription of some genes in the mother cell. Unless this intervening sequence is removed, σ^K will not be synthesized, some of the gene products required for sporulation will not be expressed, and sporulation will be blocked (see Kunkel et al., Suggested Reading).

Terminal differentiation also occurs in the formation of highly specialized cells responsible for fixing atmospheric nitrogen in some types of filamentous cyanobacteria (see Golden et al., Suggested Reading). These cells, called heterocysts, appear periodically in the filaments when the bacteria are growing under conditions of nitrogen starvation. The heterocysts never divide to form new cells and disappear when the cells are provided with a good source of nitrogen, so that they do not need a full complement of DNA and can undergo irreversible DNA rearrangements. As occurs in *B. subtilis* sporula-

tion, intervening sequences are cut out of some genes while the cells differentiate. In the cyanobacteria, these sequences are cut out of at least two genes. One is the *nifD* gene, which encodes a product required for nitrogen fixation in the heterocysts after an intervening sequence of 11 kbp is cut out.

The process of removing the intervening sequences seems to be similar in both developmental systems. Recombination between directly oriented sites bracketing the intervening sequence results in its excision. In both cases, the site-specfic recombinase required to promote recombination between the bracketing sites is encoded within the intervening sequence itself and is expressed only during differentiation. Therefore, the recombinase will be expressed only in the cells undergoing terminal differentiation, and the intervening sequence will not be lost from normally replicating cells.

The function of these intervening sequences is unclear. Even though they are removed only during development, their removal seems to play no important regulatory role in either sporulation or heterocyst differentiation. Permanently removing the intervening sequences from the chromosome, so that the σ^K and *nifD* genes are not interrupted, does not adversely affect the ability of the bacteria to sporulate or to develop heterocysts that can fix nitrogen. One possibility is that these intervening sequences are parasitic DNAs like phages or transposons. If they integrate themselves into an essential gene, the cell cannot easily delete them without killing itself. As with transposons, unless the deletion event precisely removes the intervening sequence, the gene will be inactive, disrupting an important cellular function. By excising themselves during differentiation, however, the parasitic DNA elements allow sporulation or nitrogen fixation by their hosts and so have no deleterious effects—the mark of a good parasite.

Integrases

Integrases also recognize two sequences in DNA and promote recombination between them; therefore, they are no different in principle from the site-specific recombinases that resolve cointegrates or remove intervening sequences. However, rather than remove a DNA sequence by promoting recombination between two directly repeated sequences on the *same* DNA, integrases act to integrate one DNA into another by promoting recombination between two sites on *different* DNAs.

PHAGE INTEGRASES
The best-known integrases are the Int enzymes, which are responsible for the integration of circular phage DNA into the DNA of the host to form a prophage (see the section on lysogeny in chapter 7). Briefly, the phage

integrase specifically recognizes the *attP* site in the phage DNA and the *attB* site on the bacterial chromosome and promotes recombination between them. Usually, phage integrases are extremely specific. Only the *attP* and *attB* sites will be recognized, so that the phage will only integrate at one or at most a few places in the bacterial chromosome. This type of site-specific recombination was discussed in chapter 7 and so is not repeated here.

INTEGRASES OF TRANSPOSON INTEGRONS

Integrases are also important in the evolution of some types of transposons. The first indication of this importance came from comparisons of members of some families of transposons, such as the Tn*21* family (see Figure 8.14). The three transposons in this figure are similar, except that they have different antibiotic resistance genes inserted at exactly the same site, designated *P* in the figure. The transposon In*0* has no antibiotic resistance gene inserted at the *P* site, whereas In*3* has a gene for trimethoprim resistance inserted at this site. In*2*, which is actually a transposon inserted within the Tn*21* transposon, has a gene for streptomycin and spectinomycin resistance inserted at this site. Yet other members of this family carry different genes at this site, including those for resistance to kanamycin, gentamicin, trimethoprim, chloramphenicol, and oxacillin, as well as one for mercury resistance. Presumably, In*0* is a direct descendant of

a primordial transposon that did not have an antibiotic resistance gene at the *P* site. The other transposons then evolved from this primordial transposon as different antibiotic resistance genes were inserted at the *P* site.

But how did those resistance genes come to occupy the same site on the transposon? The answer came from the discovery of a gene, homologous to the phage integrases, lying next to the *P* site on the transposons. Furthermore, the integrase encoded by this gene will promote recombination between two DNAs containing the 59-bp sequence adjacent to the integration site. Presumably, site-specific recombination between the *P* site in the transposon and similar sites in other DNAs carrying resistance genes has resulted in the incorporation of the resistance genes into the site. The advantages of this system are obvious. By carrying different resistance genes, the transposon allows the cell, and thereby itself, to survive environments containing various toxic chemicals. However, where the resistance genes originally came from remains a mystery.

Resolvases

The resolvases of transposons discussed above are another type of site-specific recombinase. These enzymes promote the resolution of cointegrates by recognizing the *res* sequences that occur in one copy in the transposon but occur in two copies in direct orientation in cointegrates. Recombination between these two *res* se-

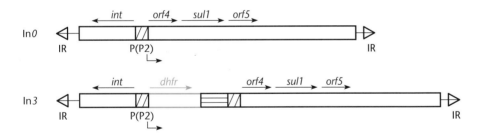

int	Phage-like integrase
sul1	Sulfonamide resistance
dhfr	Trimethoprim resistance
aadA	Streptomycin and spectinomycin resistance
◹	GTT
▤	59-bp element
P(P2)	Strong promoter

Figure 8.14 Integrons of the Tn*21* family. The transposon In*0* contains only the integrase gene and a gene for sulfonamide resistance. IR denotes a 25-bp inverted repeated sequence. In In*3*, the 3-bp GTT sequence has been the site of an integrase-mediated insertion of the gene for resistance to tetrahydrofolate reductase inhibitors, *dhfr*. In the In*2* element found inserted in Tn*21*, the GTT has been the site of insertion of the *aadA* gene, conferring resistance to streptomycin and spectinomycin. The expression of the inserted resistance genes is driven by the integron promoter P(P2). Also shown are 59-bp elements located at the 3' ends of the inserted resistance-encoding sequences. These 59-bp sequences are thought to be involved in the Int-mediated recombination reactions.

quences will excise the DNA lying between them, resolving the cointegrate.

DNA Invertases

The DNA invertases promote site-specific recombination between two closely linked sites on DNA. The main difference between the reactions promoted by DNA invertases and those catalyzed by resolvases is that two sites recognized by invertases are in reverse orientation with respect to each other whereas the sites recognized by resolvases are in direct orientation. As discussed in chapter 3 in the section on types of mutations, recombination between two sequences that are in direct orientation will delete the DNA between the two sites whereas recombination between two sites that are in inverse orientation with respect to each other will invert the intervening DNA.

The sequences acted upon by DNA invertases are called **invertible sequences**. These short sequences may carry the gene for the invertase or may be adjacent to it. Therefore, the invertible sequence and its specific invertase form an invertible cassette that sometimes plays an important regulatory role in the cell, some examples of which follow.

PHASE VARIATION IN *SALMONELLA* SPECIES

One type of invertible sequence is responsible for **phase variation** in some strains of *Salmonella*. Phase variation was discovered in the 1940s with the observation that some strains of *Salmonella* can shift from making flagella composed of one flagellin protein, H1, to making flagella composed of a different flagellin protein, H2. The shift can also occur in reverse, i.e., from making H2-type flagella to making H1-type flagella. The flagellar proteins are the strongest antigens on the surface of many bacteria, and periodically changing their flagella may help these bacteria escape detection by the host immune system.

Two features of the *Salmonella* phase variation phenomenon suggested that the shift in flagellar type was not due to normal mutations. First, the shift occurs at a frequency of about 10^{-4} per cell, much higher than normal mutation rates. Second, both phenotypes are completely reversible—the cells switch back and forth between the two types of flagella.

Figure 8.15 outlines the molecular basis for the *Salmonella* antigen phase variation (see Simon et al., Suggested Reading). A DNA invertase called the Hin invertase causes the phase variation by inverting an invertible sequence upstream of the gene for the H2 flagellin. The invertible sequence contains the invertase gene itself and a promoter for two other genes: one en-

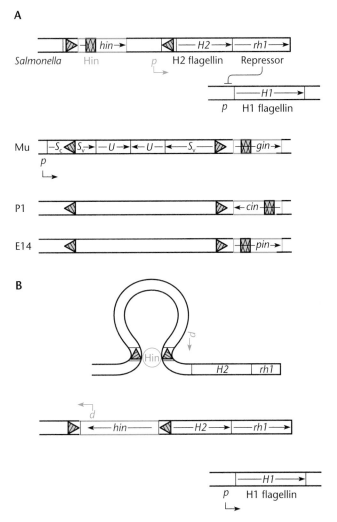

Figure 8.15 Regulation of the *Salmonella* phase variation and some other members of the family of Hin invertases. (A) Invertible sequences, bordered by blue arrows, of *Salmonella* and several phages. (B) Hin-mediated inversion. In one orientation, the H2 flagellin gene, as well as the repressor gene *rh1*, are transcribed from the promoter *p* (in blue). In the other orientation, neither of these genes is transcribed and the H1 flagellin is synthesized instead. The invertase Hin is made constitutively from its own promoter. Redrawn from G. F. Hatfull and N. D. F. Grindley, p. 362, *in* R. Kucherlapati and G. R. Smith (ed.), *Genetic Recombination*, American Society for Microbiology, Washington, D.C., 1988.

codes the H2 flagellin and another encodes a repressor of H1 gene transcription. With the invertible sequence in one orientation, the promoter will transcribe the H2 gene and the repressor gene, and only the H2-type flagellum will be expressed on the cell surface. When the sequence is in the other orientation, neither the H2 gene nor repressor gene will be transcribed, because the promoter will be facing backward. Now, however, without the repressor, the H1 gene can be transcribed; therefore, in this state, only the H1-type flagellum will be expressed on the cell surface. Clearly, the Hin DNA invertase that is encoded in the invertible sequence is expressed in either orientation, or the inversion would not be reversible.

OTHER INVERTIBLE SEQUENCES

There are a few other known examples of regulation by invertible sequences in bacteria and phages. For example, fimbria synthesis in some pathogenic strains of *E. coli* is regulated by an invertible sequence. Fimbriae are required for the attachment of the bacteria to the eukaryotic cell surface and may also be important targets of the host immune system.

Invertible sequences also exist in some phages. Some examples are the invertible regions of phages Mu and P1 and of the defective prophage e14 shown in Figure 8.15. These phages use invertible sequences to change their tail fibers. The tail fibers made when the invertible sequences are in one orientation differ from those made when the sequences are in the other orientation, broadening the host range of the phage. In phage Mu, the tail fiber genes expressed when the invertible sequence is in one orientation allow the phage to adsorb to *E. coli*, and when it is in the other orientation, the phage adsorbs to *Citrobacter* spp.

THE Hin FAMILY OF RECOMBINASES

As mentioned above, many site-specific recombinases, whether they be integrases, resolvases, or invertases, are sometimes closely related to each other. Dramatic evi-

dence for these relationships came from the discovery that the Hin invertase will invert the Mu, P1, and e14 invertible sequences and vice versa (see Van De Putte et al., Suggested Reading). Because they are related, these DNA invertases are called the Hin family of invertases, named after the first member to be discovered. The members of the Hin family of invertases are also related by sequence to the resolvases of Tn3-like transposons. It is tempting to speculate that these recombinases were all derived from the same original enzyme, which has been pressed into service in many different regulatory systems.

Topoisomerases

As discussed in chapter 1, topoisomerases are enzymes that make single- or double-stranded breaks in DNA and allow DNA strands to pass through the breaks before resealing them. Because they catalyze the breakage and joining of DNA strands, topoisomerases essentially perform a nonhomologous recombination reaction. However, since they usually rejoin the original strands after DNA passes through the break, they do not, in general, create new combinations of sequences. Only when they mistakenly join DNA strands from two different regions or different DNA molecules do they cause recombination. Because this recombination is often between sequences that have little or no sequence similarity, the recombination promoted by topoisomerases is also an example of nonhomologous recombination. Topoisomerases are sometimes responsible for mutations, including deletions and frameshifts.

SUMMARY

1. Nonhomologous recombination is the recombination between specific sequences on DNA that occurs even if the sequences are mostly dissimilar.

2. Transposition is the movement of certain DNA sequences, called transposons, from one place in DNA to another. The smallest known bacterial transposons are IS elements, which contain only the genes required for their own transposition. Other transposons carry genes for resistance to substances such as antibiotics and heavy metals. Transposons have played an important role in evolution and are useful for mutagenesis, gene cloning, and random gene fusions.

3. Composite transposons are composed of DNA sequences bracketed by IS elements. Inside-end transposition

by composite transposons can cause the deletion and inversion of neighboring DNA sequences.

4. Bacterial transposons transpose by two distinct mechanisms: replicative transposition and cut-and-paste, or conservative, transposition. In replicative transposition, which is used by such transposons as Tn3 and Mu, the entire transposon is replicated, leading to formation of a cointegrate. In cut-and-paste transposition, which is used by many IS elements and composite transposons, the transposon is cut out of the donor DNA and inserted somewhere else.

5. Site-specific recombinases are enzymes that promote recombination between certain sites on the DNA. Examples of site-specific recombinases include resolvases, integrases, *(continued)*

SUMMARY (continued)

and DNA invertases. The genes for many of these site-specfic recombinases have sequences in common, suggesting they may have been derived from a common ancestor.

6. Resolvases are site-specific recombinases encoded by replicative transposons that resolve cointegrates by promoting recombination between short *res* sequences in the copies of the transposon in the cointegrate.

7. Phage integrases promote recombination between specific sequences in bacterial DNA and a specific sequence on phage DNA, causing integration of the phage DNA into the bacterial chromosome. Integrases also play a role in the evolution of transposons by integrating antibiotic resistance genes into them.

8. DNA invertases promote nonhomologous recombination between short inverted repeats, thereby changing the orientation of the DNA sequence between them. The sequences they invert, invertible sequences, are known to play an important role in changing the host range of phages and the bacterial surface antigens to avoid host immune defenses.

9. Topoisomerases can also perform nonhomologous recombination. However, instead of creating new combinations of sequences, topoisomerases usually reattach the original strands. Only rarely do they promote an illegitimate recombination event.

QUESTIONS FOR THOUGHT

1. For those transposons that transpose replicatively (e.g., Tn3), why are there not multiple copies of the transposon around the genome?

2. How do you think that transposon Tn3 and its relatives have spread throughout the bacterial kingdom?

3. Do you think that transposons are just parasitic DNAs, or do they serve a useful purpose for the host?

4. Where do you suppose the genes that were inserted into integrons in the evolution of transposons came from?

5. If the DNA invertase enzymes are made continuously, why do the invertible sequences invert so infrequently?

PROBLEMS

1. Outline how you would use an HFT λ strain to show that some *gal* mutations of *E. coli* are due to insertion of an IS element.

2. In the experiments shown in Figures 8.12 and 8.13, what would have been observed if Tn10 transposed by a replicative mechanism?

3. The red color of *Serratia marcesens* is reversibly lost with a high frequency. Outline how you would attempt to determine if the change in pigment is due to an invertible sequence.

4. How would you prove that the Mu phage transposon probably does not encode a resolvase?

SUGGESTED READING

Bender, J., and N. Kleckner. 1986. Genetic evidence that Tn10 transposes by a nonreplicative mechanism. *Cell* 45:801–815.

Bissonnette, L., and P. H. Roy. 1992. Characterization of In0 of *Pseudomonas aeruginosa* plasmid pVS1, an ancestor of multiresistance plasmids and transposons of gram-negative bacteria. *J. Bacteriol.* 174:1248–1257.

Craig, N. L. 1996. Transposition, p. 2339–2362. *In* F. C. Neidhardt, R. Curtiss III, J. L. Ingraham, E. C. C. Lin, K. B. Low, B. Magasanik, W. S. Reznikoff, M. Riley, M. Schaechter, and H. E. Umbarger (ed.), *Escherichia coli and Salmonella: Cellular and Molecular Biology*, 2nd ed. ASM Press, Washington, D.C.

Foster, T. J., M. A. Davis, D. E. Roberts, K. Takeshita, and N. Kleckner. 1981. Genetic organization of transposon Tn10. *Cell* 23:201–213.

Francia, M. V., and J. M. Garcia Lobo. 1996. Gene integration in the *Escherichia coli* chromosome mediated by Tn21 integrase (Int21). *J. Bacteriol.* 178:894–898.

Golden, J. W., S. G. Robinson, and R. Haselkorn. 1985. Rearrangement of nitrogen fixation genes during heterocyst differentiation in the cyanobacterium Anabaena. *Nature* (London) 327:419–423.

Kanaar, R., P. Van de Putte, and N. R. Cozzarelli. 1988. Gin-mediated DNA inversion: product structure and the mechanism of strand exchange. *Proc. Natl. Acad. Sci. USA* 85:752–756.

Kunkel, B., R. Losick, and P. Stragier. 1990. The *Bacillus subtilis* gene for the developmental transcription factor σ^K is generated by excision of a dispensable DNA element containing a sporulation recombinase gene. *Genes Dev.* 4:525–535.

May, E. W., and N. L. Craig. 1996. Switching from cut-and-paste to replicative Tn7 transposition. *Science* 272:401–404.

Nash, H. A. 1996. Site-specific recombination: integration, excision, resolution, and inversion of defined DNA segments, p. 2363–2376. *In* F. C. Neidhardt, R. Curtiss III, J. L. Ingraham, E. C. C. Lin, K. B. Low, B. Magasanik, W. S. Reznikoff, M. Riley, M. Schaechter, and H. E. Umbarger (ed.), *Escherichia coli and Salmonella: Cellular and Molecular Biology*, 2nd ed. ASM Press, Washington, D.C.

Shapiro, J. A. 1979. Molecular model for the transposition and replication of bacteriophage Mu and other transposable elements. *Proc. Natl. Acad. Sci. USA* 76:1933–1937.

Simon, M., J. Zeig, M. Silverman, G. Mandel, and R. Doolittle. 1980. Phase variation: evolution of a controlling element. *Science* 209:1370–1374.

Van De Putte, P., R. Plasterk, and A. Kuijpers. 1984. A Mu *gin* complementing function and an invertible DNA region in *Escherichia coli* K-12 are situated on the genetic element e14. *J. Bacteriol.* 158:517–522.

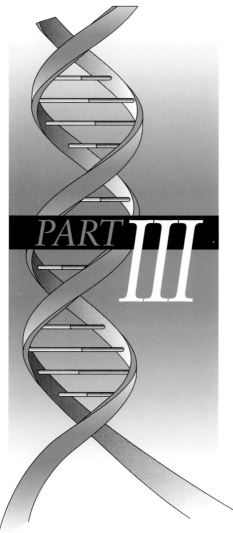

E VEN THOUGH THEY ARE THE SIMPLEST ORGANISMS on Earth, with only a few thousand genes, the cells of bacteria are nevertheless almost unimaginably complex. The few thousand gene products of bacteria interact to perform many functions required by the living cell. The task of molecular genetics is to identify the genes whose products are involved in the various functions of the cell and to determine what each of these gene products contributes to the function. The cellular functions that we will discuss are recombination, repair, and gene regulation. These functions are central to living cells and involve the DNA itself, so in a sense we are using genetics to study genetics. Also, many of these basic pathways have been found to have features in common with the analogous pathways in other organisms, including humans, so the studies with bacteria have laid the foundation for similar studies in other organisms.

Chapter 9

Molecular Basis of Recombination

EVEN TWO ORGANISMS belonging to a species that must reproduce sexually will not be genetically identical. As chromosomes are assembled into germ cells (sperm and eggs), one chromosome of each pair of homologous chromosomes is chosen at random. Consequently, it is highly unlikely that two siblings will be alike, because each of their parents' germ cells contains a mixture of chromosomes originally derived from the siblings' grandparents. In addition, the chance of two siblings being the same is even lower because of **genetic recombination**. Because of recombination, genetic information that was previously associated with one DNA molecule may become associated with a different DNA molecule, or the order of the genetic information in a single DNA molecule may be altered.

All organisms on Earth probably have some mechanism of recombination, suggesting that recombination is very important for species survival. The new combinations of genes obtained through recombination allow the species to adapt more quickly to the environment and speed up the process of evolution. Recombination can allow an organism to change the order of its own genes or move genes to a different replicon, for example, from the chromosome to a plasmid. Recombination genes also play an important role in repair of damage to DNA and in mutagenesis, topics that will be covered in the next chapter.

As we mentioned at the beginning of chapter 8, the two general types of recombination are nonhomologous, or site-specific, recombination and homologous recombination. Nonhomologous recombination occurs relatively rarely and requires special proteins that recognize specific sequences and promote recombination between them. **Homologous recombination** occurs much more often. This type of recombination can occur between any two

DNA sequences that are the same or very similar, and it usually involves the breaking of two DNA molecules in the same region, where the sequences are similar, and the joining of one DNA to the other. The result is called a **crossover**. Homology-dependent crossovers can occur between homologies as short as 23 bases, although homologies greater than 50 to 100 bp produce more frequent crossovers.

Because of its importance in genetics, we have already mentioned homologous recombination in previous chapters in discussions of deletion and inversion mutations and the integration of plasmid and phage DNA into other DNA molecules in cells. Determination of recombination frequencies also allows us to measure the distance between mutations and thus can be used to map mutations with respect to each other, as we discuss in chapters 13 and 14. Moreover, the clever use of recombination can take some of the hard work out of cloning genes and making DNA constructs.

We were able to introduce recombination before describing its molecular basis because the results of recombination can be understood without knowledge of how it works on the molecular level. Furthermore, to use recombination for genetic mapping and other types of applications, we can ignore the actual mechanisms involved and take a simple view of this process, in which two DNA molecules are broken at the same place and then are rejoined in new combinations. In this chapter, we focus on the actual mechanisms of recombination—what actually happens to the DNA molecules involved—discussing some molecular models and some of the genetic evidence that supports or contradicts those models. We also discuss the proteins involved in recombination in *Escherichia coli*, the bacterium for which recombination is best understood.

Overview of Recombination

Recombination is a remarkable process. Somehow, two enormously long DNA molecules in a cell link up and exchange sequences. Moreover, recombination usually occurs only at homologous regions of two DNAs. Thus, these regions must line up so that they can be broken and rejoined, and the long DNA molecules on either side of the point of recombination must change their configuration with respect to each other. This complicated process clearly involves many functions. But before presenting detailed models for how recombination might occur, and the functions involved, we consider the basic requirements any recombination model must satisfy and what functions to expect of gene products directly involved in recombination.

Requirement 1: Identical or Very Similar Sequences in the Crossover Region

The distinguishing feature of homologous recombination is that the deoxynucleotide sequence in the two regions of DNA where a crossover occurs must be the same or very similar, and all recombination models must start with this requirement. This prerequisite serves a very practical function in recombination. The sequence of nucleotides in molecules of DNA from different individuals of the same species will usually be almost identical over the entire lengths of the molecules, and by extension, two DNA molecules will generally share identical sequences only in the same regions. Thus, recombination usually occurs only between sites on the two DNAs that are in the same place with respect to the entire molecule. Recombination between different regions in two DNA molecules does sometimes occur because the same or similar sequences occur in more than one place in the DNAs. This type of recombination, sometimes called **ectopic** or **homeologous recombination**, gives rise to deletions, duplications, inversions, and other gross rearrangements of DNA (see chapter 3). Not surprisingly, cells have evolved special mechanisms to discourage ectopic recombination, one of which we discuss in the next chapter in the section on mismatch repair.

Requirement 2: Complementary Base Pairing between Double-Stranded DNA Molecules

Mandatory complementary base pairing between strands of the two DNA molecules ensures that recombination will occur only between sequences at the same **locus**, that is, the same place on the molecules. The point at which two double-stranded DNA molecules are held together by complementary base pairing between their strands is called a **synapse**. All recombination models must involve synapse formation. However, in double-stranded DNA, the bases are usually hidden inside the helix, where they are not available for base pairing. Separating the strands in various regions to expose the bases would be too slow to allow efficient recombination. Therefore, the cell must contain some functions that allow one strand of DNA to locate and pair with its complementary sequence even though the bases are mostly hidden from pairing inside the double helix.

Requirement 3: Recombination Enzymes

For a crossover, or true recombination, to occur between two DNA molecules, the strands of each molecule must be broken and rejoined to the corresponding strands of the other DNA molecule. Therefore, DNA endonucleases and ligases—enzymes that break DNA strands and rejoin them, respectively—are required for recombination.

Requirement 4: Heteroduplex Formation

The regions of complementary base pairing between the two DNA molecules in a synapse are called **heteroduplexes**, because the strands in these regions come from different DNA molecules. In principle, for a synapse to form, heteroduplexes need form only between two of the strands of the DNA molecules being recombined. However, evidence indicates that all strands of the two DNA molecules are involved in heteroduplex formation, an observation that must be explained by the models.

Molecular Models of Recombination

Several models have been proposed to explain recombination at the molecular level. These models include the required features of recombination discussed above and also account for additional experimental evidence. However, no single model of recombination can make an exclusive claim to the truth, and recombination may occur by different pathways in different situations. Nevertheless, these models serve as a framework for forming hypotheses that can be tested through experimentation and

will help focus thinking about recombination at the molecular level.

The Holliday Double-Strand Invasion Model

The first widely accepted model for recombination was proposed by Robin Holliday in 1964. Figure 9.1 illustrates the basic steps of the **Holliday model**. According to this model, recombination is initiated by two single-stranded breaks made simultaneously at exactly the same place in the two DNA molecules to be recombined. Then the free ends of the two broken strands cross over each other, each pairing with its complementary sequence in the other DNA molecule to form two heteroduplexes. The ends are then ligated to each other, resulting in a cruciform-like structure called a **Holliday junction** in which the two double-stranded molecules are held together by their crossed-over strands (see Figure 9.1). The formation of Holliday junctions is central to all the recombination models discussed here.

Once formed, Holliday junctions can undergo a rearrangement that changes the relationship of the strands to each other. This rearrangement is called an **isomerization** because no bonds are broken. As shown in the figure, isomerization causes the crossed strands in configuration I to uncross. A second isomerization occurs to create configuration II, where the ends of the two double-stranded DNA molecules are in the recom-

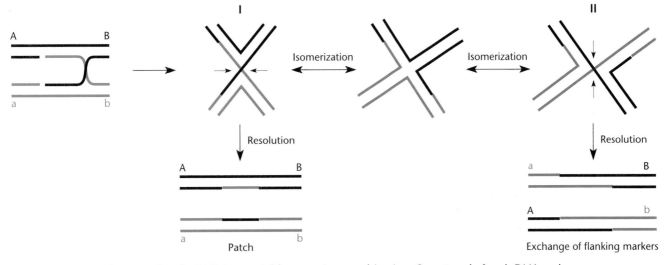

Figure 9.1 The Holliday model for genetic recombination. One strand of each DNA molecule is cut at the same position and then pairs with the other molecule to form a heteroduplex (region of blue paired with black). The strands are then ligated to form the Holliday junction. This DNA structure can isomerize between forms I and II. Cutting and ligating resolves the Holliday junction. Depending on the conformation of the junction, the flanking markers A, B, a, and b will recombine or remain in their original conformation. The product DNA molecules will contain heteroduplex patches.

binant configuration with respect to each other. This may seem surprising, but experiments with models show that the two structures I and II are actually equivalent to each other and the Holliday junction can change from one form to the other without breaking any hydrogen bonds between the bases. Hence, flipping from one configuration to the other requires no energy and should occur quickly, so that each configuration should be present approximately 50% of the time.

Once formed, the Holliday junction can be cut and religated, or **resolved**, as shown in Figure 9.1. Whether or not recombination occurs depends on the configuration of the junction at resolution. If the Holliday junction is in configuration I, the flanking DNA sequences will not recombine. However, if the Holliday junction is in the II configuration when it is resolved, DNA sequences will be exchanged, as indicated by the flanking markers shown in Figure 9.1.

Holliday junctions also move up and down the DNA by breaking and re-forming the hydrogen bonds between the bases. This process is called **branch migration** (Figure 9.2). The same number of hydrogen bonds are broken and re-formed as the cross-connection moves, so that no energy is required. However, without the expenditure of some energy, hydrogen bonds may not be broken fast enough for efficient branch migration. Specific ATP-hydrolyzing proteins seem to be required for branch migration (see the section on the Ruv proteins).

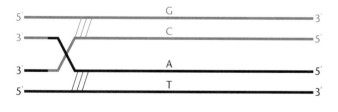

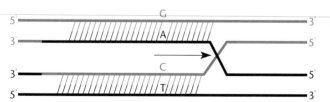

Figure 9.2 Migration of Holliday junctions. By breaking the hydrogen bonds holding the DNAs together in front of the branch and re-forming them behind, the junction migrates and extends the regions of pairing (i.e., the heteroduplexes) between the two DNAs. The heteroduplex region is crosshatched. In the example, two mismatches, GA and CT, form in the heteroduplex region because one of the DNA molecules has a mutation in this region.

As shown in Figure 9.2, branch migration in Holliday junctions can have the effect of increasing the length of the heteroduplex regions. If a heteroduplex extends to include a region of differing sequence, a mismatch will occur, possibly leading to gene conversion (discussed later in the chapter). The other models we discuss all invoke branch migration of Holliday junctions to explain the experimental evidence concerning gene conversion and the length and distribution of heteroduplexes.

The Holliday model is called a double-strand invasion model because one strand from each DNA molecule invades the other DNA molecule (Figure 9.1), explaining how heteroduplexes can form on both molecules of DNA during recombination. However, one problem with this model is that the two DNA molecules must be simultaneously cut at almost the same place to initiate recombination. But how could the two like DNA molecules line up for pairing before they are cut when the bases are hidden inside the double-stranded DNA helix and so are not free to pair with other DNA molecules? Also, if the two DNA molecules were not aligned, how could they be cut at exactly the same place? To answer these questions, Holliday proposed the existence of certain sites on DNA that are cut by special enzymes to initiate the recombination. However, there is no evidence for such sites, and recombination seems to occur more or less randomly over the entire DNA.

Despite these objections, the Holliday double-strand invasion model has served as the standard against which all other models of recombination are compared. All the most favored models involve the formation of Holliday junctions and branch migration. They mostly differ in the earlier stages, before Holliday junctions have formed.

Single-Strand Invasion Model

One way to overcome the objections to the Holliday model is to modify it with the proposal of a single-strand invasion model such as the one shown in Figure 9.3 (see Meselson and Radding, Suggested Reading). In this model, a strand of one of the two DNA molecules is cut at random and then the exposed end invades another double-stranded DNA until it finds its complementary sequence. If it finds such a sequence, it will displace one strand. The DNA polymerase will then fill the gap left by the invading strand, using the remaining strand as a template. The displaced strand is degraded, and its remaining end is ligated to the newly synthesized DNA strand on the opposite molecule to form a Holliday junction (Figure 9.3). Once the Holliday junction has formed, isomerization and resolution can produce recombinants as in the Holliday model.

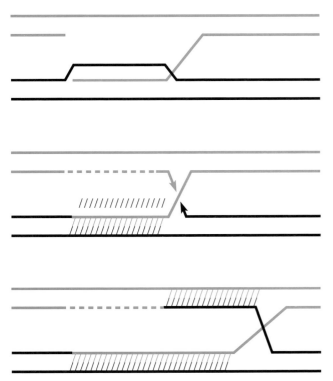

Figure 9.3   The single-strand invasion model. A single-stranded break in one of the two DNA molecules frees a single-stranded end that invades the other DNA molecule. The gap left on the cut DNA is filled by DNA polymerase (dashed line). The displaced strand on the other DNA molecule is degraded, and the two ends are joined (arrows). Initially, a heteroduplex, represented by crosshatching, will form on only one of the two DNA molecules. Branch migration will cause another heteroduplex to form on the other DNA molecule. Isomerization can recombine the flanking DNA molecules, as in the Holliday model.

By this model, a heteroduplex will at first form on only one of the two DNAs. However, once the Holliday junction forms, branch migration can create a heteroduplex on the other DNA. This accounts for the formation of heteroduplexes on both DNAs.

Double-Strand Break Repair Model

In both the single-strand and the double-strand invasion models, a single-strand break in one or both of the two DNA molecules, respectively, initiates the recombination event. A priori, it seems unlikely that a double-strand break in DNA could initiate recombination. If a single-strand break is made in DNA, one strand still holds the molecule together. However, if a double-strand break occurred, the two parts of the DNA should fall apart, a lethal event. Thus, double-strand break models were at first ruled out as being counterintuitive.

However, it now seems clear that recombination is initiated by a double-strand break in one of the two DNA molecules, at least in some situations.

The first evidence that double-strand breaks in DNA can initiate recombination came from genetic experiments with *Saccharomyces cerevisiae* (see Szostak et al., Suggested Reading). These experiments were aimed at analyzing the ability of recombination between plasmids and chromosomes to repair double-strand breaks and gaps in the plasmids by inserting the corresponding sequence from the chromosomes. However, the initiation of recombination by double-strand breaks is now known to be a general mechanism. For example, homing DNA endonucleases in bacteria, phages, and lower eukaryotes initiate recombination by making a double-strand break (see Box 9.1).

Figure 9.4 illustrates the model for this type of recombination (see Szostak et al., Suggested Reading). Both

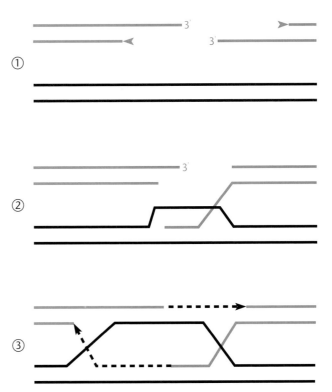

Figure 9.4   The double-strand break repair model. (1) A double-strand break in one of the two DNAs initiates the recombination event. The arrows indicate the degradation of the 5' ends at the break. (2) One 3' end, or tail, then invades the other DNA, displacing one of the strands. (3) This 3' end serves as a primer for DNA polymerase, which extends the tail until it can eventually be joined to a 5' end (black arrow). Meanwhile, the displaced strand (in black) serves as a template to fill the gap left in the first DNA (dashed lines). Two Holliday junctions form and may produce recombinant flanking DNA, depending upon how they are resolved.

BOX 9.1

Homing Enzymes, Introns, and Inteins

RNA Introns

RNA introns are sequences spliced out of mRNA before it is translated. Until recently, introns were thought to be confined to eukaryotes, but we now know that introns also exist in prokaryotes. Some of these are self-splicing introns, because they splice themselves out of the RNA without the help of proteins; therefore, in a sense, they are enzymes. Self-splicing introns were one of the first examples of an RNA enzyme, or ribozyme, to be discovered, and Tom Cech was awarded a Nobel Prize in 1989 for their discovery.

The self-splicing RNA introns are remarkably similar among bacteriophages and lower eukaryotes, which supports the speculation that gene exchange between prokaryotes and lower eukaryotes has continued throughout time. Also, the position of one such intron represents additional evidence that the organelles of higher eukaryotes are derived from symbiotic bacteria. A leucine tRNA gene of cyanobacteria (*Calothrix* spp.) contains a self-splicing intron in exactly the same place as the corresponding *leu* tRNA gene of plant chloroplasts, as though the intron has been preserved through evolution. Plant chloroplasts are generally considered to be derived from cyanobacteria, and the position of the intron supports this hypothesis.

Protein Inteins

Not only can some RNA sequences splice themselves out of RNA, but also some protein sequences, called inteins, can splice themselves out of protein. So far, the number of known inteins is limited to three examples, one from *S. cerevisiae*, one from *Archaea*, and one from the *recA* gene of the eubacterium *Mycobacterium tuberculosis*, the causative agent

of tuberculosis. These short coding sequences in the gene encode amino acid sequences that are spliced out of the polypeptide after it is made.

Homing Introns and Inteins

The DNA sequences for some RNA introns and protein inteins can move from one place in the DNA to another and so, in a sense, are transposable. However, the movement mechanism of many of these sequences is more analogous to gene conversion due to homologous recombination than to transposition. The elements encode a DNA endonuclease that makes a double-stranded break in the target sequence, and this break initiates the recombination event. The target sequences of homing introns and inteins are very specific, and these elements will transpose only into the same place in the same gene that harbors them (but on a different molecule). Therefore, their transposition is sometimes called homing. In a sense, the gene for the endonuclease moves itself, and the sequence of the intron or intein comes along for the ride through gene conversion. By associating itself with an intron or intein, the DNA endonuclease gene can hop into an essential gene without inactivating it and killing the cell.

References

Belfort, M., M. E. Reaban, T. Coetzee, and J. Z. Dalgaard. 1995. Prokaryotic introns and inteins: a panoply of form and function. *J. Bacteriol.* **177**:3897–3903.

Shub, D. A., J. M. Gott, M.-Q. Xu, B. F. Lang, F. Michel, J. Tomaschewski, J. Pedersen-Lane, and M. Belfort. 1988. Structural conservation among three homologous introns of bacteriophage T4 and group I introns of eukaryotes. *Proc. Natl. Acad. Sci. USA* **85**:1151–1155.

strands of one of the two DNA molecules participating in the recombination are broken, and the 5' ends from each break are digested by an exonuclease, leaving a gap with exposed single-stranded 3' tails. One of these tails invades the other double-stranded DNA until it finds its complementary sequence. Then DNA polymerase extends the tail along the complementary sequence, displacing the other strand with the same sequence, until it reaches the free 5' end of the invading strand and is joined to it by DNA ligase. The other free 3' end can then be used as a primer to fill in the remaining gap by using the displaced strand as a template before being joined to the free 5' end in its own strand by DNA ligase. This causes two Holliday junctions to form (Figure 9.4).

Whether recombination occurs depends on which configuration the two Holliday junctions are in when resolved. If both are in the same configuration—either I or II (Figure 9.1)—when they are resolved, no crossovers, and thus no recombination, will result. However, if the two junctions are in different configurations at resolution, recombination will occur.

In this model, heteroduplex DNA will initially form in only one DNA molecule, between the strand that does the initial invading and its complementary sequence. However, as in the single-strand invasion models, the migration of the Holliday junctions can lead to heteroduplex formation on both strands.

Molecular Basis for Recombination in *E. coli*

As with many cellular phenomena, much more is known about the molecular basis for recombination in *E. coli* than in any other organism. Numerous proteins involved in recombination in *E. coli* have been identified, and specific roles have been assigned to many of these.

chi Sites and the RecBCD Nuclease

The first step in recombination in *E. coli* is usually performed by the RecBCD nuclease. This enzyme has 3'-to-5' DNA exonuclease activities as well as a DNA endonuclease activity. Very little recombination occurs without RecBCD, which is required for the major pathway of recombination in *E. coli*.

Recombination initiated by the RecBCD nuclease is greatly stimulated by sequences in the DNA called *chi* (or Chi) **sites** (for the Greek letter χ, which looks like a crossover). These sites have the sequence 5'GCTG-GTGG3'. The lack of symmetry in this sequence (i.e., that it does not read the same on both strands in the 5'-to-3' direction) explains why *chi* sites stimulate recombination in only one direction, as we explain later.

Figure 9.5 shows how a *chi* site might stimulate the RecBCD nuclease to initiate recombination. According to this model, the RecBCD protein loads onto the DNA at one end or at an internal double-strand break. The RecBCD protein then moves into the DNA, creating a loop that its exonuclease activity immediately degrades. When the moving RecBCD enzyme encounters a *chi* site in the correct orientation, the exonuclease activity of the RecBCD protein is inhibited, so that the 3' end will no longer be degraded and a stable single-stranded 3' end will form. This single-stranded end can invade another homologous double-stranded DNA with the help of the RecA protein, forming a synapse (see the next section).

This model seems complicated but is supported by experimental evidence on *chi*-stimulated recombination. For example, it accounts for the fact that *recD* mutants are not defective in recombination, nor do *chi* sites stimulate recombination in such mutants. These mutants do not have the exonuclease activity, so the RecD subunit must be required for this activity. The model predicts that, in a *recD* mutant, the displaced strand would not be degraded. So even if the RecBCD nuclease had not passed over a *chi* site, the displaced strand would be available to invade another DNA.

This model also explains why recombination is only stimulated on the 5' side of a *chi* site. Until the RecBCD enzyme reaches a *chi* site, the displaced strand is de-

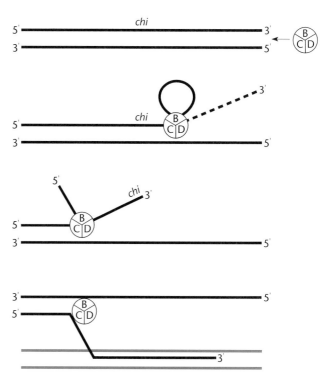

Figure 9.5 Model for promotion of recombination initiation at a *chi* site by the RecBCD enzyme. The RecBCD enzyme loads onto the DNA at a free end or at a double-strand break internal to the DNA. It then moves along the DNA, displacing a loop, and the 3' end is degraded by the exonuclease activity of the RecBCD enzyme (dotted line). When it encounters a *chi* site, the exonuclease activity is inhibited and so the 3' end is no longer degraded and is free to invade another DNA.

graded and so is not available for recombination. Only after the RecBCD protein passes completely through a *chi* site will the strand survive to invade another DNA, so that only DNA on the 5' side of the site will survive. This model is also supported by the known enzymatic activites of the RecBCD protein, as well as by electron microscopic visualization of the RecBCD protein acting on DNA. It has also received experimental support from the results of in vitro experiments with purified RecBCD nuclease and DNA containing a *chi* site (see Dixon and Kowalcykowski, Suggested Reading).

According to this model, the RecBCD enzyme can load onto DNA only at a double-strand break or end. But how and where are these breaks made? In crosses resulting from transduction and transformation, the pieces of DNA that recombine are often fairly short, so that free ends are never far away. In conjugation, the RecBCD enzyme may be able to load on the DNA at periodic discontinuities in the molecule. Evidence indicates that *chi* sites preferentially stimulate recombination close to the ends of DNA molecules, which is consistent with the model presented here.

Synapse Formation and the RecA Protein

Once the action of RecBCD has freed a single-stranded 3' end of DNA, this end must still find and invade another homologous double-stranded DNA to form a synapse. As mentioned, the search for complementary sequences in another DNA could be quite difficult, since the base pairs are on the inside of the helix, where they are not available for pairing. This process would be very slow if the double strands had to be separated each time. This is where the RecA protein plays its part. Rather than separate strands, this protein locates sequences complementary to the 3' end by creating a **triple-stranded helix** involving double-stranded DNA and the invading single strand (Figure 9.6) (see Rao et

al., Suggested Reading). The RecA protein first binds to the single strand, forcing it into an extended helical conformation. If this helical single-stranded DNA then encounters a double-stranded DNA with a homologous sequence, the single strand can pair with the double-stranded DNA in its major groove to form an extended triple-stranded helix. This structure facilitates the displacement of one strand of the double-stranded DNA, which can then pair with the strand formerly paired to the invading single strand, forming a Holliday junction.

If RecBCD and RecA are the only proteins required for formation of Holliday junctions, the Holliday junctions should form in the presence of these proteins alone. In fact, it has been demonstrated, by using purified RecBCD protein and purified RecA protein, that Holliday junctions can be formed by the concerted action of RecBCD and RecA protein on purified DNA in the presence of ATP (see Dixon and Kowalcykowski, Suggested Reading, for references).

The Ruv and RecG Proteins and the Migration and Cutting of Holliday Junctions

Once Holliday junctions have been formed by the action of RecBCD and RecA, the branches can migrate to increase the length of the heteroduplexes on both strands. The Holliday junctions can then isomerize, and cutting at the crossed strands will resolve them, forming the recombinant DNA molecules, as shown in the models.

In *E. coli,* branch migration and the resolution of Holliday junctions can be performed by the three Ruv proteins, RuvA, RuvB, and RuvC, which are encoded by adjacent genes. According to the model shown in Figure 9.7 (see Parsons et al., Suggested Reading), the products of the *ruvA* and *ruvB* genes are involved in the migration of Holliday junctions. First, the RuvA protein binds to the Holliday junction. Then RuvB binds to RuvA, and ATP cleavage drives the migration of the Holliday junction.

Early on, researchers thought that branch migration should occur spontaneously, without the need for proteins or energy. As many hydrogen bonds are formed behind a migrating branch as are broken ahead of it, so there is no net loss of hydrogen bonds. However, although spontaneous branch migration may occur, the RuvA and RuvB proteins greatly accelerate its rate in vivo. Spontaneous branch migration appears to be too slow to explain the migration of Holliday junctions during normal recombination.

As discussed earlier, the isomerization reaction that crosses and uncrosses the strands of a Holliday junction probably does not require proteins or an expenditure of energy. However, the RuvC protein resolves Holliday

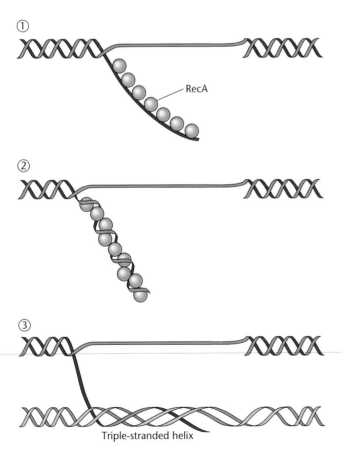

Triple-stranded helix

Figure 9.6 Synapse formation between two homologous DNAs by RecA protein. In steps 1 and 2, the RecA protein binds to the single-stranded end and forces it into an extended helical structure. In step 3, the helical single-stranded DNA can pair with a homologous double-stranded DNA in its major groove to form a stable extended triple-helix.

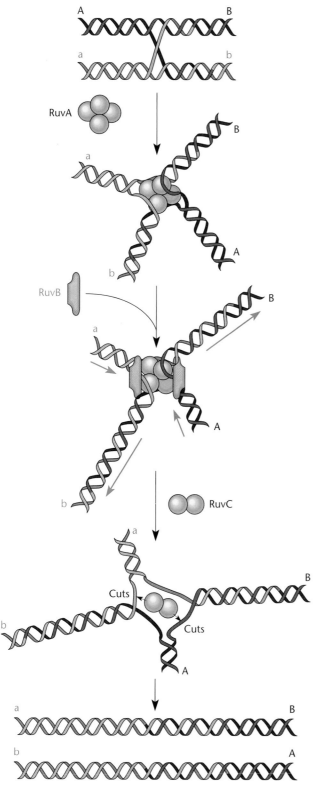

junctions by cleaving the two crossed strands as shown in Figure 9.7. As mentioned above, whether recombination occurs depends on whether the Holliday junction is in the original or the recombinant conformation at the time of the cleavage.

Support for this model of the action of RuvA, RuvB, and RuvC has come from observations of purified Ruv proteins acting on artificially synthesized structures that resemble Holliday junctions (see Parsons et al., Suggested Reading). These junctions are constructed by annealing four synthetic single-stranded DNA chains that have pairwise complementarity to each other (Figure 9.8). These synthetic structures are not completely analogous to a real Holliday junction in that they are not made from naturally occurring DNA. Rather, four single strands are synthesized that are complementary to each other in the regions shown and therefore form a cross. A Holliday junction made this way will be much more stable than a natural Holliday junction because the branch cannot migrate. Real Holliday junctions are too unstable for these experiments; they quickly separate into two double-stranded DNA molecules.

Experiments performed with such synthetic Holliday junctions indicated that purified Ruv proteins act sequentially on a synthetic Holliday junction in a manner consistent with the above model. First, RuvA protein bound specifically to the synthetic Holliday junctions, and then a combination of RuvA and RuvB caused a disassociation of the synthetic Holliday junctions, simulating branch migration in a natural DNA molecule. The dissocation required the energy in ATP to break the hydrogen bonds holding the Holliday junction together, as predicted. The RuvC protein then specifically cut the synthetic Holliday junction in two of the four strands.

Surprisingly, in spite of the important role of the Ruv proteins in recombination, mutations that inactivate a *ruv* gene do not prevent recombination. The *ruv* genes were discovered only because mutations within them increase the sensitivity of the cells to killing by UV irradiation, hence the name *ruv* (repair *UV*). The reason that these proteins are nonessential for recombination is that another protein, the RecG protein, can substitute for them. We discuss the interaction between RecG and the Ruv proteins later in this chapter in the section on genetic analysis of recombination in bacteria. Many other proteins, including those of the RecF pathway, are also involved in recombination in *E. coli*, at least under

Figure 9.7   Model for the mechanism of action of the Ruv proteins. RuvA binds to the Holliday junction. Note that the figure starts with one turn of blue-gray heteroduplex. Then RuvB binds to RuvA, and the junction migrates, deriving en- ergy from ATP cleavage. RuvC cleaves two strands of the Holliday junction to resolve the junction into separate DNA molecules. Note the three turns of blue-gray heteroduplex after junction migration.

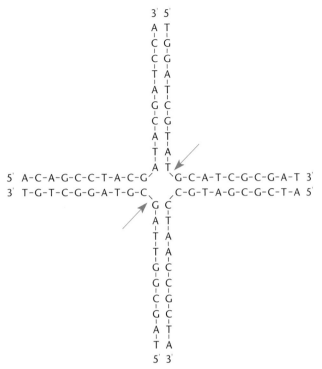

Figure 9.8 A synthetic Holliday junction with four complementary strands. The junction cannot migrate but can be disrupted by RuvA and RuvB. It can also be cut by RuvC (arrows) and other Holliday junction resolvases such as RecG and the gene *49* and gene *3* products of phages T4 and T7, respectively.

generally only small pieces of incoming donor DNA recombine with the chromosome, whereas in other organisms, two enormously long double-stranded DNA molecules of equal size usually recombine. Nevertheless, the requirements for recombination in bacteria are likely to be similar to those in other organisms. In fact, accumulating evidence supports the existence of proteins analogous to many of the recombination proteins of bacteria—including RecA analogs and Holliday junction resolvases—in higher organisms.

Phage Recombination Pathways

Many phages encode their own recombination functions, some of which can be important for the multiplication of the phage. For example, as discussed in chapter 7, some phages use recombination to make primers for replication and concatemers for packaging. Also, phage recombination systems may be important for repairing damaged phage DNA and for exchanging DNA between related phages to increase diversity. Phages may encode their own recombination systems to avoid dependence on host systems for these important functions.

Many phage recombination functions have been shown to be analogous to the recombination proteins of the host bacteria (Table 9.2), and in many cases, the phage recombination proteins were discovered before their host counterparts. As a result, studies of bacterial recombination systems have been heavily influenced by simultaneous studies of phage recombination systems.

Rec Proteins of Phages T4 and T7

Phages T4 and T7 depend upon recombination for the formation of DNA concatemers after infection (see

some conditions (Table 9.1). These other pathways and how they were discovered are likewise covered in the section on genetic analysis of recombination in bacteria.

Recombination in bacteria has some differences from recombination in most other organisms. In bacteria,

TABLE 9.1	Genes encoding recombination functions		
Gene	Mutant phenotype	Enzymatic activity	Probable role in recombination
recA	Recombination deficient	Enhanced pairing of homologous DNAs	Synapse formation
recBC	Reduced recombination	Exonuclease, ATPase, helicase, chi-specific endonuclease	Initiates recombination by separating strands cutting DNA at the *chi* site
recD	Rec+	Exonuclease	Degrades 3' ends
recF	Reduced plasmid recombination	Binds ATP and single-stranded DNA	Substitutes for RecBCD
recJ	Reduced recombination in RecBC⁻	Single-stranded exonuclease	Substitutes for RecBCD
recN	Reduced recombination in RecBC⁻	ATP binding	Substitutes for RecBCD
recO	Reduced recombination in RecBC⁻	DNA binding and renaturation	Substitutes for RecBCD
recQ	Reduced recombination in RecBC⁻	DNA helicase	Substitutes for RecBCD
recR	Reduced recombination in RecBC⁻	Binds double-stranded DNA	Substitutes for RecBCD
recG	Reduced Rec in RuvA⁻B⁻C⁻	Binds to Holliday junctions, ATPase	Migration and resolution of Holliday junctions
ruvA	Reduced recombination in RecG⁻	Binds to Holliday junctions	Migration of Holliday junctions
ruvB	Reduced recombination in RecG⁻	Binds to RuvA, ATPase	Migration of Holliday junctions
ruvC	Reduced recombination in RecG⁻	Holliday junction-specific nuclease	Resolution of Holliday junctions

TABLE 9.2	Analogy between phage and host recombination functions
Phage function	**Analogous *E. coli* function**
T4 UvsX	RecA
T4 gene *49*	RuvC, RecG
T7 gene *3*	RuvC, RecG
T4 genes *46* and *47*	RecBCD
λ ORF in *nin* region	RecO, RecR, RecF
Rac *recE* gene	RecJ, RecQ

chapter 7). Therefore, recombination functions are essential for the multiplication of these phages. Many of the T4 and T7 Rec proteins are analogous to those of their hosts. For example, the gene *49* protein of T4 and the gene *3* protein of T7 are endonucleases that resolve Holliday junctions and are representative of phage proteins discovered before their host counterparts, in this case RuvC and RecG. The gene *46* and *47* products of phage T4 may perform a reaction similar to the RecBCD protein of the host, although no evidence indicates the presence of *chi*-like sites associated with this enzyme. The UvsX protein of T4 is analogous to the RecA protein of the host.

The RecE Pathway of the *rac* Prophage

Another classic example of a phage-encoded recombination pathway is the RecE pathway encoded by the *rac* prophage of *E. coli* K-12. The *rac* prophage is integrated at 29 min in the *E. coli* genetic map and is related to λ. This defective prophage cannot be induced to produce infective phage, as it lacks some essential functions for multiplication.

The RecE pathway was discovered by isolating suppressors of *recBCD* mutations, called *sbcA* mutations (for suppressor of *BC*), that restored recombination in conjugational crosses. The *sbcA* mutations were later found to activate a normally repressed recombination function of the defective prophage *rac*. Apparently, *sbcA* mutations inactivate a repressor that normally prevents the transcription of the *recE* gene, as well as other prophage genes. When the repressor gene is inactivated, the RecE protein will be synthesized and can then substitute for the RecBCD nuclease in recombination.

The Phage λ *red* System

Phage λ also encodes recombination functions. The best characterized is the *red* system, which requires the products of adjacent λ genes *red*α and *red*β. How the products of these genes participate in recombination is not clear. The product of the *red*α gene is an exonuclease that degrades one strand of a double-stranded DNA

from the 5′ end to leave a 3′ single-stranded tail. This enzyme, called λ exonuclease, is discussed in chapter 15 in connection with the molecular biology technique of making nested deletions in cloned DNAs. The *red*β gene product is known to help the renaturation of denatured DNA and to bind to the λ exonuclease. Unlike the other recombination systems that we have discussed, the λ *red* recombination pathway does not require the RecA protein.

Interestingly, the RecE protein of *rac* and the λ *red*α exonuclease may be similar. The RecE protein needs RecA to promote *E. coli* recombination, but it does not need RecA to promote λ recombination. It is not too surprising that λ and *rac* encode similar recombination functions, since *rac* and λ are related phages.

Besides the *red* system, phage λ encodes another recombination function that can substitute for components of the *E. coli* RecF pathway (see Sawitzke and Stahl, Suggested Reading). Apparently, phages can carry components for more than one recombination pathway.

Genetic Analysis of Recombination in Bacteria

One reason we understand so much more about recombination in *E. coli* than in most other organisms is because of the relative ease of doing genetic experiments with this organism. In this section, we'll discuss some of the genetic experiments that have led to our present picture of the mechanisms of recombination in *E. coli*.

Isolating Rec⁻ Mutants of *E. coli*

As in any genetic analysis, the first step in studying recombination in *E. coli* was to isolate mutants defective in recombination. Such mutants are called **Rec⁻ mutants** and have mutations in the *rec* genes, whose products are required for recombination. Two very different approaches were used in the first isolations of Rec⁻ mutants of *E. coli*.

One approach used for selection of Rec⁻ mutants was to select them directly on the basis of their inability to support recombination (see Clark and Margulies, Suggested Reading). The idea behind this selection was that an *E. coli* strain with a mutation that inactivates a required *rec* gene should not be able to produce recombinant types when crossed with another strain.

In one experiment designed to isolate mutants defective in recombination, a Leu⁻ strain of *E. coli* was mutagenized with nitrosoguanidine. Individual strains, some of which might now also have a *rec* mutation, were then crossed separately with an Hfr strain. A Rec⁻ mutant would have produced no Leu⁺ recombinants when crossed with the Hfr strain.

To cross thousands of the mutagenized bacteria separately with the Hfr strain to find one that gave no recombinants would have been very laborious, but replica plating facilitated the crosses. A few mutant strains gave no Leu⁺ recombinants when crossed with the Hfr strain and were candidates for Rec⁻ mutants. We discuss bacterial techniques such as replica plating and Hfr crosses in chapter 14.

However, just because the mutants give no recombinants in a cross does not mean that they are necessarily Rec⁻ mutants. For instance, the mutants might have been normal for recombination but defective in the uptake of DNA during conjugation. This possibility was ruled out by crossing the mutants with an F'-containing strain instead of an Hfr strain. As discussed in chapter 5, apparent recombinant types can appear without recombination in an F' cross because the F' factor can replicate autonomously in the recipient cells; that is, it is a replicon. However, the DNA must still be taken up during transfer of the F' factor, so that if mutants are defective in DNA uptake, no apparent recombination types would appear in the F' cross. Normal frequencies of apparent recombinant types appeared when some of the mutants were crossed with F' strains; therefore, these mutants were not defective in DNA uptake during conjugation but, rather, had defects in recombination. These and other criteria were used to isolate several recombination-deficient Rec⁻ mutants.

The other approach used in the first isolations of recombination-deficient mutants of *E. coli* was less direct (see Howard-Flanders and Theriot, Suggested Reading). These isolations depended on the fact that some recombination functions are also involved in the repair of UV-damaged DNA. Therefore, using methods described in chapter 10, Howard-Flanders and Theriot isolated several repair-deficient mutants and tested them to determine if any were also deficient in recombination. Some, but not all, of these repair-deficient mutants could also be shown to be defective in recombination in crosses with Hfr strains.

COMPLEMENTATION TESTS WITH *rec* MUTATIONS

Once Rec⁻ mutants had been isolated, the number of *rec* genes could be determined by complementation tests. The original *rec* mutations defined three genes of the bacterium: *recA*, *recB*, and *recC*. The *recB* and *recC* mutants were less defective in recombination and repair than were the *recA* mutants. In fact, the RecA function is the only gene product absolutely required for recombination in *E. coli* and many other bacteria.

MAPPING *rec* GENES

The next step was to map the *rec* genes. This task may seem impossible, as a *rec* mutation causes a deficiency in recombination, but it is the frequency of recombination with known markers that reveals map position (see chapter 14). Crosses with Rec⁻ mutants can be successful only if the donor, but not the recipient, strain has the *rec* mutation. After DNA transfer, the recipient cell will retain recombination activity at least long enough for recombination to occur. Other markers are selected, and the presence or absence of the *rec* mutation is scored as an unselected marker by the UV sensitivity of the recombinants or some other phenotype due to the *rec* mutation.

Using crosses such as those described above, investigators found that *recA* mutations mapped at 51 min and *recB* and *recC* mutations mapped close to each other at 54 min on the *E. coli* genetic map. Later studies showed that the products of the *recB* and *recC* genes, as well as of the adjacent *recD* gene, comprise the RecBCD nuclease, which initiates the major recombination pathway in *E. coli* (see above). The *recD* gene was not found in the original selection, because its product is not essential for recombination under these conditions. In fact, as discussed above, mutations in the *recD* gene can stimulate recombination by preventing degradation of the displaced strand and making recombination independent of *chi* sites.

Other Recombination Genes

In addition to the *recA* and *recBCD* genes, other genes whose products participate in recombination in *E. coli* have been found (Table 9.1). Many of these genes were not found in the original selections, because inactivating them alone does not sufficiently reduce either recombination or repair in wild-type *E. coli*.

THE RecF PATHWAY

The RecF pathway in *E. coli* involves the products of the *recF*, *recJ*, *recN*, *recO*, *recQ*, and *recR* genes. The genes of the RecF pathway of recombination were not discovered in the original selections of *rec* mutations because they are not normally required for recombination. By themselves, the *recB* and *recC* mutations reduce recombination after an Hfr cross to about 1% of its normal level. Mutations in any of the genes of the RecF pathway can prevent the residual recombination that occurs in a *recB* or *recC* cell, suggesting that these genes are responsible for a minor pathway of recombination in *E. coli*.

The RecF pathway would be more efficient, except that the products of two other *E. coli* genes interfere with it. These genes, named *sbcB* and *sbcC*, were first detected because mutations in them suppress the recombination deficiency in *recB* and *recC* mutants (hence the name *sbc*, for *s*uppressor of *BC*). We know that these suppressor mutations act by allowing stimulation of the RecF pathway, because, for example, the extra recombination in a *recB sbcB* mutant was eliminated by a third mutation in a gene of the RecF pathway, for example, *recQ*. Apparently, the products of the *sbcB* and *sbcC* genes normally inhibit the RecF pathway, perhaps because they have an enzymatic activity that destroys one of the intermediates of this pathway.

But why would the *E. coli* cell seem to invest so many genes in a pathway that is normally responsible for only a small part of the total recombination in the cell, especially when the efficiency of this second pathway is limited by other gene products that the cell itself synthesizes? The answer may lie in the way the original experiments were performed, which may have diminished the role of the RecF pathway. The experiments that revealed the RecF pathway were Hfr crosses in which a large portion of the donor DNA was introduced into the recipient cell by conjugation. These experiments also put no constraints on the time required for recombination, which could have occurred over a long period. However, the RecF pathway may be very important if short pieces of DNA are introduced and recombination is allowed to proceed for only a limited time. Evidence for this hypothesis comes from transduction in *Salmonella typhimurium* (see Miesel and Roth, Suggested Reading).

THE *ruvABC* AND *recG* GENES

As discussed above, the products of the *ruv* and *recG* genes are involved in the migration and cutting of Holliday junctions. There are three adjacent *ruv* genes: *ruvA*, *ruvB*, and *ruvC*. The *ruvA* and *ruvB* genes are transcribed into a polycistronic mRNA, and the *ruvC* gene is adjacent but independently transcribed. The *recG* gene lies elsewhere in the genome.

The discovery of the role the *ruv* genes play in recombination involved some interesting genetics. The *recG* and *ruvABC* genes were not found in the original selections for recombination-deficient mutants because by themselves, mutations in these genes do not significantly reduce recombination. The *ruv* genes were found only because mutations in them can increase the sensitivity of *E. coli* to killing by UV irradiation. The *recG* gene was found because double mutants with mutations in the *recG* gene and one of the *ruv* genes are severely deficient in recombination (see Lloyd, Suggested Reading). The RecG and Ruv proteins perform the same function in recombination and so can substitute for each other. If

one of the *ruv* genes is inactive, the *recG* gene product is still available to resolve Holliday junctions and vice versa. Redundancy of function is a common explanation in genetics for why some gene products are nonessential. We discuss other examples of redundant functions elsewhere in this book.

Once the Ruv proteins had been shown to be involved in recombination, it took some clever intuition to find that their role is in the migration and resolution of Holliday junctions (see Parsons et al., Suggested Reading). The Ruv proteins were first suspected to be involved in a late stage of recombination because of a puzzling observation: even though *ruv* mutants give normal numbers of transconjugants when crossed with an Hfr strain, they give many fewer transconjugants when crossed with strains containing F' plasmids. As mentioned above, mating with F' factors should result in, if anything, more transconjugants than crosses with Hfr strains, because F' factors are replicons, which can multiply autonomously and do not rely on recombination for their maintenance.

One way to explain this observation is to propose that the Ruv proteins function late in recombination. If the Ruv proteins function late, recombinational intermediates might accumulate in cells with *ruv* mutations, thus having a deleterious effect on the the cell. If so, *recA* mutations, which block an early step in recombination by preventing the formation of synapses, might suppress the deleterious effect of the *ruv* mutations on F' crosses. This was found to be the case. A *ruv* mutation had no effect on the frequency of transconjugants in F' crosses if the recipient cells also had a *recA* mutation. Once genetic evidence supported a late role for the Ruv proteins in recombination, the process of Holliday junction resolution became the candidate for that role, since this is the last step in recombination. Biochemical experiments were then used to show that the Ruv proteins help in the migration and resolution of synthetic Holliday junctions, as described above.

DISCOVERY OF *chi* SITES

The discovery on DNA of *chi* sites that stimulate recombination by the RecBCD nuclease also required some interesting genetics. Many sites that are subject to single- or double-strand breaks are known to be "hot spots" for recombination. In some cases, such as recombination initiated by homing enzymes (see Box 9.1), it is clear that breaks at specific sites in DNA can initiate recombination. In general, however, the frequency of recombination seems to correlate fairly well with physical distance on DNA, as though recombination occurs fairly uniformly throughout DNA molecules.

It came as some surprise, therefore, to discover that the major recombination pathway for Hfr crosses in *E. coli*, the RecBCD pathway, does occur through specific sites on the DNA—the *chi* sites. Like many important discoveries in science, the discovery of *chi* sites started with an astute observation. This observation was made during studies of host recombination functions using λ phage (see Stahl et al., Suggested Reading). The experiments were designed to analyze the recombinant types that formed when the phage *red* recombination genes were deleted. Without its own recombination functions, the phage requires the host RecBCD nuclease. In addition, if the phage is also a *gam* mutant, it will not form a plaque unless it can recombine. Therefore, plaque formation by a *red gam* mutant phage λ is an indication that recombination has taken place.

The reason that *red gam* mutant phage λ cannot multiply to make a plaque without recombination is somewhat complicated. As discussed in chapter 7, phage λ cannot package DNA from genome-length DNA molecules but only from concatemers in which the λ genomes are linked end to end. Normally, the phage makes concatemers by rolling-circle replication. However, if the phage is a *gam* mutant, it cannot switch to the rolling-circle mode of replication because the RecBCD nuclease, which is normally inhibited by Gam, will somehow block the switch. Therefore, the only way a *gam* mutant phage λ can form concatemers is by recombination between the circular λ DNAs formed via θ replication. If the phage is also a *red* mutant (i.e., lacks its own recombination functions), the only way it can form concatemers is by RecBCD recombination, the major host pathway. Therefore, *red gam* mutant phage λ requires RecBCD recombination to form plaques, and the formation of plaques can be used as a measure of RecBCD recombination under these conditions.

chi mutations were discovered when large numbers of *red gam* mutant phage λ were plated on RecBCD⁺ *E. coli*. Very few phage were produced, and the plaques that formed were very tiny. Apparently, very little RecBCD recombination was occurring between the circular phage DNAs produced by θ replication. However, λ mutants that produced much larger plaques sometimes appeared. The circular λ DNAs in these mutants were apparently recombining at a much higher rate. The responsible mutations were named **chi** mutations because they increase the frequency of crossovers (crossed-over chromosomes are called *chi*asma in eukaryotes). Once the mutations were mapped and the DNA was sequenced, *chi* mutations in λ were found to be base pair changes that created the sequence 5'GCTGGTGG3' somewhere in the λ DNA. The presence of this sequence

appears to stimulate recombination by the RecBCD pathway. Since wild-type λ has no such sequence anywhere in its DNA, recombination by RecBCD is very infrequent unless the *chi* sequence is created by a mutation.

Why wild-type λ has no *chi* sequences seems obvious. Phage λ normally uses its own recombination system, and so with no need for the RecBCD enzyme, it also has no need for *chi* sites. Bacterial DNA, by contrast, has many *chi* sites—many more than would be predicted by chance alone.

Further experimentation with *chi* sites revealed several interesting properties. For example, they stimulate crossovers only to one side of themselves, the 5' side. Very little stimulation of crossovers occurs on the 3' side. Also, if only one of the two parental phages contains a *chi* site, most of the recombinant progeny will not have the *chi* site, so that the chi site itself is preferentially lost during the recombination. These properties of *chi* sites led to models for *chi* site action such as the one presented earlier in this chapter.

Gene Conversion and Other Manifestations of Heteroduplex Formation during Recombination

GENE CONVERSION

As discussed earlier in this chapter, models for recombination are based in part on the evidence concerning the formation of heteroduplexes during recombination. The first such evidence came from studies of gene conversion in fungi. Understanding this process requires some knowledge of the sexual cycles of fungi. Some fungi have long been favored organisms for the study of recombination because the spores that are the products of a single meiosis are often contained in the same bag, or ascus (see any general genetics textbook). When two haploid fungal cells mate, the two cells fuse to form a diploid zygote. Then the homologous chromosomes pair and replicate once to form four chromatids that recombine with each other before they are packaged into spores. Therefore, each ascus will contain four spores (or eight in fungi such as *Neurospora crassa*, in which the chromatids replicate once more before spore packaging).

Since both haploid fungi contribute equal numbers of chromosomes to the zygote, their genes should show up in equal numbers in the spores in the ascus. In other words, if the two haploid fungi have different alleles of the same gene, two of the four spores in each ascus should have an allele from one parent and the other two spores should have the allele from the other parent. This is called a 2:2 segegation. However, the two parental alleles sometimes do not appear in equal numbers in the spores. For example, three of the spores in a particular ascus might have the allele from one parent while the remaining spore has the allele from the other parent—a 3:1 segregation instead of the expected 2:2 segregation. In this case, an allele of one of the parents appears to have been converted into the allele of the other parent during meiosis. The term **gene conversion** originates from this phenomenon.

Gene conversion is caused by repair of mismatches created on heteroduplexes during recombination, and Figure 9.9 shows how such mismatch repair can con-

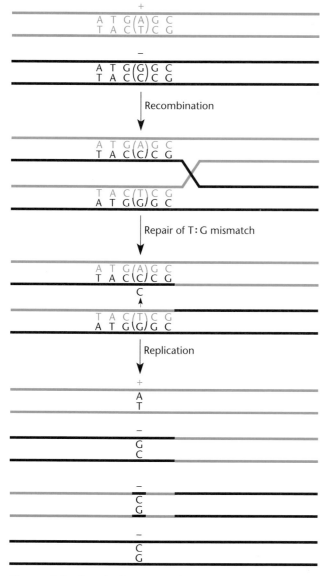

Figure 9.9 Repair of a mismatch in a heteroduplex region formed during recombination can cause gene conversion. A plus sign indicates the wild-type sequence, and a minus sign indicates the mutant sequence. See text for details.

vert one allele into the other when the DNA molecules of the two parents recombine. In the illustration, the DNAs of two parents are identical in the region of the recombination except that a mutation has changed a wild-type AT pair into a GC in one of the DNA molecules. Hence, the parents have different alleles of this gene. When the two individuals mate to form a diploid zygote and their DNAs recombine during meiosis, one strand of each DNA may pair with the complementary strand of the other DNA in this region. A mismatch will result, with a G opposite a T in one DNA and an A opposite a C in the other DNA (Figure 9.9). If a repair system changes the T opposite the G to a C in one of the DNAs, then after meiosis three molecules will carry the mutant allele sequence, with GC at this position, but only one DNA will have the wild-type allele sequence, with AT at this position. Hence, one of the two wild-type alleles has been converted into the mutant allele.

MANIFESTATIONS OF MISMATCH REPAIR IN HETERODUPLEXES IN PHAGES AND BACTERIA

Gene conversion is difficult to detect in crosses with bacteria and phages, since the products of a single recombination event are not contained in an ascus like they are in some fungi. Heteroduplexes do form during recombination in bacteria and phages, but the repair of mismatches in these structures is manifested in other ways.

Map Expansion

In prokaryotes, mismatch repair in heteroduplexes can increase the apparent recombination frequency between two closely linked markers, making the two markers seem farther apart than they really are. This manifestation of mismatch repair in heteroduplexes formed during recombination is called **map expansion** because the genetic map appears to increase in size.

Figure 9.10 shows how mismatch repair can affect the apparent recombination frequency between two markers. In the illustration, the two DNA molecules participating in the recombination have mutations that are very close to each other, so that crossovers between the two mutations to give wild-type recombinants should be very rare. However, a Holliday junction occurs nearby, and the region of one of the two mutations is included in the heteroduplex, creating mismatchs that can be repaired. If the G in the GT mismatch in one of the DNA molecules is repaired to an A, the progeny with the DNA will appear to be a wild-type recombinant. Therefore, even though the potential crossover that caused the formation of heteroduplexes did not

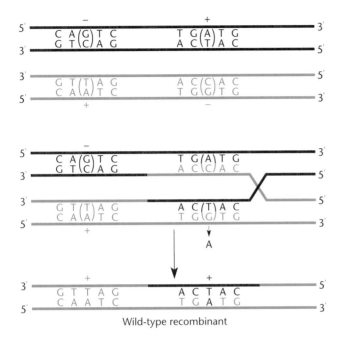

Figure 9.10 Repair of mismatches can give rise to recombinant types between two mutations. A plus sign indicates the wild-type sequence, and a minus sign indicates the mutant sequence. The position of the mutations is shown in parentheses. See the text for details.

occur in the region between the two mutations, apparent wild-type recombinants resulted. The apparent recombination due to mismatch repair might occur even when the Holliday junction is resolved so that the flanking sequences are in their original configuration; thus, a true crossover does not result. Therefore, although gene conversion and other manifestations of mismatch repair are generally associated with recombinant DNA molecules, they do not appear only in DNA molecules with obvious crossovers.

Marker Effects

Mismatch repair of heteroduplexes also can cause **marker effects**, phenomena in which two different markers at exactly the same locus show different recombination frequencies when crossed with the same nearby marker. For example, two different transversion mutations might change a UAC codon into UAA and UAG codons in different strains. However, when these two strains are crossed with another strain with a third nearby mutation, the recombination frequency between the ochre mutation and the third mutation might appear to be much lower than the recombination frequency between the amber mutation and the third mutation, even

though the amber and ochre mutations are exactly the same distance on DNA from the third mutation. Such a difference between the two recombination frequencies can be explained because the amber and ochre mutations are causing different mismatches to form during recombination and one of these may be recognized and repaired more readily by the mismatch repair system than the other.

Marker effects also occur because the lengths of single DNA strands removed and resynthesized by different mismatch repair systems will vary (see chapter 10), and the chance that mismatch repair will lead to apparent recombination depends upon the length of these sequences, or patches. As is apparent in Figure 9.10, a wild-type recombinant will only occur if mismatches due to both mutations are not removed on the same repair patch. If the patch that is removed in repairing one mismatch also removes the other mismatch, one of the parental DNA sequences will be restored and no apparent recombination will occur.

High Negative Interference

Another manifestation of mismatch repair in heteroduplexes is **high negative interference**. This has the reverse effect of interference in eukaryotes, a phenomenon in which the stiffness of the chromatids can cause one crossover to reduce the chance of another crossover nearby. In high negative interference, one crossover greatly *increases* the apparent frequency of another crossover nearby.

High negative interference is often detected during three-factor crosses with closely linked markers. In chapters 13 and 14, we discuss how three-factor crosses can be used to order three closely linked mutations. Briefly, if one parent has two mutations and the other parent a third mutation, the frequency of the different types of recombinants after the cross will depend on the order of the sites of the three mutations in the DNA. Twice as many apparent crossovers are required to produce the rarest recombinant type as are needed for the more frequent recombinant types.

However, owing to mismatch repair of heteroduplexes, the rarest recombinant type can occur much more frequently than expected. Figure 9.11 shows a three-factor cross between markers that are created by three mutations. Wild-type sequences are marked with a plus and mutant sequences with a minus. With the molecules pictured, the formation of a wild-type recombinant should require two crossovers: one between mutations 1 and 2 and another between mutations 2 and 3. Theoretically, if the two crossovers were truly

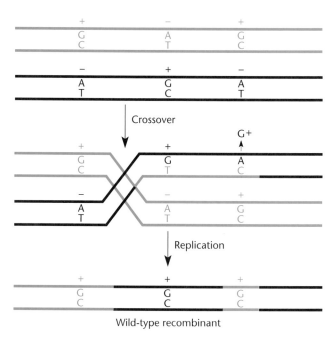

Figure 9.11 High negative interference due to mismatch repair. See the text for details.

independent, the frequency of wild-type recombinants in the three-factor cross should equal the frequency when mutation 1 is crossed with mutation 2 times the frequency when mutation 2 is crossed with mutation 3, in separate crosses. Instead, the frequency of the wild-type recombinants in the three-factor cross is often much higher than this product. As shown in Figure 9.11, if a crossover occurs between two of the mutations, a heteroduplex formed by the Holliday junction might include the region of the third mutation. Then repair of the third mutant site mismatch can give the appearance of a second nearby crossover, greatly increasing the apparent frequency of that crossover.

It is important to realize that repair of mismatches can occur wherever two DNAs are recombining. However, only if two markers are close enough together that normal crossovers are infrequent will mismatch repair in the heteroduplexes contribute significantly to apparent recombination frequencies and cause effects such as map expansion, high negative interference, and marker effects.

SUMMARY

1. Recombination is the joining of DNA strands in new combinations. Homologous recombination occurs only between two DNA molecules that have the same sequence in the region of the crossover.

2. Models of recombination can be divided into one- or two-strand invasion models and double-strand break repair models.

3. All recombination models involve the formation of Holliday junctions. The Holliday junctions can migrate and isomerize so that the crossed strands uncross and then recross in a different orientation. Holliday junctions can then be resolved by specific DNA endonucleases to give recombinant DNA products.

4. The region over which two strands, originating from different DNA molecules, are paired in a Holliday junction is called a heteroduplex.

5. Repair of mismatches on the heteroduplex DNA molecules formed as intermediates in recombination can give rise to such phenomena as gene conversion, map expansion, high negative interference, and marker effects.

6. In *E. coli*, the major pathway for recombination during conjugation and transduction is the RecBCD pathway. A model for RecBCD action has been proposed in which the RecBCD protein loads on the DNA at a double-stranded break and moves along the DNA, looping out a single strand and degrading one strand from the 3' end. If a sequence in the DNA called the *chi* sequence is encountered as the protein moves along the DNA, the exonuclease activity on RecBCD is inhibited, leaving a free 3' end that can invade other double-stranded DNAs.

7. In the absence of RecBCD, the RecF pathway can be activated. The RecF pathway requires the products of the *recF*, *recJ*, *recN*, *recO*, *recQ*, and *recR* genes. This pathway can function only if the *sbcB* and *sbcC* genes have been inactivated. Apparently, the SbcB and SbcC enzymes destroy intermediates created during recombination by the RecF pathway.

8. The RecA protein promotes synapse formation and is required for recombination by both the RecBCD and RecF pathways. The RecA protein forces a single-stranded free end of DNA into a double-stranded DNA helix, allowing it to find its complementary sequence in double-stranded DNA through formation of an elongated triple-stranded helix. In the triple-stranded helix, the single-stranded DNA pairs with its complementary strand in the double-stranded DNA through the major groove without actually separating the strands.

(continued)

SUMMARY (continued)

9. Holliday junctions can be resolved by two separate pathways in *E. coli*, the RuvABC pathway and the RecG pathway. In the first pathway, the RuvA protein binds to Holliday junctions and then the RuvB protein binds to RuvA and promotes branch migration with the energy derived from cleaving ATP. The RuvC protein is a Holliday junction-specific endonuclease that cleaves Holliday junctions to resolve recombinant products. The RecG protein is also a Holliday junction-specific endonuclease and can substitute for the RuvABC system.

10. Many phages also encode their own recombination systems. Sometimes, phage recombination functions are analogous to host recombination functions. The gene *49* product of T4 phage and the gene *3* product of T7 resolve Holliday junctions. A cryptic phage, *rac*, found in *E. coli* K-12 strains encodes a recombination system called RecE, which can be activated by *sbcA* mutations and substitute for the RecBCD system in conjugational and transductional crosses. Phage λ also encodes recombination systems, one of which can partially substitute for components of the RecF pathway. Another λ recombination system, the *red* system, is encoded by two genes, *redα* and *redβ*. The Redα protein is partially homologous to RecE of the *rac* prophage and is the λ exonuclease commonly used in molecular biology.

QUESTIONS FOR THOUGHT

1. Why do you suppose essentially all organisms have recombination systems?

2. Why do you suppose the RecBCD protein promotes recombination through such a complicated process?

3. Why are there overlapping pathways of recombination that can substitute for each other?

4. What is the real function of the RecF pathway? Why do you think the cell encodes the *sbcB* and *sbcC* gene products that interfere with the RecF pathway?

5. Why do some phage encode their own recombination systems—why not rely exclusively on the host pathways?

PROBLEMS

1. Describe how you would determine if recombinants in an Hfr cross have a *recA* mutation. Note: *recA* mutations make the cells very sensitive to mitomycin and UV irradiation.

2. How would you determine if the products of other genes participate in the *recG* pathway of migration and resolution of Holliday junctions? How would you find such genes?

3. Describe the recombination promoted by homing double-stranded nucleases to insert an intron by using the double-strand break repair model.

4. Design an experiment to determine whether recombination due to the RecBC nuclease without the RecD subunit is still stimulated by *chi*.

SUGGESTED READING

Clark, A. J., and A. D. Margulies. 1965. Isolation and characterization of recombination-deficient mutants of *Escherichia coli* K12. *Proc. Natl. Acad. Sci. USA* **62:**451–459.

Dixon, D. A., and S. C. Kowalcykowski. 1993. The recombination hotspot Chi is a regulatory sequence that acts by attenuating the nuclease activity of the *E. coli* RecBCD enzyme. *Cell* **73:**87–96.

Holliday, R. 1964. A mechanism for gene conversion in fungi. *Genet. Res.* **5:**282–304.

Howard-Flanders, P., and L. Theriot. 1966. Mutants of *Escherichia coli* defective in DNA repair and in genetic recombination. *Genetics* **53:**1137–1150.

Jones, M., R. Wagner, and M. Radman. 1987. Mismatch repair and recombination in *E. coli*. *Cell* **50:**621–626.

Kowalczykowski, S. C., D. A. Dixon, A. K. Eggleston, S. D. Lauder and W. M. Rehrauer. 1994. Biochemistry of homologous recombination in *Escherichia coli*. *Microbiol. Rev.* **58:**401–465.

Lloyd, R. G. 1991. Conjugal recombination in resolvase-deficient *ruvC* mutants of *Escherichia coli* K-12 depends on *recG*. *J. Bacteriol.* **173:**5414-5418.

Lloyd, R. G., and C. Buckman. 1985. Identification and genetic analysis of *sbcC* mutations in commonly used *recBC sbcB* strains of *Escherichia coli* K-12. *J. Bacteriol.* **164:**836–844.

Lloyd, R. G., and K. B. Low. 1996. Homologous recombination, p. 2236–2255. *In* F. C. Neidhardt, R. Curtiss III, J. L. Ingraham, E. C. C. Lin, K. B. Low, B. Magasanik, W. S. Reznikoff, M. Riley, M. Schaechter, and H. E. Umbarger (ed.), *Esherichia coli and Salmonella: Cellular and Molecular Biology*, 2nd ed. ASM Press, Washington, D.C.

Meselson, M. S., and C. M. Radding. 1975. A general model for genetic recombination. *Proc. Natl. Acad. Sci. USA* **2:**358–361.

Miesel, L., and J. R. Roth. 1996. Evidence that SbcB and RecF pathway functions contribute to RecBCD-dependent transductional recombination. *J. Bacteriol.* **178**:3146–3155.

Parsons, C. A., I. Tsaneva, R. G. Lloyd, and S. C. West. 1992. Interaction of *Escherichia coli* RuvA and RuvB proteins with synthetic Holliday junctions. *Proc. Natl. Acad. Sci. USA* **89**:5452–5456.

Rao, B. J., M. Dutreix, and C.M. Radding. 1991. Stable three-stranded DNA made by RecA protein. *Proc. Natl. Acad. Sci. USA* **88**:2984–2988.

Sawitzeke, J. A., and F. W. Stahl. 1992. Phage λ has an analog of *Escherichia coli recO*, *recR* and *recF* genes. *Genetics* **130**:7–16.

Stahl, F. W., M. M. Stahl, R. E. Malone, and J. M. Crasemann. 1980. Directionality and nonreciprocality of Chi-stimulated recombination in phage λ. *Genetics* **94**:235–248.

Szostak, J. W., T. L. Orr-Weaver, R. J. Rothstein, and F. W. Stahl. 1983. The double-strand-break repair model for recombination. *Cell* **33**:25–35.

Taylor, A. F., and G. R. Smith. 1992. RecBCD enzyme is altered upon cutting DNA at a recombination hotspot. *Proc. Natl. Acad. Sci. USA* **89**:5226–5230.

Chapter 10

DNA Repair
and Mutagenesis

THE CONTINUITY OF SPECIES from one generation to the next is a tribute to the stability of DNA. If DNA were not so stable and were not reproduced so faithfully, there could be no species. Before DNA was known to be the hereditary material and its structure determined, a lot of speculation centered around what types of materials would be stable enough to ensure the reliable transfer of genetic information over so many generations (see, for example, Schrodinger, Suggested Reading). Therefore, the discovery that the hereditary material is DNA—a chemical polymer no more stable than many other chemical polymers—came as a surprise.

Evolution has resulted in a design for the DNA replication apparatus that minimizes mistakes (see chapter 1). However, mistakes during replication are not the only threats to DNA. Since DNA is a chemical, it is constantly damaged by chemical reactions. Many environmental factors can damage this molecule. Heat can speed up spontaneous chemical reactions, leading, for example, to the deamination of bases. Chemicals can react with DNA, adding groups to the bases or sugars, breaking the bonds of the DNA, or fusing parts of the molecule to each other. Irradiation at certain wavelengths can also chemically damage DNA, which can absorb the energy of the photons. Once the molecule is energized, bonds may be broken or parts may be fused. Chemical damage can be very deleterious to cells because their DNA may not be able to replicate over the damaged area and so the cells could not multiply. Even if the damage does not block replication, replicating over the damage can cause mutations, many of which may be deleterious or even lethal. Obviously, cells need mechanisms for DNA damage repair.

To describe DNA damage and its repair, we need to first define a few terms. Chemical damage in DNA is called a **lesion**. Chemical compounds or treatments that cause lesions in DNA can kill cells and can also increase the frequency of mutations in DNA. Such chemicals that generate mutations are called **mutagenic chemicals** or **mutagens**. Some mutagens, known as **in vitro mutagens**, can be used to damage DNA in the test tube and then produce mutations when this DNA is introduced into cells. Other mutagens damage DNA only in the cell, for example by interfering with base pairing during replication. These are called **in vivo mutagens**.

In this chapter, we discuss the types of DNA damage, how each type of damage to DNA might cause mutations, and how bacterial cells repair the damage to their DNA. Many of these mechanisms seem to be universal and are shared by higher organisms including humans.

Evidence for DNA Repair

Before discussing specific types of DNA damage, we should make some general comments on the outward manifestations of DNA damage and its repair. The first question is how we even know a cell has the means to repair a particular type of damage to its DNA. One way is to measure killing by a chemical or by irradiation. The chemical agents and radiation that damage DNA also often damage other cellular constituents, including RNA and proteins. Nevertheless, cells exposed to these agents usually die as a result of chemical damage to the DNA. The other components of the cell can usually be resynthesized and/or exist in many copies, so that even if some molecules are damaged, more of the same type of molecule will be there to substitute for the damaged ones. However, a single chemical change in the enormously long chromosomal DNA of a cell can prevent the replication of that molecule and subsequently cause cell death unless the damage is repaired.

To measure cell killing—and thereby demonstrate that a particular type of cell has DNA repair systems—we can compare the survival of the cells exposed intermittently to small doses of a DNA-damaging agent with that of cells that receive the same amount of treatment continuously. If the cells have DNA repair systems, more will survive the short intervals of treatment because some of the damage will be repaired between treatments. Consequently, at the end of the experiment, fewer intermittently treated cells will have been killed than cells exposed to continuous treatment. In contrast, if DNA is not repaired during the rest periods, whether the treatment occurs at intervals or continuously should make no difference. The cells will accumulate the same amount of damage regardless of the different treatments, and the same fraction of cells should survive both regimes.

Another indication of repair systems comes from the shape of the killing curves. A killing curve is a plot of

the number of surviving cells versus the extent of treatment by an agent that damages DNA. The extent of treatment can refer to the length of time the cells are irradiated or exposed to a chemical that damages DNA or to the intensity of irradiation or the concentration of the damaging chemical.

The two curves in Figure 10.1 contrast the shapes of killing curves for cells with and without DNA repair systems. In the curve for cells without a repair system for the DNA-damaging treatment, the fraction of surviving cells drops exponentially, since the probability that each cell will be killed by a lethal "hit" to its DNA is the same during each time interval. This exponential decline gives rise to a straight line when plotted on semilog paper as shown.

The other curve shows what happens if the cell has DNA repair systems. Rather than dropping exponentially with increasing treatment, this curve extends horizontally first, creating a "shoulder." The shoulder appears because repair mechanisms repair lower levels of damage, allowing many of the cells to survive. Only with higher treatment levels, when the damage becomes so extensive that the repair systems can no longer cope with it, will the number of surviving cells drop exponentially with increasing levels of treatment.

Among the survivors of DNA-damaging agents, there may be many more mutants than before. However, it is very important to distinguish DNA damage from muta-

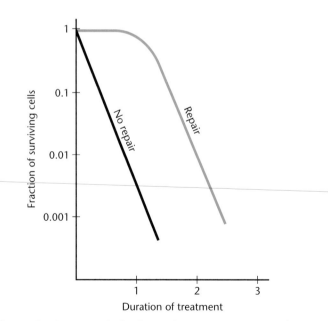

Figure 10.1 Survival of cells as a function of the time of treatment with a DNA-damaging agent. The fraction of surviving cells is plotted against the duration of treatment. A shoulder on the survival curve indicates the presence of a repair mechanism.

genesis. In particular types of cell, some types of DNA damage are mutagenic while others are not, independent of their effect on cell survival. Recall from chapter 3 that mutations are permanent heritable changes in the sequence of nucleotides in DNA. The damage to DNA caused by a chemical or by irradiation is not by itself a mutation because it is not heritable. A mutation might occur because the damage is not repaired and the replication apparatus must proceed over the damage, making mistakes because complementary base pairing does not occur properly at the site of the damage. Alternatively, mistakes might be made during attempts to repair the damage, causing changes in the sequence of nucleotides at the site of the damage. In the following sections, we discuss some types of DNA damage, how they might cause mutations, and the repair systems that can repair them. Most of what we describe is known for *Escherichia coli*, for which these systems are best understood.

Specific Repair Pathways

Different agents damage DNA in different ways, and different repair pathways operate to repair the various forms of damage. Some of these repair pathways repair only a certain type of damage, whereas others are less specific and repair many types. We first discuss examples of damage repaired by specific repair pathways.

Deamination of Bases

One of the most common types of damage to DNA is the deamination of bases. Some of the amino groups in adenine, cytosine, and guanine are particularly vulnerable and can be removed spontaneously or by many chemical agents (Figure 10.2). When adenine is deaminated, it becomes **hypoxanthine**. When guanine is deaminated, it becomes **xanthine**. When cytosine is deaminated, it becomes **uracil**.

Deamination of DNA bases is mutagenic because it results in base mispairing. As shown in Figure 10.2, hypoxanthine derived from adenine will pair with the base cytosine instead of thymine, and uracil derived from the deamination of cytosine will pair with adenine instead of guanine.

The type of mutation caused by deamination depends on which base is altered. For example, the hypoxanthine that results from deamination of adenine will pair with cytosine during replication, incorporating C instead of T at that position. In a subsequent replication, the C will pair with the correct G, causing an AT-to-GC transition in the DNA. Similarly, a uracil resulting from the deamination of a cytosine will pair with an adenine during replication, causing a GC-to-AT transition.

DEAMINATING AGENTS

Although deamination often occurs spontaneously, especially at higher temperatures, some types of chemicals react with DNA and remove amino groups from the bases. Treatment of cells or DNA with these chemicals, known as **deaminating agents,** can greatly increase the rate of mutations. Which deaminating agents are mutagenic in a particular situation depends upon the properties of the chemical.

Hydroxylamine

Hydroxylamine specifically removes the amino group of cytosine and consequently causes only GC-to-AT transitions in the DNA. However, hydroxylamine, an in vitro mutagen, cannot enter cells, so it can be used only to mutagenize purified DNA or viruses. Mutagenesis by hydroxylamine is particularly effective when the treated DNA is introduced into cells deficient in repair by the uracil-*N*-glycosylase enzyme (see chapter 1), for reasons we discuss below.

Bisulfite

Bisulfite will also deaminate only cytosine, but these cytosines must be in single-stranded DNA. This property of bisulfite has made it useful for **site-directed mutagenesis**. If the region to be mutagenized in a clone is made single stranded, the bisulfite will preferentially limit mutagenesis to this single-stranded region. Use of bisulfite for site-directed mutagenesis of cloned DNAs has been largely supplanted by oligonucleotide-directed and PCR mutagenesis (see chapter 15).

Nitrous Acid

Nitrous acid not only deaminates cytosines but also removes the amino groups of adenine and guanine (Figure 10.2). It also causes other types of damage. Because it is less specific, nitrous acid can cause both GC-to-AT and AT-to-GC transitions as well as deletions. Nitrous acid can enter some types of cells and so can be used as a mutagen both in vivo and in vitro.

REPAIR OF DEAMINATED BASES

Because base deamination is potentially mutagenic, special enzymes have evolved to remove deaminated bases from DNA. These enzymes, **DNA glycosylases**, break the glycosyl bond between the damaged base and the sugar in the nucleotide. A unique DNA glycosylase exists for each type of deaminated base and will remove only that particular base. Specific DNA glycosylases discussed in later chapters remove bases damaged in other ways.

Figure 10.3 illustrates the removal of damaged bases from DNA by DNA glycosylases. After the base has

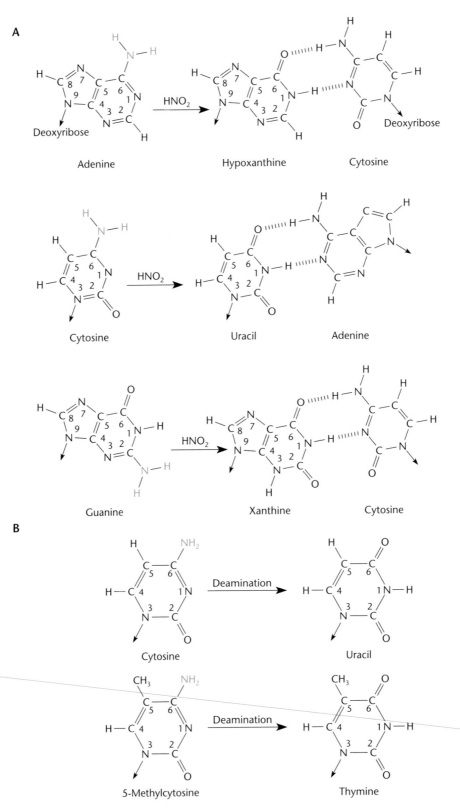

Figure 10.2 (A) Modified bases created by deaminating agents such as nitrous acid (HNO₂). Some deaminated bases will pair with the wrong base, causing mutations. (B) Spontaneous deaminations. Deamination of 5-methylcytosine will produce a thymine that is not removable by the uracil-*N*-glycosylase.

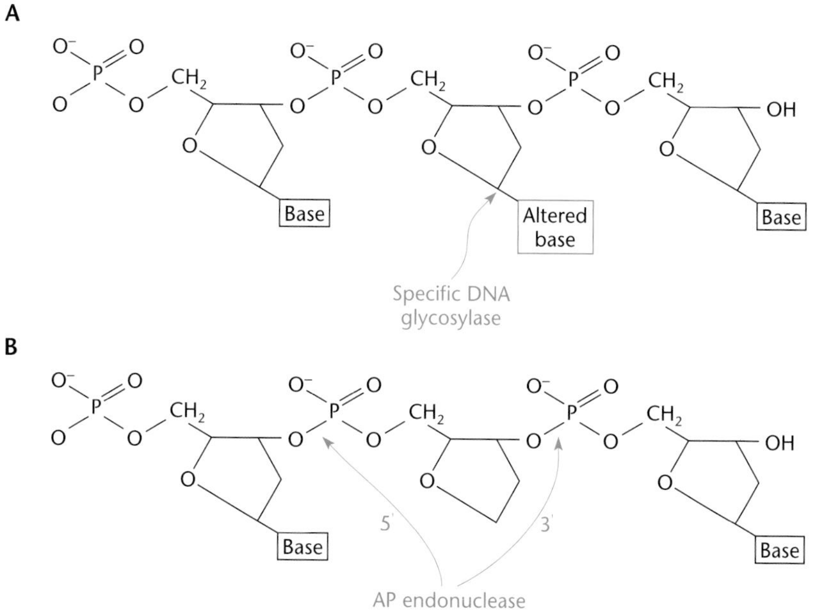

A

B

Figure 10.3 Repair of altered bases by DNA glycosylases. (A) The specific DNA glycosylase removes the altered base. (B) Most apurinic or apyrimidinic (AP) endonucleases cut the DNA backbone on the 5′ side of the apurinic or apyrimidinic site. The strand is degraded and resynthesized, and the correct base is restored (not shown).

been removed by the specific glycosylase, nucleases called **AP endonucleases** cut the sugar-phosphate backbone of the DNA next to the missing base. The AP endonuclease can cut either next to the spot from which a pyrimidine (C or T) has been removed (an *apyrimidinic* site) or next to where a purine (A or G) has been removed (an *apurinic* site). After the cut is made, the free 3′-hydroxyl end is used as a primer by the repair DNA polymerase (DNA polymerase I in *E. coli*) to synthesize more DNA, while the 5′ exonuclease activity associated with the DNA polymerase degrades the strand ahead of the DNA polymerase. In this way, the entire region of the DNA strand around the deaminated base is resynthesized and the normal base is inserted in place of the damaged one.

VERY SHORT PATCH (VSP) REPAIR OF DEAMINATED 5-METHYLCYTOSINE

Most organisms have some 5-methylcytosine bases instead of cytosines at specific sites in their DNA. These bases are cytosines with a methyl group at the 5 position on the pyrimidine ring instead of the usual hydrogen (Figure 10.2B). Specific enymes called **methyltransferases** transfer the methyl group to this position after the DNA is synthesized. The function of these 5-methylcytosines is often obscure, but we know that they sometimes help protect DNA against cutting by restriction endonucleases and may help regulate gene expression in higher organisms.

The sites of 5-methylcytosine in DNA are often hot spots for mutagenesis, because deamination of 5-methylcytosine yields thymine rather than uracil (Figure

10.2B), and thymine in DNA will not be recognized by the uracil-N-glycosylase since it is a normal base in DNA. These thymines in DNA will be opposite guanines and so could in principle be repaired by the methyl-directed mismatch repair system (see below and chapter 1). However, in a GT mismatch created by a replication mistake, the mistakenly incorporated base can be identified because it is in a newly replicated strand, as yet unmethylated by Dam methylase, whereas the GT mismatches created by the deamination of 5-methylcytosine are generally not in newly synthesized DNA. Repairing the wrong strand will cause a GC-to-AT transition in the DNA.

In *E. coli* K-12, most of the 5-methylcytosine in the DNA occurs in the second C of the sequence 5′CCWGG3′/3′GGWCC5′, where the middle base pair (W) is generally either AT or TA. The second C in this sequence is methylated by an enzyme called DNA cytosine methylase (Dcm) to give C^{m}CAGG/GGTC^{m}C. Because of the mutation potential, *E. coli* K-12 has evolved a special repair mechanism for deaminated 5-methylcytosines that occur in this sequence. This repair system specifically removes a thymine whenever it appears as a TG mismatch in this sequence.

Because a small region, or "patch," of the DNA strand containing the T is removed and resynthesized during the repair process, the mechanism is called **very short patch (VSP) repair**. In VSP repair, the Vsr endonuclease, the product of the *vsr* gene, binds to a TG mismatch in the C^{m}CAGG/GGTTC sequence and makes a break next to the T. The T is then removed and the strand resynthesized by the repair DNA polymerase

(DNA polymerase I), which inserts the correct C (see Hennecke et al., Suggested Reading).

The Vsr repair system is very specialized and will repair TG mismatches only in the sequence shown above. If methylated C's did not occur in this particular sequence, this repair system would be useless. The *vsr* gene is immediately downstream of the gene for the Dcm methylase, ensuring that cells that inherit the gene to methylate the C in $C^mCAGG/GGTC^mC$ will also usually inherit the ability to repair the mismatch correctly if it is deaminated. While only *E. coli* K-12 has been shown to have this particular repair system, many other organisms have 5-methylcytosine in their DNA, and we expect that similar repair systems will be found in these organisms.

Damage Due to Reactive Oxygen

Although molecular oxygen (O_2) is not damaging to DNA, other, more reactive forms of oxygen are very damaging. The reactive forms of oxygen have more electrons than molecular oxygen and include superoxide radicals, hydrogen peroxide, and hydroxyl radicals.

These forms of oxygen may be produced by normal cellular reactions or by cytochromes, such as P-450, which are involved in detoxification of chemicals by the liver. Alternatively, they can arise as a result of environmental factors, including UV irradiation and chemicals such as the herbicide paraquat.

Because the reactive forms of oxygen normally appear in cells, all aerobic organisms must contend with the resulting DNA damage and have evolved elaborate mechanisms to remove these chemicals from the cellular environment. In bacteria, some of these systems are induced by the presence of the reactive forms of oxygen, and these genes encode enzymes such as superoxide dismutases, catalases, and peroxide reductases, among others, which help destroy the reactive forms. These systems also include genes that encode repair enzymes that help repair the oxidative damage to DNA caused by the reactive forms. The accumulation of this type of damage may be responsible for the increase in cancer rates with age and for many age-related degenerative diseases (see Box 10.1).

BOX 10.1

Oxygen: the Enemy Within

To respire, all aerobic organisms, including humans, must take up molecular oxygen (O_2). At normal temperatures, molecular oxygen reacts with very few molecules. However, some of it is converted into more reactive forms such as superoxide radicals (O_2^-), hydrogen peroxide (H_2O_2), and hydroxyl radicals (·OH). These reactive molecules can be formed by cytochromes used for normal aerobic metabolism or for detoxification in the liver. They are also formed in specialized cells that use them in lysosomes to kill invading bacteria. They can also be produced by externally added chemicals such as the herbicide paraquat, which has been used to spray marijuana fields in Central and South America. These forms of oxygen will react with many cellular molecules, including DNA, and so can damage DNA and cause mutations.

Cells have evolved many enzymes to help reduce this damage. These include catalases that reduce reactive oxygen molecules as they form and repair enzymes such as exonucleases and glycosylases that remove the damaged bases from DNA before they can cause mutations.

Accumulation of DNA damage due to reactive oxygen has been linked to many degenerative diseases such as cancer, arthritis, cataracts, and cardiovascular disease. It has been estimated that a rat at 2 years of age has about 2 million DNA lesions per cell, and some types of human cells have been shown to accumulate DNA damage with age. The synthesis of these active forms of oxygen helps explain why some compounds (such as asbestos) that are not themselves mutagenic or chronic infections can increase the rate of cancer. The reactive forms of oxygen that are synthesized in response to these conditions by macrophages may be the real mutagens (see Box 10.4).

Obviously, any mechanism for reducing the levels of these active forms of oxygen should increase longevity and reduce the frequency of many degenerative diseases. Fruits and vegetables produce antioxidants, including asorbic acid (vitamin C), tocopherol (vitamin E), and carotenes such as β-carotene (found in large amounts in carrots), that destroy these molecules. Consumption of adequate amounts of fruits and vegetables that contain these compounds may greatly reduce the rate of cancer and many degenerative diseases.

References

Ames, B. N., M. K. Shigenaga, and T. M. Hagen. 1993. Oxidants, antioxidants and the degenerative diseases of aging. *Proc. Natl. Acad. Sci. USA* **90:**7915–7922.

Farr, S. B., and T. Kogama. 1991. Oxidative stress responses in *Escherichia coli* and *Salmonella typhimurium*. *Microbiol. Rev.* **55:**561–585.

8-oxoG

One of the most mutagenic lesions in DNA caused by reactive oxygen is the oxdized base **7,8-dihydro-8-oxoguanine (8-oxoG or GO)** (Figure 10.4). This base appears frequently in DNA and often mispairs with adenine. Because of the mutagenic potential of 8-oxoG, cells have evolved many mechanisms for avoiding the resultant mutations.

The *mut* Genes

Studies with certain of the *mut* genes of *E. coli* have confirmed the mutagenic potential of 8-oxoG. Bacteria with a mutation in a *mut* gene suffer higher than normal rates of spontaneous mutagenesis. This fact led to the identification of several *mut* genes in *E. coli*. In other chapters, we mentioned some *E. coli mut* genes, including the *mutD* gene, which encodes the DNA Pol III editing function, and the *mutS*, *mutL*, and *mutH* genes, which encode components of the methyl-directed mismatch repair system. In chapter 14, we discuss the isolation of *mut* mutants of *E. coli* as an example of bacterial genetic analysis.

The products of three of the *E. coli mut* genes, *mutM*, *mutT*, and *mutY*, are devoted exclusively to preventing

A

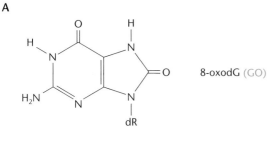

8-oxodG (GO)

B

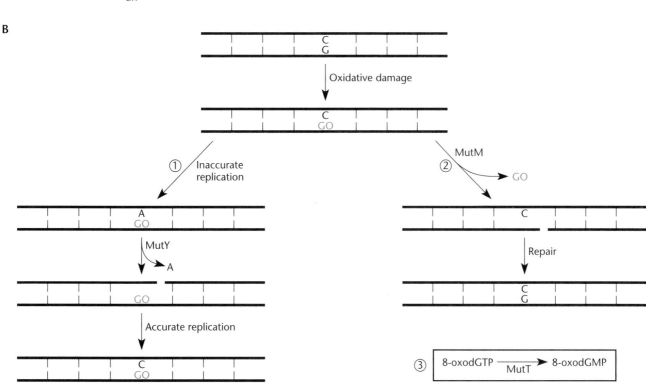

Figure 10.4 (A) Structure of 8-oxodG. (B) Mechanisms for avoiding mutagenesis due to 8-oxoG (GO). In pathway 1, an A mistakenly incorporated opposite 8-oxoG is removed by a specific glycosylase (MutY), and the strand is degraded and resynthesized with the correct C. In pathway 2, the 8-oxoG is itself removed by a specific glycosylase (MutM), and the strand is degraded and resynthesized with a normal G. In a third pathway, the 8-oxodG is prevented from entering DNA by a specific phosphatase (MutT) that degrades the triphosphate 8-oxodGTP to the monophosphate 8-oxodGMP. Adapted from M. L. Michaels and J. H. Miller, *J. Bacteriol.* **174:**6321–6325, 1992.

mutations due to 8-oxoG and are important components of the defense against this type of oxidative damage. The effects of mutations in these genes are additive in the sense that the rate of spontaneous mutations is higher if two or all three of the *mut* genes are mutated than if only one of them is mutated. Moreover, the generally high rate of spontaneous mutagenesis in *mutM*, *mutT*, and *mutY* mutants supports the conclusions that internal oxidation of DNA is an important source of spontaneous mutations and that 8-oxoG, in particular, is a very mutagenic form of damage to DNA. How the gene products MutM, MutT, and MutY prevent mutations due to 8-oxoG lesions is quite well understood (see Michaels et al., Suggested Reading). We discuss each separately.

MutM

The MutM enzyme is an *N*-glycosylase that specifically removes the 8-oxoG base from the deoxyribose sugar in DNA (Figure 10.4). This repair pathway functions like other *N*-glycosylase repair pathways we have discussed except that the depurinated strand is cut by the AP endonuclease activity of MutM itself, degraded by an exonuclease, and resynthesized by DNA polymerase I (Figure 10.3).

The MutM protein is present in larger amounts in cells that have accumulated reactive oxygen, because the *mutM* gene is part of a regulon induced in response to oxidative stress. We discuss regulons in more detail in chapter 12.

MutY

The MutY enzyme is also a specific *N*-glycosylase. However, rather than removing 8-oxoG directly, the MutY *N*-glycosylase specifically removes adenine bases that have been mistakenly incorporated opposite an 8-oxoG in DNA (Figure 10.4). Repair synthesis then introduces the correct C to prevent a mutation, as with other *N*-glycosylase-initiated repair pathways.

In vitro, the MutY enzyme will also recognize a mismatch that results from accidental incorporation of an A opposite a normal G and will remove the A. However, its major role in avoiding mutagenesis in vivo seems to be to prevent mutations due to 8-oxoG. As evidence, mutations that cause the overproduction of MutM completely suppress the mutator phenotype of *mutY* mutants (see Michaels et al., Suggested Reading). The interpretation of this result is as follows. If a significant proportion of all spontaneous mutations in a *mutY* mutant resulted from misincorporation of A's opposite normal G's, excess MutM should have no effect on the mutation rate, because removal of 8-oxoG should not affect this type of mispairing. However, the fact that excess MutM suppresses mutagenesis in *mutY* mutants suggests that it functions to ensure that very little 8-oxoG will persist in the DNA to mispair with A.

MutT

The MutT enzyme operates by a very different mechanism (see Figure 10.4): it prevents 8-oxoG from entering the DNA in the first place. The reactive forms of oxygen can oxidize not only guanine in DNA to 8-oxoG but also the base in dGTP to form 8-oxodGTP. Without MutT, 8-oxodGTP will be incorporated into DNA by DNA polymerase, which cannot distinguish 8-oxodGTP from normal dGTP. The MutT enzyme is a phosphatase that specfically degrades 8-oxodGTP to 8-oxodGMP so that it cannot be used in DNA.

Genetics of 8-oxoG Mutagenesis

Genetic evidence obtained with *mutM*, *mutT*, and *mutY* mutants is consistent with these functions for the products of the genes. First, these activities explain why the effects of mutations in these genes are additive. If *mutT* is mutated, more 8-oxodGTP will be present in the cell to be incorporated into DNA, increasing the spontaneous mutation rate. If MutM does not remove these 8-oxoG's from DNA, spontaneous mutation rates will increase even further. If MutY does not remove some of the A's that mistakenly pair with the 8-oxoG's, the spontaneous mutation rate will be higher yet.

Mutations in the *mutM*, *mutY*, and *mutT* genes also increase the frequency of only some types of mutations, which again can be explained by the activities of these enzymes. For example, the enzyme functions explain why *mutM* and *mutY* mutations increase only the frequency of GC-to-AT transversion mutations. If MutM does not remove 8-oxoG from DNA, then mispairing of the 8-oxoG's with A's can lead to these transversions. Moreover, GC-to-TA transversions will occur if MutY does not remove the mispaired A's opposite the 8-oxoG's in the DNA. By contrast, while *mutT* mutations can increase the frequency of GC-to-TA transversion mutations, they can also increase the frequency of TA-to-GC transversions. This is possible because some of the 8-oxodGTP present in the cell owing to the *mutT* mutation may enter the DNA incorrectly by pairing with an A (instead of a C).

Alkylation

Alkylation is another common type of damage to DNA. Both the bases and the phosphates in DNA can be akylated. The responsible chemicals, known as **alkylating agents**, add alkyl groups (CH_3, CH_3CH_2, etc.) to the bases or phosphates. Some examples of alkylating agents are ethyl methanesulfonate (EMS; nitrogen mustard gas), methyl methanesulfonate (MMS), and *N*-methyl-*N*'-nitro-*N*-nitrosoguanidine (nitrosoguanidine, NTG, or MNNG).

Many reactive groups of the bases can be attacked by alkylating agents. The most reactive are the N^7 of guanine and the N^3 of adenine. These nitrogens can be alkylated by EMS or MMS to yield methylated or ethylated bases such as N^7-methylguanine and N^3-methyladenine. Alkylation of the bases at these positions can severely alter their pairing with other bases, causing major distortions in the helix.

Some akylating agents, such as nitrosoguanidine, can also attack other atoms in the rings, including the O^6 of guanine and the O^4 of thymine. The addition of a methyl group to these atoms makes O^6-methylguanine (Figure 10.5) and O^4-methylthymine, respectively. Altered bases with an alkyl group at these positions are particularly mutagenic because the helix is not significantly distorted, so that the lesions cannot be repaired by the more general repair systems discussed below. However, the altered base will often mispair, producing a mutation. In this section, we discuss the repair systems specific to these types of alkylated bases.

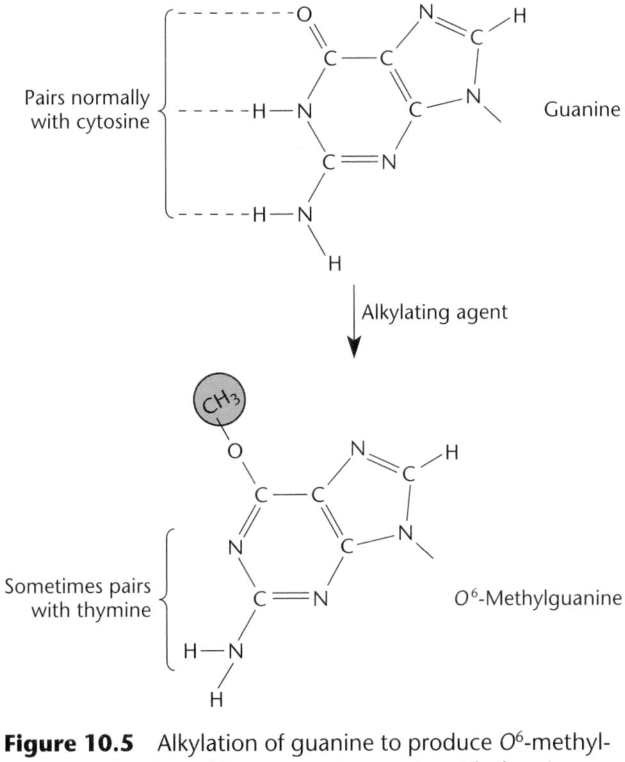

Figure 10.5 Alkylation of guanine to produce O^6-methylguanine. The altered base sometimes pairs with thymine, causing mutations.

SPECIFIC N-GLYCOSYLASES

Alkylated bases can be removed by specific *N*-glycosylases. The repair pathways involving these enzymes work in the same way as other *N*-glycosylase repair pathways in that first the alkylated base is removed by the specific *N*-glycosylase and then the apurinic or apyrimdinic DNA strand is cut by an AP endonuclease. Exonucleases degrade the cut strand, which is then resynthesized by DNA polymerase I. In *E. coli*, an *N*-glycosylase that removes the bases 3-methyladenine and 3-methylguanine has been identified. This enzyme, which is encoded by the *alkA* gene, is induced as part of the adaptive response (see below).

METHYLTRANSFERASES

The cell can also use special proteins to repair bases damaged from alkylation of the O^6 carbon of guanine and the O^4 carbon of thymine. These proteins, called methyltransferases, directly remove the alkyl group at these positions by transferring the group from the altered base in the DNA to themselves. They are not true enzymes because they do not catalyze the reaction but, rather, are consumed during it. Once they have transferred a methyl or other alkyl group to themselves, they are inactive and will be eventually degraded. That the cell is willing to sacrifice an entire protein molecule to repair a single O^6-methylguanine or O^4-methylthymine lesion is a tribute to the mutagenic potential of these lesions.

THE ADAPTIVE RESPONSE

Many of the genes whose products, including specific *N*-glycosylases and methytransferases, repair alkylation damage in *E. coli* are part of the **adaptive response**. The products of these genes are normally synthesized in small amounts but are produced in much greater amounts if the cells are exposed to an alkylating agent. The name "adaptive response" comes from early evidence suggesting that *E. coli* "adapted" to damage caused by alkylating agents. If *E. coli* cells are briefly treated with an alkylating agent such as nitrosoguanidine (NTG), they will be better able to survive subsequent treatments with alkylating agents. We now know that the cell adapts to the alkylating agent by inducing a number of genes whose products are involved in repairing alkylation damage to DNA.

Some of the the genes induced as part of the adaptive response include *ada*, *aidB*, *alkA*, and *alkB*. The product of the *ada* gene is a methyltransferase that removes methyl or ethyl groups from O^6-methylguanine or O^4-methylthymine or from the phosphate groups in the backbone of the DNA. The Ada protein is also the regulatory protein that regulates transcription of the other genes of the adaptive response, as we discuss below. The *alkA* gene product is a an N-glycosylase that removes many alkylated bases, including 3-methyladenine and 3-methylguanine, from DNA (see above). The products of the *aidB* and *alkB* genes are unknown but must be involved in alkylation repair because *aidB* and *alkB* mutants are much more sensitive to killing by alkylating agents.

Regulation of the Adaptive Response

How does treatment with an alkylating agent cause the concentration of some of the enzymes involved in repairing alkylation damage to increase from a few to many thousands of copies? The regulation is achieved through the state of methylation on the Ada protein, a methyltransferase, as shown in Figure 10.6. After alkylation damage occurs, methyl groups can be transferred to amino acids at two different positions on the Ada protein. The amino acid at one position accepts a methyl group from either O^6-methylguanine or O^4-methylthymine, and the amino acid at the other position accepts methyl groups from any phosphate in the DNA backbone. Methylation at either position will prevent further acceptance of methyl groups at that position. Transfer of a methyl group from an alkylated base to the Ada C terminus inactivates the Ada protein for additional acceptance of methyl or other groups (i.e., suicide inactivation in Figure 10.6). However, methyl groups transferred from the phosphate backbone to the Ada N terminus convert the Ada protein into a transcriptional activator. In this state, the Ada protein will activate transcription of its own gene, the *ada* gene, as well as the other genes of the adaptive response. We have discussed transcriptional activators in chapter 7 in connection with the CI and CII proteins of phage λ, and we discuss them again in much more detail in chapter 11.

Pyrimidine Dimers

One of the major sources of natural damage to DNA is UV irradiation due to sun exposure. Every organism that is exposed to sunlight must have mechanisms to repair UV damage to its DNA. The conjugated-ring structure of the bases in DNA causes them to strongly absorb light in the UV wavelengths. The absorbed photons energize the bases, causing their double bonds to react with other nearby atoms and hence forming additional chemical bonds. These chemical bonds result in abnormal linkages between the bases in the DNA and other bases or between bases and the sugars of the nucleotides.

One common type of UV irradiation damage is the **pyrimidine dimer**, in which the rings of two adjacent

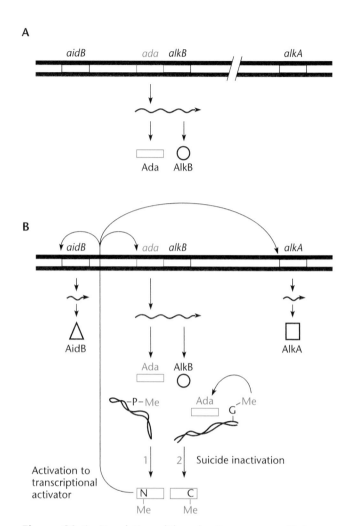

Figure 10.6 Regulation of the adaptive response. Only a few copies of the Ada protein normally exist in the cell. After damage due to alkylation, the Ada protein, a methyltransferase, transfers alkyl groups from damaged DNA phosphates to itself, converting itself into a transcriptional activator (1), or from a damaged base, inactivating itself (2). See the text for details.

pyrimidines become fused (Figure 10.7). In one of the two possible dimers, the 5-carbon atoms and 6-carbon atoms of two adjacent pyrimidines are joined to form a **cyclobutane ring**. In the other type of dimer, the 6-carbon of one pyrimidine is joined to the 4-carbon of an adjacent pyrimidine to form a **6-4 lesion**.

PHOTOREACTIVATION OF CYCLOBUTANE DIMERS

Because the cyclobutane-type pyrimidine dimer due to UV irradiation is so common, a special type of repair system called **photoreactivation** has evolved to repair it. The photoreactivation repair system separates the fused

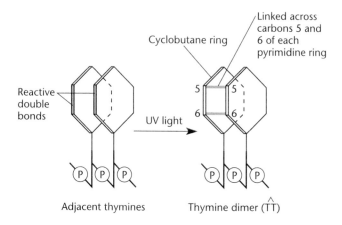

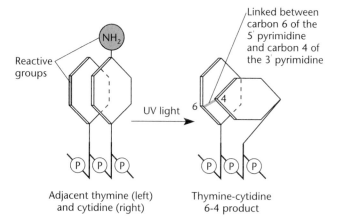

Figure 10.7 Two common types of pyrimidine dimers caused by UV irradiation. In the top panel, two adjacent thymines are linked through the 5- and 6-carbons of their rings to form a cyclobutane ring. In the bottom panel, a 6-4 dimer is formed between the 6 carbon of a cytosine and the 4 carbon of a thymine 3' to it.

bases of the cyclobutane pyrimidine dimers rather than replacing them. This mechanism is named "photoreactivation" because this type of repair occurs only in the presence of visible light.

In fact, photoreactivation was the first DNA repair system to be discovered. In the 1940s, Albert Kelner observed that the bacterium *Streptomyces griseus* was more likely to survive UV irradiation in the light than in the dark. Photoreactivation is now known to exist in most organisms on earth, with the important exception of placental mammals like ourselves. Don't think that when you are soaking up rays the damaging effects of UV irradiation are being repaired just because you are also being exposed to visible light. You don't have a photoreactivation system, although you might have a repair system derived from it.

The mechanism of action of photoreactivation is shown in Figure 10.8. The enzyme responsible for the repair is called **photolyase**. This enzyme, which contains a reduced flavin adenine dinucleotide ($FADH_2$) group that will absorb light of wavelengths between 350 and 500 nm, binds to the fused bases. Absorption of light then gives photolyase the energy it needs to separate the fused bases.

There is some evidence that the photoreactivating system may also help repair pyrimidine dimers even in the dark. By binding to pyrimidine dimers, it may help make them more recognizable by the nucleotide excision repair system discussed below.

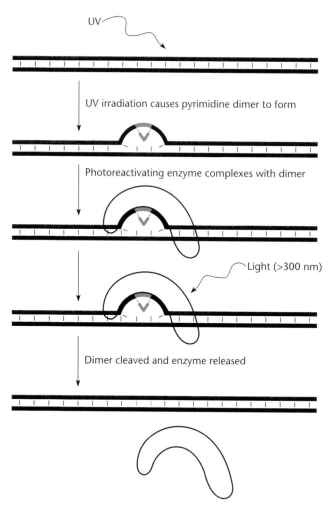

Figure 10.8 Photoreactivation. The photoreactivating enzyme (photolyase) binds to cyclobutane pyrimidine dimers (blue) even in the dark. Absorption of light by the photolyase causes it to cleave the bond between the two pyrimidines, restoring the bases to their original form. Adapted with permission from E. C. Friedberg, *DNA Repair*, W. H. Freeman and Co., New York, 1985.

N-GLYCOSYLASES SPECIFIC TO PYRIMIDINE DIMERS

There are also specific N-glycosylases that recognize and remove pyrimidine dimers. This repair mechanism is similar to the mechanisms for deaminated and alkylated bases discussed above and involves AP endonucleases and the removal and resynthesis of strands of DNA containing the dimers.

General Repair Mechanisms

As mentioned, not all repair mechanisms in cells are specific for a certain type of damage to DNA. Some types of repair systems can repair many different types of damage. Rather than recognizing the damage itself, these repair systems recognize distortions in the DNA structure caused by improper base pairing.

The Methyl-Directed Mismatch Repair System

One of the major pathways for avoiding mutations in *E. coli* is the **methyl-directed mismatch repair system**. We have already mentioned the mismatch repair system in chapter 1, in connection with lowering the rate of mistakes made during the replication of DNA, and in chapter 9, in connection with gene conversion. The methyl-directed mismatch repair system is not specific and can repair essentially any damage that causes a slight distortion in the DNA helix, including mismatches, frameshifts, incorporation of base analogs, and some types of alkylation damage that cause only minor distortions in the DNA helix. Alkylation that causes more significant distortions is repaired by other systems including the nucleotide excision repair discussed later. In general, DNA damage that causes only minor distortions of the helix is repaired by the methyl-directed mismatch repair system whereas damage that causes more significant distortions is repaired by other pathways.

BASE ANALOGS

Base analogs are chemicals that resemble the normal bases in DNA. Because they resemble the normal bases, these analogs can sometimes be converted into a deoxynucleoside triphosphate and enter DNA. Incorporation of a base analog can be mutagenic because the analog often pairs with the wrong base, leading to base pair changes in the DNA. Figure 10.9 shows two base analogs, 2-aminopurine (2-AP) and 5-bromouracil (5-BU). 2-AP resembles adenine, except that it has the amino group at the 2 position rather than at the 6 position. This analog sometimes incorrectly pairs with cytosine. 5-BU, which resembles thymine except for a

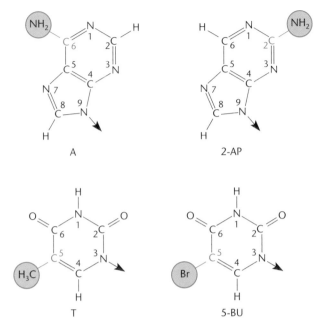

Figure 10.9 Base analogs 2-aminopurine (2-AP) and 5-bromouracil (5-BU). The amino groups that are at different positions in adenine (A) and 2-AP are circled in blue, as are the methyl group in thymine (T) and the bromine group in 5-BU.

bromine atom instead of a methyl group at the 5 position, sometimes incorrectly pairs with guanine.

Figure 10.10 shows how mispairing by a base analog might cause a mutation. In the illustration, the base analog 2-AP has entered a cell and been converted into the nucleoside triphosphate. The deoxyribose 2-aminopurine triphosphate is then incorporated during synthesis of DNA and sometimes pairs with cytosine in error. Which type of mutation occurs depends on when the 2-AP mistakenly pairs with cytosine. Even if the 2-AP enters the DNA by mistakenly pairing with C instead of T, in subsequent replications it will usually pair correctly with T. This will cause a GC-to-AT transition mutation. However, if it is incorporated properly by pairing with T but pairs mistakenly with C in subsequent replications, an AT-to-GC transition mutation will result.

FRAMESHIFT MUTAGENS

Another type of damage repaired by the methyl-directed mismatch repair system is the incorporation of frameshift mutagens, which are usually planar molecules of the acridine dye family (see chapter 1). These chemicals are mutagenic because they intercalate between bases in the same strand of the DNA, thereby increasing the distance between the bases and preventing them from aligning properly with bases on the other strand. The frameshift mutagens include acridine dyes such as 9-aminoacridine, proflavin, and ethidium bromide, as well as some aflatoxins made by fungi.

Figure 10.11 illustrates a model for mutagenesis by a frameshift mutagen. Intercalation of the dye forces two of the bases apart, causing the two strands to slip with respect to each other. One base will thus be paired with the base next to the one with which it previously paired. This slippage is most likely to occur where a base pair in the DNA is repeated, for example in a string of AT or GC base pairs. Whether a deletion or addition of a base pair occurs depends upon which strand slips, as shown in Figure 10.11. If the dye is intercalated into the template DNA prior to replication, the newly synthesized strand might slip and incorporate an extra nucleotide. However, if it is incorporated into the newly synthesized strand, the strand might slip backward, leaving out a base pair in subsequent replication.

MECHANISM OF METHYL-DIRECTED MISMATCH REPAIR

As discussed briefly in chapter 1, the methyl-directed mismatch repair system requires the products of the

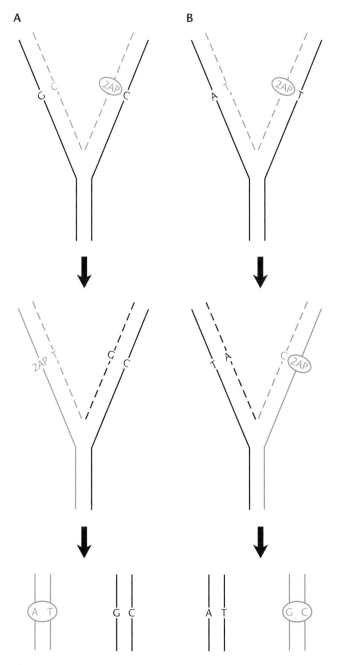

Figure 10.10 Mutagenesis by incorporation of the adenine analog 2-AP into DNA. The 2-AP is first converted to the deoxyribose nucleoside triphosphate and then inserted into DNA. (A) Analog incorrectly pairs with a C during its incorporation into the DNA strand. (B) Analog is incorporated correctly opposite a T but mispairs with C during subsequent replication. The mutation is circled in both panels.

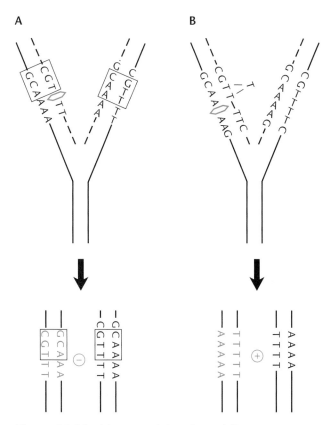

Figure 10.11 Mutagenesis by a frameshift mutagen. Intercalation of a planar acridine dye molecule between two bases in a repeated sequence in DNA forces the bases apart and can lead to slippage. (A) The dye inserts itself into the new strand, removing a base pair (–). (B) The dye comes into the old strand, adding a base pair (+).

mutS, mutL, and *mutH* genes. Like the *mut* genes whose products are involved in avoiding 8-oxoG, these *mut* genes were found in a search for mutations that increase spontaneous mutation rates (see chapter 14). The products of these three genes were predicted to participate in the same repair pathway because of the observation that double mutants did not exhibit higher spontaneous mutation rates than cells with mutations in each of the single genes alone. In other words, the effect of inactivating two of these *mut* genes by mutation is not additive. Generally, mutations that affect steps of the same pathway do not have additive effects.

As discussed in chapter 1, another gene product that participates in the methyl-directed mismatch repair system is the product of the *dam* gene. The *dam* gene encodes the Dam methylase, which methylates adenine in the sequence GATC/CTAG. This enzyme methylates the DNA at this sequence only after it is synthesized, so that the new strand of DNA will be temporarily unmethylated after the replication fork moves on. By having the *mut* gene products repair only the newly synthesized strand, the cell usually avoids mutagenesis, because most mistakes are made during synthesis of new strands. If the old strand were degraded and resynthesized with the new strand as a template, the damage would be fixed as a mutation and would be passed on to future generations.

It is mysterious how the mismatch repair system can use the state of methylation of the GATC/CTAG sequence to direct itself to the newly synthesized strand, even though the nearest GATC/CTAG sequence is probably some distance from the alteration. Figure 10.12 presents a model that may explain this mechanism. First, the MutS protein binds to the alteration in the DNA that is causing a minor distortion in the helix (an AC mismatch in the figure). Then two molecules of the MutH protein bind to MutS, and the DNA slides through the complex in both directions, resulting in a loop containing the alteration and held together by the MutS-MutH complex as shown. When one of the MutH subunits in the complex reaches the nearest hemimethylated GATC/CTAG sequence in either the 5' or 3' direction, it cuts the unmethylated strand, which is then degraded. DNA polymerase III then resynthesizes a new strand, using the other strand as the template and removing the cause of the distortion. The function of MutL in this repair process is not clear, but we do know that it is bound to the complex formed by MutS and MutH, as shown in Figure 10.12 (see Modrich, Suggested Reading). It is somewhat surprising that DNA polymerase III, the replication DNA polymerase, is used for this repair. Most other repair reactions use DNA polymerase I or II, the normal repair enzymes.

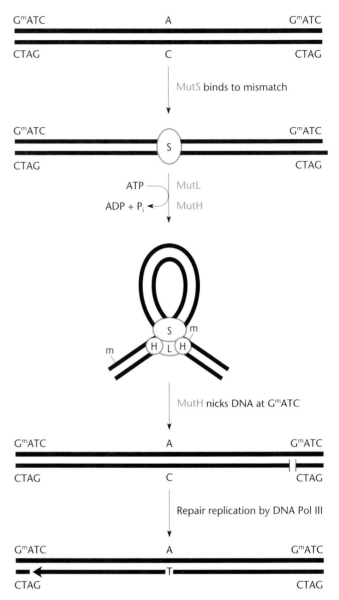

Figure 10.12 Model for repair, by the methyl-directed mismatch repair system, of slight distortions in the helix as a result of mismatches and of incorporation of base analogs and frameshift mutagens into DNA. The MutS protein binds to the distortion (in the example, this is due to incorporation of a C opposite an A). The MutL and MutH proteins then bind together, and the DNA is drawn through the complex in both directions until it reaches a hemimethylated GATC sequence. The MutH protein nicks the DNA in the unmethylated strand, which is then degraded back over the mismatch. Resynthesis of a new strand by DNA polymerase III restores the correct base. Adapted with permission from P. Modrich, *Annu. Rev. Genet.* **25**:229–253, 1991.

GENETIC EVIDENCE FOR METHYL-DIRECTED MISMATCH REPAIR

Genetic experiments support general models of methyl-directed mismatch repair like that presented in Figure 10.12. Probably the most convincing evidence for the role of methylation in directing the repair system came from experiments with heteroduplexes of λ DNA (see Pukkila et al., Suggested Reading). We have discussed the synthesis and use of heteroduplex DNAs for experiments in other chapters. For example, in chapter 8 we showed how they can be used to distinguish replicative from cut-and-paste transposition.

To determine the role of methylation in directing the mismatch repair system, heteroduplex λ DNAs were prepared in which the two strands came from phage with different mutations—we shall call them phage mutant 1 and phage mutant 2. The DNA from phage mutant 2 was unmethylated because it was propagated on *E. coli* with a *dam* mutation. One of the phage was also propagated in medium with heavy isotopes, as in the Meselson and Stahl experiment (see chapter 1), so that the heteroduplex DNAs could be purified later. After the DNA from both mutants was purified, the DNA strands of both were separated by heating and then were renatured in the presence of each other. The heteroduplex DNAs were then purified on density gradients. Because mutant 1 and mutant 2 carried different mutations, the heteroduplex DNA had mismatches at the sites of those mutations. These DNAs were transfected into cells, the phage progeny were plated, and their genotypes were tested to see which mutation they had inherited and which of the two strands had been preferentially repaired. The progeny phage exhibited predominantly the genotype of mutant 1, which had the methylated DNA. In a similar experiment, mutant 1 was propagated on the *dam* mutant so that it would have the unmethylated DNA. In this case, the progeny phage that arose from the heteroduplex DNA had the genotype of mutant 2; therefore, once again, the methylated DNA sequence was preferentially preserved. These results indicate that the sequence of the unmethylated strand is usually the one that is repaired to match the sequence of the methylated strand.

Other evidence supporting the role of methylation in directing the mismatch repair system came from genetic studies of 2-aminopurine (2-AP) sensitivity in *E. coli* (see Glickman and Radman, Suggested Reading). These experiments were based on the observation that a *dam* mutation makes *E. coli* particularly sensitive to killing by 2-AP, as expected from the model. As mentioned, cells incorporate 2-AP indiscriminately into their DNA because they cannot distinguish it from the normal base adenine. The methylated mismatch repair system re-

pairs DNA containing 2-AP because the incorporated 2-AP causes a slight distortion in the helix. Since the 2-AP is incorporated into the newly synthesized strand during the replication of the DNA, the strand containing the 2-AP will be normally transiently unmethylated, so that the 2-AP-containing strand will be repaired, removing the 2-AP. In a *dam* mutant, however, neither strand will be methylated, and so the mismatch repair system cannot tell which strand was newly synthesized and may try to repair both strands. If two 2-APs are incorporated close to each other, the mismatch repair system may try to simultaneously remove the two mismatches by repairing the opposite strands. However, cutting two strands at sites opposite each other will result in a double-strand break in the DNA, which may kill the cell.

One prediction based on this proposed mechanism for the sensitivity of *dam* mutants to 2-AP is that the toxicity of 2-AP in *dam* mutants should be reduced if the cells also have a *mutL*, *mutS* or *mutH* mutation. Without the products of the mismatch repair enzymes, the DNA will not be cut on either strand, much less on both strands simultaneously. The cells may suffer higher rates of mutations but at least they will survive. This prediction was fulfilled. Double mutants that have both a *dam* mutation and a mutation in one of the three *mut* genes were much less sensitive to 2-AP than were mutants with a *dam* mutation alone. Furthermore, *mutL*, *mutH*, and *mutS* mutations can be isolated as suppressors of *dam* mutations on media containing 2-AP. In other words, if large numbers of *dam* mutant *E. coli* cells are plated on medium containing 2-AP, the bacteria that survive are often double mutants with the original *dam* mutation and a spontaneous *mutL*, *mutS*, or *mutH* mutation.

THE ROLE OF THE MISMATCH REPAIR SYSTEM IN PREVENTING HOMEOLOGOUS AND ECTOPIC RECOMBINATION

As discussed in chapter 3, some DNA rearrangements, such as deletions and inversions, are caused by recombination between similar sequences in different places in the DNA. This is called **ectopic recombination,** or "out-of-place" recombination. Many sequences at which ectopic recombination occurs are similar but not identical. Also, recombination between DNAs from different species often occurs between sequences that are similar but not identical. In general, recombination between similar but not identical sequences is called **homeologous recombination.**

The mismatch repair system reduces ectopic and other types of homeologous recombination. As evidence, recombination between similar but unrelated bacteria such as *E. coli* and *Salmonella typhimurium* is greatly enhanced if the recipient cell has a *mutL*, *mutH*, or *mutS* mutation. Also, the frequency of deletions and other types of DNA rearrangements is enhanced among bacteria with a *mutL*, *mutH*, or *mutS* mutation. Because the sequences at which the homeologous recombination occurs are not identical, the heteroduplexes formed during recombination between different regions in the DNA or between the same region from different species often contain many mismatches that bind MutS and the other proteins of the mismatch repair system. Recent evidence suggests that the mismatch repair system may inhibit homeologous and ectopic recombination by interfering with the ability of RecA to form synapses where there are extensive mismatches (see Worth et al., Suggested Reading).

MISMATCH REPAIR SYSTEMS IN OTHER ORGANISMS

All organisms probably have mismatch repair systems that help reduce mutagenesis. In fact, deficiencies in a human mismatch repair system have been linked to human cancer (see Box 10.2). However, most eukaryotes and even most types of bacteria do not have a Dam methylase and so cannot use GATC methylation to direct the mismatch repair system to the newly synthesized strand. In fact, some organisms such as yeasts have very little methylation of any type in the bases in their DNA. Some other mechanism is apparently used to direct the mismatch repair system to the newly synthesized strand.

Nucleotide Excision Repair

One of the most important general repair systems in cells is **nucleotide excision repair**. This type of repair is very efficient and seems to be common to most types of organisms on Earth. It is also relatively nonspecific and will repair many different types of damage. Because of its efficiency and relative lack of specificity, the nucleotide excision repair system is very important to the ability of the cell to survive damage to its DNA.

The nucleotide excision repair system is relatively nonspecific because, like the mismatch repair system, it recognizes the distortions in the normal DNA helix that results from damage, rather than the chemical structure of the damage itself. However, nucleotide excision repair recognizes only major distortions in the helix; it will not repair lesions such as base mismatches, O^6-methylguanine, O^4-methylthymine, 8-oxoG, or base analogs, all of which result in only minor distortions and must be repaired by other repair systems. Nucleotide excision repair will, however, repair almost all the types of damage caused by UV irradiation, including

cyclobutane dimers, 6-4 lesions, and base-sugar cross-links. Nucleotide excision repair will also collaborate with recombination repair to remove cross-links formed between the two strands by some chemical agents such as psoralens, *cis*-diamminedichloroplatinum (cisplatin), and mitomycin (see below).

The excision repair system can be distinguished from most other types because the DNA containing the damage is actually excised from the DNA. For example, after cells are irradiated with UV light, short pieces of DNA (oligodeoxynucleotides) containing pyrimidine dimers and the other types of damage induced by UV appear in the medium.

Cancer and Mismatch Repair

Cancer is a multistep process that is initiated by mutations in oncogenes and other tumor-suppressing genes. It is not surprising, therefore, that DNA repair systems form an important line of defense against cancer. There is recent evidence that people with a mutation in a mismatch repair gene are much more likely to develop some types of cancer, including cancer of the colon, ovary, uterus, and kidney. This disease, called HNPCC (for hereditary nonpolyposis colon cancer), results from mutations in a gene called *hMSH2* (for human *mutS* homolog 2). This gene was first suspected to be involved in mismatch repair because people who inherited the mutant gene showed a higher frequency of short insertions and deletions that should normally be repaired by the mismatch repair system (see the text). People with this hereditary condition were found to have inherited a mutant form of a gene homologous to the *mutS* gene of *E. coli*. Moreover, the *hMSH2* gene can cause increased spontaneous mutations when expressed in *E. coli*, probably because it interferes with the normal mismatch repair system. Like its analog, MutS, the *hMSH2* gene product may bind to mismatches but then does not interact properly with MutL and MutH. In any case, the mismatch repair system in humans seems to play an important role in defending against cancer.

References

Fishel, R., M. K. Lescoe, M. R. S. Rao, N. G. Copeland, N. A. Jenkins, J. Garber, M. Kane, and R. Kolodner. 1993. The human mutator gene homolog MSH2 and its association with hereditary nonpolyposis colon cancer. *Cell* **75**:1027–1038.

Kolodner, R. 1996. Biochemistry and genetics of eukaryotic mismatch repair. *Genes Dev.* **10**:1433–1442.

MECHANISM OF NUCLEOTIDE EXCISION REPAIR

Because nucleotide excision repair is such an important line of defense against some types of DNA-damaging agents, including UV irradiation, mutations in the genes whose products are required for this type of repair can make cells much more sensitive to these agents. In fact, mutants defective in excision repair were identified because they were killed by much lower doses of irradiation than was the wild type. Table 10.1 lists the *E. coli* genes whose products are required for nucleotide excision repair. The products of some of these genes, such as *uvrA*, *uvrB*, and *uvrC*, are involved only in excision repair, while the products of others, including the *polA* and *uvrD* genes, are also required for other types of repair.

How these gene products participate in excision repair is illustrated in Figure 10.13. The products of the *uvrA*, *uvrB*, and *uvrC* genes interact to form what is called the **UvrABC endonuclease** (or, sometimes, the UvrABC exc*i*nuclease). The function of these gene products is to make a nick close to the damaged nucleotide, causing it to be excised.

In more detail, two copies of the UvrA protein and one copy of the UvrB protein form a complex that binds nonspecifically to DNA even if it is not damaged. This complex then migrates up and down the DNA until it hits a place where the helix is distorted because of DNA damage (in the illustration, because of a thymine dimer). The complex will then stop, the UvrB protein will bind to the damage, and the UvrA protein will leave, being replaced by UvrC. The binding of UvrC protein to UvrB causes UvrB to cut the DNA about 4 nucleotides 3' of the damage. The UvrC protein then cuts the DNA 7 nucleotides 5' of the damage. Once the DNA is cut, the UvrD helicase removes the oligonucleotide containing the damage and the DNA polymerase I resynthesizes the strand that was removed, using the complementary strand as a template.

TABLE 10.1	Genes involved in the UvrABC endonuclease repair pathway
Gene	Function of gene product
uvrA	DNA-binding protein
uvrB	Loaded by UvrA to form a DNA complex; nicks DNA 3' of lesion
uvrC	Binds to UvrB-DNA complex; nicks DNA 5' of lesion
uvrD	Helicase II; helps remove damage-containing oligonucleotide
polA	Pol I; fills in single-strand gap
lig	Ligase; seals single-strand nick

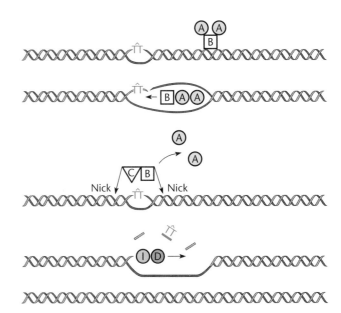

Figure 10.13 Model for nucleotide excision repair by the UvrABC endonuclease. See the text for details. A, UvrA; B, UvrB; C, UvrC; D, UvrD; I, Pol I.

INDUCTION OF NUCLEOTIDE EXCISION REPAIR

Although the genes of the excision repair system are almost always expressed at low levels, *uvrA*, *uvrB*, and *uvrD* are expressed at much higher levels after the DNA has been damaged. This is a survival mechanism that ensures that larger amounts of the repair proteins will be synthesized when they are needed. Because they are inducible by DNA damage, the *uvr* genes that are induced by DNA damage fall into a class of genes known as the *din* genes (for *damage inducible*), which includes *recF*, *recA*, *umuC*, and *umuD*. Many *din* genes, including *uvrA*, *uvrB*, and *uvrD*, are part of the SOS regulon (see later sections).

CAN EXCISION REPAIR BE DIRECTED?

Damage in some regions of the DNA presents a more immediate problem for the cell than does damage in other regions. For example, pyrimidine dimers in transcribed regions of DNA will block not only replication of the DNA but also transcription of RNA from the DNA. It makes sense for the cell to repair the damage that occurs in transcribed genes first, so that these genes can be transcribed and translated into proteins.

Recent evidence suggests that a mechanism that directs the nucleotide excision repair system preferentially to transcribed regions of the DNA may exist (see Box 10.3). A protein called transcription repair coupling factor (TRCF) displaces RNA polymerase molecules that are stalled at the site of the DNA damage because they

BOX 10.3

Transcription-Repair Coupling

DNA damage in transcribed regions presents a special problem for the cell because the RNA polymerase can stall at the damage, interfering with both expression of the genes and repair of the damage. Recent evidence indicates that DNA damage that occurs within transcribed regions of the DNA and in the transcribed strand will be repaired preferentially by the nucleotide excision repair system. This is called transcription-repair coupling. Some of the evidence for transcription-repair coupling is that mutations occur more frequently when the nontranscribed strand of DNA in a particular region is the one damaged, as expected if damage in this strand were not repaired by the relatively mistake-free excision repair system. This enhanced repair of transcribed strands requires that the gene be actively transcribed and suggests that the RNA polymerase molecule blocked at the site of the DNA damage helps direct the UvrB protein to the damage. The protein that seems to be directly involved in directing the UvrB protein to stalled RNA polymerase is TRCF (transcription-repair coupling factor), the product of the *mfd* gene in *E. coli*. TRCF may bind to the DNA with a stalled RNA polymerase and remove it to get it out of the way of the UvrABC endonuclease. TRCF may then remain bound to the DNA and direct the UvrAB proteins to the damage, accelerating the rate of repair.

Human cells have a counterpart to the TRCF that may play a similar role. Deficiencies in the human analog of TRCF may be the cause of the human genetic disease known as Cockayne's syndrome.

References

Selby, C. P., and A. Sancar. 1993. Transcription-repair coupling and mutation frequency decline. *J. Bacteriol.* **175:**7509–7514.

Selby, C. P., and A. Sancar. 1993. Molecular mechanism of transcription-repair coupling. *Science* **260:**53–58.

Eventually, the successive rounds of DNA replication and cell division dilute out the damage so that it is no longer a problem. Because these systems do not actually repair damage, they might be more properly called DNA-damage-tolerating systems.

One damage-tolerating system is named either **postreplication repair**, because it operates only after the replication fork has moved over the damage, or **recombination repair**, because it requires the recombination functions of the cell. Figure 10.14 shows a model for how recombination repair occurs and how it allows the cell to tolerate DNA damage. In this example, UV irradiation has caused a thymine dimer to form in the DNA ahead of the replication fork. The thymine dimer cannot base pair properly, and so the replication fork stalls. Replication may reinitiate further along, leaving a gap opposite the thymine dimer. However, this is not a permanent solution to the problem. The gap will result in a break in the DNA the next time a replication fork tries to replicate over this region. However, RecA protein can bind to the single-stranded gap, form a triple-stranded region between one daughter DNA helix and the single-stranded damaged region, and promote recombination. As a result, a good strand from the double-stranded daughter DNA helix moves opposite the thymine dimer. The gap remaining can then be filled in by DNA polymerase. This process can be repeated each time the replication fork moves past DNA damage, allowing the chromosome to replicate without ever actually repairing the thymine dimer. Eventually, as the cells continue to replicate their DNA and divide, the damage will be diluted out, so that only very few cells will carry any of the original damage. This model is based on early evidence for recombination repair that indicated that gaps occurred opposite the DNA damage after replication and that recombination eventually placed a good DNA strand opposite the damage (see Rupp et al., Suggested Reading).

The other type of damage-tolerating system is SOS mutagenic repair. We discuss this system later in the chapter.

cannot transcribe over the damage. The TRCF protein may then help the UvrA and UvrB proteins bind to the damage and the UvrA protein to leave, thereby increasing the rate of repair.

Postreplication or Recombination Repair

All of the repair systems we have discussed so far correct the damage to DNA. However, the two systems discussed next allow the cell to tolerate damage but do not actually repair it. Instead, the damage remains in the DNA but is prevented from interfering with replication.

REPAIR OF INTERSTRAND CROSS-LINKS IN DNA
Many chemicals such as light-activated psoralens, mitomycin, cisplatin, and ethyl methanesulfonate can form **interstrand cross-links** in the DNA, in which two bases in the opposite strands of the DNA are covalently joined to each other (hence the prefix "inter," or "between"). Interstrand cross-links present special problems and cannot be repaired by either nucleotide excision repair or recombination repair alone. Cutting both strands of the DNA by the UvrABC endonuclease would cause double-strand breakage and death of the cell. Also, recombination repair by itself cannot repair

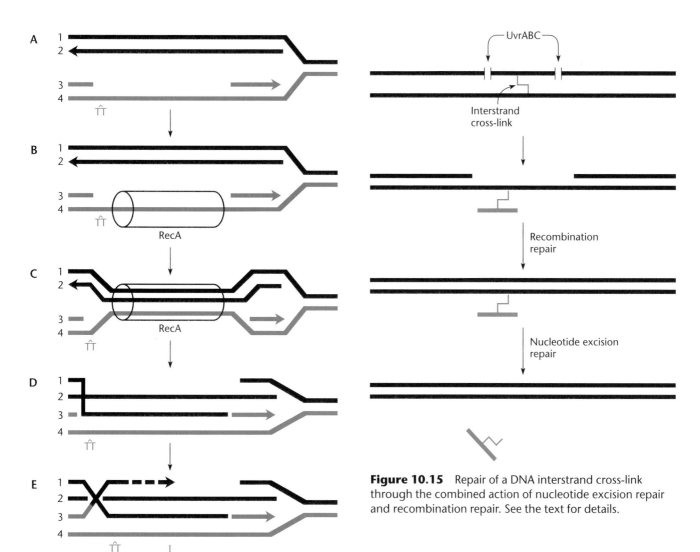

Figure 10.14 Model for postreplicative recombination repair. See the text for details.

Figure 10.15 Repair of a DNA interstrand cross-link through the combined action of nucleotide excision repair and recombination repair. See the text for details.

DNA cross-links, because the cross-link prevents the replication fork from separating the strands.

Although DNA cross-links cannot be repaired by either excision or recombination repair alone, they can be repaired by a combination of recombination repair and nucleotide excision repair, as shown in Figure 10.15. In the first step, the UvrABC endonuclease makes nicks in one strand on either side of the interstrand cross-link, as though it were repairing any other type of damage. This

will leave a gap opposite the DNA damage, as shown in the figure. In the second step, the gap is widened by the 5' exonuclease activity of DNA polymerase I. In the third step, recombination repair replaces the gap with a good strand from the other daughter DNA in the cell. The DNA damage is now confined to only one strand and is opposite a good strand; therefore, it can be repaired by the nucleotide excision repair pathway. Notice that this repair is possible only when the DNA has already replicated and there are already two copies of the DNA in the cell, since the replication apparatus cannot proceed past the interstrand cross-link. However, in fast-growing cells, most of the DNA does exist in more than one copy (see chapter 1).

SOS Inducible Repair

As mentioned above, DNA damage leads to the induction of genes whose products are required for DNA repair. In this way, the cell is better able to repair the damage and survive.

The first indication that some repair systems are inducible came from the work of Jean Weigle on the reactivation of UV-irradiated λ (see Weigle, Suggested Reading). He irradiated phage λ and tested their survival by plating them on *E. coli*. He observed that more irradiated phage survived when plated on *E. coli* cells that had themselves been irradiated than when plated on unirradiated *E. coli* cells. Because the phage were restored to viability, or reactivated, by being plated on preirradiated *E. coli* cells, this phenomenon was named **Weigle reactivation or W-reactivation.** Apparently, a repair system was being induced in the irradiated cells that could repair the damage to the λ DNA when this DNA entered the cell.

THE SOS RESPONSE

Earlier in the chapter, we mentioned a class of genes induced after DNA damage, called the *din* genes, which includes genes that encode products that are part of the excision repair and recombination repair pathways. The products of other *din* genes help the cell survive DNA damage in other ways. For example, some *din* gene products transiently delay cell division until the damage can be repaired, and some allow the cell to replicate past the DNA damage.

Many *din* genes are regulated by the the **SOS response,** so named because this mechanism rescues cells from severe DNA damage. Genes under this type of control are called the **SOS genes.**

Figure 10.16 illustrates how genes under SOS control are induced. The SOS genes are normally repressed by a protein called the LexA repressor, which binds to sites called operators upstream of the SOS genes and prevents their transcription. However, after DNA damage, the LexA repressor cleaves itself, an action known as **autocleavage,** thereby inactivating itself and allowing transcription of the SOS genes. This is reminiscent of the induction of phage λ described in chapter 7, in which the CI repressor is also cleaved following DNA damage. The two mechanisms are in fact remarkably similar, as we discuss later.

The reason the LexA repressor no longer binds to DNA after it is cleaved is also well understood. Each LexA polypeptide has two separable domains. One, the **dimerization domain,** binds to another LexA polypeptide to form a dimer (hence its name), and the other, the **DNA binding domain,** binds to the DNA operators upstream of the SOS genes and blocks their transcription. The LexA protein will bind to DNA only if it is in the dimer state. The point of cleavage of the LexA polypeptide is between the two domains, and autocleavage separates the DNA binding domain from the dimerization

domain. The DNA binding domain cannot dimerize and so by itself cannot bind to DNA and block transcription of the SOS genes.

Figure 10.16 also presents a model to answer the next question: why is the LexA repressor cleaved only after DNA damage? After DNA damage, single-stranded DNA appears in the cell owing to excision repair and postreplication repair. This single-stranded DNA binds to RecA protein, which then binds to LexA and causes it to autocleave. This feature of the model—that LexA cleaves itself in response to RecA binding, rather than being cleaved by RecA—is supported by experimental evidence. When heated under certain conditions, even in the absence of RecA, LexA will eventually cleave. This result indicates that RecA acts as a **coprotease** to facilitate LexA autocleavage.

The RecA protein thus plays a central role in the induction of the SOS response. It senses the single-stranded DNA that accumulates in the cell as a by-product of attempts to repair damage to the DNA and then causes LexA to cleave itself, inducing the SOS genes. We have already discussed another activity of RecA, the recombination function involved in synapse formation.

We can speculate why the RecA protein might have two different roles, the one in recombination and the other in repair. It must bind to single-stranded DNA to promote synapse formation during recombination, and so it must be a sensor of single-stranded DNA in the cell. It is always present in large enough amounts to bind to all the LexA repressor and quickly promote its autocleavage. The activated RecA coprotease also promotes the autocleavage of the UmuD protein involved in SOS mutagenesis (see below) and the λ repressor (see chapter 7).

As mentioned, the regulation of the SOS response through cleavage of the LexA repressor is strikingly similar to the induction of λ through cleavage of the CI repressor. Like the LexA repressor, the λ repressor must be a dimer to bind to the operator sequences in DNA and repress λ transcription. Each λ repressor polypeptide consists of an N-terminal DNA binding domain and a C-terminal dimerization domain. Autocleavage of the λ repressor stimulated by the activated RecA coprotease also separates the DNA binding domain and dimerization domain, preventing the binding of the λ repressor to the operators and allowing transcription of λ lytic genes. The sequences of amino acids around the sites at which the LexA and λ repressors are cleaved are also similar. However, the cell would not have evolved the SOS system to help λ prophage induce itself and escape in the event of irreversible damage to the cell DNA.

A

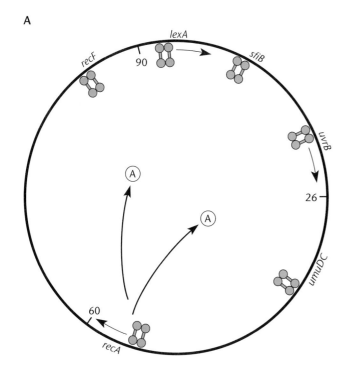

B

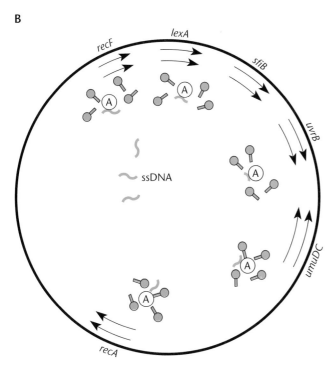

Figure 10.16 Regulation of the SOS response regulon in *E. coli*. (A) A number of genes around the *E. coli* chromosome are normally repressed by the binding of a LexA dimer (barbell structure) to their operators. Some SOS genes are expressed at low levels, as indicated by single arrows. (B) After DNA damage, the single-stranded (ss) DNA that accumulates in the cell binds to RecA (circled A), activating its co-

Rather, it is more likely that the cell evolved the SOS induction mechanism to help *itself* survive DNA damage by inducing the SOS *din* genes after DNA damage. By modeling its own repressor after the LexA repressor, the λ prophage allows the SOS regulatory system of the host to induce it following DNA damage, thereby allowing λ to escape a doomed host cell.

SOS INDUCIBLE MUTAGENESIS

At least one of the repair systems encoded by the SOS genes appears to be very mistake prone, and its operation often causes mutations. As we show, this system, known as **mutagenic repair**, allows the replication fork to proceed over damaged DNA so that the molecule can be replicated and the cell can survive. This mechanism may be a last resort that operates only when the DNA damage is so extensive that it cannot be repaired by other, less mistake-prone mechanisms.

The first indications that a mistake-prone pathway for UV damage repair is inducible in *E. coli* came from the same studies that showed that repair itself is inducible (see Weigle, Suggested Reading). In addition to measuring the survival of UV-irradiated λ when they were plated on UV-irradiated *E. coli*, Weigle counted the number of clear-plaque mutants among the surviving phage. (Recall from chapter 7 that lysogens will form in the center of wild-type phage λ plaques, making the plaques cloudy, but mutants that cannot lysogenize form clear plaques.) There were more mutants among the surviving phage if the bacterial hosts had been UV irradiated prior to infection than if they had not been irradiated. This inducible mutagenesis was named **Weigle mutagenesis** or just **W-mutagenesis**. Later studies showed that the inducible mutagenesis results from induction of the SOS genes; it will not occur without RecA and the cleavage of the LexA repressor. Thus, the inducible mutagenesis Weigle observed is called **SOS mutagenesis**.

Determining Which Repair Pathway Is Mutagenic

Although Weigle's experiments showed that one of the inducible UV damage repair systems in *E. coli* is mistake prone and causes mutations, they did not indicate

protease activity. The activated RecA binds to LexA, causing LexA to cleave itself. The cleaved repressor can no longer bind to the operators of the genes, and the genes are induced as indicated by two arrows. The approximate positions of some of the genes of the SOS regulon are shown.

which system was responsible. A genetic approach was used to answer this question (see Kato et al., Suggested Reading). To detect UV-induced mutations, the experimenters used the reversion of a *his* mutation. Their basic approach was to make double mutants with both a *his* mutation and a mutation in one or more of the genes of each of the repair pathways. The repair-deficient mutants were then irradiated with UV light and plated on media containing limiting amounts of histidine, so that only His⁺ revertants could multiply to form a colony. Under these conditions, each reversion to *his⁺* will result in only one colony, making it possible to measure directly the number of *his⁺* reversions that have occurred. This number, divided by the total number of surviving bacteria, will give an estimation of the susceptibility of the cells to mutagenesis by UV light.

The results of these experiments led to the conclusion that recombination repair does not seem to be mistake prone. While *recB* and *recF* mutations reduced the survival of the cells exposed to UV light, the number of *his⁺* reversions per surviving cell was no different from that of cells lacking mutations in their *rec* genes. Also, nucleotide excision repair does not seem to be mistake prone. Addition of a *uvrB* mutation to the *recB* and *recF* mutations made the cells even more susceptible to killing by UV light but also did not reduce the number of *his⁺* reversions per surviving cell.

However, *recA* mutations did seem to prevent UV mutagenesis. While these mutations made the cells extremely sensitive to killing by UV light, the few survivors did not undergo additional mutations. We now know that two genes, *umuC* and *umuD*, must be induced for mutagenic repair, and *recA* mutations prevent their induction. The RecA protein may also be directly involved in SOS mutagenesis (see below).

Mechanism of Induction of SOS Mutagenesis

The *umuC* and *umuD* genes, whose products are required for mutagenesis after UV irradiation and some other types of DNA damage, were identified in a search for mutants of *E. coli* that are not mutagenized by UV irradiation (see chapter 14). The genes were named *umu* for UV-induced *mu*tagenesis. These two genes are adjacent in the *E. coli* genome and are cotranscribed in the direction *umuD-umuC* from a promoter immediately upstream of the *umuD* gene.

Figure 10.17 illustrates that there are two reasons why UV mutagenesis is inducible in *E. coli*. First, like other SOS genes, the *umuC* and *umuD* genes are normally repressed by LexA and so are transcribed only after the DNA of the cell is damaged and LexA is autocleaved. However, UmuD is also autocleaved. When sin-

gle-stranded DNA binds to RecA and promotes the autocleavage of LexA, it also promotes the autocleavage of UmuD. Only the cleaved form of UmuD (called UmuD') is active for mutagenesis.

The autocleavage of UmuD may require larger quantities of activated RecA coprotease than does cleavage of LexA, ensuring that SOS mutagenesis will occur only when DNA damage is greater than that needed to induce the other SOS genes. Requiring higher levels of DNA damage to induce mutagenic repair than to induce the other repair pathways may ensure that mutagenic repair occurs only as a last resort, as we suggested at the beginning of this section.

Mechanism of Mutagenesis by the UmuC and UmuD Proteins

Figure 10.17 also illustrates how the UmuC and UmuD' proteins might help the cell survive DNA damage but in the process cause mutations. According to this model, the UmuC and UmuD' proteins allow the cell to tolerate the damage by permitting the DNA polymerase to replicate over the lesion. This process is called **replicative bypass** and allows the chromosome to replicate to completion and the cells to divide. However, the damaged bases do not pair properly, and mistaken base pairing can lead to mutations.

Figure 10.18 shows a more specific model of how replicative bypass works. The reason DNA polymerase cannot normally replicate over damage forms the foundation for this model. Normally, the DNA polymerase would stop at regions of DNA damage because of the action of the editing functions (see chapter 1). Since the damaged bases cannot pair properly, the entering bases would be perceived as wrong and the editing functions would stop the DNA polymerase. Under normal conditions, the β protein is part of this editing mechanism and forms a circular clamp holding the DNA polymerase on the DNA. The UmuC and UmuD' proteins may substitute for the β protein to form a more relaxed clamp that allows the DNA polymerase to read through some types of DNA damage.

While there is little direct evidence in support of this model, there is some indirect evidence. For instance, artificially overproducing the UmuC and UmuD' proteins, even in the absence of induction of the SOS response by DNA damage, causes increased rates of spontaneous mutagenesis. This finding suggests that the normal editing functions do not work properly when UmuC and UmuD' are present. Furthermore, UmuD and the β protein of *E. coli* share partial homology, as might be expected if UmuD substitutes for β in its clamping role. UmuC and UmuD also share homology with the bacte-

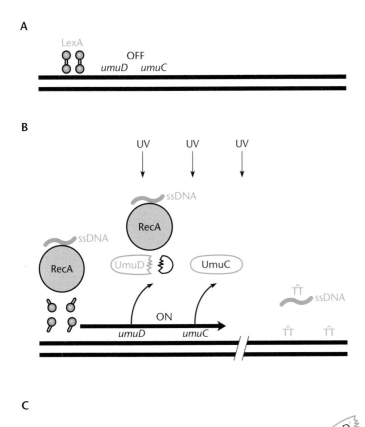

Figure 10.17 Regulation of SOS mutagenesis in *E. coli*. (A) Before DNA damage occurs, the LexA protein represses the transcription of SOS genes including the *umuDC* operon. (B) After DNA damage, the activated RecA protein promotes the autocleavage of LexA, causing the induction of the SOS genes including the *umuDC* operon. The activated RecA protein also promotes the autocleavage of UmuD to its active form, UmuD'. (C) The UmuD' and UmuC proteins then permit the DNA polymerase to replicate over DNA damage. However, random mistakes, here G's inserted opposite T's, may be made (see Figure 10.18 for a more detailed model). ss, Single-stranded.

riophage T4 protein 45, which is known to perform a β-like function, forming a clamp that holds T4 DNA polymerase on the DNA, as well as proteins 44/62, which are thought to aid in the formation of this clamp (see Kong et al., Suggested Reading, and chapter 7). Nevertheless, proof of the relaxed clamping model for mutagenesis by UmuC and UmuD' will require more direct experimental evidence.

Another Role for RecA in SOS Mutagenesis?

The UmuC and UmuD' proteins may not act alone to cause mutations after induction of the SOS response. Some evidence suggests that the RecA protein may play a more direct role in the mutagenesis besides the one it plays in the cleavage of LexA and UmuD. First, RecA⁻ mutants of *E. coli* are not mutagenized by UV light, even if the cells constitutively express UmuC and UmuD', the cleaved active form of UmuD. However, RecA⁻ mutants are so sensitive to UV light that these ex-

periments are difficult to interpret. Another line of evidence is that some *recA* mutations prevent mutagenesis after DNA damage but do not inactivate the other functions of RecA; the RecA protein in these mutants retains its coprotease activity and allows cleavage of LexA and UmuD as well as remaining proficient in recombination. It has been proposed that RecA may play a direct role in mutagenesis by helping UmuC and UmuD' promote the putative replicative bypass of DNA lesions.

ROLE OF MUTAGENIC REPAIR

In the discussion above, we assume that SOS mutagenic repair is a last-ditch attempt by the cell to survive extensive damage to its DNA. However, the actual function of SOS mutagenic repair remains a mystery. If this explanation for the existence of mutagenic repair were correct, then *umuC* and *umuD* mutants that lack mutagenic repair should be much more sensitive than wild-type bacteria to being killed by agents that extensively

A

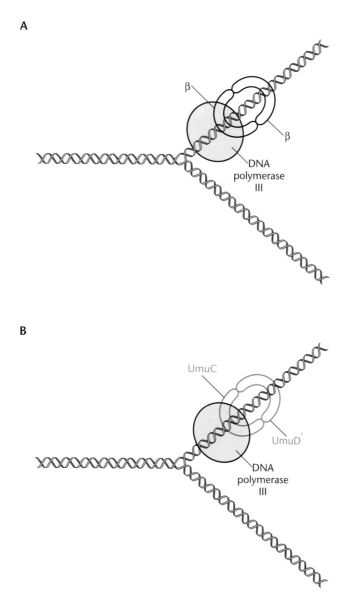

B

Figure 10.18 Speculative model for SOS mutagenesis by UmuC and UmuD' proteins. (A) Two copies of the β-subunit of the DNA polymerase III holoenzyme normally form a circular clamp holding the enzyme on the DNA and preventing it from moving over base mismatches until the mismatches have been removed by the editing function. (B) After induction of SOS mutagenesis, the UmuC and UmuD' proteins form an alternative clamp that replaces the β subunits and, possibly with the help of RecA, allows the DNA polymerase III to replicate over the damage.

damage DNA. However, mutations that inactivate the *umuC* and/or *umuD* gene make *E. coli* only slightly more sensitive to killing by UV irradiation and other DNA-damaging agents. One way to explain this puzzling observation is to propose that UmuC and UmuD protect against types of DNA-damaging agents other

than those that have been tested. Another possibility is that UmuC and UmuD offer more protection against DNA damage under conditions different from those that exist in the laboratory.

OTHER FUNCTIONS ANALOGOUS TO UmuC AND UmuD

Whatever the real role of UmuC and UmuD, most types of bacteria seem to have similar mechanisms. Genes analogous to *umuC* and *umuD* have now been found in many bacteria including the gram-positive *Streptomyces* spp. In addition, some naturally occurring plasmids carry analogs of the *umuC* and *umuD* genes. The best studied of these genes are the *mucA* and *mucB* genes of plasmid R46, of which pKM101 is a derivative (see chapter 5). The products of these plasmid genes can substitute for UmuC and UmuD' in mutagenic repair of *E. coli* and *Salmonella* spp. Because the pKM101 plasmid makes *Salmonella* spp. more sensitive to mutagenesis by many types of DNA-damaging agents, it has been introduced into the *Salmonella* strains used for the Ames test (see Box 10.4).

Other *din* Genes

As mentioned above, the SOS genes are only a subset of the *din* genes, many of which have been identified by making random transcriptional fusions with transposons such as Mu*dlac* (see chapter 14). Fusions to the promoters of these genes make more of the reporter gene product after DNA damage. Most *din* genes are expressed constitutively but are induced to higher levels of expression after DNA damage. The functions of many of these genes have yet to be determined.

The *groEL* and *groES* genes are examples of *din* genes that are not repressed by LexA and so are not SOS genes. The GroEL and GroES proteins are chaperones that help the folding of proteins and are also induced after heat shock (see chapter 12). In addition, these proteins play an indirect role in SOS mutagenesis, perhaps by helping to fold UmuC so that it interacts properly with UmuD'.

Summary of Repair Pathways in *E. coli*

Table 10.2 shows the repair pathways we have discussed and some of the genes whose products participate in each pathway. Some pathways, such as photoreactivation and most base excision pathways, have evolved to repair specific types of damage to DNA. Some, such as VSP repair, amend damage only in certain sequences. Others, such as mismatch repair and nucleotide excision repair, are much more general and will repair essentially

BOX 10.4

The Ames Test

It is now well established that cancer is initiated by mutations in genes including oncogenes and tumor-suppressing genes. Therefore, chemicals that cause mutations are often carcinogenic for humans. Yet many new chemicals are being used as food additives or otherwise come into contact with humans, and each of these chemicals must be tested for its carcinogenic potential. However, such testing in animal models is expensive and time-consuming.

Because many carcinogenic chemicals damage DNA, they are mutagenic for bacteria as well as humans. Therefore, bacteria can be used in initial tests to determine if chemicals are apt to be carcinogenic. The most widely used of these tests is the Ames test, developed by Bruce Ames and his collaborators. This test uses revertants of *his* mutations of *Salmonella* spp. to detect mutations. The chemical is placed on a plate lacking histidine and on which has been spread a His⁻ mutant of *Salmonella*. If the chemical can revert the *his* mutation, a ring of His⁺ revertant colonies will appear around the chemical on the plate. A number of different *his* mutations must be used because different mutagens cause different mutations and they all have preferred sites of mutagenesis, or hot spots. Also, the test is made more sensitive by eliminating the non-mutagenic nucleotide excision repair system with a *uvrA* mutation and introducing the pKM101 plasmid containing the *mucA* and *mucB* genes. These genes are analogs of *umuC* and *umuD* and so increase mutagenesis (see text).

Some chemicals are not mutagenic themselves but can be converted into mutagens by enzymes in the mammalian liver. To detect these mutagens, we can add a liver extract from rats to the plates and spot the chemical on the extract. If the extract converts the chemical into a mutagen, His⁺ revertants will appear on the plate.

Reference
McCann, J., and B. N. Ames. 1976. Detection of carcinogens as mutagens in the Salmonella/microsome test assay of 300 chemicals: discussion. *Proc. Natl. Acad. Sci. USA* **73**:950–954.

TABLE 10.2	Genetic pathways for damage repair	
Repair mechanism	**Genetic loci**	**Function**
Methyl-directed mismatch repair	*dam*	DNA adenine methylase
	mutS	Mismatch recognition
	mutH	Endonuclease that cuts at hemimethylated sites
	mutL	Interacts with MutS and MutH
	uvrD (*mutU*)	Helicase
Very short patch repair	*dcm*	DNA cytosine methylase
	vsr	Endonuclease that cuts at 5′ side of T in TG mismatch
"GO" (guanine oxidizations)	*mutM*	Glycosylase that acts on GO
	mutY	Glycosylase that removes A from A:GO mismatch
	mutT	8-oxodGTP phosphatase
Alkyl	*ada*	Alkyltransferase and transcriptional activator
	alkA	Glycosylase for alkylpurines
Nucleotide excision	*uvrA*	Component of UvrABC
	urvB	Component of UvrABC
	uvrC	Component of UvrABC
Base excision	*xthA*	AP endonuclease
	nfo	AP endonuclease
Photoreactivation	*phr*	Photolyase
Recombination repair	*recA*	Strand exchange
	recBCD	Helicase and exonuclease
	recF	Recombination function
	ssb	Single-stranded DNA-binding protein
SOS system	*recA*	Coprotease
	lexA	Repressor
	umuDC	Involved in translesion synthesis and SOS mutagenesis

any damage to DNA, provided that it causes a distortion in the DNA structure.

The separation of repair functions into different pathways is in some cases artificial. Some repair genes are inducible, and the repair enzymes themselves can play a role in their induction as well as in the induction of genes in other pathways. For example, the RecBCD nuclease is involved in recombination repair but can also help make the single-stranded DNA that activates the RecA coprotease activity after DNA damage to induce SOS functions. The RecA protein is required for both recombination repair and induction of the SOS functions and may also be directly involved in SOS mutagenesis. Needless to say, the overlap of the functions of the repair gene products in different repair pathways has complicated the assignment of roles in these pathways.

Bacteriophage Repair Pathways

The DNA genomes of bacteriophages are also subject to DNA damage, either when the DNA is in the phage particle or when it is replicating in a host cell. Not surprisingly, many phages encode their own DNA repair enzymes. In fact, the discovery of some phage repair pathways preceded and anticipated the discovery of the corresponding bacterial repair pathways. By encoding their own repair pathways, phages avoid dependence on host pathways and repair proceeds more efficiently than it might with the host pathways alone.

The repair pathways of phage T4 are perhaps the best understood. This large phage encodes at least seven different repair enzymes that help repair DNA damage due to UV irradiation, and some of these enzymes are also involved in recombination (see chapter 9). Table

TABLE 10.3   Bacteriophage T4 repair enzymes

Repair enzyme	Host analog
DenV	UV endonuclease of *M. luteus*
UvsX,Y	RecA
UvsW	Unknown
gp46/47 exonuclease	RecBCD recombination repair
gp49 resolvase	RuvABC recombination repair

10.3 lists the function of some T4 gene products and the homologous bacterial enzymes. The phage also uses some of the corresponding host enzymes to repair damage to its DNA.

One of the most important functions for repairing UV damage in phage T4 is the product of the *denV* gene (see Dodson and Lloyd, Suggested Reading), which is both an N-glycosylase and a DNA endonuclease. The N-glycosylase activity specifically breaks the bond holding one of the pyrimidines to its sugar in pyrimidine cyclobutane dimers of the *cis-syn* type (Figure 10.7). The endonuclease activity then cuts the DNA just 3' of the pyrimidine dimer, and the dimer is removed by the exonuclease activity of the host cell DNA polymerase I. This enzyme thus works very differently from the UvrABC endonuclease of *E. coli* and, in a sense, combines both the N-glycosylase and AP endonuclease activities of some other repair pathways. The bacterium *Micrococcus luteus* has a similar enzyme. The endonuclease activity of the DenV protein will also function independently of the N-glycosylase activity and will cut apurinic sites in DNA as well as heteroduplex loops caused by short insertions or deletions. Because it cuts next to pyrimidine dimers in DNA, the purified DenV endonuclease of T4 is often used to determine the persistence of pyrimidine dimers after UV irradiation.

SUMMARY

1. All organisms on Earth probably have mechanisms to repair damage to their DNA. Some of these repair systems are specific to certain types of damage, while others are more general and repair any damage that makes a significant distortion in the DNA helix.

2. Specific DNA glycosylases remove some types of damaged bases from DNA. Specific DNA glycosylases are known that remove uracil, hypoxanthine, some types of alkylated bases, 8-oxoG, and any A mistakenly incorporated opposite 8-oxoG in DNA. After the damaged base is removed by the specific glycosylase, the DNA is cut by an AP endonuclease and the strand is degraded and resynthesized to restore the correct base.

3. The positions of 5-methylcytosine in DNA are particularly susceptible to mutagenesis, because deamination of 5-methylcytosine produces thymine instead of uracil, and thymine will not be removed by the uracil-N-glycosylase. *E. coli* has a special repair system, called VSP repair, that recognizes the thymine in the thymine-guanine mismatch at the site of 5-methylcytosine and removes it, preventing mutagenesis.

4. The photoreactivation system uses an enzyme called the photolyase to specifically separate the bases of one type of pyrimidine dimers created during UV irradiation. The photolyase binds to pyrimidine dimers in the dark but requires visible light to separate the fused bases.

(continued)

SUMMARY (continued)

5. Methyltransferase enzymes remove the methyl or ethyl group from certain aklylated bases and phosphates and transfer it to themselves. In *E. coli*, these specific methyltransferases are inducible as part of the adaptive response. Methylation of the Ada protein converts it into a transcriptional activator for its own gene as well as for some other repair genes involved in repairing alkylation damage.

6. The methyl-directed mismatch repair system recognizes mismatches in the DNA and removes and resynthesizes one of the two strands, restoring the correct pairing. In *E. coli*, the newly synthesized strand is degraded and resynthesized because it has not yet been methylated by the Dam methylase.

7. The nucleotide excision repair pathway encoded by the *uvr* genes of *E. coli* removes many types of DNA damage that cause gross distortions in the DNA helix. The UvrABC endonuclease cuts on both sides of the DNA damage, and the entire oligonucleotide including the damage is removed and resynthesized.

8. Recombination repair does not actually repair the damage but helps the cell tolerate it. Replication proceeds past the damage, leaving a gap opposite the damaged bases. Recombination with the other daughter DNA in the cell then puts a good strand opposite the damage, and the process can continue. This type of repair in *E. coli* requires the recombination functions RecBCD, RecF, and RecA.

9. A combination of excision and recombination repair may remove interstrand cross-links in DNA. The excision repair system cuts one strand, and the single-stranded break is enlarged by exonucleases to leave a gap opposite the damage. Recombination repair can then transfer a good strand to a position opposite the damage. Excision repair can then remove the damage, since it is now confined to one strand. Interstrand cross-links can be repaired only if there are two or more daughter chromosomes in the cell.

10. The SOS regulon includes many genes that are induced following DNA damage. The genes are normally repressed by the LexA repressor, which is cleaved after DNA damage. The cleavage is triggered by RecA bound to single-stranded DNA fragments that accumulate following DNA damage.

11. SOS mutagenesis is due to the the products of genes *umuC* and *umuD*, which are induced following DNA damage. These proteins may allow the DNA polymerase to proceed over the damage in the template, often causing mutations.

12. The UmuD protein also cleaves itself upon binding the activated RecA coprotease. The cleaved form of UmuD, called UmuD', is the active form for UV mutagenesis.

13. The RecA protein may play a direct role in SOS mutagenesis, helping the UmuC and UmuD' proteins to allow the DNA polymerase to replicate over damage in the DNA.

QUESTIONS FOR THOUGHT

1. Some types of organisms, for example, yeasts, do not have methylated bases in their DNA and so would not be expected to have a methyl-directed mismatch repair system. Can you think of any other ways besides methylation that a mismatch repair system could be directed to the newly synthesized strand during replication?

2. Why do you suppose so many pathways exist to repair some types of damage in DNA?

3. Why do you think the SOS mutagenesis pathway exists if it contributes so little to survival after DNA damage?

PROBLEMS

1. Outline how you would determine if a bacterium isolated from the gut of a marine organism at the bottom of the ocean has a photoreactivation DNA repair system.

2. How would you show in detail that the mismatch repair system preferentially repairs the base in a mismatch on the strand unmethylated by Dam methylase? Hint: Make heteroduplex DNA of two mutants of your choice.

3. Outline how you would determine if the photoreactivating system is mutagenic.

4. Outline how you would determine if the nucleotide excision repair system of *E. coli* can repair damage due to aflatoxin B. Hint: Use a *uvr* mutant.

5. How would you determine if the RecA protein has a direct role in SOS mutagenesis independent of its role in cleaving LexA and UmuD?

6. Outline how you would determine if the *recN* gene is a member of the SOS regulon.

SUGGESTED READING

Dodson, M. L., and R. S. Lloyd. 1989. Structure-function studies of the T4 endonuclease V repair enzyme. *Mutat. Res.* 218:49–65.

Friedberg, E. C., G. C. Walker, and W. Siede. 1995. *DNA Repair and Mutagenesis*. American Society for Microbiology, Washington, D.C.

Glickman, B. W., and M. Radman. 1980. *Escherichia coli* mutator mutants deficient in methylation instructed DNA mismatch correction. *Proc. Natl. Acad. Sci. USA* **77:**1063–1067.

Hennecke, F., H. Kolmer, K. Brundl, and H. J. Fritz. 1991. The *vsr* gene product of *E. coli* K-12 is a strand- and sequence-specific DNA mismatch endonuclease. *Nature* (London) **353:**776–778.

Kato, T., R. H. Rothman, and A. J. Clark. 1977. Analysis of the role of recombination and repair in mutagenesis of *Escherichia coli* by UV irradiation. *Genetics* **87:**1–18.

Kong, X.-P., R. Onrust, M. O'Donnell, and J. Kuriyan. 1992. Three-dimensional structure of the β-subunit of *E. coli* DNA polymerase III holoenzyme. *Cell* **69:**425–437.

Lin, J.-J., and A. Sancar. 1992. (A)BC exinuclease: the *Escherichia coli* nucleotide excision repair enzyme. *Mol. Microbiol.* **6:**2219–2224.

Michaels, M. L., C. Cruz, A. P. Grollman, and J. H. Miller. 1992. Evidence that MutY and MutM combine to prevent mutations by an oxidatively damaged form of guanine in DNA. *Proc. Natl. Acad. Sci. USA* **89:**7022–7025.

Modrich, P. 1991. Mechanism and biological effects of mismatch repair. *Annu. Rev. Genet.* **25:**229–253.

Nohni, T., J. R. Battista, L. A. Dodson, and G. C. Walker. 1988. RecA-mediated cleavage activates UmuD for mutagenesis: mechanistic relationship between transcriptional derepression and postranslational activation. *Proc. Natl. Acad. Sci. USA* **85:**1816–1820.

Pukkila, P. J., J. Petersson, G. Herman, P. Modrich, and M. Meselson. 1983. Effects of high levels of adenine methylation on methyl directed mismatch repair in *Escherichia coli*. *Genetics* **1044:**571–582.

Rupp, W. D. 1996. DNA repair mechanisms, p. 2277–2294. *In* F. C. Neidhardt, R. Curtiss III, J. L. Ingraham, E. C. C. Lin, K. B. Low, B. Magasanik, W. S. Reznikoff, M. Riley, M. Schaechter, and H. E. Umbarger (ed.), *Escherichia coli and Salmonella: Cellular and Molecular Biology*, 2nd ed. ASM Press, Washington, D.C.

Rupp, W. D., C. E. I. Wilde, D. L. Reno, and P. Howard-Flanders. 1971. Exchanges between DNA strands in ultraviolet irradiated *E. coli*. *J. Mol. Biol.* **61:**25–44.

Schrodinger, E. 1944. *What is Life? The Physical Aspect of the Living Cell*. Cambridge University Press, Cambridge.

Teo, I., B. Sedgwick, M. W. Kilpatrick, T. V. McCarthy, and T. Lindahl. 1986. The intracellular signal for induction of resistance to alkylating agents in *E. coli*. *Cell* **45:**315–324.

Weigle, J. J. 1953. Induction of mutation in a bacterial virus. *Proc. Natl. Acad. Sci. USA* **39:**628–636.

Worth, L., Jr., S. Clark, M. Radman, and P. Modrich. 1994. Mismatch repair proteins MutS and MutL inhibit RecA-catalyzed strand transfer between diverged DNAs. *Proc. Natl. Acad. Sci. USA* **91:**3238–3241.

Chapter 11

Regulation of Gene Expression

T HE DNA OF A CELL will contain thousands to hundreds of thousands of genes depending on whether the organism is a relatively simple single-celled bacterium or a complex multicellular eukaryote like a human. All of the features of the organism are due, either directly or indirectly, to the products of these genes. However, all the cells of an organism do not always look or act the same, even though they share essentially the same genes. Even the cells of a single-celled bacterium can look or act differently depending upon the conditions in which the cells find themselves, because the genes of cells are not always expressed at the same levels. The process by which the expression of genes is turned on and off at different times and under different conditions is called the **regulation of gene expression**.

Cells regulate the expression of their genes for many reasons. A cell may only express the genes that it needs in a particular environment so that it does not waste energy making RNAs and proteins it does not need at that time. Or the cell may turn off genes whose products might interfere with other processes going on in the cell at the time. Cells also regulate their genes as part of developmental processes, such as embryogenesis and sporulation.

As described in chapter 2, the expression of a gene moves through many stages, any one of which offers an opportunity for regulation. First, RNA is transcribed from the gene. Even if RNA is the final gene product, that molecule may require further processing to be active. If the final product of the gene is a protein, the mRNA synthesized from the gene might have to be processed before it can be translated into protein. Then the protein might need to be further processed or transported to its final location to be active. Even after the gene product is synthesized in its final form, its activ-

265

ity might also be modulated under certain environmental conditions.

By far the most common type of regulation occurs at the first stage, when RNA is made. Genes that are regulated at this level are said to be **transcriptionally regulated**. This form of gene regulation seems the most efficient, since synthesizing mRNA that will not be translated seems wasteful. Yet not all genes are transcriptionally regulated, at least not exclusively. Examples abound in which the expression of a gene is regulated even after RNA synthesis.

Any regulation that occurs after the gene is transcribed into mRNA is called **posttranscriptional regulation**. There are many types of posttranscriptional regulation; the most common is **translational regulation**. If a gene is translationally regulated, the mRNA may be continuously transcribed from the gene but its translation is sometimes inhibited.

Transcriptional Regulation in Bacteria

Thanks to the relative ease of doing genetics with bacteria, transcriptional regulation in bacteria is better understood than that in other organisms and has served as a framework for understanding transcriptional regulation in eukaryotic organisms. There are important differences between the mechanisms of transcriptional regulation in bacteria and higher organisms, many of which relate to the presence of a nuclear membrane. Nevertheless, many of the strategies used are similar throughout the biological world, and many general principles have been uncovered through studies of bacterial transcriptional regulation.

The Bacterial Operon

The concept of an **operon** is central to hypotheses about bacterial transcriptional regulation. Bacterial genes are often arranged so that more than one gene can be transcribed into the same polycistronic mRNA (see chapter 2). In such cases, the genes are said to be **cotranscribed**. Cotranscription of more than one gene into a polycistronic mRNA seems to be unique to bacteria and their phages and affects the types of translational initation regions (TIR) used on the mRNA. In eukaryotes, generally the first AUG codon in an mRNA is used to initiate translation so that only one polypeptide can be encoded by each mRNA. However, bacterial TIRs are much more complex. As discussed in chapter 2, Shine-Dalgarno and other sequences help define the TIR. Because of its distinct structure, a bacterial TIR will be recognized wherever it appears in the mRNA, and more than one TIR can be recognized in the same mRNA.

A bacterial operon is the region on the DNA that includes genes cotranscribed into the same mRNA plus all of the adjacent *cis*-acting sequences required for transcription of these genes, including the genes' promoter as well as operators and other sequences involved in regulating the transcription of the genes. Because the genes of an operon are all transcribed from the same promoter and use the same regulatory sequences, all the genes of an operon can be transcriptionally regulated simultaneously.

In this chapter, we discuss the regulation of some bacterial operons. The examples were chosen both because of their historical importance and because they illustrate many of the mechanisms used in transcriptional regulation. Where possible, we describe the genetic experiments used to uncover the regulatory mechanisms.

Repressors and Activators

Before discussing our examples of transcriptional regulation in bacteria, we give a brief overview of the types of regulation known to occur and define some of the terms that describe the factors involved. The transcription of bacterial operons is regulated by the products of **regulatory genes**, which are often proteins called **repressors** or **activators**. These regulatory proteins bind close to the operon's promotor and regulate transcription from the promoter. Sometimes, a regulatory protein can play dual roles and can also perform an enzymatic reaction in the pathway encoded by the operon. Because they bind to DNA, repressors and activators often have the helix-turn-helix motif shared by many DNA-binding proteins (see Box 11.1).

Repressors and activators work in opposite ways, as illustrated in Figure 11.1A. Repressors bind to sites called **operators** and turn *off* the promoter, thereby preventing transcription of the genes of the operon. Activators, in contrast, bind to **activator sites** and turn *on* the promoter, thereby facilitating transcription of the operon genes.

Negative and Positive Regulation

Another important concept is the distinction between **negative** and **positive regulation** by regulatory proteins. If a regulatory protein in its active state turns *off* the expression of the operon, the operon is said to be *negatively* regulated by the regulatory protein. If the regulatory protein in its active state turns *on* the operon, the operon is *positively* regulated by the regulatory protein. An operon regulated by a repressor is therefore negatively regulated, because the presence of the active repressor prevents transcription of that operon. In con-

The Helix-Turn-Helix Motif of DNA-Binding Proteins

Proteins that bind to DNA, including repressors and activators, often share similar structural motifs determined
by the interaction between the protein and the DNA helix.
One such motif is the helix-turn-helix motif. A region of approximately 7 to 9 amino acids forms an α-helical structure
called helix 1. This region is separated by about 4 amino
acids from another α-helical region of 7 to 9 amino acids
called helix 2. The two helices are at approximately right
angles to each other, hence the name helix-turn-helix.
When the protein binds to DNA, helix 2 lies in the major
groove of the DNA double helix while helix 1 lies crosswise
to the DNA, as shown in the figure. Because they lie in the
major groove of the DNA double helix, the amino acids in
helix 2 can contact and form hydrogen bonds with specific
bases in the DNA. Thus, a DNA-binding protein containing
a helix-turn-helix motif recognizes and binds to specific regions on the DNA. Many DNA-binding proteins exist as
dimers and bind to inverted repeated DNA sequences. In
such cases, the two polypeptides in the dimer are arranged
head to tail so that the amino acids in helix 2 of each
polypeptide can make contact with the same bases in the
inverted repeats.

In the absence of structural information, the existence of
a helix-turn-helix motif in a protein can often be predicted
from the amino acid sequence, since some sequences of
amino acids cause the polypeptide to assume an α-helical
form and the bent region between the two helices usually
contains a glycine. The presence of a helix-turn-helix motif in
a protein helps identify it as a DNA-binding protein.

Reference

Steitz, T. A., D. H. Ohlendorf, D. B. McKay, W. F. Anderson, and
B. W. Matthews. 1982. Structural similarity in the DNA-binding do-
mains of catabolite gene activator and *cro* repressor proteins. *Proc.
Natl. Acad. Sci. USA* **79**:3097–3100.

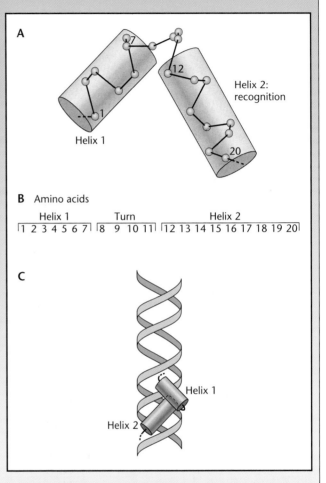

Helix-turn-helix motifs of DNA-binding proteins. (A) The
structure of a helix-turn-helix domain. (B) The number of
amino acids in the helix-turn-helix domain of the CAP protein. (C) The interactions of helix 1 and helix 2 with double-
stranded DNA.

trast, an operon regulated by an activator is positively
regulated, because in its active state the activator protein
turns on transcription of the operon under its control.

Activator and repressor proteins usually bind to different regions of the DNA (Figure 11.1B). Activators usually bind upstream of the −35 sequence of the promoter,
where they can make contact with the RNA polymerase
bound to the promoter. Repressors often bind to the promoter region itself, or at least very close to it, and thereby
block access by RNA polymerase to the promoter.

Some regulatory proteins can be both repressors and
activators, depending upon the situation. The λ repres-

sor discussed in chapter 7 is an example. It represses
transcription from the the p_L and p_R promoters but activates the p_{RM} promoter. The binding site on the DNA for
the regulatory protein often changes when the protein
shifts from being an activator to a repressor. We discuss
more examples of bacterial regulatory proteins that are
both repressors and activators later in the chapter.

Inducers and Corepressors

Whether a regulatory protein is active sometimes depends on whether it is bound to a small molecule. Small
molecules that bind to proteins and change their proper-

A

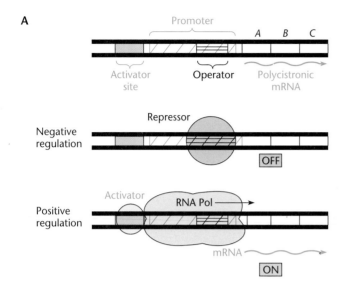

B

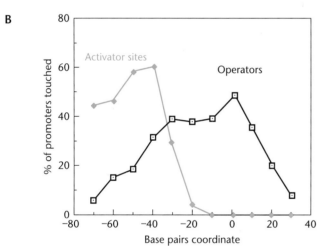

Figure 11.1 (A) The two general types of transcriptional regulation. In negative regulation, a repressor binds to an operator and turns the operon off. In positive regulation, an activator protein binds upstream of the promoter and turns the operon on. (B) Graph showing the usual locations of activator sites relative to operators. Activator sites are usually farther upstream. Each datum point indicates the middle of the known region on the DNA where a regulatory protein binds. Zero (0) on the *x* axis marks the start point of transcription. Adapted from J. Collado-Vides, B. Magasanik, and J. D. Gralla, *Microbiol. Rev.* **55:**371–394, 1991.

ties are called **effectors.** An effector that binds to a repressor or activator and thereby initiates transcription of an operon is called the **inducer** of the operon. In contrast, an effector that binds to a repressor and causes it to block transcription is called a **corepressor.**

The activity of regulatory proteins is not necessarily altered only by binding to small-molecule effectors.

Some repressors and activators are covalently altered under some conditions, for example, by methylation or phosphorylation.

Genetic Evidence for Negative and Positive Regulation

Negatively and positively regulated operons behave very differently in genetic tests. One difference is in the effect of mutations that inactivate the regulatory gene for the operon. If an operon is negatively regulated, a mutation that inactivates the regulatory gene will allow transcription of the operon genes, even in the absence of inducer. If the regulation is positive, mutations that inactivate the regulatory gene will prevent transcription of the genes of the operon, even in the presence of the inducer. A mutant in which the genes of an operon are always transcribed, even in the absence of inducer, is called a **constitutive mutant.** Constitutive mutations are much more common with negatively than with positively regulated operons because any mutation that inactivates the repressor will result in the constitutive phenotype. With positively regulated operons, a constitutive phenotype can be caused only by changes that do not inactivate the activator protein but alter it so that it can activate transcription without binding to the inducer. Such changes tend to be rare.

Complementation tests reveal another difference between negatively and postively regulated operons. Constitutive mutations of a negatively regulated operon are often recessive to the wild type. This is because any normal repressor protein in the cell encoded by a wild-type copy of the gene will bind to the operator and block transcription, even if the repressor encoded by the mutant copy of the gene in the same cell is inactive. In contrast, constitutive mutations in a solely positively regulated operon should be dominant to the wild type (see chapter 14). A mutant activator protein that is active without inducer bound might activate transcription even in the presence of a wild-type activator protein.

Negative Regulation

The *E. coli lac* Operon

The classic example of negative regulation is regulation of the *E. coli lac* operon, which encodes the enzymes responsible for the utilization of the sugar lactose. The experiments of François Jacob and Jacques Monod and their collaborators on the regulation of the *E. coli lac* genes are excellent examples of the genetic analysis of a biological phenomenon in bacteria (see Jacob and Monod, Suggested Reading). Although these experiments were performed in the late 1950s, only shortly after the discovery of the structure of DNA and the existence of mRNA, they still stand as the framework with which all other studies of gene regulation are compared.

GENETICS OF THE *lac* OPERON

When Jacob and Monod began their classic work, it was known that the enzymes of lactose metabolism are **inducible** in that they are expressed only when the sugar lactose is present in the medium. If no lactose is present, the enzymes are not made. From the standpoint of the cell, this is a sensible strategy, since there is no point in making the enzymes for lactose utilization unless there is lactose available for use as a carbon and energy source.

To understand the regulation of the lactose genes, Jacob and Monod first isolated many mutations affecting lactose metabolism and regulation, which fell into two fundamentally different groups. Some mutants were unable to grow with lactose as the sole carbon and energy source and so were Lac⁻. Other mutants made the lactose-metabolizing enzymes whether or not lactose was present in the medium and so were constitutive mutants.

To analyze the regulation of the *lac* genes, Jacob and Monod needed to know which of the mutations affected *trans*-acting gene products—either protein or RNA—involved in the regulation and how many different genes these mutations represented. They also wished to know if any of the mutations were *cis* acting (affecting sites on the DNA involved in regulation).

To answer these questions, they needed to perform complementation tests, which require that the organisms be diploid, with two copies of the genes being tested. Bacteria are normally haploid, with only one copy of each of their genes, but are "partial diploids" for any genes carried on an introduced prime factor. Recall that a prime factor is a plasmid into which some of the bacterial chromosomal genes have been inserted (see chapter 5). By introducing prime factors carrying various mutated *lac* genes into cells with different mutations in the chromosomal *lac* genes, Jacob and Monod performed complementation tests on each of their *lac* mutations. Their methods depended upon the type of mutation being tested.

COMPLEMENTATION TESTS WITH *lac* MUTATIONS

Whether a particular *lac* mutation is dominant or recessive was determined by introducing an F′ factor carrying the wild-type *lac* region into a strain with the *lac* mutation in the chromosome. If the partial diploid bacteria are Lac⁺ and can multiply to form colonies on minimal plates with lactose as the sole carbon and energy source, the *lac* mutation is recessive. If the partial diploid cells are Lac⁻ and cannot form colonies on lactose minimal plates, the *lac* mutation is dominant. Jacob and Monod discovered that most *lac* mutations are recessive to the wild type and so presumably inactivate genes whose products are required for lactose utilization.

The question of how many genes are represented by recessive *lac* mutations could be answered by performing pairwise complementation tests between different *lac* mutations. Prime factors carrying the *lac* region with one *lac* mutation were introduced into a mutant strain with another *lac* mutation in the chromosome (Figure 11.2). In this kind of experiment, if the partial diploid cells are Lac⁺, the two recessive mutations can complement each other and are members of *different* complementation groups or genes. If the partial diploid cells are Lac⁻, the two mutations cannot complement each other and are members of the *same* complementation group or gene. Jacob and Monod found that most of the *lac* mutations sorted into two different complementation groups, which they named *lacZ* and *lacY*. We now know of another gene, *lacA*, which was not discovered in their original selections because its product is not required for growth on lactose.

cis-Acting *lac* Mutations

Not all *lac* mutations affect diffusible gene products and can be complemented. Immediately adjacent to the *lacZ* mutations are other *lac* mutations that are much rarer and have radically different properties. These mutations

Merodiploid	Phenotype	Interpretation
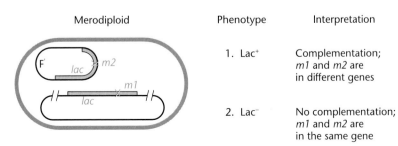	1. Lac⁺	Complementation; *m1* and *m2* are in different genes
	2. Lac⁻	No complementation; *m1* and *m2* are in the same gene

Figure 11.2 Complementation of two recessive *lac* mutations. One mutation is in the chromosome, and the other is in a prime factor. If the two mutations complement each other, the cells will be Lac⁺ and will grow with lactose as the sole carbon and energy source. The mutations will not complement if they are in the same gene or if one affects a regulatory site or is polar. See text for more details.

cannot be complemented to allow the expression of the *lac* genes on the same DNA, even in the presence of good copies of the *lac* genes. Recessive mutations that cannot be complemented are *cis* acting and presumably affect a site on DNA rather than a diffusible gene product like RNA or protein (see Table 13.1).

To show that a *lac* mutation is *cis* acting, i.e., affects only the expression of genes on the same DNA where it occurs, we could introduce an F' factor containing the potential *cis*-acting *lac* mutation into cells containing either a *lacZ* or a *lacY* mutation in the chromosome (Figure 11.3). *trans*-acting gene products encoded by the F' factor *lacZ* or *lacY* genes would complement the chromosomal *lacY* or *lacZ* mutations, respectively. However, if the resulting phenotypes are Lac⁻, the *lac* mutation in the F' factor must prevent expression of both LacZ and LacY proteins from the F' factor. The mutation in the F' factor is therefore *cis* acting.

As we discuss below, one type of the *cis*-acting *lac* mutations affects *lacp* and is a promoter mutation that prevents transcription of the *lacZ* and *lacY* genes by changing the binding site for RNA polymerase on the DNA. Another type of *cis*-acting mutation is a polar mutation in *lacZ* that prevents the transcription of the downstream *lacY* gene.

Lac⁻ Mutants with Dominant Mutations

Some Lac⁻ mutants have mutations that affect diffusible gene products but are not recessive. Such *lac* mutations will make the cell Lac⁻ and unable to use lactose even if there is another good copy of the lactose operon in the cell, either in the chromosome or in the F' factor. These dominant mutations are called *lacI*ˢ mutations, for superrepressor mutations. As we show below, these mutants have mutations that change the repressor so that it can no longer bind the inducer.

COMPLEMENTATION TESTS WITH CONSTITUTIVE MUTATIONS

As mentioned, some *lac* mutations do not make the cells Lac⁻ but, rather, make them constitutive, so that they express the *lacZ* and *lacY* genes even in the absence of

the inducer lactose. In complementation tests between constitutive mutations, partial diploids are made in which either the chromosome or the F' factor, or both, carry a constitutive mutation. The partial diploid cells are then tested to determine whether they express the *lac* genes constitutively or only in the presence of the inducer. If the partial diploid cells express the *lac* gene in the absence of the inducer, the constitutive mutation is dominant. However, if the partial diploid cells express the *lac* genes only in the presence of the inducer, the mutations are recessive. Using this test, Jacob and Monod found that some of the constitutive mutations, which could be recessive or dominant, affect a *trans*-acting gene product, either protein or RNA. Complementation between the recessive constitutive mutations revealed they are all in the same gene, which these investigators named *lacI* (Figure 11.4A).

cis-Acting *lacO*ᶜ Mutations

A rarer constitutive mutation is *cis* acting, allowing the constitutive expression of the *lacZ* and *lacY* genes from the DNA that has the mutation, even in the presence of a wild-type copy of the *lac* DNA. Jacob and Monod named these *cis*-acting constitutive mutations *lacO*ᶜ mutations, for *lac* operator-constitutive mutations. Figure 11.4B shows the partial diploid cells used in these complementation tests.

trans-Acting Dominant Constitutive Mutations

Some *lacI* mutations, called *lacI*⁻ᵈ mutations, are dominant, making the cell constitutive for expression of the *lac* operon even in the presence of a good copy of the *lacI* gene. These *lacI*⁻ᵈ mutations are possible because the LacI polypeptides form a homotetramer. A mixture of normal and defective subunits can be nonfunctional, causing the constitutive LacI⁻ phenotype. Hence, the *lacI*⁻ᵈ mutations are *trans* dominant.

JACOB AND MONOD OPERON MODEL

On the basis of this genetic analysis, Jacob and Monod proposed their **operon model** for *lac* gene regulation (Figure 11.5). The *lac* operon includes the genes *lacZ*

Figure 11.3 The *lacp* mutations cannot be complemented and are *cis* acting. A *lacp* mutation in the prime factor will prevent expression of any of the other *lac* genes on the prime factor, so that a *lac* mutation in the chromosome will not be complemented. Partial diploid cells will be Lac⁻.

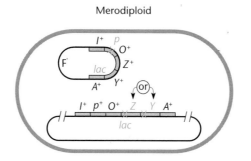

Merodiploid   Phenotype   Interpretation

Lac⁻   No complementation; the *cis*-acting *lacp* mutation prevents expression of *lacZ*, *lacY*, and *lacA*

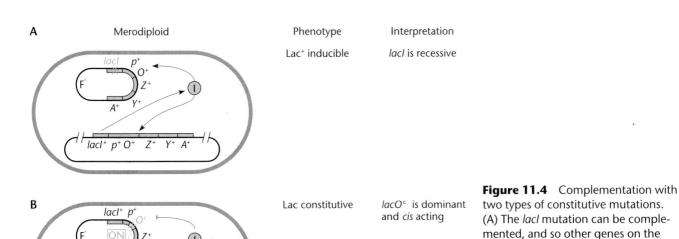

Figure 11.4 Complementation with two types of constitutive mutations. (A) The *lacI* mutation can be complemented, and so other genes on the prime factor will be inducible in the presence of a wild-type copy of the *lacI* gene. (B) In contrast, *lacO^c* mutations cannot be complemented by a wild-type *lacO* region in the chromosome and so are dominant.

and *lacY*. These genes, known as the **structural genes** of the operon, encode the enzymes required for lactose utilization. The *lacZ* gene product is a β-galactosidase, which cleaves lactose to form glucose and galactose, which can then be used by other pathways. The *lacY* gene product is a permease that allows lactose into the cell. The operon also includes the *lacA* gene. The *lacA* gene product is a transacetylase, whose function is un-

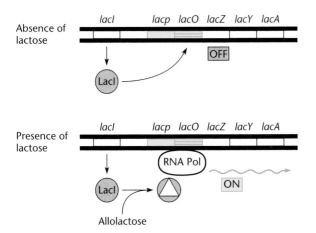

Figure 11.5 The Jacob and Monod model for negative regulation of the *lac* operon. In the absence of the inducer, lactose, the LacI repressor will bind to the operator region, preventing transcription of the other genes of the operon by RNA polymerase (RNA Pol). In the presence of lactose, the repressor can no longer bind to the operator, allowing transcription of the *lacZ*, *lacY*, and *lacA* genes. Allolactose, rather than lactose, is the true inducer molecule.

known. This enzyme was originally thought to be also encoded by the *lacY* gene.

The most important part of this model explains why the structural genes are expressed only in the presence of lactose. The product of the *lacI* gene is a repressor protein. In the absence of lactose, this repressor binds to the operator sequence (*lacO*) close to the promoter and thereby prevents the RNA polymerase from binding to the promoter and blocks the transcription of the structural genes. In contrast, when lactose is available, the inducer binds to the repressor and changes its conformation so that it can no longer bind to the operator sequence. The RNA polymerase can then bind to *lacp* and transcribe the *lacZ*, *lacY*, and *lacA* genes. The LacI repressor is very effective at blocking transcription of the structural genes of the operon. In the absence of repressor, transcription is about 1,000 times more active than in its presence.

It is worthwhile to emphasize how the Jacob and Monod operon model explains the behavior of mutations that affect the regulation of the *lac* enzymes. Mutants with *lacZ* and *lacY* mutations are Lac⁻ because they do not make an active β-galactosidase or permease, respectively, both of which are required for lactose utilization. These mutations are clearly *trans* acting, because they are recessive and can be complemented. An active β-galactosidase or permease made from another DNA in the same cell can provide the missing enzyme and allow lactose utilization.

The behavior of *lacp* mutations is also explained by the model. Jacob and Monod proposed that *lacp* mutations change the sequence on the DNA to which the

RNA polymerase binds. The RNA polymerase normally binds to *lacp* and moves through *lacZ*, *lacY*, and *lacA*, making an RNA copy of these genes. This explains why *lacp* mutations are *cis* acting; if the site on DNA at which RNA polymerase initiates transcription is changed by a mutation so that it no longer binds RNA polymerase, the *lacZ*, *lacY*, and *lacA* genes on that DNA will not be transcribed into mRNA, even in the presence of a good copy of the *lac* region elsewhere in the cell.

Their model also explains the behavior of the two constitutive mutations: *lacI* and *lacO^c*. The *lacI* mutations affect a *trans*-acting function, because they inactivate the repressor protein that binds to the operator and prevents transcription. The LacI repressor made from a functional copy of the *lacI* gene anywhere in the cell can bind to the operator sequence and block transcription in *trans*. However, the *lacO^c* mutations change the sequence on DNA to which the LacI repressor binds to block transcription. The LacI repressor cannot bind to this altered *lacO* sequence, even in the absence of lactose. Therefore, the RNA polymerase is free to bind to the promoter and transcribe the structural genes. The *lacO^c* mutations are *cis* acting because they allow the constitutive expression of the *lacZ*, *lacY*, and *lacA* genes from the same DNA, even in the presence of a good copy of the *lac* operon elsewhere in the cell.

The behavior of superrepressor *lacI^s* mutations is also explained by their model. These are mutations that change the repressor molecule so that it can no longer bind the inducer. The mutated repressor will bind to the operator even in the presence of inducer, making the cells permanently repressed and phenotypically Lac⁻. The fact that this type of mutation is dominant over the wild type is also explained. The mutated repressors will repress transcription of any *lac* operon in the same cell, even in the presence of inducer, and so they will make the cell Lac⁻ even in the presence of a good *lac* operon, either in the chromosome or in an F' factor.

The *lac* genes provide a good example of what we mean by "operon." As we stated earlier, an operon includes all the genes that are transcribed into the same mRNA plus any adjacent *cis*-acting sites that are involved in the transcription or regulation of transcription of the genes. The *lac* operon of *E. coli* consists of the three structural genes, *lacZ*, *lacY*, and *lacA*, which are transcribed into the same mRNA, as well as the *lac* promoter from which these genes are transcribed. It must also include the the *lac* operator, since this is a *cis*-acting regulatory sequence involved in regulating the transcription of the structural genes. However, the *lac* operon does *not* include the gene for the repressor, *lacI*. The *lacI* gene is adjacent to the *lacZ*, *lacY*, and *lacA* genes and

regulates their transcription, but it is not transcribed onto the same mRNA as the structural genes. Moreover, its product is *trans* acting rather than *cis* acting.

UPDATE ON THE REGULATION OF THE *lac* OPERON

The operon model of Jacob and Monod has survived the passage of time. In 1965, it earned them a Nobel Prize, which they shared with Andre Lwoff. Because of its elegant simplicity, the operon model for the regulation of the *lac* genes of *E. coli* serves as the paradigm for understanding gene regulation in other organisms.

Over the years, the operon model has undergone a few refinements. As mentioned, Jacob and Monod did not know of the existence of the *lacA* gene and thought that *lacY* encoded the transacetylase. Also, most of the mutations that Jacob and Monod defined as *lacp* were not promoter mutations but, rather, strong polar mutations in *lacZ* that prevent the transcription of all three structural genes (*lacZ*, *lacY*, and *lacA*). Later studies also revealed that the true inducer that binds to the LacI repressor is not lactose itself but, rather, allolactose, a metabolite of lactose. In most experiments, an analog of allolactose called isopropyl-β-D-thiogalactoside (IPTG) is used as the inducer because it is not metabolized by the cells.

The most significant alteration to the Jacob and Monod model was the discovery that the LacI repressor can bind to not just one but three operators, called o_1, o_2, and o_3 (Figure 11.6). The operator closest to the promoter, o_1, seems to be the most important for repressing transcription of *lac*. However, deleting both o_2 and o_3 diminishes repression as much as 50-fold.

Why does the *lac* operon have more than one operator, especially since one of the operators (o_3) is so far upstream of the promoter that it seems unlikely that it could block binding of the RNA polymerase to that site? One idea is that the LacI repressor binds to two operators simultaneously (Figure 11.6), bending the DNA—and the promoter—between them. The bent promoter might not be able to bind RNA polymerase or might not undergo the changes in structure required for the initiation of transcription. As we discuss below, the *ara* and *gal* operons also contain multiple operators that may also bend the DNA in the promoter region.

CATABOLITE REPRESSION OF THE *lac* OPERON

In addition to being under the control of its own specific repressor, the *lac* operon is regulated through **catabolite repression. Catabolites** are carbon-containing molecules that are used to build other molecules. The catabolite repression system ensures that the genes for lactose uti-

A

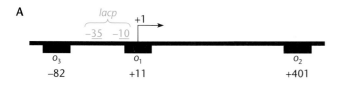

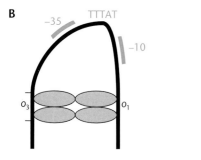

B

Figure 11.6 Locations of the three operators in the *lac* operon and a possible model for how LacI repressor binding to these operators may help repress the operon. Repressor (solid ellipses) bound to o_1 and o_3, or o_2 and o_3, could bend the DNA in the promoter region and help prevent RNA polymerase binding to the promoter. The AT-rich region may facilitate bending. Redrawn from H. Choy and S. Adhya, *in* F. C. Neidhardt, R. Curtiss III, J. L. Ingraham, E. C. C. Lin, K. B. Low, B. Magasanik, W. S. Reznikoff, M. Riley, M. Schaechter, and H. E. Umbarger, (ed.), *Escherichia coli and Salmonella: Cellular and Molecular Biology,* 2nd ed., p. 1291, ASM Press, Washington, D.C., 1996.

lization will not be expressed if a better carbon and energy source such as glucose is available. The name "catabolite repression" is a misnomer since the expression of operons under catabolite control requires a transcriptional activator, the catabolite activator protein (CAP), and the small molecule effector, cyclic AMP (cAMP). We defer a detailed discussion of the mechanism of catabolite repression until chapter 12, since this is a type of global regulation.

STRUCTURE OF THE *lac* CONTROL REGION

Figure 11.7 illustrates the structure of the *lac* control region in detail, showing the nucleotide sequences of the *lac* promoter and operators as well as the region to which the CAP protein binds. The *lac* promoter is a typical σ^{70} bacterial promoter with the characteristic −10 and −35 regions (see chapter 2). One of the operators to which the LacI repressor binds (o_1) actually overlaps the mRNA start site (+1 in Figure 11.7) for transcription of *lacZ, lacY,* and *lacA.* Although not shown, the other *lacO* operator sequences lie nearby. The CAP-binding site that enhances initation by RNA polymerase in the absence of glucose is just upstream of the promoter, as shown.

EXPERIMENTAL USES OF THE *lac* OPERON

Because it is the best understood of the bacterial regulatory systems, the *lac* genes and regulatory regions have found many uses in molecular genetics. We list a few below.

The LacI Repressor

The LacI repressor protein has served as a model for the interaction of proteins with DNA. This protein is very interesting because it has more than one function: it must be able to bind to DNA at the operator sequences as well as bind the inducer, either allolactose or IPTG. The recent success in crystallizing the protein has given us a better picture of the structure of this protein. The active repressor is a homotetramer, with four identical polypeptide subunits encoded by the *lacI* gene, so that the individual polypeptides must bind to each other. Different regions or domains of LacI participate in these various functions. In chapter 14, we discuss genetic exeriments that helped identify the domains of the LacI repessor involved in its various functions. Finally, the *lac* repressor has been used in many bacteria other than *E. coli,* and has even been used in eukaryotic cells. The LacI repressor protein is very stable and can be used to block transcription by binding to the *lac* operator sequence in any cells in which it can be synthesized. Expression vectors in which the *lac* repressor is used to repress transcription of genes have even been used in mammalian cells (see chapter 15 for a description of expression vectors).

The *lac* Promoter

The *lac* promoter or its derivatives have been used to transcribe cloned genes in many expression vectors (see chapter 15). This promoter offers many advantages in these expression vectors. The *lac* promoter is fairly strong, allowing high levels of transcription of a cloned gene. The *lac* promoter is also inducible, which makes it possible to clone genes whose products are toxic to the cell. The cells can be grown in the absence of inducer, so that the cloned gene will not be transcribed. Only when the cells reach a high density is the inducer IPTG added and the cloned gene transcribed. Even if the toxic protein kills the cells, some of the protein will usually be synthesized before the cell dies.

The derivatives of the *lac* promoter used in expression vectors retain some of the desirable properties of the *lac* operon but have additional features. For example, the mutated promoter *lacUV5* is no longer sensitive to catabolite repression and so will be active even if glucose is present in the medium (see chapter 12).

A hybrid *trp-lac* promoter called the *tac* promoter has also been widely used. The *tac* promoter has the ad-

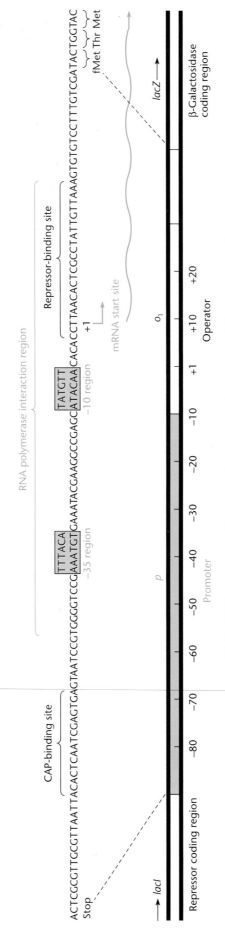

Figure 11.7 DNA sequence of the promoter and operator regions of the *lac* operon. See the text for details. The entire region is only 100 bp long. Only the o₁ operator sequence is shown.

vantages that it is even stronger than the *lac* promoter but retains its inducibility. It is also insensitive to catabolite repression.

The *lacZ* Reporter Gene

The *lacZ* gene is widely used as a reporter gene in translational and transcriptional fusions. In many ways, this gene is the quintessential reporter gene and has been used in cells ranging from bacteria to fruit fly to human cells. The product of the *lacZ* gene, β-galactosidase, is easily detected by colorimetric assays using substrates such as X-Gal (5-bromo-4-chloro-3-indolyl-β-D-galactopyranoside) and ONPG (o-nitrophenyl-β-D-galactopyranoside) and is an unusually stable protein. The fact that this polypeptide is very large can be a disadvantage or advantage depending upon the situation.

PROSPECTUS

The *lac* operon is one of the simplest regulatory systems known, and so it is fortunate that it was one the first to be chosen to study. The relatively simple regulation of the *lac* operon encouraged attempts to understand other types of regulation. As discussed later, regulation of most other operons is more complicated and would have been even more difficult to understand if the *lac* operon had not been available as a point of reference. In the next sections, we discuss the regulation of some other representative bacterial operons.

The *E. coli gal* Operon

The operon of *E. coli* involved in the utilization of the sugar galactose, the *gal* operon, is another classic example of negative regulation. Figure 11.8 shows the organization of the genes in this operon. The products of three structural genes, *galE*, *galT*, and *galK*, are required for the utilization of galactose and convert galactose into glucose, which can then enter the glycolysis pathway.

The specific reactions catalyzed by each of the enzymes of the *gal* pathway appear in Figure 11.9. The GalK gene product is a kinase that phosphorylates galactose to make galactose-1-phosphate. The product of the *galT* gene is a transferase that transfers the galactose-1-phosphate to UDPglucose, displacing the glucose to make UDPgalactose. The released glucose can then be used as a carbon and energy source. The GalE gene product is an epimerase that converts UDPgalactose to UDPglucose to continue the cycle. It is not clear why *E. coli* cells use this convoluted pathway to convert galactose to glucose so that the latter can be used as a carbon and energy source. However, many organisms, including both plants and animals, use this pathway.

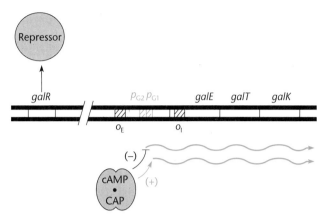

Figure 11.8 Structure of the galactose operon of *E. coli*. The *galE*, *galT*, and *galK* genes are transcribed from two promoters, p_{G1} and p_{G2}. The CAP protein with cyclic AMP (cAMP) bound turns on p_{G1} and turns off p_{G2}, as shown. There are also two operators, $galo_E$ and $galo_I$. The repressor genes are some distance away, as indicated by the broken line. Only the *galR* repressor gene is shown.

Unlike the genes for lactose utilization, not all the genes for galactose utilization are closely linked in the *E. coli* chromosome. The *galU* gene, whose product synthesizes UDPglucose, is located in a different region of the chromosome. Also, the genes for the galactose permeases, which are responsible for transporting galactose into the cell, are not part of the *gal* operon. Furthermore, unlike the *lacI* gene, which is adjacent to the *lac* operon, the regulatory genes for the *gal* operon do not reside near this operon. The scattering of the genes for galactose metabolism may reflect the fact that galactose not only serves as a carbon and energy source but also plays other roles. For example, the UDPgalactose synthesized by the *gal* operon donates galactose to make polysaccharides for lipopolysaccharide and capsular synthesis.

$$Galactose + ATP \xrightarrow[GalK]{} galactose\text{-}1\text{-}PO_4$$

$$Galactose\text{-}1\text{-}PO_4 + UDPglucose \xrightarrow[GalT]{} UDPgalactose + glucose$$

$$UDPgalactose \xrightarrow[GalE]{} UDPglucose$$

$$UTP + glucose\text{-}1\text{-}PO_4 \xrightarrow[GalU]{} UDPglucose$$

Figure 11.9 Pathway for galactose utilization in *E. coli*. See the text for details.

REGULATION OF THE *gal* OPERON
Everything about the *gal* operon comes in twos. There are two promoters and two operators. There are even two repressors, encoded by different genes. Either of these repressors can repress the *gal* operon, although one is more effective than the other.

The Repressors: GalR and GalS
The two repressors in control of the *gal* operon are GalR and GalS, encoded by the *galR* and *galS* genes, respectively. GalR was discovered first because mutations in the *galR* cause constitutive expression of the *gal* operon. However, the *gal* operon was also subject to other regulation. If the GalR repressor were solely responsible for regulating the *gal* operon, mutations that inactivate the *galR* gene should result in the same level of *gal* expression whether galactose is present or not. Yet some regulation of the *gal* operon could be observed, even in *galR* mutants. When galactose was added to the medium in which *galR* mutant cells were growing, more of the enzymes of the *gal* operon were made than if the cells were growing in the absence of galactose. The product of another gene, *galS*, was responsible for the residual regulation. As evidence, double mutants with mutations that inactivate both *galR* and *galS* are fully constitutive. Later studies showed that the product of the *galS* gene is also a repressor that negatively regulates the *gal* operon.

The GalS and GalR repressor proteins are closely related, and they both bind the inducer galactose. Even so, they may play somewhat different roles. The GalR repressor is responsible for most of the repression of the *gal* operon in the absence of galactose. The GalS repressor plays only a minor role in regulating the *gal* operon but solely controls the genes of the galactose transport system, which transports galactose into the cell. The reason for this two-tier regulation is unclear but also may be related to the diverse roles of galactose in the cell (see above).

The Two *gal* Operators
There are also two operators in the *gal* operon. One is upstream of the promoters, and the other is internal to the first gene, *galE* (Figure 11.8). The two operators are named o_E and o_I for operator *external* to the *galE* gene and *internal* to the *galE* gene, respectively. The discovery of the o_I operator involves some interesting genetics, so we discuss it in some detail.

1. Isolating gal *operator mutants.* The first mutant with an o_I mutation was isolated as part of a collection of constitutive mutants of the *gal* operon (see Irani et al.,

Suggested Reading). These mutants are easier to isolate in strains with superrepressor $galR^s$ mutations than in wild-type *E. coli*. The $galR^s$ mutations are analogous to $lacI^s$ mutations. The superreppressor mutation will make a strain Gal⁻ and uninducible because galactose cannot bind to the mutated repressor. Therefore, *E. coli* with a $galR^s$ mutation cannot multiply to form colonies on plates containing only galactose as the carbon and energy source. However, a constitutive mutation that inactivates the $GalR^s$ repressor or changes the operator sequence will prevent the mutant repressor from binding to the operator and allow the cells to use the galactose and multiply to form a colony. Thus, if bacteria with a $galR^s$ mutation are plated on medium with galactose as the sole carbon and energy source, only constitutive mutants will multiply to form a colony. However, most of the constitutive mutants isolated this way will have mutations in the *galR* gene that inactivate the $GalR^s$ repressor rather than operator mutations, since the operator is by far the smaller target. Many *galR* mutants would have to be screened before a single operator mutant was found. Therefore, to make this method practicable for isolating consitutive mutants with operator mutations, the frequency of *galR* mutants must be decreased until it is not too much higher than that of constitutive mutants with mutations in the operator sequences.

One way to reduce the frequency of *galR* mutants is to use a strain that is a partial diploid for (has two copies of) the $galR^s$ gene. Then, even if one $galR^s$ gene is inactivated by a mutation, the other $galR^s$ gene will continue to make the $GalR^s$ protein, making the cell phenotypically Gal⁻. Only two independent mutations, one in each $galR^s$ gene, can make the cell constitutive. Therefore, since two independent *galR* mutations should be no more frequent than single operator mutations, cells with operator mutations should now be a significant fraction of the constitutive mutants. Moreover, constitutive mutants with operator mutations can be distinguished from the double mutants with mutations in both the $galR^s$ genes. The operator mutations will map in the *gal* operon, unlike mutations in either copy of the $galR^s$ gene.

Accordingly, a partial diploid was constructed that had one copy of the $galR^s$ gene in the normal position and another copy in a specialized transducing λ phage integrated at the λ attachment site (see chapter 7). When this strain was plated on medium containing galactose as the sole carbon and energy source, a few Gal⁺ colonies arose from constitutive mutants. The mutations in two of these constitutive mutants mapped in the region of the *gal* operon and so were presumed to be operator mutations. When the DNA of the two mutants was sequenced, it was discovered that one mutation had changed a base pair in the known operator region upstream of the promoters (o_E), as expected. However, the other operator mutation had changed a base pair in the *galE* gene, suggesting that a sequence in that gene—o_I— also functions as an operator. Furthermore, this mutation occurred in a sequence homologous to the 15 bp making up the known operator. In fact, 12 bp of this sequence is identical in o_I and o_E. Moreover, the mutation in the *galE* gene was *cis* acting for constitutive expression of the *gal* operon, one of the criteria for an operator mutation.

2. *Escape synthesis of the Gal enzymes.* The mutations in the *galE* gene cause *cis*-acting constitutive expression of the *gal* genes, suggesting that the o_I sequence functions as an operator. However, the constitutive mutants could be explained in other ways.

If the sequence in the *galE* gene is truly an operator, it would bind the GalR repressor protein. To test this, the experiment illustrated in Figure 11.10 was performed. First, DNA containing the *galE* gene is cloned into a multicopy plasmid. When this multicopy plasmid is transformed into a cell, that cell will contain many copies of the *galE* gene. Then, if the sequence in the *galE* gene does bind the repressor, these extra copies should bind most of the GalR protein in the cell, leaving

too little to completely repress expression of the *gal* operon. Thus, the cells would appear to be constitutive mutants. This general method is called **titration**, and the enzymes of the *gal* operon are synthesized through **escape synthesis**, so named since the operon is "escaping" the effects of the repressor. In the actual experiment, cells containing many copies of the *galE* gene region did exhibit a partially constitutive phenotype. In contrast, multiple copies of mutant DNA with the putative operator mutation (i.e., o_I^c) do not cause escape synthesis of the Gal enzymes. This result was interpreted to confirm the presence in *galE* of a second binding site for repressor, which is inactivated by the mutation.

3. *Why does the* gal *operon have two operators?* There are two general hypotheses for why the *gal* operon has two operators. According to one, the two operators function independently to block transcription of the *gal* operon. The other proposes that the operators cooperate to block transcription.

Genetic evidence supports the second general hypothesis—the two operators cooperate to block transcription. If the two operators functioned independently, the effect of mutations in both operators would be additive; in other words, the level of expression of the genes in the operon when both operators were mutated would be the sum of the levels of operon expression when each of the operators was mutated separately. However, genetic ex-

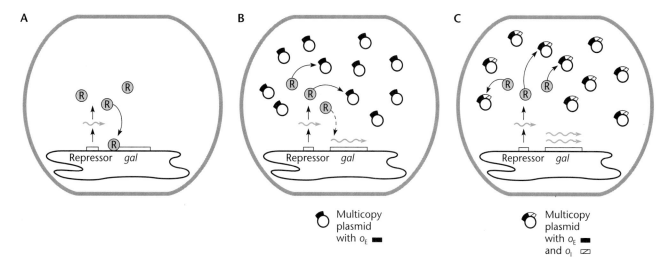

Figure 11.10 Escape synthesis of the enzymes of the galactose operon caused by additional copies of the operator regions. Clones of the operator regions in a multicopy plasmid dilute out the repressor, inducing the operon even in the absence of galactose. (A) The cell contains the normal number of operators and the operon is not induced. (B) The multicopy plasmid contains only o_E, and the operon is only partially induced. (C) The multicopy copy plasmid contains both o_E and the *galE* gene containing o_I, and the operon is fully induced.

periments showed that the level of expression in the double mutant is higher than the sum of the expression levels in each single mutant. This observation indicated cooperation between the two operators.

Figure 11.11 shows a model for this cooperation. Repressor molecules bound to the two operators interact with each other to bend the DNA of the promoter that is between the two operators, helping to prevent initiation of RNA synthesis at the promoter. A similar model had been proposed for the *ara* operon, and another was proposed later for the extra *lac* operators (Figure 11.6).

This model leads to a specific prediction: the spacing in the DNA between the two operators should be very important for the repression. As illustrated in Figure 11.12, the significance of the spacing is related to the structure of DNA. A region of DNA looks different depending on the face of the molecule. For example, on one face you might see the major groove plus one of the two strands, while on another you might see the minor groove, and so on. The two repressor molecules presumably recognize the same face of both operator sequences, since the two operator sequences are almost identical. Furthermore, if the two repressors bound at the two operators are to interact, they must be bound on the same side of the DNA; otherwise the DNA between the two operators would have to twist for the

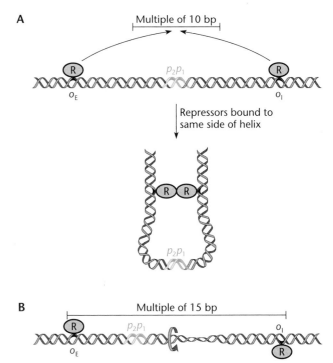

Figure 11.12 Repressor molecules (circled Rs) bound to the two operators of the *gal* operon can interact more easily if they are bound on the same side of the helix. (A) Multiples of 10 bp separating o_E and o_I allow R binding to the same side of the helix. (B) Arrows denote twisting of the molecule that is necessary when the operators are a multiple of 15 bp apart. See the text for details.

two repressor molecules to interact with each other, as shown in the figure. Twisting double-stranded DNA over such short lengths is difficult because of the stiffness of the molecule (see chapter 1). Thus, the two operators must be a multiple of 10 bp apart, since the same face of the DNA comes around only once about every 10 bp, the approximate length of a helical turn of the DNA (see chapter 1). If the two operators are more or less than a multiple of 10 bp apart, for example, 15 bp, the repressors bound to the two operators would be on opposite sides of the DNA and the DNA would have to twist for the two repressor molecules to interact, as indicated by the arrow in Figure 11.12. Sequencing has shown that the two operators in fact are normally a multiple of 10 bp apart, a spacing consistent with the model. However, this could have been just a coincidence.

In an experiment to determine the effect of the spacing on repression, the two operators were moved farther apart by inserting extra DNA sequences between them. The results were dramatic and strongly supported the model. The two operators still functioned normally if

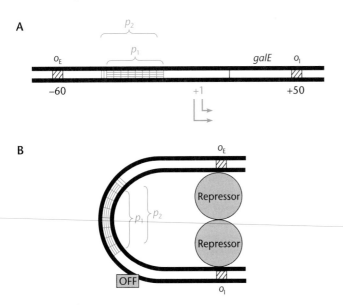

Figure 11.11 A model for the regulation of the galactose operon by two operator sequences. (A) In the presence of galactose, the repressor is not bound to the operators, and the operon is on. (B) In the absence of galactose, a dimer repressor molecule binds simultaneously to both operators, bending the DNA and preventing binding of the RNA polymerase to the promoter region, and the operon is off.

they were moved farther apart, but only if the spacing was increased in multiples of 10 bp. If the spacing was changed to a multiple of 15 bp, the cells were partially constitutive.

Two Gal Promoters and Catabolite Repression of the *gal* Operon

As mentioned, the *gal* operon also has two promoters called p_{G1} and p_{G2} (Figure 11.11). The *gal* operon may have two promoters because the enzymes are needed even when a better carbon source is available, since they are involved in making polysaccharides as well as in utilizing galactose. Like *lac*, the *gal* operon is regulated by catabolite repression so that the transcription of the *gal* genes is repressed if a better carbon source such as glucose is available, but only one of the two *gal* operon promoters is repressed. The other promoter is active even in the presence of glucose and continues making the Gal enzymes. We discuss the differential regulation of the two *gal* promoters in more detail in chapter 12 under catabolite repression.

Regulation of Biosynthetic Operons: Aporepressors and Corepressors

The enzymes encoded by the *lac*, *gal*, and *ara* operons are involved in degrading compounds to obtain catabolites in order to build other molecules. Consequently, these operons are called **catabolic operons** or **degradative operons**. Not all operons are involved in degrading compounds, however. The enzymes encoded by some operons synthesize compounds needed by the cell, such as nucleotides, amino acids, and vitamins. These operons are called **biosynthetic operons**.

The regulation of a biosynthetic operon is essentially opposite to that of a degradative operon. The enzymes of a biosynthetic pathway should *not* be synthesized in the presence of the end product of the pathway, since the product is already available and energy should not be wasted in synthesizing more. However, the mechanisms by which degradative and biosynthetic operons are regulated operate in essentially the same way. Biosynthetic operons can be regulated negatively by repressors. If the genes of a biosynthetic operon are expressed in the absence of the regulatory gene product, even if the compound is present in the medium, the operon is negatively regulated. Biosynthetic operons that are negatively regulated are constitutively expressed if the genes are expressed even if the compound is *present* in the medium.

The terminology we use to describe the negative regulation of biosynthetic operons differs somewhat from that used for catabolic operons, despite shared principles. The effector that binds to the repressor and allows it to bind to the operators is called the corepressor. A repressor that negatively regulates a biosynthetic operon is not active in the absence of the corepressor and in this state is called the **aporepressor**. However, once the corepressor is bound, the protein is able to bind to the operator and so is now called the **repressor**.

THE *trp* OPERON OF *E. COLI*

The tryptophan (*trp*) operon of *E. coli* is the classic example of a biosynthetic operon that is negatively regulated by a repressor. The enzymes encoded by the *trp* operon (Figure 11.13) are responsible for synthesizing the amino acid L-tryptophan, which is a constituent of most proteins and so must be synthesized if none is available in the medium. The products of five structural genes in the operon are required to make tryptophan from chorismic acid. These genes are transcribed from a single promoter, p_{TRP}, shown in Figure 11.13. The *trp* operon is negatively regulated by the TrpR repressor

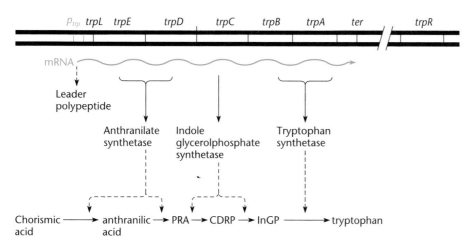

Figure 11.13 Structure of the tryptophan biosynthetic (*trp*) operon of *E. coli*. The structurural genes *trpEDCBA* are transcribed from the promoter p_{trp}. Upstream of the structural genes is a short coding sequence for the leader peptide called *trpL*. The *trpR* repressor gene is unlinked, as shown by the broken line. PRA, phosphoribosyl anthranilate; CDRP, 1-(o-carboxyphenylamino)-1-deoxyribulose-5-phosphate.

protein, whose gene, like the *gal* repressor gene, is unlinked to the rest of the operon.

Figure 11.14 shows the model for the regulation of the *trp* operon by the TrpR repressor. By binding to the operator, the TrpR repressor can prevent transcription from the p_{TRP} promoter. However, the TrpR repressor can bind to the operator only if the corepressor tryptophan is *present* in the medium. The tryptophan binds to the TrpR aporepressor protein and changes its conformation so that it *can bind* to the operator.

Autoregulation of the *trpR* Gene

The TrpR repressor negatively regulates not only the transcription of the *trp* operon but also the transcription of its own gene, *trpR*. In the absence of the TrpR protein, the transcription of the *trpR* gene is about five times higher than in its presence. When the product of a gene regulates the expression of its own gene, it is called **autoregulation**.

It is perhaps surprising that the TrpR repressor would negatively regulate the transcription of its own gene, since there would be less repressor in the cell when tryptophan is present than when tryptophan is absent. It would seem advantageous to have more repressor present when tryptophan is present to better repress the *trp* operon. One possible answer to this riddle is that through negative autoregulation of transcription of the

trpR gene, the cell ensures that repression can be established more quickly if tryptophan suddenly appears in the medium.

Isolation of *trpR* Mutants

Like other negatively regulated operons, constitutive mutations of the *trp* operon are quite common and most map in *trpR*, inactivating the product of the gene. Mutants with constitutive mutations of the *trp* operon can be obtained by selecting for mutants resistant to the tryptophan analog 5-methyltryptophan in the absence of tryptophan. This tryptophan analog also binds to the TrpR repressor and acts as a corepressor. However, 5-methyltryptophan cannot be used in place of tryptophan to make active proteins. Therefore, in the presence of the analog, the *trp* operon will not be induced even in the absence of tryptophan and the cells will starve for this amino acid. Only constitutive mutants that continue to express the genes of the operon in the presence of 5-methyltryptophan can multiply to form colonies on plates with this analog but without tryptophan.

Other Types of Regulation of the *trp* Operon

The *trp* operon is also subject to a completely different type of regulation called **attenuation**. This type of regulation is discussed later in the chapter. Also, as in many biosynthetic pathways, the first enzyme of the *trp* path-

Figure 11.14 Negative regulation of the *trp* operon by the TrpR repressor. See the text for details.

A   In the absence of tryptophan

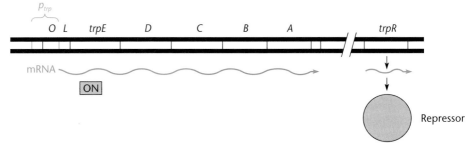

B   In the presence of tryptophan

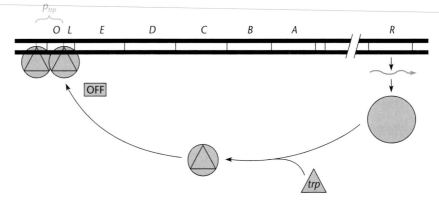

way is subject to **feedback inhibition** by the end product of the pathway, tryptophan. We also return to feedback inhibition later in the chapter.

Positive Regulation

In the first part of the chapter, we covered the classic examples of negative regulation by repressors, but many operons are regulated positively by activators. An operon under the control of an activator protein will be transcribed only in the presence of that protein with the inducer bound (Figure 11.1). We devote the next portion of the chapter to positive regulation in bacteria.

The *E. coli* L-*ara* Operon

The L-*ara* operon was the first example of positive regulation in bacteria to be discovered. The L-*ara* operon, usually called the *ara* operon, is responsible for the utilization of the five-carbon sugar L-arabinose. The genes of this operon are responsible for converting L-arabinose into D-xylulose-5-phosphate, which can be used by other pathways. The *E. coli* bacterium can also utilize D-arabinose, an isomer of L-arabinose, but the enzymes for D-arabinose utilization are encoded by a different operon, which lies elsewhere in the chromosome.

Figure 11.15A illustrates the structure of the *ara* operon. Three structural genes in the operon, *araB*, *araA*, and *araD*, are transcribed from a single promoter, p_{BAD}. Upstream of the promoter is the activator region, *araI*, where the activator protein AraC binds to activate transcription in the presence of L-arabinose, and the CAP site, at which the catabolic activator protein binds. There are also two operators, *araO₁* and *araO₂*, at which the AraC protein binds to repress transcription. The *araC* gene that encodes the regulatory protein is also shown. This gene is transcribed from the promoter p_C in the opposite direction from *araBAD*, as shown by the arrows in the figure. As described below, the AraC protein is a positive activator of transcription. As such, it is a member of a large family of activator proteins (see Box 11.2).

GENETIC EVIDENCE FOR POSITIVE
REGULATION OF THE *ara* OPERON
Early genetic evidence indicated that the *lac* and *ara* operons are regulated by very different mechanisms (see Englesberg et al., Suggested Reading). One observation was that loss of the regulator proteins results in different phenotypes. For example, deletions and nonsense mutations in the the *araC* gene—mutations that presumably inactivate the protein product of the gene—

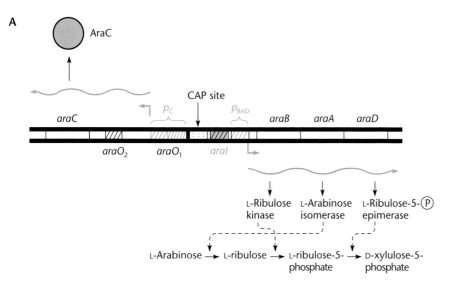

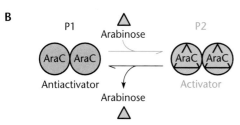

Figure 11.15 (A) Structure and function of of the L-arabinose operon of *E. coli*. (B) Binding of the inducer L-arabinose converts the AraC protein from an antiactivator P1 form to an activator P2 form. See the text for details.

BOX 11.2

Families of Activators

As the sequences of the genes for more and more bacterial repressors and activators become known, it is becoming increasingly obvious that some of them are related to each other, even though they regulate operons of very different functions and respond to different effectors. For example, the *E. coli* activators LysR and CysB, as well as the *S. typhimurium* activator MetR and the ClcR and TfdR/S activators of chlorocatechol-degrading operons on plasmids of *Pseudomonas* spp. and *Alcaligenes* spp., respectively, are all closely related to each other and are members of the LysR family of activators. The AraC family of activators includes many other activator proteins including the XylS activator protein of the Tol plasmid of *P. putida*, described in this chapter. Members of another family of activators are related to the NtrC activator that activates transcription of the *glnA* operon involved in nitrogen regulation (discussed in chapter 12). The NtrC family of activators has been studied most extensively and includes the XylR activator of the *tol* operon and the DmpR activator that activates an operon involved in phenol degradation in *Pseudomonas* sp. strain CF600. These activators function only with σ^{54}-type promoters and seem to be organized in distinguishable domains. The domain of the activator protein that either binds the inducer or is phosphorylated is at the N terminus and the DNA-binding domain is at the C terminus. The middle region of the polypeptide contains a region that interacts with RNA polymerase and has an ATPase activity required for activation, perhaps by allowing dimerization of the activator. Recent work with the DmpR activator indicates that the N-terminal inducer-binding domain masks the ATPase activity unless the inducer is bound to the N-terminal domain (Shingler and Pavel). All of the activators in this family may function in a similar way, with the only difference being that the ATPase domain can be unmasked either by binding an effector to the N-terminal domain or by phosphorylating it.

Hybrid activators, made by fusing the C-terminal DNA-binding domain of one type of activator protein to the N-terminal inducer-binding domain of another activator protein, provide a graphic demonstration of the commonality of activator action. Sometimes, such hybrid activators can still activate the transcription of an operon, but which operon is activated by the hybrid activator depends on the source of the carboxy-terminal DNA-binding domain, while which inducer induces the operon depends on the source of the N-terminal inducer-binding domain. This leads to a situation where an operon is induced by the inducer of a different operon. It is intriguing to think that all activator proteins may have evolved from a few different precursor proteins through simple changes in their effector binding and, to some extent, their DNA-binding regions yet continue to activate the RNA polymerase by the same basic mechanism.

References

Henikoff, S., G. W. Haughn, J. M. Calvo, and J. C. Wallace. 1988. A large family of bacterial activators. *Proc. Natl. Acad. Sci. USA* **85**:6602–6606.

Parek, M. R., S. M. McFall, D. L. Shinabarger, and A. M. Chakrabarty. 1994. Interaction of two LysR-type regulatory proteins CatR and ClcR with heterologous promoters: functional and evolutionary implications. *Proc. Natl. Acad. Sci. USA* **91**:12393–12397.

Schell, M. A. 1993. Molecular biology of the LysR family of transcriptional regulators. *Annu. Rev. Microbiol.* **47**:597–626.

Shingler, V., and H. Pavel. 1995. Direct regulation of the ATPase activity of the transcriptional activator DmpR by aromatic compounds. *Mol. Microbiol.* **17**:505–513.

lead to a "superrepressed" phenotype in which the genes of the operon are not expressed, even in the presence of the inducer arabinose. Recall that deletion or nonsense mutations in the regulatory gene of a negatively regulated operon such as *lac* result in a constitutive phenotype.

Another difference between *ara* and negatively regulated operons is in the frequency of constitutive mutants. Mutants that constitutively express a negatively regulated operon are relatively common because any mutation that inactivates the repressor gene will cause constitutive expression. However, mutants that constitutively express *ara* are very rare, which suggests that

mutations that result in the constitutive phenotype do not inactivate *araC*.

Isolating Constitutive Mutations of the *ara* Operon

Because constitutive mutations of the *ara* operon are so rare, special tricks are required to isolate them. One method for isolating rare constitutive mutations in *araC* uses the anti-inducer D-fucose. This anti-inducer binds to the AraC protein and prevents it from binding L-arabinose, thereby preventing induction of the operon. As a consequence, wild-type *E. coli* cannot multiply to form colonies on agar plates containing D-fucose with L-arabinose as the sole carbon and energy source. Only mutants that constitutively express the genes of the *ara* operon will form colonies under these conditions.

A more clever trick for isolating constitutive mutations in the *ara* operon depends on the presence of the other *E. coli* operon responsible for the utilization of D-arabinose, the isomer of L-arabinose. The enzymes produced by the L- and D-*ara* operons cannot use each other's isomer, with one exception. The product of the *araB* gene—the ribulose kinase enzyme of the L-*ara* operon pathway, which phosphorylates L-ribulose as the second step of the pathway—will also phosphorylate D-ribulose, so that the L-*ara* kinase can substitute for that of the D-*ara* operon. Nevertheless, *E. coli* mutants that lack the D-*ara* kinase cannot multiply to form a colony on plates containing only D-arabinose as a carbon and energy source, because D-arabinose is not an inducer of the L-*ara* operon. Therefore, constitutive mutants with mutations of the L-*ara* operon can be isolated merely by plating D-*ara* kinase-deficient mutants on agar plates containing D-arabinose as the sole carbon and energy source.

A MODEL FOR THE POSITIVE REGULATION OF THE *ara* OPERON
The contrast in phenotypes between mutations that inactivate the *lacI* and *araC* genes led to an early model for the regulation of the *ara* operon. According to this early model, the AraC protein can exist in two states, called P1 and P2. In the absence of the inducer, L-arabinose, the AraC protein is in the P1 state and inactive. If L-arabinose is present, it binds to AraC and changes the protein conformation to the P2 state. In this state, AraC binds to the DNA at the site called *araI* (Figure 11.15A) in the promoter region and activates transcription of the *araB*, *araA*, and *araD* genes.

This early model explained some, but not all, of the behavior of the *araC* mutations. It explained why mutations in *araC* that cause the constitutive phenotype are rare but do occur at a very low frequency. According to this model, these mutations, called *araC^c* mutations, change AraC so that it is permanently in the P2 state, even in the absence of L-arabinose, and thus the operon will always be transcribed. Such mutations would be expected to be very rare because only a few amino acid changes in the AraC protein could specifically change the conformation of the AraC protein to the P2 state.

AraC IS NOT JUST AN ACTIVATOR
One prediction of this model for regulation of the *ara* operon is that *araC^c* mutations should be dominant over the wild-type allele in complementation tests. If

AraC acts solely as an activator, partial diploid cells that have both an *araC^c* allele and the wild-type allele would be expected to constitutively express the *araB*, *araA*, and *araD* genes. In other words, *araC^c* mutations should be dominant over the wild type, since the mutant AraC in the P2 state should activate transcription of *araBAD*, even in the presence of wild-type AraC protein in the P1 state.

The prediction of the model was tested with complementation. An F' factor carrying the wild-type *ara* operon was introduced into cells with an *araC^c* mutation in the chromosome. Figure 11.16 illustrates that the partial diploid cells were inducible, not constitutive, indicating that *araC^c* mutations were recessive rather than dominant. This observation was contrary to the prediction of the model. Therefore, the model had to be changed.

Figure 11.17 illustrates a detailed model recently proposed to explain the recessiveness of *araC^c* mutations (see Johnson and Schleif, Suggested Reading). In this model, the P1 form of the AraC protein that exists in the absence of arabinose is not simply inactive but takes on a new identity as an antiactivator (Figure 11.15B). The P1 state is called an antiactivator rather than a repressor because it does not repress transcription like a classical repressor but, rather, acts to prevent activation by the P2 state of the protein. In the P1 state, the AraC protein preferentially binds to the operator *araO_2* and another site, *araI_1*, bending the DNA between the two sites. Because AraC in the P1 form preferentially binds to *araO_2*, it cannot bind to *araI_2* and activate transcription from the *p_{BAD}* promoter. In the presence of L-arabinose, however, the AraC protein changes to the P2 form and now preferentially binds to *araI_1* and *araI_2*.

This model explains why the *araC^c* mutations are recessive to the wild type in complementation tests (Figure 11.16), because AraC^c in the P2 form can no longer bind to *araI_1* and *araI_2* to activate transcription as long as wild-type AraC in the P1 state is already bound to *araO_2* and *araI_1*. It also explains the behavior of certain deletion mutations, known as the Englesberg deletions, which we have not mentioned yet. These deletions remove the *araO_2* region but leave the *araI_1* and *araI_2* regions intact. In this case, *araC^c* mutations are no longer recessive to the wild-type allele of *araC*. Without *araO_2* to bind to, the AraC protein in the P1 form seems unable to antiactivate transcription of the operon.

AUTOREGULATION OF AraC
The AraC protein not only regulates the transcription of the *ara* operon but also negatively autoregulates its own

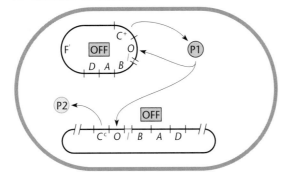

A Absence of arabinose

P1 form binds to operators of both operons, preventing *araBAD* expression

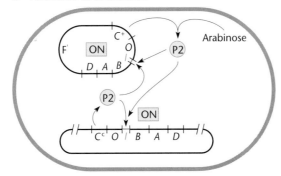

B Presence of arabinose

P2 form and/or the AraC^c (=P2) form binds to *I* (induction) sites of both operons, activating *araBAD* transcription

Figure 11.16 Recessiveness of *araC^c* mutations. See text for details.

transcription. Like TrpR, the AraC protein seems to repress its own synthesis, so that less AraC protein is synthesized in the absence of arabinose than in its presence. However, if the concentration of AraC becomes too high, its synthesis will again be repressed.

Figure 11.17 includes a model for the autoregulation of AraC synthesis. In the absence of arabinose, the interaction of two AraC monomers bound at *araO$_2$* and *araI$_1$* bend the DNA in the region of the *araC* promoter p_C, thereby inhibiting transcription from this promoter. In the presence of arabinose, the AraC protein will no longer be bound to *araO$_2$*, and so the promoter will no longer be bent and transcription from p_C will occur. However, if the AraC concentration becomes too high, the excess AraC protein binds to the operator *araO$_1$*, preventing further transcription of *araC* from the P_C promoter.

CATABOLITE REGULATION OF THE L-*ara* OPERON

The *ara* operon is also regulated through catabolite repression, so the genes for arabinose utilization will not be expressed if the medium contains a better carbon source. The CAP protein that regulates the transcrip-

tion of genes subject to catabolite repression is also a positive activator, like the AraC protein. By binding to the CAP-binding site shown in Figure 11.15A, the CAP protein may help open the loop of DNA created when AraC binds to *araO$_2$* and *araI$_1$*. Opening the loop may prevent AraC from binding to *araO$_2$* and *araI$_1$*, facilitating the binding of AraC to *araI$_1$* and *araI$_2$* and the activation of transcription from p_{BAD}. Thus, the absence of glucose or another carbon source better than arabinose may enhance the transcription of the *ara* operon.

The *E. coli* Maltose Operons

Other well-studied positively regulated operons in bacteria include those for the utilization of the sugar maltose in *E. coli*. Figure 11.18 shows that the genes for maltose transport and metabolism are organized in four clusters at 36, 75, 80, and 91 min on the *E. coli* genetic map (see Schwarz, Suggested Reading). The operon at 75 min has two genes, *malQ* and *malP*, whose products are involved in converting maltose and other polymers into glucose and glucose-1-phosphate once maltose and the polymers are in the cytoplasm. This cluster also includes the regulatory gene *malT*. The *malS* gene at 80

A Absence of L-arabinose

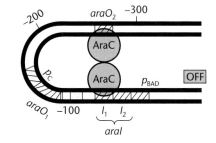

B Presence of L-arabinose

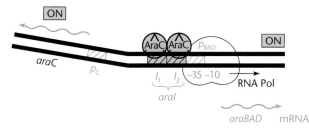

C

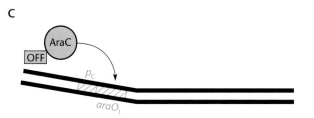

Figure 11.17 A model to explain how AraC can be a positive activator of the *ara* operon in the presence of arabinose and an antiactivator in the absence of this compound, as well as how AraC can negatively autoregulate transcription of its own gene. (A) In the absence of arabinose, AraC molecules in the P1 state preferentially bind to *araI*$_1$ and *araO*$_2$, preventing any P2 form from binding to *araI*$_2$ and activating transcription from p_{BAD}, thereby shutting off the operon. Bending of the DNA between the two sites may also inhibit transcription of the *araC* gene by inhibiting transcription from the p_C promoter. (B) In the presence of arabinose, AraC shifts to the P2 state and preferentially binds to *araI*$_1$ and *araI*$_2$. AraC bound to *araI*$_2$ activates transcription from p_{BAD}. (C) If the AraC concentration becomes too high, it will also bind to *araO*$_1$, thereby repressing transcription from its own promoter p_C.

min encodes an enzyme that breaks down polymers of maltose. The other cluster, at 91 min, has two operons whose gene products can transport maltose into the cell. An operon at 36 min encodes enzymes that degrade polymers of maltose.

Although they allow the cell to use maltose as a carbon source, the more significant function of the products of these operons is probably to enable the cell to

transport and degrade polymers of maltose called **maltodextrins**. These compounds are products of the breakdown of starch molecules, which are very long polysaccharides stored by cells to conserve energy. The sugar maltose is itself a disaccharide composed of two glucose residues with a 1–4 linkage, and the enzymes of the *malP-malQ* operon can break the maltodextrins down into maltose and then into glucose-1-phosphate, which can enter other pathways. Some bacteria, including species of *Klebsiella*, excrete extracellular enzymes that degrade long starch molecules and allow the bacteria to grow on starch as the sole carbon and energy source. *E. coli* lacks some of the genes needed to degrade starch to maltodextrins; therefore, in nature it probably depends on neighboring microorganisms to break the starch down to the maltodextrins so that it can use them as a carbon and energy source.

THE MALTOSE TRANSPORT SYSTEM

Most of the protein products of the *mal* operons are involved in transporting maltodextrins and maltose through the outer and inner membranes (Figure 11.19). The product of the *lamB* gene resides in the outer membrane, where it can bind maltodextrins in the medium. This protein forms a large channel in the outer membrane through which the maltodextrins can pass. LamB is not required for growth on maltose, probably because maltose is small enough to pass through the outer membrane without its help. The LamB protein in the outer membrane also serves as the cell surface receptor for phage λ, and so the gene name is derived from the phage name (*lamB* from lambda). Mutants of *E. coli* resistant to λ have mutations in the *lamB* gene.

Once through the outer membrane, the MalS protein in the periplasm may have to degrade large maltose polymers into smaller ones before they can be transported through the inner membrane. Then the smaller polymer of maltose binds to MalE in the periplasm between the outer and inner membranes. The MalF, MalG, and MalK genes then transport the maltodextrin through the inner membrane.

EXPERIMENTAL USES OF THE *mal* GENES

Because they have been studied so extensively, the *mal* operons have been put to much use in molecular genetics. The ability of the MalE protein, sometimes called the maltose-binding protein (MBP), to tightly bind maltose has made it useful in some biotechnological applications. In some cloning vectors, the *malE* gene is fused to other proteins, making it possible to purify the fusion protein on maltose affinity columns. Because the products of the *E. coli mal* operons are involved in trans-

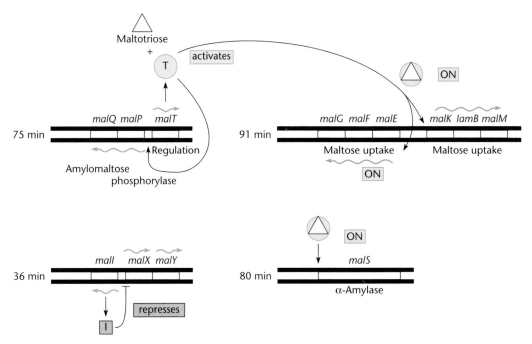

Figure 11.18 The maltose operons in *E. coli*. The MalT activator protein regulates both operons at 75 min and the operons at 80 and 91 min on the *E. coli* map. Another operon at 36 min is also induced by maltose.

porting long molecules into the cell, the *mal* operons have been of particular interest in studies of large-molecule transport systems. Also, many of the proteins encoded by the operons are localized in the inner and outer membranes or the periplasmic space. These proteins must be transported into or through the inner membrane and so serve as models for the study of protein transport through cellular membranes. We shall discuss such experiments in more detail in chapter 14.

REGULATION OF THE *mal* OPERONS

The regulation of the *mal* operons is also illustrated in Figure 11.18. The inducer of the *mal* operons is **maltotriose**, which is composed of three molecules of glucose held together by the maltose linkage. Maltotriose can be synthesized from maltose brought into the cell by some of the enzymes encoded by the operons. Also, the cell normally contains polymers of maltose that were synthesized in the cell from glucose (and so do not need

Figure 11.19 Function of the genes of the maltose regulon in the transport and processing of maltodextrins in *E. coli*. The LamB protein binds maltodextrins and transports them across the outer membrane. The MalE protein in the periplasmic space then passes them through a pore in the cytoplasmic or inner membrane formed by the MalF, MalB, and MalK proteins. Once in the cytoplasm, the maltodextrins and maltose are degraded by MalP and MalQ to glucose-1-phosphate and glucose, respectively. These compounds can then be converted into glucose-6-phosphate for use as energy and carbon sources. Taken from M. Schwartz, *in* F. C. Neidhardt, R. Curtiss III, J. L. Ingraham, E. C. C. Lin, K. B. Low, B. Magasanik, W. S. Reznikoff, M. Riley, M. Schaechter, and H. E. Umbarger (ed.), *Escherichia coli and Salmonella: Cellular and Molecular Biology*, 2nd ed., p. 1484, ASM Press, Washington, D.C., 1987.

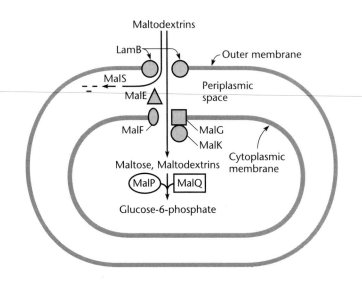

to be transported in) and can be broken down into the inducer, maltotriose. The enzymes that degrade maltose polymers therefore play an indirect role in regulating the operons.

The genes in all three clusters are regulated by an activator encoded by the *malT* gene in the first cluster. The MalT activator protein specifically binds the inducer, maltotriose, and activates transcription of the operons. The activation of the genes of the *mal* regulon by MalT involves the DNA wrapping around many copies of the protein, thereby changing the DNA conformation. The MalT activator is also a member of a large family of activators that includes SoxS, which regulates genes that relieve oxygen toxicity.

The genetic analysis of the regulation of the *mal* operons has been complicated because maltose polymers are natural components of the cell and play many roles including protecting the cell against high osmolarity. In such situations, caution must be exercised in concluding that constitutive mutations are located in regulatory genes. For example, preliminary genetic evidence suggested that the MalK protein is a repressor of the *mal* operons, since *malK* mutants appear to be constitutive for the expression of the other genes of the operons. However, the MalK protein may normally degrade maltotriose in cells that have this enzyme. As a result, the levels of this inducer are higher in *malK* mutants than in wild-type cells, and so *malK* mutants are not truly constitutive. Obviously, understanding the regulation of complex interacting pathways like those involved in maltose metabolism is not straightforward and requires extensive biochemical as well as genetic experimentation.

The *tol* Operons

Many of the bacterial operons of soil bacteria involved in the degradation of cyclic hydrocarbons are also subject to positive regulation. Cyclic hydrocarbons are based on the conjugated ring structure of benzene. Many do not exist naturally and are pesticides and other industrially important manufactured chemicals. Moreover, many of the manufactured chemicals are also chlorinated. Chlorinated compounds were very rare until the advent of modern chemistry, yet despite this short time interval, some types of bacteria have evolved enzymatic pathways to degrade some of these compounds. We have taken advantage of this quick adaptation by using bacteria and other microorganisms to remove contaminating chemicals in a process called **bioremediation**. Understanding the regulation of operons involved in degrading cyclic compounds may lead to yet more rational approaches to bioremediation of toxic waste.

The *tol* operons in plasmid pWWO, originally isolated from the soil bacterium *Pseudomonas putida*, encode enzymes that degrade toluene and the closely related compound xylene. Toluene itself is not chlorinated and has presumably always existed in nature, but operons related to *tol* have been discovered that degrade similar compounds that are chlorinated, including chlorinated catechols.

Figure 11.20 diagrams the *tol* operons. The plasmid contains two, separated by a few thousand base pairs of DNA. The first operon encodes the enzymes of the "upper pathway," which converts toluene into benzoate; the other operon encodes enzymes of the "lower pathway," which breaks the ring of benzoate and degrades it to intermediates of the tricarboxylic acid (TCA) cycle to be used for energy and to make carbon-containing compounds.

REGULATION OF THE *tol* OPERONS

The regulation of the *tol* operons is also illustrated in Figure 11.20. The regulation of the upper and lower *tol* operons is both coordinated and independent. Both operons must be coordinately turned on if toluene is present in the medium, since both operons are required to degrade toluene to TCA cycle intermediates. However, only the lower operon should be turned on if only

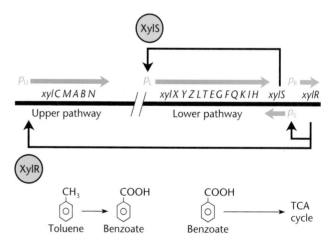

Figure 11.20 The structure of the *tol* operons of the Tol plasmid pWWO of *P. putida*. The upper pathway converts toluene (or xylene) to benzoate. The lower encodes a meta cleavage pathway that converts the benzoate to intermediates of the TCA cycle. The two operons are separated by a few thousand base pairs of DNA, as indicated by the broken line. The promoters activated by each of the activator proteins XylS and XylR are indicated by the arrows and the direction of mRNA transcription is indicated by the blue arrows. p_U is the promoter for the upper operon; p_L is the promoter for the lower operon.

benzoate is present in the medium, since there is no need to induce the upper pathway under these conditions.

The coordinate regulation of the upper and lower *tol* operons in the presence of toluene is achieved through two activators, one of which activates transcription of the other's gene. The activator for the upper operon is XylR, a member of the NtrC family of activators, and the activator for the lower pathway is XylS, a member of the LysR family of activators (see Box 11.2). In the presence of toluene (or xylene), the XylR activator activates transcription of the upper operon from the promoter called p_U. Toluene is thus degraded into benzoate, but not enough benzoate is produced to bind to XylS and activate transcription of the lower operon. However, the XylR activator also activates transcription of the *xylS* gene from the promoter p_S. At higher concentrations, XylS can activate transcription of the lower operon even without benzoate being bound. If, however, high levels of benzoate are present in the medium, the benzoate will bind to XylS and induce the lower pathway; much less XylS is required for this type of activation. This regulatory interaction ensures that both the upper and lower pathways will be induced if toluene is present but only the lower pathway will be induced in the presence of benzoate alone.

The regulation of the *tol* operons and their regulatory genes is similar to the regulation of the genes of other catabolic pathways in bacteria. For instance, the XylR protein also regulates its own transcription; that is, it is transcriptionally autoregulated, like AraC. Therefore, in the presence of toluene, more XylR will be present to activate transcription of the other genes. Another similarity is that the promoters p_U and p_L are recognized by an RNA polymerase containing the alternate sigma factor σ^{54}, the same sigma factor used to transcribe the nitrogen-regulated genes discussed in chapter 12. Why these operons should use an alternate sigma factor is not known, but the expression of operons in bacteria as distantly related as the nitrogen-regulated genes in *E. coli* and the toluene-degrading operons in *P. putida* share this feature.

GENETICS OF THE *tol* OPERONS

The above picture of the organization of the genes of the *tol* operons came from molecular genetic experiments. These studies of the organization and regulation of the *tol* operons were greatly aided by the fact that plasmid pWWO carrying the *tol* operons is a broad-host-range, self-transmissible plasmid that can transfer itself from the original *P. putida* strain into *E. coli*, where more sophisticated genetic tests have been developed. Once in *E. coli*, the plasmid could be easily muta-

genized with transposon Tn5 and then transferred back into *P. putida* to determine whether a particular transposon insertion inactivates a gene required for growth on toluene. Insertions that inactivate *tol* genes could then be located by restriction endonuclease mapping as discussed in chapter 16. In this way, the maps of the *tol* operons shown in Figure 11.20 were obtained.

Molecular genetic tests also yielded the picture for the regulation of the *tol* operons outlined above. The XylS protein was first identified as a positive activator of the lower pathway because clones that did not include the *xylS* gene failed to express the genes of the lower pathway, even in the presence of benzoate. Similarly, clones of the upper pathway that excluded the *xylR* gene did not express the upper-pathway operon, and *xylS* was transcribed at a lower rate if *xylR* was missing. These observations implicated XylR as a positive activator of the upper operon and the *xylS* gene. Clones that overexpress XylS turn on the lower pathway, even in the absence of benzoate, leading to the model that XylS can activate transcription of the lower pathway in the absence of benzoate, provided that XylS is present at a high concentration.

USING SELECTIONAL GENETICS TO BROADEN THE RANGE OF INDUCERS OF THE *tol* LOWER OPERON

A promising avenue of research is to use genetic selections to alter known pathways so that they can use alternate substrates. For example, while the *tol* lower pathway can degrade some substituted forms of benzoate, including 3-methylbenzoate and 4-methylbenzoate, and allow the cell to use them as carbon and energy sources, it cannot use other derivatives of benzoate such as 4-ethylbenzoate. This particular substrate cannot be used because the second enzyme of the pathway cannot use it as a substrate and because it does not function as an inducer of the operon. Even if the enzymes encoded by an operon can degrade a derivative of the normal substrates for the operon, the enzymes will not be present if the derivative does not function as an inducer of the pathway.

Gene fusion techniques were used to select *xylS* mutants that can use 4-ethylbenzoate and other derivatives of benzoate as an inducer of the lower *tol* operon (see Ramos et al., Suggested Reading). These studies could be performed in *E. coli* because the XylS protein can activate transcription of the lower operon even in *E. coli*. In these experiments, the promoter p_L for the lower operon was fused to a tetracycline resistance gene on a plasmid so that the tetracycline resistance gene would not be transcribed unless the p_L promoter was activated.

This plasmid was then used to transform *E. coli* containing a second compatible plasmid expressing the XylS protein. When a particular derivative of benzoate caused XylS to activate transcription from the p_L promoter, the cells became tetracycline resistant (Tet^r) and grew on plates containing tetracycline. However, when the derivative did not induce the operon, the cells remained tetracycline sensitive (Tet^s). As expected, benzoate made the cells Tet^r. Moreover, some derivatives of benzoate, including 2-chlorobenzoate, functioned well as inducers, making the cells Tet^r. However, other derivatives, such as 4-ethylbenzoate and 2,4-dichlorobenzoate, were not inducers and the cells remained Tet^s.

Selectional genetics was used to try to isolate mutants with altered XylS proteins in which 4-ethylbenzoate or similar noninducing derivatives could function as inducers. The bacteria described above containing the two plasmids were mutagenized, and large numbers were spread on plates containing tetracycline and a potential inducer, for example, 4-ethylbenzoate. Most of the bacteria did not multiply to form a colony, but a few colonies of Tet^r mutants appeared. Some of these were constitutive mutants that were Tet^r even in the absence of inducer, and so they were discarded. However, some had mutations changing the XylS protein so that it could use the new inducer. Mutants with *xylS* mutations that allowed induction by 4-ethylbenzoate were separated from other, unwanted types by isolating the *xylS*-containing plasmid and transforming it into new bacteria containing the *tet* fusion. Only if the mutation was in the plasmid containing the *xylS* gene, and therefore presumably in the *xylS* gene itself, were transformants Tet^r in the presence of ethylbenzoate. The mutation was shown to be in the *xylS* gene and not somewhere else in the plasmid by recloning the *xylS* gene into a new plasmid and showing that this plasmid also confers the Tet^r phenotype in the presence of 4-ethylbenzoate. Finally, the mutated *xylS* gene could be sequenced to determine what amino acid changes in the XylS protein can allow it to use 4-ethylbenzoate as an inducer. The success of these experiments with 4-ethylbenzoate and other benzoate derivatives revealed that the inducer specificity of activator proteins can sometimes be changed by simple mutations. This may be the origin of families of activators (see Box 11.2).

Regulation by Attenuation of Transcription

In the above examples, the transcription of an operon is regulated through the initiation of RNA synthesis at the promoter of the operon. However, this is not the only known means of regulating operon transcription. Another mechanism is the **attenuation of transcription**. Unlike repressors and activators, which turn on or off transcription from the promoter, the attenuation mechanism works by terminating transcription—which begins normally—before the RNA polymerase reaches the first structural gene of the operon. The classic examples of regulation by attenuation are the *his* and *trp* operons of *E. coli*. Closely related mechanisms regulate such *E. coli* biosynthetic operons as leucine (*leu*), phenylalanine (*phe*), threonine (*thr*), and isoleucine-valine (*ilv*) and the *Bacillus subtilis* tRNA synthetase genes (see Box 11.3). In this section, we discuss the regulation of the *trp* operon by attenuation.

Genetic Evidence for Attenuation

As discussed earlier in the chapter, the *trp* operon is negatively regulated by the TrpR repressor protein. However, early genetic evidence suggested that this is not the only type of regulation for *trp*. If the *trp* operon were regulated solely by the TrpR repressor, the levels of the *trp* operon enzymes in a *trpR* mutant would be the same in the absence and the presence of tryptophan. However, even in a *trpR* null mutant, the expression of these enzymes is higher in the absence of tryptophan than in its presence, indicating that the *trp* operon is subject to another regulatory system.

Early evidence suggested that tRNA^{Trp} plays a role in the regulation of the *trp* operon in the absence of TrpR (see Landick and Yanofsky and Morse and Morse, Suggested Reading). Mutations in the trytophanyl-tRNA synthetase (the enzyme responsible for transferring tryptophan to tRNA^{Trp}) and mutations in the structural gene for the tRNA^{Trp}, as well as mutations in genes whose products are responsible for modifying the tRNA^{Trp} increase the expression of the operon. All these mutations presumably lower the amount of aminoacylated-tRNA^{Trp} in the cell, suggesting that this other regulatory mechanism is not sensing the amount of free tryptophan in the cell but, rather, the amount bound to the tRNA^{Trp}.

Other evidence suggested that the region targeted by this other type of regulation is not the promoter but a region downstream of the promoter called the **leader region**, or *trpL* (Figures 11.14 and 11.21). Deletions in this region, which lies between the promoter and *trpE*, the first gene of the operon, eliminate the regulation. Double mutants with both a deletion mutation of the leader region and a *trpR* mutation are completely constitutive for expression of the *trp* operon. Deletions of the leader region are also *cis* acting and affect only the expression of the *trp* operon on the same DNA. Later

Regulation by Attenuation: the Aminoacyl-tRNA Synthetase Genes of *Bacillus subtilis*

Several bacterial operons are now known to be regulated by some form of attenuation and antitermination of transcription. Many of these use mechanisms different from those described for the *trp* and *his* operons. Some types of attenuation control use *trans*-acting proteins to regulate transcription through termination signals. Examples include the *bgl* operon in *E. coli* and the *trp* operon in *Bacillus subtilis*. In these operons, regulatory proteins bind to the leader RNA and prevent the formation of secondary structures, thereby enhancing or reducing transcription termination at downstream termination signals.

One of the most dramatic examples of regulation through attenuation is the regulation of the transcription of the genes for the aminoacyl-tRNA synthetase genes in *B. subtilis*. Bacteria synthesize higher levels of aminoacyl-tRNA synthetases in response to amino acid deprivation. The higher levels of the synthetases presumably allow more efficient attachment of the amino acids to their cognate tRNAs.

In *B. subtilis*, the synthetase genes are regulated by attenuation of transcription. In high concentrations of the amino acid, transcription of the synthetase gene often terminates in the leader region, so that less synthetase is synthesized. If the amino acid for that synthetase is limiting, however, transcription terminates less often in the leader sequence, and more synthetase is made.

Whether or not transcription terminates in the leader sequence of each synthetase gene is determined by the relative levels of the unaminoacylated cognate tRNA for that synthetase. At high levels of the amino acid, most of the tRNA will have its amino acid attached (i.e., be aminoacylated). However, at low concentrations of the amino acid, more of the tRNA will lack its amino acid (i.e., be unaminoacylated). The anticodon of the unaminoacylated tRNA can bind to a strategically placed codon for that amino acid upstream of the transcription terminator in the leader region. This converts the region of the codon into an antiterminator, allowing increased transcription of the synthetase gene. In this way, the synthetase genes can all be regulated by the same mechanism, but each synthetase gene will respond only to levels of its own cognate amino acid.

Reference
Grundy, F. J., and T. M. Henkin. 1993. tRNA as a positive regulator of transcription antitermination in *B. subtilis*. *Cell* **74**:475–482.

evidence indicated that transcription terminated in this leader region in the presence of tryptophan because of an excess of aminoacylated-tRNA^{Trp}. Because the regulation seemed to be able to stop, or attenuate, transcription that had already initiated at the promoter, it was called attenuation of transcription, in agreement with an analogous type of regulation already discovered for the *his* operon.

MODEL FOR REGULATION OF THE *trp* OPERON BY ATTENUATION
Figures 11.21 and 11.22 illustrate a current model for regulation of the *trp* operon by attenuation (see Yanofsky and Crawford, Suggested Reading). According to this model, the percentage of the tRNA^{Trp} that is aminoacylated (i.e., has tryptophan attached) determines which of several alternative secondary-structure hairpins will form in the leader RNA. Figure 11.21 shows that four RNA regions can form three different hairpins. Recall from chapter 2 that that the secondary structure of an RNA, or a hairpin, results from complementary pairing between the bases in RNA transcribed from inverted repeated sequences.

Whether or not transcription termination occurs depends on whether the attenuation mechanism senses relatively low or high levels of tryptophan. The *trpL* region, which contains two adjacent *trp* codons, provides the signal. The *trp* codons are there to allow the ribosome to test the water before the RNA polymerase is allowed to plunge into the structural genes of the operon. If levels of tryptophan are low, the levels of tryptophanyl-tRNA^{Trp} (tRNA^{Trp} with tryptophan attached) will also be low. When a ribosome encounters one of the *trp* codons, it will temporarily stall, unable to insert the amino acid. This stalled ribosome in the *trpL* region therefore communicates that the tryptophan concentration is low and that transcription should continue (Figure 11.22).

Figures 11.21 and 11.22 also show how the hairpins operate in attenuation. Four different regions in the *trpL* leader RNA—regions 1, 2, 3, and 4—can form three different hairpins, 1:2, 2:3, and 3:4, as shown in Figure 11.21. The formation of hairpin 3:4 causes RNA polymerase to terminate transcription because this hairpin is part of a factor-independent transcription termination signal (see chapter 2).

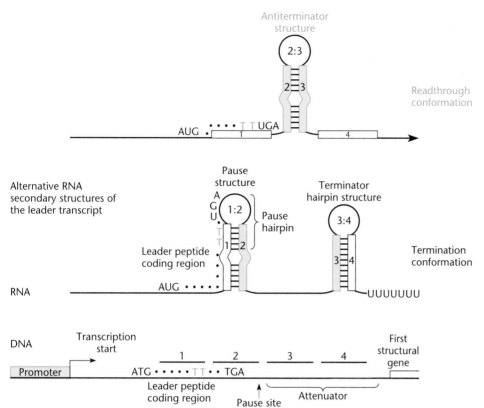

Figure 11.21 Structure and relevant features of the leader region of the *trp* operon involved in regulation by attenuation. TT (in blue) indicates the two *trp* codons in the leader region. See the text for details. Redrawn from R. Landick, C. Turnbough, Jr., and C. Yanofsky, *in* F. C. Neidhardt, R. Curtiss III, J. L. Ingraham, E. C. C. Lin, K. B. Low, B. Magasanik, W. S. Reznikoff, M. Riley, M. Schaechter, and H. E. Umbarger (ed.), *Escherichia coli and Salmonella: Cellular and Molecular Biology,* 2nd ed., ASM Press, Washington, D.C., 1996.

Whether hairpin 3:4 forms is determined by the dynamic relationship between ribosomal translation of the *trp* codons in the *trpL* region and the progress of RNA polymerase through the *trpL* region (also illustrated in Figure 11.22). After RNA polymerase initiates transcription at the promoter, it moves through the *trpL* region to a site located just after region 2, where it pauses. The hairpin formed by mRNA regions 1 and 2 is an important part of the signal to pause. The pause is short, probably less than 1 s, but it ensures that a ribosome has time to load on the mRNA before the RNA polymerase proceeds to region 3. The moving ribosome may help release the paused RNA polymerase by colliding with it.

The process of ribosome translation through the *trp* codons of *trpL* then determines whether hairpin 3:4 will form, causing termination, or 2 will pair instead with 3, preventing formation of the 3:4 hairpin. Region 3 will pair with region 2 if the ribosome stalls at the *trp* codons because of low tryptophan concentrations (Figure 11.22D). If the ribosome does not stall at the *trp* codons, it will continue until it reaches the UGA stop codon at the end of *trpL*. By remaining at the stop codon while region 4 is synthesized, the ribosome will prevent hairpin 2:3 from forming. Therefore hairpin 3:4 can form and terminate transcription (Figure 11.22C).

GENETIC EVIDENCE FOR THE MODEL

No model is satisfactory unless it is supported by experimental evidence. The existence and in vivo functioning of hairpin 2:3 were supported by the phenotypes produced by mutation *trpL75*. This mutation, which changes one of the nucleotides and prevents pairing of two of the bases holding the hairpin together, should destabilize the hairpin. In the *trpL75* mutant, transcription terminates in the *trpL* region, even in the absence of tryptophan, consistent with the model that formation of hairpin 2:3 normally prevents formation of hairpin 3:4.

That translation of the leader peptide from the *trpL* region is essential to the regulation is supported by the phenotypes of mutation *trpL29*, which changes the AUG start codon of the leader peptide to AUA, preventing initiation of translation. In *trpL29* mutants, termination also occurs even in the absence of tryptophan. The model also explains this observation as long as we can assume that the RNA polymerase paused at hairpin coding sequence 1:2 will eventually move on, even without a translating ribosome to nudge it, and will eventually transcribe the 3:4 region. Without a ribosome stalled at the *trp* codons, however, hairpin 1:2 will persist, preventing the formation of hairpin 2:3. If hairpin 2:3 does not form, hairpin 3:4 will form and transcription will terminate.

A RNA Pol pauses at 1:2 pause site

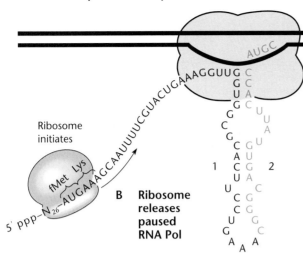

Ribosome
initiates

B **Ribosome releases paused RNA Pol**

Figure 11.22 Details of regulation by attenuation in the *trp* operon. (A) RNA polymerase pauses after transcribing regions 1 and 2. (B) A ribosome has time to load on the mRNA and begin translating, eventually reaching the RNA polymerase and bumping it off the pause site. (C) In the presence of tryptophan, the ribosome translates through the *trp* codons and prevents the formation of hairpin 2:3, thereby allowing the formation of hairpin 3:4, which is part of a transcription terminator. Transcription terminates. (D) In the absence of tryptophan, the ribosome stalls at the *trp* codons, and hairpin 2:3 forms, preventing the formation of hairpin 3:4 and allowing transcription to continue through the terminator. See the text for more details.

C Attenuation in presence of Trp

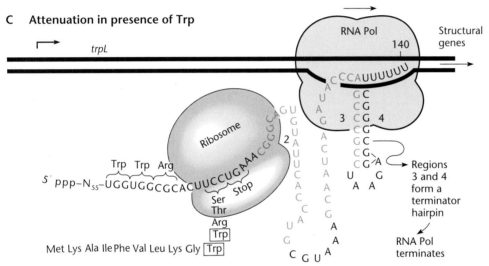

D Transcription elongation in absence of Trp

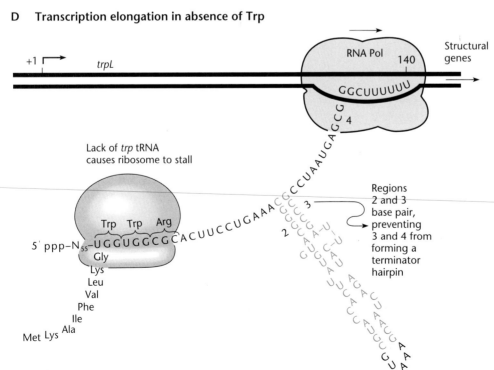

One final prediction of the model is that stopping translation at other codons in *trpL* should also relieve attenuation. The codon immediately downstream of the second tryptophan codon in the *trpL* region is for arginine. Starving the cells for arginine also prevents attenuation of the *trp* operon, fulfilling this prediction of the model.

Feedback Inhibition

Biosynthetic pathways are not regulated solely through transcriptional regulation of their operons; they are also often regulated by feedback inhibition of the enzymes once they are made. In feedback inhibition, the end product of a pathway binds to the first enzyme of the pathway, inhibiting its activity. Feedback inhibition is common to many types of biosynthetic pathways and is a more sensitive and rapid mechanism for modulating the amount of the end product than is transcriptional regulation, which responds only slowly to changes in the concentration of the end product of the pathway.

Tryptophan Operon

The Trp biosynthetic pathway is subject to feedback inhibition. Tryptophan binds to the first enzyme of the Trp synthesis pathway, anthranilate synthetase, and inhibits it, thereby blocking the synthesis of more tryptophan. We can isolate mutants defective in feedback inhibition of tryptophan biosynthesis by using the tryptophan analog 5-methyltryptophan to study this process. At high concentrations, 5-methyltryptophan will bind to anthranilate synthetase in lieu of tryptophan and inhibit the enzyme's activity, starving the cells for tryptophan. Only mutants defective in feedback inhibition because of a missense mutation in the *trpE* gene that prevents the binding of tryptophan (and 5-methyltryptophan) to the anthranilate synthetase enzyme will multiply to form a colony.

A similar method was described earlier for isolating constitutive mutants with mutations of the *trp* operon, but selection of constitutive mutants requires lower concentrations of 5-methyltryptophan. If the concentration of this analog is high enough, even constitutive mutants will be starved for tryptophan.

Isoleucine-Valine Operon

Feedback inhibition is also responsible for valine sensitivity of *E. coli*. If *E. coli* cells are presented with high concentrations of valine, they will die, as long as isoleucine is not provided in the medium. The reason is that valine and isoleucine are synthesized by the same pathway, encoded by the *ilv* (isoleucine-valine) operon. The first enzyme of the pathway, acetohydroxy acid synthase, is feedback inhibited by valine, so that if the concentration of valine is high, the cells can make neither valine nor isoleucine. The cells will then starve for isoleucine unless this amino acid is provided in the medium. Such a situation seldom occurs in nature, since degraded proteins are the usual source of amino acids and since isoleucine and valine are two of the most common amino acids and so are present in most proteins.

While most *E. coli* strains are valine sensitive, mutants that are valine resistant are easily isolated by plating *E. coli* in the presence of high concentrations of valine with no isoleucine. Any colonies that arise are due to the multiplication of valine-resistant mutants. These mutants are about as frequent as mutants resistant to the tryptophan analog 5-methyltryptophan, and a priori one might assume they had the same molecular basis, in this case, an altered acetohyroxy acid synthase that is still active but is no longer feedback inhibited by valine. However, mutations to valine resistance are often revertants of a mutation that normally inactivates a gene for another acetohydroxy acid synthase that is not feedback inhibited by valine and so performs the first step in the synthesis of isoleucine, regardless of the valine concentration.

SUMMARY

1. Regulation of gene expression can occur at any stage in the expression of a gene. If the amount of mRNA synthesized from the gene differs under different conditions, the gene is transcriptionally regulated. If the regulation occurs after the mRNA is made, the gene is posttranscriptionally regulated. A gene is translationally regulated if the mRNA is made but not always translated at the same rate.

2. In bacteria, more than one gene is sometimes transcribed into the same mRNA. Such a cluster of genes, along with their adjacent controlling sites, is called an operon.

3. The regulation of operon transcription can be negative, positive, or a combination of the two. If a protein blocks the transcription of the operon, the operon is negatively regulated and the regulatory protein is a repressor. If a protein is required for transcription of an operon, the operon is positively regulated and the regulatory protein is an activator.

4. If an operon is negatively regulated, mutations that inactivate the regulatory gene product will result in constitutive mutants in which the operon genes are always expressed. If

(continued)

SUMMARY (continued)

the operon is positively regulated, mutations that inactivate the regulatory protein will cause permanent repression of the expression of the operon. Therefore in general, constitutive mutations are much more common with negatively regulated operons than with positively regulated operons.

5. Sometimes the same protein can be both a repressor and an activator in different situations, which complicates the analysis of the regulation.

6. The regulation of transcription of bacterial operons is often achieved through small molecules called effectors, which bind to the repressor or activator protein, changing its conformation. If the presence of the effector causes the operon to be transcribed, it is often called an inducer; if its presence blocks trancription of the operon, it is called a corepressor. The substrates of catabolic operons are usually inducers, whereas the end products of biosynthetic pathways are usually corepressors.

7. The regions on DNA to which repressors bind are called operators. Some repressors seem to act by binding to two operators simultaneously, bending the DNA between them and inactivating the promoter.

8. The regions to which activator proteins bind are called activator sequences. The activator proteins seem to activate transcription by binding directly to the RNA polymerase at the promoter. Interaction with the activator protein changes the conformation of the RNA polymerase so that it can initiate transcription from the promoter.

9. Some regulatory proteins seem to bind to different regions on DNA depending upon whether the effector is present, in this way acting as both repressors and activators.

10. Some operons are transcriptionally regulated by a mechanism called attenuation. In operons regulated by attenuation, transcription will begin on the operon but then terminate after a short leader sequence has been transcribed if the enzymes encoded by the operon are not needed. Attenuation of biosynthetic and degradative operons is sometimes determined by whether certain codons in the leader sequence are translated. Pausing of the ribosome at these codons can cause secondary structure changes in the leader mRNA, leading to termination of transcription by RNA polymerase before it reaches the first gene of the operon.

11. The activity of biosynthetic operons is often not regulated solely through the transcription of the genes of the operon. It may also be regulated through reversible regulation of the activity of the enzymes of the pathway. This reversible regulation, called feedback inhibition, usually results from binding of the end product of the biosynthetic pathway to the first enzyme of the pathway.

QUESTIONS FOR THOUGHT

1. Why do you suppose both negative and positive mechanisms of transcriptional regulation are used to regulate bacterial operons?

2. Why are regulatory protein genes sometimes autoregulated?

3. What advantages or disadvantages are there to regulation by attenuation? Would you expect operons other than amino acid biosynthetic/degradative operons to be regulated by attenuation? Why or why not?

PROBLEMS

1. Outline how you would isolate a *lacI*^s mutant in *E. coli*.

2. Is the AraC protein in the P1 or P2 state with D-fucose bound?

3. The *phoA* gene of *E. coli* is turned on only if phosphate is limiting in the medium. What kind of genetic experiments would you do to determine whether the *phoA* gene is positively or negatively regulated?

4. Outline how you would use 5-methyltryptophan to isolate constitutive mutants of the *trp* operon. Then use these to isolate feedback inhibition mutants.

SUGGESTED READING

Choy, H., and S. Adhya. 1996. Negative control, p. 1287–1299. *In* F. C. Neidhardt, R. Curtiss III, J. L. Ingraham, E. C. C. Lin, K. B. Low, B. Magasanik, W. S. Reznikoff, M. Riley, M. Schaechter, and H. E. Umbarger (ed.), *Escherichia coli and Salmonella: Cellular and Molecular Biology*, 2nd ed. ASM Press, Washington, D.C.

Englesberg, E., C. Squires, and F. Meronk. 1969. The arabinose operon in *Escherichia coli* B/r: a genetic demonstration of two functional states of the product of a regulator gene. *Proc. Natl. Acad. Sci. USA* **62**:1100–1107.

Irani, M. H., L. Orosz, and S. Adhya. 1983. A control element within a structural gene: the *gal* operon of *Escherichia coli*. *Cell* **32**:783–788.

Jacob, F., and J. Monod. 1961. Genetic regulatory mechanisms in the synthesis of proteins. *J. Mol. Biol.* **3**:318–356.

Johnson, C. M., and R. F. Schleif. 1995. *In vivo* induction kinetics of the arabinose promoters in *Escherichia coli. J. Bacteriol.* **177**:3438–3442.

Landick, R., C. Turnbough, Jr., and C. Yanofsky. 1996. Transcription attenuation, p. 1263–1286. *In* F. C. Neidhardt, R. Curtiss III, J. L. Ingraham, E. C. C. Lin, K. B. Low, B. Magasanik, W. S. Reznikoff, M. Riley, M. Schaechter, and H. E. Umbarger (ed.), *Escherichia coli and Salmonella: Cellular and Molecular Biology*, 2nd ed. ASM Press, Washington, D.C.

Landick, R., and C. Yanofsky. 1987. Transcription attenuation, p. 1276–1301. *In* F. C. Neidhardt, R. Curtiss III, J. L. Ingraham, E. C. C. Lin, K. B. Low, B. Magasanik, W. S. Reznikoff, M. Riley, M. Schaechter, and H. E. Umbarger (ed.), *Escherichia coli and Salmonella: Cellular and Molecular Biology*, 2nd ed. ASM Press, Washington, D.C.

Morse, D. E., and A. N. C. Morse. 1976. Dual control of the tryptophan operon is mediated by both tryptophanyl-tRNA synthetase and the repressor. *J. Mol. Biol.* **103**:209–226.

Oxender, D. L., G. Zurawski, and C. Yanofsky. 1979. Attenuation in the *Escherichia coli* tryptophan operon. Role of RNA secondary structure invoving the tryptophan codon region. *Proc. Natl. Acad. Sci. USA* **76**:5524–5528.

Ramos, J. L., C. Michan, F. Rojo, D. Dwyer, and K. Timmis. 1990. Signal-regulator interactions: genetic analysis of the effector binding site of *xylS*, the benzoate-activated positive regulator of *Pseudomonas* Tol plasmid *meta*-cleavage pathway operon. *J. Mol. Biol.* **211**:373–382.

Schleif, R. 1996. Two positively regulated systems, *ara* and *mal*, p. 1300–1309. *In* F. C. Neidhardt, R. Curtiss III, J. L. Ingraham, E. C. C. Lin, K. B. Low, B. Magasanik, W. S. Reznikoff, M. Riley, M. Schaechter, and H. E. Umbarger (ed.), *Escherichia coli and Salmonella: Cellular and Molecular Biology*, 2nd ed. ASM Press, Washington, D.C.

Schwarz, M. 1987. The maltose operon, p. 1482–1502. *In* F. C. Neidhardt, R. Curtiss III, J. L. Ingraham, E. C. C. Lin, K. B. Low, B. Magasanik, W. S. Reznikoff, M. Riley, M. Schaechter, and H. E. Umbarger (ed.), *Escherichia coli and Salmonella: Cellular and Molecular Biology*, 2nd ed. ASM Press, Washington, D.C.

Summers, A. O. 1992. Untwist and shout: a heavy metal-responsive transcriptional regulator. *J. Bacteriol.* **174**:3097–3101.

Yanofsky, C., and I. P. Crawford. 1987. The tryptophan operon, p. 1453–1472. *In* F. C. Neidhardt, R. Curtiss III, J. L. Ingraham, E. C. C. Lin, K. B. Low, B. Magasanik, W. S. Reznikoff, M. Riley, M. Schaechter, and H. E. Umbarger (ed.), *Escherichia coli and Salmonella: Cellular and Molecular Biology*, 2nd ed. ASM Press, Washington, D.C.

Global Regulatory Mechanisms

BACTERIA MUST BE ABLE TO ADAPT to a wide range of environmental conditions to survive. Nutrients are usually limiting, so bacteria must be able to protect themselves against starvation until an adequate food source becomes available. Different environments also vary greatly in the amount of water or in the concentration of solutes, so bacteria must also be able to adjust to desiccation and differences in osmolarity. Temperature fluctuations are also a problem for bacteria. Unlike humans and other warm-blooded animals, bacteria cannot maintain their own cell temperature and so must be able to function over wide ranges of temperature.

Mere survival is not enough for a species to prevail, however. The species must compete effectively with other organisms in the environment. Competing effectively might mean being able to efficiently use scarce nutrients or taking advantage of plentiful ones to achieve higher growth rates and thereby become a higher percentage of the total population of organisms in the environment. Moreover, different compounds may be available for use as carbon and energy sources. The bacterium may need to choose the carbon and energy source it can use most efficiently and ignore the rest, so that it does not waste energy making extra enzymes.

Conditions not only vary in the environment, but their changes can be abrupt. The bacterium may have to adjust the rate of synthesis of its cellular constituents quickly in response to the change in growth conditions. For example, different carbon and energy sources allow different rates of bacterial growth. Different growth rates require different rates of synthesis of cellular macromolecules such as DNA, RNA, and proteins, which in turn require different concentrations of the components of the cellular macromolecular synthesis machinery such as ribosomes, tRNA, and RNA polymerase. Moreover, the relative rates of synthesis of the different cellular components

297

must be coordinated so that the cell does not accumulate more of some components than it needs.

Adjusting to major changes in the environment requires regulatory systems that simultaneously regulate numerous operons. These systems are called **global regulatory mechanisms**. Often in global regulation, a single regulatory protein controls a large number of operons, which are then said to be members of the same **regulon**.

Table 12.1 lists some global regulatory mechanisms known to exist in *Escherichia coli*. If the genes are under the control of a single regulatory gene (and so are members of the same regulon), the regulatory gene is also listed. We have discussed some examples of regulons in previous chapters. For example, all the genes under the control of the TrpR repressor, including the *trpR* gene itself, are part of the TrpR regulon. The Ada regulon comprises the adaptive response genes, including those encoding the methyltransferases that repair alkylation damage to DNA (see chapter 10); all of these genes are under the control of the Ada protein. Similarly, the SOS genes induced after UV irradiation and some other types of DNA-damaging treatments are all under the control of the same protein, the LexA repressor, and so are part of the LexA regulon. In other cases, the molecular basis of the global regulation is less well understood and may involve a complex interaction between several cellular signals.

In this chapter, we discuss what is known about how some global regulatory mechanisms operate on the molecular level and describe some of the genetic experiments that have contributed to this knowledge. The ongoing studies on the molecular basis of global regulatory mechanisms represent one of the most active areas of research involving bacterial molecular genetics; therefore, new developments will probably have occurred in many of these areas by the time you read this book.

Catabolite-Sensitive Operons

One of the largest global regulatory systems in bacteria coordinates the expression of genes involved in carbon and energy source utilization. All cells must have access to high-energy, carbon-containing compounds, which they degrade to generate ATP for energy and smaller molecules needed for cellular constituents. Smaller molecules resulting from the metabolic breakdown of larger molecules are called **catabolites**.

In times of plenty, bacterial cells may be growing in the presence of several different carbon and energy sources, some of which can be used more efficiently than others. Energy must be expended to synthesize the enzymes needed to metabolize the different carbon sources, and the utilization of some carbon compounds requires

more enzymes than does the utilization of others. By making only the enzymes for the carbon and energy source that yields the highest return, the cell will get the most catabolites and energy, in the form of ATP, for the energy it expends. The mechanism for ensuring that the cell will preferentially use the best carbon and energy source available is called **catabolite repression**, and operons subject to this type of regulation are **catabolite sensitive**. The name "catabolite repression" originates from the fact that cells growing in better carbon sources have more catabolites, which seem to repress the transcription of operons for the utilization of poorer carbon sources. However, as we shall see, the name "catabolite repression" is a misnomer because, at least in some of the regulatory systems in *E. coli*, the genes under catabolite control are *activated* when poorer carbon sources are the only ones available. Catabolite repression is sometimes called the **glucose effect** because glucose, which yields the highest return of ATP per unit of expended energy, usually strongly represses operons for other carbon sources. To use glucose, the cell need only convert it to glucose-6-phosphate, which can enter the glycolytic pathway. Thus, glucose is the preferred carbon and energy source for most types of bacteria.

Figure 12.1 illustrates what happens when *E. coli* cells are growing in a mixture of glucose and galactose. The cells first use the glucose, and only after it is depleted do they begin using the galactose. When the glucose is gone, the cells stop growing briefly while they synthesize the enzymes for galactose. Growth then resumes but at a slightly lower rate.

cAMP and the cAMP-Binding Protein

Most bacteria and lower eukaryotes are known to have systems for catabolite repression. The best understood is the **cyclic AMP (cAMP)**-dependent system of *E. coli* and other enteric bacteria. cAMP is like AMP (see chapter 2), with a single phosphate group on the ribose sugar, but the phosphate is attached to both the 5'-hydroxyl and the 3'-hydroxyl groups of the sugar, thereby making a circle out of the phosphate and sugar. Only *E. coli* and other closely related enterics seem to use this cAMP-dependent system. Other bacteria may use an entirely different system, which does not involve cAMP. Even *E. coli* has a second, cAMP-independent system for catabolite repression, which we discuss later.

REGULATION OF cAMP SYNTHESIS

Catabolite regulation in *E. coli* is achieved through fluctuation in the levels of cAMP, which vary inversely with the levels of cellular catabolites. In other words, cellular concentrations of cAMP are higher when catabolite levels are lower, the situation that prevails when the bacte-

TABLE 12.1	Some *E. coli* global regulatory systems			
System	Response	Regulatory gene(s) (protein[s])	Category of mechanism	Some genes and operons regulated
A. Nutrient limitation				
Carbon	Catabolite repression	*crp* (CAP, also called CRP)	DNA-binding activator or repressor	*lac, ara, gal, mal,* and numerous other C source operons
	Control of fermentative vs oxidative metabolism	*cra* (*fruR*) (CRA)	DNA-binding activator or repressor	Enzymes of glycolysis, Krebs cycle
Nitrogen	Response to ammonia limitation	*rpoN* (NtrA)	Sigma factor (σ^{54})	*glnA* (GS) and operons for amino acid degradation
		ntrBC (NtrBC)	Two-component system	
Phosphorus	Starvation for inorganic orthophosphate (P_i)	*phoBR* (PhoBR)	Two-component system	38 genes, including *phoA* (bacterial alkaline phosphatase) and *pst* operon (P_i uptake)
B. Growth limitation				
Stringent response	Response to lack of sufficient aminoacylated-tRNAs for protein synthesis	*relA* (RelA), *spoT* (SpoT)	(p)ppGpp metabolism	rRNA, tRNA
Stationary phase	Switch to maintenance metabolism and stress protection	*rpoS* (RpoS)	Sigma factor (σ^s)	>40 genes, including *otsBA* (trehalose synthesis) and *dps* (DNA-binding protein)
Oxygen	Response to anaerobic environment	*fnr* (Fnr)	CAP family of DNA-binding protein	>31 transcripts, including *narGHJI* (nitrate reductase)
		arcAB (ArcAB)	Two-component system	>20 genes, including *cob* (cobalamin synthesis)
C. Stress				
Osmoregulation	Response to abrupt osmotic upshift	*kdpDE* (KdpD, KdpE)	Two-component system	*kdpFABC* (K^+ uptake system)
	Adjustment to osmotic environment	*rpoS* (RpoS)	Sigma factor (σ^s)	>16 genes, including *osmB* (an outer membrane lipoprotein) (and genes regulated in stationary phase above)
		micF	Antisense RNA	*ompF* (porin)
Oxygen stress	Protection against reactive oxygen species	*soxS* (SoxS)	AraC family of DNA-binding proteins	>10 genes, including *sodA* (superoxide dismutase) and *micF* (antisense RNA regulator of *ompF*)
		oxyR (OxyR)	LysR family of DNA-binding proteins	>10 genes, including *katG* (catalase)
Heat shock	Tolerance to abrupt temperature increase	*rpoH* (RpoH)	Sigma factor (σ^{32})	>20 hsps (heat shock proteins), including *dnaK, dnaJ,* and *grpE* (chaperones), and *lon, clpP, clpX,* and *hflB* (proteases)
		rpoE (RpoE)	Sigma factor (σ^{24})	>10 proteins, including *rpoH* (σ^{32}) and *degP* (a periplasmic protease)
pH shock	Tolerance of acidic environment	*cadC* (CadC)	ToxR-related DNA-binding protein	Several genes, including *cadBA* (amino acid decarboxylase)

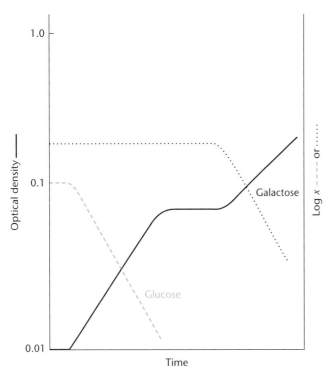

Figure 12.1 Growth of *E. coli* in a mixture of glucose and galactose. The concentration of the sugars in the medium is shown as the dashed lines. The optical density, a measure of cell density, is shown in black. Only after the cells deplete all the glucose will they begin to use the galactose, and then only after a short lag while they induce the *gal* operon (plateau in optical density). They then grow more slowly on the galactose. See the text for details.

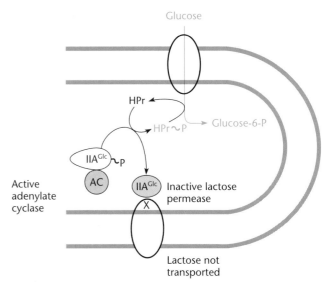

Figure 12.2 Exogenous glucose inhibits both cAMP synthesis and the uptake of other sugars, such as lactose. Activation of adenylate cyclase (AC) requires that the protein IIA^{Glc} be phosphorylated (IIA^{Glc}~P). In the presence of glucose, the HPr protein transfers the phosphate from IIA^{Glc}~P to glucose. The unphosphorylated IIA^{Glc} protein also inhibits other permeases such as the lactose permease. See the text for details.

ria are growing in a relatively poor carbon source such as lactose or maltose.

The synthesis of cAMP is controlled through the regulation of the activity of adenylate cyclase. This enzyme, which makes cAMP from ATP, is more active when cellular concentrations of catabolites are low and less active when catabolite concentrations are high. The adenylate cyclase enzyme is associated with the inner membrane and is the product of the *cya* gene.

Figure 12.2 outlines the current picture of the regulation of adenylate cyclase activity. An important factor in the regulation is the phosphoenolpyruvate (PEP)-dependent sugar phosphotransferase system (PTS), which, as the name implies, is responsible for transporting certain sugars, including glucose, into the cell. One of the protein components of the PTS system, named IIA^{Glc}, can exist in either an unphosphorylated (IIA^{Glc}) or a phosphorylated (IIA^{Glc}~P) form. The IIA^{Glc}~P form activates adenylate cyclase to make cAMP. However, the IIA^{Glc} protein is dephosphorylated when one of the sugars it is involved in transport of, such as glucose, is present in the

medium. Another component of PTS, the phosphotransferase named HPr (for histidine protein), transfers the phosphate from IIA^{Glc}~P to the sugar as the sugar is transported. Less of the IIA^{Glc}~P form will then exist to activate the adenylate cyclase, and cAMP levels will drop.

The unphosphorylated form of IIA^{Glc} also inhibits other sugar-specific permeases that transport sugars such as lactose (Figure 12.2). Therefore, less of these other sugars will enter the cell if glucose or another better carbon source is available, and so less inducer will be present to initiate synthesis from their respective operons (see chapter 11).

THE CATABOLITE ACTIVATOR PROTEIN CAP

The mechanism by which cAMP turns on catabolite-sensitive operons in *E. coli* is quite well understood. The cAMP binds to the protein product of the *crp* gene, which is an activator of transcription of catabolite-sensitive operons. This activator protein goes by two names, CAP (for catabolite gene activator protein) and CRP (for cAMP receptor protein). We will call it CAP. CAP with cAMP bound (CAP-cAMP) functions like other activator proteins discussed in chapter 11 in that it interacts with the RNA polymerase to activate transcription from promoters for operons under its control, including *lac, gal, ara,* and *mal*. These operons are all members of the **CAP regulon**, or the catabolite-sensitive regulon (Table 12.1). However, the mechanism of CAP-cAMP regulation varies. As discussed below, CAP can function not only as an activator but also as a repressor, depending on where it binds relative to the promoter (see the section on regulation of other catabolite-sensitive operons by CAP-cAMP, below).

REGULATION OF *lac* BY CAP-cAMP

A detailed model for CAP activation of transcription from the *lac* promoter is shown in Figure 12.3. This model incorporates much of what has been learned from in vitro experiments and genetic studies about the requirements for CAP activation. Upstream of the promoter is a short sequence called the **CAP-binding site**, which is similar in all catabolite-sensitive operons and so can be easily identified. CAP can bind to this site only when it is bound to cAMP, so that this site will be occu-

pied only when cAMP levels are high. RNA polymerase binds very weakly to the *lac* promoter in the absence of CAP and cAMP at the CAP-binding site. However, the cAMP-CAP bound on this site will make contact with an incoming RNA polymerase, facilitating its binding to the promoter and thereby making the *lac* promoter much stronger.

Recent observations shed light on the details of CAP activation. As illustrated in Figure 12.3, CAP bends the DNA when it binds to the CAP-binding site, although the significance of this bending is unknown. Also, the region of the RNA polymerase with which CAP makes contact to activate transcription is now known. The CAP protein contacts the carboxy-terminal portion of the α subunit of RNA polymerase (see Zou et al. and Tang et al., Suggested Reading). Recall from chapter 2 that the RNA polymerase of *E. coli* and most other eubacteria consists of five subunits, two of which are identical α subunits. This region of the RNA polymerase interacts with some other activator proteins as well as with DNA sequences called UP elements, which lie upstream of some very strong promoters including those for rRNA genes. UP elements do not exist upstream of catabolite-sensitive promoters, but when CAP is bound at its binding site, it may serve as a substitute. We discuss UP elements later in the section on rRNA regulation.

REGULATION OF OTHER CATABOLITE-SENSITIVE OPERONS BY CAP-cAMP

The positioning of the CAP-binding sequence relative to the promoter is different in operons other than *lac*. In the *ara* operon, the CAP-binding site is far upstream, with the AraC-binding site between it and the promoter (see Figure 11.15). In this case, the CAP protein presumably binds too far from the promoter to make physical contact with the incoming RNA polymerase, and so the CAP-cAMP complex may activate RNA polymerase indirectly, perhaps through an interaction with the AraC protein.

Other operons in the CAP regulon, such as *gal*, are less sensitive to catabolite repression than others. The *gal* operon is not totally repressed by glucose in the medium because it has two promoters, p_{G1} and p_{G2} (see Figure 11.8). p_{G2} does not require CAP for its activation. In fact, CAP-cAMP represses this promoter, probably because it binds to the −35 sequence of the promoter. As discussed in chapter 11, the p_{G2} promoter permits some expression of the *gal* operon even in the presence of glucose. This low level of expression is necessary to allow the synthesis of cell wall components that include galactose, since the UDP-galactose synthesized by the operon serves as the donor of galactose in biosynthetic reac-

A

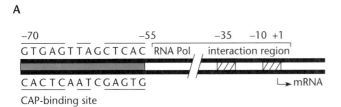

B   The CAP protein bends DNA

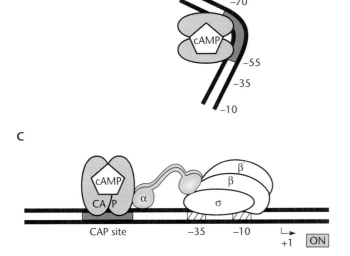

Figure 12.3 Mechanism of activation of the *lac* promoter by the CAP-cAMP complex. (A) Sequence of the CAP-binding site upstream of the promoter. (B) Binding of CAP at this site bends the DNA. (C) The bound CAP-cAMP complex makes contact with the α subunit of the RNA polymerase at the promoter, activating the initiation of transcription.

tions. However, the level of expression of the *gal* operon from p_{G2} is not high enough for the cells to grow well on galactose as a carbon and energy source.

RELATIONSHIP OF CATABOLITE REPRESSION TO INDUCTION

An important point about CAP regulation of catabolite-sensitive operons is that it occurs in addition to any other regulation to which the operon is subject (Figure 12.4). Two conditions must be met before catabolite-sensitive operons can be transcribed: better carbon sources such as glucose must be absent, and the inducer of the operon must be present. Take the example of the *lac* operon. If no carbon sources better than lactose are available, cAMP levels will be high and CAP-cAMP will bind upstream of the *lac* promoter to activate transcription. However, even at high cAMP levels, the *lac* operon will not be transcribed unless the inducer allolactose is

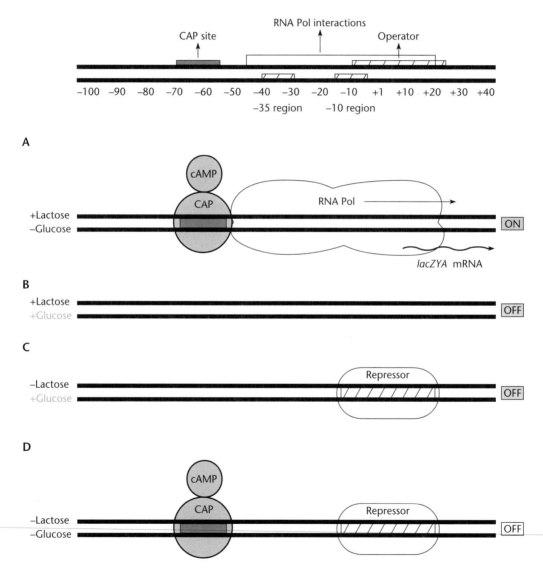

Figure 12.4 Regulation of the lactose operon by both glucose and the inducer lactose. (A) The operon will be ON only in the absence of glucose and the presence of lactose, which is converted into the inducer, allolactose. (B and C) The operon will be OFF in the presence of glucose whether or not lactose is present, because the CAP-cAMP complex will not be bound to the CAP site. (D) The operon will be OFF if lactose is not present, even if glucose is also not present, because the repressor will be bound to the operator. The relative positions of the CAP-binding site, operator, and promoter are shown. The entire regulatory region covers about 100 bp of DNA.

also present. In the absence of inducer, the LacI repressor will be bound to the operator and prevent the RNA polymerase from binding to the promoter and transcribing the operon (see chapter 11).

Genetic Analysis of Catabolite Regulation in *E. coli*

The above model for the regulation of catabolite-sensitive operons is supported by both genetic and biochemical analysis of catabolite repression in *E. coli*. This analysis has involved the isolation of mutants defective in the global regulation of all catabolite-sensitive operons, as well as mutants defective in the catabolite regulation of specific operons.

ISOLATION OF *crp* AND *cya* MUTATIONS

According to the model presented above, mutations that inactivate the *cya* and *crp* genes for adenylate cyclase and CAP, respectively, should prevent transcription of all the catabolite-sensitive operons. In these mutants, there is no CAP with cAMP attached to bind to the promoters. In other words, *cya* and *crp* mutants should be Lac⁻, Gal⁻, Ara⁻, Mal⁻, and so on. In genetic terms, *cya* and *crp* mutations are **pleiotropic**, because they cause many phenotypes, i.e., the inability to use many different sugars as carbon and energy sources.

The fact that *cya* and *crp* mutations should prevent cells from using several sugars was used in the first isolations of *crp* and *cya* mutants (see Schwartz and Beckwith, Suggested Reading). The selection was based on the fact that colonies of bacteria will turn tetrazolium salts red while they multiply, provided that the pH remains high. However, bacteria that are fermenting a carbon source will give off organic acids such as lactic acid that will lower the pH, preventing the conversion to red. As a consequence, wild-type *E. coli* cells growing on a fermentable carbon source form white colonies on tetrazolium-containing plates, whereas mutant bacteria that cannot use the fermentable carbon source will use a different carbon source and so will form red colonies. Some of these red-colony-forming mutants might have *crp* or *cya* mutations, although most would have mutations that inactivate a gene within the operon for the utilization of the fermentable carbon source. Thus, without a way to increase the frequency of *crp* and *cya* mutants among the red-colony-forming mutants, many red-colony-forming mutants would have to be tested to find any with mutations in either *cya* or *crp*.

For these experiments, the investigators reasoned that they could increase the frequency of *crp* and *cya* mutants by plating heavily mutagenized bacteria on tetrazolium agar containing *two different* fermentable sugars, for example, lactose and galactose or arabinose

and maltose. Then, to prevent the utilization of both sugars, either two mutations, one in each operon, or a single mutation, in *cya* or *crp*, would have to occur. Since mutants with single mutations should be much more frequent than mutants with two independent mutations, the *crp* and *cya* mutants should be a much higher fraction of the total red-colony-forming mutants growing on two carbon and energy sources. Indeed, when the red-colony-forming mutants that could not use either of the two sugars provided were tested, most of them were found to be either deficient in adenylate cyclase activity or lacking a protein, now named CAP, later shown to be required for the activation of the *lac* promoter in vitro.

PROMOTER MUTATIONS THAT AFFECT CAP ACTIVATION

Three types of mutations have been isolated in the *lac* promoter. Those belonging to class I change the CAP-binding site so that CAP can no longer bind to it. The *lac* promoter mutation L8 is an example (Figure 12.5). By preventing the binding of CAP-cAMP upstream of the promoter, this mutation weakens the *lac* promoter. As a result, the *lac* operon is expressed poorly, as measured by β-galactosidase activity, even when cells are growing in lactose and cAMP levels are high. However, the low level of expression of the *lac* operon is less affected by the carbon source and is not reduced much more if glucose is added and cAMP levels drop.

Class II mutations change the −35 region of the RNA polymerase-binding site so that the promoter is less active even when cAMP levels are high. However, with this type of mutation, the residual expression of the *lac* operon will still be sensitive to catabolite repression. Consequently, the amount of β-galactosidase synthesized when cells are growing in the presence of lactose plus glucose while cAMP levels are low will be smaller than the amount synthesized when the cells are growing in the presence of a poorer carbon source plus lactose and cAMP levels are high.

A third, very useful mutated *lac* promoter, class III, was found by isolating apparent Lac⁺ revertants of class I mutations such as p_{L8} or of *cya* or *crp* mutations. One such mutation is called p_{lacUV5}. This mutant promoter is almost as strong as the wild-type *lac* promoter but no longer requires cAMP-CAP for activation. As shown in Figure 12.5, the *lacUV5* mutation changes a 2-bp stretch of the −10 region of the *lac* promoter, so that the sequence reads TATAAT instead of TATGTT. This mutant −10 sequence more closely resembles the sequence of a consensus σ⁷⁰ promoter (see chapter 2). Some expression vectors use the *lacUV5* promoter rather than the wild-type *lac* promoter, so that the promoter can be

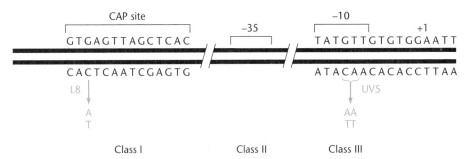

Figure 12.5 Mutations in the *lac* regulatory region that affect activation by CAP. The class I mutation L8 changes the CAP-binding site so that CAP can no longer bind and the promoter cannot be turned on even in the absence of glucose (high cAMP). The class III mutation UV5 changes 2 bp in the –10 sequence of the promoter so that the promoter no longer requires activation by CAP and the operon can be induced even in the presence of glucose (low cAMP). No class II mutations are shown. The changes in the sequence in each mutation appear in blue.

induced even if the bacteria are growing in glucose-containing media.

INTERACTION OF RNA POLYMERASE WITH CAP

Biochemical experiments first identified the carboxy terminus of the α subunit as a region of the RNA polymerase that contacts CAP (see Igarashi and Ishihama, Suggested Reading). In these experiments, the α subunits of RNA polymerase were synthesized in vitro by using a clone of the *rpoA* gene, the gene for the α subunit. If the region of the *rpoA* gene encoding the carboxy terminus of the α subunit is deleted from the clone, by methods described in chapter 15, an RNA polymerase containing the truncated α subunits can be assembled and is still active for transcription from most promoters. However, the defective RNA polymerase cannot initiate transcription from catabolite-sensitive promoters, even in the presence of CAP-cAMP. These experiments therefore suggested that CAP interacts with the carboxy-terminal portion of the α subunit of RNA polymerase to activate transcription from catabolite-sensitive promoters.

Genetic experiments confirmed this role of the carboxy-terminal region of the α subunit and identified some of the amino acids important for the interaction (see Tang et al., Suggested Reading). The investigators mutagenized the entire *rpoA* gene on a plasmid clone and introduced the mutagenized clone into cells with a wild-type *rpoA* gene in their chromosome. The *rpoA* gene was cloned in a multicopy plasmid, so that most of the α subunit synthesized in these cells would have come from the plasmid. Consequently, most RNA polymerase molecules would have had at least one α subunit from the mutagenized clone. The investigators then se-

lected mutants defective in the utilization of two sugars, lactose and ribose, by methods similar to those used for the original isolation of *crp* and *cya* mutants (see above). They found such mutants, which presumably had amino acid changes in the α subunit of their RNA polymerase that prevented its interaction with CAP. To prove that this was the case, the investigators isolated the plasmid from some of the mutants and introduced it into two different strains of *E. coli*. One strain had the wild-type *lac* operon in its chromosome, and the other had the *lac* operon with a *lac* promoter with both the L8 and UV5 mutations discussed above, which thus did not require CAP for activation. The mutations in *rpoA* prevented synthesis of the β-galactosidase product of the *lacZ* gene in the strain with the wild-type *lac* promoter but not in the strain with the mutated *lac* promoter. When the mutations in the *rpoA* gene were located by DNA sequencing, they were found to change amino acids in the carboxy terminus of the α subunit.

Uses of cAMP in Other Organisms

The use of cAMP to regulate catabolite-sensitive operons seems unique to enteric bacteria such as *E. coli*. Other bacteria use different mechanisms to regulate catabolite-sensitive operons (see Box 12.1). Many bacteria do not appear to make cAMP at all, whereas in others the levels of cAMP do not vary depending on the carbon sources available. Furthermore, the CAP-cAMP-mediated catabolite regulatory system is not the only system used by *E. coli* and other enteric bacteria (see Box 12.1).

In eukaryotes, cAMP does not regulate carbon source utilization but does have many other uses, including the regulation of G proteins and roles in cell-to-

BOX 12.1

cAMP-Independent Catabolite Repression

Recent evidence indicates that not all catabolite repression in *E. coli* and other enteric bacteria is mediated by cAMP. One mechanism of cAMP-independent catabolite repression involves the Cra protein, named for its function as a *catabolite repressor/activator* (sometimes called FruR, for fructose repressor). The Cra protein is encoded by the *cra* gene and is a DNA-binding protein similar to LacI and GalR. Cra was discovered during the isolation of mutants with mutations that suppress *ptsH* mutations of *E. coli* and *S. typhimurium*. The *ptsH* gene encodes the HPr protein that phosphorylates many so-called PTS sugars, including glucose, during transport (Figure 12.2). Therefore, *ptsH* mutants cannot use these sugars, because phosphorylation is the first step in the glycolytic pathway. The *cra* mutations suppress *ptsH* mutations and allow growth on PTS sugars by allowing the constitutive expression of the fructose catabolic operon, one of whose genes encodes a protein that can substitute for HPr. The *cra* mutants were found to be pleiotropic in that they are unable to synthesize glucose from many substrates, including acetate, pyruvate, alanine, and citrate. However, they demonstrate elevated expression of genes involved in glycolytic pathways.

The pleiotropic phenotype of *cra* mutants suggested that Cra functioned as a global regulatory protein, activating the transcription of some genes and repressing the synthesis of others. As for other regulation proteins, whether Cra activates or represses transcription depends on where it binds to the operon. If its binding site is upstream of the promoter, it activates transcription of the operon; if its binding site overlaps or is downstream of the promoter, it represses transcription. In either case, Cra will come off the DNA if it binds the sugar catabolites fructose-1-phosphate and fructose-1,6-bis-phosphate that are present in high concentrations during growth in the presence of sugars such as glucose. What effect this will have on the transcription of a particular operon depends on whether Cra functions as a repressor or activator of that operon. If it functions as a repressor, the transcription of the operon will increase; if it functions as an activator, the transcription of the operon will decrease—a process called antiactivation. In general, the Cra protein functions as a repressor of operons whose products are involved in alternate pathways for sugar catabolism, such as the Embden-Meyerhof and Entner-Doudoroff pathways, so that the transcription of these genes will increase when glucose and other good carbon sources are available. In contrast, it usually activates operons whose products are involved in synthesizing glucose from acetate and other metabolites (gluconeogenesis), and so it will not activate the transcription of these operons if glucose is present. Why make glucose if some is already available in the medium?

Catabolite repression has also been studied extensively in *Bacillus subtilis*. In this bacterium, catabolite-sensitive genes are not regulated by cAMP levels. In many ways, catabolite regulation in *B. subtilis* is analogous to the cAMP-independent Cra pathway in *E. coli*. Numerous genes for carbon source utilization are regulated by a protein named CcpA (catabolite control protein A), which is also a member of the LacI and GalR family of regulators. Furthermore, CcpA seems to repress some genes and activate others.

References

Henkin, T. M. 1996. The role of CcpA transcriptional regulator in carbon metabolism in *Bacillus subtilis*. *FEMS Microbiol. Lett.* **135**:9–15.

Saier, M. H., Jr., and T. M. Ramseier. 1996. The catabolite repressor/activator (Cra) protein of enteric bacteria. *J. Bacteriol.* **178**:3411–3417.

cell communication. In view of the general importance of cAMP in various biological systems, it is somewhat surprising that, among the bacteria, only the enteric bacteria seem to use this nucleotide to regulate carbon source utilization.

Regulation of Nitrogen Assimilation

Nitrogen is a component of many biological molecules, including nucleotides, vitamins, and amino acids. Thus, all organisms must have a source of nitrogen atoms for growth to occur. For bacteria, possible sources include ammonia (NH_3) and nitrate (NO_3^-), as well as nitrogen-containing organic molecules such as nucleotides and amino acids. Some bacteria can even use atmospheric nitrogen (N_2) as a nitrogen source, a capability that makes them apparently unique on Earth (see Box 12.2).

NH_3 is the preferred source of nitrogen for most types of bacteria. All biosynthetic reactions involving the addition of nitrogen either use NH_3 directly or transfer nitrogen in the form of an NH_2 group from glutamate and glutamine, which in turn are synthesized by directly adding NH_3 to α-ketoglutarate and glutamate, respectively.

Thus, because NH_3 is directly or indirectly the source of nitrogen in biosynthetic reactions, most other forms

Nitrogen Fixation

Some bacteria can use atmospheric nitrogen (N_2) as a nitrogen source by converting it to NH_3 in a process called *nitrogen fixation,* which appears to be unique to bacteria. However, N_2 is a very inconvenient source of nitrogen. The bond holding the two nitrogen atoms together must be broken, which is very difficult. Bacteria that can fix nitrogen include the cyanobacteria and members of the genera *Klebsiella, Azotobacter, Rhizobium,* and *Azorhizobium.* These organisms play an important role in nitrogen cycles on Earth.

Some types of nitrogen-fixing bacteria, including members of the genera *Rhizobium* and *Azorhizobium,* are symbionts that fix N_2 in nodules on the roots or stems of plants and allow the plants to live in nitrogen-deficient soil. In return, the plant furnishes nutrients and an oxygen-free atmosphere in which the bacterium can fix N_2. This symbiosis therefore benefits both the bacterium and the plant. An active area of biotechnology is the use of N_2-fixing bacteria as a source of natural fertilizers.

The fixing of N_2 requires the products of many genes, called the *nif* genes. In free-living nitrogen-fixing bacteria such as *Klebsiella* spp., the fixing of nitrogen requires about 20 *nif* genes arranged in about eight adjacent operons. Some of the *nif* genes encode the nitrogenase enzymes directly responsible for fixing N_2. Other *nif* genes encode proteins involved in assembling the nitrogenase enzyme and in

regulating the genes. Plant symbiotic bacteria also require many other genes whose products produce the nodules on the plant (*nod* genes) and whose products allow the bacterium to live and fix nitrogen in the nodules (*fix* genes).

Because atmospheric nitrogen is such an inconvenient source of nitrogen, the genes involved in N_2 fixation are part of the Ntr regulon and are under the control of the NtrC activator protein. In *Klebsiella pneumoniae,* in which the regulation of the *nif* genes has been studied most extensively, the phosphorylated form of NtrC does not activate all eight operons involved in N_2 fixation operons directly. Instead, the phosphorylated form of NtrC activates the transcription of another activator gene, *nifA,* whose product is directly required for the activation of the eight *nif* operons. In addition, in the presence of oxygen, the *nif* operons are negatively regulated by the product of the *nifL* gene, which senses oxygen. Because the nitrogenase enzymes are very sensitive to oxygen, the bacteria can fix nitrogen only in an oxygen-free (anaerobic) environment, such as exists in the nodules on the roots of plants.

References

Fischer, H.-M. 1994. Genetic regulation of nitrogen fixation in rhizobia. *Microbiol. Rev.* **58:**352–386.

Gussin, G. N., C. W. Ronson, and F. M. Ausubel. 1986 Regulation of nitrogen fixation genes. *Annu. Rev. Genet.* **20:**567–591.

---

of nitrogen must be reduced to NH_3 before they can be used in biosynthetic reactions. This process is called **assimilatory reduction** of the nitrogen-containing compounds, because the nitrogen-containing compound converted into NH_3 will be introduced, or assimilated, into biological molecules. In another type of reduction, **dissimilatory reduction,** oxidized nitrogen-containing compounds such as NO_3^- are reduced when they serve as electron acceptors in anaerobic respiration (in the absence of oxygen). However, the compounds are generally not reduced all the way to NH_3 in this process, and the nitrogen is not assimilated into biological molecules. Here, we discuss only the assimilatory uses of nitrogen-containing compounds. The genes whose products are required for anaerobic respiration are members of a different regulon (Table 12.1).

Pathways for Nitrogen Assimilation

Enteric bacteria use different pathways to assimilate nitrogen depending on whether NH_3 concentrations are

low or high (Figure 12.6). When NH_3 concentrations are low, for example when the nitrogen sources are amino acids, which must be degraded for their NH_3, an enzyme named **glutamine synthetase (GS)**, the product of the *glnA* gene, adds the NH_3 directly to glutamate to make glutamine. Most of this glutamine is then converted to glutamate by another enzyme, **glutamate synthase,** which removes an NH_2 group from glutamine and adds it to α-ketoglutarate to make two glutamates. These glutamates can, in turn, be converted into glutamine by glutamine synthetase. Because the NH_3 must all be routed by glutamine synthetase to glutamine when NH_3 concentrations are low, the cell needs a lot of the glutamine synthetase enzyme under these conditions. We shall return to the significance of this later.

If NH_3 concentrations are high because the medium contains NH_3 in the dissolved form of NH_4OH, the nitrogen is assimilated through a very different pathway. In this case, the enzyme **glutamate dehydrogenase** adds the NH_3 directly to α-ketoglutarate to make glutamate.

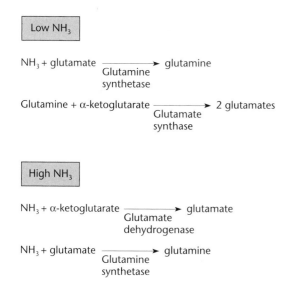

Figure 12.6 Pathways for nitrogen assimilation in *E. coli* and other enteric bacteria. With low NH_3 concentrations, the glutamine synthetase (GS) enzyme adds NH_3 directly to glutamate to make glutamine. Glutamate synthase can then convert the glutamine plus α-ketoglutarate into two glutamates, which can reenter the cycle. In the presence of high NH_3 concentrations, the NH_3 is added directly to α-ketoglutarate by the glutamate dehydrogenase to make glutamate, which can be subsequently converted to glutamine by GS.

Some of the glutamate is subsequently converted into glutamine by glutamine synthetase. Much less glutamine is required for protein synthesis and biosynthetic reactions than for assimilation of limiting nitrogen from the medium. Therefore, cells need much less glutamine synthetase when growing in high concentrations of NH_3 than when growing in low concentrations.

The reaction catalyzed by glutamate dehydrogenase when NH_3 concentrations are high is very efficient, which is why bacteria that have a glutamate dehydrogenase enzyme prefer NH_3 as a nitrogen source. The rapid assimilation of nitrogen under these conditions allows them to multiply rapidly. However, a few bacteria, including *Bacillus subtilis*, lack glutamate dehydrogenase. These bacteria prefer glutamine to NH_3 as a nitrogen source and cannot assimilate NH_3 as quickly.

REGULATION OF NITROGEN ASSIMILATION PATHWAYS BY THE Ntr SYSTEM

The operons for nitrogen utilization are part of a nitrogen-regulated regulon (the **Ntr system**). Ntr regulation ensures that the cell will not waste energy making enzymes for the use of nitrogen sources such as amino acids or nitrate when NH_3 is available. In this section, we discuss what is known about how the Ntr global

regulatory system works. As usual, geneticists led the way, by identifying the genes whose products are involved in the regulation, so that a role could eventually be assigned to each one. The Ntr regulatory system is remarkably similar in all genera of gram-negative bacteria, including *Escherichia*, *Salmonella*, *Klebsiella*, and *Rhizobium*, but with important exceptions, some of which we point out. Early indications are that this system may be different in gram-positive bacteria, although work on Ntr in these bacteria is in its early stages.

Regulation of the *glnA-ntrB-ntrC* Operon by a Signal Transduction Pathway

Since cells need more glutamine synthetase while growing at low NH_3 concentrations than when growing at high NH_3 concentrations, the expression of the *glnA* gene, which encodes glutamine synthetase, must be regulated according to the nitrogen source that is available. This gene is part of an operon called *glnA-ntrB-ntrC*. The products of the two other genes in the operon, NtrB and NtrC (*n*itrogen *r*egulator B and C), are involved in regulating the operon. (These proteins are also called NR_{II} and NR_I, respectively, but we use the Ntr names in this chapter.) Because the *ntrB* and *ntrC* genes are part of the same operon as *glnA*, their products are also synthesized at higher levels when NH_3 concentrations are low.

Figure 12.7 illustrates the regulation of the *glnA-ntrB-ntrC* operon in detail. In addition to NtrB and NtrC, the proteins GlnD and P_{II} participate in regulation of the operon. These four proteins form a **signal transduction pathway**, in which news of the available nitrogen source is passed, or transduced, from one protein to another until it gets to its final destination, the regulator protein.

The first indication of the available nitrogen sources is the ratio of glutamine to α-ketoglutarate in the cell. This ratio will be high when the cells are growing in high NH_3 concentrations and low when the cells are growing in low NH_3 concentrations. This ratio influences the activity of the GlnD protein, an enzyme that can either add or remove UMP from the P_{II} protein. Glutamine stimulates GlnD to remove UMP from P_{II}, while α-ketoglutarate stimulates it to add UMP to P_{II}. As a consequence, when NH_3 concentrations are low, the P_{II} protein will exist mostly as P_{II}-UMP, and when concentrations are high, it will exist as just P_{II}.

NtrB, the third protein in the pathway, is an **autokinase**. As an autokinase, it adds a phosphate (P) to itself to make NtrB~P. This process is known as **autophosphorylation**.

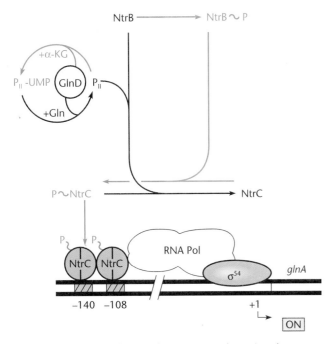

Figure 12.7 Regulation of Ntr operons by a signal transduction pathway in response to NH_3 levels. At low NH_3 concentrations, the reactions shown in blue will predominate. At high NH_3 concentrations, the reactions shown in black will predominate. See the text for details.

NtrC is the fourth protein in the pathway. This protein is a transcriptional activator, which, when phosphorylated, will activate transcription from Ntr-type promoters. In low concentrations of NH_3, phosphates can be transferred from NtrB to NtrC to make NtrC~P, which will activate transcription from the promoters. However, in high NH_3, the P_{II} protein without UMP attached can bind to NtrB and somehow interfere with the phosphorylation of NtrC, as well as promote the removal of phosphate from NtrC~P. Therefore, as a result of this signal transduction pathway, NtrC will be unphosphorylated when NH_3 concentrations are high, and the Ntr promoters will not be activated. When NH_3 concentrations are low, NtrC will exist as NtrC~P. Hence, the Ntr promoters, including that for the *glnA-ntrB-ntrC* operon, will be activated, more mRNA will be made from the operon, and more glutamine synthetase will be synthesized.

NtrB and NtrC: a Two-Component Sensor-Response Regulator System

The NtrB and NtrC proteins form a two-component system in which NtrB is a **sensor** and NtrC is a **response regulator**. Such protein pairs are now known to be common in bacteria, and the corresponding members of

such pairs are remarkably similar to each other (see Box 12.3). Typically, one protein of the pair "senses" an environmental parameter and phosphorylates itself at a histidine. This phosphoryl group is then passed on to an aspartate in the second protein. The activity of this second protein, the response regulator, depends on whether it is phosphorylated. Many response regulators are transcriptional regulators. In chapter 6, we discussed another example of a sensor-response regulator pair, ComP and ComA, involved in the development of transformation competence in *B. subtilis*, and later in this chapter we describe other pairs.

The Other Ntr Operons

The other operons besides *glnA-ntrB-ntrC* that are activated by NtrC~P depend on the type of bacterium and the other nitrogen sources they can use. In general, operons under the control of NtrC~P are those involved in using poorer nitrogen sources. For example, genes for the uptake of the amino acids glutamine in *E. coli* and histidine and arginine in *Salmonella typhimurium* are under the control of NtrC~P. An operon for the utilization of nitrate as a nitrogen source in *Klebsiella pneumoniae* is activated by NtrC~P, but neither *E. coli* nor *S. typhimurium* has such an operon.

In some bacteria, the Ntr genes are not regulated directly by NtrC~P but are under the control of another gene product whose transcription is activated by NtrC~P. For example, operons for amino acid degradative pathways in *Klebsiella aerogenes* use σ^{70} promoters that do not require activation by NtrC~P. However, they are indirectly under the control of NtrC~P because transcription of the gene for their transcriptional activator, *nac*, is activated by NtrC~P. The nitrogen fixation genes of *K. pneumoniae* are similarly under the indirect control of NtrC~P because this protein activates transcription of the gene for their activator protein, *nifA* (see Box 12.2).

Another Ntr Regulatory Pathway?

The need for so many steps in the regulation of the Ntr system is not clear, but it may reflect the central role that nitrogen utilization plays in the physiology of the cell. Other pathways may intersect with the various steps of the signal transduction pathway for nitrogen utilization, so that expression of the genes can be coordinated with many other cellular functions. In support of this idea, some Ntr regulation occurs even in *ntrB* mutants (see the section on genetics of Ntr regulation below). At least one other pathway besides the one involving NtrB must lead to the phosphorylation and dephosphorylation of NtrC in response to changes in the nitrogen source.

BOX 12.3

Sensor-Response Regulator Two-Component Systems

Many regulation mechanisms require that the cell sense changes in the external environment and change the expression of its genes or the activity of its proteins accordingly. Examples of such regulation include motility in response to chemical attractants and induction of pathogenesis operons upon entry into a suitable host. The mechanisms regulating many of these systems are remarkably similar. They often involve two proteins. The first in the pair, an autokinase sensor protein, is usually a membrane protein. The second protein, the response regulator protein, receives the phosphoryl group from the first. Some response regulator proteins, such as NtrC or OmpR, are activators of transcription. Others activate another cellular function; for example, CheY activates flagellar movement. The phosphorylation of the response regulator modulates its regulatory activity (see figure).

Not only do many sensor and response regulator pairs of proteins work in remarkably similar ways, but also they share considerable amino acid sequence homology, as though they were evolutionarily derived from each other. The sensor proteins have a conserved region near their carboxy terminus called the transmitter domain. This region includes the histidine that is phosphorylated. Whatever their role, the response regulators are similar in their N-terminal region, the so-called receiver domain, which includes the aspartate that is phosphorylated. The remainder of the protein can differ depending on its function, although different subfamilies of response regulators share regions of high homology in other parts of the protein, including the helix-turn-helix motifs of transcriptional regulators.

The similarities between these systems suggest that under some conditions the sensor of one system may phosphorylate the response regulator of a different system. Such cross-regulation could provide communication between different regulatory systems and so coordinate the different regulatory cascades. The extent to which cross-regulation occurs in vivo is not yet known.

References

Hoch, J. A., and T. J. Silhavy (ed.). 1995. *Two-Component Signal Transduction.* ASM Press, Washington, D.C.

Parkinson, J. S., and E. C. Kofoid. 1992. Communication modules in bacterial signaling proteins. *Annu. Rev. Genet.* **26:**71–112.

Stock, J. B., A. J. Ninfa, and A. M. Stock. 1989. Protein phosphorylation and regulation of adaptive responses in bacteria. *Microbiol. Rev.* **53:**450–490.

Wanner, B. L. 1992. Is cross-regulation by phosphorylation of two-component response regulator proteins important in bacteria? *J. Bacteriol.* **174:**2053–2058.

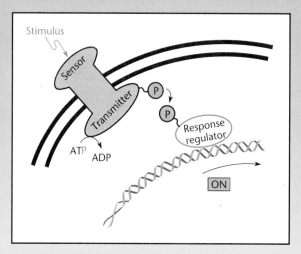

Sensor-response regulator two-component systems. An input signal triggers a conformational change in the input (sensor) domain and a corresponding conformational change in the transmitter domain of the sensor protein. This causes the transmitter domain to phosphorylate itself and donate the phosphoryl group to the receiver domain of the response regulator. Phosphorylation of its receiver domain will cause the output domain of the response regulator to undergo a conformational change, leading to the output signal. In many cases this involves transcriptional regulation of specific genes.

TRANSCRIPTION OF THE *glnA-ntrB-ntrC* OPERON

The *glnA-ntrB-ntrC* operon is transcribed from three promoters, only one of which is NtrC~P dependent. The positions of the three promoters and the RNAs made from each are shown in Figure 12.8. Of the three promoters, only the p_2 promoter is activated by NtrC~P and is responsible for the high levels of glutamine synthetase and NtrB and NtrC synthesis in low NH_3. We defer discussion of the other two promoters, p_1 and p_3, until later.

The p_2 promoter is immediately upstream of the *glnA* gene, and RNA synthesis initiated at this promoter continues through all three genes, as shown in Figure 12.8. However, some transcription terminates at a transcriptional terminator located between the *glnA* and *ntrB* genes, so that much less NtrB and NtrC is made than glutamine synthetase.

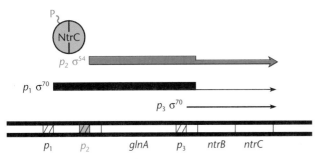

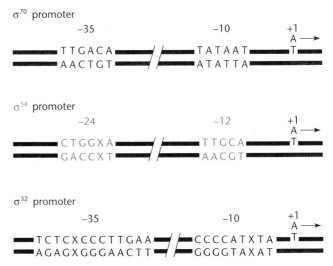

Figure 12.8 The *glnA-ntrB-ntrC* operon of *E. coli*. Three promoters service the genes of the operon. The arrows show the mRNAs that are made from each promoter. The blue arrow indicates the mRNA expressed from the nitrogen-regulated promoter, p_2. The thickness of the lines indicates how much RNA is made from each promoter in each region. See the text for details.

Figure 12.9 Sequence comparison of the promoters recognized by RNA polymerase holoenzyme carrying the normal sigma factor (σ^{70}), the nitrogen sigma factor (σ^{54}), and the heat shock sigma factor (σ^{32}). Instead of consensus sequences at −10 and −35 bp, with respect to the RNA start site, the σ^{54} promoter has consensus sequences at −12 and −24 bp. The σ^{32} promoter has consensus sequences at approximately −10 and −35 bp, but they are different from the consensus sequences of the σ^{70} promoters. X indicates that any base pair can be present at this position. +1 is the start site of transcription.

The "Nitrogen Sigma" σ^{54}

The p_2 promoter and other Ntr-type promoters activated by NtrC~P are unusual in terms of the RNA polymerase holoenzyme that recognizes them. Most promoters are recognized by the RNA polymerase holoenzyme with σ^{70} attached, but the Ntr-type promoters are recognized by holoenzyme with σ^{54} attached (see chapter 2 for a discussion of sigma factors). As shown in Figure 12.9, promoters that are recognized by the σ^{54} holoenzyme look very different from promoters recognized by the σ^{70} holoenzyme. Unlike the typical σ^{70} promoter, which has RNA polymerase-binding sequences at −35 and −10 bp upstream of the RNA start site, the σ^{54} promoters have very different binding sequences at −24 and −12 bp. Because promoters for the genes involved in Ntr regulation are recognized by RNA polymerase with σ^{54}, this sigma factor was first named the "nitrogen sigma" and the gene for σ^{54} was named

rpoN (Table 12.2). However, σ^{54}-type promoters have been found in operons unrelated to nitrogen utilization, including in the flagellar genes of *Caulobacter* spp. and some promoters of the toluene-biodegradative operons of the *Pseudomonas putida* Tol plasmid (see chapter 11). Interestingly, all of the known σ^{54}-type promoters require activation by an activator protein, which always belongs to the NtrC family (see chapter 11, Box 11.2).

TABLE 12.2	Genes for nitrogen regulation		
Gene	Alternate name	Product	Function
glnA		Glutamine synthetase	Synthesize glutamine
glnB		P$_{II}$, P$_{II}$-UMP	Inhibit phosphatase of NtrB, activate adenylyltransferase
glnD		Uridylyltransferase (UTase)/Uridylyl-removing enzyme (UR)	Transfer UMP to and from P$_{II}$
glnE		Adenylyltransferase (ATase)	Transfer AMP to glutamine synthetase
glnF	*rpoN*	σ^{54}	RNA polymerase recognition of promoters of Ntr operons
ntrC	*glnG*	NtrC, NtrC-PO$_4$	Activator of promoters of Ntr operons
ntrB	*glnL*	NtrB, NtrB-PO$_4$	Autokinase, phosphatase; phosphate transferred to NtrC

NtrC

The polypeptide chains of NtrC-type activators have the basic structure shown in Figure 12.10. A DNA-binding domain that recognizes the σ^{54}-type promoter lies at the carboxy-terminal end of the polypeptide. A regulatory domain that either binds inducer or is phosphorylated is present at the amino-terminal end. The region of the polypeptide responsible for transcriptional activation is in the middle. This region has an ATP-binding domain and an ATPase activity that cleaves ATP to ADP. As mentioned in chapter 11 (Box 11.2), the N-terminal domain somehow masks the middle domain for activation unless the N-terminal domain has bound inducer or has been phosphorylated.

Detailed Model for Activation of Ntr-Type Promoters

The NtrC-activated promoters, including p_2 of the *glnA-ntrB-ntrC* operon, are unusual in that NtrC binds to an **upstream activator sequence (UAS)**, which lies more than 100 bp upstream of the promoter. For most prokaryotic promoters, the activator protein binding sequences are adjacent to the promoter (see Figure 11.1B). Activation at a distance, such as occurs with the NtrC-activated promoters, is much more common in eukaryotes, where many examples are known.

Figure 12.10 also shows one detailed model for how NtrC~P activates transcription from p_2 and other Ntr-activated promoters. NtrC oligomers bind strongly to the UAS only when the protein is in the NtrC~P form. Because the UAS has an inverted repeat and one NtrC~P polypeptide can bind to each copy of the repeated sequence, two dimers of NtrC~P bind to the UAS. The DNA bends between the UAS and the promoter so that the two molecules of NtrC~P bound at the UAS make contact with both the RNA polymerase and the promoter. Like other activators, NtrC~P must touch the RNA polymerase before the RNA polymerase can initate transcription. However, σ^{54}-type promoters may require that the activator protein also help separate the strands of the DNA at the promoter, which the RNA polymerase cannot do by itself. This may be the func-

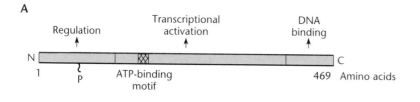

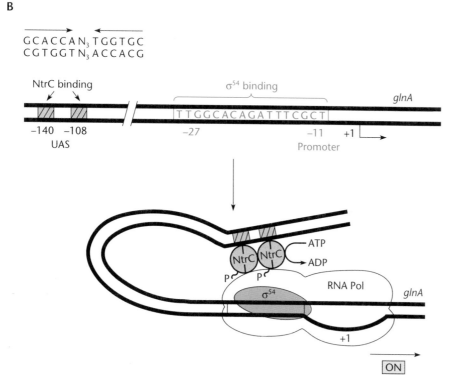

Figure 12.10 Model for the activation of the p_2 promoter by phosphorylated NtrC protein. (A) Functions of the various domains of NtrC. P denotes a phosphate. (B) Two dimers of phosphorylated NtrC bind to the inverted repeats in the upstream activator sequence (UAS). The DNA is bent between the UAS and the promoter, allowing contact between NtrC and the RNA polymerase bound at the promoter more than 100 bp downstream.

tion of the ATPase activity of the NtrC protein. Cleaving ATP to ADP may generate the energy required to separate the strands of the DNA in the promoter as shown.

This is only one of many models devised to explain the activation. Another model proposes that phosphorylation of NtrC allows the formation of higher NtrC multimers, which, in conjunction with the binding to the UAS, activates the ATPase activity of NtrC~P, thereby activating the promoter. However, in neither model is it clear how ATP cleavage to ADP helps separate the strands of DNA at the promoter.

Function of the Other Promoters of the *glnA-ntrB-ntrC* Operon

As mentioned, the p_2 promoter is only one of three promoters that service the *glnA-ntrB-ntrC* operon (Figure 12.8). The other promoters are p_1 and p_3. The p_1 promoter is further upstream of *glnA* than is p_2, and most of the transcription that initiates at the p_1 promoter terminates at the transcription termination signal just downstream of *glnA*, as shown in Figure 12.8. The p_3 promoter is between the *glnA* and *ntrB* genes and services the *ntrB* and *ntrC* genes.

The p_1 and p_3 promoters use σ^{70} and so do not require NtrC~P for their activation. In fact, they are repressed by NtrC~P, so they will not be used if NH_3 concentrations are low. The function of these promoters is presumably to ensure that the cell will have some glutamine synthetase and NtrB and NtrC when NH_3 concentrations are high. Unless glutamine is provided in the medium, the cell must have glutamine synthetase to make glutamine for use as the NH_2 group donor in some biosynthetic reactions and in protein synthesis. The cell must also have some NtrB and NtrC in case conditions change suddenly from a high-NH_3 to a low-NH_3 environment, in which the products of the Ntr genes are suddenly needed.

ADENYLYLATION OF GLUTAMINE SYNTHETASE

Regulating the transcription of the *glnA* gene is not the only way the activity of glutamine synthetase is regulated in the cell. This activity is also modulated by the adenylylation of (transfer of AMP to) a specific tyrosine in the enzyme by an adenylyltransferase (ATase) enzyme when NH_3 concentrations are high. The adenylylated form of the glutamine synthetase enzyme is less active and is also much more susceptible to feedback inhibition by glutamine than is the unadenylylated form (see chapter 11 for an explanation of feedback inhibition). This makes sense, considering how glutamine synthetase plays different roles when NH_3 concentra-

tions are high and when they are low. When NH_3 concentrations are high, the primary role of the glutamine synthetase is to make glutamine for protein synthesis, which requires less enzyme; the little enzyme activity that remains should be feedback inhibited to ensure that the cells will not accumulate too much glutamine. When NH_3 concentrations are low, more enzyme is required. In this situation, the enzyme should not be feedback inhibited, because its major role is to assimilate nitrogen.

The state of adenylylation of the glutamine synthetase enzyme is also regulated by GlnD and the state of the P_{II} protein. When the cells are in low NH_3 so that the P_{II} protein has UMP attached (Figure 12.7), the adenylyltransferase will remove AMP from glutamine synthetase. When the cells are in high NH_3, so that the P_{II} protein does not have UMP attached, the P_{II} protein will bind to and stimulate the adenylyltransferase to add more AMP to the glutamine synthetase.

The P_{II} protein may not be the only way to stimulate the deadenylylation of glutamine synthetase in *E. coli*, however. Recent work has suggested that another protein, GlnK, may form GlnK-UMP in response to ammonia deprivation. The GlnK-UMP protein can also stimulate the adenylyltransferase to adenylylate glutamine synthetase when NH_3 concentrations are high (see van Heeswijk et al., Suggested Reading). Clearly, many pathways interact to regulate nitrogen source utilization by bacteria.

Coordination of Catabolite Repression, the Ntr System, and the Regulation of Amino Acid Degradative Operons

Not only must bacteria sometimes use one or more of the 20 amino acids as a nitrogen source, but also they must sometimes use amino acids as carbon and energy sources. The type of amino acids that different bacteria can use varies. For example, *E. coli* can use only alanine, proline, arginine, and tryptophan as carbon and nitrogen sources, whereas some *Salmonella* species can also use proline, arginine, and histidine. Like the sugar-utilizing and biosynthetic operons, the amino acid-utilizing operons not only have their own specific regulatory genes, so that they are transcribed only in the presence of their own inducer, but often are also part of larger regulons. As discussed in this section, they are often under Ntr regulation and so will not be induced in the presence of their inducer while a better nitrogen source such as NH_3 is in the medium. In some bacteria, these operons are also under the control of the catabolite regulon and will not be expressed in the presence of better carbon sources, such as glucose.

Sometimes multiple levels of operon control can be a disadvantage to the bacterium. For example, *Salmonella typhimurium* cells will starve for nitrogen if they are growing in the presence of histidine plus glucose but no other nitrogen source. The glucose apparently prevents CAP from activating the promoter for the *his* operon, so that the cells cannot use the histidine as a nitrogen source. Glucose itself contains no nitrogen, and so although the cells can use it as a carbon source, they will starve for nitrogen.

In addition to their potential as nitrogen, carbon, and energy sources, amino acids are necessary for other purposes, including protein synthesis and, in the case of proline, osmoregulation. Therefore, the use of amino acids as carbon and nitrogen sources can present strategic problems for the cell. The way in which all these potentially conflicting regulatory needs are resolved is bound to be complicated.

Genetic Analysis of Nitrogen Regulation in Enteric Bacteria

The present picture of nitrogen regulation in bacteria began with genetic studies. Most of this work was first done in *K. aerogenes*, although some genes were first found in *S. typhimurium* or *E. coli*.

The first indications of the central role of glutamine and glutamine synthetase in Ntr regulation came from the extraordinary number of genes that, when mutated, could affect the regulation of the glutamine synthetase or give rise to an auxotrophic growth requirement for glutamine (see Magasanik, Suggested Reading). The *gln* genes were originally named *glnA*, *glnB*, *glnD*, *glnE*, *glnF*, *glnG*, and *glnL* (Table 12.2). These genes are not lettered consecutively because, as often happens in genetics, genes presumed to exist because of a certain phenotype were later found to not exist or to be the same as another gene, and so their letters were retired. Sorting out the various contributions of the *gln* gene products to arrive at the model for nitrogen regulation outlined above was a remarkable achievement. It took many years and required the involvement of many people. In this section, we describe how the various *gln* genes were first discovered and how the phenotypes caused by mutations in the genes led to the model.

THE *glnB* GENE
The *glnB* gene encodes the P_{II} protein. The first *glnB* mutations were found among a collection of *K. aerogenes* mutants with the Gln⁻ phenotype, the inability to multiply without glutamine in the medium. Mutations in *glnB* apparently can prevent the cell from making enough glutamine for growth. However, early genetic

evidence indicated that these *glnB* mutations do not exert their Gln⁻ phenotype by inactivating the P_{II} protein. As evidence, transposon insertions and other mutations that should totally inactivate the *glnB* gene (null mutations) do not result in the Gln⁻ phenotype. In fact, null mutations in *glnB* are intragenic suppressors of the Gln⁻ phenotype of the original *glnB* mutations (see chapter 3 for a discussion of the different types of suppressors).

We now know that the original *glnB* mutations do not inactivate P_{II} but, rather, change the binding site for UMP so that UMP cannot be attached to it by GlnD. This should have two effects. P_{II} without UMP will bind to NtrB~P and prevent phosphorylation of NtrC. Therefore, even in low NH_3 concentrations, little glutamine will be synthesized. By itself, however, this effect does not explain the Gln⁻ phenotype, since the *glnA* gene can also be transcribed from the p_1 promoter, which does not require NtrC~P for activation (see above). It is the second function of P_{II}—the stimulation of the adenylyltransferase—that causes the Gln⁻ phenotype. In the mutants, enough P_{II} without UMP attached accumulates to stimulate the adenylyltransferase to the extent that glutamine synthetase will be too heavily adenylated to synthesize enough glutamine for growth. This also explains why null mutations in *glnB* do not cause the Gln⁻ phenotype. By inactivating P_{II} completely, null mutations prevent the P_{II} protein from stimulating the adenylyltransferase so that less glutamine synthetase is adenylylated and enough glutamine is synthesized for growth.

THE *glnD* GENE
The *glnD* gene was also discovered in a collection of mutants with mutations that cause the Gln⁻ phenotype. However, in this case, experiments showed that null mutations in *glnD* cause the Gln⁻ phenotype. Furthermore, null mutations in *glnB* suppress the Gln⁻ phenotype of null mutations in *glnD*. These observations are consistent with the above model. Since the GlnD protein is the enzyme that transfers UMP to P_{II}, null mutations in *glnD* should behave like the original *glnB* mutations and prevent UMP attachment to P_{II} but leave the P_{II} protein intact. This will make the cells Gln⁻ for the reasons given above. Null mutations in *glnD* are suppressed by null mutations in *glnB* because the absence of GlnD protein does not matter if the cell contains no P_{II} protein to bind to the adenylyltransferase and stimulate the attachment of AMP to glutamine synthetase. The glutamine synthetase without AMP attached will remain active and synthesize enough glutamine for growth.

THE *glnL* GENE

The *glnL* gene was later renamed *ntrB* to better reflect its function. Mutations in this gene were first discovered not because they cause the Gln⁻ phenotype but because they are extragenic suppressors of *glnD* and *glnB* mutations. These mutations do not inactivate NtrB completely; instead, they leave NtrB intact and prevent the binding of P_{II} so that NtrB will transfer its phosphate to NtrC regardless of the presence of P_{II}-UMP. Mutants with null mutations in *ntrB* also retain some regulation of the *glnA-ntrB-ntrC* operon. As previously mentioned, this evidence supports the hypothesis of an alternative NtrB-independent pathway of NtrC phosphorylation (see the section on another Ntr regulatory pathway, above).

THE *glnG* GENE

The *ntrC* gene, originally called *glnG*, was discovered because mutations in it can suppress the Gln⁻ phenotype of *glnF* mutations. The *glnF* gene encodes the nitrogen sigma σ⁵⁴ (see below). Further work showed that the suppressors of *glnF* are null mutations in *ntrC* because transposon insertions and other mutations that inactivate the *ntrC* gene also suppress *glnF* mutations. Null mutations in *ntrC* do not cause the Gln⁻ phenotype, because the *glnA* gene can also be transcribed from the p_1 promoter, which does not require NtrC for activation. They do, however, prevent the expression of other Ntr operons that require NtrC~P for their activation, many of which do not have alternative promoters.

Some mutations in *ntrC* do cause the Gln⁻ phenotype, however. These presumably are mutations that change NtrC so that it can no longer activate transcription from p_2, but it retains the ability to repress transcription from p_1 because it can be phosphorylated.

THE *glnF* GENE

As mentioned, the glnF gene, now renamed *rpoN*, encodes the nitrogen sigma, σ⁵⁴. The gene was also discovered in a collection of Gln⁻ mutants. Without σ⁵⁴, the p_2 promoter cannot be used to transcribe the *glnA-ntrB-ntrC* operon. By itself, inactivation of p_2 would not be enough to cause the Gln⁻ phenotype, since the *glnA* gene can also be transcribed from the p_1 promoter, which does not require σ⁵⁴. However, the NtrC~P form of NtrC will repress the p_1 promoter if the cells are growing in low NH₃ concentrations (see above). In high NH₃ concentrations, the p_1 promoter will not be repressed, but the small amount of glutamine synthetase synthesized will be heavily adenylylated, preventing the synthesis of sufficient glutamine for growth. This interpretation of the Gln⁻ phenotype of *rpoN* mu-

tations is consistent with the fact that null mutations in *ntrC* suppress the Gln⁻ phenotype of *rpoN* mutations as discussed above. Without NtrC~P present to repress the p_1 promoter, sufficient glutamine synthetase will be synthesized from the p_1 promoter, even in low NH₃ concentrations.

Regulation of Porin Synthesis

Most bacteria can survive outside a eukaryotic host. Because they are free living, they often find themselves in media of changing solute concentrations and consequently variable osmolarity. The osmotic pressure is normally higher inside the inner membrane than outside it. This pressure would cause the bacterium to swell, but the rigid cell wall keeps the cell from expanding. Even the cell wall is not invincible, however, and bacteria must keep the difference in osmotic pressure inside and outside the cell from becoming too great. The ability to monitor osmolarity can be important to bacteria for a second reason: bacteria also sometimes sense changes in their external environment by the changes in osmolarity. In fact, one way in which pathogenic bacteria sense that they are inside a host, and induce their virulence genes, is by the much higher osmolarity inside the host (see the section on global regulation in pathogenic bacteria, below). The systems by which bacterial cells sense these changes in osmolarity and adapt are global regulatory mechanisms, and many genes are involved.

Much is known about how bacteria respond to media of different osmolarities. One way they regulate the differences in osmotic pressure across the membrane is by excreting or accumulating K⁺ ions and other solutes such as proline and glycine betaine. Gram-negative bacteria such as *E. coli* have the additional problem of maintaining an equal osmotic pressure across the outer membrane. They achieve this in part by synthesizing oligosaccharides in the periplasmic space to balance solutes in the external environment.

One of the major mechanisms by which gram-negative bacteria balance osmotic pressure across the outer membrane is by synthesizing pores to let solutes into and out of the periplasmic space. These pores are composed of outer membrane proteins called **porins**. To form pores, these proteins trimerize in the outer membrane.

The two major porin proteins in *E. coli* are OmpC and OmpF. Pores composed of OmpC are slightly smaller than those composed of OmpF, and the size of the pores can affect the rate at which solutes pass through the pores and thus confer protection under some conditions. For example, the smaller pores, composed of OmpC, may prevent the passage of some toxins such as the bile salts in the intestine. The larger pores, composed of OmpF, may allow more rapid passage of solutes and so confer an advantage in dilute aqueous environments. Accordingly, *E. coli* cells growing in a medium of high osmolarity, such as the human intestine, have more OmpC than OmpF, whereas *E. coli* cells growing in a medium of low osmolarity, such as dilute aqueous solutions, have less OmpC than OmpF.

Other environmental factors besides osmolarity can alter the ratio of OmpC to OmpF. This ratio increases at higher temperatures or when the cell is under oxidative stress due to the accumulation of reactive forms of oxygen (see chapter 10). The ratio also increases when the bacterium is growing in the presence of organic solvents such as ethanol or some antibiotics. Presumably, the smaller size of OmpC pores limits the passage of many toxic chemicals into the cell. The reason for the temperature effect is more obscure, however. There is no obvious reason why the size of the pores should help defend against higher temperatures. One possibility is that the bacterium normally uses a temperature increase as an indication that it has passed from the external environment into the intestine of a warm-blooded vertebrate host, its normal habitat. It then must synthesize mostly OmpC-containing pores to keep out toxic materials such as bile salts, as mentioned above. While not yet completely understood, the regulation of OmpC and OmpR porin synthesis in *E. coli* has served as a model for systems that allow the cell to sense the external environment and adjust their gene expression accordingly; therefore, we shall discuss this subject in some detail.

Genetic Analysis of Porin Regulation

As in the genetic analysis of any regulatory system, the first step in studying the osmotic regulation of porin synthesis in *E. coli* was to identify the genes whose products are involved in the regulation. The isolation of mutants defective in the regulation of porin synthesis was greatly aided by the fact that the some of the porin proteins also serve as receptors for phages and bacteriocins, so that mutants that lack a particular porin will be resistant to a given phage or bacteriocin. This fact offers an easy selection for mutants defective in porin synthesis. Only mutants that lack a certain porin in the outer membrane will be able to form colonies in the presence of the corresponding phage or bacteriocin.

Using such selections, investigators isolated mutants that had reduced amounts of the porin protein, OmpF, in their outer membrane. These mutants were found to have mutations in two different loci, which were named *ompF* and *ompB*. Mutations in the *ompF* locus can completely block OmpF synthesis, whereas mutations

in *ompB* only partially prevent its synthesis. The quantitative difference in the effect of mutations in the two loci suggested that *ompF* is the structural gene for the OmpF protein and that an *ompB*-encoded protein(s) is required for the expression of the *ompF* gene. Complementation experiments showed that the *ompB* locus actually consists of two genes, *envZ* and *ompR*. Using *lacZ* fusions to *ompF* to monitor the transcription of the *ompF* gene (see chapter 2), investigators confirmed that EnvZ and OmpR are required for optimal transcription of the *ompF* gene (see Hall and Silhavy, Suggested Reading).

EnvZ AND OmpR: A SENSOR-RESPONSE REGULATOR PAIR OF PROTEINS

The *envZ* and *ompR* genes were cloned and sequenced by methods such as those discussed in chapter 15. Similarities in amino acid sequence between EnvZ and OmpR and other sensor-response regulators, including NtrB and NtrC, suggested that these proteins are also a sensor-response regulator pair of proteins. Like many sensor proteins, the EnvZ protein is an inner membrane protein, with its N-terminal domain in the periplasm and its C-terminal domain in the cytoplasm (see the figure in Box 12.3). Current evidence indicates the N-terminal domain of EnvZ apparently senses an unknown signal in the periplasm that indicates that the osmolarity is low and transfers this information to the cytoplasmic domain. The information is then transferred to the OmpR protein, a transcriptional regulator that regulates transcription of the porin genes. Like the NtrB protein, the EnvZ protein is known to be autophosphorylated, and its phosphate is also known to be transferred to OmpR.

An Early Model for *ompC* and *ompF* Regulation
Because of the similarity of NtrB and NtrC to EnvZ and OmpR, respectively, it was tempting to speculate that EnvZ and OmpR might regulate the transcription of *ompC* and *ompF* by the following simple model. According to this model, the EnvZ protein would be an autokinase that phosphorylates itself when the osmolarity is high. This phosphate is then transferred to OmpR. The phosphorylated form of OmpR then activates transcription of *ompC*, whereas the unphosphorylated form of OmpR activates transcription of *ompF*. Therefore, the relative rates of transcription of *ompC* and *ompF* would depend on the state of phosphorylation of OmpR.

Genetic Evidence Fails To Support a Simple Model
Several lines of genetic evidence indicate that the phosphorylated form (OmpR~P) does not simply activate

transcription of *ompC* while the unphosphorylated form of OmpR activates the *ompF* gene. The regulation is more complicated; therefore, at this point, we will outline some of the genetic evidence.

1. Mutant phenotypes. Table 12.3 lists several relevant phenotypes. According to the simple model, mutations that totally inactivate the EnvZ protein (*envZ* null mutations) should completely prevent the phosphorylation of OmpR under any conditions, freezing it in the form that activates transcription of *ompF*. Therefore, *envZ* null mutations would be predicted to completely prevent the transcription of *ompC* but allow high-level transcription of *ompF*. However, the evidence shows that while *envZ* null mutations do prevent the transcription of *ompC*, they allow only very limited transcription of *ompF* (shown as OmpF^{+-} in Table 12.3; see Slauch et al., Suggested Reading). The EnvZ protein is apparently required to activate the transcription of *both* porin genes, perhaps because some OmpR~P is required to activate the transcription of both genes.

2. Constitutive mutations in ompR. One type of constitutive mutation, called *ompR2*(Con) in Table 12.3, prevents the expression of *ompC* but causes the constitutive expression of *ompF*, even when coupled with a null mutation in *envZ*. A second type of constitutive mutation, called *ompR3*(Con), causes the constitutive expression of *ompC* but prevents the expression of *ompF*. The existence of these constitutive mutations could have been predicted from the simple model. In analogy to the AraC activator (see chapter 11), the OmpR activator may exist in two forms (see chapter 11), one when it is active and another when it is not. The *ompR3*(Con) mutations could change OmpR into the form in which it normally exists when phosphory-

lated, even without phosphorylation. It could then activate the transcription of *ompC* but not *ompF*.

3. Diploid analysis. The behavior of the *ompR*(Con) mutations in complementation tests is hard to reconcile with the simple model. According to the simple model, in partial diploids carrying both a constitutive allele and a wild-type allele (*envZ*$^+$, *ompR3*(Con)/*envZ*$^+$, *ompR*$^+$ in Table 12.3), both *ompC* and *ompF* would be predicted to be expressed at high levels, even at low osmolarity, because the mutant OmpR protein should activate the transcription of *ompC* whereas the wild-type OmpR protein should activate transcription of *ompF*. In genetic terms, the constitutive mutations should be dominant over the wild-type allele for the expression of *ompC* but recessive for the expression of *ompF* under conditions of low osmolarity. Instead, as shown in Table 12.3, *ompF* is not expressed in the partial diploids, even in media of low osmolarity, although *ompC* is constitutively expressed.

A Current Model

A current model for how EnvZ and OmpR regulate the transcription of the porin genes incorporates these and other observations about the regulation. The EnvZ protein is known to have both phosphotransferase and phosphatase activities, allowing it to both donate a phosphoryl group to and remove one from OmpR. Whether the phosphorylated or unphosphorylated form of OmpR predominates depends on whether the phosphotransferase or phosphatase activity of EnvZ is most active. Under conditions of high osmolarity, the phosphotransferase activity predominates and the levels of phosphorylated OmpR (OmpR~P) are high. Under conditions of low osmolarity, the phosphatase activity predominates, and most of the OmpR is unphosphorylated.

To explain how *envZ* null mutations can prevent the optimal transcription of both genes, we must propose that the phosphorylated form of OmpR is required to activate transcription of both the *ompC* and *ompF* genes. But how can the relative levels of phosphorylated OmpR regulate the transcription of *ompC* and *ompF* so that higher levels of OmpR~P allow the transcription of one gene to predominate whereas lower levels allow transcription of the other to predominate? One solution to this question is to propose the existence of multiple binding sites for OmpR~P upstream of the promoters for the *ompF* and *ompC* genes (see Pratt et al., Suggested Reading). Some of these sites bind OmpR~P more tightly than others; in other words, some are high-affinity sites whereas others are low-affinity sites. By binding to these sites, the OmpR~P protein can either

TABLE 12.3	Phenotypes of *envZ* and *ompR* mutations
Genotype	**Phenotype**
envZ$^+$ *ompR*$^+$	OmpC$^+$ OmpF$^+$
envZ$^+$ *ompR1*	OmpC$^-$ OmpF$^-$
envZ(null) *ompR*$^+$	OmpC$^-$ OmpF^{+-a}
envZ$^+$ *ompR2*(Con)	OmpC$^-$ OmpF$^+$ (low osmolarity)
	OmpC$^-$ OmpF$^+$ (high osmolarity)
envZ(null) *ompR2*(Con)	OmpC$^-$ OmpF$^+$ (low osmolarity)
	OmpC$^-$ OmpF$^+$ (high osmolarity)
envZ$^+$ *ompR3*(Con)	OmpC$^+$ OmpF$^-$ (low osmolarity)
	OmpC$^+$ OmpF$^-$ (high osmolarity)
envZ$^+$ *ompR3*(Con)/*envZ*$^+$ *ompR*$^+$	OmpC$^+$ OmpF$^-$ (low osmolarity)
	OmpC$^+$ OmpF$^-$ (high osmolarity)

^{a}+ – indicates that OmpF levels are reduced but not eliminated.

activate or repress transcription from the promoters, depending on the position of the binding site relative to the promoter (see Figure 11.1). Upstream of the *ompF* gene are both high- and low-affinity sites, while upstream of the *ompC* gene are only low-affinity sites. Binding of OmpR~P to the high-affinity sites upstream of *ompF* will activate its transcription, but binding of OmpR~P to the low-affinity sites upstream of *ompF* will repress its transcription. OmpR~P binding to the low-affinity sites upstream of *ompC* will activate the transcription of *ompC*. Hence, when the osmolarity is low, so that the cell contains less OmpR~P, the OmpR~P will be bound only to the high-affinity sites upstream of *ompF* and activate the transcription of this gene. However, when osmolarity is high and thus the concentration of OmpR~P is high, the OmpR~P protein will be bound to both the high- and low-affinity sites. By binding to the low-affinity sites upstream of *ompF*, OmpR~P will repress *ompF* transcription, but its binding to the low-affinity sites upstream of *ompC* will activate the transcription of *ompC*. This model is consistent with the dual role of other regulatory proteins, such as AraC, which can be either repressors or activators depending on where they are bound relative to the promoter (see chapter 11). However, refining the model will require more evidence.

The EnvZ protein is a transmembrane protein that can communicate information about what is happening outside the cell to the internal regulatory pathways. Therefore, studies with this system should further our understanding of how cells sense the external environment. How the EnvZ protein achieves this feat should tell us much about how information is transferred across cellular membranes in general.

MicF

As mentioned, the OmpC/OmpF ratio increases not only when the osmolarity increases but also when the temperature increases or when toxic chemicals including organic solvents such as ethanol are in the medium. A different mechanism is responsible for this regulation. It involves a regulatory RNA called MicF. The *micF* gene is immediately upstream of the *ompC* gene and is transcribed in the opposite direction from *ompC*.

There are many examples of regulatory RNAs in bacteria, including RNA I, which regulates the replication and incompatibility of ColE1 plasmids (see chapter 4). Other regulatory RNAs inhibit the translation of mRNAs such as that for the transposase of the transposon Tn*10*. In general, regulatory RNAs have sequences of nucleotides that are complementary to the sequences of a target RNA, so that they can inhibit the function of the target RNA by base pairing with it. The double-stranded region thus formed may either mask important sequences in the target RNA or create a substrate for a cellular RNase and thereby cause the target RNA to be degraded. Regulatory RNAs that can base pair with an mRNA and thereby inhibit its translation are called **antisense RNAs** by analogy to the antisense strand of DNA, which has the complementary sequence of the sense, or coding, strand of DNA (see chapter 1).

The MicF RNA is an antisense RNA to the mRNA for OmpF. It is not complementary to the entire OmpF mRNA but only to a short sequence including the translational initiation region (TIR), which contains the Shine-Dalgarno sequence and the AUG initiation codon. By forming double-stranded RNA with the TIR, MicF can prevent access by ribosomes to the TIR and hence translation of OmpF. As a consequence, *E. coli* cells containing higher concentrations of MicF RNA, the situation that prevails when they are growing at higher temperatures or in media containing organic solvents or certain antibiotics, will make less OmpF protein and therefore have a higher OmpC to OmpF ratio.

The cellular levels of MicF RNA increase under certain conditions because the promoter for the *micF* gene contains binding sites for many transcriptional activators. Each seems to work independently and activates transcription from the *micF* promoter under its own particular set of conditions. For example, the transcriptional activator SoxS activates transcription from the *micF* promoter when the cell is under oxidative stress. Another, MarA, activates transcription of *micF* when weak acids or some antibiotics are present. Even OmpR may activate the transcription of *micF*, allowing another level of regulation of OmpF synthesis by osmolarity. Other proteins of unknown function may also bind close to the *micF* promoter and activate or repress transcription of the *micF* gene. The regulation of porin synthesis by osmolarity and other environmental factors is obviously central to cell survival, which is why it is so complicated and involves so many interacting systems.

Regulation of Virulence Genes in Pathogenic Bacteria

The virulence genes of pathogenic bacteria represent another type of global regulon. Virulence genes allow pathogenic bacteria to adapt to their eukaryotic hosts and cause disease. Most pathogenic bacteria express their virulence genes only in the eukaryotic host; somehow, the conditions inside the host turn on the expression of these genes. Virulence genes can be identified because mutations that inactivate them render the bacterium nonpathogenic but do not affect its growth outside the host. We discuss some examples of the regulation of virulence regulons in this section.

Diphtheria

Diphtheria is caused by the bacterium *Corynebacterium diphtheriae*, a gram-positive bacterium that colonizes the human throat. It is spread from human to human through aerosols created by coughing or sneezing. The colonization of the throat by itself results in few symptoms. However, strains of *C. diphtheriae* that harbor a prophage named β (see chapter 7) produce diphtheria toxin, which is responsible for most of the symptoms. Excreted from the bacteria in the throat, the toxin enters the bloodstream, where it does its damage.

DIPHTHERIA TOXIN

The diphtheria toxin is a member of a large group of A-B toxins, so named because they have two subunits, A and B. In most A-B toxins, the A subunit is an enzyme that damages host cells and the B subunit helps the A subunit enter the host cell by binding to specific cell receptors. The two parts of the diphtheria toxin are first synthesized from the *tox* gene as a single polypeptide chain, which is cleaved into the two subunits A and B as it is excreted from the bacterium. These two subunits are held together by a disulfide bond until they are translocated into the host cell, where the disulfide bond is reduced and broken, releasing the individual A subunit into the cell.

The action of the diphtheria toxin A subunit on eukaryotic cells is quite well understood. The A subunit enzyme specifically ADP-ribosylates (adds ADPribose to) a modified histidine amino acid of the translation elongation factor EF-2 (called EF-G in bacteria; see chapter 2). The ADP-ribosylation of the translation factor blocks translation and kills the cell. Interestingly, the opportunistic pathogen *Pseudomonas aeruginosa* makes a toxin that is identical in action to the diptheria toxin, although it has a somewhat different sequence.

Regulation of the *tox* Gene

Like many virulence genes, the *tox* gene is turned on only under the conditions present in the infected organism. One factor that characterizes the eukaryotic environment is the lack of available iron. All of the iron in the body is tied up in other molecules, such as transferrins and hemoglobin. The bacterium senses this lack of free iron, concludes that it is in the eukaryotic host, and turns on the *tox* gene and other genes required for pathogenesis.

Interestingly, even though the *tox* gene encoding the toxin of *C. diphtheriae* is carried on the lysogenic phage β, it is regulated by the product of a chromosomal gene called *dtxR*, as shown in Figure 12.11 (see Schmidt and Holmes, Suggested Reading). The DtxR protein is a transcriptional repressor that binds to the operator region for the *tox* gene only in the presence of ferrous ions (Fe^{2+}) and some other divalent cations. Therefore, the *tox* gene is expressed only when ferrous ions are limiting, the condition that prevails in the eukaryotic host. This is an interesting example of close cooperation between normal chromosomal genes of a bacterium and genes encoded by a foreign DNA element.

IRON REGULATION

Iron is not only a signal of the internal eukaryotic environment but also an essential nutrient for bacteria. Thus, iron limitation presents a problem for all bacteria but for pathogenic bacteria in particular. To multiply in a eukaryotic host, pathogenic bacteria must extract the iron from the transferrins and other proteins to which it is bound. For this purpose, *C. diphtheriae* and many other bacteria use siderophores, which are small molecules excreted from the cells. Siderophores bind Fe^{2+} more tightly than do other molecules and so can wrest Fe^{2+} from them. They then are transported back into the bacterial cell with their "catch" of Fe^{2+}.

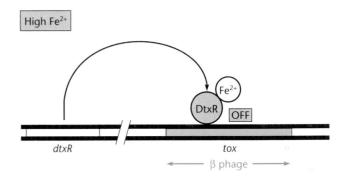

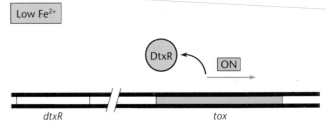

Figure 12.11 Regulation of the *tox* gene of the *Corynebacterium diphtheriae* prophage β. The DtxR repressor protein binds to the operator for the *tox* gene only in the presence of ferrous ions (Fe^{2+}).

The genes for making the siderophores and a high-efficiency transport system for iron are also regulated by the DtxR protein in *C. diphtheriae*. Therefore, these components are synthesized only when iron is limiting.

THE Fur REGULON OF *E. COLI*

The DtxR protein of *C. diphtheriae* is highly homologous to a protein called Fur, which is responsible for iron regulation in *E. coli*. Over 30 *E. coli* genes, including toxin genes in pathogenic strains of *E. coli*, are regulated by Fur. Like DtrX, Fur is an iron-dependent repressor that binds to operators only in the presence of Fe^{2+}. Genes highly homologous to the *E. coli fur* repressor gene and the *dtxR* gene of *C. diphtheriae* have been found in many other bacteria, including *P. aeruginosa*, *Yersinia pestis*, and *Vibrio cholerae*, and presumably also encode iron-dependent repressors.

Cholera

The disease cholera is another well-studied example of the global regulation of virulence genes. *V. cholerae*, the causative agent, is a gram-negative bacterium that is spread through water contaminated with human feces. The disease continues to be a major health problem worldwide, with periodic outbursts, especially in countries with poor sanitation. When ingested by a human, *V. cholerae* colonizes the small intestine, where it synthesizes cholera toxin, which acts on the mucosal cells to cause a severe form of diarrhea. Like the diphtheria toxin, the cholera toxin is encoded by a lysogenic phage, in this case a relative of the single-stranded DNA phage M13 (see chapter 7). Other virulence determinants are the flagellum that allows the bacterium to move in the mucosal layer of the small intestine and pili that allow it to stick to the mucosal surface.

CHOLERA TOXIN

The mechanism of action of the cholera toxin has been the subject of intense investigation, in part because of what it reveals about the normal action of eukaryotic cells. The cholera toxin is composed of two subunits, CtxA and CtxB. Like diphtheria toxin, the CtxA subunit of cholera toxin is an ADP-ribosylating enzyme. However, rather than ADP-ribosylating an elongation factor for translation, the cholera toxin ADP-ribosylates a mucosal cell membrane protein called Gs, which is part of a signal transduction pathway that regulates the activity of the adenylcyclase enzyme that makes cAMP. The ADP-ribosylation of Gs causes cAMP levels to rise and alters the activity of transport systems for sodium and chloride ions. The loss of sodium and chloride ions causes loss of water from the cells, resulting in severe diarrhea.

Regulation of Cholera Toxin Synthesis

The *ctxA* and *ctxB* genes encoding the cholera toxin are part of a large regulon including as many as 20 genes. In addition to the *ctx* genes, the genes of this regulon include those encoding pili, colonization factors, and outer membrane proteins related to osmoregulation. Although some of these genes, including the *ctx* genes, are carried on the prophage, others, including the pilin genes, are carried on the bacterial chromosome. The transcription of the genes of this regulon are activated only under conditions of high osmolarity and in the presence of certain amino acids, conditions that may mimic those in the small intestine. Here we describe three genes involved in *ctx* regulation.

1. ToxR. The cholera virulence regulon is under the control, either directly or indirectly, of the activator protein ToxR, the product of the *toxR* gene (see DiRita, Suggested Reading). The ToxR protein combines the functions related to both proteins of the two-component sensor and response regulator type of system in the same polypeptide. That is, it is involved in both sensing and transcriptional activation. Like EnvZ, the ToxR polypeptide traverses the inner membrane so that the carboxy-terminal part of the protein is in the periplasm, where it can sense the external environment. However, the amino-terminal part is in the cytoplasm, where, like OmpR, it can activate the transcription of genes under its control.

2. ToxS. The ToxR protein is not activated by phosphorylation, unlike orthodox response regulator proteins. Clues to its activation lie with another protein, ToxS. The *toxS* gene is immediately downstream of *toxR* in the same operon, and the gene was discovered because mutations that inactivate it also prevent expression of the genes of the ToxR regulon. The ToxS protein is also anchored in the inner membrane, but a large domain sticks out into the periplasm.

A model for how ToxS might activate ToxR is shown in Figure 12.12. The ToxR protein is a monomer, composed of only one polypeptide encoded by the *toxR* gene, when the bacterium is outside the human body. However, ToxR must be a dimer before it can activate transcription of the cholera toxin genes and other genes of the regulon under its control. After ToxS receives a signal that the bacterium is in the small intestine, this protein helps ToxR to dimerize. Presumably, the portion of the ToxS protein that is in the periplasm interacts with the carboxy-terminal portions of two ToxR polypeptides, which are also in the periplasm, helping to bring the two polypeptides together. The dimerized

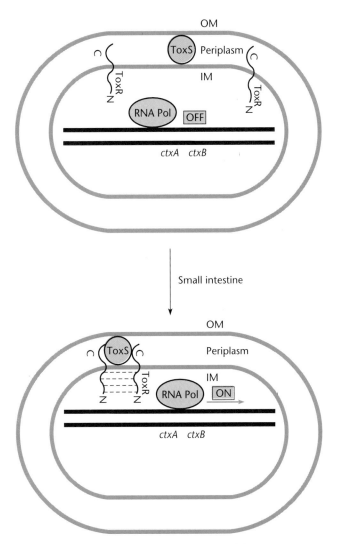

Figure 12.12 Model for the regulation of the toxin genes of *Vibrio cholerae* through dimerization of the ToxR protein by the ToxS protein. The ToxR polypeptides traverse the inner membrane (IM), with the carboxy terminus (C) in the periplasm and the amino terminus (N) in the cytoplasm. The ToxS protein is in the periplasm and helps dimerize the two ToxR polypeptides when the bacterium enters the small intestine. The cytoplasmic N-terminal ends of the dimerized ToxR polypeptides can then activate RNA polymerase (RNA Pol) to transcribe the cholera toxin genes (*ctxA* and *ctxB*). OM, outer membrane.

ToxR protein then activates transcription of the cholera toxin genes and other genes of the regulon. A question that this model leaves unanswered is why the ToxS protein helps dimerize ToxR only in the small intestine or under conditions that mimic those in the small intestine.

Strong evidence that this model is at least partially correct has come from experiments in which the car-

boxy-terminal portion of the ToxR protein was fused to PhoA. The PhoA protein is an alkaline phosphatase enzyme and itself normally dimerizes in the periplasm. The rationale behind the experiment was that the PhoA portion of the fusion protein, while dimerizing itself, may also dimerize the ToxR portion, even without the presence of ToxS. The fusion protein was indeed active and the *ctx* genes were transcribed, even in the absence of ToxS protein, as though dimerization of ToxR were sufficient to activate transcription.

Other evidence suggests that dimerization, by itself, may not be enough to activate the ToxR protein. Some feature related to the membrane anchoring of ToxR may also be required. If dimerization were sufficient, attaching other dimerization domains, such as the ones from the CI repressor protein of λ phage, to the cytoplasmic domain of the ToxR protein should cause ToxR to dimerize in the cytoplasm and activate transcription. However, fusion proteins composed of ToxR and other dimerization domains are still inactive unless the ToxR protein retains its transmembrane domain. Perhaps the ToxR protein must be at least partly in the membrane to be active.

3. ToxT. As mentioned, *V. cholerae* has many virulence genes besides the toxin genes that are also considered part of the ToxR regulon, since *toxR* mutations prevent their expression. However, most of the ToxR regulon genes are not directly controlled by the ToxR protein but, rather, are controlled by an intermediary, the ToxT protein, which is a member of the AraC family of activators (see Box 11.2). Transcription of the *toxT* gene is activated by ToxR, and the transcription of the other genes is then activated by ToxT. Therefore, this is a regulatory cascade. The ToxS protein senses the external environment and dimerizes ToxR, which in turn activates the transcription of the *toxT* gene, whose product then activates the transcription of the other genes of the regulon.

CLONING THE *toxR* GENE

The current level of understanding of the virulence regulon in *V. cholerae* would not have been possible without clones of the *toxR* gene (see Miller and Mekalanos, Suggested Reading). When this work was begun, little was known except that mutations in a gene of *V. cholerae* called *toxR* caused the cells to produce little toxin. This suggested that the *toxR* gene encoded a positive regulatory protein that was required to activate transcription of the toxin genes.

For convenience, the cloning was attempted in *E. coli* rather than in *V. cholerae*. As discussed in chapter 4, many cloning vectors have been developed for *E. coli*, and cloning techniques are much more advanced for this species than for other bacteria. The investigators had reason to hope that the *toxR* gene would be expressed and function in *E. coli*, since *V. cholerae* and *E. coli* are fairly closely related and since most *V. cholerae* genes that had been tested are expressed well in *E. coli*.

Their strategy was to screen a library (see chapter 15) for clones that expressed a protein that could activate transcription from the promoter for the toxin genes, the *ctxA* promoter, in *E. coli*. They first made a partial *Sau*3A library of wild-type *V. cholerae* DNA in an *E. coli* plasmid cloning vector (see chapter 15). This library was then transformed into a strain of *E. coli* harboring a lysogenic phage carrying a transcriptional fusion of *ctxA* to a reporter gene, *lacZ*. The *ctxA* and *lacZ* genes were arranged so that the *lacZ* gene would be transcribed only from the *ctxA* promoter. The investigators then plated the transformants on agar containing X-Gal and observed whether the cells synthesized β-galactosidase, the product of the *lacZ* gene. As mentioned in other chapters, X-Gal turns blue if it is cleaved by β-galactosidase. Most transformants made only faintly blue colonies. However, 2 of about 5,000 transformants made deep blue colonies. The clones in these transformants were presumed to contained the *toxR* gene, which was activating transcription from the *ctxA* promoter.

Further confirmation that the clones contained the *toxR* gene came from complementation tests in which plasmids containing the clones were mobilized into *toxR* mutants of *V. cholerae* with a ColE1-derived plasmid cloning vector, so that it could be mobilized by the F-plasmid Tra functions (see chapter 5). The narrow-host-range ColE1-derived plasmids can nevertheless replicate to some extent in *V. cholerae*, since, as mentioned, *V. cholerae* is closely related to *E. coli*. Because the plasmid clones complemented *toxR* mutations in *V. cholerae* to allow normal synthesis of the toxin, they presumably contained the *toxR* gene.

With clones of the *toxR* gene, it was also possible to construct PhoA fusions to investigate the membrane topology of the ToxR protein (see chapter 14). These studies revealed that the N-terminal part of the ToxR protein is in the cytoplasm while the C-terminal part is in the periplasm. Some of these fusions exhibited ToxR activation of transcription, independent of ToxS, inspiring the model for activation by dimerization discussed above.

Whooping Cough

Another well-studied disease involving global regulation of virulence genes is whooping cough, caused by the gram-negative bacterium *Bordetella pertussis*. Whooping cough is mainly a childhood disease and is character-

ized by uncontrolled coughing, hence the name. The bacteria colonize the human throat and are spread through aerosols resulting from the coughing. Effective vaccines have been developed, but the disease continues to kill hundreds of children worldwide, mainly in areas where the vaccines are not available.

In spite of their very different symptoms, the diseases caused by *V. cholerae* and *B. pertussis* have a similar molecular basis. *B. pertussis* also makes a complex A-B toxin (pertussis toxin) that is in some ways remarkably similar to the cholera toxin. The pertussis toxin has six subunits, although only two of them are identical. One of the subunits (S1) is the enzyme, while the others (S2 to S5) are involved in adhesion to the mucosal surface of the throat. The pertussis toxin is also similar to the cholera toxin in that it ADP-ribosylates a G protein in a signal transduction pathway involved in deactivating the adenylate cyclase, leading to elevated levels of cAMP. However, rather than causing a loss of water from the cells, the elevated cAMP levels seem to cause an increase in mucus production, presumably because the pertussis toxin attacks cells of the throat rather than of the small intestine.

In addition to pertussis toxin, *B. pertussis* also synthesizes a number of other toxins. One is an adenylate cyclase enzyme that enters host cells and presumably directly increases intracellular cAMP levels by synthesizing cAMP. This observation supports the importance of increased cAMP levels to the pathogenesis of the bacterium, although the contribution of this adenylate cyclase to the symptoms is unknown. Other known toxins include one that causes necrotic lesions on the skin of mice and a cytotoxin that is not a protein but a peptidoglycan fragment and that kills ciliated cells of the throat. Other virulence factors are involved in the adhesion of the bacterium to the mucosal layer. The pertussis toxin itself may play a role in such adhesion.

REGULATION OF PERTUSSIS VIRULENCE GENES

Like the virulence genes of *C. diphtheriae* and *V. cholerae*, many of the virulence genes of *B. pertussis* are presumably expressed only when the bacterium enters the eukaryotic host. The regulation of the virulence genes of *B. pertussis* is achieved by a sensor-response regulator pair of proteins encoded by linked genes, *bvgA* and *bvgS* (for *b*ordetella *v*irulence *g*enes); *bvgS* encodes the sensor, and *bvgA* encodes the transcriptional activator.

The BvgS-BvgA system is similar to many other sensor-response regulators in that the BvgS protein is a transmembrane protein, with its N terminus in the periplasm and its C terminus in the cytoplasm, allowing it to communicate information from the external environment across the membrane to the inside of the cell. Also, like many other sensor proteins that work in two-component systems, the BvgS protein is autophosphory-lated in response to some signal from the external environment and donates this phosphate to the BvgA protein, which then activates transcription of the virulence genes. One difference between BvgS-BvgA and other sensor-response regulator pairs is that in the BvgS sensor protein, other amino acids are phosphorylated in addition to the histidine that is autophosphorylated in the other sensors. The significance of these other sites of phosphorylation for the activation of the pertussis virulence genes is not clear.

In vitro, the signal transduction pathway to transcribe the pertussis toxin gene can be activated by growing the bacteria in media with low nicotinamide and magnesium concentrations and possibly low iron concentrations, conditions that presumably mimic those in the throat. Also, expression of the pertussis toxin is highest at 37°C, the temperature of the human body. The BvgA protein may be both a transcriptional activator and a repressor, since the signal transduction pathway seems to activate some virulence genes and repress others. Also, BvgA may activate other regulatory genes that in turn activate additional genes, making a sort of regulatory cascade, like that in the *V. cholerae* system.

CLONING THE *bvgA-bvgS* OPERON

As with the ToxR regulon, the results described above would not have been possible without clones of the *bvg* gene region. The *bvg* locus, where the two genes lie, was first found by Tn*5* transposon insertion mutations that prevent the synthesis of many of the virulence factors of *B. pertussis* (see Stibitz et al. [1988], Suggested Reading). By cloning the kanamycin resistance gene on Tn*5* and the flanking chromosomal sequences, the investigators constructed a clone of the *bvg* region. This clone was then used as a probe in colony hybridizations to screen a library made from wild-type *B. pertussis* DNA to identify bacteria containing the wild-type *bvg* locus. We discuss this general method for cloning genes from gram-negative bacteria other than *E. coli* in detail in chapter 16.

When the *bvg* locus was sequenced, it was found to contain an operon with two open reading frames (ORFs), one of which encoded a protein with sequence homology to sensor autokinases and so was presumably a gene encoding a sensor autokinase. The other ORF encoded a protein with a DNA-binding motif and other features of a transcriptional activator and so presumably encoded one of these proteins. The genes were later named *bvgS* and *bvgA,* respectively.

Clones of the *bvg* genes facilitated the genetic analysis of the activation mechanism. Some mutations in *bvgS* cause the constitutive expression of the virulence genes, even in high concentrations of nicotinamide and magnesium. Also, deletions of the N-terminal periplasmic domain of the protein forestall the requirement for conditions of low nicotinamide and magnesium concentrations to activate the regulon. Apparently, these mutations change BgvS so that it no longer requires a signal from the environment to be autophosphorylated.

PHASE VARIATION OF *BORDETELLA* VIRULENCE

B. pertussis cells spontaneously lose the ability to express many of their virulence factors at a frequency of about 1 in 10^6, which is fairly high for spontaneous mutations. The loss of virulence is reversible, and some bacteria also regain the ability to make the virulence factors at a fairly high frequency.

The switch from virulence to nonvirulence is often called **phase variation**, by analogy to the phenomenon in *Salmonella typhimurium* (see chapter 8). Like *S. typhimurium* cells, which periodically express different flagellin proteins, *B. pertussis* cells may undergo phase variation to avoid the host immune system. Many of the virulence factors, such as the proteins for flagella and pili, are strong antigens, and the reversible loss of the ability to make these virulence factors may help the bacteria establish themselves in the host.

Figure 12.13 illustrates the mechanism of phase variation in *B. pertussis*. Rather than result from inversion of a DNA segment as in *S. typhimurium,* phase variation in *B. pertussis* is caused by a mutation that inactivates the *bvgS* gene, thereby blocking synthesis of all the virulence factors (see Stibitz et al. [1989], Suggested Reading). These mutations add a GC base pair in a string of six GC base pairs in the *bvgS* gene, causing a frameshift and inactivating the product of the *bvgS* gene. The *bvgS* mutations are reversible because the activity of the gene can be restored by a second mutation that removes a GC base pair somewhere in the same string of GC base pairs, restoring the original sequence of the gene. The sequence around the string of GCs in *bvgS* may make the region a hot spot for frameshift mutations, thereby increasing the frequency of the phase shift.

The Heat Shock Regulon

The heat shock regulon is one of the most extensively studied regulons in bacteria and other organisms. One of the major challenges facing cells is to survive abrupt changes in temperature. To adjust to abrupt increases,

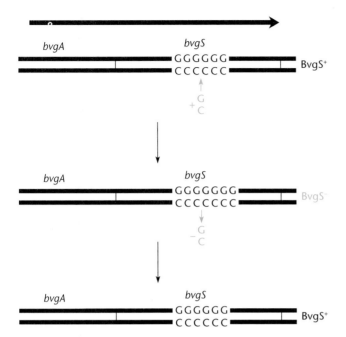

Figure 12.13 Operon structure of the *bvg* locus of *Bordetella pertussis* and the molecular basis for phase variation. The toxin genes of *B. pertussis* are regulated by a two-component system consisting of the sensor (BvgS) and the response regulator (BvgA). The phase variation that causes the reversible loss of virulence determinants results from a frameshift mutation in a string of six GC base pairs in the *bvgS* gene. Adapted from J. F. Miller, S. A. Johnson, W. J. Black, D. T. Beattie, J. J. Mekalanos, and S. Falkow, *J. Bacteriol.* **174**:970–979, 1992.

cells induce at least 30 different genes encoding proteins called the **heat shock proteins (Hsps)**. The concentrations of these proteins quickly increase in the cell after a temperature upshift and then slowly decline, a phenomenon known as the **heat shock response**. Besides being induced by abrupt increases in temperature, the heat shock genes are induced by other types of stress, such as the presence of ethanol and other organic solvents in the medium or DNA damage. Therefore, these proteins are more of a general stress response rather than a specific response to an abrupt increase in temperature. Nevetheless, the name heat shock has stuck, and we will use it here.

Unlike most shared cellular processes, the heat shock response was observed in cells of higher organisms long before it was seen in bacteria. Some of the Hsps are remarkably similar in all organisms on Earth and presumably play similar roles in protecting cells against heat shock (see Box 12.4). The mechanism of regulation of the heat shock response may also be similar in organisms ranging from bacteria to higher eukaryotes (see Craig and Gross, and Mager and de Kruijff, Suggested Reading).

Heat Shock Response in *E. coli*

The molecular basis for the heat shock response is better understood in *E. coli* than in any other organism. In *E. coli,* the genes for the Hsps form two regulons, the **heat shock regulons**. The major one, the σ^{32} regulon, includes about 30 genes encoding 30 different Hsps. The minor regulon, the σ^E regulon, encodes only three proteins, and we do not discuss it in detail here.

The functions of many Hsps are known (see Table 12.1). Most Hsps play roles during the normal growth of the cell and so are always present at low concentrations, but after a heat shock, their rate of synthesis increases markedly and then slowly declines to normal levels.

Some Hsps, including GroE, DnaK, DnaJ, and GrpE, are chaperones that direct the folding of newly translated proteins (see chapter 2). The names of these proteins do not reflect their function but, rather, how they were orginally discovered. For example, DnaK and

---

BOX 12.4

Evolutionary Conservation of Heat Shock Proteins

All types of cells that have been studied undergo similar changes in response to an abrupt increase in temperature or exposure to other types of stress such as organic solvents. After being stressed, the synthesis of a few proteins, the heat shock proteins, is greatly enhanced, while the synthesis of other proteins is reduced. Consequently, essentially all the protein synthesis machinery of the cell is devoted to making the heat shock proteins.

One of the remarkable features of the heat shock proteins is their evolutionary conservation. For example, the DnaK protein of *E. coli* is about 50% homologous to heat shock proteins of similar size (~70 kDa; Hsp70) found in humans and fruit flies. Furthermore, the Hsp70 proteins may play a similar role in regulating the heat shock response in all organisms.

References
Allen, S. P., J. O. Pollazi, J. K. Gierse, and A. M. Easton. 1992. Two novel heat shock genes encoding proteins produced in response to heterologous protein expression in *Escherichia coli. J. Bacteriol.* **174**:6938–6947.

Bardwell, J. C. A., and E. A. Craig. 1984. Major heat-shock gene of *Drosophila* and *Escherichia coli* heat-inducible *dnaK* gene are homologous. *Proc. Natl. Acad. Sci. USA* **81**:848–852.

Zeilstra-Ryalls, J., O. Fayet, and C. Georgopoulous. 1991. The universally conserved GroE (Hsp60) chaperonins. *Annu. Rev. Microbiol.* **45**:301–325.

DnaJ were found because they affect the stability of a protein required for DNA replication, but they themselves have little to do with replication. Chaperones may help the cell survive a heat shock by binding to proteins denatured by the abrupt rise in temperature and either helping them to refold properly or targeting them for destruction.

Other Hsps, including Lon and Clp, are proteases, which may degrade proteins that are so badly denatured by the heat shock that they are irreparable and so are best degraded before they poison the cell. Some other Hsps are proteins normally involved in protein synthesis, including special aminoacyltransferases that are induced after a heat shock. The function of this type of Hsp in protecting the cell after a temperature rise is not clear.

Knowing that many Hsps are involved in helping proteins fold properly or in destroying denatured proteins helps explain the transient nature of the heat shock response. Immediately after the temperature increases, the concentrations of salts and other cellular components are not appropriate for protein stability at the higher temperature, leading to massive protein unfolding. Later, after the temperature has been elevated for some time, the internal conditions will have had time to adjust, and so proteins will not continue to be denatured and the increased numbers of chaperones and other Hsps will no longer be necessary. Hence, the synthesis of the Hsps declines.

Genetic Analysis of the *E. coli* Heat Shock Regulon

As with other regulatory systems, the analysis of the heat shock response was greatly aided by the discovery of mutants with defective regulatory genes. The first of such mutants was found in a collection of temperature-sensitive mutants. It was later shown to be unable to induce the Hsps after a shift to high temperature (see Zhou et al., Suggested Reading). This mutant, which failed to make the regulatory gene product, made it possible to clone the regulatory gene by complementation (see chapter 15 for methods). A library of wild-type *E. coli* DNA was introduced into the mutant strain, and clones that permitted the cells to survive at high temperatures were isolated. When the sequence of cloned gene for the regulatory Hsp was compared with that of other proteins, it was found to encode a new type of sigma factor, σ^{32}, and was named the heat shock sigma. The RNA polymerase holoenzyme with σ^{32} attached recognizes promoters for the heat shock genes that are different from the promoters recognized by the normal σ^{70} and the nitrogen sigma, σ^{54} (Figure 12.9). The gene for

the heat shock sigma was named *rpoH* for RNA polymerase subunit *h*eat shock.

Regulation of σ^{32} Synthesis

Normally, very few copies of σ^{32} exist in the cell. However, immediately after an increase in temperature from 30 to 42°C, the amount of σ^{32} in the cell increases 15-fold. This increase in concentration leads to a significant rise in the rate of transcription of the heat shock genes, since they are transcribed from σ^{32}-type promoters. Understanding how the heat shock genes are turned on after a heat shock requires an understanding of how this increase in the amount of σ^{32} occurs.

An abrupt increase in temperature might increase the amount of σ^{32} through several mechanisms. One possibility is that the *rpoH* gene for σ^{32} is transcriptionally autoregulated. According to this hypothesis, the cell would normally contain a small amount of σ^{32}, which is somehow activated after heat shock, and this activated σ^{32} would direct more of the RNA polymerase to the *rpoH* gene, leading to the synthesis of more σ^{32}, and so forth. For this hypothesis to be correct, the *rpoH* gene would have to be strongly transcriptionally regulated. However, although the rate of transcription of the *rpoH* gene increases slightly after a heat shock, it does not increase enough to explain the large rise in σ^{32} levels. Moreover, if the *rpoH* gene is transcriptionally autoregulated, at least one of the promoters servicing the *rpoH* gene should be of the type that uses the heat shock sigma. However, of the four different promoters from which *rpoH* is transcribed, none are recognized by σ^{32}. Three promoters are used by RNA polymerase with the normal σ^{70}, and another is probably used by RNA polymerase with another type of sigma factor, σ^{E} (also called σ^{24}), which is more active at higher temperatures, perhaps explaining the slight increase in *rpoH* transcription after a heat shock.

Because the transcription of the *rpoH* gene does not increase significantly after heat shock, the amount of σ^{32} in the cell must be posttranscriptionally regulated. In fact, immediately after temperature upshift, σ^{32} stability increases markedly and the translation rate of the *rpoH* mRNA increases 10-fold.

DnaK: THE CELLULAR THERMOMETER?
Recent evidence indicates that at least some of the posttranscriptional regulation of σ^{32} levels after a heat shock is due to the protein chaperones DnaK, DnaJ, and GrpE. These chaperones normally bind to nascent proteins in the process of being synthesized and help them to fold properly (see chapter 2). Under heat shock conditions, they can also bind to denatured proteins and help them refold.

The protein-binding ability of these Hsps may allow them to indirectly regulate their own synthesis and that of the other Hsps. According to this hypothesis, at least one of these Hsps is a "cellular thermometer" that senses the change in temperature and induces the transcription of the heat shock genes. The Hsp DnaK is the prime suspect for the cellular thermometer, and so we will assume that it is responsible, although DnaJ and GrpE could also be involved (see Craig and Gross, Suggested Reading).

Figure 12.14 shows a model for how the protein-binding ability of the DnaK chaperone may indirectly

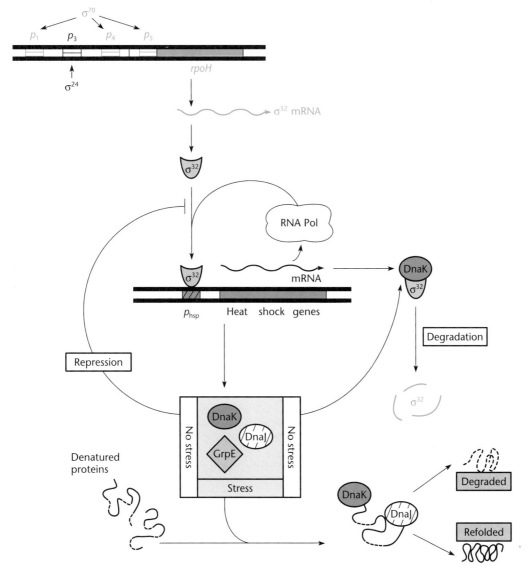

Figure 12.14 Role of DnaK in the induction of the heat shock genes after cells are exposed to an abrupt increase in temperature or other type of stress. The σ^{32} with DnaK bound is susceptible to a protease and is quickly degraded. In addition, σ^{32} with DnaK bound is less active for the initiation of transcription from the heat shock promoters. After an abrupt increase in temperature, many other proteins are denatured, and DnaK binds to these to help them refold properly. This action frees σ^{32}, stabilizing it and making it more active for transcription initiation. When the cell adjusts to the higher temperature and DnaK accumulates to the point where some is again available to bind to σ^{32}, the activity of σ^{32} in the cell again drops, and the transcription of the heat shock genes returns to basal levels. Adapted from W. H. Mager and A. J. J. de Kruijff, *Microbiol. Rev.* **59**:506–531, 1995.

regulate the synthesis of the heat shock proteins. One of the proteins to which DnaK binds is σ^{32} (see Liberek et al., Suggested Reading). By binding to σ^{32}, the DnaK protein regulates the transcription of the heat shock genes in two ways. First, it affects the stability of σ^{32}. The σ^{32} protein with DnaK bound is more susceptible to a cellular protease than is free σ^{32}. Therefore, DnaK can cause σ^{32} to be degraded almost as soon as it is made, keeping the amount of σ^{32} at low levels and therefore reducing the transcription of the heat shock genes. Second, DnaK inhibits the activity of σ^{32}. With DnaK protein bound, σ^{32} may be less active in transcription, either because it is less able to bind to RNA polymerase or because the complex of RNA polymerase, σ^{32}, and DnaK is less able to bind to the heat shock promoters. By inhibiting the activity of σ^{32}, the DnaK protein will lower the transcription of the heat shock genes.

According to the model, the release of DnaK inhibition of σ^{32} activity is responsible for the increase in transcription of the heat shock genes after an abrupt increase in temperature. However, this seems to be backward. The concentration of DnaK increases after a heat shock, but its effects on σ^{32} activity are all negative. How could the increase in a protein that inhibits the activity of σ^{32} lead to an increase in the transcription of the heat shock genes?

The answer to this dilemma lies in the chaperone role of DnaK; that is, in addition to binding to the σ^{32} protein, DnaK binds to denatured proteins to help them refold properly. As mentioned, salt concentrations and other conditions in the cell influence the stability of proteins, and the optimal conditions for protein stability may change with the temperature. If the temperature increases too rapidly, the cellular conditions may not have time to adjust, and many proteins may be denatured. In this case, most of the DnaK protein in the cell will be commandeered to help renature the unfolded proteins, leaving less DnaK available to bind to σ^{32}. The σ^{32} protein will then be more stable and will accumulate in the cell. It will also be more active, increasing the transcription of the heat shock genes, including the *dnaK* gene itself.

This model also explains the transient nature of the heat shock response, in which the concentration of the Hsps slowly declines after a temperature increase. When enough DnaK has accumulated to bind to all the unfolded proteins and internal conditions have adjusted so that no more proteins are denatured at the higher temperature, extra DnaK will once again be available to bind to and inhibit σ^{32}, leading to the observed drop in the rate of synthesis of the Hsps.

The σ^{32} protein may also be **translationally autoregulated**, in other words, able to repress its own translation.

Generally, such proteins bind to their own TIR in the mRNA, thereby blocking access by ribosomes. After an upshift in temperature, the translational repression of σ^{32} synthesis is less complete but then returns, suggesting that DnaK (and DnaJ) may be involved. The translational repression of σ^{32} synthesis may also involve secondary structures that are formed in the σ^{32} mRNA at lower temperatures and removed at higher temperatures.

In addition to sensing a shift in temperature indirectly, through the presence of denatured proteins in the cell, the DnaK protein might be able to sense the temperature more directly, since its ATPase is much more active at higher temperatures. It is interesting how the protein binding and other activities of some of the Hsps may be used to regulate their own synthesis as well as that of other Hsps.

Regulation of Ribosome and tRNA Synthesis

To compete effectively in the environment, cells must make the most efficient use possible of the available energy. One of the major ways in which cells conserve energy is by regulating the synthesis of their ribosomes and tRNAs, so that they make only enough to meet their needs. More than half of the RNA made at any one time comprises rRNA and tRNA. Moreover, each ribosome is composed of about 50 different proteins, and there are about as many different tRNAs.

The number of ribosomes and tRNA molecules needed by the cell varies greatly, depending on the growth rate. Fast-growing cells require many ribosomes and tRNA molecules to maintain the high rates of protein synthesis required for fast cellular growth. Cells growing more slowly, either because they are using a relatively poor carbon and energy source or because some nutrient is limiting, need fewer ribosomes and tRNAs. For example, a rapidly growing *E. coli* cell contains as many as 70,000 ribosomes, but slowly growing cells have fewer than 20,000. As with most of the global regulatory systems we have discussed, the regulation of ribosome and tRNA synthesis is much better understood for *E. coli* than any other organism on earth. Even in *E. coli*, however, some major questions remain unanswered. In this section, we confine our discussion to *E. coli*, with occasional references to other bacteria when information is available.

Ribosomal Proteins

Ribosomes are composed of both proteins and RNA (see chapter 2). The ribosomal proteins are designated by the letter L or S, to indicate whether they are from the *large* (50S) or *small* (30S) subunit of the ribosome, followed by a number for the particular protein. Thus, protein L11 is protein number 11 from the 50S subunit of the ribosome, whereas protein S9 is protein number 9 from the 30S subunit. The gene names begin with *rp*, for ribosomal protein, followed by a lower case *l* or *s* to indicate whether the protein product resides in the large or small subunit. Another capital letter designates the specific gene. For example, the gene *rplK* is ribosomal protein gene K encoding the L11 protein; note that K is the 11th letter of the alphabet. Similarly, *rpsL* encodes the S12 protein; *L* is the 12th letter of the alphabet.

MAPPING RIBOSOMAL PROTEIN GENES

A total of 54 different genes encode the 54 polypeptides that compose the *E. coli* ribosome, and mapping these genes was a major undertaking. Some ribosomal protein genes were mapped simply by mapping mutations that caused resistance to antibiotics such as streptomycin, which binds to the ribosomal protein S12, blocking the translation of other genes (see chapter 2). More complex techniques involving specialized transducing phages and DNA cloning (see chapters 15 and 16) were needed to map the other ribosomal protein genes. Often, clones containing these genes were identified because they synthesize a particular ribosomal protein in coupled transcription-translation systems (see chapter 2).

The final map revealed some intriguing aspects to the organization of the ribosomal protein genes in the chromosome of *E. coli*. Rather than being randomly scattered around the chromosome, the 54 genes are organized into large clusters of operons, with the largest at 73 and 90 min in the *E. coli* genome. Furthermore, these operons also often contain genes for other components of macromolecular synthesis, including subunits of RNA polymerase, tRNAs, and genes for proteins of the DNA replication apparatus.

In the cluster shown in Figure 12.15, four genes for tRNAs, *thrU, tryD, glyT,* and *thrT*, are followed by *tufB*, a gene for the translation elongation factor EF-Tu. These five genes constitute one operon; they are all co-transcribed into one long precursor RNA, from which the individual tRNAs are cut out later. The next operon in the cluster contains two genes for ribosomal proteins, *rplK* and *rplA*, and a third operon has four genes, *rplJ* and *rplL* (encoding two more ribosomal proteins) and *rpoB* and *rpoC* (encoding the β and β' subunits of the RNA polymerase, respectively).

Several hypotheses have been proposed to explain why genes involved in macromolecular synthesis would be clustered in *E. coli* DNA. First, the products of these genes must all be synthesized in large amounts to meet

Figure 12.15 Arrangement of a gene cluster in *E. coli* encoding ribosomal proteins and other gene products involved in macromolecular synthesis. The cluster contains three operons, transcribed in the direction shown by the arrows. The genes *thrU, tryU, glyT,* and *thrT* all encode tRNAs. The *tufB* gene encodes translation elongation factor EF-Tu. The genes *rplK, rplA, rplJ,* and *rplL* encode proteins of the large subunit of the ribosome. The genes *rpoB* and *rpoC* encode subunits of the RNA polymerase.

the cellular requirements. Some clusters are near the origin of replication, *oriC*, and cells growing at high growth rates have more than one copy of the genes near this site (see chapter 1), which allows higher rates of synthesis of the gene products. Other possible reasons have to do with the structure of the bacterial nucleoid (see chapter 1). Genes clustered together will probably be on the same loop of the nucleoid. A loop for the macromolecular synthesis genes might be relatively large and extend out from the core of the nucleoid to allow RNA polymerase and ribosomes to gain easier access to the genes. A third possible explanation is that the genes for macromolecular synthesis must be coordinately regulated with the growth rate and that their assembly in clusters of operons facilitates their coordinate regulation, although why this should be the case is not clear.

REGULATION OF THE SYNTHESIS OF RIBOSOMAL PROTEINS

The regulation of ribosomal protein synthesis is best understood in *E. coli*. However, there is every reason to believe that the regulation is similar in all bacteria. The ribosomal proteins and rRNAs are synthesized independently and then assembled into mature ribosomes. Nevertheless, there is never an excess of either free ribosomal proteins or free rRNA in the cell, suggesting that their synthesis is somehow coordinated. Either the rate of ribosomal protein synthesis is adjusted to match the rate of synthesis of the rRNA, or vice versa. As it turns out, the rate of protein synthesis is adjusted to match the rate of rRNA synthesis.

Like the synthesis of the heat shock sigma (σ^{32}), the synthesis of the ribosomal proteins is translationally autoregulated. The ribosomal proteins bind to TIRs in the mRNA and repress their own translation. Rather than each ribosomal gene of the operon translationally regulating itself independently, the protein product of only one of the genes of the operon is designated as responsible for regulating the translation of all the ribosomal proteins encoded by the operon. This designated protein can also bind to free rRNA in the cell, which, as we discuss later, is what coordinates the synthesis of the ribosomal proteins with the synthesis of rRNA.

Figure 12.16 illustrates this regulation with the relatively simple operon *rplK-rplA*. Of the two proteins encoded by the operon, L1 is the one believed to function in regulation of the operon. The L1 protein can bind to both free rRNA and the TIR on the *rplK* mRNA. By binding to the *rplK* TIR and thereby inhibiting the translation of this gene, the L1 protein simultaneously inhibits the translation of its own gene, since the two genes are translationally coupled (see chapter 2). However, if the cell contains free rRNA, the L1 protein will preferentially bind to it, allowing translation of L11 and L1 to resume. The same thing happens to the other ribosomal proteins, and so their synthesis also resumes. When there is no longer any free rRNA in the cell—because it is all taken up by the ribosomes—free L1 protein will begin to accumulate, bind to the TIR for the *rplK* gene, and once again repress the translation of itself and the L11 protein.

The protein in each operon designated to be the regulator may have been chosen because it normally binds to rRNA during the assembly of the ribosome and so already has an rRNA-binding ability. In at least some cases, the sequence to which the designated protein binds in the rRNA may be mimicked in the TIR for the first gene in the mRNA.

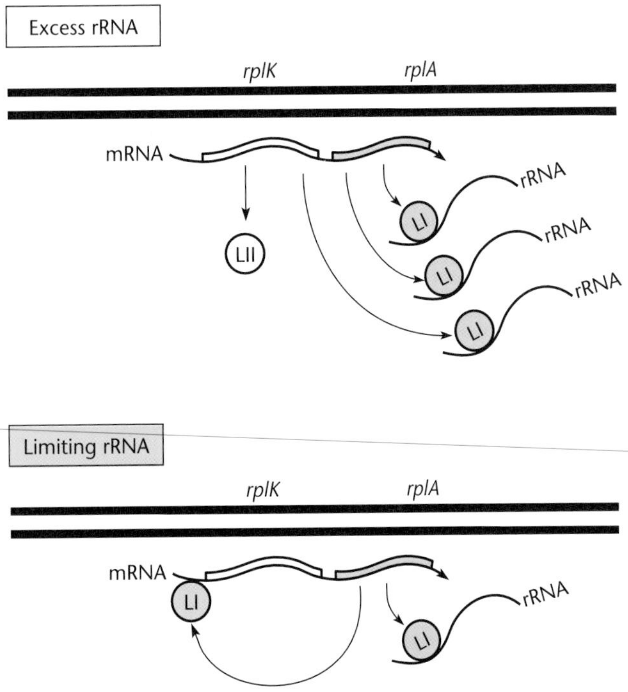

Experimental Support for the Translational Autoregulation of Ribosomal Proteins

Evidence that the synthesis of ribosomal proteins is translationally autoregulated came from a series of experiments called **gene dosage experiments**, in which the number of copies, or dosage, of a gene in the cell is increased and the effect of this increase on the rate of synthesis of the protein product of the gene is determined (see Yates and Nomura, 1980 and 1981, Suggested Reading). Figure 12.17 illustrates the principle behind a gene dosage experiment. If the gene is not autoregulated, the rate of synthesis of the gene product should be approximately proportional to the number of copies of the gene. However, if the gene is autoregulated, the product of the gene should repress its own synthesis, and the rate of synthesis of the gene product should not increase, regardless of the number of gene copies.

In the actual experiment, specialized transducing phage carrying operons encoding one or more ribosomal proteins were integrated at the normal phage attachment site (see chapter 7). As a result, the cell contained two copies of these ribosomal protein genes,

Figure 12.16 Translational autoregulation of the ribosomal proteins, as illustrated by the *rplK-rplA* operon shown in Figure 12.15. See the text for details.

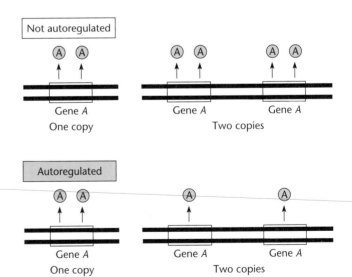

Figure 12.17 A gene dosage experiment to determine if a gene is autoregulated. The number of copies of the gene for the A protein is doubled. If the gene is not autoregulated, twice as much of protein A will be made. If gene *A* is autoregulated, the same amount of protein A will be made.

one at the normal position and another copy at the site of integration of the transducing phage DNA. The same amount of the ribosomal proteins was synthesized with two copies of the ribosomal protein genes as with one, indicating that the synthesis is autoregulated.

The next step was to determine if the autoregulation occurs at the level of transcription or translation. If the synthesis of the ribosomal proteins were transcriptionally autoregulated, the rate of transcription of the ribosomal genes would not increase when the number of copies of the genes increased. However, if the autoregulation occurred only at the level of translation, the rate of gene transcription would increase. The investigators found that the rate of transcription approximately doubled when the gene dosage was doubled, indicating that the autoregulation did not occur transcriptionally. If genes are not transcriptionally autoregulated, they must be translationally autoregulated. Therefore, the ribosomal proteins are capable of repressing their own translation.

The next step was to determine whether all the proteins in the operon independently repress their own translation or whether some of the proteins in each operon are responsible for regulating translation of themselves as well as the others. To answer this question, the investigators systematically deleted genes for some of the proteins encoded by each operon in the transducing phage and evaluated the effect of their absence on the synthesis of the other proteins. Deleting most of the genes in each operon had no effect on the synthesis of the proteins encoded by the other genes. However, when one particular gene in each operon was deleted, the rate of synthesis of the other proteins doubled. Therefore, this one protein was known to repress the translation of itself and the other proteins encoded by the same operon.

rRNA and tRNA Regulation

As discussed in chapter 2, the 16S, 23S, and 5S rRNAs are synthesized together as a long precursor RNA, often with intermingled tRNA sequences. After synthesis, the long precursor RNA is processed into the individual rRNAs and tRNAs. Every ribosome contains one copy of each of the three types of rRNAs, and the synthesis of the three rRNAs as part of the same precursor RNA ensures that all three will be made in equal amounts.

Each cell has tens of thousands of ribosomes, requiring the synthesis of large amounts of rRNA. To meet this need, bacteria have evolved many ways to increase the output of their rRNA genes. For example, many bacteria have more than one copy of the gene for rRNA. *E. coli* and *S. typhimurium* have seven copies of the

genes for the rRNAs. Many bacteria also have very strong promoters for their rRNA genes, so that as many as 50 RNA polymerase molecules can be transcribing each rRNA operon simultaneously, almost as many as a DNA of this length will hold. The RNA polymerase molecules initiate trancription and start down the operon, one immediately after another. Some of the rRNA promoters are so strong because they have a sequence called the UP element upstream of the promoter. This sequence enhances initiation of transcription from the promoter by interacting with the carboxyl terminus of the α subunit of RNA polymerase, much like CAP and some other activator proteins, which interact with the polymerase to activate transcription (see Rao et al., Suggested Reading).

The rate of rRNA gene transcription is also increased because rRNA operons have antitermination sequences lying just downstream of the promoter and in the spacer region between the 16S and 23S coding sequences. These antitermination sequences reduce pausing by RNA polymerase and prevent termination at ρ-dependent transcription termination sites (chapter 2), allowing the synthesis of a complete rRNA to be completed in a shorter time (see Condon et al., Suggested Reading).

The rRNA and tRNA genes in *E. coli* seem subject to at least two types of regulation. In the first type, **stringent control**, rRNA and tRNA synthesis ceases when cells are starved for an amino acid. In the second, **growth rate regulation**, the rate of rRNA and tRNA synthesis decreases when cells are growing slowly on a poor carbon source. As we discuss next, these two types of regulation may have some features in common.

STRINGENT CONTROL

Protein synthesis requires all 20 amino acids. A cell is said to be **starved** for an amino acid when the amino acid is missing from the medium and the cell cannot make it. The ribosomes of the cell will then stall whenever they encounter a codon for the missing amino acid because it will not be available for insertion into the growing polypeptide chain.

In principle, rRNA synthesis can continue in cells starved for an amino acid, since RNA does not contain amino acids. However, in *E. coli*, and probably other types of cells, the synthesis of rRNA and tRNA ceases when an amino acid is lacking. This coupling of the synthesis of rRNA and tRNA to the synthesis of proteins is called stringent control. Stringent control saves energy; there is no point in making ribosomes and tRNA if one or more of the amino acids are not available for protein synthesis.

Synthesis of ppGpp during Stringent Control

The stringent control of rRNA synthesis results from the accumulation of an unusual nucleotide, **guanosine tetraphosphate (ppGpp)**. A similar nucleotide, **guanosine pentaphosphate (pppGpp)**, also accumulates under these conditions, and some is subsequently converted to ppGpp. These nucleotides are made by transferring two phosphates from ATP to the 3′ hydroxyl of GDP or GTP, respectively. These nucleotides were originally called magic spots I and II (MSI and MSII) because they show up as a distinct spots during some types of chromatography (see Cashel and Gallant, Suggested Reading). All evidence indicates that the two nucleotides have similar effects, so we refer to them collectively as ppGpp.

1. RelA. Figure 12.18 shows a model for how amino acid starvation stimulates the synthesis of ppGpp. The nucleotides are made by an enzyme called RelA (for *re-laxed control gene A*), which is bound to the ribosome. When *E. coli* cells are starved for an amino acid (lysine in the example), the tRNAs for that amino acid, tRNA^Lys, are uncharged. The uncharged tRNA will not bind EF-Tu (see chapter 2) and so will not enter the ribosome. Consequently, when a ribosome moving along an mRNA encounters a codon for that amino acid (the codon AAA in the example), the ribosome will stall. If the ribosome stalls long enough, an uncharged tRNA may eventually enter the A site of the ribosome even though it is not bound to EF-Tu. Uncharged tRNA entering the A site stimulates the RelA protein on the ribosome to synthesize ppGpp.

The intracellular levels of ppGpp during amino acid starvation are also regulated by the SpoT (for magic spot) protein, which is the product of the *spoT* gene. The SpoT protein normally degrades ppGpp, but its degradation activity is inhibited after amino acid starvation, leading to more accumulation of ppGpp. Therefore, after amino acid starvation, the cellular concentration of ppGpp is determined both by the activation of the ppGpp synthesis activity of RelA and the inhibition of the ppGpp-degrading activity of SpoT. Surprisingly, as we discuss later, the *spoT* gene product not only can degrade ppGpp but also may synthesize ppGpp during growth rate regulation.

2. Isolating relA *mutants.* Evidence that the accumulation of ppGpp is responsible for stringent control came from the behavior of *relA* mutants of *E. coli* that lack the RelA enzyme activity. These mutants do not accumulate ppGpp after amino acid starvation and also do not shut off rRNA and tRNA synthesis. Because rRNA

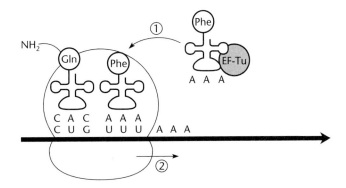

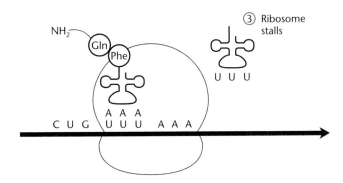

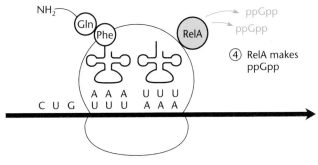

Figure 12.18 A model for synthesis of ppGpp after amino acid starvation. Cells are starved for the amino acid lysine. The tRNA^Lys will then have no lysine attached, and EF-Tu cannot bind to a tRNA that is not aminoacylated. A ribosome moving along the mRNA will then stop when it arrives at a codon for lysine (AAA), because it has no aminoacylated-tRNA to translate the codon. However, if the ribosome remains stalled at the codon long enough, a tRNA^Lys (anticodon UUU in the example) will bind to the A site of the ribosome even though EF-Tu is not bound. This binding will cause RelA to synthesize ppGpp.

synthesis and tRNA synthesis are not stringently coupled to protein synthesis in *relA* mutants, strains with *relA* mutations are called **relaxed strains**, which is the orgin of the gene name.

As is often the case, the first *relA* mutant was isolated by chance. A mutant strain of *E. coli* was observed to have a difficult time recovering after amino acid starvation, whereas the wild-type parent could start growing almost immediately after the amino acid was restored. A later study showed that rRNA and tRNA synthesis continued after this mutant was starved for an amino acid, and so this mutant was called a relaxed mutant.

The relative inability of the original *relA* mutant to recover from amino acid starvation suggested a way of enriching for *relA* mutants (see Fiil and Frieson, Suggested Reading). This procedure is based on the fact that growing cells are killed by ampicillin but cells that are not growing will survive (see chapter 14 for methods). *E. coli* cells that were auxotrophic for an amino acid were mutagenized, washed, and resuspended in a medium without the amino acid, so that they stopped growing. After a long period of incubation, the amino acid was restored and ampicillin was added to the medium. The ampicillin was removed shortly thereafter, and the process was repeated. After a few cycles of this treatment, the bacteria were plated. A high percentage of the surviving bacteria that multiplied to form colonies were *relA* mutants, as evidenced by the fact that most of the surviving strains continued to synthesize rRNA and tRNA after amino acid starvation.

How Does ppGpp Block rRNA and tRNA Synthesis?

The mechanism by which ppGpp blocks rRNA and tRNA synthesis is still not clearly established. Most evidence suggests that ppGpp blocks the initiation of transcription of the genes for these RNAs by acting directly on their promoters. Mutations in the promoters for rRNA and tRNA genes can render them insensitive to ppGpp. The presence of ppGpp can also inhibit the binding of RNA polymerase to rRNA promoters in vitro and can inhibit in vitro rRNA synthesis.

However, other evidence suggests that ppGpp does not block initiation but, rather, the elongation of transcription. In support of this hypothesis is the fact that transcription of the rRNA genes from promoters that are normally not sensitive to stringent control, including the T7 or λ phage promoters, is still sensitive to ppGpp. Thus, the target of ppGpp may not be exclusively the promoter of the rRNA genes, and ppGpp may also block rRNA transcription after it is under way. Why different experimental methods produce apparently conflicting results on the mechanism by which ppGpp blocks rRNA and tRNA synthesis is not clear.

Although ppGpp seems to block rRNA and tRNA synthesis, it also stimulates the transcription of some other genes, including the *his* operon and some other biosynthetic operons. Failure to adequately express these biosynthetic operons in the absence of ppGpp makes *relA spoT* double mutants auxotrophic for several growth substances. High levels of ppGpp may also decrease the rate of transcription and translation and lower the translation error rates of many other genes. More experiments are needed to determine how ppGpp affects transcription and whether all the effects are related.

GROWTH RATE REGULATION OF RIBOSOMAL AND tRNA SYNTHESIS

As mentioned above, cells growing fast in rich medium have many more ribosomes and a higher concentration of tRNA than do cells growing slowly in poor medium. This regulation of rRNA and tRNA synthesis is called growth rate regulation. Growth rate regulation seems to depend on the number of free untranslating ribosomes. If the number of ribosomes exceeds that needed for translation, some of the ribosomes will remain free of mRNA. These free ribosomes in the cell seem to inhibit the synthesis of rRNA and tRNA, thereby ensuring that the cell will contain no more ribosomes and tRNA than are needed for translation at a particular growth rate. However, how free ribosomes repress the synthesis of rRNA and tRNA is an area of controversy.

Role of ppGpp in Growth Rate Regulation

Some evidence indicates that the nucleotide ppGpp may also be responsible for growth rate regulation of tRNA and rRNA synthesis. The levels of ppGpp are higher in bacteria growing more slowly in a poor carbon source than they are in bacteria growing faster in a good carbon source. However, we know that the ppGpp that accumulates when cells are growing in a poor carbon source is not synthesized by the RelA enzyme, since *relA* mutations do not prevent the increase in ppGpp levels observed when cells are growing in a poor carbon source. Nor do *relA* mutations prevent the growth rate regulation of tRNA and rRNA synthesis.

It has been known for a long time that cells have more ribosomes and other components of the translational machinery when they are growing faster than when they are growing more slowly. The determination of growth rates is the most central global regulatory system of all and, accordingly, must be coordinated with many of the other cell functions. Whatever the roles of free ribosomes and ppGpp turn out to be in this regulation, growth rate regulation remains one of the major unanswered questions in studies of bacterial physiology and molecular genetics.

SUMMARY

1. The coordinate regulation of a large number of genes is called global regulation. Operons that are regulated by the same regulatory protein are part of the same regulon.

2. In catabolite repression, the operons for the use of alternate carbon sources cannot be induced when a better carbon and energy source, such as glucose, is present. In *E. coli* and other enteric bacteria, this catabolite regulation is achieved, in part, by cAMP, which is made by adenylate cyclase, the product of the *cya* gene. When the bacteria are growing in a poor carbon source such as lactose or galactose, the adenylate cyclase is activated and cAMP levels are high. When the bacteria are growing in a good carbon source such as glucose, cAMP levels are low. The cAMP acts through a protein called CAP, the product of the *crp* gene. CAP is a transcriptional activator, which, with cAMP bound, will activate transcription of catabolite-sensitive operons such as *lac* and *gal*.

3. Bacterial cells induce different genes depending on the nitrogen sources available. Genes that are regulated through the nitrogen source are called Ntr genes. Most bacteria, including *E. coli*, prefer NH_3 as a nitrogen source and will not transcribe genes for using other nitrogen sources when growing in NH_3. Glutamine concentrations are low when NH_3 concentrations are low. A signal transduction pathway is then activated, culminating in the phosphorylation of NtrC. This signal transduction pathway begins with the GlnD protein, a uridylyl transferase, which is the sensor of the glutamine concentration in the cell. At low concentrations of glutamine, GlnD transfers UMP to the P_{II} protein, inactivating it. However, at high concentrations of glutamine, the GlnD protein removes UMP from P_{II}. The P_{II} protein without UMP attached can bind to NtrB, somehow preventing the transfer of phosphate to NtrC and causing the removal of phosphates from NtrC. The phosphorylated NtrC protein activates the transcription of the *glnA* gene, the gene for glutamine synthetase, as well as the *ntrB* gene and its own gene, *ntrC*, since they are part of the same operon as *glnA*. It also activates the transcription of operons for using other nitrogen sources.

4. The NtrB and NtrC proteins form a sensor-response regulator pair and are highly homologous to other sensor-kinase/response-regulator pairs in bacteria.

5. NtrC-regulated promoters of *E. coli* and other enterics require a special sigma factor called σ^{54}, which is also used to transcribe the flagellar genes and some biodegradative operons in other types of bacteria.

6. The cell also regulates the activity of glutamine synthetase by adenylylating the glutamine synthetase enzyme. The enzyme is highly adenylylated at high glutamine to α-ketoglutarate ratios, which makes it less active and subject to feedback inhibition.

7. One of the ways that bacteria adjust to changes in the osmolarity of the medium is by changing the ratio of their porin proteins, which form pores in the outer membrane through which solutes can pass to equalize the osmotic pressure on both sides of the membrane. The major porins of *E. coli* are OmpC and OmpF, which make pores of different sizes, thereby allowing the passage of different-sized solutes. The relative amounts of the OmpC and OmpF change in response to changes in the osmolarity of the medium. The *ompC* and *ompF* genes in *E. coli* are regulated by a sensor-response regulator pair of proteins, EnvZ and OmpR, which are similar to NtrB and NtrC. The EnvZ protein is an inner membrane protein with both kinase and phosphatase activities that, in response to a change in osmolarity, can transfer a phosphoryl group to or remove one from OmpR, a transcriptional acivator. The state of phosphorylation of OmpR affects the relative rates of transcription of the *ompC* and *ompF* genes.

8. The ratio of OmpF to OmpC porin proteins is also affected by an antisense RNA named MicF. A region of the MicF RNA can base pair with the translation initiation region of the OmpF mRNA and block access by ribosomes, thereby inhibiting OmpF translation. The *micF* gene is regulated by a number of transcriptional regulatory proteins, including SoxS, which induces the oxidative stress regulon, and MarA, which induces genes involved in excluding toxic chemicals and antibiotics from the cell.

9. The virulence genes of pathogenic bacteria are also global regulons and are normally trancribed only when the bacterium is in its vertebrate host.

10. The diphtheria toxin gene, *dtxR*, encoded by a prophage of *Corynebacterium diphtheriae*, is turned on only when iron is limiting, a condition mimicking that in the host. The *dtxR* gene is regulated by a chromosomally encoded repressor protein, DtxR, which is similar to the Fur protein involved in regulating the genes of iron availability pathways in *E. coli* and other enteric pathogens.

11. The toxin genes of *Vibrio cholerae* are also carried on a prophage and are regulated by a transcriptional activator, ToxR. The ToxR protein traverses the inner membrane and is activated by being dimerized by a periplasmic protein, ToxS, in response to conditions that mimic those in the host. The ToxR protein also activates transcription of the gene for another transcriptional activator, ToxT, which activates transcription of other virulence genes.

12. The virulence genes of *Bordetella pertussis* are regulated by a sensor-response regulator pair of proteins, BvgS and BvgA. Phase variation in *B. pertussis* results from frameshift mutations in *bvgS*.

(continued)

SUMMARY (continued)

13. Bacteria induce a set of proteins called the heat shock proteins in response to an abrupt increase in temperature. Some of the heat shock proteins are chaperones, which assist in the refolding of denatured proteins. Others are proteases, which degrade denatured proteins. The heat shock response is common to all organisms, and the heat shock proteins have been highly conserved throughout evolution.

14. The promoters of the heat shock genes are recognized by RNA polymerase holoenzyme with an alternative sigma factor called the heat shock sigma, or σ^{32}. The amount of this sigma factor markedly increases following a heat shock, leading to increased transcription of the heat shock genes. The increase in σ^{32} following a heat shock involves DnaK, a chaperone that is one of the heat shock proteins.

15. The genes encoding the ribosomal proteins, rRNAs, and tRNAs are part of the largest regulon in bacteria, with hundreds of genes that are coordinately regulated. A large proportion of the cellular energy goes into making the rRNAs, tRNAs, and ribosomal proteins; therefore, regulating these genes saves the cell considerable energy.

16. The synthesis of rRNA and proteins is coordinated by coupling the translation of the ribosomal protein genes to the amount of free rRNA that is not yet in a ribosome. The ribosomal protein genes are organized into operons, and one ribosomal protein of each operon plays the role of translational repressor. The same protein also binds to free rRNA, so that when there is excess rRNA in the cell, all of the repressor protein binds to the free rRNA and none is available to repress translation.

17. The synthesis of rRNA and tRNA following amino acid starvation is inhibited by guanosine tetraphosphate (ppGpp), synthesized by an enzyme associated with the ribosome called RelA. All types of bacteria contain ppGpp, and so the regulation may be universal. However, it is not yet clear how higher levels of ppGpp inhibit transcription of the genes for rRNA and tRNA.

QUESTIONS FOR THOUGHT

1. Why do you suppose that proteins involved in gene expression (i.e., transcription and translation) are among the heat shock proteins?

2. Why do you think genes for the utilization of amino acids as a nitrogen source are not under Ntr regulation in *Salmonella* spp. but they are in *Klebsiella* spp.?

3. Why are the corresponding sensor and response regulator genes of the various two-component systems so similar to each other?

4. Why is the enzyme responsible for ppGpp synthesis in response to amino acid starvation different from the one responsible for ppGpp synthesis during growth rate control? Why might SpoT be used to degrade ppGpp made by RelA after amino acid starvation but be used to synthesize it during growth rate control?

PROBLEMS

1. You have isolated a mutant of *E. coli* that cannot use either maltose or arabinose as a carbon and energy source. How would you determine if your mutant has a *cya* or *crp* mutation or whether it is a double mutant with mutations in both the *ara* operon and a *mal* operon?

2. What would you expect the phenoptypes of the following mutations to be?

a. a *glnA* (glutamine synthetase) mutation.

b. an *ntrB* mutation.

c. an *ntrC* mutation.

d. a *glnD* null mutation that inactivates the UTase so that P_{II} has no UMP attached.

e. a constitutive *ntrC* mutation that changes the NtrC protein so that it no longer needs to be phosphorylated to be active.

f. a *dnaK* mutation.

g. a *dtrR* mutation of *Corynebacterium diphtheriae*.

h. a *relA spoT* double mutation.

3. How would you show that the toxin gene of a pathogenic bacterium is not a normal chromosomal gene but is carried on a prophage not common to all the bacteria of the species?

4. Explain how you would use gene dosage experiments to prove that the heat shock sigma (σ^{32}) gene is not transcriptionally autoregulated.

5. Explain how you would show which of the ribosomal proteins in the *rplJ-rplL* operon is the translational repressor.

SUGGESTED READING

Bartlett, M. S., and R. L. Gourse. 1994. Growth rate-dependent control of the *rrnB p*₁ core promoter in *Escherichia coli*. *J. Bacteriol.* **176:**5560–5563.

Brissette, R. E., K. Tsung, and M. Inouye. 1991. Intramolecular second-site revertants to the phosphorylation site mutation in *ompR*, a kinase-dependent transcriptional activator in *Escherichia coli. J. Bacteriol.* **173:**3749–3755.

Brunel, C., P. Romby, C. Sacerdot, M. de Smit, M. Gaffe, J. Dondon, J. van Duin, B. Ehresmann, C. Ehresmann, and M. Springer. 1995. Stabilized secondary structure at a ribosomal binding site enhances translational repression. *J. Mol. Biol.* **253:**277–290.

Cashel, M., and J. Gallant. 1969. Two compounds implicated in the function of the RC gene in *Escherichia coli*. *Nature* (London) **221:**838–841.

Condon, C., C. Squires, and C. L. Squires. 1995. Control of rRNA transcription in *Escherichia coli*. *Microbiol. Rev.* **59:**623–645.

Craig, E., and C. A. Gross. 1991. Is Hsp70 the cellular thermometer? *Trends Biochem. Sci.* **16:**135–140.

Csonka, L. N. 1989. Physiological and genetic responses of bacteria to osmotic stress. *Microbiol. Rev.* **53:**121–147.

DiRita, V. J. 1992. Coordinate expression of virulence genes by ToxR in *Vibrio cholerae*. *Mol. Microbiol.* **6:**451–458.

Fiil, N., and J. D. Frieson. 1968. Isolation of relaxed mutants of *Escherichia coli. J. Bacteriol.* **95:**729–731.

Hall, M. N., and T. J. Silhavy. 1981. Genetic analysis of the *ompB* locus in *Escherichia coli* K-12. *J. Mol. Biol.* **151:**1–15.

Henngge-Aronis, R. 1996. Back to log phase: σ^s as a global regulator of osmotic control of gene expression in *Escherichia coli*. *Mol. Microbiol.* **21:**887–893.

Hernandez, V. J., and H. Bremer. 1991. *Escherichia coli* ppGpp synthetase II activity requires *spoT*. *J. Biol. Chem.* **266:**5991–5999.

Higgens, D. E., and V. J. DiRita. 1996. Genetic analysis of the interaction between *Vibrio cholerae* transcription activator ToxR and *toxT* promoter DNA. *J. Bacteriol.* **178:**1080–1087.

Igarashi, K., and A. Ishihama. 1991. Bipartite functional map of *E. coli* RNA polymerase α subunit: involvement of the C-terminal region in transcription activation by cAMP-CRP. *Cell* **65:**1015–1022.

Kolter, R., D. A. Siegele, and A. Tormo. 1993. The stationary phase of the bacterial life cycle. *Annu. Rev. Microbiol.* **47:**855–874.

Liberek, K., T. P. Galitski, M. Zyliez, and C. Georgopoulos. 1992. The DnaK chaperon modulates the heat shock response of *E. coli* by binding to the σ^32 transcription factor. *Proc. Natl. Acad. Sci. USA* **89:**3516–3520.

Magasanik, B. 1982. Genetic control of nitrogen assimilation in bacteria. *Annu. Rev. Genet.* **16:**135–168.

Mager, W. H., and A. J. J. de Kruijff. 1995. Stress-induced transcriptional activation. *Microbiol. Rev.* **59:**506–531.

Miller, J. F., J. J. Mekalanos, and S. Falkow. 1989. Coordinate regulation and sensory transduction in the control of bacterial virulence. *Science* **243:**916–922.

Miller, V. L., and J. J. Mekalanos. 1984. Synthesis of cholera toxin is positively regulated at the transcriptional level by *toxR*. *Proc. Natl. Acad. Sci. USA* **81:**3471–3475.

Nagai, H., H. Yuzawa, and T. Yura. 1991. Interplay of two *cis*-acting mRNA regions in translational control of σ^32 synthesis during the heat shock response of *Escherichia coli*. *Proc. Natl. Acad. Sci. USA* **88:**10515–10519.

Neidhardt, F. C., R. Curtiss III, J. L. Ingraham, E. C. C. Lin, K. B. Low, B. Magasanik, W. S. Reznikoff, M. Riley, M. Schaechter, and H. E. Umbarger (ed.). 1996 *Escherichia coli and Salmonella: Cellular and Molecular Biology*, 2nd ed., vol. 1. ASM Press, Washington, D.C.

Nomura, M., J. L. Yates, D. Dean, and L. E. Post. 1980. Feedback regulation of ribosomal protein gene expression in *Escherichia coli*: structural homology of ribosomal RNA and ribosomal protein mRNA. *Proc. Natl. Acad. Sci. USA* **77:**7084–7088.

Pratt, L. A., W. Hsing, K. E. Gibson, and T. J. Silhavy. 1996. From acids to *osmZ*: multiple factors influence synthesis of the OmpF and OmpC porins in *Escherichia coli*. *Mol. Microbiol.* **20:**911–917.

Rao, L., W. Ross, A. Appelman, T. Gaal, S. Leirmo, P. J. Schlax, M. T. Record, and R. L. Gourse. 1994. Factor independent activation of *rrnB p*₁: an "extended" promoter with an upstream element that dramatically increases promoter strength. *J. Mol. Biol.* **235:**1421–1435.

Saier, M. H., Jr. 1989. Protein phosphorylation and allosteric control of inducer exclusion and catabolic repression by the bacterial phosphoenolpyruvate:sugar phosphotransferase system. *Microbiol. Rev.* **53:**109–120.

Schmidt, M., and R. K. Holmes. 1993. Analysis of diphtheria toxin repressor-operator interactions and the characterization of mutant repressor with decreasing binding activity for divalent metals. *Mol. Microbiol.* **9:**173–181.

Schwartz, D., and J. R. Beckwith. 1970. Mutants missing a factor necessary for the expression of catabolite-sensitive operons in *E. coli*, p. 417–422. *In* J. R. Beckwith and D. Zipser (ed.), *The Lactose Operon*. Cold Spring Harbor Laboratory Press, Cold Spring Harbor, N.Y.

Slauch, J. M., S. Garrett, D. E. Jackson, and T. J. Silhavy. 1988. EnvZ functions through OmpR to control porin gene expression in *Escherichia coli* K-12. *J. Bacteriol.* **170:**439–441.

Stibitz, S., W. Aaronson, D. Monack, and S. Falkow. 1989. Phase variation in *Bordetella pertussis* by frameshift mutation in a gene for a novel two-component system. *Nature* (London) **338:**266–269.

Stibitz, S., A. A. Weiss, and S. Falkow. 1988. Genetic analysis of a region of *Bordetella pertussis* chromosome encoding filamentous hemagglutinin and the pleiotropic regulatory locus *vir. J. Bacteriol.* **170:**2904–2913.

Tang, H., K. Severinov, A. Goldfarb, D. Fenyo, B. Chait, and R. H. Ebright. 1994. Location, structure, and function of the target of a transcriptional activator protein. *Genes Dev.* 8:3058–3067.

van Heeswijk, W. C., S. Hoving, D. Molenaar, B. Stegeman, D. Kahn, and H. V. Westerhoff. 1996. An alternative P_{II} protein in the regulation of glutamine synthetase in *Escherichia coli*. *Mol. Microbiol.* 21:133–146.

Yates, J. L., and M. Nomura. 1980. *E. coli* ribosomal protein L4 is a feedback regulatory protein. *Cell* 21:517–522.

Yates, J. L., and M. Nomura. 1981. Localization of the mRNA binding sites for ribosomal proteins. *Cell* 24:243–249.

Zhou, Y. N., N. Kusukawa, J. W. Erickson, C. A. Gross, and T. Yura. 1988. Isolation and characterization of *Escherichia coli* mutants that lack the heat shock sigma factor σ^{32}. *J. Bacteriol.* 170:3640–3649.

Zou, C., N. Fujita, K. Igarashi, and A. Ishihama. 1992. Mapping the cAMP receptor protein contact site on the alpha subunit of *Escherichia coli* RNA polymerase. *Mol. Microbiol.* 6:2599–2605.

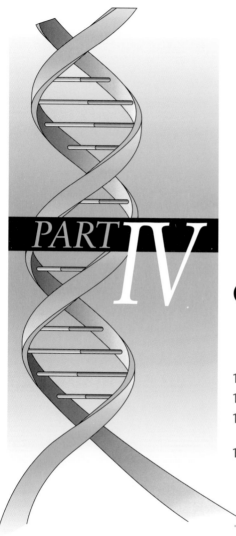

Genes in Practice

*I*N PREVIOUS PARTS, we reviewed many of the facts about bacterial cells that have been revealed through molecular genetic studies. However, we have not gone into detail about how molecular genetic data are assembled and analyzed. In this last part, we discuss the analysis of genetic data in some detail and give some examples of classical genetic analysis in phages and bacteria. We also review some of the techniques of recombinant DNA technology and how these techniques can be applied to the cloning and study of bacterial genes. Finally, we give some up-to-date examples of how the molecular genetics of bacteria is now being used to understand the world around us and to develop technologies to improve our lives. It is our hope that with the background assimilated in the previous parts, you will be able to imagine yourself actually participating in these developments.

Chapter *13*

Genetic Analysis in Phages

GREGOR MENDEL PERFORMED the first genetic analysis of a cellular function almost 150 years ago, when he crossed wrinkled peas with smooth peas and counted the number of progeny of each type. The methods of genetic analysis have become considerably more sophisticated since then and offer one of the most powerful approaches to understanding how organisms function at the cellular and molecular level. The first information about many basic cellular and developmental functions often comes from a genetic analysis of the function.

There are many advantages to genetic analysis. This approach requires few assumptions and can be applied to a function and an organism about which little to nothing is known. Genetic analysis is also one of the only ways to determine how many gene products are involved in a function and to obtain a preliminary idea of the role that the gene products play in the function. Through suppressor analysis, it also offers one of the only ways to ascertain which gene products interact with each other in performing the function.

As discussed in the Introduction, a classical genetic analysis always follows the same basic pattern. First, mutants in which the function is altered are isolated. Then the responsible mutations are mapped and complementation experiments are performed to determine how many gene products are involved. Finally, attempts are made to determine, from the properties of each of the mutants, the contribution of each mutated gene and region to the function.

Phages are ideal for genetic analysis (see Introduction). They have short generation times and are haploid. Mutant strains can be stored for long periods and resurrected only when needed. Phages multiply clonally in plaques, and large numbers can be propagated on plates or in small volumes of liquid media. Different phage strains can be easily crossed with each other, and the progeny can be readily analyzed. Because of these advantages, phages were central to the development of molecular genetics, and important genetic principles such as recombination, complementation, suppression, and *cis-trans* mutations are most easily demonstrated with phages. In this chapter, we give a few examples of the use of classical genetic analysis to study gene structure and function in phages. However, most of the basic genetic principles presented here are the same for genetic studies of all organisms, including humans.

Steps in a Genetic Analysis with Phages

The first step in any classical genetic analysis of a function is to isolate mutants that are altered in that function. Without such mutants, we cannot perform a genetic analysis of the function. A priori, we have no way of knowing what kinds of mutants are possible or all the mutations that can give a particular mutant phenotype. Even if the mutant phenotype we are interested in is possible to obtain, the mutations that produce that type of mutant may be very rare. However, part of the fun of genetics is to try to predict what types of mutations are possible and what effects the mutations will have on the organism.

Phages, like all viruses, are just genes wrapped in a protein and/or membrane coat. The genes of a virus are not expressed until it infects a cell. Therefore, genetic experiments with viruses require that cells be infected with the virus of interest. Hence, the nucleic acid genome, whether DNA or RNA, is injected into the cell and the genes are expressed, leading to the synthesis of more viruses.

Infection of Cells

As discussed in chapter 7, each type of virus can infect and multiply in only a few types of cells. An appropriate host cell must have specific receptors to which the virus can bind and thereby enter the cell. Such a cell is called a **sensitive cell** for that virus. For the virus to complete its life cycle, not only must the cells be sensitive to the virus, but they must also be able to provide all the necessary machinery for the virus to express its genes and replicate its nucleic acid genome, be it DNA or RNA.

In principle, infecting sensitive cells with a virus is simple enough. The virus and sensitive cells need only be mixed with each other. Some cells and viruses will collide at random, leading to virus infection. However, what percentage of the cells will be infected depends upon the concentration of viruses and cells. If the

viruses and cells are very concentrated, they will more often collide with each other to initiate an infection than if they were more dilute.

MULTIPLICITY OF INFECTION

The efficiency of infection is affected not only by the concentration of viruses and cells but also by the ratio of viruses to cells, the **multiplicity of infection** (MOI). For example, if 2.5×10^9 phage are added to 5×10^8 bacteria, there will be $2.5 \times 10^9/5 \times 10^8 = 5$ viruses for every cell, and the MOI is 5. If only 2.5×10^8 viruses had been added to the same number of bacteria, the MOI would have been 0.5.

The MOI can be either high or low. If the number of viruses greatly exceeds the number of cells, the cells are infected at a **high MOI**. Conversely, a **low MOI** indicates that the cells outnumber the viruses. To illustrate, an MOI of 5 is considered to be high; there are five times as many viruses as bacteria. An MOI of 0.5 is low; there is only one virus for every two bacteria.

The MOI is important because it determines what percentage of the cells will be infected by a virus. At a high MOI, most of the cells will be infected by a virus. At a low MOI, many of the cells will remain uninfected. However, not all the cells will be infected even at a very high MOI. There are two reasons for this. First, infection by viruses is never 100% efficient. On its surface, each cell may have only one or a very few receptors for the virus, and a virus can infect a cell only if it happens to bind to one of these receptors. Second, the number of viruses infecting each cell will follow a normal distribution; therefore, statistical variation dictates that not all the cells will be infected. At an MOI of 5, for example, some cells will be infected by five viruses—the MOI— but some will be infected by six viruses, some by four, some by none, and so on. Even at the highest MOIs, some cells will remain uninfected.

The minimum fraction of cells that will escape infection due to statistical variation can be calculated by using the Poisson distribution, which is an approximation to the normal distribution (see chapter 3). According to the Poisson distribution, the probability of a cell receiving no viruses and remaining uninfected (P_0) is at least e^{-MOI}, since the MOI is the average number of viruses per cell. If MOI = 5, then $P_0 = e^{-5} = \sim0.0067$; i.e., at least 0.67% of the cells will remain uninfected. At an MOI of only 1, e^{-1}, or at least ~37%, of the cells will remain uninfected. In other words, at most ~63% of the cells will be infected at an MOI of 1. Even this is an overestimation of the fraction of cells that will be infected, since, as mentioned, some of the viruses will never actually infect a cell.

Virus Crosses

Once the mutations to be tested are chosen for a genetic analysis, the mutated DNAs of two members of the same species must be put together into the same cell. This is called **crossing**. If the DNAs of the two different organisms are in a cell at the same time, the genes of both mutant strains can be expressed and the two DNAs can recombine with each other. In a eukaryotic organism, crosses are usually performed by mating the two organisms to form zygotes that can develop into the mature organism. In viruses, crosses are performed by infecting cells with different strains of the virus at the same time.

To cross two strains of a virus, at least some of the cells in a culture must be infected by both strains. The Poisson distribution can be used to calculate the maximum fraction of cells that will be infected by both mutant viruses at a given MOI of each. If MOI = 1 for each mutant virus is used for the infection, then at least e^{-1} or ~0.37 of the cells will be uninfected by each virus strain and at most $1 - 0.37 = 0.63$ of the cells will be infected with each mutant strain of the virus. Since the chance of being infected with one strain is independent of the chance of being infected by the other strain, at most $0.63 \times 0.63 \approx 0.40$, or ~40%, of the cells will be infected by *both* virus strains at an MOI of 1. You can see that only when both virus strains have a high MOI will most of the bacteria be infected by both strains.

Recombination and Complementation Tests with Viruses

The two basic concepts in classical genetic analysis are **recombination** and **complementation**. The type of information derived from these tests is completely different. In recombination, the *DNA* of the two parent organisms is assembled in new combinations, so that the progeny have DNA sequences from both parents. In complementation, the *gene products* synthesized from two different DNAs interact in the same cell to produce a phenotype.

Recombination Tests

The principles of recombination as a genetic technique are the same for all organisms, but they are most easily illustrated with viruses. Figure 13.1 gives a simplified view of what happens when two DNA molecules from different strains of the same virus recombine. The two mutant virus strains infecting the cell are almost identical, except that one has a mutation at one end of the DNA and the other has a mutation at the other end. The sequences of the two DNAs therefore differ only at the

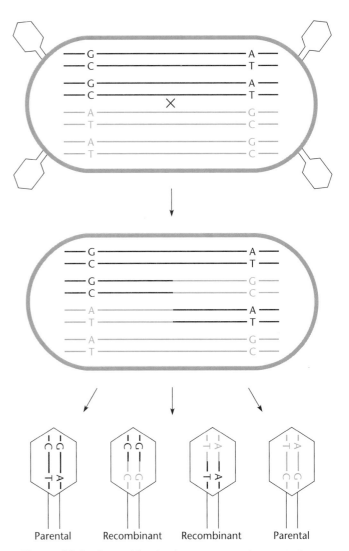

Figure 13.1 Recombination between two virus mutations. The two different mutant parent viruses infect the same permissive host cell and their DNA replicates. Crossovers occur in the region between the two mutations, giving rise to recombinant types that are unlike either parent virus. Only the positions of the mutated base pairs are shown. The DNA of one parent virus is shown in black, and the other is shown in blue.

ends, the sites of the two mutations. As described in chapter 9, recombination occurs by means of a crossover between the sites of two mutations. In Figure 13.1, a crossover in the middle of the DNA molecule between the sites of the two mutations can yield two new types of recombinant DNA molecules: those with neither mutation and those with both mutations. Progeny viruses that have packaged the DNAs with these new DNA sequences are called **recombinant types** because they are unlike either parent. Progeny viruses that have packaged a DNA molecule with only one mutation at one end or the other are called **parental types** because they are like the original viruses that infected the cell. The appearance of recombinant types tells us that recombination has occurred.

In Figure 13.1, one strain had one mutation and the other strain had the other mutation. However, one strain could have had both mutations and the other strain could have had neither mutation. In that case, the recombinant types would have only one or the other of the two mutations and the parental types would have either both mutations or neither mutation.

RECOMBINATION FREQUENCY

The closer together the regions of sequence difference are to each other in the DNA, the less room there is between them for a crossover to occur. Therefore, the frequency of recombinant-type progeny is a measure of how far apart the mutations are in the DNA of the virus. This number is usually expressed as the recombination frequency. In general, independent of the type of organism involved in the cross, the **recombination frequency** is defined as the number of recombinant progeny divided by the total progeny produced in the cross. When the recombination frequency is expressed as a percentage, it is called the **map unit**. For example, the regions of two mutations in the DNA give a recombination frequency of 0.01 if 1 in 100 of the progeny are recombinant types. The regions of the two mutations are then $0.01 \times 100 = 1$ map unit apart.

Different organisms differ greatly in their recombination activity; therefore, map distance is only a relative measure, and a map unit represents a different physical length of DNA for different organisms. Also, recombination frequency can indicate the proximity of two mutated regions only when the mutations are not too far apart. If they are far apart, two crossovers will often occur between them, reducing the apparent recombination frequency. Note that while one crossover between the regions of two mutations will create recombinant types, two crossovers will re-create the parental types. In general, odd numbers of crossovers will produce recombinant types and even numbers of crossovers will re-create parental types.

Complementation Tests

Rather than measure the frequency with which recombination occurs between the regions of mutations, complementation measures the interaction between gene products synthesized from different DNAs in the same cell. One question that complementation tests can answer is whether two mutations inactivate the same or different functional units (usually genes) on DNA. If two mutations complement each other, they are usually in different functional units or genes. If they do not complement each other, they are usually in the same functional unit or gene. Therefore, complementation cannot be used to measure distance on DNA. Of course, mutations that are in the same gene are often closer together than mutations that are in different genes.

The process of assessing complementation is similar to that of measuring recombination. In organisms other than viruses, two mutant strains of the same species with different mutations that cause the same phenotype are crossed to make cells that contain DNA from both organisms. Now, however, rather than testing the progeny organisms to see if any are recombinant types, we test the phenotypes of the progeny organisms. If the phenotype is changed by the presence of both DNAs, the mutations complement each other. If the organism exhibits the phenotype of a parental strain, the mutations do not complement each other. Table 13.1 summarizes the interpretation of complementation tests.

Complementation is also most easy to demonstrate with viruses, although the nomenclature and concepts are the same for all organisms. For complementation tests with viruses, cells are infected simultaneously by different strains of a particular virus. The course of the infection then reveals whether the mutations in the two strains complement each other and so are likely to be in different genes.

In the example presented in Figure 13.2, two strains of a virus have different mutations that prevent multiplication in cells that can serve as hosts of the wild-type virus. In other words, these strains are host range mutants of the virus. The cells in which the virus mutants cannot multiply are the nonpermissive hosts, whereas the cells in which the mutant viruses can multiply are the permissive hosts (see chapter 7). Infection of the nonpermissive cells with either mutant strain of the virus alone gives basically the same phenotype: the mutation in the strain will prevent multiplication of the virus. However, if the nonpermissive cells are infected

TABLE 13.1	Interpretation of complementation tests
Test result	**Possible explanations**
x and y complement	Mutations are in different genes Intragenic complementation has occurred[a]
x and y do not complement	Mutations are in the same gene One of the mutations is dominant One of the mutations affects a regulatory site or is polar

[a]See the text for an explanation of intragenic complementation. This is a less likely explanation than the mutations being in different genes.

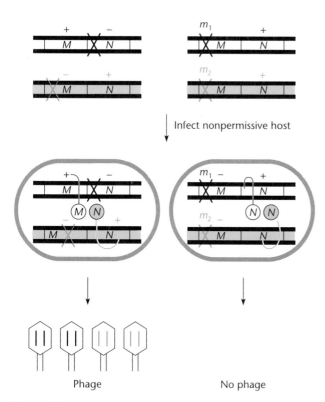

| Infect nonpermissive host

Phage   No phage

Figure 13.2 Complementation tests between virus mutations. Viruses with different mutations infect the same host cell, in which neither mutant virus can mulitiply. (Left) The mutations, represented by the minus signs, in different genes (M and N). Each mutant virus will synthesize the gene product that the other one cannot make; complementation will occur, and new viruses will be produced. (Right) The mutations (minus signs) prevent the synthesis of the M gene product. There is no complementation, the mutants cannot help each other multiply, and no viruses are produced.

simultaneously with the two mutant strains, the outcome of the infection might be altered if the mutations are in different genes. If this is the case (left side of Figure 13.2), each DNA furnishes one of the needed gene products, and so the two mutations will complement each other and both mutant viruses can multiply. If, however, the two mutations are in the same gene (right side of Figure 13.2), neither DNA will furnish one of the gene products, and so the mutations will not complement each other and neither mutant strain will multiply.

Note the difference in the method for how recombination and complementation tests are performed with virus strains that have the same two mutations that prevent multiplication in a particular host. In a recombination test, the *permissive* host cells are infected with the two strains. The virus is allowed to multiply before the *genotype* of the progeny viruses is tested by plating the

progeny on the nonpermissive host. However, in the complementation test, the *nonpermissive* host cells are infected with both mutant strains. Only if the mutations complement each other will the viruses multiply to form more mutant viruses. If complementation occurs, most of these progeny viruses will still be the parental types, which nevertheless were produced in the nonpermissive host because their mutant defects were complemented.

INTRAGENIC COMPLEMENTATION
Very rarely, complementation can occur between certain mutations within the same gene. This phenomenon, called **intragenic complementation** (*intra* means within), usually occurs only between mutations in a gene whose protein product is a homomultimer, that is, one made up of more than one subunit encoded by the same gene (see chapter 2). When the protein is made in cells that have two copies of the gene with different mutations, the multimeric protein will sometimes be assembled from the different mutant polypeptides. By chance, some of these mixed multimers might be active, even if multimers made up of only one mutant subunit or the other are inactive. Thus, the two mutations will seem to complement each other, even though they are in the same gene. However, intragenic complementation is rare, occurring between only certain mutations, even in genes that encode multimeric proteins.

POLARITY AND TRANSLATIONAL COUPLING
Polarity and translational coupling can also complicate the interpretation of complementation experiments because they sometimes cause mutations in different genes to not complement each other. Although the effects of polarity and translational coupling are similar, their molecular basis is very different (see chapter 2). However, in both mechanisms, a nonsense mutation, or an insertion or frameshift that causes a nonsense codon to be encountered during translation of one gene, can prevent transcription or translation of a downstream gene normally transcribed into the same mRNA. In this case, the mutation in the upstream gene will not complement mutations in the downstream gene, because it prevents expression of the downstream gene. Mutations such as nonsense mutations that disrupt translation of the upstream gene can affect complementation tests this way.

RECESSIVE VERSUS DOMINANT MUTATIONS
Complementation tests can also reveal whether a mutation is **recessive** or **dominant** to the normal form of the gene, or the **wild-type allele**. Whether a mutation is recessive or dominant to the wild type depends on whether it shows its phenotype in the presence of the

wild-type form of the gene. The wild-type phenotype will prevail if both recessive and normal alleles of the gene are present. In contrast, the mutant phenotype will prevail even in the presence of a normal allele of the gene when the mutation is dominant.

Most mutations are recessive because mutations usually inactivate a gene product, either RNA or protein. In cells with a normal and a mutated copy of the gene, the needed product can be furnished by the normal copy, restoring the normal or wild-type phenotype.

A few types of mutations are dominant. Mutations can be dominant for many reasons. For instance, the mutant gene product may be inactive but nevertheless able to interfere with the function of the normal gene product. If the product of the gene is a homomultimer, a protein made of normal and mutated subunits may be inactive. Alternatively, a dominant mutation may create a gene product that acts in a way that the normal gene product cannot, such as one that performs its role in an unregulated way.

Complementation tests can be used to determine if two mutations are in the same or different genes only if both mutations are recessive. If one of the mutations is dominant, it will express its phenotype in the presence of another mutation, regardless of whether the other mutation is in a different gene.

CIS- VERSUS TRANS-ACTING ELEMENTS

Complementation tests can also reveal whether a mutation affects a *trans*-acting function or a *cis*-acting site. As discussed in previous chapters, mutations that affect a protein or RNA product of a gene affect a **trans-acting** function (*trans* means on the other side of), because the gene product can move or diffuse throughout the cell, while mutations that are **cis acting** often change a site on the DNA rather than a diffusible gene product. Mutations that affect a *trans*-acting function can usually be complemented by the normal or wild-type allele, since the protein or RNA product of the gene can be made by the wild-type allele of the gene in another DNA in the same cell. *cis*-acting mutations cannot be complemented, because they affect the DNA even in the presence of a normal copy of the region. *cis*-acting mutations can manifest themselves in many ways, as we discuss later in this chapter.

Experiments with the *r*II Genes of Phage T4

We illustrate the basic principles of virus genetics by using the *r*II genes of phage T4. Experiments with the *r*II genes of T4 were responsible for many early develop-

ments in molecular genetics, including the discovery of nonsense codons, the definition of the nature of the genetic code, and the discovery of gene divisibility. Because of their historical importance, the *r*II genes deserve equal status with Mendel's peas, although to this day, the function of these genes is unknown. As we discuss, this does not prevent a genetic analysis.

The *r*II genes have an unconventional name because they were discovered and studied before the convention of using three-letter names for genes became established. The name *r*II means "rapid lysis mutants type II." Phage with this mutation cause the infected cells to lyse more quickly than the normal (r^+) phage, a property that *r*II phage share with the rapid-lysis mutants types I and III. The rapid-lysis phenotype of these mutants on *Escherichia coli* B indicator bacteria causes them to form hard-edged, clear plaques. In contrast, the plaques formed by r^+ phage have fuzzy edges because of a phenomenon called lysis inhibition, which delays lysis of the infected cells. The hard-edge, clear-plaque phenotype makes it easy to distinguish rapid-lysis mutants from the wild type (Figure 13.3).

A property of *r*II mutants that distinguishes them from the other types of rapid-lysis mutants and makes them particularly desirable for genetic analysis is that they are host range mutants. They can multiply in all strains of *E. coli* in which the wild type can multiply except those that carry the λ prophage (*E. coli* K12λ or

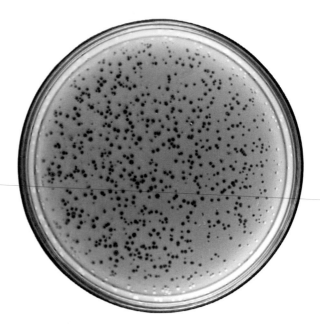

Figure 13.3 Plaques of phage T4. Most plaques are fuzzy edged, but some due to *r*II or other *r*-type mutants have hard edges because of rapid lysis of the host cells.

Kλ; see chapter 7). As we discuss, this property greatly facilitates complementation tests with *r*II mutants and makes possible the detection of even very rare recombinant types.

Complementation Tests with *r*II Mutants

In about 1950, Seymour Benzer and others realized the potential of using the *r*II genes of T4 to determine the detailed structure of genes. The first question asked was how many genes, or **complementation groups**, are represented by *r*II mutations? For an answer, numerous phage with *r*II mutations were isolated, and pairwise complementation tests were performed in which two different *r*II mutants infected the nonpermissive host, *E. coli* Kλ. When the two *r*II mutations complemented each other, phage were produced. These complementation tests revealed that all *r*II mutations could be sorted into two complementation groups, or genes, which were named *r*IIA and *r*IIB. Presumably, *r*IIA and *r*IIB encode different polypeptides, both of which are required for multiplication in *E. coli* Kλ.

Recombination Tests with *r*II Mutants

The next step in the genetic analysis was to perform recombination tests between *r*II mutations to determine the location of the *r*IIA and *r*IIB genes with respect to each other and to order the mutations within these genes. Recombination between *r*II mutations was measured by infecting permissive *E. coli* B with two different *r*II mutants and allowing the phage to multiply. The progeny phage were then plated to measure the frequency of recombinant types.

If recombination can occur between the two *r*II mutations, two different recombinant-type progeny would appear—double mutants with both *r*II mutations, and wild-type, or *r*⁺, recombinants with neither mutation. The recombinant types with both mutations are difficult to distinguish from the parental types because all *r*II mutants make *r*-type plaques on *E. coli* B and do not form plaques on *E. coli* Kλ. However, the *r*⁺ recombinants with neither *r*II mutation are easy to detect because they can multiply and form plaques on *E. coli* Kλ. Therefore, when the progeny of the cross are plated on *E. coli* Kλ, the number of plaques that appear equals the number of *r*⁺ recombinants. As discussed earlier, the recombination frequency equals the *total* number of recombinant types divided by the total progeny of the cross. We can assume that about half of the recombinants are not being detected because they are double-mutant recombinants, and so the total number of recombinant types is the number of *r*⁺ recombinants

multiplied by 2. Meanwhile, all the progeny should form plaques on *E. coli* B. Hence, the recombination frequency between the two *r*II mutations is twice the number of phage that form plaques on *E. coli* Kλ divided by the number of phage that form plaques on *E. coli* B.

Before the number of progeny phage that can form plaques on *E. coli* Kλ and *E. coli* B can be counted, however, a certain practical problem must be overcome. There will be many fewer *r*⁺ recombinant progeny than total progeny, so that if the cross is plated directly on the two types of bacteria, the number of phage that form plaques on *E. coli* Kλ will be much smaller than the number of phage that form plaques on *E. coli* B. There may be only a few plaques on *E. coli* Kλ and millions of plaques on *E. coli* B—too many to count. To overcome this problem, the phage progeny must be serially diluted by different amounts (see Introduction) before they are plated on each of the bacteria. The dilutions should be adjusted so that there are about 100 to 500 plaques on both the Kλ and B plates, which is not too many plaques to count but is large enough to give a reasonably low statistical error. The number of plaques on each type of indicator bacteria is then multiplied by the dilution factor to give the total number of progeny phage of each type.

To illustrate, let us cross an *r*II mutant that has the mutation *r168* with another *r*II mutant that has the mutation *r131*. We infect *E. coli* B with the two mutants and incubate the infected cells to allow the phage to multiply. To determine the number of *r*⁺ recombinants, we dilute the phage by a factor of 10^5 and plate on the *E. coli* Kλ indicator bacteria. To determine the total number of progeny, we dilute the phage by a factor of 10^7 and plate on the *E. coli* B indicator. After incubating the plates overnight, we observe 108 plaques on the *E. coli* Kλ plate and 144 plaques on the *E. coli* B plate.

From the equation for recombination frequency (RF),

$$RF = \text{total recombinant progeny/total progeny}$$
$$= \frac{2 \times 108 \times 10^5}{144 \times 10^7} = 1.5 \times 10^{-2} = 0.015$$

the recombination frequency between the two mutations is 0.015. If we want to express this in map units, we multiply the recombination frequency by 100 to give 1.5. The two mutations, *r168* and *r131*, are therefore 1.5 map units apart.

Using such crosses, Benzer began to determine the distance between some of his *r*IIA and *r*IIB mutations. He found that *r*IIA mutations are close to *r*IIB mutations, suggesting that the *r*IIA and *r*IIB genes are adjacent in the T4 DNA. He then went on to order a large number of mutations within the *r*IIA and *r*IIB genes.

Ordering *r*II Mutations by Three-Factor Crosses

Measuring the recombination frequency between two mutations can give an estimate of how close together the two mutations are in the DNA. However, to determine the relative order of mutations from such crosses, it is necessary to measure recombination frequencies very accurately. For example, let us say we have three *r*IIA mutations, *r21*, *r3*, and *r12*, and we want to use recombination frequencies to determine their order in the *r*IIA gene. When we cross *r12* with *r3,* we obtain a recombination frequency of approximately 0.01. When we cross *r21* with *r12*, we also obtain a recombination frequency of approximately 0.01. When we cross *r3* with *r21*, we get a recombination frequency of approximately 0.02. From these data alone, we suspect that *r12* is between *r3* and *r21* and that the order of the three mutations is

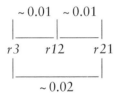

However, with crosses such as these, even a small error could cause an over- or underestimation of the recombination frequency and give the wrong order. The recombination frequency could be underestimated if, for example, the MOIs of the two mutant phages used for one of the infections were not exactly the same or if the cells lysed prematurely in one or more of the crosses.

Three-factor crosses offer a less ambiguous method for ordering mutations. In this method, a mutant strain that has two mutations is crossed with another strain that has the third mutation. The number of wild-type recombinants is then determined. In such a cross, the number of crossovers required to make a wild-type recombinant will depend on the order of the three mutations, and the more crossovers required to make a wild-type recombinant, the less frequent that recombinant type will be.

In Figure 13.4, a three-factor cross is being used to order the three *r*II mutations *r21*, *r3*, and *r12*. First, a double mutant is constructed with the *r21* and *r3* mutations. This double mutant is then crossed with a mutant that had only the *r12* mutation. If the order is *r3-r21-*

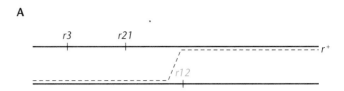

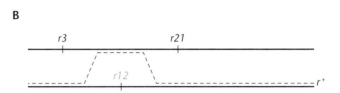

Figure 13.4 A three-factor cross to map *r*II mutations. See the text for details.

r12, only one crossover between *r21* and *r12* is required to make an *r*+ recombinant and the frequency of *r*+ recombinants should be about 0.01, the frequency of recombination between *r21* and *r12*. The same is true if the order is *r21-r3-r12*, where, again, only one crossover is required to make *r*+ recombinants. If, however, the order is *r3-r12-r21*, with the *r12* mutation in the middle, as we suspect from the earlier two-factor crosses, two crossovers are required and the frequency of *r*+ recombinants will be much lower than 0.01.

Theoretically, if the two crossovers were independent, the frequency of double crossovers should be the product of the frequencies of each of the single crossovers, or about 0.01 × 0.01 = 0.0001, which is only 1/100 of the frequency of the single crossover. However, because of high negative interference (see chapter 9), crossovers close to each other in the DNA are not truly independent, and one crossover will greatly increase the likelihood of what appears to be a second crossover nearby, making the frequency of apparent double crossovers much higher than predicted. Nevertheless, the frequency of double crossovers will generally be much lower than the frequencies of the individual crossovers, permitting the use of three-factor crosses to unambiguously order mutations.

ORDERING MUTATIONS BY DELETION MAPPING

As part of his genetic analysis of the structure of the *r*II genes, Benzer wanted to determine how many sites there are for mutations in *r*IIA and *r*IIB and to determine whether all sites within these genes are equally mutable or whether some are preferred. To find these answers, he turned to **deletion mapping**. The principle behind this method is that if a phage with a point mutation is crossed with another phage with a deletion mutation, no wild-type recombinants will appear when the point mutation lies within the deleted region. It is much easier to determine whether there are any r^+ recombinants at all than it is to carefully measure recombination frequencies. Therefore, deletion mapping offers a conveneient way to map large numbers of mutations.

For this approach, Benzer needed deletion mutations extending known distances into the *r*II genes. Some of the *r*II mutants he had already isolated had the properties of deletion mutations (see chapter 3). First, the *r*II mutations in these particular mutants did not revert, as indicated by the lack of plaques due to r^+ revertant phage, even when very large numbers of *r*II mutant phage were plated on *E. coli* Kλ. Second, these mutations did not map at a single position, or point, as would base pair changes or frameshift mutations. They did not give r^+ recombinants when crossed with many different *r*II mutations, at least some of which gave r^+ recombinants when crossed with each other, and so must have been at different positions in the *r*II genes.

To determine the extent of the deleted regions in some of these *r*II mutants, Benzer crossed them with some of the point mutations he had mapped by using three-factor crosses. When the deletion gave no r^+ recombinants in a cross with a particular point mutation, he concluded that the deletion must extend at least as far as the point mutation. He thus was able to estimate the endpoints of the deletions in the rII genes fairly accurately.

Figure 13.5 shows a set of Benzer deletions that are particularly useful for mapping *r*IIA mutations. These lengthy deletions begin somewhere outside of *r*IIB and remove all of that gene, extending various distances into *r*IIA. One deletion, *r1272*, extends through the entire *r*II region, completely removing both *r*IIA and *r*IIB.

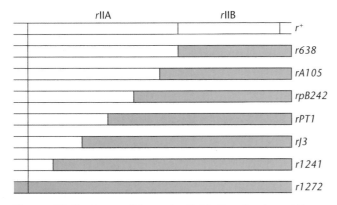

Figure 13.5 Some of Benzer's *r*II deletions in phage T4. The deletions remove all of *r*IIB and extend various distances into *r*IIA. The shaded bars show the region deleted in each of the mutations.

Armed with deletions with known endpoints, Benzer quickly localized the position of any new *r*II point mutation by crossing the mutant phage with other phage that each had one of these deletions. For example, a point mutation gave r^+ recombinants with the deletion *rA105* but not with *rpB242*; this point mutation must lie in the short region between the end of *rA105* and the end of *rpB242*. Therefore, with no more than seven crosses, a mutation in an *r*II mutant could be localized to one of seven segments of the *r*IIA gene. The position of the mutation could be located more precisely through additional crosses between phage with the point mutation and phage with short deletions or other point mutations that are located within this segment.

MUTATIONAL SPECTRA

The numerous point mutations within the *r*IIA and *r*IIB genes that Benzer found by deletion mapping included spontaneous as well as induced mutations. Figure 13.6 illustrates the map locations of some of the spontaneous mutations. Spontaneous mutations can occur everywhere in the *r*IIA gene, but Benzer noted that some sites are "hot spots," in which many more mutations occur than at other sites. Mutagen-induced mutations also have hot spots, which differ from each other and from those of spontaneous mutations (data not shown).

The tendency of different mutagens to mutate some sites much more frequently than others has practical consequences in genetic analysis. If Benzer had studied only spontaneous mutations in the *r*II genes, almost 30% of these would have been at B5, the major hot spot for spontaneous mutations. Therefore, to obtain a random collection of mutations in a gene, we should isolate not only spontaneous mutations but also induced ones.

Methods were recently developed that allow essentially random mutagenesis of selected regions of DNA. These methods involve the use of special oligonucleotide primers for site-specific mutagenesis and PCR mutagenesis. We discuss some of them in chapter 15.

The *r*II Genes and the Nature of the Genetic Code

Of all the early experiments with the T4 *r*II genes, some of the most elegant were those that revealed the nature of the genetic code. These were conducted by Francis Crick and his collaborators (see Crick et al., Suggested Reading). These experiments not only have great historical importance but also are a good illustration of classical genetic principles and analysis.

At the time Crick and his collaborators began these experiments, he and Watson had used the X-ray diffraction data of Rosalind Franklin and Maurice Wilkins

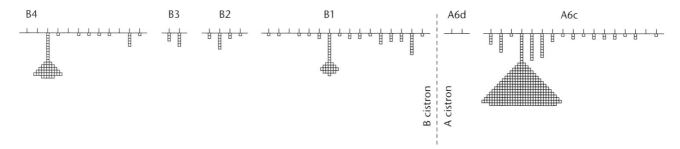

Figure 13.6 Mutational spectrum for spontaneous mutations in a short region of the *r*II genes. Each small box indicates one mutation observed at that site. Large numbers of boxes at a site indicate hot spots, where spontaneous mutations often occur. Adapted with permission from S. Benzer, *Proc. Natl. Acad. Sci. USA* **47**:403–415, 1961.

and the biochemical data of Erwin Chargaff and others to solve the structure of DNA (see Suggested Reading, *The Path to the Double Helix,* in chapter 1). This structure indicated that the sequence of bases in DNA determines the sequence of amino acids in protein. However, the question remained of how the sequence of bases is read. For example, how many bases in DNA encode each amino acid? Does every possible sequence of bases encode an amino acid? Is the code "punctuated," with each code word demarcated, or does the cell merely begin reading at the beginning of a gene and continue to the end, reading a certain number of bases each time? The ease with which the *r*II genes of T4 could be manipulated made this system the obvious choice to answer these questions.

The experiments of Crick et al. were successful for two reasons. First, the extreme N-terminal region of the *r*IIB polypeptide, the so-called B1 region, is nonessential for activity of the *r*IIB protein. Consequently, the B1 region can be deleted or all the amino acids it encodes can be changed without affecting the activity of the polypeptide. Note that this is not normal for proteins. Most proteins cannot tolerate extensive amino acid changes in any region.

The second reason for the success of these experiments is that acridine dyes specifically induce frameshift mutations by causing the removal or addition of a base pair in DNA (see chapter 10). This conclusion required a leap of faith at the time. The mutations caused by acridine dyes are usually not leaky but are obviously not deletions because they map as point mutations and revert. Also, the frequency of these revertants increases when the phage are propagated in the presence of acridine dyes themselves but not when they are propagated in the presence of base analogs, which at the time were suspected to cause only base pair changes. It was reasoned that if acridine dye-induced mutations could not be reverted by base pair changes, the mutations induced

by acridine dyes themselves could not be base pair changes. This evidence that acridine dyes cause frameshift mutations may seem flimsy in retrospect, yet it was convincing enough to Crick et al. that they proceeded with their experiments on the nature of the genetic code.

INTRAGENIC SUPPRESSORS OF A FRAMESHIFT MUTATION IN *r*II B1

The first step in their analysis was to induce a frameshift mutation in the *r*II B1 region by propagating cells infected with the phage in the presence of the acridine dye proflavin. Crick and his colleagues named their first *r*II mutation FC0 for "Francis Crick Zero." The FC0 mutation prevents T4 multiplication in *E. coli* Kλ because it inactivates the *r*IIB polypeptide. That an acridine-induced mutation in the region encoding the nonessential B1 portion of the B polypeptide can inactivate the *r*IIB polypeptide in itself suggests the mutations are unusual, since, as mentioned above, merely changing an amino acid in the B1 region should not inactivate the gene.

ISOLATING SUPPRESSOR MUTATIONS OF FC0

The next step in the Crick et al. analysis was to isolate suppressor mutations of FC0. As discussed in chapter 3, a suppressor mutation restores the function of a mutated gene product but occurs in a DNA sequence different from the original mutation. To isolate suppressors of FC0, Crick et al. merely plated large numbers of FC0 mutant phage on *E. coli* Kλ. A few plaques due to phenotypically *r*+ phage appeared. These phage could either have been revertants of the original FC0 mutation or have had two mutations, the original FC0 mutation plus a suppressor.

To determine which of the *r*+ phage had suppressor mutations, Crick et al. applied the classic genetic test for suppression. In such a test, an apparent revertant is crossd with the wild type. If any of the progeny are recombinant types with the mutant phenotype, the interpretation is that the mutation had not reverted but had been suppressed by a second site mutation that restored the wild-type phenotype.

Figure 13.7 illustrates the principle behind this test as applied to the apparent revertants of *r*II mutants of T4. The apparent wild-type revertant is crossed with the wild-type phage. If the mutation has reverted, all of the progeny will be *r*⁺ and there will be no *r*II mutant recombinants. In contrast, if the mutation has been suppressed, the suppressing mutation can be crossed away from the FC0 mutation, and *r*II mutant recombinant types will appear that cannot multiply in *E. coli* Kλ. In the test of Crick et al., most of the apparent *r*⁺ revertants of FC0 gave some *r*II mutant recombinants when crossed with the wild type; therefore, the FC0 mutation in these apparent revertants was being suppressed rather than reverted. Moreover, there were very few *r*II mutant recombinants, and so the suppressing mutations must have been very close to the original FC0 mutation, presumably also in the B1 region of the *r*IIB gene. Only crossovers between the regions of the FC mutation and the suppressing mutation will give rise to rII mutant recombinants.

ISOLATING THE SUPPRESSOR MUTATIONS BY THEMSELVES

A double mutant with both the FC0 mutation and the suppressor mutation is *r*⁺ and will multiply in *E. coli* Kλ. But would a mutant with a suppressor mutation alone be phenotypically *r*II or *r*⁺? If the suppressor mu-

A   Reversion

B   Suppression

Figure 13.7 Classical genetic test for suppression. Phages that have apparently reverted to wild type (*r*⁺) are crossed with wild-type *r*⁺ phage. (A) If the mutation has reverted, all of the progeny will be *r*⁺. (B) If the mutation has been suppressed by another mutation, x, there will be some *r*II mutant recombinants among the progeny.

tations by themselves produce a phenotypically *r*II mutant, presumably some of the *r*II mutant recombinants obtained by crossing the suppressed FC0 mutant with wild-type T4 were single-mutant recombinants with only the suppressor mutation. This can be easily tested. If a recombinant is phenotypically *r*II because it has the suppressor mutation rather than the FC0 mutation, it should give some *r*⁺ recombinants when crossed with the FC0 single mutant. Some of the *r*II mutant recombinant phages did give some *r*⁺ recombinants when crossed with FC0 mutants, indicating that these phages have an *r*II mutation, presumably the suppressor mutation, different from the FC0 mutation.

ISOLATING SUPPRESSOR-OF-SUPPRESSOR MUTATIONS

Because the suppressor mutations of FC0 by themselves make the phage *r*IIB⁻ and prevent multiplication on *E. coli* Kλ, the next question was whether the suppressor mutations of FC0 could be suppressed by "suppressor-of-suppressor" mutations. As with the original FC0 mutation, when Crick et al. plated large numbers of T4 with a suppressor mutation on Kλ, they observed a few plaques. Most of these resulted from second-site suppressors. Moreover, these suppressor-of-suppressor mutations were *r*IIB⁻ when isolated by themselves. This process could be continued indefinitely.

FRAMESHIFT MUTATIONS AND IMPLICATIONS FOR THE GENETIC CODE

To explain these results, Crick and his collaborators proposed the model shown in Figure 13.8. According to this model, FC0 is a frameshift mutation that alters the reading frame of the *r*IIB gene by adding or removing a base pair so that all the amino acids inserted in the protein from that point on are wrong. This explains how the FC0 mutation can inactivate the *r*IIB polypeptide, even though it occurs in the nonessential B1 region of the gene.

The suppressors of FC0 are also frameshift mutations in the *r*IIB1 region. The suppressors either remove or add a base pair, depending on whether FC0 adds or removes a base pair, respectively. As long as the other mutation has the opposite effect to FC0, an active *r*IIB polypeptide will often be synthesized (see Figure 13.8).

IMPLICATIONS FOR THE CODE

The results of these experiments had several implications for the genetic code.

The Code Is Unpunctuated

At the time, it was not known if something demarcated where a code word in the DNA begins and ends. Con-

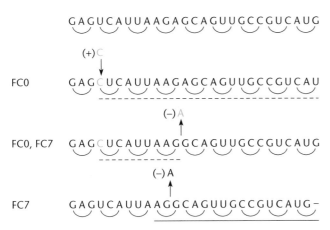

Figure 13.8 Frameshift mutations and suppression. The FC0 frameshift is caused by the addition of 1 bp, which alters the reading frame and makes an *r*II mutant phenotype. FC0 can be suppressed by another mutation, FC7, that deletes 1 bp and restores the proper reading frame and the *r*+ phenotype. The FC7 mutation by itself confers an *r*II mutant phenotype. Regions translated in the wrong frame are underlined.

sider a language in which all the words have the same number of letters. If there were spaces between the words, we could always read the words of a sentence correctly because we would know where one word ended and the next word began. However, if the words did not have spaces between them, the only way we would know where the words began and ended would be to count the letters. If a letter were left out or added to a word, we would read all the following words wrong. This is what happens when a base pair is added to or deleted from a gene. The remainder of the gene is read wrong; therefore, the code must be unpunctuated.

The Code Is Three Lettered
The experiments of Crick et al. also answered the question of how many letters are in each word of this language; i.e., how many bases in DNA are being read for each amino acid inserted in the protein? At the time, there were theoretical reasons to believe that the number is larger than two. Since DNA contains four "letters"—or bases (A, G, T and C)—only $4 \times 4 = 16$ possible amino acid code words could be made out of only two of these letters. However, at least 20 amino acids were known to be inserted into proteins (the known number is now 22 [see chapter 2]), so a two-letter code would not yield enough code words for all the amino acids. However, three bases per code word results in $4 \times 4 \times 4 = 64$ possible code words, plenty to encode all of the amino acids.

The assumption that the code is three lettered was testable. The reading frame of a three-letter code would not be altered if 3 bp was added or removed in the *r*IIB1 region. Continuing with the letter analogy, an extra word would then be put in or left out but all the other words would be read correctly. Therefore, in the B1 region, if three suppressors of FC0 or three suppressors of suppressors were combined in the same phage DNA, a complete new code word would be added to or subtracted from the molecule and the correct reading frame would be restored. Thus, the phage should be *r*+ and multiply in *E. coli* Kλ. Experimental results were consistent with this hypothesis, indicating that 3 bp in DNA encodes each amino acid inserted into a protein.

The Code Is Redundant
The results of these experiments also indicated that the code is redundant; that is, more than one word codes for each amino acid. Crick et al. reasoned that if the code were not redundant, most of the code words, i.e., $64 - 20 = 44$, would not encode an amino acid; then a ribosome translating in the wrong frame would almost immediately encounter a code word that does not encode an amino acid, and translation would cease. The fact that most combinations of suppressors with FC0 and with suppressors of suppressors restored the *r*+ phenotype indicated that most of the possible code words do encode an amino acid.

Nonsense Code Words and Termination of Translation
Although their evidence indicated that most code words encode an amino acid, it also indicated that not all of them do. If all possible words signified an amino acid, the entire *r*IIB1 region should be translatable in any frame and a functional polypeptide would result, provided that the correct frame was restored before the translation mechanism entered the remainder of the *r*IIB gene. However, if not all the words encode an amino acid, a "forbidden" code word that does not encode an amino acid might be encountered during translation in a wrong frame. In this situation, not all combinations of suppressors and suppressors of suppressors would restore the *r*+ phenotype. Crick et al. observed that some combinations of suppressors and suppressors of suppressors did cause forbidden code words to be encountered in the *r*IIB region. However, other combinations in the same region resulted in a functional *r*IIB polypeptide. For example, in Figure 13.9, a nonsense codon (UAA) is encountered when a region is translated in the +1 frame because 1 bp was removed. However, no nonsense codons are encoun-

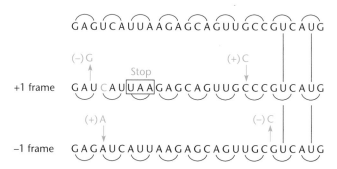

Figure 13.9 Frameshift suppression and nonsense codons. The nonsense codon UAA is encountered in the +1 frame due to deletion of 1 bp in the DNA. Translation will terminate even if the correct translational frame is restored farther downstream. In the –1 frame, due to addition of 1 bp, no nonsense codons are encountered. A downstream deletion restores the correct reading frame, and the active polypeptide is translated.

tered when 1 bp is added and the same region is translated in the –1 frame.

Even more convincing evidence that some code words do not encode an amino acid came from experiments with the deletion *r1589* (see Benzer and Champe, Suggested Reading). The deletion *r1589* removes much of *r*IIA and the nonessential *r*IIB1 region, eliminating the nonsense codon or codons that are normally at the end of the *r*IIA gene and the translation initiation region of *r*IIB. This deletion mutation thereby causes translation initiated at *r*IIA to proceed into *r*IIB, resulting in a fusion protein in which the N terminus comes from *r*IIA and the rest of the protein comes from *r*IIB. Since most of *r*IIA but only the nonessential B1 region of *r*IIB is deleted, this fusion protein has *r*IIB activity but not *r*IIA activity, as can be demonstrated by complementation tests.

Although the fusion protein does not require the *r*IIA portion for *r*IIB activity, Benzer and Champe found that some base pair change mutations in the *r*IIA region prevented *r*IIB activity. These base pair changes presumably occurred in one of the nonsense codons that stopped translation in the *r*IIA region. Other base pair change mutations that did not disrupt *r*IIB activity were presumably missense mutations, which resulted in insertion of the wrong amino acid in the rIIA portion of the fusion protein but did not stop translation.

Benzer and Champe also found that even the presumed nonsense mutations did not prevent *r*IIB activity in some strains of *E. coli* and so were in a sense ambivalent. We now know that these "permissive" strains of *E. coli* are nonsense suppressor strains with mutations in tRNA genes that allow readthrough of one or more of the nonsense codons (see chapter 3).

POSTSCRIPT ON THE CRICK ET AL. EXPERIMENTS

The experiments of Crick and his collaborators laid the groundwork for the subsequent deciphering of the genetic code by Marshall Nirenberg and his colleagues, who assigned an amino acid to each of the 61 3-base sense codons. Other researchers later used reversion and suppression studies to determine that the nonsense codons are UAG, UAA, and UGA.

Isolating Duplication Mutations of the *r*II Region

Our final example of the genetic manipulation of the *r*II genes of T4 is the isolation of tandem duplication mutations of the *r*II region (see, for example, Symonds et al., Suggested Reading). These experiments help contrast the differences between complementation and recombination and also illustrate some of the genetic properties of tandem duplication mutations.

The isolation of tandem duplication mutations of the *r*II region depended on the properties of two deletions in this region, the aforementioned *r1589* deletion and another deletion, *r638* (Figure 13.10). As mentioned, the *r1589* deletion removes the B1 region of the *r*IIB gene as well as part of *r*IIA. Phage with this deletion are phenotypically *r*IIA⁻ *r*IIB⁺. The deletion mutation *r638* deletes all of *r*IIB but does not enter *r*IIA, so that phage with this deletion are *r*IIA⁺ *r*IIB⁻. Because one deleted DNA makes the product of the *r*IIA gene and the other makes the the product of the *r*IIB gene, the two deletion mutations can complement each other. However, they cannot recombine to give *r*⁺ recombinants, because they overlap, both deleting the B1 region of the *r*IIB gene.

Even though recombination should not occur between the two deletions to give *r*⁺ recombinants, when

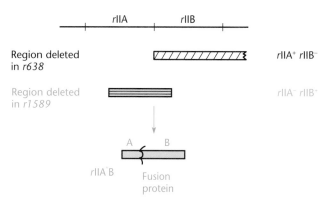

Figure 13.10 The *r*II deletions *r638* and *r1589*. An *r*IIAB fusion protein is made in a strain with *r1589*. See the text for details.

E. coli B is infected simultaneously with the two deletion mutants and the progeny are plated on *E. coli* Kλ, a few rare plaques due to apparent *r*⁺ phage arise. These phenotypically *r*⁺ phage have tandem duplications of the *r*II region (Figure 13.11A). Each copy of the *r*II region has a different deletion mutation, and these mutations complement each other to give the r⁺ phenotype.

Figure 13.11A also illustrates how these tandem duplication mutations might arise. Sometimes, while the DNA is replicating, recombination mistakenly occurs between two short directly repeated regions on either side of the *r*II region. Such mistaken crossovers are rare because repeated sequences in DNA, when they exist at all, are usually very short. However, once such a crossover occurs, one of the recombinant-type phage will have a duplicate of the *r*II region, both copies of which, in the example, have the *r1589* mutation. If a phage with such a duplication then infects the same cell as a phage with the *r638* deletion, one of the copies of the *r*II region can recombine with the DNA of the other parent and the *r638* deletion will replace one of the *r1589* deletions. This second recombination will occur very frequently, because of the extensive homology between the duplicated regions. The phage that package this DNA will then have two copies of the *r*II region, one copy with the *r1589* deletion and the other copy with the *r638* deletion. In subsequent infections, the two deletion mutations can complement each other to make the phage phenotypically *r*⁺. These phage are diploid for the *r*II regions and surrounding genes because they have two copies of these genes.

The salient property of tandem duplication mutations is that they are very unstable, because recombination anywhere in the long duplicated region will destroy the duplication (see chapter 3). These *r*⁺ phage exhibit this instability. If the *r*⁺ phage with the putative duplication are propagated in *E. coli* B, where there is no selection for phage with the duplication, a very high percentage of the progeny phage will be unable to multiply in *E. coli* Kλ.

Figure 13.11B illustrates why the duplications are unstable. A crossover between either of the duplicated segments, x or y, will cause the intervening sequences to be deleted and one copy of the duplication to be lost. The resulting phage, some of which have the *r1589* deletion whereas others have the *r638* deletion, are called **haploid segregants**. The phage are considered haploid because they have lost one copy of the duplicated region. The term "segregants" is used rather than recombinants because the recombination that destroys the duplication occurs spontaneously while the phage is multiplying and does not require crosses. The x and y

regions are usually quite long and identical in sequence, so that haploid segregants appear quite frequently.

The two haploid types segregate at a characteristic frequency for each duplication mutation. As we can see from the duplication shown in Figure 13.11B, a crossover in region x will yield the *r1589* haploid whereas a crossover in region y will yield the *r638* haploid. Therefore, which haploid type will segregate at the highest frequency will depend upon which region—x or y—is longer. If x is longer than y, the *r1589* single mutant will segregate more frequently. However, if x is shorter than y, *r638* will segregate more frequently.

Constructing the Genetic Linkage Map of a Phage

A picture that shows many of the genes of a phage and how they are ordered with respect to each other is known as the **genetic linkage map** of the phage, so named because it shows the proximity or linkage of the genes to each other. This linkage is determined by genetic crosses. Physical methods for mapping DNA, discussed in chapter 15, give rise to a physical map, which can often be correlated with the genetic map.

To be informative, a genetic linkage map should include many if not most of the genes of the phage. The *r*II genes of T4 could be identified because mutations in these genes cause the phage to make *r*-type plaques on *E. coli* B and prevent its mutiplication in *E. coli* Kλ. However, *r*IIA and *r*IIB are only two of the more than 200 genes of phage T4. To identify other genes of the phage, we need to examine other phenotypes.

Conditional-lethal temperature-sensitive and nonsense mutations can be used to identify any gene that is essential for multiplication of a phage. If a phage has a mutation to the amber codon (UAG) in an essential gene, it will multiply and form a plaque only on a permissive host with an amber suppressor tRNA (see chapter 7). Phage with a temperature-sensitive mutation in an essential gene will multiply and form plaques at a lower (permissive) temperature but not at a higher (nonpermissive) temperature. Because such mutations can be isolated in any essential gene, nonsense and temperature-sensitive mutations can be used to identify many of these genes and to construct a more complete genetic linkage map of a phage.

The first step in constructing the conditional-lethal map of a phage is to isolate a large number of temperature-sensitive and nonsense mutations of the phage by first mutagenizing the phage with various mutagens and then plating the surviving phage on suppressing bacteria at the permissive temperature. Then plaques are picked and the phage is tested for multiplication on nonsuppressing bacteria and at the nonpermissive temperature. Phage that cannot form plaques on the nonsuppressing bacteria or at the nonpermissive temperature have non-sense mutations or temperature-sensitive mutations, respectively, in essential genes.

Identifying Phage Genes by Complementation Tests

Once a large collection of mutations of the phage have been assembled, the mutations can be placed into com-

plementation groups or genes. As discussed above, two mutations that do not complement each other are probably in the same complementation group or gene. Complementation tests are done under conditions where neither mutant can multiply. For example, to test for complementation between two different amber mutations, nonsuppressing bacteria are infected with the two mutants simultaneously and the progeny are plated on amber-suppressing bacteria to determine how many progeny phage are produced. To test for complementa-

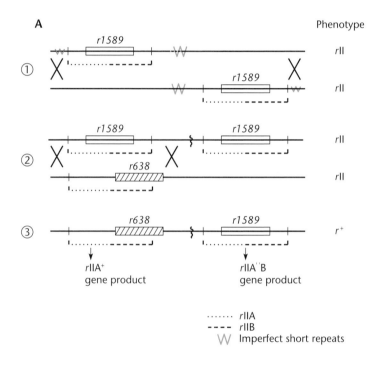

A

Phenotype

① *rII*

rII

② *rII*

rII

③ *r⁺*

rIIA⁺ gene product

rIIA⁻B gene product

······· *rIIA*
---- *rIIB*
W Imperfect short repeats

Figure 13.11 (A) Model for how tandem duplications in the *rII* region form. Recombination between short repeated sequences flanking the *rII* region may occur at a low frequency, giving rise to a duplication of one or more genes. This duplication can recombine with the other deletion strain in the region of one of the repeated sequences, giving rise to a duplication in which one copy of *rIIB* has the *r1589* deletion and the other copy has the *r638* deletion. The two deletions complement each other, so that the phage is phenotypically *r⁺*. (B) Tandem duplications are unstable because recombination between the duplicated regions can destroy the duplication. Which deletion mutation remains in the haploid segegant depends upon where the recombination occurs. (1) Recombination between the y duplicated segments gives rise to a haploid segregant with only the *r638* deletion. (2) Recombination between the two duplicated regions designated x gives rise to a haploid segregant with only the *r1589* deletion. In this example, *r1589* haploid segegants will be about three times more frequent than *r638* segregants because x is approximately three times as long as y.

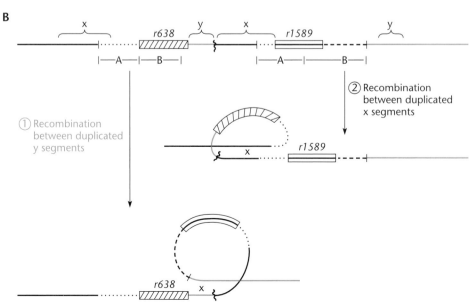

B

tion between a temperature-sensitive and an amber mutation, nonsuppressing bacteria are infected at the high (nonpermissive) temperature. The progeny phage are then plated on amber-suppressing bacteria at the permissive temperature to see how many phage are produced. Note that a temperature-sensitive and an amber mutation could be in the same gene; in other words, they could be allelic. Even though they are different types of mutations, if they are allelic, they will not complement each other.

Each time we find a mutation that complements all the other mutations in the collection, we have found a new gene. Once two mutations have been found that do not complement each other, only one of the two mutations need be used for further complementation tests. If many of the complementation groups are represented by only a single mutation, many other essential genes are probably not yet represented by any mutations. More mutants must then be isolated to identify most of the essential genes. Eventually, more and more of the new mutations will sort into one of the previously identified complementation groups and the collection of mutations in essential genes will be almost complete.

Mapping Phage Genes

Once most of the genes of the phage have been identified, representative mutations in each of the genes can be mapped with respect to each other. To measure the frequency of recombination between two mutations, cells are infected under conditions that are permissive for both mutations. For example, to cross a temperature-sensitive mutation with an amber mutation, amber suppressor cells are infected at the low (permissive) temperature. After phage are produced, the progeny phage are plated under conditions permissive for both mutations, to measure the total progeny, and under conditions that are nonpermissive for both mutations, to measure the number of wild-type recombinants. Hence, if an amber mutant strain is crossed with a temperature-sensitive mutant strain, the progeny should be plated on nonsuppressor bacteria at the high (nonpermissive) temperature to measure the amber of wild-type recombinants and at the low (permissive) temperature on amber-suppressing bacteria to measure the total progeny. Under the former conditions, only the wild-type recombinants, which lack both mutations, can multiply to form plaques. From the frequency of wild-type recombinants, the map distance between the mutations and therefore the genes they are in can be calculated by using the equation for recombination frequency given above.

Mutations in closely linked genes may need to be ordered on the DNA by using three-factor crosses. The de-

tails of the method depend on the type of mutations being mapped. Take the example of ordering amber mutations in two genes, *am*66 and *am*231, with respect to a temperature-sensitive mutation, *ts*21, in a third gene. First, the double mutant with both the *am*66 and *ts*21 mutations could be constructed by crossing phage strains with each of the mutations and identifying a progeny strain that has both an amber mutation and a temperature-sensitive mutation. Mutant strains with both these mutations will not form plaques on either nonsuppressor bacteria at the permissive temperature or suppressor bacteria at the nonpermissive temperature. This double mutant is then crossed with a single mutant having only the *am*231 mutation. The *am*⁺ recombinants are then tested for the *ts*21 mutation by picking the plaques from nonsuppressing bacteria at the permissive temperature and testing them at the nonpermissive temperature. If the order is *am*66-*am*231-*ts*21, most of the *am*⁺ recombinants should have the *ts*21 mutation and so should be temperature sensitive. If the order is *am*231-*am*66-*ts*21, most of the *am*⁺ recombinants should not have the *ts*21 mutation and so should not be temperature sensitive.

Nonsense and temperature-sensitive mutations can be used to identify only essential genes of the phage, or at least genes whose products are essential for multiplication in a particular type of host bacterium. The methods for finding nonessential genes can be more difficult, and they differ with each gene. Nonessential genes are also more difficult to map, since recombinants are usually more difficult to identify.

Once we have identified numerous phage genes and ordered them with respect to each other, we can begin the difficult task of trying to determine how the product of each gene functions during phage multiplication. This would involve experiments such as those described in chapter 7, to identify gene products involved in DNA replication, transcriptional regulatory genes, and so on.

Genetic Linkage Maps of Some Phages

Genetic linkage maps have been determined for several commonly used phages. Chapter 7 presents some of these maps and describes the function of some of the gene products.

One noticeable feature of most phage genetic maps is that genes whose products must physically interact, such as the products involved in head or tail formation, tend to be clustered together. This clustering may allow recombination between closely related phages without disruption of gene function. If the genes whose products must physically interact were not close to each other, recombination would tend to separate the genes and could give rise to inviable phage. For example, if the head genes were not clustered, recombination with the DNA of another related phage would often replace some of the head genes of the first phage with the corresponding head genes from the other phage. If the head of the phage could not be assembled from this mixture of head proteins, the phage would be inviable. Therefore, to maintain the advantage of new combinations of genes, evolution has left genes whose products must physically interact clustered together, so that recombination between other genes will not lead to inviable phage. This hypothesis has received support recently from the structure of phages related to λ. These phages seem to be made up of regions, or cassettes, from different phages assembled by recombination.

Factors That Determine the Form of the Linkage Map

We can see from the maps in chapter 7 that some phages have a linear genetic linkage map while others have a circular map. This form does not necessarily correlate with the linearity or circularity of the phage DNA. Some phages with circular DNA have a linear linkage map, while some with linear DNA have a circular map. To understand how the genetic maps arise, we need to review how the DNA of phages replicates and how it is packaged into phage heads.

PHAGE λ

Phage λ has a linear genetic map, even though the DNA forms a circle after it enters the cell. Phage λ has a linear map because its concatemers are cleaved at *cos* sites before being packaged into the phage head. As an illustration, consider a cross between two phages with mutations in the *A* and *R* genes at the ends of the phage DNA (see the λ map in chapter 7). Even though different parental alleles of the *A* and *R* genes can be next to each other in the concatemers prior to packaging, these alleles will be separated when the DNA is cut at the *cos* site during packaging of the DNA. Therefore the *A* and *R* genes will appear to be unlinked in genetic crosses and the genetic map will be linear, with the *A* and *R* genes at its ends. For this reason, all types of phages with unique *pac* or *cos* sites have linear genetic linkage

maps with the ends defined by the position of the *pac* or *cos* site.

PHAGE T4

Phage T4 has a circular genetic map even though its DNA never forms a circle. T4 has no unique *pac* site, and the DNA is packaged by a headful mechanism from long concatemers (Figure 13.12). Consequently, the T4 phage DNAs in different phage heads do not have the same ends but, rather, are cyclic permutations of each other. Therefore, genes that are next to each other in the concatemers will still be together in most of the phage heads and so will appear linked in crosses, producing a circular map.

PHAGE P22

Phage P22 is a phage of *Salmonella* spp. It is closely related to λ and replicates by a similar mechanism. However, unlike λ, it has a circular linkage map. The difference is that P22 begins packaging at a unique *pac* site like λ but then packages a few genomes by a processive headful mechanism like T4.

PHAGE P1

Phage P1 also replicates as a circle and packages by a headful mechanism; nevertheless, it has a linear map because of a very active site-specific recombination system called *cre-lox* that promotes recombination at a particular site in the DNA. The function of this site-specific recombination system may be to promote the formation of concatemers during lytic development. Because recombination at this site is so frequent, genetic markers on either side of the site appear to be unlinked, giving rise to a linear map terminating at the *cre-lox* site.

Genetic Experiments with Phage λ

Phage λ has been one of the major model systems for genetic analysis, and we probably know more about this phage and its interaction with its host, *E. coli*, than we know about any other genetic system. Phage λ is also ideal for illustrating genetic concepts such as intragenic complementation and *cis*- and *trans*-acting mutations. The following are some examples of how genetic

Figure 13.12 T4 DNA headful packaging. Packaging of DNA longer than a single genome equivalent gives rise to repeated terminally redundant ends and circularly permuted genomes.

analysis has been used to study the various functions of phage λ.

Genetics of λ Lysogeny

Phage λ is capable of lysogeny, and the plaques of λ are cloudy in the middle because of the growth of immune lysogens in the plaque (Figure 13.13). Mutants of λ that cannot form lysogens are easily identified by their clear plaques. These "clear-plaque mutants" have mutations in genes whose products are required for the phage to form lysogens.

Complementation tests revealed how many genes are represented by clear-plaque mutants. Now, however, rather than asking whether two mutants can help each other to multiply, we ask whether two mutants can help each other to form a lysogen, since this is the function of the genes represented by clear-plaque mutants. Cells are infected by two different clear-plaque mutants simultaneously, and the appearance of lysogens is monitored. Lysogens can be recognized by their immunity to infection by the phage, which allows them to form colonies in the presence of the phage. One way to perform this test is to mix one of the mutant phages with the bacteria and streak the mixture on a plate. The other mutant is then streaked at right angles to the first streak. If bacteria grow in the region where the two streaks cross, immune lysogens are forming because some cells are simultaneously infected with both mutants and the two mutations are complementing each other. Such comple-

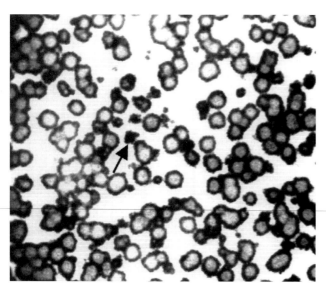

Figure 13.13 Phage λ plaques. Note the cloudy center, giving the plaques a fried-egg appearance. Some plaques are clear (indicated by an arrow), having arisen from clear-plaque mutants.

mentation tests revealed three complementation groups or genes to which the clear-plaque mutations belonged: cI, cII, and cIII. In addition to mutations in the clear-plaque genes, mutations in the *int* gene can also prevent the formation of stable lysogens, although Int⁻ mutants make somewhat cloudy plaques. In this case, the λ DNA, while not integrated into the chromosome, may nevertheless make the cells transiently immune.

Further genetic tests revealed different roles for CI, CII, and CIII in lysogeny. Mutations in the cII and cIII genes can be complemented to form lysogens, and these lysogens can harbor a single prophage with a cII or cIII mutation. However, lysogens harboring a single prophage with a cI mutation are never seen. Apparently, the cI mutation in the phage must be complemented by another mutation to maintain the phage in the lysogenic state. This observation led to the idea that CII and CIII are required to form lysogens but are not necessary to maintain the lysogenic state once a lysogen has formed. The CI protein, on the other hand, is required to form a lysogen and to maintain the lysogenic state.

Because they can be complemented, the cI, cII, and cIII mutations must affect *trans*-acting functions, either proteins or RNAs required to form lysogens. Another set of mutations, called the *vir* mutations, also prevent lysogeny and cause clear-plaque formation but cannot be complemented and so are *cis* acting. These *vir* mutations allow the mutant phage to multiply and form clear plaques even on λ lysogens. DNA sequencing has revealed that phage with *vir* mutations are double mutants with mutations in the $o_R{}^1$ and $o_R{}^2$ sequences. These mutations change the o_R operator so that it can no longer bind the CI repressor, thereby preventing lysogeny (see chapter 7).

Genetics of the λ CI Repressor

The λ repressor product of the cI gene has served as a model for proteins with separable domains. The first indication that the CI repressor has separable domains came from genetic experiments that demonstrated intragenic complementation between temperature-sensitive mutations in the cI gene (see Lieb, Suggested Reading). As discussed, complementation usually occurs only between mutations in different genes, and intragenic complementation is possible only if the protein product of the gene is a multimer composed of more than one identical polypeptide encoded by that gene (Table 13.1).

Figure 13.14 illustrates the experiments that demonstrated intragenic complementation by some temperature-sensitive mutations in the cI gene. Lysogenic cells containing a prophage with one cI temperature-sensitive mutation were heated to the nonpermissive temperature

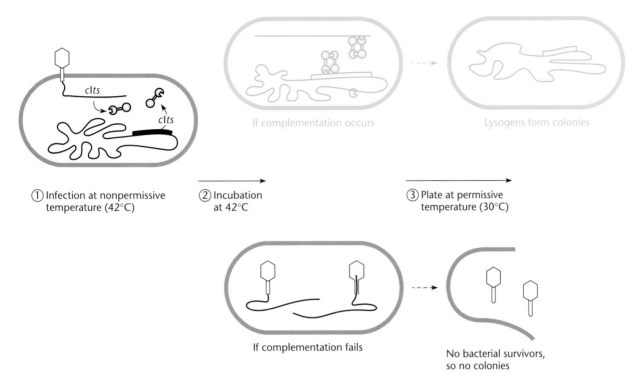

Figure 13.14 An experiment to show intragenic complementation in the cI gene of λ. See the text for details.

and infected with a phage carrying a different *cI* temperature-sensitive mutation. At this temperature, phages with either mutation will invariably kill the cell because the repressor will be inactivated so that λ cannot form a lysogen. However, if the two mutations complement each other to form an active repressor, a few cells may become lysogens and survive.

The results clearly demonstrated intragenic complementation between some of the mutations in the *cI* gene. In particular, some mutations in the amino (N)-terminal part of the polypeptides, which we now know to be involved in DNA binding, complement some mutations in the carboxy (C)-terminal part, which we now know to be involved in dimer formation. Apparently, in some cases, dimers can form if only one of the two polypeptides has a mutation in the C-terminal domain. Furthermore, the dimer can sometimes bind to DNA if only one of the two polypeptides in the dimer has a mutation in the N-terminal domain. The ability to form active repressor out of two mutant polypeptides is what leads to intragenic complementation.

trans- versus *cis*-Acting Functions in λ DNA Replication

As with other replicons, both *cis*- and *trans*-acting functions are required for the initiation of λ DNA replica-

tion. The genetic requirements for λ DNA replication (see chapter 7) include the *trans*-acting protein products of the λ *O* and *P* genes. The products of these genes are *trans* acting because they encode protein products, although *O* mutations are weakly *cis* acting because the *O* protein may bind preferentially to an *ori* region on the same DNA. Mutations in the *ori* region are *cis* acting because they affect a site on the DNA rather than a *trans*-acting gene product. In addition, the *ori* region must be transcribed to be activated, so that mutations that inactivate the promoter used to transcribe the *ori* region are *cis* acting.

The *O* and *P* genes of λ were found because amber mutations in these genes prevented λ DNA replication after infection of a nonpermissive host. Since most mutations in these genes could be complemented, the genes were assumed to encode *trans*-acting gene products, either protein or RNA required for replication. In a mixed infection between phage with an *O* or *P* mutation and wild-type λ, both the *O* or *P* mutant and the λ wild type will replicate their DNA and produce progeny. The *ori* region was found to lie within gene *O*, because of *cis*-acting mutations in gene *O* that reduced λ DNA replication. In a mixed infection between the wild-type phage and λ with one of these *O* mutations, only the wild-type DNA will replicate normally, and so most of the progeny phage produced are wild type. It would not be possible to isolate phages with mutations that completely inactivate the *ori* region, since such DNA would not be able to replicate, even in the presence of a wild-type helper phage. Therefore, only mutations that partially inactivate the *ori* region can be isolated under normal circumstances.

Isolation of λ *nut* Mutations

The last genetic analysis we discuss is the isolation and mapping of *nut* mutations of λ. As discussed in chapter 7, the *nut* sites of λ are the N utilization sites required for antitermination of transcription, and they allow the RNA polymerase to proceed through transcription termination sites and into the λ genes beyond. Our present picture of how these *nut* sites function is that they are sequences on the mRNA to which the antiterminator protein N can bind. Once bound to the RNA, the N protein can then bind to the RNA polymerase, making the RNA polymerase insensitive to transcription termination (see Figure 7.10). Two *nut* sites, *nutL* and *nutR*, act to antiterminate transcription from the p_L and p_R promoters, respectively, thereby allowing transcription of genes to the left and the right of the *cI* gene (see Figure 7.9 for a genetic map of λ). According to the model, mutations in the DNA coding sequence for one of these *nut* sites could prevent the binding of the N protein to the mRNA, thereby causing transcription to stop at the next transcription termination site and preventing transcription of downstream genes.

The first *nut* mutations to be isolated were in *nutL* (see Salstrom and Syzbalski, Suggested Reading). It seemed easier to isolate *nutL* mutants than *nutR* mutants because a *nutR* mutation should prevent the transcription of essential genes of the phage, including the O and P genes, as well as other genes whose products are required for phage development and plaque formation (see λ map, Figure 7.9). Therefore, mutations that completely inactivate *nutR* should be lethal. In contrast, transcription termination at t_L^1 due to lack of antitermination by *nutL* will prevent only the transcription of genes to the left of the terminator, including *gam*, *red*, *int*, and *xis*, none of which are required for multiplication of the phage to form plaques. However, even though *nutL* mutations should not be lethal, they might be very rare. The *nut* sequences in DNA may be very short, consisting of only a few base pairs, and only mutations that changed one of these base pairs would inactivate the *nut* site. Selecting rare mutations requires a positive selection. As we discuss in more detail in the next chapter on isolating bacterial mutants, **positive selection** means establishing conditions under which only

the desired mutant and not the wild type can multiply. In this case, it meant finding conditions under which only phages with a mutation that inactivates the *nutL* site can form plaques whereas wild-type λ cannot.

The positive selection used to isolate *nutL* mutations is illustrated in Figure 13.15. The selection is based on the observation that, for unknown reasons, wild-type λ cannot multiply in *E. coli* lysogenized by another phage, P2, because the *gam* and *red* gene products of the infecting λ can interact somehow with the *old* gene product of the P2 prophage and kill the cell. However, mutants of λ that fail to make both the Gam and Red proteins can multiply and form plaques on a P2 lysogen. Therefore, mutations that inactivate *nutL* and thereby prevent the transcription of both the *gam* and *red* genes should result in phage able to form plaques on *E. coli* lysogenic for P2. Isolating *nutL* mutants of λ should therefore be easy: just plate millions of mutagenized λ on a P2 lysogen, and any plaques that form may be due to λ with *nutL* mutations.

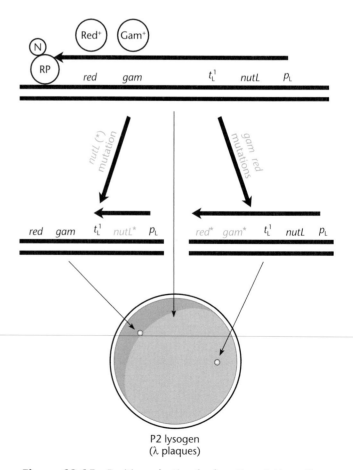

Figure 13.15 Positive selection for λ *nutL* mutations. See the text for details. RP, RNA polymerase.

Unfortunately, *nutL* mutants are not the only type of mutant that can also form plaques under these conditions. For example, λ with point mutations in both their *gam* and *red* genes or with a deletion mutation that simultaneously inactivates both the *red* and *gam* genes will also multiply and form plaques on a P2 lysogen. Fortunately, these types of mutants should not be much more common than *nutL* mutants. Even though *nutL* mutants may be rare, mutants with two point mutations, one in *gam* and one in *red*, will also be rare. Also, deletion mutations that inactivate both *red* and *gam* will be rare if the λ phage have been mutagenized with a mutagen that causes only point mutations (see chapter 10).

Even though some of their mutants were other types, the investigators could distinguish λ mutants with *nutL* mutations from those with *gam* and *red* mutations by genetic mapping experiments, since *gam* and *red* mutations should map to the left of the t_L^1 terminator while *nutL* mutations should map to the right of the t_L^1 terminator (Figure 13.15). Potential *nutL* mutations were mapped in a manner analogous to deletion mapping (Figure 13.16). In this case, however, specialized transducing phage in which bacterial genes were substituted for some phage genes were used instead of deletion mutants (see chapter 7 for a discussion of specialized transducing phage). The *E. coli* genes that replaced the λ genes included the *bio* operon, which encodes the pathway for the vitamin biotin. As shown in Figure 13.16, a Red⁻ Gam⁻ mutant of λ is crossed with a phage λ in which a *bio* substitution includes the *red* and *gam* genes. The appearance of Red⁺ Gam⁺ recombinants indicates that the mutation that makes the phage Red⁻ Gam⁻ must lie outside the substituted region. As discussed in chapter 9, only Red⁺ Gam⁺ recombinants can form concatemers and therefore plaques on *recA E. coli* mutants. Therefore, even very rare Red⁺ Gam⁺ recombinants can be detected by plating the cross on *recA E. coli* mutants.

If the mutation lies upstream of the *red* and *gam* genes, it could not be either a double point mutation in *red* and *gam* or a deletion that removes both *red* and *gam*. Other types of mutations upstream of *red* and *gam* besides *nutL* mutations that prevent *gam* and *red* transcription may also be possible, but these can be distinguished from *nutL* mutations by other criteria. For example, mutations that completely inactivate the *N* gene (null mutations in *N*) will cause termination at t_L^1, but they should not allow the phage to form plaques, because they will also cause premature termination of rightward transcription from the p_R promoter. Leaky *N* mutations might reduce the transcription of *gam* and *red* enough to allow plaques to form on a P2 lysogen

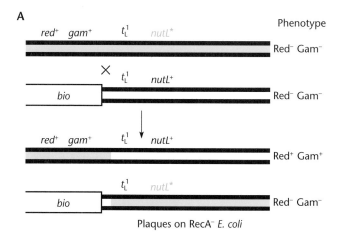

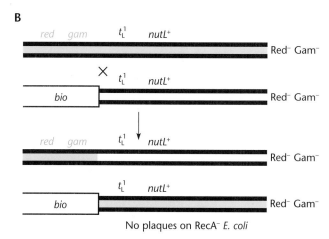

Figure 13.16 Using λp*bio*-substituted phage to map *nutL* mutations. If the mutation that makes λ phenotypically Gam⁻ Red⁻ maps outside the substituted region, Gam⁺ Red⁺ recombinants will arise that can multiply in RecA⁻ *E. coli*. Only the region of the *nutL* region and the *gam* and *red* genes is shown.

and might also allow sufficient *O* and *P* transcription for a plaque to form. However, such *N* mutations might also be rare and could be distinguished from *nutL* mutations because they are *trans* acting rather than *cis* acting, like *nutL* mutations. Note that *nut* mutations are *cis* acting even though they affect a site on the RNA rather than a site on the DNA, because the mutation affects transcription termination only from the same DNA.

Once *nutL* mutations had been isolated, investigators located the region of the mutation through DNA sequencing by comparing the sequence to the known sequence of wild-type λ in this region. The mutations changed the sequence shown as *nutL* in Figure 7.11. An identical sequence was found downstream of the promoter p_R that transcribes genes to the right in λ. This is the *nutR* sequence at which N acts to allow transcription of the genes to the right in λ, including O, P, and Q, required for lytic development. As discussed in chapter 7, the *nutL* and *nutR* sequences of λ are similar to sequences found in some other genes, including the rRNA genes of bacteria. In some cases, these other sequences have also been shown to function as antiterminators. Nevertheless, antiterminators were discovered and characterized first in phage λ.

SUMMARY

1. Bacteriophages are ideal for illustrating the basic principles of classical genetics, including complementation and recombination.

2. Recombination tests are a measure of distance on DNA and can be used to locate mutations on DNA. The recombination frequency, defined as the number of recombinant types divided by the total progeny, is a measure of the distance between two mutations.

3. Complementation is used to determine if two mutations affect the same function. In general, two mutations will complement each other only if the mutations are in different genes.

4. Complementation can also be used to determine if a mutation affects a *trans*-acting product (protein or RNA) or a *cis*-acting site on DNA. If a mutation can be complemented, it is affecting a *trans*-acting function; if it is recessive but cannot be complemented, it usually is affecting a *cis*-acting site in the DNA or RNA. For example, mutations that affect the O and P proteins of λ affect *trans*-acting functions, while mutations that affect the *ori* region of λ, the promoters, or *nut* sites affect *cis*-acting sites.

5. The order of mutations in the DNA can be determined by three-factor crosses. In this approach, a double mutant constructed with two of the mutations is crossed with a single mutant containing the third mutation. The frequency of the various recombinant types allows an unambiguous ordering of the regions of the three mutations.

6. Deletion mapping is a convenient way to localize point mutations. The mutant with the point mutation being mapped is crossed with a mutant with a deletion mutation with known endpoints in the gene. The appearance of wild-type recombinants indicates that the point mutation lies outside the deleted region.

7. Phages are convenient for illustrating the classical genetic concepts of reversion and suppression. A mutation has reverted if the function of the gene has been restored by changing the mutated base pair, restoring the original DNA sequence of the gene or at least the function of the gene product. A mutation has been suppressed if a mutation somewhere else restores the function. Suppressor mutations give clues to the interaction of different gene products and different regions within the same gene product.

PROBLEMS

1. You have two *rII* mutations, *r231* and *r686*. You cross them by infecting *E. coli* B. After 1 h, you lyse the infected cells and plate the progeny phage on *E. coli* B and on *E. coli* K-12λ. On *E. coli* B, a 1 in 10^8 dilution gives you 200 plaques. On *E. coli* K-12λ, a 1 in 10^6 dilution gives you 100 plaques. What is the recombination frequency between *r231* and *r686*?

2. You want to use the deletion mutations of Benzer to map the point mutation *r231*. You cross *r231* with all of the deletions separately but obtain r^+ recombinants only with the *r1241* deletion. Where is *r231* located?

3. The *rII* mutation *r736* gives a recombination frequency of 0.1% with *r26* but a recombination frequency of 1.2% with *r686*. If *r736* and *r26* complement each other, should *r736* and *r866* also complement? Why or why not?

4. *r26* and *r686* are both point mutations. How would you construct and identify a double mutant having both the *r26* and the *r686* mutation?

5. You cross your double mutant with a single mutant having the point mutation *r736*. Only 0.005% of the progeny are r^+

recombinants. Do you expect the order of the three mutations to be *r686-r736-r26* or *r686-r26-r736*? Why?

6. To order the three genes *A*, *M*, and *Q* in a previously uncharacterized phage you have isolated, you cross a double mutant having an amber mutation in gene *A* and a temperature-sensitive mutation in gene *M* with a single mutant having an amber mutation in gene *Q*. About 90% of the AM^+ recombinants that can form plaques on the nonsuppressor host are temperature sensitive. Is the order *Q-A-M* or *A-Q-M*? Why?

7. The N-terminal region of the *rIIB* gene is nonessential for activity and can be deleted or the amino acids changed without affecting activity. However, you observe that the mutation *r291* in this region abolishes *rIIB* function. What possible kinds of mutations could *r291* be?

8. A stock of virus can be subjected to titer determination (i.e., the concentration of viruses can be determined) by counting virus particles under the electron microscope. How would you determine the effective MOI for a virus (i.e., the

fraction of viruses that actually infect a cell) under a given set of conditions?

9. Phage T1 packages DNA from concatamers beginning at a unique *pac* site and then packaging by a processive headful mechanism, cutting about 6% longer than a genome length each time. However, it packages a maximum of only three headfuls from each concatamer. Would you expect T1 to have a linear or circular genetic map? Draw a hypothetical T1 map.

10. How would you show that a *nutL* mutation of λ is *cis* acting?

SUGGESTED READING

Benzer, S. 1961. On the topography of genetic fine structure. *Proc. Natl. Acad. Sci. USA* **47:**403–415.

Benzer, S., and S. P. Champe. 1962. A change from nonsense to sense in the genetic code. *Proc. Natl. Acad. Sci. USA* **48:**1114–1121.

Crick, F. H. C., F. R. S. Leslie Barnett, S. Brenner, and R. J. Watts-Tobin. 1961. General nature of the genetic code for proteins. *Nature* (London) **192:**1227–1232.

Echols, H., and H. Murialdo. 1978. Genetic map of bacteriophage lambda. *Microbiol. Rev.* **42:**577–591.

Fluck, M. M., and R. H. Epstein. 1980. Isolation and characterization of context mutations affecting the suppressibility of nonsense mutations. *Mol. Gen. Genet.* **177:**615–627.

Friedman, D.I ., M. F. Baumann, and L. S. Baron. 1976. Cooperative effects of bacterial mutations affecting λ N gene expression. *Virology* **73:**119–127.

Lieb, M. 1976. λ mutants: intragenic complementation and complementation with a *cI* promoter mutant. *Mol. Gen. Genet.* **146:**291–297.

Nomura, M., and S. Benzer. 1961. The nature of deletion mutants in the *rII* region of phage T4. *J. Mol. Biol.* **3:**684–692.

Salstrom, J. S., and W. Szybalski. 1978. Coliphage λ *nutL*: a unique class of mutations defective in the site of N product utilization for antitermination of leftward transcription. *J. Mol. Biol.* **124:**195–222.

Studier, F. W. 1969. The genetics and physiology of bacteriophage T7. *Virology* **39:**562–574.

Symonds, N., P. Vander Ende, A. Dunston, and P. White. 1972. The structure of *rII* diploids of phage T4. *Mol. Gen. Genet.* **116:**223–238.

Genetic Analysis in Bacteria

I N THE LAST CHAPTER, we discussed how phages are ideal for genetic analysis; however, bacteria are also excellent genetic subjects because of the ease with which bacterial mutants can be isolated and the responsible mutations mapped.Other advantages offered by bacteria include the formation of discrete colonies on agar plates in which each bacterium is a clone of the original bacterium that multiplied to form the colony. Bacteria are also haploid and so have only one allele of each gene. This feature makes isolating mutants and identifying recombinants easier in bacteria than in diploid organisms. Most importantly, because bacteria are haploid and able to multiply asexually to form discrete colonies on plates, we can apply **selectional genetics**, an approach that allows us to isolate even very rare types of mutants and recombinants.

In this chapter, we first discuss the general principles of applying genetic analysis to bacteria and how data are analyzed. We then give some specific examples of genetic analysis in bacteria.

Isolating Bacterial Mutants

We mentioned in chapter 13 that the isolation of mutants altered in a particular function is the first step in any classical genetic analysis of function and that there is no way of knowing from the outset which mutants can be obtained or what type of mutations will result in a given mutant phenotype. However, predicting possible mutations and their effects on the organism is one of the more enjoyable aspects of genetics.

The procedures for isolating mutants differ for bacteria and phages, and many more types of bacterial mutants are possible. In this section, we describe the steps involved in isolating bacterial mutants.

To Mutagenize or Not To Mutagenize?

Mutations can occur spontaneously or be induced. **Spontaneous mutations** arise during the normal multiplication of the organism. **Induced mutations** are caused by deliberately adding mutagenic chemicals or by irradiating the cells with UV light.

Both spontaneous and induced mutations have advantages in a genetic analysis. To decide whether to mutagenize the cells and which mutagen to use, we must first ask how frequent the mutations are likely to be. Spontaneous mutations are usually much rarer than induced mutations and so are more difficult to isolate. On the other hand, mutants containing spontaneously arising mutations are less likely to contain multiple mutations, which can confuse the analysis.

To isolate very rare types of mutants or ones for which there is no good selection, we might have to use a mutagen. However, since organisms differ in their susceptibility to various mutagens, we might have to test many different mutagens to find one that is effective in the organism being studied.

One major advantage of mutagens is that they sometimes cause a particular type of mutation. For example, acridine dyes cause only frameshift mutations and base analogs cause only base pair changes. Therefore, the use of a particular mutagen may make it possible to restrict the mutations to the type desired. In contrast, spontaneous mutations can be base pair changes, frameshifts, deletions, and so on: all types of mutations are represented among spontaneous mutations.

Isolating Independent Mutations

Mutants defective in a function should be as representative as possible of all the mutations that cause the phenotype. If the strains in a collection of mutants carry many different mutations, we can get a better idea of how many genes can be mutated to give the phenotype and how many types of mutations can cause the phenotype.

There are two ways to ensure that the strains of a collection of mutants have different mutations. One is to use different mutagens. All mutagens have preferred hot spots and tend to mutagenize some sites rather than others (see chapter 13). If mutants are all obtained with the same mutagen, many of them will have mutations in the same hot spot, but mutants obtained with different mutagens will tend to have different mutations. Another way to increase the variety of mutations is to avoid picking **siblings**, which are organisms that are descendents of the same original mutant. Two sibling mutants will have the same mutation. The best way to avoid picking siblings is to isolate the mutants from different cultures, all started from nonmutant bacteria. If two mutants arose in different cultures, they must have arisen independently and hence could not be siblings.

Selecting Mutants

Even after mutagenesis, mutants are rare and still must be found among the myriad of bacteria that remain normal for the function. The process of finding the mutants is called **screening** and usually involves finding **selective conditions** under which either the mutant or the wild type will not be able to multiply to form a colony. Agar plates or media with selective conditions are called **selective plates** or **selective media**, respectively.

Screening for mutants is usually the most creative part of a genetic analysis. One must anticipate the phenotypes that mutations in the genes for a particular function might cause. This is where the geneticist earns her or his pay, as guessing the possibly phenotypes can be difficult if the function is incompletely characterized. For example, what do you imagine would be the phenotype of mutants defective in protein transport through the membrane? We discuss examples of these and other mutants later in the chapter.

Selections can be either positive or negative. In a **positive selection**, selective conditions are chosen under which the mutant but not the wild type can multiply. In a **negative selection**, selective conditions are used under which the wild type but not the mutant can grow. Screening for mutants is much easier with positive selections. In a sense, negative selections are not really selections at all, because the selective conditions are being used to *screen* for the mutants rather than to eliminate all other cells. Nevertheless, we shall use the common terms in this discussion.

ISOLATING MUTANTS BY NEGATIVE SELECTIONS

Most mutants, for example those that are auxotrophic, temperature sensitive, or DNA repair deficient, can be isolated only by using negative selections because the majority of an organism's gene products help it to multiply. Therefore, mutations that inactivate a gene product are more likely to make the organism *unable* to multiply under a given set of conditions than *able* to multiply under conditions that prevent multiplication of the wild-type cell.

To isolate mutants by negative selection, the bacteria are first plated under **nonselective** conditions, in which both the mutant and the wild type can multiply. When the colonies have developed, some of the bacteria in each colony are transferred to a selective plate to determine which colonies contain bacteria that cannot multiply under those conditions. Once such a colony has been identified, the mutant bacteria can be retrieved from the original nonselective plate.

Replica Plating

Because of the general rarity of mutants, many colonies usually have to be screened when using a negative selection. **Replica plating** can be used to streamline this process. As illustrated in Figure 14.1, a few hundred bacteria are spread on a nonselective plate and the plate is incubated to allow colonies to form. A replica is made of this plate by inverting the plate and pressing it down

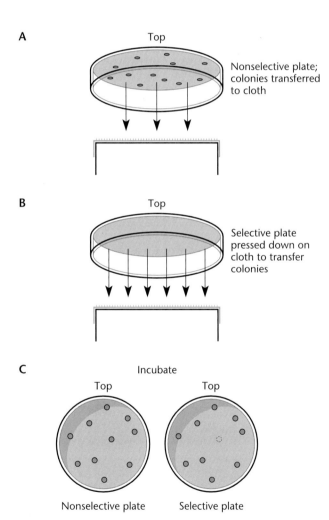

Figure 14.1 Replica plating. (A) A few hundred bacteria are spread on a nonselective plate, and the plates are incubated to allow colonies to form. The plate is then inverted over velveteen cloth to transfer the colonies to the cloth. (B) A second plate is then inverted and pressed down over the same cloth and incubated. (C) Both plates after incubation. The dotted circle indicates the position of a colony missing from the selective plate. See the text for details.

over a piece of fuzzy cloth, such as velveteen. Then a plate containing the selective medium is inverted and pressed down over the same cloth so that the colonies are transferred from the cloth to the selective plate.

After the selective plate has been incubated, it can be held in front of the original nonselective plate to identify colonies that did not reappear on the second, selective plate. The missing colonies presumably contain descendants of a mutant bacterium that are unable to multiply on the second, selective plate. The mutant bacteria can then be taken from the colony on the original, nonselective plate.

Enrichments

If the mutant being sought is rare, finding it by negative selection can be very laborious, even with replica plating. No more than about 500 bacteria can be spread on a plate and still give discrete colonies. So, for example, if the mutant occurs at a frequency of 1 in 1 million, more than 2,000 plates might have to be replicated to find a single mutant!

Many fewer colonies need to be screened if the frequency of mutants is first increased through **mutant enrichment**. This method depends on the use of antibiotics such as ampicillin and bromodeoxyuridine (BUdR) that kill growing but not nongrowing cells. Ampicillin inhibits cell wall synthesis and causes a growing bacterial cell to "grow out of its skin" and lyse. In contrast, BUdR is incorporated into replicating DNA in lieu of thymidine, and BUdR-containing DNA is much more sensitive to UV light than is normal DNA. If the cells are not growing, their DNA will not be replicating, BUdR will not be incorporated, and the cells can survive a dose that would kill BUdR-containing cells.

To enrich for mutants that cannot grow under a particular set of selective conditions, we place the population of mutagenized cells under selective conditions in which the desired mutants will not multiply but also will not die. Meanwhile, the nonmutant wild-type cells will continue to multiply. The antibiotic—either ampicillin or BUdR—is then added to kill any multiplying cells. The cells are then filtered or centrifuged to remove the antibiotic and transferred to nonselective conditions. The mutant cells will have survived preferentially and will have become a higher percentage of the population. However, no enrichment is 100% effective, and the surviving bacteria must still be plated and colonies must be tested for the mutant phenotype. Yet even if the enrichment only makes the mutant 100 times more frequent, only 1/100 as many colonies and therefore 1/100 as many plates must be replicated to find a mutant after an enrichment. In the example given above, we could replicate 20 plates instead of 2,000. Enrichments have been a great boon to selective genetics.

Unfortunately, enrichments cannot be applied to all types of mutants. Some mutants are killed by the selective conditions and so cannot be enriched by these procedures. To be enriched, the mutant must resume multiplying after it is removed from the antibiotic and the selective conditions.

POSITIVE SELECTIONS

In a positive selection, selective conditions are established in which the mutant but not the original or wild-type bacterium can grow. This type of selection includes those for antibiotic resistance and phage resistance. Mutations that revert or suppress auxotrophic mutations can also often be isolated by positive selection because these mutations allow the mutant to multiply under the selective conditions. Figure 14.2 illustrates how a positive selection can be used to isolate cells revertant for a *his* mutation. These cells would no longer require histidine for growth.

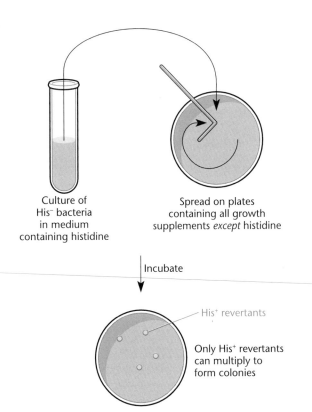

Figure 14.2 Positive selection of a His⁺ revertant. A His⁻ mutant bacterium is plated on minimal media with all the growth requirements except histidine. Any colonies that form after the plate is incubated are due to His⁺ revertants that can multiply without histidine in the medium.

As mentioned earlier, positive selections are preferable to negative selections because it is so much easier to isolate mutants by positive selection. Billions of bacteria can be placed on a single selective agar plate, and only the mutants that can multiply under those conditions will form a colony. Even if the billions of bacteria contained only one mutant bacterium that could multiply on the plate, a colony would appear and mutant bacteria could be purified. In the Introduction, we compared this selective ability to finding a single human from among the entire human population of the Earth. But a positive selection of a bacterial mutant takes place on a single petri plate about 5 in. in diameter!

Genetic Mapping of Mutations

Once a collection of mutants with the desired phenotype has been assembled, the next step is to genetically map the responsible mutations. The methods to use for mapping depend on the means of genetic exchange available for the type of bacteria being studied. If a generalized transducing phage has been isolated for the bacteria, transduction might be used. If the bacteria are naturally transformable, transformation might yield mapping data. If a self-transmissible plasmid has been isolated for the bacteria, mapping by Hfr conjugation might be possible.

The procedures for interpretation of genetic data usually differ in bacteria and other organisms and depend on the method of gene exchange that is being used. Therefore, before we discuss genetic mapping in bacteria, we present a few definitions and explain a few concepts. Most of these words and concepts are used throughout genetics, but some are specific to bacteria and even to a particular mapping method.

Genetic Markers

The DNAs of two individuals of the same species are usually almost identical in sequence over their entire length. However, sequence differences sometimes occur at specific sites in the DNA. These differences could occur if one of the individuals carries a mutation or has a transposon or another DNA element inserted in its DNA. Differences could also occur because of natural strain variation. If these differences in sequence cause phenotypic differences in the two individuals, the position of the difference in the DNA can be mapped genetically. A DNA sequence variation that can be used for genetic mapping is called a **genetic marker**.

In classical genetic mapping, the position of one genetic marker is determined with respect to other genetic markers. To map the markers, we need to cross an individual of a species that has one or more genetic markers with another individual of the same species that has different genetic markers. The frequencies of the various recombinant types are then analyzed to determine the map positions of the markers. How the recombinants are detected and which phenotypes are scored depend on what is convenient. Nevertheless, what is being mapped is a genetic marker, a position on the DNA of the species where a difference in DNA sequence can cause a specific phenotype.

DONOR AND RECIPIENT

To genetically map markers, we have to cross organisms that differ in the markers. Crossing bacteria is a unique process in the biological world. In crosses with most organisms, viruses and humans included, both parents participate equally in the cross. The progeny organisms of the cross are as likely to have DNA from one parent as the other. However, in crosses between two bacteria, whether due to transformation, transduction, or conjugation, only part of the DNA of one bacterium, the donor, is transferred to the other bacterium: the recipient. Consequently, in a bacterial cross, only the *recipient* can become a recombinant, and a recombinant type is a recipient in which one or more of the **recipient alleles** or sequences has been replaced by the corresponding **donor alleles** or sequences.

As an example, consider a cross between a recipient that has a *hisG* mutation, making it require histidine (His⁻), and a His⁺ donor. After the cross, the His⁺ recipient bacteria may be recombinant for the *hisG* marker. In the recombinant, the wild-type *hisG*⁺ allele of the donor has replaced the mutant *hisG* allele of the recipient. In contrast, when the donor has the *hisG* mutation and the recipient is His⁺, the recombinant types for the *hisG* marker will be His⁻ recipient bacteria. In this case, the *hisG* mutant allele of the donor has replaced the wild-type nonmutant *hisG*⁺ allele of the recipient. Note again that only the recipient bacterium can become a recombinant, because the donor bacterium contributes only a piece of its DNA to the cross.

SELECTED AND UNSELECTED MARKERS

Recombinants are usually quite rare in bacterial crosses, so they must often first be isolated by positive selection. We usually select for one of the markers first and then test the recombinants for the other markers. The marker chosen to be the selected one is called the **selected marker**, and the other markers are the **unselected markers**.

Convenience dictates which marker will be the selected one and whether the wild-type or mutant phenotype will be selected. For example, if we choose *hisG* as our selected marker, it is much easier to select recombinants if the recipient has the *hisG* auxotrophic mutation and the donor has the wild-type *hisG*⁺ allele. We then select the recombinants for the *hisG* marker by plating the cross on medium without histidine, where only the His⁺ recombinants can multiply. For other types of markers, it might be easier if the donor has the mutant allele and the recipient has the wild-type allele. For example, if the selected marker is a mutation that confers resistance to an antibiotic such as rifampin (*rif*), it may be more convenient to have the resistance mutation in the donor rather than in the recipient. Then recombinants for the selected marker can be selected on plates containing the antibiotic. Nevertheless, even though the allele being selected may differ for convenience, it is the donor allele that is selected in each case, whether the donor has the mutant or the wild-type sequence for the region.

Once the donor allele for one marker has been selected, the recombinants can be tested for other, unselected, markers. It makes much less difference for the unselected markers which allele is in the donor and which is in the recipient, since recombinants for the selected marker must be individually tested for the unselected markers anyway. However, strains recombinant for the unselected markers are still those that received the allele of the donor. How data from such crosses are interpreted depends upon the type of cross being performed. In the next sections, we discuss how genetic evidence is analyzed by each of the mapping methods.

Mapping by Hfr Crosses

The easiest way to map a genetic marker on the entire bacterial chromosome is by Hfr crosses. Recall from chapter 5 that an Hfr strain usually has a copy of a self-transmissible plasmid integrated somewhere in its chromosome. If the integrated plasmid attempts to transfer itself, at least part of the bacterial chromosome will also be transferred, starting from the *oriT* site in the integrated plasmid and continuing in one direction around the chromosome. The Hfr strain from which the DNA is transferred is the donor strain, and the plasmidless strain into which the DNA is transferred is the recipient strain. Recipient bacteria that have received some DNA from the donor are **transconjugants**. If a region of the chromosome is transferred, recombinants can form for a marker in that region, where the recipient DNA in the region of the marker will be replaced by the donor region.

Hfr crosses can be used to map markers because the DNA on one side and closest to the site of integration of the plasmid will be transferred first, followed by DNA farther away. The last DNA to be transferred will be that closest to the plasmid on the other side. Therefore, it is possible to tell the order of markers in the chromosome by measuring the frequency of recombinants for the markers after Hfr crosses.

THE GRADIENT OF TRANSFER

The **gradient of transfer** is a concept based on the fact that the farther a genetic marker is from the origin of transfer in an Hfr cross, the less the likelihood that it will be transferred to the recipient. Transfer could cease because the cells are constantly moving owing to Brownian motion and thus two cells will sometimes break apart. Alternatively, transfer could stop because of a single-strand break in the bacterial chromosome that occurs before the marker region. These periodic interruptions of mating lead to an exponential decay in the transfer frequency of markers farther away from the origin of transfer.

To use the gradient of transfer to map a genetic marker by Hfr crosses, a donor Hfr bacterium and a recipient bacterium are mixed and the mating is allowed to proceed long enough that the entire chromosome *could* be transferred, even though it seldom will be. The mating mixture is then plated under conditions selective for one of the markers. Recombinants for this marker are then tested for the other, unselected, markers. The data on the frequency of recombinants for the various unselected markers are then used to determine the map position of the unknown markers.

Table 14.1 lists some data for a typical Hfr cross, and Figures 14.3 through 14.5 illustrate how these data might be analyzed. In this example, the marker with the unknown position is due to a mutation, *rif-8*, that confers resistance to the antibiotic rifampin. The Hfr strain used for the cross is PK191, which requires proline because of a small deletion including the *proC* gene Δ(*lac-pro*). The recipient strain has mutations that make it require histidine (*hisG1*), arginine (*argH5*), and tryptophan (*trpA3*). It also has the *rif-8* mutation being mapped. The positions of all the markers except the *rif-8* marker are known (see the *Escherichia coli* genetic map in Figure 14.4).

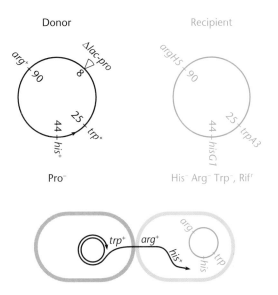

Plate: Arginine plus tryptophan: His⁺
Tryptophan plus histidine: Arg⁺
Arginine plus histidine: Trp⁺

Figure 14.3 Mapping by Hfr crosses. The phenotypes and positions of the mutations in the genetic maps of the donor and recipient bacteria are shown. The chromosome will be transferred from the donor to the recipient, starting at the position of the integrated self-transmissible plasmid (arrowhead). The plating media used to select the markers are also shown.

To map markers by the gradient-of-transfer method, we mix the Hfr strain with the recipient strain and incubate the mixture for a sufficient time to permit transfer of the entire chromosome (more than 100 min for *E. coli* at 37°C). The mating mixture is then plated under conditions in which *neither* the donor *nor* the recipient but *only* the recombinants being selected can multiply to form a colony; otherwise the parent bacteria would grow up and cover the plates, making the detection of recombinants impossible. Plating under conditions where the donor cannot multiply is known as **counterselecting the donor**. In this case, the donor can be counterselected by omitting proline from the plates, since the donor requires proline for growth. To plate under conditions where only recipients that are recombinant for a particular marker can multiply, the selective plates should contain two of the three amino acids required by the recipient but lack the one that corresponds to the selected marker (e.g., if the selected marker is *hisG*, the medium should lack histidine). Then only recipient bacteria that are recombinant for that marker will be able to multiply and form a colony on the plates.

To illustrate, assume that the marker being selected is the *hisG* marker. The mating mixture is then plated on

TABLE 14.1	Typical results of an Hfr cross			
Selected marker	**Percent recombinant for unselected markers**			
	hisG	*trpA*	*argH*	*rif*
hisG		1	7	6
trpA	33		29	31
argH	28	12		89

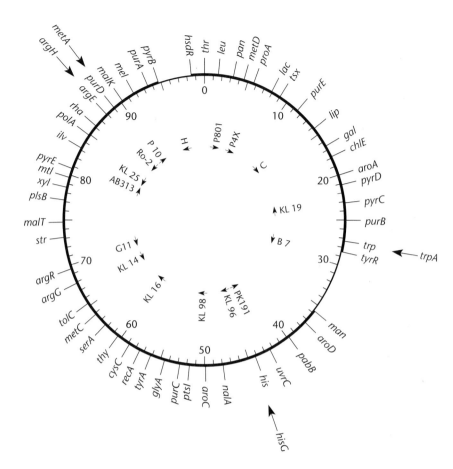

Figure 14.4 Partial genetic linkage map of *E. coli* showing the positions (large arrows) of the known markers used for the Hfr gradient of transfer in Figure 14.5. The small arrows indicate the position of integration of the F plasmid in some Hfr strains, including PK191 (located near the position of *hisG* at 44 min). In each of these Hfr strains, the chromosomal DNA will be transferred beginning from the tip of the arrow. Adapted from B. J. Bachmann, K. B. Low, and A. L. Taylor, *Bacteriol. Rev.* **40**:116–167, 1976.

plates lacking histidine. On these plates, only the His⁺ recombinants of the recipient can multiply to form a colony. Note that the His⁺ recombinants will form colonies on these plates whether or not they are also Arg⁺ or Arg⁻, Trp⁺ or Trp⁻, or Rifʳ or Rifˢ if the plates contain arginine and tryptophan and lack rifampin. Similarly, to select the *argH* marker, the mating mixture is plated on minimal plates plus tryptophan and histidine. To select the *trpA* marker, it is plated on arginine plus histidine.

TESTING FOR UNSELECTED MARKERS

After transconjugants recombinant for one of the markers have been selected, they are tested to determine if they are also recombinant for one or more of the unselected markers. The transconjugants are first colony purified on the same type of plate medium used to select them, to remove any contaminating donors and nonrecombinant recipients. Then the transconjugants are tested by picking some bacteria from each colony with a loop or toothpick and transferring them onto each of the other selective plates to determine how many colonies are also recombinant for one or more of the other unselected markers.

In Table 14.1, one of the markers has been selected, and the percentage of these recombinants that are also recombinant for each of the unselected markers is given. Remember that the recipient is recombinant for a marker when it has the allele of the donor. Thus, for example, the recombinants for the *argH* marker are Arg⁺ but the recombinants for the *rif* marker are rifampin sensitive.

The first step in analyzing the data by gradient of transfer is to decide which known markers must have come into the recipient *before* the unknown marker. Data obtained from the crosses in which the selected marker came in after the unknown marker cannot be used because the frequency of recombinants for the unknown marker will not vary with its map position. As shown in Table 14.1, if the selected marker, such as the *trpA* marker, is transferred late, all of the recombinants will have also received the regions of the *argH* and *hisG* markers, which are transferred ahead of *trpA* by the Hfr strain PK191. In the map in Figure 14.4, the position of the integrated plasmids in some Hfr strains including PK191 is shown as an arrow. The DNA, starting at the point marked by the arrowhead, is transferred into a recipient cell as though it were an arrow being shot into

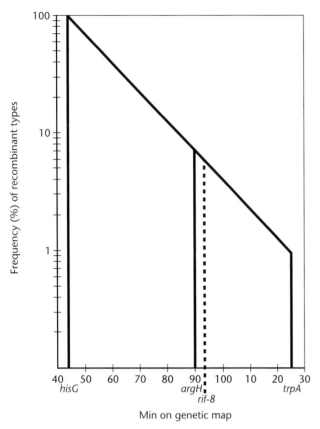

Figure 14.5 Mapping by gradient of transfer during an Hfr cross. The ordinate shows the frequency of each unselected marker with *his* as the selected marker. The abscissa is the distance in minutes from the selected marker. The dashed line shows an estimate of the position of *rif-8* based on the percentage of Rif⁵ recombinants from the data in Table 14.1.

the recipient. The frequencies of recombinants for the *hisG* and *argH* markers (~33 and 29%, respectively) are almost the same even though they are far apart on the genetic map. Apparently, the frequency of recombinants for any genetic marker, independent of its position on the genetic map, is almost the same (~30%) when the selected marker is transferred later. Referring to the data in Table 14.1 again, it seems likely that the region of the *rif* marker also transferred in before the *trpA* marker since its recombination frequency when the *trpA* marker is selected is 31%. This information does not help localize *rif-8*—most of the chromosome is transferred in before the *trpA* marker. We need to perform another cross and select a marker that is transferred before the *rif* marker. The *hisG* marker is probably transferred before the *rif* marker, because *hisG* is one of the first regions to be transferred in this cross and the fre-

quency of recombinants for the *rif* marker is much less than 30%. Therefore, we shall use the data obtained by selecting the *hisG* marker to determine the map position of *rif-8*.

PLOTTING THE GRADIENT OF TRANSFER
Now that we have decided which marker to use in the analysis, we can determine what the recombination frequency should be for any marker that comes a certain distance after the selected marker by drawing a standard curve on semilog paper. In Figure 14.5, the frequencies of transconjugants that are recombinant for the known *argH* and *trpA* markers are plotted against the known distances of these markers from the *hisG* marker (consult the map in Figure 14.4). This curve appears as a straight line on semilog paper since, as mentioned, the frequency of transmission of a particular marker falls off exponentially the farther it is from the selected marker. The frequency of transmission of the unknown *rif* marker is then placed on this line, reading down to determine what map position would give rise to this transmission frequency. These data place the *rif-8* mutation at approximately 90 min, close to the *argH* marker.

The placement of the *rif-8* mutation close to *argH5* is also supported by the results when *argH* is the selected marker. A very high percentage (89%) of the recombinants selected for *argH⁺* are also rifampin sensitive and so recombinant for the *rif* marker. Apparently, few crossovers occurred between the regions of the *argH* and *rif* markers when the *argH* region of the donor replaced the *argH* region of the recipient, indicating that the two markers are very closely linked. If markers are much farther apart than this, so many crossovers occur between the two markers that such genetic linkage is not apparent.

A CAVEAT
The interpretation of mapping data from Hfr crosses can be complicated if the marker being mapped is too close to the marker used to counterselect the donor. In the example, if the *rif-8* mutation is very close to the *proC* mutation of the Hfr donor, there will be very few crossovers between the *proC* mutation and the *rif-8* mutation, so that most of the transconjugants recombinant for the *rif* marker will also be Pro⁻ and will not grow on plates selective for proline prototrophy. Therefore, in this case, most of the transconjugants would be rifampin sensitive, no matter which marker was selected. Accordingly, to get a reliable map position for an unknown marker, it may be necessary to use a variety of markers to counterselect the donor and to use

various Hfr strains with the self-transmissible plasmid integrated at different positions. Detailed protocols for use of Hfr strains are given in K. B. Low, Suggested Reading.

CONCLUSIONS

Because Hfr crosses are the easiest way to map markers on the entire bacterial chromosome, distances on bacterial genetic maps are often measured according to the time it takes for an Hfr strain to transfer the DNA into a recipient strain starting from an arbitrary position on the chromosome. The circular genetic map of *E. coli* has 100 min, the length of time it takes for an Hfr strain of *E. coli* to transfer the entire chromosome in rich medium at 37°C. The zero position on the *E. coli* genetic map is the site of insertion of the self-transmissible F plasmid in one of the original Hfr strains, named Hfr Hayes. Similar Hfr crosses have been used to map the chromosomes of many other gram-negative bacteria by methods discussed in chapter 16.

Mapping by Transduction or Transformation

In the example described above, data from a single Hfr cross pointed to the approximate position of the *rif-8* mutation when a priori this mutation could have been anywhere on the entire *E. coli* chromosome. Hfr crosses are very useful for locating the approximate map position of markers. However, for more precise mapping, we need to use techniques such as generalized transduction and transformation.

As explained in chapter 7, generalized transduction involves a phage that mistakenly packages the DNA of the donor bacterium during an infection and injects it into the recipient cell in the next infection. The recipient is then called the transductant. Chapter 6 described transformation, in which naked DNA from the donor bacterium is taken up by a recipient, which is then called a transformant. Because the mapping data obtained from transductional and transformational crosses are analyzed similarly, we discuss them together.

ANALYZING THE DATA OF TRANSDUCTIONAL AND TRANSFORMATIONAL CROSSES

As in Hfr crosses, in transduction and transformation only a piece of the chromosome of the donor is transferred to the recipient, so that only recipient bacteria can become recombinants. Also, a recipient bacterium becomes recombinant for a marker if the recipient allele is replaced by the corresponding allele from the donor. However, unlike the procedure for Hfr crosses, the donor bacteria need not be counterselected in transductional or transformational crosses because they will

have been removed from the phage or their DNA will have been extracted from them. The only memorial to the donor bacteria is their DNA, which is broken in pieces and either packaged in a phage heads or left free in the medium.

The method of analyzing genetic data obtained from transduction or transformation differs from that used in Hfr crosses. In contrast to the lengths of DNA transferred by Hfr donors, the pieces of transferred DNA are usually quite short, so that both ends enter the recipient cell almost simultaneously and there is no gradient of transfer. Hence, mapping by transduction or transformation is usually based on whether markers can be carried on the same piece of DNA and how often recombination occurs between markers. If the regions of two markers are not close enough together to be carried on the same piece of DNA, they will almost never be brought into the cell together, since transduction and transformation are generally inefficient. Markers that are close enough together to be carried on the same piece of DNA during transduction or transformation are said to be **cotransducible** or **cotransformable**, respectively, and these attributes are evidence that two markers are closely linked genetically. In fact, the higher the percentage of the transductants that are recombinant for both markers, the closer together the two markers are likely to be. The percentage of the total transductants or transformants selected for one marker that are also recombinant for the other marker is called the **cotransduction** or **cotransformation frequency** between the two markers, respectively. In principle, the cotransduction frequency between two markers is a constant for any two markers and should be independent of which of the two markers is selected and whether it is the mutant or wild-type form of the gene that is selected. A cross with the selected and unselected markers reversed is called a **reciprocal cross**.

In this section, we illustrate how mapping data from transductional crosses is interpreted by using an actual example. Similar reasoning would apply to transformational crosses.

MAPPING BY COTRANSDUCTION FREQUENCIES

In the example shown in Figure 14.6, phage P1 is being used for transduction to further refine the mapping of the *rif-8* mutation. In particular, we want to know the map position of the *rif-8* mutation relative to the known map position of the nearby *argH5* mutation.

The first step is to determine if the regions of the *argH* and *rif* markers are close enough to each other on the DNA to be cotransducible. We select one of the markers and then determine whether any of the recom-

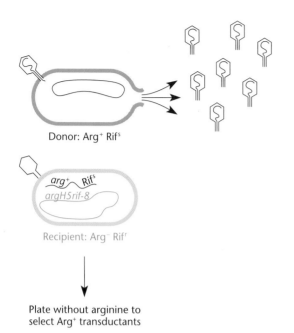

Plate without arginine to
select Arg⁺ transductants

Figure 14.6 Cotransduction of two genetic markers. The regions of the *argH* and *rif-8* mutations are close enough together that both regions can be carried on a piece of DNA fitting into a phage head. After transduction, approximately 37% of the Arg⁺ transductants are also rifampin sensitive. See the text for details.

binants for that marker are also recombinant for the other marker. For practical reasons, it is easier to use the *argH* marker, rather than the *rif* marker, as the selected marker. It is also much easier to select Arg⁺ transductants than Arg⁻ transductants, since the former need only be plated on minimal plates without arginine. Thus, we shall select the *argH* marker by using a donor bacterium that is Arg⁺ and a recipient that has the *argH5* mutation. The rifampin resistance mutation, the unselected marker, can be in either the donor or the recipient.

In the experiment illustrated in Figure 14.6, the transducing phage are grown on donor cells that are wild type for the *argH* and *rif* genes so are phenotypically Arg⁺ Rifˢ. These phage are then used to infect recipient calls that have the *argH* and rifampin resistance mutations and so are Arg⁻ Rifʳ. The Arg⁺ transductants are then selected by plating the infected cells on minimal plates without arginine, and the purified transductants are tested for the unselected *rif* marker. In the Arg⁺ transductants that are also rifampin sensitive, the region of the donor DNA that contains the *rif* marker will have replaced the corresponding region of the recipient. In the example, 37% of the Arg⁺ transductants are also

Rifˢ. Thus, the two markers are cotransducible, and the cotransduction frequency is about 37%. Note that if we did the reciprocal cross and grew the phage on a donor with the *rif-8* mutation and used it to transduce an *argH5* mutant, selecting for rifampin-resistant transductants and then testing for the *argH* marker, the result should be about the same. About 37% of the Rifʳ transductants should be Arg⁺, again indicating that the *rif* and *argH* markers are 37% cotransducible.

We can estimate how close together on the *E. coli* DNA the *argH* and *rif* markers would have to be so that they could be cotransducible by P1 phage. The chromosome is 100 min long, and the P1 phage head only holds about 2% of that length; therefore, the two markers must be less than 2 min apart. Translating this distance into base pairs of DNA, the *E. coli* chromosome is about 4.5 million × 10⁶ bp long, and the P1 phage head will hold only 0.02 times 4.5×10^6, or 90,000 bp. Thus, to be cotransducible, two markers in the DNA must be less than 90,000 bp apart.

ORDERING THREE MARKERS BY COTRANSDUCTION FREQUENCIES

As mentioned, the closer together two markers are in the DNA, the more likely they are to be carried in the same phage head, and the higher their cotransduction frequency. Therefore, cotransduction frequencies can also be used to determine which markers are closest to each other on the DNA and therefore to determine the order of markers. To illustrate, we shall use cotransduction to order the *argH* and *rif* markers with respect to another marker in this region. The third marker is due to a mutation in the *metA* gene, whose product is required to make methionine. We have already determined that the *argH* and *rif* markers are about 37% cotransducible. By doing similar transduction experiments, we determine that the *argH* and *metA* markers are about 20% cotransducible and the *rif* and the *metA* markers are almost 80% cotransducible. Hence, the *metA* marker appears to be closer to the *rif* marker than either is to the *argH* marker. We also know that the *rif* marker is somewhat closer to the *argH* marker (37%) than the *metA* marker is to the *argH* marker (20%). From these data, we conclude that the *rif* and *metA* markers must be on the same side of *argH* and that the order of the three markers is *argH–rif-8–metA*.

ORDERING MUTATIONS BY THREE-FACTOR CROSSES

A careful determination of cotransduction frequencies can reveal the order of markers in the DNA. However, **three-factor crosses** offer a less ambiguous way to determine marker order. In chapter 13, we discussed three-factor crosses in connection with mapping phage markers. As in a phage cross, in a transductional or transformational cross, the number of crossovers required to make a particular recombinant type depends on the order of the three markers. Now, however, the sequence of the donor DNA must replace the chromosomal DNA sequence of the recipient in the region of the marker. This requires at least two crossovers for the following reasons.

In recombination between a short piece of linear DNA and the chromosome, odd numbers of crossovers (one, three, five, etc.) will break the chromosome and be lethal, so that any viable recombinant types must originate from an even number of crossovers (two, four, six, etc.). Therefore, the smallest number of crossovers required to make a recombinant type is two, and the rarest recombinant type requires a minimum of four crossovers. Which recombinant type is rarest therefore reveals the order of the three markers. We also discuss this reasoning in chapter 16.

To illustrate the ordering of bacterial markers by transductional three-factor crosses, we again use the example of the *argH*, *metA*, and *rif* markers. In our example, the donor has the *metA15* mutation and the recipient has the *argH5* and *rif-8* mutations. After the cross, the *argH* marker was selected and the Arg⁺ transductants were tested for the *metA* and *rif* markers. Four recombinant types are possible in a cross of this type. Table 14.2 lists representative data for the frequency of each of these four recombinant types.

As mentioned, the most important of the four recombinant types for determining the map order is the rarest, because this is usually the type that requires four crossovers. Therefore, in our example, the Arg⁺ Met⁻ Rifʳ recombinant type, with only two representatives, probably requires the four crossovers. As shown in Figure 14.7, the markers must be ordered *argH–rif-8–metA* to produce this recombinant type. In contrast, only two crossovers are required with an order of *argH–metA–rif-8*. We conclude that the order is probably *argH–rif-8–metA*.

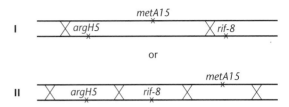

Figure 14.7 The number of crossovers required to make the rarest recombinant type, Arg⁺ Rifʳ Met⁻, with two different orders of the three markers. Since order II, *argH–rif-8–metA*, requires four crossovers, this is probably the order of the three markers. See the text for details.

This order is also consistent with the cotransduction frequencies of the various markers calculated from the data in Table 14.2. Recombinants that are either Arg⁺ Met⁺ Rifˢ or Arg⁺ Met⁻ Rifˢ are cotransductants for the *argH* and *rif* markers. Since there were 22 + 15 = 37 of these, the cotransduction frequency of the *argH* and *rif* markers from this cross was 37/100 = 37%, which is within the experimental error of the cotransduction frequency obtained in the two-factor cross discussed above. The cotransductants for the *argH* and *metA* markers are either Arg⁺ Met⁻ Rifˢ or Arg⁺ Met⁻ Rifʳ. There are only 15 + 2 = 17 of these, for a cotransduction frequency of only 17%, again close enough to the data from the two-factor cross. Therefore, both the cotransduction frequencies and the three-factor-cross data support the conclusions that the *rif-8* marker is closest to the *argH* marker and the order of the genetic markers in the chromosome is *argH–rif-8–metA*.

USING TRANSDUCTION AND TRANSFORMATION TO CONSTRUCT MUTANT STRAINS

Often, strains with the right combination of mutations needed for a cross are not available and must be constructed. Transduction and transformation are very useful for constructing such strains. For example, we may want a donor strain that is Rifˢ Arg⁺ Met⁺. This strain differs from the donor we used above only in that it is Met⁺ rather than Met⁻. To obtain such a strain, we could transduce our donor above with a donor that is Met⁺ and select for Met⁺ transductants.

USING TRANSPOSON INSERTIONS AS GENETIC MARKERS TO CONSTRUCT MUTANT STRAINS

Transposon insertions often make very convenient genetic markers for mapping or moving mutations into other strains by transduction or transformation, particularly if the mutations do not have easily selectable phenotypes. Strains of *E. coli* have been constructed with Tn*10* or Tn*5* transposons inserted around the genetic

TABLE 14.2	Typical transductional data from a three-factor cross
Recombinant phenotype	**No. of recombinants**
Arg⁺ Met⁺ Rifʳ	61
Arg⁺ Met⁺ Rifˢ	22
Arg⁺ Met⁻ Rifʳ	2
Arg⁺ Met⁻ Rifˢ	15

map, and these strains have found many uses in genetic research. Each of the transposon insertions in these strains is cotransducible with at least two other transposon insertions on either side, so that any marker in *E. coli* is cotransducible with at least one of the transposon insertions.

The use of these transposon-containing strains to move mutation *x* from one strain to another requires two transductions. First, a transposon-containing strain that contains the transposon inserted close to the region where mutation *x* occurred is obtained in the mail and used as a donor to transduce the new mutant strain, and the antibiotic resistance encoded by the transposon is selected. If the site of the transposon insertion and the region of mutation *x* are cotransducible, some of the transductants will retain the mutation, but some will not. In the second transduction, one of the original transductants that retained mutation *x* is used as a donor to transduce the mutation into other strains and the antibiotic resistance on the transposon is again selected. Some of these other strains should now have mutation *x*. Note that for these applications, the transposon does not transpose. The transposon is merely being used as a genetic marker and is recombined into the chromosome of the recipient bacteria by recombination between homologous surrounding sequences (**flanking sequences**). The transposon might transpose sometimes, but the frequency of transposition is much lower than the frequency of recombination.

If no strain with a transposon inserted close to the region of mutation *x* is available, it may be necessary to make one. In this case, the first step is to use transposon mutagenesis (see chapter 16) to randomly mutagenize the strain of bacteria carrying mutation *x* and then select for the antibiotic resistance gene on the transposon. These antibiotic-resistant mutants will all have the transposon inserted somewhere in their chromosome. The antibiotic-resistant insertion mutants are pooled and used as donors to transduce another strain that is *x*⁺. The transductants are then tested for the phenotype caused by mutation *x*. Any transductants that have the mutation probably have the transposon inserted close by. This position can be confirmed by using one of them as a donor to transduce another strain. A high percentage of the antibiotic-resistant transductants should now have mutation *x*. After a transposon insertion that is close to the mutation has been isolated, the antibiotic resistance gene on the transposon can be used to map mutation *x* or move it into other strains. Note that this application does depend upon the ability of a transposon to transpose, but only in the first step, transposon mutagenesis. Subsequent steps rely on homologous re-

combination. We discuss transposon mutagenesis and its applications in more detail in chapter 16.

Mapping Other Types of Markers

The same mapping techniques used to map markers due to mutations in the chromosome can also be used to map other types of markers.

MAPPING SUPPRESSOR MUTATIONS
As discussed in chapter 3, suppressor mutations are second-site mutations that restore the wild-type phenotype lost as the result of a mutation. In other words, these mutations suppress the effect of another mutation. Suppressors can be either **intragenic** (if they lie in the same gene) or **extragenic** (**intergenic**) (if they lie in a gene different from that containing the original mutation (see chapter 13).

There are many reasons for mapping suppressors. For example, extragenic suppressors often provide clues about other gene products that interact with the gene product in question. Similarly, intragenic suppressors can provide clues to the parts of the protein product of a gene that interact with other parts of the protein.

Mapping suppressor mutations is much like mapping other types of mutations. The major difference is that both the donor and recipient usually must have the original mutation, particularly if the suppressor mutation causes no phenotype when it is by itself. Then recombinants that are wild-type for the suppressor locus will show the original mutant phenotype because they still have the original mutation but not the suppressor.

Like other types of mutations, extragenic suppressor mutations can be localized by Hfr crosses and then the map position can be further refined by transductional or transformational crosses. To illustrate, take the example of a *trp* nonsense mutation suppressed by a nonsense suppressor (a tRNA mutation [see chapter 3]) elsewhere in the genome. By mapping the nonsense suppressor mutation, we will locate the gene for the tRNA that has been mutated so that it responds to the nonsense codon. The first step could be an Hfr cross in which the donor was an Hfr strain with the original *trp* mutation and the recipient was a strain carrying both the original *trp* mutation and the suppressor mutation. Various markers around the genome are selected and tested for the Trp⁻ phenotype. Recombinants that have had the region with the suppressor mutation replaced by the corresponding wild-type sequence will be Trp⁻, since both the donor and the recipient have the same *trp* mutation. The data are then analyzed as for other mutations being mapped by Hfr crosses.

Intragenic suppressor mutations are always closely linked to the site of the original mutation and so are usually more difficult to map. By using modern techniques, it may be easier just to sequence the gene to find the site of the suppressor mutation (see chapter 15).

MAPPING CLONED GENES

You may have cloned a gene (see chapter 15), but you do not know its position in the chromosome. A phenotype of some sort is needed to map the gene by using genetic techniques, but it is not certain which phenotype(s) if any, that mutations in the gene would cause. In such cases, the cloned gene can be mapped by being crossed back into the chromosome along with a marker to use for the mapping. The marker might be a gene for antibiotic resistance in the cloning vector, or it might be a resistance gene introduced into the clone itself. The marker can then be mapped by methods described above to locate the region in the chromosome that was cloned. The methods for crossing cloned genes back into the chromosome are discussed in chapter 16.

Complementation Tests in Bacteria

As we discussed in chapter 13 regarding phages, only complementation tests can reveal whether two mutations are in the same or different genes or show how many genes are represented by a collection of mutations. Complementation tests can also provide needed information about the type of mutation being studied, whether the mutations are dominant or recessive, and whether they affect a *trans*-acting function or a *cis*-acting site on the DNA.

Complementation tests require that two different alleles of a gene or genes be introduced into the same cell. Because bacteria are normally haploid, with only one allele for each gene, they can, at best, be made diploid for only small regions of the chromosome. The case is different for diploid organisms, whose homologous chromosomes can carry different alleles of all genes, and for viruses, different strains of which can simultaneously infect cells (see chapter 13). Bacteria diploid for a small region of their chromosome are called **partial diploids** or **merodiploids**.

Selecting Prime Factors

One way to make partial diploids with bacteria is with **prime factors**. As discussed in chapter 5, prime factors are plasmids that contain part of the bacterial chromosome, and if this chromosomal DNA includes the region of the mutations being studied, complementation tests can be performed.

Chapter 11 included an account of how prime factors were used to complement *lac* mutations, which led to the operon model for *lac* gene regulation. However, this chapter did not discuss how such prime factors were obtained. Prime factors appear rarely, and so the power of selectional genetics must be used to isolate them.

SELECTING PRIME FACTORS ON THE BASIS OF EARLY TRANSFER OF MARKERS

In chapter 5, we described how prime factors arise from Hfr strains when a plasmid is incorrectly excised from the chromosome, carrying part of the chromosome with it. A prime factor is usually quite small compared with the chromosome and usually contains the chromosomal genes that flanked it while it was integrated in the Hfr strain. For these reasons, the prime factor will transfer some markers more efficiently than the Hfr strain will.

Figure 14.8 shows how we take advantage of these features to select transconjugants containing a prime factor carrying a certain region of the chromosome. In this case, the region includes the *proA* gene of *E. coli*. The selection begins with an Hfr strain that has the F plasmid integrated close to the wild-type *proA*⁺ gene of

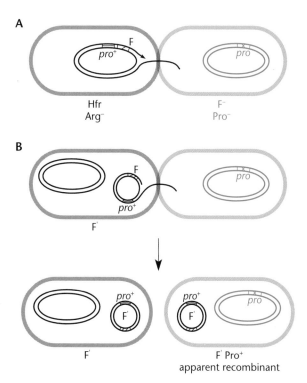

Figure 14.8 Selecting prime factors on the basis of early transfer of a marker. The hatched region indicates F DNA. See the text for details.

E. coli. However, the plasmid is oriented so that the *proA* gene will be transferred late, if at all, by the Hfr strain. Meanwhile, in a few cells, the F' factor will have been excised, and the region of the *proA* marker may have been excised with it and so will be transferred early and efficiently. The prime factor can be isolated from these few cells by mixing the Hfr strain with recipient cells that have a *proA* mutation and mating for only only a short time. The few recipient cells that become Pro+ will probably contain F' factors that include the *proA* gene. These apparently recombinant strains will be partial diploids containing two *proA* genes, one in the F' plasmid and the other in the chromosome. They will be Pro+ because the *proA+* gene in the plasmid complements the *proA* mutant gene in the chromosome. The partial diploid cells can be distinguished from true recombinants caused by an Hfr cross because the partial diploids contain the entire plasmid sequence and so will themselves be able to produce pili and will therefore be sensitive to male-specific phage (see chapters 5 and 7); they will also transfer markers on the F' factor, including the *proA* marker, into other bacteria with high efficiency.

SELECTING PRIME FACTORS BECAUSE THEY ARE REPLICONS

The bacterial genes in a prime factor are part of a replicon, and so can replicate and be transferred independently of the chromosome. In contrast, genes received during an Hfr mating are lost unless they recombine with the chromosome, but recombination between incoming DNA and the chromosome will not occur if the recipient cells have a *recA* mutation.

Therefore, we can select for prime factors by mating the donor cells with *recA* mutant recipients. No recombinants should be obtained in a cross between an Hfr strain and a *recA* mutant recipient, but a few *recA* mutant cells will become apparent recombinants for a marker in such a mating. These cells will most probably contain prime factors (Figure 14.9).

This second method of isolating prime factors is generally preferred because after transfer, the prime factors will be in *recA* mutant cells, where they are more stable and will not be destroyed by recombination with the chromosome. Of course, this latter method is applicable only if *recA* mutants are available for the type of bacteria being studied.

One problem with prime factors, especially if they are very large, is that they tend to be unstable and often suffer large deletions or are lost spontaneously (i.e., cured) as the cells multiply. Consequently, cells containing prime factors usually must be grown under conditions that select for the genes on the prime factor.

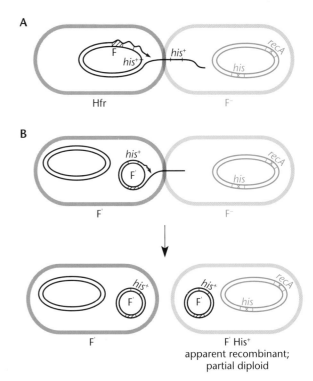

Figure 14.9 Selecting a prime factor by mating into a RecA⁻ recipient. The prime factor is a replicon and so does not need to recombine with the chromosome to be maintained. Any apparent recombinants presumably contain a prime factor. The hatched region indicates F DNA.

Once a prime factor carrying the region of interest has been isolated, it can be mated into other strains for complementation tests. This was how the prime factors used for the complementation tests in previous chapters were obtained.

Complementation by Specialized Transducing Phages

The use of prime factors is not the only way to make partial diploids in bacteria for complementation tests. As discussed in chapter 7, lysogenic phages called specialized transducing phages also carry regions of the bacterial chromosome. The bacterial DNA can be introduced into the phage either by DNA cloning techniques (see chapter 15) or by selecting transducing phages that have excised incorrectly from the chromosome during induction (see chapter 7). Lysogenic phages are advantageous for complementation in that they usually exist in only a single copy in the chromosome. This feature can be important, because too many copies of the gene can distort the effects of complementation. Also, specialized transducing phage-containing strains tend to be quite stable and can often be easily induced to isolate large quantities of the complementing gene.

Complementation from Plasmid Clones

Complementation tests can also be performed with chromosomal DNA cloned in plasmid cloning vectors. The disadvantage of this method is that plasmid cloning vectors usually exist in relatively high copy numbers and, as mentioned, too many copies of a gene can distort the effects of the complementation. Another problem is that a cloned gene is often transcribed from a plasmid promoter instead of its own promoter, and so too much of the gene product may be made. In some cases, **suppressive effects** can occur, in which unrelated genes can appear to complement mutations if the cloned gene exists in too many copies or if too much of the gene product is synthesized.

In addition to these problems, designing plasmid cloning vectors for complementation tests involves a catch-22 situation. On one hand, low-copy-number plasmids, which would be more appropriate for complementation experiments, are usually too difficult to detect and isolate for use as cloning vectors. On the other hand, high-copy-number plasmid cloning vectors can distort the results of complementation tests. This problem can sometimes be circumvented through the use of vectors with two origins, one of each type of plasmid. Under some conditions, only the low-copy-number origin would be used, whereas under others, the high-copy-number origin would be functional, facilitating isolation of the cloning vector.

Examples of Genetic Analysis in Bacteria

We conclude this chapter with some examples of basic genetic analysis in bacteria. We chose these particular examples because they have historical importance for bacterial genetics and/or, in our opinion, they illustrate particularly well the application of genetic analysis to the study of gene structure and function in bacteria.

Deletion Mapping of *lacI* Missense Mutations

The most efficient way to map a large number of mutations in a particular gene is by deletion mapping. We discussed deletion mapping of *r*II mutations in phage T4 in the previous chapter, and the principles are the same here: no wild-type recombinants will appear in crosses between a mutant with a point mutation and a mutant with a deletion mutation if the point mutation lies within the deleted region.

As we discussed in chapter 11, the *lac* operon encodes the enzymes responsible for the utilization of lactose. The operon is regulated by a repressor, encoded by the *lacI* gene, that binds to operator regions and pre-

vents transcription of the structural genes of the operon. However, when the inducer allolactose or one of its analogs binds to the LacI repressor, LacI can no longer bind to the operators, and transcription ensues. Missense mutations in *lacI* result in various phenotypes (Table 14.3). For example, *lacI* mutations may inactivate the repressor by preventing binding to the operators, preventing tetramer formation, or both. The *lacI* mutations can be subdivided into two groups: recessive and dominant. Any *lacI* mutation that completely inactivates the repressor so that it no longer binds to the operators is recessive to the wild type and is called simply a *lacI* mutation. However, some *lacI* mutations inactivate the operator-binding domain of the repressor but do not affect the tetramerization domain and so cannot form tetramers. These *lacI* mutations can be dominant to the wild type if mixed tetramers, made up of both wild-type and mutant subunits, cannot bind to the operators. They are therefore called *lacI*⁻ᵈ mutations, where the *d* stands for dominant. Other dominant *lacI* mutations, called *lacI*ˢ, result in permanent repression even in the presence of inducer. These mutations prevent the binding of the inducer to the repressor. With some mutations, called *lacI*ʳᶜ, the situation is reversed. These mutations change the repressor so that it binds to the operators only in the *presence* of inducer.

The position in the *lacI* gene of these mutations may reveal information about which regions of the LacI protein are involved in its various functions. Accordingly, numerous *lacI* mutations have been isolated and mapped to determine their position in the gene (see Miller, Schmeissner, et al., and Miller and Schmeissner, Suggested Reading). In the following sections, we describe how these mutations were isolated and how this mapping was done.

ISOLATION OF *lacI* DELETIONS
Before a large number of *lacI* missense mutations could be mapped, a set of deletion mutations extending various distances into the *lacI* gene was needed (Figure 14.10). The *tonB* gene of *E. coli* was used for these

TABLE 14.3	Types of *lacI* mutations	
Mutation	**Function affected**	**Phenotype**
lacI	Operator binding or tetramer formation	Constitutive; recessive
lacI⁻ᵈ	Operator binding	Constitutive; dominant
*lacI*ˢ	Inducer binding	Permanently repressed; dominant
*lacI*ʳᶜ	Conformational change after inducer binding	Repressed only with inducer bound; dominant

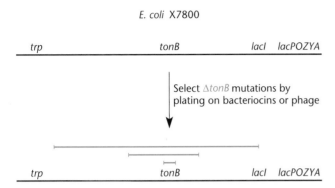

E. coli X7800

| trp | tonB | lacI lacPOZYA |

Select Δ*tonB* mutations by
plating on bacteriocins or phage

| trp | tonB | lacI lacPOZYA |

Figure 14.10 Selecting *tonB* deletions that extend into
the *lacI* gene in *E. coli* X7800. Mutant bacteria that have
tonB deletion mutations extending into *lacI* will form blue
colonies on X-Gal plates in the absence of inducer.

selections because mutations that inactivate this gene
make the cells resistant to phage T1 and some types of
bacteriocins, offering a positive selection for *tonB* mutants. In fact, mutants of *E. coli* resistant to phage T1
were used in the earliest studies of inheritance in bacteria (see chapter 3, the Luria and Delbruck and the Newcombe experiments).

Normally, the *lacI* gene is not near the *tonB* gene;
therefore, the investigators had to design a special strain
of *E. coli* in which *lacI* is repositioned next to *tonB*. The
genomic structure of this strain, *E. coli* X7800, is dia-
grammed in Figure 14.10. These bacteria were then selected for two phenotypes—resistance to phage T1 and
constitutive Lac expression—in order to isolate the
tonB mutants with deletion mutations exending into
lacI. One property of deletion mutations is that they can
inactivate more than one gene simultaneously (see chapter 3), and, conversely, most mutants in which more
than one gene has been inactivated have deletion mutations. Therefore, mutants that are both phage T1 resistant and mutant for *lacI* most probably have deletion
mutations extending from *tonB* to *lacI*, some of which
may end within the *lacI* gene.

To isolate mutants that are both T1 resistant and mutant for *lacI*, the investigators plated *E. coli* X7800 on
X-Gal plates containing T1 phage and X-Gal but lacking allolactose or its analogs. Any bacterium that forms
a colony on these plates must be T1 resistant, and any
blue colonies must have originated from a mutant constitutive for *lac* operon expression. The β-galactosidase
product of the *lacZ* gene will cleave the X-Gal, making
the colony blue, but X-Gal itself is not an inducer of the
lac operon, and so only constitutive Lac mutants will
turn blue without the inducer present. The endpoints of
the deletions in the *lacI* gene were then mapped by

crossing them with a few *lacI* point mutations that had
been mapped previously by three-factor crosses.

ISOLATING *lacI* MISSENSE MUTATIONS

The next step was to isolate a large number of the different types of *lacI* missense mutations. So that later
mapping would be easier, these *lacI* mutations were isolated in an F′ factor rather than in the chromosome. F′
lac-pro, the F′ factor used, contains the *lac* genes as well
as the wild-type *proB* gene as a selectable marker. This
prime factor was maintained in a strain that had the
chromosomal *lac* genes deleted, so that any *lacI* mutations would occur in the F′ factor.

The different types of *lacI* missense mutations required different selection procedures. To isolate mutants
with constitutive *lacI* mutations, the bacteria containing
the F′ factor were mutagenized and mutants that form
blue colonies on X-Gal plates in the absence of inducer
were selected. As discussed, under these conditions,
only constitutive *lacI* mutants will express the *lacZ* gene
and make blue colonies.

Isolating strains with *lacI*s mutations was somewhat
more difficult. These mutants form colorless colonies on
X-Gal plates, even with inducer, but *lacZ* mutants also
form colorless colonies under these conditions and are
much more frequent. However, unlike most other mutations that make the colonies colorless, *lacI*s mutations
are dominant over the wild type. They will make a
strain colorless even if there is a wild-type *lac* operon in
the chromosome. Therefore, the mutagenized F′ factor
was mated into a strain with a functional *lac* operon in
the chromosome and the transconjugants were plated
on X-Gal plates with the inducer. Under these conditions, many of the cells that form colorless colonies will
have a *lacI*s mutation in the F′ factor.

MAPPING *lacI* POINT MUTATIONS

Figure 14.11 shows how, once a collection of *lacI* mutations with different properties was assembled, they could
be mapped. The F′ factor containing the particular point
mutation to be mapped, as well as the wild-type *proB*
gene, was crossed into derivatives of *E. coli* X7800 that
each contained one of the *lacI* deletion mutations in the
chromosome. After selection for *proB*, the partial
diploid strains were grown to allow possible recombination between the mutant *lacI* gene on the F′ factor and
the partially deleted *lacI* gene in the chromosome. The
presence of any *lacI*+ recombinants in the population
was evidence that the *lacI* mutation in the F′ factor lay
outside the deleted region in the chromosome. However,
LacI+ recombinants are rare and must be selected from
among the majority of the bacteria, which remain LacI−.

Step 1

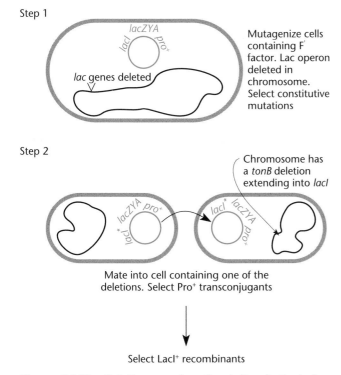

Mutagenize cells containing F' factor. Lac operon deleted in chromosome. Select constitutive mutations

Step 2

Chromosome has a *tonB* deletion extending into *lacI*

Mate into cell containing one of the deletions. Select Pro⁺ transconjugants

Select LacI⁺ recombinants

Figure 14.11 Deletion mapping of mutations in the *lacI* gene of *E. coli*. Step 1: cells lacking a chromosomal *lac* region but with an F' factor containing this region are mutagenized. Step 2: the F' prime factor is mated into a strain with one of the *lacI* deletions isolated, as in Figure 14.10, and Pro⁺ bacteria are selected. A few *lacI*⁺ recombinants will appear if the deleted region in the chromosome does not extend into the region of the point mutation in the prime factor.

The method used to select LacI⁺ recombinants depended on the type of mutation being mapped.

Mapping *lacI* Mutations

To detect *lacI*⁺ recombinants in crosses with mutants that had *lacI* point mutations, the *lac* deletion mutant strains were made mutant for *galE* by methods such as those described above for moving mutations into strains. The investigators could then use phenyl-β-D-galactoside (Pgal) and the fact that *galE* mutants are killed by galactose (see chapter 3) to select *lacI*⁺ recombinants. The selection was based on the facts that the β-galactosidase product of the *lacZ* gene can cleave the galactose off Pgal and kill *galE* mutant strains and that Pgal is not an inducer of the *lac* genes. In the absence of inducer, only strains that are mutants constitutive for the expression of the *lacZ* gene will cleave Pgal and kill themselves. As a consequence, if Pgal is added to the partial-diploid *galE* mutant strain, all the bacteria that are constitutive for *lac* gene expression will be

killed and only *lacI*⁺ recombinants will multiply to form a colony. Therefore, when the transconjugants in step 2 of Figure 14.11 were plated on agar containing Pgal, the appearance of colonies on the plates was evidence that the *lacI* point mutation lies outside the deleted region.

Mapping *lacI*ˢ Mutations

The selection method outlined above could not be used to detect *lacI*⁺ recombinants with *lacI*ˢ mutant F' plasmids. The *lacI*ˢ mutations are dominant, and so the *lacZ* gene will not be expressed to make β-galactosidase to cleave Pgal and kill the *galE* mutant cells, even in partial diploids with a *lacI* deletion in the chromosome. Therefore, the transconjugants will grow on the Pgal plates regardless of whether recombination with the chromosome occurs, rendering the detection of rare *lacI*⁺ recombinants impossible.

One way to detect possible *lacI*⁺ recombinants with *lacI*ˢ mutations depends on the fact that the partial diploids with a *lacI*ˢ mutation in the F' factor and a *lacI* deletion in the chromosome cannot use lactose because the *lacZ* and *lacY* genes cannot be induced. Therefore, if the transconjugants are plated on minimal plates with lactose as the sole carbon and energy source, cells in which the *lacI*ˢ gene in the F' factor has recombined with the *lacI* deletion in the chromosome will become inducible and Lac⁺ and will multiply to form a colony. However, another type of recombinant also will exhibit the Lac⁺ phenotype. In this recombinant, both the F' factor and the chromosome have the *lacI* deletion, because the *lacI* deletion has been transferred to the F' factor by recombination. This latter type of recombinant can occur even if the *lacI*ˢ mutation lies in the deleted region. Therefore, to determine whether the *lacI*ˢ mutation lies outside the deleted region, only Lac⁺ recombinants that are *lacI*⁺ should be counted. In the experiment, the two types of Lac⁺ recombinants were distinguished by replicating the colonies onto X-Gal plates in the presence and absence of the inducer isopropyl-β-D-thiogalactopyranoside (IPTG). Only *lacI*⁺ recombinants should form blue colonies in the presence of IPTG and colorless colonies in the absence of IPTG. The *lacI* recombinants should form blue colonies on both types of plates.

IMPLICATIONS OF THE MAP POSITION OF DIFFERENT TYPES OF *lacI* MUTATIONS FOR THE STRUCTURE OF THE LacI REPRESSOR

The location of the various types of mutations in the *lacI* gene suggested that the LacI repressor protein is divided into domains much like the λ repressor. The fact that *lacI⁻ᵈ* mutations mostly map in the region encoding the N terminus of the polypeptide suggested that the N-terminal domain probably binds to DNA. Other types of *lacI* mutations are scattered throughout the gene, suggesting that the domain of the protein involved in tetramer formation is less well defined. The *lacIˢ* mutations are somewhat scattered throughout the gene but tend to be clustered in a few places. Some of these sites presumably form the binding site for the inducer, although some may also cause more general conformational changes that interfere with inducer binding. This analysis of the structure of the LacI repressor is still under way and serves as a prototype for mutational analyses of proteins. Because the three-dimensional structure of the LacI repressor protein is now known, structural studies can now confirm or disprove the conclusions of this elegant genetic analysis.

Isolating Tandem Duplications of the *his* Operon in *Salmonella typhimurium*

The process of selecting tandem duplication mutations is a good example of the genetic analysis of chromosome structure in bacteria. As discussed in chapters 3 and 13, tandem duplication mutations can occur by recombination between directly repeated sequences, causing the duplication of the DNA between the repeated sequences. However, once they form, tandem duplication mutations are usually very unstable because recombination anywhere within the duplicated segments can destroy the duplication.

Most long tandem duplication mutations do not cause easily detectable phenotypes, because they do not inactivate any genes. Even the genes in which the mistaken recombination occurred to create the duplication exist in a functional copy at the other end of the duplication. Therefore, special methods must be used to select bacteria with duplication mutations.

Transduction has been used to select tandem duplications of the *his* region of *Salmonella typhimurium* (see Anderson, et al., Suggested Reading). The selection used by Anderson et al. depends on the properties of two deletion mutations in the *his* region, Δ*his*2236 and Δ*his*2527 (see Figure 14.13). These deletion mutations complement each other because one ends in *hisC* and the other ends in *hisB*. However, because the endpoints of the two deletion mutations are very close to each other, crossovers between them will occur infrequently.

P22 transducing phage were propagated on a strain with one of the deletion mutations and used to transduce a strain with the other deletion mutation, as shown in Figure 14.12. The His⁺ tranductants were then selected by plating on minimal plates without histidine.

Step 1   Grow phage P22 in *S.typhimurium* with one *his* deletion

Step 2   Use phage to transduce bacteria with the other *his* deletion

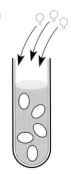

Step 3   Plate on medium without histidine; incubate

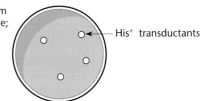

His⁺ transductants

Figure 14.12   Using transduction to select bacteria with duplication mutations of the *his* region of *Salmonella typhimurium*. The P22 transducing phage are grown on bacteria with one deletion and used to transduce bacteria with the other deletion. The transductants are plated on medium without histidine to select for the His⁺ phenotype. See the text for details and conclusions.

A few His⁺ transductants arose, even though there should be very little recombination between the deletions. Moreover, many of the His⁺ transductants that arose were unusual. Most were very unstable, spontaneously giving off His⁻ segregants at a high frequency when they multiplied in the presence of histidine. Also, some of the His⁺ transductants were slimy, or mucoid, in appearance, but the His⁻ segregants had lost this mucoidy. On the basis of these observations, these investigators concluded that unstable His⁺ transductants had tandem duplications of the *his* region (Figure 14.13), in which one copy has the Δ*his*2236 deletion whereas the other has the Δ*his*2527 deletion. The two deletion mutations can complement each other, making the transductants His⁺.

These duplications probably arose by recombination between directly repeated sequences in some of the recipient bacteria while they were multiplying (Figure

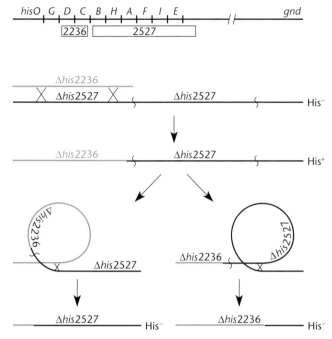

Figure 14.13 Mechanism for generation and destruction of *his* duplications. One of the two copies of the *his* region of bacteria with a preexisting duplication (in the example, Δ*his2527*) recombines with incoming transducing DNA carrying the other deletion (in the example, Δ*his2236*) to replace one copy of the duplicated region with the corresponding region of the donor. The two deletions complement each other to make the bacteria His⁺. The duplication can be destroyed by looping out one of the two duplicated segments, giving rise to His⁻ segregants with one or the other of the two deletions.

14.13). In a bacterium that contains a duplicated *his* operon with one deletion mutation, a *his* operon with the other deletion, brought into the cell by a transducing phage, can replace one of the *his* regions of the recipient cell. This type of His⁺ transductant, although rare, might be more frequent than His⁺ transductants produced from recipients that do not have a duplicated *his* operon, since in the latter case, His⁺ transductants would arise only from the cells in which recombination had occurred between the two very closely linked deletion mutations.

The unusual properties of the His⁺ transductants observed in the experiment are explained by the duplication structure. As with other tandem duplication mutations, the His⁺ phenotype of these transductants gives rise to His⁻ segregants because recombination anywhere in the duplicated regions will leave only one copy of the *his* region, with one or the other of the deletions (Figure 14.13). These cells are called haploid segregants because they have only one copy of the duplicated re-

gion and occur spontaneously without the need for genetic crosses (see chapter 13). The mucoidy of some of the His⁺ transductants can also be explained. The duplicated region in some of the His⁺ transductants may contain another gene that makes colonies appear mucoid when twice as much of the gene product is synthesized.

A *recA* MUTATION HELPS STABILIZE THE DUPLICATIONS

If the instability of the putative duplications results from recombination between the repeated segments, then *recA* mutations, which prevent homologous recombination, should stabilize the duplications. To test this hypothesis, the investigators introduced a *recA* mutation into some cells containing putative duplications by means of Hfr crosses, with the selection for a *serA* marker closely linked to *recA*. Transconjugants that had the *recA* mutation were identified by their sensitivity to UV light (see chapter 10). The tandem duplication in these *recA* recombinants was now stable and did not give His⁻ haploid segregants even when grown in media with histidine.

DETERMINING THE LENGTH OF TANDEM DUPLICATIONS

These same investigators also used genetic experiments to determine the length of the segment duplicated in some of the strains, especially to see whether the duplicated region ever extended as far as the *metG* gene, about 2 min (~100,000 bp) away from the *his* region. For this experiment, the investigators selected *his* duplications in a strain that also had a *metG* mutation. Strains containing these duplications were propagated without histidine in the medium to maintain the duplication and then transduced a second time with phage propagated on a *metG*⁺ donor. Met⁺ transductants were selected. The reasoning was that if the *metG* gene is included in the duplicated region, there should be two copies of the *metG* gene in the cell, and only one of the two *metG* genes in the recipient need have been replaced by the wild-type *metG*⁺ gene of the donor to give Met⁺ transductants (Figure 14.14). If only one of the two copies of *metG* has been transduced to *metG*⁺, when the transductants with one *metG*⁺ and one *metG* gene are grown in media containing histidine (and methionine), some of the His⁻ haploid segregants should also be Met⁻, depending on whether they lose the wildtype or the mutant *metG*. The results of this work revealed that many duplications of the *his* operon did include the *metG* gene. In fact, some duplications extended much farther, even as far as the *aroD* gene, which is 10 min, or 10% of the entire genome, away from *his*.

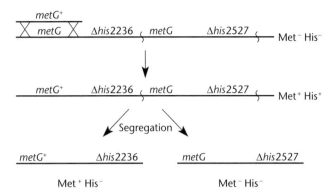

Figure 14.14 Determining if the duplicated segment in a *his* duplication extends as far as the *metG* gene. A duplication is made as before but in a strain with a *metG* mutation. A strain with the duplication to be tested is transduced with phage grown on *metG*+ bacteria. If the *metG* gene is in fact duplicated, one of the two copies of the *metG* gene will be transduced to *metG*+ and the strain will spontaneously give rise to Met⁻ segregants.

FREQUENCY OF SPONTANEOUS DUPLICATIONS

These investigators also attempted to estimate the frequency of spontaneous tandem duplication mutations in a growing population of bacteria. They did this by comparing the frequency of tranductants when the donor and recipient bacteria had different deletion mutations with normal transduction frequencies, when only the recipient had one of the deletions. In the first case, most of the recipient cells that were transduced must have had a tandem duplication of the *his* region, whereas in the second case, most of the recipients will have had only one copy of the *his* region. The investigators estimated that duplications occur as frequently as once every 10^4 times a cell divides, which is hundreds of times greater than normal mutation frequencies. Apparently, spontaneous duplications in bacteria occur quite often during cell multiplication and can occur over very large regions of the chromosome. However, because they cause few phenotypic changes, most of these large duplications have little effect on the organism.

Genetic Analysis of Repair Pathways

Genetic analysis can also be applied to understanding DNA repair pathways, and much of what is known about the repair of DNA damage and mutagenesis has come from genetic experiments. In this section, we discuss the genetic experiments that helped elucidate some of the repair pathways discussed in chapter 10.

ISOLATION OF *mut* MUTANTS

As discussed in chapter 10, the products of the *mut* genes of *E. coli* and other bacteria help to lower sponta-

neous mutation rates. In *E. coli*, the *mut* genes include the genes of the mismatch repair system (*mutS*, *mutH*, and *mutL*), the genes for repair of oxidation damage due to 8-oxodeoxyguanine lesions (*mutM*, *mutY*, and *mutT*), and a gene encoding the DNA Pol III holoenzyme editing function (*mutD*). Some of these same gene products also act to prevent mutations caused by some mutagens.

The *mut* genes were discovered because mutations in these genes increase spontaneous mutation rates. The resulting phenotype is often referred to as "mutator." A common method for detecting mutants with abnormally high mutation rates is **colony papillation**. This scheme is based on the fact that all of the descendants of a bacterium with a particular mutation will be mutants of the same type. Thus, a growing colony will contain **papillae**, or sectors composed of mutant bacteria of various types. A *mut* mutation will increase the frequency of papillation for many types of mutants, so that the phenotype used in a colony papillation test is purely a matter of convenience.

In *E. coli*, Lac⁺ revertants of Lac⁻ mutations are an obvious choice for these tests. Lac⁻ mutants growing on plates containing X-Gal will give rise to colorless colonies, but any Lac⁺ revertants appearing in the growing colony will give rise to blue papillae. As shown in Figure 14.15, with higher than normal rates of spontaneous mutagenesis because of a *mut* mutation, the colony will have more blue papillae than normal. This is how many of the *mut* genes we have discussed were detected.

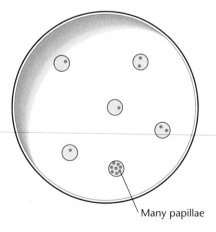

Figure 14.15 Colonies due to *mut* mutants have more papillae. A *lacZ* mutant was plated on X-Gal-containing medium. Revertants of the *lacZ* mutation will produce blue sectors or papillae. A *mut* mutant will give more blue papillae than normal owing to increased spontaneous mutation frequencies (arrow).

GENETIC ANALYSIS OF INDUCIBLE MUTAGENESIS: ISOLATION OF *umuC* AND *umuD* MUTATIONS

The early experiments by Jean Weigle, described in chapter 10, showed that UV mutagenesis is inducible in *E. coli*. He observed that UV-irradiated phage λ suffer many more mutations when they infect cells that had been UV-irradiated prior to infection than when they infect unirradiated cells. Later experiments by Kato et al. (chapter 10, Suggested Reading) showed that most of the mutations induced by UV light were not due to mistakes made by the recombination or excision repair pathways. If these pathways were mistake prone, mutations that inactivated a gene in the pathway should reduce the frequency of mutants among the surviving bacteria. This is not the case. Instead, evidence indicated that these repair pathways make very few mistakes when they are repairing the damage in DNA due to UV irradiation. Repair by these pathways does, however, markedly increase the number of cells that survive the irradiation.

Once it was established that most of the mutagenesis after UV irradiation can be attributed to an inducible pathway different from recombination and nucleotide excision repair, the next step was to identify the genes of this pathway. Mutations that inactivate a *mut* gene or another repair pathway gene *increase* the rate of spontaneous mutations or mutagen-induced mutations because normally the gene products repair damage before it can cause mistakes in replication. However, as noted above, mutations that inactivate gene products of a mutagenic or mistake-prone repair pathway should have the opposite effect, *decreasing* the rate of at least some induced mutations. Because the repair pathway itself is mutagenic, mutations should be less frequent if the repair pathway does *not* exist than if it *does* exist. Cells with a mutation in a gene of the mistake-prone repair pathway may be less likely to survive DNA damage, but there should be a lower percentage of newly generated mutants among the survivors.

The first *umuC* and *umuD* mutants were isolated in two different laboratories by essentially the same method—reversion of a *his* mutation to measure mutagenesis—but we shall describe the one used by Kato and Shinoura (see Suggested Reading). The basic strategy was to treat *his* mutants with a mutagen that induces DNA damage similar to that caused by UV irradiation and to identify mutant bacteria in which fewer *his*+ revertants occurred. These *his* mutants would have a second mutation that inactivated the mutagenic repair pathway.

Figure 14.16 illustrates the first part of the experiment in detail. A *his* point mutant of *E. coli* was heavily

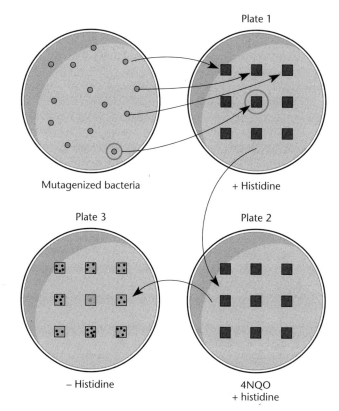

Figure 14.16 Detection of a mutant defective in mutagenic repair. Colonies of mutagenized *his* bacteria are picked individually from a plate and patched onto a new plate containing histidine. This plate (plate 1) is then replicated onto a plate containing 4NQO (plate 2) to induce DNA damage similar to that induced by UV irradiation. The 4NQO-containing plate is then replicated onto another plate with limiting amounts of histidine (plate 3). After incubation, mottling of a patch caused by many His+ revertants indicates that the bacterium that made the colony on the original plate was capable of mutagenic repair. The colony circled in blue on the original plate may have arisen from a mutant deficient in mutagenic repair, because it gives fewer His+ revertants when replicated onto plate 3.

mutagenized to try to induce mutations in the genes of the mutagenic repair pathway, and individual colonies of the bacteria were patched with a loop onto plates with medium containing histidine. The plates were incubated until patches due to bacterial growth first appeared; each plate was then replicated onto another plate containing 4-nitroquinoline-1-oxide (4NQO), which causes DNA damage similar to that caused by UV irradiation. After the patches developed, this plate was replicated onto a third plate containing limiting amounts of histidine. Most bacteria formed patches with a few regions of heavier growth due to His+ revertants, indicating they were being mutagenized by the

4NQO. However, a few bacteria formed patches with very few areas of heavier growth and therefore had fewer His⁺ revertants. The bacteria in the patches were candidates for descendants of mutants that could not be mutagenized by 4NQO and presumably UV irradiation.

The mutations that prevented mutagenesis by UV irradiation mapped to a number of genes, including *recA* and *lexA*. These mutations presumably prevent the induction of all the SOS genes including those for mutagenic repair. The *recA* mutations presumably inactivate the coprotease activity of the RecA protein that binds single-stranded DNA and causes the autocleavage of the LexA repressor, inducing transcription of the SOS genes (see chapter 10). The *lexA* mutations in all probability change the LexA repressor protein so that it cannot be cleaved, presumably because one of the amino acids around the site of cleavage has been altered. This is a special type of *lexA* mutation called a *lexA(ind)* mutation. If the LexA protein is not cleaved following UV irradiation of the cells, the SOS genes, including the genes for mutagenic repair, will not be induced.

Most important for this discussion, however, are mutations in a third locus at 25 min on the *E. coli* map. Complementation tests between mutations at this locus revealed two genes at this site required for UV mutagenesis, later named *umuC* and *umuD* for UV mutagenesis C and D. Later experiments also showed that these genes are transcribed into the same mRNA and so form an operon, *umuDC*, in which the *umuD* gene is transcribed first. Experiments with *lacZ* fusions revealed that the *umuDC* operon is inducible by UV light and is an SOS operon, since it is under the control of the LexA repressor (see Bagg et al., Suggested Reading).

Experiments To Show that Only *umuC* and *umuD* Must Be Induced for SOS Mutagenesis

Just because *umuC* and *umuD* are inducible by DNA damage and required for SOS mutagenesis does not mean that they are the *only* genes that must be induced for this pathway. Might other genes that must be induced for SOS mutagenesis have been missed in the mutant selections? Several investigators sought an answer to this question (see Sommer et al., Suggested Reading). Their experiments used a *lexA(ind)* mutant, which, as mentioned, should permanently repress all the SOS genes. They then mutated the operator site of the *umuDC* operon so that these genes would be expressed constitutively and would no longer be under the control of the LexA repressor. Under these conditions, the only SOS gene products that should be present are the *umuC* and *umuD* genes, since all other such SOS genes will be permanently repressed by the LexA(*ind*) repressor. In

addition, a shortened form of the *umuD* gene was used. This altered gene synthesizes only the carboxy terminal UmuD' fragment that is the active form for SOS mutagenesis (see chapter 10). With the altered *umuD* gene, the RecA coprotease is not required for UmuD to be autocleaved to the active form.

The experiments showed that UV irradiation induced mutations in the *lexA(ind)* mutants that expressed UmuC and UmuD' constitutively, indicating that *umuC* and *umuD* are the only genes that *need* to be induced for SOS mutagenesis. However, this result does not entirely eliminate the possibility that other SOS gene products are involved. As discussed in chapter 10, the RecA protein may also be directly required for UV mutagenesis. The *recA* gene is induced to higher levels following UV irradiation but apparently is present in large enough amounts for UV mutagenesis even without induction. The GroEL and GroES proteins are also required for UV mutagenesis, but the *groEL* and *groES* genes are not under the control of the LexA repressor and so are expressed even in the *lexA(ind)* mutants.

Genetic Analysis of Protein Transport

The final examples of genetic analysis in bacteria are the studies of protein transport and protein topology in membranes. About one-fifth of the proteins made in a bacterium do not remain in the cytoplasm but are transported into or through the surrounding membranes. The process of passing proteins out of the cytoplasm is called **protein export**. Proteins that are passed into or through the membranes are **secretory proteins**, and the process of transporting proteins through the membranes is known as **protein secretion**. Proteins that remain in either the inner or outer membrane are **membrane proteins**, while those that remain in the periplasmic space are **periplasmic proteins**. Proteins that are passed out of the cell into the surrounding environment are **extracellular proteins**.

We have already discussed some exported proteins. For example, the β-lactamase enzyme that makes the cell resistant to penicillin resides in the periplasm, and so must be secreted through the inner membrane. The protein disulfide isomerases also reside in the periplasm and form disulfide linkages in some periplasmic or extracellular enzymes when they are in the periplasm. The *tonB* gene product of *E. coli* must also pass through the inner membrane to its final destination in the outer membrane, where it participates in transport processes and serves as a receptor for some phages and colicins.

Genetic analysis of protein secretion in bacteria has revealed general principles common to all cells. Indeed, many aspects of protein transport are similar in bacteria and eukaryotes. Secreted proteins of both bacteria and eukaryotes have signal sequences (see below). Recently, analogs to Srps (signal recognition particles), small organelles composed of both proteins and RNA that are known to play a role in the secretion of proteins in eukaryotic cells, were found in bacterial cells. However, protein secretion clearly differs in bacteria and in eukaryotes. Protein secretion in eukaryotes involves organelles such as the rough endoplasmic reticulum and the Golgi apparatus, whereas bacteria do not have such organelles. These organelles may be necessitated by the larger size of eukaryotic cells (see Introduction).

USE OF THE *mal* GENES TO STUDY PROTEIN TRANSPORT

The products of many of the genes of the *mal* operons of *E. coli* (discussed in chapter 11) are membrane or periplasmid proteins and so are transported. In this section, we discuss how the *mal* genes have been used to identify the important features of proteins that allow them to be secreted as well as to identify other functions required for the secretion of many proteins.

Isolating Mutations That Affect the Signal Sequence of the MalE and LamB Proteins

The products of the *lamB* and *malE* genes reside in the outer membrane and periplasm of *E. coli*, respectively (see Figure 11.19). To reach their final destination, these proteins must be secreted through the inner membrane. Like most secreted proteins, LamB and MalE are first made as precursor polypeptides with approximately 25 extra amino acids at their N-terminal ends. These amino acids, known as the **signal sequence**, are cleaved off as the polypeptides traverse the inner membrane.

Gene fusion techniques and selectional genetics were used to determine which amino acids in the signal sequence of MalE and LamB were important for the secretion of these proteins (see Bassford and Beckwith, Suggested Reading). Mutations that cause amino acid changes in a short sequence like a signal sequence are rare and require a positive selection. The method capitalized on translational fusions that joined the N terminus of the MalE or LamB protein, including the signal sequence, to the LacZ protein, which is normally a cytoplasmic protein (see chapter 2 for an explanation of translational fusions). The N-terminal signal sequence will direct the fusion protein to the membrane, and the transport machinery will attempt to export it. However, for unknown reasons, the large fusion protein will become trapped in the membrane, killing the cells, perhaps by causing them to lyse.

The sensitivity of cells to transport of the fusion proteins offers a means of positively selecting transport-defective mutants (Figure 14.17). Since the fusion protein is made in large amounts only in the presence of the inducer maltose, which turns on transcription from the *mal* promoter (see chapter 11), bacteria containing these fusions are maltose sensitive (Mals) and are killed by the addition of maltose to the growth medium. However, bacteria with mutations that prevent the transport

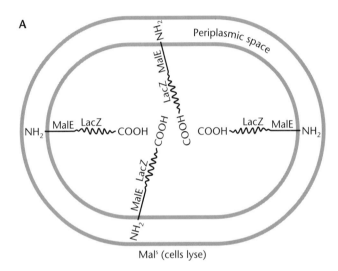

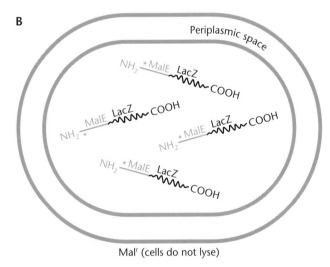

Figure 14.17 Model for the maltose sensitivity (Mals) of cells containing a *malE-lacZ* gene fusion. (A) In Mals cells, the presence of maltose induces the synthesis of the fusion protein, which cannot be transported completely through the membrane and so lodges in the memebrane, causing the cell to lyse. (B) In Malr cells, a mutation in the region encoding the signal sequence (asterisk) prevents transport of the fusion protein into the membrane.

of the fusion proteins will be resistant to maltose (Mal^r) and will survive. Therefore, any colonies that form when bacteria containing the gene fusions are plated on maltose-containing plates may be due to mutant bacteria that no longer can transport the fusion proteins, provided that they still make the fusion protein in large amounts.

Some of the Mal^r mutants may have mutations that change the coding region for the signal sequence in the MalE or LamB portion of the fusion protein so that the signal sequence no longer functions in transport. Then the fusion protein will no longer enter the membrane and kill the cell, and the cell will be resistant to maltose. Determining which amino acids in the signal sequence have been changed in these Mal^r mutants should reveal which amino acids in the signal sequence are required for signal sequence function. These mutations can be distinguished from other mutations that cause defects in secretion by their map positions. Signal sequence mutations should map in the *malE-lacZ* fusion gene, while mutations that affect other proteins required for transport should map elsewhere in the *E. coli* genome. Furthermore, mutations that change the signal or some other sequence in the MalE or LamB part of the fusion protein should specifically affect the transport of the fusion protein and not affect the transport of other membrane proteins.

On the basis of this selection, a number of signal sequence mutations were isolated and later sequenced to determine which amino acid changes could prevent the function of the signal sequence. Many of these were changes from hydrophobic to charged amino acids that might interfere with the insertion of the signal sequence into the hydrophobic membrane.

ISOLATING MUTATIONS IN *sec* GENES

The *mal* genes were also used to select mutants with mutations in genes whose products are part of the protein transport machinery (see Oliver and Beckwith, Suggested Reading). The products of these genes, named the *sec* genes for protein *sec*retion, are required for the transport of some or maybe even all of the membrane, periplasmic, and exported proteins. Therefore, unlike signal sequence mutations, which affect the transport of only one protein, *sec* mutations cause defects in the transport of many proteins into or through the membranes and map in many places on the genome.

A different, somewhat more sensitive selection was used to isolate *sec* mutants rather than signal sequence mutations. This selection was based on the observation that cells containing a *malE-lacZ* fusion do make some MalE-LacZ fusion proteins in the absence of maltose in the medium, but not enough to kill the cells. The amount of protein made is not enough to make the cells Lac$^+$ and able to multiply with lactose as the sole carbon and energy source, even though the MalE-LacZ fusion protein retains the β-galactosidase activity of the LacZ protein, which is the only enzyme required for growth on lactose (see chapter 11). The investigators reasoned that the cells may not exhibit the β-galactosidase activity because the fusion protein is being transported to the periplasmic space. Furthermore, the normally cytoplasmic LacZ polypeptide may be inactivated in the periplasm because its cysteines are cross-linked by the periplasmic disulfide isomerases.

The fact that transport of the fusion protein inactivates its β-galactosidase activity, thereby preventing growth on lactose, offers a positive selection for *sec* mutants. Any mutation that prevents the secretion of the fusion protein into the membrane should make the cells Lac$^+$ and able to form colonies on lactose minimal plates, because the MalE-LacZ fusion protein should remain in the cytoplasm and retain its β-galactosidase activity. Therefore, to isolate *sec* mutants, these investigators could merely plate cells containing the *malE-lacZ* fusion on minimal plates containing lactose but no maltose. They identified six different *sec* genes this way and named them *secA*, *secB*, *secD*, *secE*, *secF*, and *secY*.

The function of the products of the *sec* genes in protein transport is the subject of current investigations. The product of the *secB* gene is a chaperone (see chapter 2) that probably acts by preventing the premature folding of membrane proteins before they can be translocated through the membrane (Figure 14.18). By binding to a protein while it is being synthesized, SecB keeps the signal sequence at the amino terminus exposed until the protein is ready to enter the inner membrane. In the absence of SecB, the signal sequence may be folded into the interior of the protein, where it is not available to guide the protein into the inner membrane. Somewhat surprisingly, the *secB* gene product is not absolutely essential for growth. Mutants with null mutations in *secB* can multiply to form colonies, especially in poor medium. Either some protein translocation occurs, even without SecB, or none of the proteins that require SecB for translocation are absolutely required for cell growth.

Another *sec* gene that has been extensively studied is *secA*. The SecA protein is an ATPase that binds to the inner surface of the inner membrane and probably binds proteins to be secreted into the membrane. Unlike SecB, the SecA protein is essential for growth and so null mutations in the *secA* gene are lethal.

A

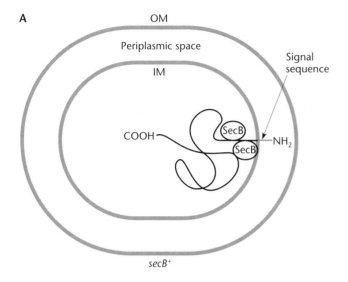

secB⁺

B

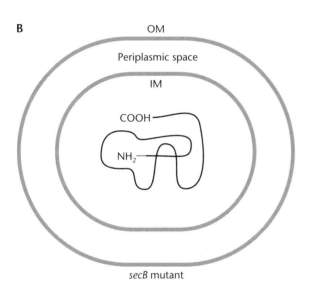

secB mutant

Figure 14.18 Function of SecB in protein transport. (A) The SecB protein (black) prevents premature folding of a protein, keeping the signal sequence (blue) exposed so that it can enter the membrane after protein synthesis is complete. OM, outer membrane; IM, inner membrane. (B) In a *secB* mutant, the signal sequence folds into the interior of the protein and so the protein cannot be transported.

Not all exported proteins or signal sequences require the *sec* genes to be translocated through the membranes of gram-negative bacteria. Some transported proteins have evolved their own specific transport systems rather than depending on the general secretory system used by most proteins (see Box 14.1). For this reason, the name "secreted" protein is often reserved for proteins that are transported by the general secretory system.

STRUCTURE AND TOPOLOGY OF TRANSMEMBRANE PROTEINS

Rather than being completely buried in the membrane, most membrane proteins have regions that are exposed to the cytoplasm, the periplasm, and/or the external environment. These exposed regions allow the membrane protein to pass information from the external environment to the interior of the cell or from one cellular component to another. Proteins that are exposed at both surfaces of a membrane are called **transmembrane proteins**, and the regions of the polypeptide that traverse the membrane from one surface to the other are called the **transmembrane domains**. We discussed transmembrane proteins in chapter 12 in connection with two-component sensor kinase response regulator systems in which the sensor kinase protein is exposed to both the periplasm and the cytoplasm so that it can communicate information from the periplasm to the response regulator protein in the cytoplasm, which then regulates gene expression.

Some transmembrane proteins traverse the membrane many times. The transmembrane domains alternate with those exposed to the cytoplasm (**cytoplasmic domains**) and the periplasm (**periplasmic domains**). The term **membrane topology** refers to how the different sections of the protein are distributed in the membrane.

Gene fusions to the alkaline phosphatase gene (*phoA*) of *E. coli* have been used to identify *E. coli* genes that encode transmembrane proteins and to study the membrane topology of inner membrane proteins of *E. coli*. The alkaline phosphatase product of the *phoA* gene is a "scavenger enzyme," which can obtain phosphates for cellular reactions from larger phosphate-containing molecules that cannot be easily transported into cells. To fulfill this role, the alkaline phosphatase of *E. coli* resides in the periplasmic space, where it can obtain phosphates from molecules that are too large to be easily transported. The alkaline phosphatase removes the phosphates for transport, and the rest of the molecule remains outside.

Alkaline phosphatase is active only in the periplasm and not in the cytoplasm, possibly because it is a homodimer in which the two identical polypeptide products of the *phoA* gene are held together by disulfide bonds between their cysteines. The enzyme must be in this dimer form to be active. The disulfide bonds may form only in the periplasm because, as mentioned above, the protein disulfide isomerase enzymes may be present only in the periplasm.

The fact that the PhoA protein is active only when it is transported into the periplasmic space has made the *phoA* gene very useful as a reporter gene (see Hoffman and Wright, Suggested Reading). For such experiments,

BOX 14.1

Specific Protein Translocation Systems

Some bacterial proteins are exported into the external medium. Examples include nucleases, amylases, and proteases involved in degrading long molecules so that they can be transported into the cell. Some toxins and other pathogenicity factors are also exported. For a protein to be exported into the medium from a gram-negative bacterium, it must be translocated through both the inner and outer membranes. Most exported proteins use the general transport system and have a signal sequence to get them through the inner membrane, but they must use special systems to get them through the outer membrane. For example, the cholera toxin CtxA exported by *Vibrio cholerae* has a signal sequence and probably uses the normal secretion system to get through the inner membrane but then needs to coat itself with a specific protein, CtxB, to get through the outer membrane. Once in the extracellular medium, it can stimulate the adenylate cyclase of animal cells, causing extreme diarrhea.

Some exported proteins have evolved their own specific systems to get through both the inner and outer membranes of gram-negative bacteria. Many of these secreted proteins are involved in pathogenicity in both plants and animals. Examples include the alpha-hemolysins of *Escherichia coli*, which cause dysentery; the adenylate cyclase/hemolysin of *Bordetella pertussis*, which causes whooping cough; and the metalloprotease system of the plant pathogen *Erwinia chry-*

santhemi. These exported proteins do not have signal sequences that are cleaved off; instead they may have sequences closer to their carboxy terminus that are recognized by the transport systems. The transport systems probably form some sort of channel from the cytoplasm through the inner membrane, periplasm, and outer membrane through which the protein can be exported. Although the transport systems are quite specific, sometimes the transport proteins from one system can substitute for the proteins of another system. There is a common structure to the transport systems in that two transport proteins are in the inner membrane and one is in the outer membrane.

References
Koronakis, V., E. Koronakis, and C. Hughes. 1989. Isolation and analysis of the C-terminal signal directing export of *Escherichia coli* hemolysin across both bacterial membranes. *EMBO J.* **8:**545–605.

Sauvonnet, N., and A. P. Pugsley. 1996. Identification of two regions of *Klebsiella oxytoca* pullanase that together are capable of promoting β-lactamase secretion by the general secretory pathway. *Mol. Microbiol.* **22:**1–7

Thomas, W. D., Jr., S. P. Wagner, and R. A. Welch. 1992. A heterologous membrane protein domain fused to the C-terminal ATP-binding domain of HylB can export *Escherichia coli* hemolysin. *J. Bacteriol.* **174:**6771–6779.

Wandersman, C., and P. Delepelaire. 1990. TolC, an Escherichia coli outer membrane protein required for hemolysin secretion. *Proc. Natl. Acad. Sci. USA* **87:**4776–4780.

the carboxy terminal part of PhoA without its signal sequence is fused to the N terminus of another protein. Only cells in which the signal sequence of the other protein transports the PhoA portion of the hybrid protein to the periplasmic space will have alkaline phosphatase activity.

Other features have made the *phoA* gene of *E. coli* a convenient reporter gene. The PhoA enzyme is easily assayed by using the chromogenic compound 5-bromo-4-chloro-3-indolylphosphate (XP), which turns blue when phosphate is removed by alkaline phosphatase. Furthermore, the alkaline phosphatase is an unusually stable protein and is active even when fused to other proteins, as long as the PhoA portion of the fusion protein can dimerize.

Identifying Genes for Transported Proteins Using *phoA* Fusions

Some transposons have been engineered to generate random gene fusions by transposon mutagenesis (see chapter 16). These transposons contain a reporter gene that

will be expressed only if the transposon hops into an expressed gene in the correct orientation. One such transposon, Tn*phoA*, was developed to identify genes whose protein products are transported into or through the inner membrane (see Gutierrez et al., Suggested Reading). Tn*phoA* has the *phoA* reporter gene inserted so that it does not have its own promoter or translation initiation region and lacks its own signal sequence coding region. A fusion protein with PhoA fused to another protein will be synthesized whenever the transposon hops into an expressed open reading frame in the right reading frame, but the fusion protein will have alkaline phosphatase activity only when Tn*phoA* integrates into a gene whose protein product is translocated and so can furnish a signal sequence to transport the fusion protein.

So that the genes for secreted proteins could be identified, cells were mutagenized by Tn*phoA* and the transposon insertion mutants were plated on XP agar. Mutant bacteria that form blue colonies under these conditions have the transposon in a gene whose protein product is normally secreted into or through the inner membrane. The gene can then be mapped and/or cloned using methods discussed in chapters 15 and 16.

Use of *phoA* Fusions To Study the Topology of Inner Membrane Proteins

The transmembrane domains of a polypeptide can often be distinguished from the periplasmic or cytoplasmic domains merely by the primary sequence of the gene. Because the transmembrane domains are embedded in the hydrophobic membrane, they are composed mostly of hydrophobic amino acids such as phenylalanine and tyrosine, which makes them more soluble in hydrophobic environments. The periplasmic and cytoplasmic domains have a larger number of charged amino acids, such as arginine, lysine, or glutamate, which make them more soluble in the ionic environments outside the membrane.

Translational fusions to the *phoA* gene have been used to determine the membrane topology of transmembrane proteins (see San Milan et al., Suggested Reading) and hence discover which regions may be in contact with the periplasm and, by extension, the external environment and which regions could be in contact with other proteins in the cytoplasm.

For this analysis, translational fusions are made between the *phoA* gene and different portions of the transmembrane protein gene, resulting in fusion proteins composed of PhoA and different lengths of the N terminus of the transmembrane protein. Whether the fusion protein has alkaline phosphatase activity will depend on whether the domain of the transmembrane protein to which PhoA is fused is in the periplasm or the cytoplasm.

In Figure 14.19, the transmembrane protein traverses the inner membrane twice, with domains x and z in the periplasm and domain y in the cytoplasm. Because the PhoA part of the fusion protein will be active only if it is in the periplasm, cells in which the *phoA* gene was fused to the coding region for regions x or z of the transmembrane protein will make blue colonies on XP plates. However, cells in which the N terminus ending in region y is fused to PhoA will make colorless colonies.

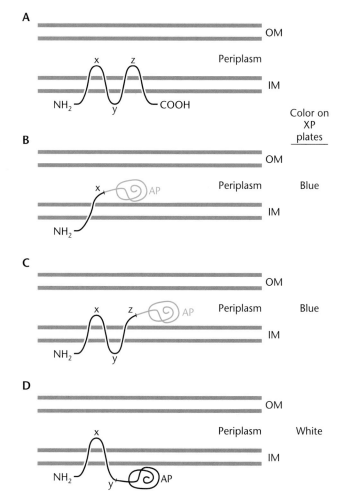

Figure 14.19 Using *phoA* fusions to determine the membrane topology of a transmembrane protein. (A) A transmembrane protein. (B and C) A fusion that joins the transmembrane protein to alkaline phosphatase (AP) at x or z leaves the alkaline phosphatase in the periplasm where it will be active. The bacteria form blue colonies on XP plates. (D) The transmembrane protein is fused at y to alkaline phosphatase, leaving the alkaline phosphatase in the cytoplasm, where it is inactive. The bacteria form white colonies on XP plates.

SUMMARY

1. Genetic analysis of bacteria has been a powerful tool to aid the understanding of many complex biological phenomena. The possibility of doing selectional genetics with single-celled organisms has made bacteria particularly amenable to genetic analysis. Also, the fact that bacteria are haploid greatly facilitates the selection of mutants. However, the mechanism of genetic exchange in bacteria affects how genetic data are interpreted.

2. Any genetic analysis depends on the existence of genetic markers in which the sequence is changed and the pheno-

(continued)

SUMMARY (continued)

type is different from the normal or wild type. Useful markers can be due to mutations, transposon and other types of insertions, or even natural strain differences.

3. Mutations can be either spontaneous or induced. To increase the frequency of spontaneous mutants, cultures of cells can be transferred a few times to increase the frequency of all types of mutants in the population. Mutagenesis has the advantages of increasing the frequency of mutants and offering better control over the type of mutations because different types of mutagens cause different types of mutations.

4. Mutants can be detected by either positive or negative selection. In positive selection, the mutant, but not the original bacterium, will multiply to form a colony. In negative selection, the mutant cannot multiply whereas the original type can multiply. Procedures such as replica plating have been devised to facilitate the detection of mutants by negative selection.

5. Enrichment procedures can be used to increase the frequency of some types of mutants when only negative selection is possible. Enrichments depend on antibiotics, such as ampicillin or bromodeoxyuridine, that preferentially kill growing cells .

6. Genetic mapping in bacteria can be done by conjugation, transduction, or transformation. How the data are analyzed depends on the method used. In a bacterial cross, only a small region of the DNA is transferred from one bacterium (the donor) to the other (the recipient). Therefore, only the recipient can become recombinant.

7. Because recombinants are usually rare in bacterial crosses, first one marker is selected (the selected marker) and then the recombinants are tested for the other markers (the unselected markers).

8. Complementation tests in bacteria are complicated by the fact that bacteria are usually haploid. It is therefore nec-

essary to make partial diploids of the region being studied. Partial diploids can be made by using prime factors, specialized transducing phages, or plasmid cloning vectors to introduce cloned copies of the region.

9. Proteins that are located in the inner or outer membranes or periplasmic space of the cell must be transported through the membranes after synthesis to reach their final destination and be active. Fusions to the *mal* genes of *E. coli* have been useful for determining the genetic requirements for protein transport.

10. Transported proteins usually have short sequences of about 25 amino acids, called signal sequences, at their N termini; these sequences insert first into the inner membrane as the polypeptide is synthesized. The signal sequence is removed by specific peptidases as the polypeptide is translocated through the membrane.

11. The products of the *sec* genes are proteins that help transport other proteins through membranes. Some are chaperones that prevent the premature folding of transported proteins.

12. The Tn*phoA* transposon has been used to identify genes whose protein products are transported. The *phoA* gene on the transposon lacks its own signal sequence, so that the PhoA enzyme will be transported to the periplasmic space only if it is fused to a transported protein.

13. The membrane topology of a protein refers to how it is distributed in the membrane. Transmembrane proteins traverse the inner membrane one or more times, with some domains in the cytoplasm and some domains in the periplasm. The sections of the polypeptide chain that lie in the membrane are the transmembrane domains. Because the PhoA portion of a fusion protein will be active only if it is in the periplasmic space, translational fusions to the alkaline phosphatase *phoA* gene of *E. coli* have been useful for analyzing membrane protein topology in *E. coli*.

PROBLEMS

1. List in your own words the advantages and disadvantages of bacteria for genetic analysis.

2. Why is it necessary to isolate mutants from different cultures to be certain of getting independent mutations?

3. In a collection of cultures, all started from a few wild-type bacteria, which cultures are most likely to have the most independent mutations, those with a few mutant bacteria or those with an exceptionally large number of mutant bacteria? Why?

4. Design a positive selection for each of the following types of mutants. Discuss what kind of selective plates and/or conditions you would use.

a. Mutants resistant to the antibiotic coumermycin.

b. Revertants of a *trp* mutation that makes cells require tryptophan.

c. Double mutants with a suppressor mutation that relieves the temperature sensitivity due to another mutation in *dnaA*.

d. Mutants with a suppressor in *araA* that relieves the sensitivity to arabinose due to a mutation in *araD*.

e. Mutants with a mutation in *suA* (the gene for the transcription terminator protein Rho) that relieves the polarity of a *hisC* mutation on the *hisB* gene. Hint: make a partial diploid and do complementation.

5. Design an enrichment procedure for reversible, temperature-sensitive mutants with mutations in genes whose products are required for cell growth.

6. You have crossed the *E. coli* Hfr *hisB4 recA1* strain with a *thyA8* strain and plated it on minimal plates with no growth supplements. Almost 80% of the *thy*⁺ recombinants are very UV sensitive. Where is the *recA* marker located in the chromosome?

7. You have mutagenized *E. coli metB1* (90 min) *serA5* (63 min) *strA7* (73 min) with the transposon Tn*5*, and you want to know where the transposon has inserted. You cross the mutant strain with an Hfr strain that is streptomycin sensitive and that transfers counterclockwise from 0 min (Figure 14.4) and has a *hisG2* mutation (44 min) that makes it require histidine. After incubation, you plate the cells on minimal plates plus tryptophan, histidine, and streptomycin to select the *metB* marker and counterselect the donor. After purifying 100 of the Met⁺ transconjugants, you test them for the other markers. You find that 85 of them are His⁺, 33 are Ser⁺, and 80 are kanamycin resistant. Where is the transposon inserted?

8. You wish to determine if the order of three markers is *metB1–argH5–rif-8* or *argH5–metB1–rif-8*, so you do a three-factor cross. The donor for the transduction has the *rif-8* mutation, and the recipient has the *metB1* and *argH5* mutations. You select the *argH5* marker by plating on minimal plates plus methionine, purify 100 of the Arg⁺ transductants, and test for the other markers. You find that 17 are Arg⁺ Met⁺ Rif ʳ, 20 are Arg⁺ Met⁺ Rifˢ, 60 are Arg⁺ Met⁻ Rifˢ, and 3 are Arg⁺ Met⁻

Rifʳ. What is the cotransduction frequency of the *argH* and *metB* markers? Of the *argH* and *rif-8* markers? What is the order of the three markers deduced from the three-factor cross data? Are the results consistent?

9. You have isolated a His⁻ mutant of *E. coli* that you suspect has a nonsense mutation because it is suppressed by intergenic suppressors. You want to map one of the suppressors. To do this, you use an Hfr strain that has the original *his* mutation and transfers clockwise from 30 min on the *E. coli* map. As recipient, you use a strain that has the suppressor as well as the *his* mutation and an *argG* mutation. You find that 80% of the Arg⁺ recombinants are His⁻. Where is the suppressor mutation located?

10. Are nonsense suppressor mutations dominant or recessive? Why?

11. You have isolated a constitutive mutant of the *trp* operon of *E. coli*. When you introduce the wild-type *trpR* repressor gene on an F′ factor into cells with the mutant *trpR* gene, it is still constitutive. Describe how you would determine if it has a *cis*-acting operator mutation or if it has a dominant repressor *trpRᵈ* mutation.

12. The product of the *envZ* gene is a transmembrane protein in the inner membrane of *E. coli*. Describe how you would determine which of the regions of the EnvZ protein are in the periplasm and which are in the cytoplasm.

13. Outline how you would isolate suppressors of mutations in the signal sequence coding region of the *malE* gene.

SUGGESTED READING

Anderson, R. P., C. G. Miller, and J. R. Roth. 1976. Tandem duplications of the histidine operon observed following generalized transduction in *Salmonella typhimurium*. *J. Mol. Biol.* 105:201–218.

Bagg, A., C. J. Kenyon, and G. C. Walker. 1981. Inducibility of a gene product required for UV and chemical mutagenesis in *Escherichia coli*. *Proc. Natl. Acad. Sci. USA* 78:5749–5753.

Bassford, P., and J. Beckwith. 1979. *Escherichia coli* mutants accumulating the precursor of a secreted protein in the cytoplasm. *Nature* (London) 277:538–541.

Friedman, D. I., M. F. Baumann, and L. S. Baron. 1976. Cooperative effects of bacterial mutations affecting λ N gene expression. *Virology* 73:119–127.

Gutierrez, C., J. Barondess, C. Manoil, and J. Beckwith. 1987. The use of transposon Tn*phoA* to detect genes for cell envelope proteins subject to a common regulatory stimulus. *J. Mol. Biol.* 195:289–297.

Hill, C. W., J. Foulds, L. Soll, and P. Berg. 1969. Instability of a missense suppressor resulting from a duplication of genetic material. *J. Mol. Biol.* 39:563–581.

Hoffman, C. S., and A. Wright. 1985. Fusions of secreted proteins to alkaline phosphatase: an approach for studying protein secretion. *Proc. Natl. Acad. Sci. USA* 82:5107–5111.

Kato, T., and Y. Shinoura. 1977. Isolation and characterization of mutants of *Escherichia coli* deficient in induction of mutations by ultraviolet light. *Mol. Gen. Genet.* 156:121–131.

Low, K. B. 1991. Conjugational methods for mapping with Hfr and F-prime strains. *Methods Enzymol.* 204:43–62.

Low, K. B. 1996. Genetic mapping, p. 2511–2517. *In* F. C. Neidhardt, R. Curtiss III, J. L. Ingraham, E. C. C. Lin, K. B. Low, B. Magasanik, W. S. Reznikoff, M. Riley, M. Schaechter, and H. E. Umbarger (ed.), *Escherichia coli and Salmonella: Cellular and Molecular Biology*, 2nd ed. ASM Press, Washington, D.C.

Miller, J. H. 1992. *A Short Course in Bacterial Genetics*. Cold Spring Harbor Laboratory, Cold Spring Harbor, N.Y.

Miller, J. H., and U. Schmeissner. 1979. Genetic studies of the *lac* repressor X. Analysis of missense mutations in the *lacI* gene. *J. Mol. Biol.* 131:223–248.

Oliver, D. B., and J. Beckwith. 1981. *E. coli* mutant pleiotropically defective in the export of secreted proteins. *Cell* 25:2765–2772.

Pugsley, A. P. 1993. The complete general secretory pathway in gram-negative bacteria. *Microbiol. Rev.* 57:50–108.

San Milan, J. L., D. Boyd, R. Dalbey, W. Wickner, and J. Beckwith. 1989. Use of *phoA* fusions to study the topology of the *Escherichia coli* inner membrane protein leader peptidase. *J. Bacteriol.* **171**:5536–5541.

Schatz, G., and B. Dobberstein. 1996. Common principles of protein translocation across membranes. *Science* **271**:1519–1526.

Schatz, P. J., and J. Beckwith. 1990. Genetic analysis of protein export in *Escherichia coli. Annu. Rev. Genet.* **24**:215–248.

Schmeissner, U., D. Ganem, and J. H. Miller. 1977. Genetic studies of the *lac* repressor II. Fine structure deletion map of the *lacI* gene, and its correlation with the physical map. *J. Mol. Biol.* **109**:303–326.

Sommer, S., K. Knezevic, A. Bailone, and R. Devoret. 1993. Induction of only one SOS operon, *umuDC,* is required for SOS mutagenesis in *E. coli. Mol. Gen. Genet.* **239**:137–144.

Yanofsky, C., D. R. Helinski, and B. D. Maling. 1961. The effects of mutation on the composition and properties of the A protein of *Escherichia coli* tryptophan synthetase. *Cold Spring Harbor Symp. Quant. Biol.* **26**:11–23.

Recombinant DNA Techniques and Cloning Bacterial Genes

E XTENSIVE KNOWLEGE, METICULOUSLY ASSEMBLED, about macromolecular synthesis and other functions in bacterial cells has made possible the techniques of molecular genetics and recombinant DNA, two premier examples of the value of basic research. Bacterial molecular genetics was one of the cornerstones of these developments, and the molecular genetic and recombinant DNA techniques orginally developed to study bacteria and their phages are now being used for many applications in biology and medicine, including the analysis of gene and chromosome structure in higher organisms, the cloning of genes from humans and plants to synthesize proteins of medical and biotechnological importance, and the identification of genes responsible for many human genetic diseases, to give just a few examples.

Most books on molecular genetics emphasize the applications of these techniques to the study of eukaryotic cells. However, molecular genetics and recombinant DNA can also be used to study bacterial cells. In some respects, bacterial analysis by these methods requires special techniques. However, before discussing the special applications, we review some basic methods. Our treatment is by no means intended to be comprehensive; entire books have been written on these topics. The examples chosen here are only intended to give the genetic and biological background for some of the methods. For more comprehensive treatments and for specific protocols, we refer the reader to manuals containing the details of recombinant DNA techniques, commonly known as cloning manuals. Two popular cloning manuals are listed in Suggested Reading at the end of this chapter.

Restriction Endonucleases and Recombinant DNA

Modern molecular biology and recombinant DNA technology would not have been possible without the discovery of restriction endonucleases in bacteria. Some of their uses are discussed in this chapter, and others are mentioned in Box 15.1.

Like many important developments in science, the discovery of restriction endonucleases came from some seemingly humble observations (see Linn and Arber, Suggested Reading, for a review). Phage propagated on one strain of *Escherichia coli* could not subsequently multiply on a second strain of *E. coli*. However, a very few phage did survive, and these could multiply normally in the second strain. Yet if they were again propagated in the first strain, the phage would again lose the ability to multiply in the second strain. This phenome-

non obviously does not result from mutations in the phage DNA that allow it to multiply in the second strain. The ability to grow on the second strain is not permanently inheritable, since it can be lost when the phage are propagated on the first strain. Rather, a reversible modification allows the phage to multiply in the second strain. Because it is only the DNA of phages that generally enters the infected cell, the modification was surmised to occur in the phage DNA. We now know that the phage DNA is being methylated at certain sequences, which prevents restriction endonucleases from cutting at, or close to, those sequences. The methylation enzymes and restriction endonucleases that do the cutting were together named **restriction modification systems.** Restriction modification systems have been discovered in most types of bacteria, and similar enzymes have been found in some lower eukaryotes.

BOX 15.1

Applications of Restriction Endonucleases and Methyltransferases

Restriction modification systems have many uses in addition to physical mapping and cloning. Some of these applications take advantage of the methylating function of restriction modification systems. The methyltransferases of restriction systems usually methylate the 5-carbon of the cytosine ring, the 4-nitrogen of the cytosine ring, or the 6-nitrogen of the adenine ring, thereby protecting the DNA from cutting by the restriction component.

Methyltransferases are useful when two restriction systems that cut at the same sequence (called isoschizomers) protect the DNA by methylating different bases in this sequence. For example, the isoschizomers HpaII and MspI both cut in the sequence CCGG/GGCC, but the MspI methyltransferase protects the DNA by methylating the first C in the sequence whereas the HpaII methyltransferase methylates the second C. As a consequence, MspI but not HpaII will cut DNA if the second C in the sequence CCGG is methylated.

This difference between the two enzymes has been exploited in the analysis of the methylation patterns of the DNA of higher eukaryotes, in which there is a high correlation between the state of methylation of a gene and the activity of the gene in a particular tissue. In higher eukaryotes, most of the 5-methylcytosine in DNA is in the sequence CG. If the DNA of a region or gene is cut by MspI but not by HpaII, the DNA is methylated.

Most restriction endonucleases will not cut DNA if their recognition sequence is methylated. However, some types of

restriction endonucleases will cut their recognition sequence only if it *is* methylated. Examples of these enzymes, called methyl-dependent restriction endonucleases, include DpnI from *Streptococcus* spp. and the *mcrA, mcrB, mcrC,* and *mrr* gene products of *E. coli* K-12. Note that methyl-dependent restriction endonucleases do not have a corresponding methyltransferase function like other restriction enzymes. The DNA would be cut and the cell killed as a result of methylation of the recognition sequence.

Methyl-dependent restriction endonucleases have been used to study the fate of DNA in vivo. For example, DpnI can be used to determine if transforming DNA has entered a cell and replicated. The A's in the sequence GATC/CATG are methylated in *E. coli* by the Dam methylase (see chapters 3 and 13) and must be methylated before DpnI will cut this sequence. Therefore, if DNA prepared in *E. coli* is used to transform another type of cell that lacks a Dam methylase, whether the DNA can be cut by DpnI depends on whether the transforming DNA has replicated. Newly replicated DNA will not be methylated and so cannot be cut. Thus, we can determine what percentage of the DNA has entered the cell and replicated.

Reference

Wilson, G. G., and N. E. Murray. 1991. Restriction and modification systems. *Annu. Rev. Genet.* 25:585–627.

Biological Role of Restriction Modification Systems

Restriction modification systems have proven very useful in molecular genetics, but why do bacteria make them? They presumably did not evolve so that humans could someday make use of them! They must confer some advantage to the bacterium. One probable use is to prevent infection by phages, since incoming phage DNA will be cut if it is not properly modified. Many phages have evolved antirestriction systems to help them overcome host restriction modification systems. However, other restriction modification systems do not appear to function in phage restriction. Some (eight-hitters [see below]) cut DNA so infrequently that they will not cut the DNA of most phages even once.

Some restriction modification systems may serve as addiction systems (see chapter 4, Box 4.3). Addiction systems encoded by some plasmids will kill host cells if they are cured of the plasmids. A restriction modification system could serve this purpose if its methylating activity is less stable than its cutting activity. If the cell is cured of the genes for the restriction modification system so that no more can be synthesized, the cutting activity will outlast the methylating activity. Newly replicated chromosomal DNA will then be unmodified and so will be cut by the cutting activity, and the cell will die. Most probably, no single reason can explain the existence of all restriction modification systems, and different types of restriction modification systems may play different roles in the cell.

Types of Restriction Modification Systems

The restriction modification systems have been divided into three general groups, types I, II, and III.

TYPE I

The type I restriction modification systems were the first to be discovered. The type I enzymes are composed of three subunits: HsdS, which recognizes a specific DNA sequence; HsdM, which methylates the DNA; and HsdR, the restriction endonuclease that cuts the DNA. Each of these subunits will function only when all three are together in a complex, so that the restriction endonuclease activity will not cut DNA unless the methylating activity is also present. Moreover, the methylation and cutting activities of type I enzymes are closely coupled; the cutting reaction requires both ATP and S-adenosylmethionine, the donor of methyl groups, both of which are also required for the methylation reaction. Consequently, DNA will often be methylated by a type I restriction modification system rather than be cut, so that not all the sites will be cut. Furthermore, the recog-

nition sequences are quite long, with no recognizable features such as symmetry, and the enzyme usually does not cut inside the recognition sequence but, rather, some distance from it. These and other properties limit the usefulness of type I enzymes in molecular biology and recombinant DNA technology (see below).

TYPE III

The type III restriction modification systems are closely related to the type I enzymes. They also have limited usefulness, for many of the same reasons. Like type I enzymes, type III enzymes are also composed of a complex, although this complex consists of only two subunits, HsdM and HsdR. The cleavage reaction also requires S-adenosylmethionine. Type III enzymes recognize a symmetric sequence in the DNA and cleave approximately 5 bp to one side of this sequence.

TYPE II

Most of the useful restriction modification systems are of type II. Type II enzymes are much simpler than the others, and the cutting reaction does not have any special requirements. These enzymes have both an HsdM modification and an HsdR restriction endonuclease subunit, but the two subunits work separately, so that the restriction endonuclease by itself can cut the DNA. Most importantly, however, type II enzymes recognize a defined, usually symmetric, sequence and cut either within the recognition site or a defined distance from it. Many of them also make a staggered break in the DNA, as we discuss later. Table 15.1 lists the sequences recognized by the restriction endonucleases of some type II systems and the places in the recognition site where the cuts occur.

Four-Hitters, Six-Hitters, and Eight-Hitters

Type II restriction endonucleases are often grouped by the length of the sequence they recognize. For example, Sau3A1 is a **four-hitter** because it recognizes a 4-bp sequence (Table 15.1), whereas EcoRI, BamHI, and PstI

TABLE 15.1	Recognition sequences of restriction endonucleases
Enzyme	Recognition sequence[a]
Sau3A	*GATC/CTAG*
BamHI	G*GATCC/CCTAG*G
EcoRI	G*AATTC/CTTAA*G
PstI	CTGCA*G/G*ACGTC
HindIII	A*AGCTT/TTCGA*A
SmaI	CCC*GGG/GGG*CCC
NotI	GC*GGCCGC/CGCCGG*CG

[a]Asterisks indicate where the endonucleases cut.

are **six-hitters** and *Not*I is an **eight-hitter**. In general, four-hitters will cut a DNA into smaller pieces than six-hitters will. The reason is simple. By chance alone, the shorter recognition sequences will, in general, occur more often in the DNA than the longer recognition sequences will.

The number of times, on average, that a restriction endonuclease with a particular length recognition sequence should cut a DNA is quite easy to calculate. Since four bases (G, C, A, or T) are possible at each position in the recognition sequence, the recognition site for a four-hitter restriction endonuclease should occur approximately every $4 \times 4 \times 4 \times 4 = 256$ bp; a six-hitter site should occur about every $4 \times 4 \times 4 \times 4 \times 4 \times 4 = 4{,}096$ bp; and sites for an eight-hitter should occur only once every $4 \times 4 \times 4 \times 4 \times 4 \times 4 \times 4 \times 4 = 65{,}536$ bp, or only a few times in an entire bacterial genome.

Although the simple calculation above gives an estimate of how often particular restriction sites should occur in DNA, this value is only the *average* interval between recognition sites for a particular restriction endonuclease. Even with a four-hitter, sites can be, by chance, thousands of base pairs apart or only a few base pairs apart. Also, the recognition sequences for a given restriction endonuclease may occur more often or less often than calculated because these sequences may play special roles in the DNA or happen to occur on an IS element, the coding sequence for an rRNA, or another sequence that is repeated many times in the genome. Furthermore, the frequency of sites depends on the base composition of both the recognition sequence and the DNA being cut. In general, if the frequency of GC and AT base pairs in the recognition sequence and the overall base pair composition are approximately equal, the recognition sequences will occur more often than if they are very different. For example, the sequence CCGG/GGCC, which is recognized by the four-hitter *Hpa*II, will occur more frequently in GC-rich DNAs such as those from the genera *Streptomyces* and *Pseudomonas* than in AT-rich DNAs like that from phage T4.

Physical Mapping with Restriction Endonucleases

Because restriction endonucleases cut only at or near specific sequences in DNA, they can be used to obtain a physical map of the DNA of an organism. A **physical map** of DNA shows the actual distance in base pairs between different positions in the molecule, in contrast to the genetic linkage maps shown in chapters 13 and 14, which show only approximate and relative distances, based on recombination frequencies, between genetic markers.

The first physical map to be constructed by using restriction endonucleases was that of the animal virus simian virus 40 (SV40) (see Danna and Nathans, Suggested Reading). Physical maps for many bacteria, viruses, and lower eukaryotes have now been constructed, and maps for some higher eukaryotes, including humans, are under construction.

The size of **restriction fragments**, or the fragments of DNA left after cutting by a particular restriction endonuclease, form the basis of physical maps. The distance between sites for a particular restriction endonuclease determines the size of the restriction fragments—the farther apart the recognition sequences in a particular region, the larger the restriction fragment from that region.

We determine the size of restriction fragments by a method called **gel electrophoresis**. The DNAs are applied to a gel made of agarose or acrylamide, and the gel is exposed to an electric field. Because they are negatively charged, the DNA fragments will migrate toward the positive electrode, with the smaller fragments migrating faster than the larger fragments. We determine the size of a restriction fragment by comparing the distance it has moved in a given time with the distance moved by fragments of known size on the same gel.

Figure 15.1 shows how such a gel will look. The distance moved by each fragment is an exponential function of its size, so that if we plot distance moved by the fragments against their size on semilog paper, we obtain a more or less straight line for the marker fragments. The size of the restriction fragments can then be estimated from their position on that line. We shall return to a more detailed description of this method, with actual examples, in the next chapter.

MAPPING RESTRICTION SITES

The size of the restriction fragments alone will not reveal the position of the recognition sites for a given restriction endonuclease but will reveal only their distance from each other on the DNA molecule. Additional steps are required to determine the position of the sites along the DNA.

One method for determining the position of the sites is to use two different restriction endonucleases and to determine the position of the recognition sites for each of them relative to the other. Figure 15.2 illustrates the principles behind this method. The DNA is cut with one of the restriction endonucleases, the fragments are separated by gel electrophoresis, and their sizes are determined. Each of the fragments is then removed from the gel and cut with the other restriction endonuclease. Finally, the sizes of these fragments are determined. This

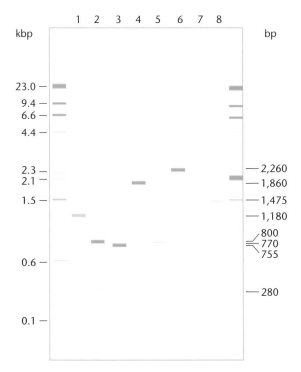

Figure 15.1 Agarose gel electrophoresis of fragments of DNA. Smaller fragments will migrate faster on the gel so will move farther in the same amount of time. The outside lanes are marker DNAs of known size for comparison. Adapted from E. L. Rosey, M. J. Kennedy, D. K. Petrella, R. G. Ulrich, and R. J. Yancey, Jr., *J. Bacteriol.* **177:**5959–5970, 1995.

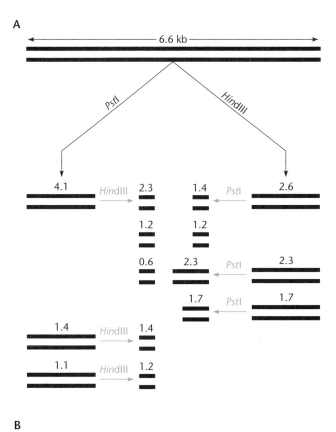

Figure 15.2 A method for mapping the restriction sites on a DNA fragment. The original fragment of 6.6 kb (6.6 × 10^3 bp) contains two recognition sites for the restriction endonuclease *Hin*dIII and two sites for *Pst*I. (A) The fragment is digested with *Hin*dIII and *Pst*I separately, and the isolated fragments are digested with the other restriction endonuclease (in blue). (B) From the size of the fragments, the order of the sites must have been as shown.

process is repeated by starting with the other restriction endonuclease. As shown in Figure 15.2, we can determine which fragments overlap in the DNA by comparing the sizes of the final fragments produced after cutting with both enzymes. By aligning all the adjacent fragments, we obtain a physical map showing the positions of the recognition sites for the two enzymes.

Southern Blot Hybridizations To Map Restriction Sites

Many phage and plasmid DNAs are short enough that a limited number of restriction fragments can be obtained when they are cut with a six-hitter restriction endonuclease. With such relatively simple DNAs, physical maps can be constructed by using six-hitter restriction endonucleases and methods such as the one discussed above. However, bacterial and eukaryotic DNAs are much longer than virus and plasmid DNAs, and cutting with a six-hitter will leave thousands of restriction fragments, too many to separate by gel electrophoresis. For physical maps of regions of such long DNAs, methods such as **Southern blot hybridization** are necessary. This

method is named after its discoverer, Ed Southern, and related methods were given nonsensical names—Northern blots, Western blots, and so on. Southern blots allow the detection of restriction fragments from a certain DNA region in the presence of large numbers of fragments from other regions that would otherwise obscure them.

Figure 15.3 shows how a Southern blot hybridization is performed, and Figure 15.4 shows some typical data. The DNA is isolated from the cells and digested with restriction endonucleases before being applied to an agarose gel and subjected to an electric field, as before. The DNA on the gel is then denatured (the strands of the DNA are separated), and the single strands are

Step 1   DNA isolation and digestion

Step 2   DNA electrophoresis

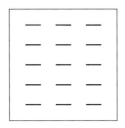

Step 3   DNA transfer

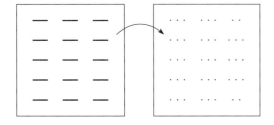

Step 4   DNA hybridrization

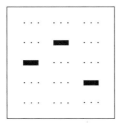

Figure 15.3 Method of Southern blot hybridization. In step 1, DNA is isolated and digested with a restriction endonuclease. In steps 2 and 3, after electrophoresis (step 2), the DNA is transferred and fixed to a filter (step 3). In step 4, the filter is hybridized with a probe. Only bands complementary to the probe appear as dark bands because the signal detection procedure reveals the radioactivity or reactive chemical in the probe.

transferred to a membrane, called a **blot,** where they are fixed by drying or UV irradiation. Only single-stranded DNA will stick to the filter or hybridize to other DNAs, which becomes important in our discussion later.

Once the blot is made, a small fragment of the DNA, called the **probe,** is added to it. The probe has been labeled either with radioisotopes or by a chemical method so that it can be detected after the hybridization. Each band that contains DNA complementary to the probe

will hybridize with it. These bands can then be identified by autoradiography or other methods, depending on how the probe was labeled. For a detailed protocol for Southern blot hybridizations, consult one of the cloning manuals.

Southern blot hybridizations can be used to determine which restriction fragments overlapped each other. A DNA fragment obtained with one restriction endonuclease is labeled and used as a probe on a filter blot of a gel obtained by digesting the DNA with a different restriction endonuclease. Only those fragments that overlap in the DNA will have complementary sequences and so will hybridize, provided that the probe does not contain sequences that are repeated elsewhere in the DNA. Southern blot hybridizations have many other uses in molecular genetics, and we shall refer to this method often.

Ordering the Clones in a Library
Another physical mapping method is to make a library of the DNA of the organism by using partial digests (see below for what is meant by a library). Then the clones in the library can be ordered by Southern blot hybridizations to determine which clones overlap. Once the clones have been put in order, the restriction sites within each of the much smaller clones can be mapped as before, since the clones are much smaller than the bacterial DNA itself. When the process is completed, the sites for some restriction endonucleases will have been mapped for the entire genome of the organism.

Correlating Genetic and Physical Maps
We often want to know where genes located by genetic methods lie on physical maps. One way to correlate genetic and physical maps is to map the sites of mutations with respect to known restriction sites. If the site of a mutation that inactivates a gene can be ascertained on the physical map, the position of that gene will be known. We can then locate the exact position of the gene on the map by correlating open reading frames (ORFs) identified by DNA sequencing with protein products, as discussed in chapter 2.

Some mutations make relatively large changes in the physical map. For example, those due to insertion of a transposon or a deletion or inversion are relatively easy to locate because they change the location of restriction sites (see chapter 16 for an example). However, it is usually difficult to determine the position of a point mutation such as a base pair change by physical methods because point mutations do not change the size of restriction fragments. For this purpose, other methods such as complementation and marker rescue

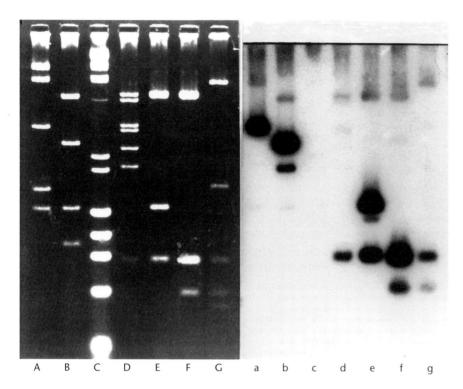

Figure 15.4 Results of a Southern blot hybridization. Lanes A to G show the total DNA in each lane. Lanes a to g show the bands that hybridize to a specific probe. Markers are shown in lane C. Reprinted from C. Kao and L. Snyder, *J. Bacteriol.* **170**:2056–2062, 1988.

must be used in combination with physical mapping (see chapter 16).

Restriction Fragment Length Polymorphisms

Although individuals within a species have basically the same genes arranged in basically the same genetic map, there are usually subtle differences in the sequence of deoxynucleotides in their DNA. These differences are referred to as **polymorphisms**, and in genetics this term usually refers to the many forms that a gene can take within a species. People have taken advantage of polymorphisms in breeding programs for plants and animals since prehistoric times. By breeding two individuals from different strains of the same species, the breeder makes a new strain, perhaps with desirable properties.

Because the different forms of a gene reflect differences in the DNA sequence, the positions of recognition sites for restriction endonucleases may vary. Between members of the same species, these variations are called **restriction fragment length polymorphisms (RFLPs)**. Generally, Southern blot hybridizations are used to detect RFLPs within a certain region of the DNA, with DNA from that region serving as the probe.

Because physical maps of restriction sites can be used to distinguish different members of the same species, this technique has many current applications, such as the unambiguous identification of the source of a DNA molecule. For example, when a particular gene in two

parent organisms differs in the location of some restriction sites, the restriction pattern of DNA from the progeny of a cross can tell us from which parent the progeny received a particular gene. This fact has been important for tracking genetic diseases and mapping the sites of mutations causing those diseases. Evidence derived from restriction fragment length polymorphisms is also now being used to identify individuals in legal actions, including paternity suits and criminal trials. Polymorphisms in DNA sequences between individuals of the same species can also be identified by techniques such as the polymerase chain reaction (PCR; see below).

Joining DNA Fragments Cut with Restriction Endonucleases

Recombinant DNA technology has been one of the major technological advances in the latter part of this century. In this process, two different DNAs are joined in the test tube, in analogy to what happens in the living cell during genetic recombination. The properties of some of the type II restriction endonucleases have made them particularly useful for making recombinant DNA.

STICKY ENDS

Some type II restriction endonucleases leave single-stranded ends after cutting (see above), a property that makes them quite useful in recombinant DNA applications. These nucleases leave single-stranded ends because they cut in specific **symmetric sequences**, which read the same in the 5'-to-3' direction on both strands, and they cut between the same two bases in the sequence on each strand. When the two deoxynucleotides between which the restriction endonuclease cuts are not exactly in the middle of the recognition sequence, the resulting break will be staggered, leaving single-stranded, complementary ends (Figure 15.5). Because the endonuclease always cuts at the same sequence, the single-stranded end created at one site will be complementary to those created at any other site by the same enzyme, as illustrated in Figure 15.5. Because these ends can base pair with each other and hold the ends together, they are called **sticky ends**. Two DNA molecules held together at their sticky ends can then be joined by DNA ligase to form continuous double-stranded DNA (Figure 15.5). If the two joined DNA molecules were

not adjacent in the original molecule because they came from different regions or different sources, the DNA will be recombinant DNA.

5' and 3' Overhangs
Depending on where the restriction endonuclease cuts the recognition sequence, the single-stranded ends will terminate in a 5' phosphate or in a 3' hydroxyl. Most restriction endonucleases are like *Eco*RI and *Bam*HI in that they cut in the 5' side of the recognition sequence and leave a 5'-terminated end or a **5' overhang**. However, a few restriction endonucleases including *Pst*I and *Nsi*I leave a 3'-hydroxyl terminated end or a **3' overhang**. The fact that different types of restriction endonucleases leave different overhangs can be capitalized upon in some DNA manipulations, such as making blunt ends and constructing deletions in vitro, as we discuss below.

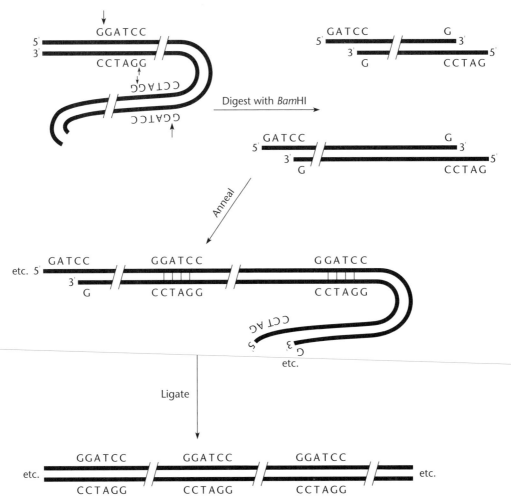

Figure 15.5 Creation of sticky complementary ends by cutting with a restriction endonuclease. The two single-stranded ends can pair with each other, and the nicks can be sealed by DNA ligase.

Compatible Sticky Ends

Some restriction endonucleases recognize different sequences but leave the same sticky ends. For example, the restriction endonuclease *Bgl*II recognizes the sequence AGATCT/TCTAGA and cuts between A and G on each strand, leaving the sticky end GATC. The restriction endonuclease *Bam*HI cuts at the somewhat different sequence GGATCC/CCTAGG (Table 15.1) but also leaves GATC sticky ends. If two restriction endonucleases leave the same sticky ends, they are said to be **compatible**. The compatible restriction endonucleases *Bam*HI, *Bgl*II, *Sau*3A, and *Bcl*I all leave the sticky end GATC. The restriction endonucleases *Xho*I and *Sal*I are compatible because they both leave the sticky end TCGA.

Although the sticky ends generated by cutting with compatible restriction endonucleases will pair with each other, in general, the hybrid sites formed when the two ends are ligated to each other will not be cut by either enzyme. As shown in Figure 15.6, the hybrid site does not match the recognition site of either endonuclease.

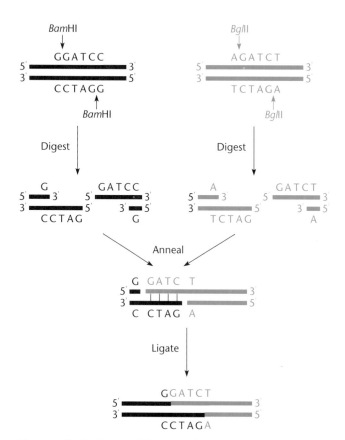

Figure 15.6 Compatible restriction endonucleases. *Bam*HI and *Bgl*II leave the same sticky ends, which can pair with each other. The religated ends leave a hybrid sequence, which cannot be cut by either enzyme.

BLUNT ENDS

Not all the type II restriction endonucleases make staggered breaks in the DNA. Some, such as *Hpa*I, *Eco*RV, and *Sma*I, cut the two strands exactly opposite each other, leaving flush or **blunt ends**.

One advantage that blunt enders offer in making recombinant DNA is that ends left by one blunt ender are essentially compatible with ends left by any other, since these enzymes do not leave single-stranded ends that might not be complementary. Therefore, the ends left by all blunt enders can be joined to all other blunt ends. The drawback is that blunt-end ligations are usually much less efficient than sticky-end ligations, as we might expect, since during blunt-end ligation, there is no annealing of single-stranded ends to hold the pieces together during ligation.

CONVERTING STICKY ENDS TO BLUNT ENDS

Sometimes an experiment requires that two incompatible ends be joined. This can be accomplished by first converting the single-stranded sticky ends to blunt ends in one of two general ways. The methods that are available depend on the type of overhangs left by the restriction endonuclease (Figure 15.7). If the restriction endonuclease leaves a 5′ overhang, DNA polymerase can use the recessed 3′-hydroxyl end as a primer and the single strand as a template to synthesize DNA out to the 5′ end of the molecule. Only a fragment (the Klenow fragment) of *E. coli* DNA polymerase I is usually used for filling-in reactions, because the entire DNA polymerase I enzyme has a 5′-exonuclease activity that can degrade into the double-stranded DNA. Alternatively, single-strand-specific nucleases, such as S1 nuclease and mung bean nuclease, can be used to remove either 5′ or 3′ overhangs. However, the S1 and mung bean nucleases sometimes

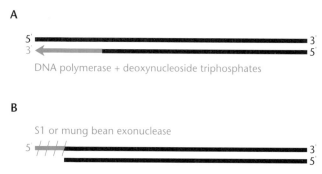

Figure 15.7 Two methods for converting a sticky end into a blunt end. (A) The α (Klenow) fragment of *E. coli* DNA polymerase I is used to fill in the ends. (B) The single-stranded ends are removed with mung bean or S1 single-strand-specific exonuclease.

continue degrading into the double-stranded DNA, creating short deletions.

DNA Cloning with Restriction Endonucleases

In biology, to **clone** an organism means to obtain from an individual a population of organisms that are all genetically identical to that individual and each other (see Introduction). In molecular biology, to clone a DNA means to obtain a population of DNA molecules that are all descended from, and identical to, an individual DNA molecule. In this section, we discuss how restriction endonucleases can be used in DNA cloning and a few tricks to aid this process.

Cloning Vectors

A cloned piece of DNA must be replicated many times. Most pieces of DNA are not replicons in cells and so cannot replicate autonomously to make copies of themselves. These DNAs must be joined to another DNA that is a replicon, for example, a plasmid or virus DNA. Plasmids, virus DNAs, and other DNAs that have been specially designed for the replication of other DNAs are called **cloning vectors**. We have already mentioned some commonly used plasmid and phage cloning vectors in chapters 4 and 7.

Cloning DNA into a Cloning Vector

Figure 15.8 illustrates the steps in cloning a piece of DNA into a cloning vector. In the example, the cloning vector is a circular DNA, either a phage DNA or a plasmid. In the first step, the vector DNA and the DNA to be cloned are cut with compatible restriction endonucleases. The cloning vector should have only one recognition site for the enzyme that cuts it, so that it will be cut in only one place. After being cut, the DNAs are mixed and DNA ligase is added. Sometimes, the two ends of the cut vector join with the ends of the DNA to be cloned, and a **recombinant DNA** forms. This recombinant DNA will be able to replicate if it is introduced by transformation, with plasmid DNA, or transfection, with phage DNA, into a bacterium in which the cloning vector can replicate (see chapter 6). After the DNA replicates, the transformed or transfected cell will contain many copies, or clones, of the original DNA fragment.

Tricks for Optimizing Cloning

Usually, in a ligation such as the one shown in Figure 15.8, the cloning vector will merely recyclize without picking up a DNA insert. Recyclizing requires only that

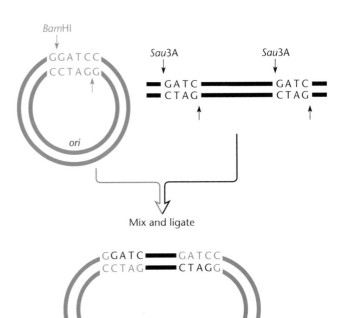

Figure 15.8 DNA cloning. The compatible restriction endonucleases *Sau*3A and *Bam*HI were used to clone a piece of DNA into a cloning vector. The DNA to be cloned was cut with *Sau*3A and ligated into a cloning vector cut with *Bam*HI. The piece of DNA inserted into the cloning vector cannot replicate by itself, since it lacks an *ori* region; however, once it is inserted into the cloning vector, it will replicate each time the cloning vector replicates.

the two ends of the same DNA collide with each other, whereas to pick up another DNA, the ends of the cloning vector must collide with the ends of the other molecule, which happens less often, particularly if the DNA solution is dilute. A number of tricks have been devised to minimize the recyclization of the vector during cloning so that the plasmids or phages in most of the transformants will have a DNA insert. Some of these tricks are mentioned below.

USING COMPATIBLE RESTRICTION ENDONUCLEASES
As mentioned, the hybrid recognition sites generated when the recognition sites of two different but compatible restriction endonucleases are ligated to each other cannot be cut with either enzyme (Figure 15.6). This fact can sometimes be used to eliminate transformants in which the cloning vector has merely recyclized without picking up a piece of the DNA being cloned. In this method, the cloning vector and the DNA to be cloned are cut with different but compatible restriction en-

donucleases. After ligation, the restriction endonuclease used to cut the vector DNA is added to the ligation mix. The restriction nuclease will cut the vector DNAs that have recyclized without picking up an insert but not the vectors that have cyclized with an inserted piece of DNA, since the latter will have hybrid sites. When the ligation mix is then used to transform competent bacteria, any transformants should contain the vector with the cloned DNA if the bacterium is like wild-type *E. coli,* which cannot be efficiently transformed with linear DNA (see chapter 6).

DOUBLE DIGESTS

Another way to increase the frequency of tranformants that contain cloned DNA is the double-digest method (Figure 15.9). Both the cloning vector and the DNA to be cloned are simultaneously cut with two incompatible restriction endonucleases. Under these conditions, the vector will be cut in two places—with a piece excised—and cannot recyclize without the missing piece. The DNA to be cloned, cut with the same two restriction endonucleases, is then added at a concentration high enough relative to the cloning vector that the vec-

tor will be more likely to cyclize by ligating to this added DNA.

DIRECTIONAL CLONING

One of the major advantages of cloning by double digests is that a piece of foreign DNA can be inserted in the cloning vector in only one orientation. Thus, in Figure 15.9, the *Eco*RI end of the fragment can ligate only to the *Eco*RI end of the cloning vector and the *Pst*I ends can ligate only to each other. This is called directional cloning. The orientation in which a gene is inserted in a cloning vector can be important for expressing the gene from a promoter and/or a translational initiation region (TIR) in an expression vector (see below).

Cloning Bacterial Genes

One of the major applications of recombinant DNA technology is to clone specfic genes of an organism. These might be genes that encode gene products useful in medicine or biotechnology or simply gene products that might be relevant to our understanding of cell function. To clone a particular gene, we must identify it from among the thousands or even millions of other genes of the organism. In this section, we discuss some of the methods for cloning and identifying genes.

Constructing DNA Libraries

The first step in cloning a gene of any organism is usually to construct a **DNA library** of the organism, which is a collection of DNA clones that includes all, or at least almost all, the DNA sequences of the organism. First, the DNA is cut with a restriction endonuclease, and the pieces are ligated into a cloning vector cut with a compatible enzyme. The mixture is then transformed or transfected into cells, and the transformants or plaques are pooled. If the collection is large enough, every DNA sequence of the organism will be represented somewhere in the pooled clones, and the library is complete.

The number of clones required to make a complete DNA library of an organism depends on the complexity of the DNA of that organism. For example, if we make a library of *E. coli* DNA cut with *Eco*RI, we will need about $4.5 \times 10^6/4 \times 10^3 \approx 1,100$ different clones, since *E. coli* DNA contains approximately 4.5×10^6 bp and a six-hitter like *Eco*RI cuts the DNA about once every 4,000 bp. In contrast, a library of λ DNA should require only about $5 \times 10^4/4 \times 10^3 \approx 13$ clones, since λ DNA has only about 50,000 bp. An important point is that these are minimum estimates of the number of clones required to make a library of the DNA of the or-

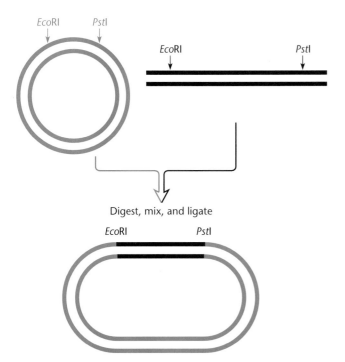

Figure 15.9 Double digests. The cloning vector cannot recyclize because it has been cut with *Eco*RI and *Pst*I, which do not leave compatible sticky ends. The cloning vector can cyclize only if it picks up another piece of DNA with *Eco*RI- and *Pst*I-cut ends, either the piece cut out of it or a piece cut from another DNA.

ganism; not all clones will be equally represented in the library because of random statistical fluctuation. Also, some pieces may be easier to clone, for example, because they are smaller or contain no genes whose products are toxic to the cell.

PARTIAL DIGESTS

Six-hitter restriction enodnucleses are popular for making libraries because they cut DNA into convenient-sized pieces that on average contain only a few genes. However, there are two potential problems with making libraries with six-hitter endonucleases. First, the pieces of DNA in the library will vary greatly in size. The distribution of restriction sites is random, so that some restriction fragments may be only hundreds of base pairs long while others may be tens of thousands of base pairs long. Second, the gene being cloned may contain one or more recognition sites for the restriction endonuclease being used. Then the gene will be split between different restriction fragments, and none of the clones in the library will contain the complete gene.

To avoid these problems, the library can be made by using a **partial digest** of the DNA with a four-hitter. In this method, only some—not all—of the recognition sequences in the DNA are cut. We accomplish this by keeping the concentration of the restriction enzyme low or incubating the reaction for only a short time.

In a standard partial digest, we use a four-hitter enzyme, such as *Sau*3A, and allow the digestion to proceed until the average-size fragment is the size of the clone desired in the library, say approximately 4,000 bp. The pieces are then ligated into a cloning vector cut with a compatible six-hitter enzyme such as *Bam*HI. The library should then have random pieces from every region of the DNA, with any particular gene intact in at least some of the pieces, even if there are multiple *Sau*3A sites in the gene, since not all the *Sau*3A sites will be cut, and different ones will be cut to make different pieces.

THE KOHARA LIBRARY OF *E. COLI* DNA

Libraries have already been constructed for many organisms. For example, an *E. coli* DNA library has been constructed with a λ phage cloning vector. This library is named the Kohara library after the head of the laboratory where it was made (see Kohara et al., Suggested Reading). The clones in this commercially available library are fixed on filters in the order of their appearance in the genome of *E. coli* for ease of use in physically mapping cloned genes by methods such as those described above. Complete, ordered libraries will soon become available for many other types of organisms.

Identifying Clones in a Library

Once a library of the DNA of an organism has been constructed, it is necessary to identify the clones that carry the gene of interest. Several methods are available; their use depends on the situation.

PLATE HYBRIDIZATIONS

Plate hybridizations offer a way of identifying clones containing genes for which a complementary probe is available. The method can be used regardless of whether the cloning vector used to make the library is a phage, in which case the library will be housed in plaques, or a plasmid, for which bacterial colonies will house the library. The method of plate hybridizations illustrated in Figure 15.10 is in principle no different from the Southern blot hybridizations illustrated in Figure 15.4, except that the DNA transferred to filters is from phage plaques or bacterial colonies on plates rather than being in bands on gels. Plates containing several plaques or colonies, each with a different clone from the library, are overlaid with a filter, which picks up some of the phage or bacteria. The phage or bacteria on the filter are then treated with alkali to release and denature their DNA. The filter containing the spots of denatured DNA is then immersed in hybridization buffer with a chemically or radioactively labeled probe that has sequences complementary to sequences in the gene being cloned. Only DNA from a colony or plaque that contains that gene sequence will hybridize to the probe and become detectable. Colonies or plaques containing the hybridizing clones will occupy the corresponding positions on the original plate, from which they can be picked. Detailed protocols for plate hybridizations are given in most cloning manuals.

Plate hybridization will succeed only when no sequences complementary to the probe occur in the cloning vector or in the chromosome of the bacterium in which the library has been constructed. This problem is particularly troublesome when the cloned gene is from the same bacterium in which the library has been constructed, for example, if an *E. coli* gene is cloned in *E. coli*. In this case, all the chromosomal DNA in the colonies (with a plasmid vector) or all the DNA in the bacterial lawn (with a phage vector), would hybridize to the probe. Even so, it is sometimes possible to identify clones, since the cloned gene in a multicopy vector will exist in more copies than in the chromosome, so that the signal from colonies or plaques containing the gene may be stronger than the background signal.

MAKING PROBES

Identifying a clone by plate hybridization requires a probe that is complementary to at least part of the gene

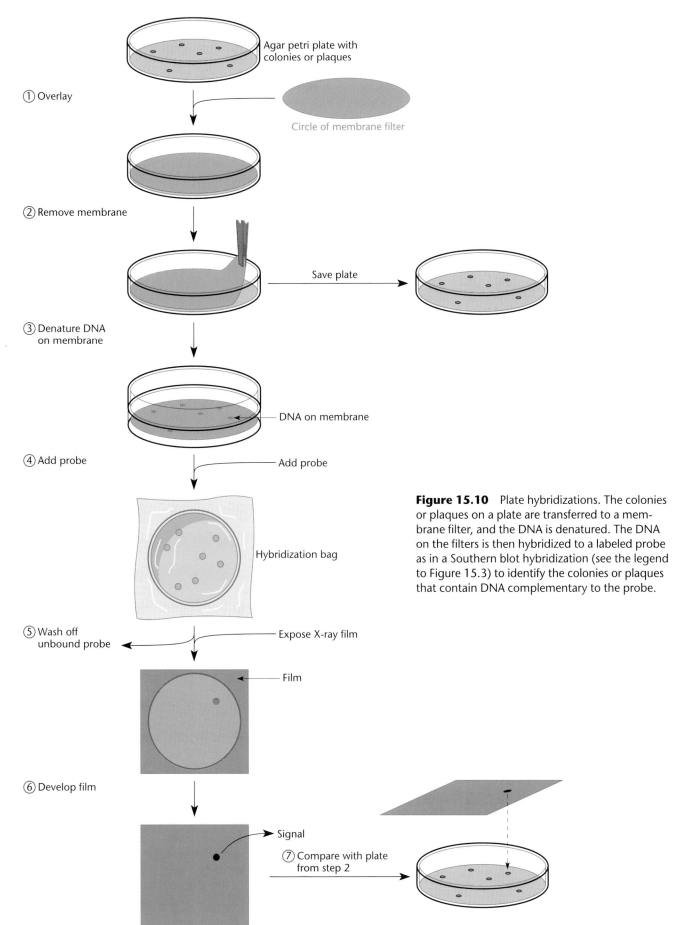

① Overlay

Agar petri plate with colonies or plaques

Circle of membrane filter

② Remove membrane

Save plate

③ Denature DNA on membrane

④ Add probe

DNA on membrane

Add probe

Hybridization bag

⑤ Wash off unbound probe

Expose X-ray film

Film

⑥ Develop film

Signal

⑦ Compare with plate from step 2

Figure 15.10 Plate hybridizations. The colonies or plaques on a plate are transferred to a membrane filter, and the DNA is denatured. The DNA on the filters is then hybridized to a labeled probe as in a Southern blot hybridization (see the legend to Figure 15.3) to identify the colonies or plaques that contain DNA complementary to the probe.

405

being cloned. Probes can either be natural DNAs or RNAs or be constructed from oligonucleotides, which are chemically synthesized short pieces of single-stranded DNA. Whatever the source, a probe used for filter hybridizations should generally be longer than about 10 bases to ensure efficient hybridization. We can obtain a complementary probe in a number of ways.

Heterologous Probes
Sometimes, someone has already cloned the same type of gene we want to clone but from a different organism. These two genes might have quite similar sequences in at least some regions if the two organisms are closely related or if the gene appears to have been highly conserved in evolution. If the genes do share sequence homology, we can probably use the gene from the related organism, or part of it, as a **heterologous probe**, so named because it comes from a different organism. Unless the two organisms are very closely related, the probe will not be 100% complementary to the gene being cloned. Nevertheless, the sequences might be similar enough for hybridization.

We have no way of predicting with certainty whether corresponding genes from different organisms are similar enough in sequence to hybridize. However, some genes are more highly conserved between species than are others. For example, genes encoding the protein part of cytochromes or components of the protein synthetic machinery are usually more highly conserved than genes for biosynthetic pathways, such as the synthesis of an amino acid. Also, some regions of a gene might be more highly conserved than other regions. In general, the sequences for the most important parts of proteins such as the active centers of enzymes tend to be more conserved than less important regions. We discuss cloning with heterologous probes in more detail in chapter 16.

Synthetic Oligonucleotide Probes
Another way to make a complementary probe is to chemically synthesize it on the basis of the amino acid sequence of the protein. This method is possible if the protein has been purified but the gene for the protein has not been cloned. We sequence part of the protein, usually the N terminus or a short peptide cut from it, and then chemically synthesize a piece of DNA that could code for that part of the protein.

Degenerate Probes
Even if the amino acid sequence of part of a protein is known, designing a probe with exactly the same sequence as a region of the gene of interest is usually impossible. More than one codon often encodes the same amino acid, and we have no way of knowing which codon stipulated each amino acid in the protein. To try to overcome this problem, we can choose a region of the protein where the redundancy of the code does not allow too many possibilities (Figure 15.11). Preferred codon usage (see chapter 2) can also help guide the choice of oligonucleotide probes. Within species, some codons are used more than others for the same amino acid; this is particularly true among organisms with compositions high in AT or GC. In organisms whose DNAs are rich in AT base pairs, for example, the codon that ends in an A or U will tend to be used for a particular amino acid.

Even with these adjustments, we usually cannot be certain which codons encode the region of the protein. Hence, we may have to make a **degenerate probe**, which is a mixture of oligonucleotides with different sequences based on the possible codons. Some of these nucleotide chains should have the correct sequence, even if most do not. Figure 15.11 illustrates the making of a degenerate probe.

IDENTIFYING CLONES BY IMMUNOLOGICAL TECHNIQUES
We can sometimes identify clones containing a certain gene by using antibodies to the protein the gene encodes. For example, the genes in the library may have been randomly fused to another gene. The procedure is much like that used in plate hybridizations with oligonucleotide probes, except that the proteins rather

Figure 15.11 Designing a degenerate probe for a gene from the partial sequence of the protein product of the gene. Because of redundancy in the code, more than one codon is possible for each amino acid. The region with the least redundancy is chosen to make the probe.

Amino acid sequence	NH$_2$— — —Trp—Pro—Met—Asp—Glu—Phe—Trp—Val—

Possible codons								
	UGG	CCA	AUG	GAC	GAA	UUC	UGG	GUA
		CCC		GAU	GAG	UUU		GUC
		CCU						GUU
		CCG						GUC

Synthetic probes:
17 nucleotides long
8 possible sequences

AUG CA$_\mathrm{U}^\mathrm{C}$ GA$_\mathrm{G}^\mathrm{A}$ UU$_\mathrm{U}^\mathrm{C}$ UGG GU than the DNA from the colonies are transferred to filters and the filters are treated with the labeled antibody to identify clones containing the protein. Cloning vectors such as the phage cloning vector λgt11, which fuses genes to the *lacZ* gene, are commonly used for such applications.

IDENTIFYING CLONES BY GENETIC TECHNIQUES

Genetic methods have the advantage over the previously discussed methods in that they do not depend upon making a probe. Consequently, we do not need to know the actual function of the gene product or have a related gene from another organism from which to make a heterologous probe. All that is needed to identify clones genetically is a phenotype for mutants with mutations in the gene to be cloned.

Identifying Clones by Marker Rescue Recombination

A clone with a particular gene can sometimes be identified through the ability of the gene to recombine with a mutant chromosome to give wild-type recombinants. This process is sometimes called **marker rescue** because the marker, or mutation, in the chromosome has been "rescued" by the piece of cloned DNA.

Figure 15.12 illustrates this technique. In this example, a *thyA* mutant of *E. coli* is being used to identify plasmid clones containing the *thyA* gene of *E. coli*. The product of *thyA* is thymidylate synthetase, which is used to make dTMP from dUMP; therefore, a *thyA* mutant cannot synthesize thymine and will not grow on plates lacking it.

In the first step, an *E. coli* plasmid library made from wild-type *E. coli* is introduced into the *thyA* mutant, and the antibiotic resistance gene on the cloning vector is selected for. The plates should also contain thymine, so that all the transformants can form a colony. Each transformant colony will contain a clone of the library, one or more of which should contain part or all of the *thyA* gene. The colonies are then replicated onto a minimal-medium plate containing all the growth requirements except thymine. Any bacteria growing at the position of the original colony may be due to Thy⁺ recombinants arising in the colony as the result of recombination between the *thyA* gene in the clone and the *thyA* gene in the chromosome (Figure 15.12B). Note that for Thy⁺ recombinants to arise, the entire *thyA* gene need not be on the clone, only the region corresponding to the *thyA* mutation in the chromosome. Once clones containing at least part of the *thyA* gene have been identified, the plasmid containing the clone

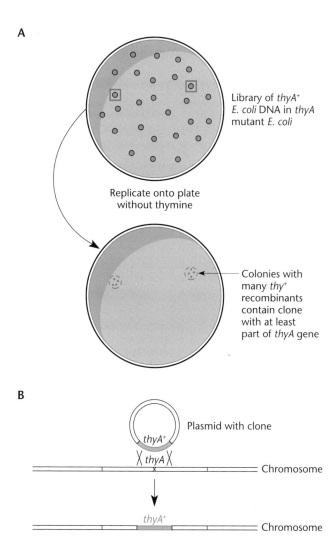

Figure 15.12 Use of marker rescue to identify a clone containing at least part of the *thyA* gene of *E. coli*. (A) The boxed colonies correspond to Thy⁺ recombinants grown on a replicated plate. See the text for details. (B) The *thyA⁺* gene on the cloning vector recombines with the *thyA* mutant gene on the chromosome to produce Thy⁺ recombinants.

should be purified from the bacteria in the corresponding colony on the original plate, where most of the plasmids will not have recombined with the chromosome and the clone will be intact in most of the bacteria. Clones should not be taken from the plate without thymine, where the only bacteria growing are ones in which the clone in the plasmid has already been altered by recombination with the chromosome. In marker rescue cloning, the selective plate is used only to identify colonies in which some marker rescue recombination has occurred, not to isolate the plamid containing the clone.

Complementation Cloning

The other genetic method for identifying clones is based on their ability to complement mutations in the chromosome. Figure 15.13 shows how complementation cloning is different from marker rescue cloning. Unlike in the latter method, in complementation cloning, the entire gene must be included in the clone and the gene must be expressed from the clone into its protein product. Also, complementation cloning does not require recombination between the clone and the corresponding region in the chromosome. In fact, the cloning is often performed in strains that are RecA⁻ to preclude recombination (see chapter 10).

To clone the *thyA* gene by complementation as shown in Figure 15.13, a library made with *thyA⁺ E. coli* DNA is used to transform a *thyA* mutant *E. coli* that is also *recA*, and an antibiotic resistance gene on the plasmid cloning vector is selected. Now, however, the selective plates lack thymine. Expression of the *thyA* gene from the clone, either from the promoter for the gene itself or from a promoter in the cloning vector, will allow the cells to multiply to form a colony. Therefore,

only cells transformed by a plasmid cloning vector with an intact *thyA* gene will multiply to form a colony. Complementation cloning has the advantage that *all* of the transformed cells will multiply, not just the few cells rescued by recombination. The colonies containing the clone of the gene can be picked directly from the original selective plate, in the example the plate without thymine.

Because complementation cloning requires that in general, the only clones that will be identified are those that contain the entire gene as well as a functional promoter to transcribe the gene, fewer clones will be identified by complementation than by marker rescue. Also, because the gene on the clone must be expressed, complementation can generally be used only to identify clones in the same or a closely related organism that has similar TIRs and other physiological properties. Therefore, complementation cloning usually must be performed in the host of origin, where the promoter and translation initation site of the gene will be recognized, so that the gene will be expressed. We discuss these issues in more detail in chapter 16.

Locating Genes on Cloned DNA

Clones usually contain many genes besides the one being cloned. Therefore, once the desired clone has been identified, the gene must be localized on it.

TRANSPOSON MUTAGENESIS

By randomly inserting a transposon into a clone, we can locate a particular gene on that clone. We identify transposon insertion mutations that inactivate the gene by testing for complementation, since a transposon insertion will inactivate the cloned gene, which will no longer complement a mutation in the chromosome. The site of such inactivating insertions in the clone will reveal the location of the gene. Transposon insertions are relatively easy to map in plasmid DNAs by physical methods, which we describe in the next chapter.

One caveat to this method is that transposon insertion mutations are often polar. A polar mutation between a gene and its promoter can prevent expression of the gene, even when the mutation is not in the ORF for that gene (see chapter 2). If polarity occurs, transposon insertions can inactivate the expression of a gene without necessarily being in the gene itself; instead, it can be upstream of the gene in the same transcription unit.

SUBCLONING

A method for locating a gene on a clone and removing the excess DNA is subcloning. To **subclone** means to clone smaller fragments of DNA containing the gene

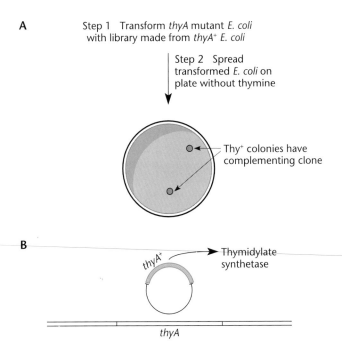

Figure 15.13 Identification of clones of the *thyA* gene of *E. coli* by complementation. (A) A *thyA* mutant of *E. coli* is transformed by a library of DNA from *thyA⁺ E. coli,* and the transformants are selected directly on plates lacking thymidine but containing the antibiotic to which the cloning vector confers resistance. (B) Thy⁺ transformants that contain a clone of the *thyA* gene synthesize the thymidylate synthetase, thus complementing the mutation in the chromosome.

from the larger DNA fragments. The clone is digested with restriction endonucleases, and smaller restriction fragments from the original large fragment are cloned into a cloning vector. The smaller clones containing the gene of interest can then be identified by reapplying the same method used to identify the larger clones containing the gene. For example, if the clone had been identified through hybridization to a probe, smaller clones containing the gene would also hybridize to the probe. Alternatively, if the original, larger clone had been identified by complementation, the smallest piece of DNA that can still complement the chromsomal mutation could be selected. This smaller subclone will probably contain very little extra DNA beyond what is required to express the gene.

Making Deletions with Exonucleases

Subcloning usually depends on the presence of convenient restriction sites in the larger clone, but when the clone contains none, portions of the cloned DNA could be deleted with DNA exonucleases, which are enzymes that remove nucleotides from the ends of DNA molecules. Commonly used exonucleases are *Bal* 31, exonuclease III (Exo III), and λ exonuclease.

The standard procedure for creating a deletion by using an exonuclease is illustrated in Figure 15.14. First the DNA is cut with a restriction endonuclease close to the region to be deleted. Then the DNA is digested with the exonuclease, as shown. After the digestion, the ends left on the DNA are ligated to each other. The DNA that was digested by the exonuclease will then have been deleted from the original clone. The longer the exonuclease was allowed to act, the longer will be the region deleted.

Each type of exonuclease has advantages and disadvantages for making subclones. Exonuclease III is particularly advantageous because it can be used to digest DNA in only one direction from a restriction site, so that the cloning vector will not be digested—only the excess cloned DNA. Exonuclease III will digest one strand of double-stranded DNA from the 3' end, but it usually will not digest single-stranded DNA from either end or double-stranded DNA from the 5' end. Therefore, it will digest DNA from an end with a 5' overhang, since the 3' end is recessed and therefore double stranded, but not from a 3' overhang, which would be single stranded.

Figure 15.15 illustrates this method. A plasmid containing a cloned DNA to be deleted is cut with *Eco*RI and *Pst*I, with the *Eco*RI site closest to the cloned DNA. Exonuclease III will degrade the cloned DNA because it ends in a single-stranded 5' end without entering the

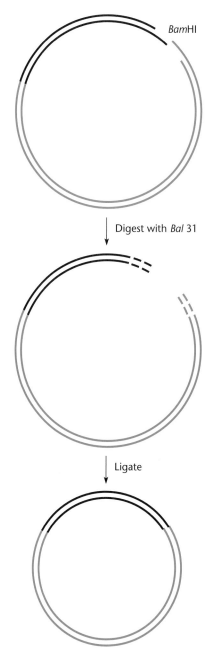

Figure 15.14 Deletion of part of a clone with *Bal* 31 exonuclease. In the example, the DNA is cut with *Bam*HI, for which there is a unique site next to the insert DNA. Parts of both the insert and the cloning vector will be deleted when the digested DNA is religated.

cloning vector, which ends in a single-stranded 3' overhang, as shown. After the one strand is digested, the other strand can be removed with a single-strand-specific nuclease such as S1 or mung bean nuclease to leave a blunt end for religating.

Figure 15.15 Deletion in only one direction with Exo III. The exonuclease will digest only from the *Eco*RI-cut end with a 5' overhang and not from the *Pst*I-cut end with a 3' overhang. The remaining single-stranded DNA is then removed with the single-strand-specific nuclease S1, and the DNA is religated. Only the cloned DNA and not the cloning vector will be deleted.

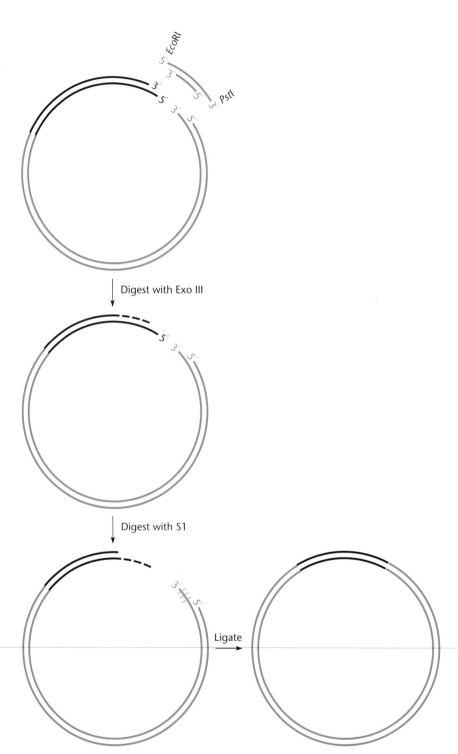

Applications Involving Cloned Genes

Cloned genes have numerous uses in biological studies. For example, they allow us to sequence the genes or make expression vectors for specific proteins. In this section, we briefly describe a few of these manipulations. For more detailed treatments, consult one of the cloning manuals.

DNA Sequencing

Once we have cloned the smallest DNA fragment containing an unknown gene, we can sequence the fragment. The DNA sequence may reveal the exact starting and ending points of the ORF that encodes the protein product of the gene. This information will be important if we decide to make an expression vector from the cloned gene (see below). The DNA sequence of the ORF, by reference to the genetic code, will also tell us the amino acid sequence of the protein product. It is much easier to sequence DNA than to sequence RNA or protein; therefore, by far the easiest way to determine a sequence of amino acids is to sequence the gene encoding it. Furthermore, any similarities in the sequence of the cloned gene to other known genes may point to possible biochemical functions for the protein product of the gene.

DNA can be sequenced by two general methods: **Maxam-Gilbert sequencing** or **Sanger dideoxy sequencing**. Since dideoxy sequencing has become generally more popular, we discuss only this method.

Sequencing by the dideoxy method is based on the chain-terminating property of the dideoxynucleotides. As discussed in chapter 1, the dideoxynucleotides are like the normal deoxynucleotides except that they have hydrogens instead of hydroxy groups at the 3' positions of the sugar. The dideoxynucleotides can be phosphorylated to give the dideoxynucleoside triphosphates, which can be incorporated into DNA by DNA polymerase. However, because it lacks a hydroxyl group at the 3' position, an incorporated dideoxynucleotide cannot be joined to the 5' phosphate of the next deoxynucleotide, and the growing DNA chain will terminate.

In a dideoxy sequencing experiment, four separate polymerizing reactions are run. Each of the four reaction mixtures contains a small amount of one of the dideoxynucleoside triphosphates (ddTTP, ddGTP, ddATP, or ddCTP) mixed with the normal deoxynucleoside triphosphates. A short DNA primer complementary to a known sequence adjacent to the unknown DNA sequence is annealed to the DNA, and DNA polymerase is added. Each time the DNA polymerase encounters a base in the template DNA that is complementary to the dideoxynucleotide used in that reaction, there is a chance that the dideoxynucleotide instead of the normal deoxynucleotide will be incorporated into the chain. As soon as a dideoxynucleotide is incorporated, the growth of the chain will be terminated and the chain will end in the dideoxynucleotide. Since each reaction mixture contains a different dideoxynucleoside triphosphate, each of the four reactions will produce a set of shortened DNA chains of different lengths determined by the position of the complementary nucleotides in the template DNA. Therefore, the sequence of the template DNA can be determined by measuring the lengths of the shortened chains in each of the reactions.

To analyze the data from dideoxy sequencing reactions, we first determine which of the four reaction mixtures contains the shortest chain with a measurable length. If the shortest chain that can be measured was produced in the reaction with ddT, then an A must be in the template DNA closest to the site where the primer DNA annealed. If the next-shortest chain came from the reaction with ddG, then a C must be next to this A. If the reaction mixture containing ddC produced the third-shortest chain, then the sequence in the template DNA must read ACG. The fourth-shortest chain is found, the fourth nucleotide in the template DNA is ascertained, and this process continues until the cloned DNA is sequenced.

In practice, it is possible to sequence only about 200 to 300 bases of a cloned DNA in a single set of four dideoxy sequencing reactions. A longer region must be broken up into smaller regions before it can be sequenced. The sequences of the smaller regions can then be assembled into the sequence of the larger region.

Several methods allow us to break up long regions into shorter ones of less than 300 bases. For example, we could break up the longer region into subclones and determine the sequence of the original clone from that of the subclones, provided that we know the order of the subclones. If the subclones come from overlapping regions of the larger clone, they can often be ordered according to the overlapping sequences. Another way to subdivide a larger clone for sequencing is to delete the sequence progressively inward from the end adjacent to the site where the primer hybridizes, using exonucleases such as *Bal* 31 (see above). When the DNA polymerase polymerizes from the primer hybridized to a deleted clone, it will be making a chain complementary to a sequence farther into the original clone. A third way is to first sequence about 300 bases in from one end. When this sequence is known, another primer can be designed complementary to the far end of the sequenced region, and this primer can be used to sequence the next region, and so forth. In this way, the entire sequence can be "walked through," one 300-base step at a time. The methods of DNA sequencing are being automated, and larger and larger DNAs, including entire bacterial chromosomes, have been sequenced (see Box 15.2).

Expression Vectors

One reason for cloning and sequencing a gene is to make large amounts of its protein product. To do so, we generally need to clone the gene into an **expression vector**. This is a vector that contains a strong bacterial promoter and TIR from which the gene will be expressed. Different expression vectors have different properties.

BOX 15.2

Bacterial Genomics

Several efforts, known as genome projects, are under way to sequence the DNA of entire organisms. The genomic DNAs of some viruses were the first to be sequenced. Even this was not an easy task. Virus DNAs range from only a few thousand to hundreds of thousands of base pairs in length.

Because viruses are not truly alive and depend upon the host for most of their functions, they tell us little about which genes are required for life. Such information can be obtained only by sequencing cellular DNA. Considering the length of genomic DNAs, this is a daunting task. The genomic DNA from even a simple bacterium is almost a millimeter long, and the DNA from a single human cell is almost a meter (or yard) long!

There are several reasons to sequence the entire DNA of a living organism. The sequences of more and more genes have been entered into databases, and proteins with the same function often have sequences, called motifs, in common (see chapter 2). Therefore, from the sequence of the gene alone, it is often possible to search the databases to find similar proteins of known function. This may make it possible to guess the function of many of the gene products of the organism. A number of insights into evolution may emerge. For example, there may be interesting patterns in how genes of similar function are organized in different species. Also, the minimum number of genes required to make an organism might be determined by comparing the total genes of many different organisms. Only the genes shared by all types of organisms would be absolutely required to make a living organism.

Because bacteria are the simplest living organisms on Earth, we would expect that bacterial genomes would be the first to be sequenced. In fact, at the time of this writing, the complete genomic sequences of two bacteria, *Haemophilus influenzae* (see Fleischmann et al., below) and *Mycoplasma genitalium*, have been determined, and more bacterial genome sequences are forthcoming. These two bacteria were chosen because their genomes are particularly small.

They are obligate parasites and so can multiply only in a human host; therefore, they do not require some functions required of a free-living organism.

The strategy for sequencing the genomes of these bacteria was to first obtain a random library of DNA fragments of about 2,000 bp. Many of these clones were then sequenced by standard methods. The sequences were introduced into the computer to determine which clones contained identical sequences and so overlapped. The computer could then order the overlapping clones, which are called contigs because they are contiguous in the genome. Repetitive DNA elements, such as IS elements, presented a problem in that they registered as false contigs; these elements had to be eliminated. Of course, there were also gaps in the original sequence, which had to be filled in by looking for contigs of clones on either side of a gap or by using the sequences of the ends of clones on either side of the gap to design primers for PCR amplification of the missing region. However, all the gaps were eventually filled in, and the entire sequence was determined. This work greatly benefited from automated sequencing techniques and improved computer technology for searching sequences and ordering contigs, both of which grew out of the better-financed Human Genome Project. Even with such advances, however, the projects still required an enomous amount of work, and dozens of researchers were involved, as is evident from the Fleischmann et al. citation.

Reference

Fleischmann, R. D., M. D. Adams, O. White, R. A. Clayton, E. F. Kirkness, A. R. Kerlavage, C. J. Bult, J.-F. Tomb, B. A. Dougherty, J. M. Merrick, K. McKenney, G. Sutton, W. Fitzhugh, C. Fields, J. D. Gocayne, J. Scott, R. Shirley, L.-I. Liu, A. Glodek, J. M. Kelley, J. F. Weldman, C. A. Phillips, T. Spriggs, E. Hedblom, M. D. Cotton, T. R. Utterback, M. C. Hanna, D. T. Nguyen, D. M. Saudek, R. C. Brandon, L. D. Fine, J. L. Fritchman, J. L. Fuhrmann, N. S. M. Geoghagen, C. L. Gnehm, L. A. McDonald, K. V. Small, C. M. Fraser, H. O. Smith, and J. C. Venter. 1995. Whole-genome random sequencing and assembly of *Haemophilus influenzae* Rd. *Science* **269**:496–512.

pUC PLASMIDS

The prototype expression vectors are the pUC vectors (Figure 15.16). The pUC plasmids are very small, with only 2,700 bp, have a very high copy number of 30 to 50, and have the easily selectable ampicillin resistance (Amp^r) gene. These plasmids also have a **multiple cloning site** containing the recognition sequences for many different restriction endonucleases. This multiple cloning

site is immediately downstream of the strong *lac* promoter on the plasmid, called p_{lac} in the figure, so that any DNA cloned into one of the restriction sites in the polyclonal region of the pUC vector will be transcribed from the *lac* promoter. If the cloned DNA contains a gene in the correct orientation, the gene will be transcribed into mRNA in *E. coli*. The *lac* promoter is also inducible and will be turned on only if an inducer, such as IPTG or lac-

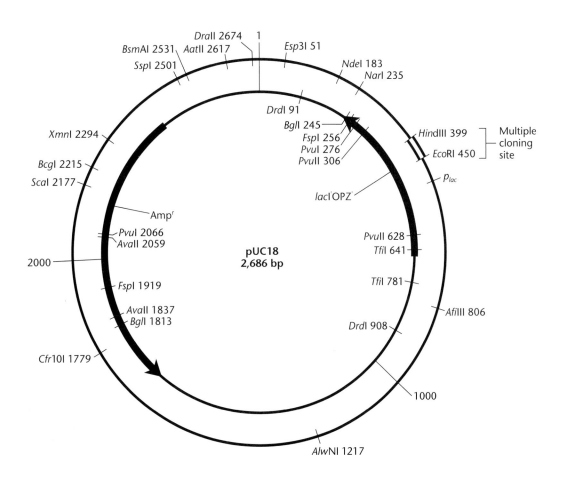

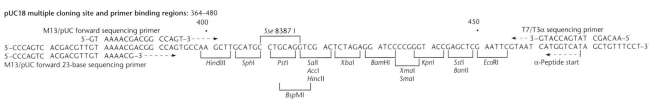

Figure 15.16 A pUC expression vector. A gene cloned into one of the restriction sites in the multiple-cloning site will almost invariably disrupt the coding sequence for the *lacZ* α-peptide (see chapter 4). If it is inserted in the correct orientation, the gene will be transcribed from the *lac* promoter called *P$_{lac}$* in the figure. If the ORF for the gene is in the same reading frame as the ORF for the *lacZ* coding sequence, the gene will also be translated from the *lacZ* TIR, and the N-terminal amino acids of *lacZ* will be fused to the polypeptide product of the gene. Adapted with permission from the *BRL-Gibco Catalog*, p. 766, BRL-Gibco, Gaithersburg, Md., 1992.

tose, is added. Thus, the cells can be propagated before the synthesis of the gene product is induced, a feature that is particularly desirable if the gene product is toxic to the cell. Genes cloned into one of the polyclonal restriction sites in the *lacZ* gene can also be translated from the *lacZ* TIR on the plasmid, provided that there are no intervening nonsense codons and the gene is cloned in the same reading frame as the upstream *lacZ* sequences.

Making proteins by fusing them to other proteins encoded by the cloning vector has disadvantages for some applications. The fusion protein may not be active or may have other undesirable properties. Expression vectors have been designed that allow a foreign gene to be translated from the TIR on the vector without fusing the protein to peptide sequences on the vector. These plasmids do not have a complete bacterial translation initiation site upstream of the cloning site but have only a Shine-Dalgarno sequence (see chapter 2). If a foreign gene is cloned into one of these vectors so that the Shine-Dalgarno sequence on the vector is within 10 to 15 bp of the first codon of the gene (usually AUG), it will probably be translated correctly in *E. coli*. Methods such as PCR have enabled the construction of subclones with the initiator codon of the gene the correct distance from the Shine-Dalgarno sequence on the cloning vector (see below). This general approach has been used to synthesize in *E. coli* several foreign proteins that are identical to those made in the original organism.

pET VECTORS

The promoters and TIRs of phages are often very strong and allow the expression of large amounts of protein from a cloned gene. These vectors are also useful for making specific RNAs in vitro to use as substrates for enzymatic reactions and as hybridization probes, as we discuss here.

The pET vectors are a family of expression vectors that use the T7 phage RNA polymerase and T7-specific promoters (see chapter 7) to express foreign genes in *E. coli* (see Figure 15.17). The pET vectors capitalize on the fact that some phages, such as T7, encode their own RNA polymerase and that this polymerase will recognize only the phage promoters. Therefore, the phage promoter on the vector will be used only if the phage RNA polymerase is present.

The general strategy for using a pET vector is illustrated in Figure 15.18. To provide a source of phage T7 RNA polymerase, *E. coli* strains that contain the gene for T7 RNA polymerase, gene 1 of the phage, have been constructed. The T7 phage gene 1 is cloned downstream of the *lac* promoter in the chromosome of these strains, so that the phage polymerase gene will be transcribed from the *lac* promoter. In these strains the T7 RNA polymerase will be synthesized only if an inducer of the *lac* promoter, such as IPTG, is added. The newly synthesized T7 RNA polymerase will then transcribe the foreign gene in the pET plasmid.

TAG VECTORS

Some of the cloning vectors have been engineered so that the protein being expressed will be fused to another protein, called a **tag**. Examples of commonly used tags are strings of histidines, glutathione-*S*-transferase, MalE, or some other protein that can be easily purified on an affinity column. The tag vectors are also constructed so that the coding sequence for an amino acid sequence cleaved by a specific protease, such as thrombin or factor X, is inserted between the coding sequence for the tag and the gene being expressed. After purifica-

tion, the tag protein can be cleaved off with the specific protease, leaving at most a few extra amino acids attached to the N terminus of the purified protein. Such **tag vectors** are particularly useful for purifying proteins for which the gene has been cloned but for which the activity is not known or for which there is no convenient assay.

RIBOPROBES

As mentioned, the pET vectors and other phage promoter-based expression vectors also are useful for making specific RNA in vitro to use as probes for hybridization experiments (**riboprobes**) and for making RNA substrates for processing reactions. The purified vector DNA containing the clone is cut downstream of the clone with a restriction endonuclease and then used as a template for in vitro RNA synthesis with the phage RNA polymerase and the four ribonucleoside triphosphates. The RNA polymerase will run off the DNA at the cut end, so that only the clone will be transcribed into RNA. If the clone is in the opposite orientation with respect to the phage promoter, RNA will be made on the noncoding, or template, strand of the clone DNA (see chapter 2). This probe will then hybridize to the mRNA from the gene, a procedure used in some methods for quantifying mRNA. For riboprobe construction, one of the nucleoside triphosphates is often radioactive or chemically labeled, so that the RNA made from the clone will be labeled. Phage T7 and some other phage RNA polymerases are available from many biochemical supply companies.

Site-Specific Mutagenesis

Once a gene for a protein or other DNA has been cloned and sequenced, the effect of changing specific amino acids in the protein can be studied by changing specific nucleotides. The various methods for changing the sequence of short regions of DNA are all known as **site-specific mutagenesis** because, instead of subjecting all the DNA in the cell to mutagens, the mutagenesis is restricted to one site, the cloned DNA, and maybe even to only 1 bp in the cloned DNA.

Site-specific mutagenesis is often useful for changing a specific amino acid in a protein to see what effect this change has on the activity of the protein. Alternatively, the cloned DNA can be mutagenized so that amino acids will be changed at random in the protein. In combination with some sort of selection, this more random mutagenesis may allow us to find an altered clone that makes an altered protein with some desired property, even though we may not have been able to predict which amino acid changes in the original protein would confer that property.

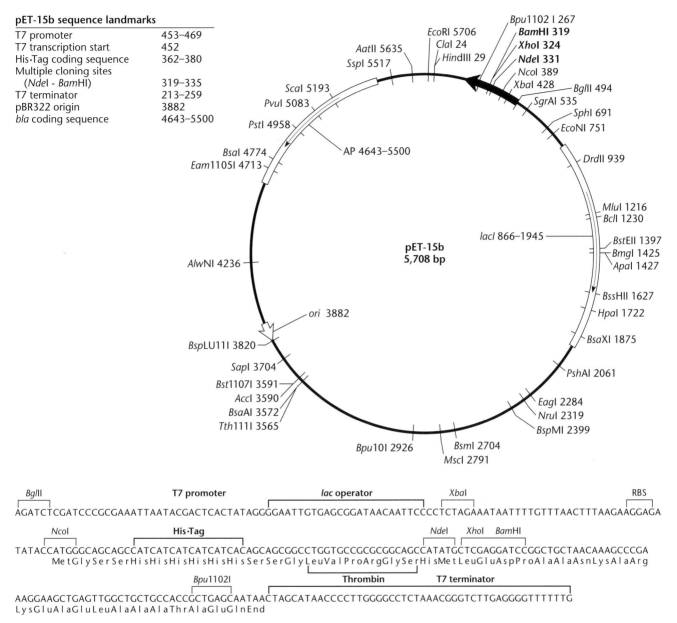

pET-15b sequence landmarks

T7 promoter	453–469
T7 transcription start	452
His·Tag coding sequence	362–380
Multiple cloning sites	
(NdeI - BamHI)	319–335
T7 terminator	213–259
pBR322 origin	3882
bla coding sequence	4643–5500

Figure 15.17 A pET expression vector. A gene cloned into the BamHI, XhoI, or NdeI site will be transcribed from the T7 phage promoter on the plasmid. If the cloned gene is in frame with the TIR (RBS in the figure), the gene will be translated with the amino acids shown in the figure fused to the N terminus of the polypeptide. In this particular pET vector, the upstream coding sequence includes the codons for a string of histidines that serve as a tag, allowing the purification of the fusion protein on affinity columns. The His tag can then be removed by cutting with the specific protease thrombin at the site indicated. Adapted with permission from the *Novagen Catalog,* p. 108, Novagen, Madison, Wis., 1997.

CHEMICAL MUTAGENESIS

One way to do site-specific mutagenesis is to treat the cloned DNA with chemicals that react with DNA in vitro, such as hydroxylamine, nitrous acid, formic acid, hydrazine, and bisulfate. The chemically treated DNA is then introduced into cells. Chemical mutagenesis is useful for making random mutations in cloned DNA but is less useful for making predetermined specific changes in the DNA sequence. Also, because chemical mutagens have specfic hot spots, or preferred sites of mutagenesis,

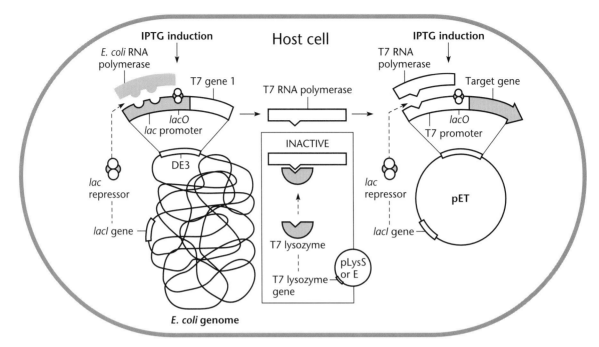

Figure 15.18 Strategy for regulating the expression of genes cloned into a pET vector. The gene for T7 RNA polymerase (gene 1) is inserted into the chromosome of *E. coli* and transcribed from the *lac* promoter; therefore, it will be expressed only if the inducer IPTG is added. The T7 RNA polymerase will then transcribe the gene cloned into the pET vector. If the protein product of the cloned gene is toxic, it may be necessary to further reduce the transcription of the cloned gene before induction. The T7 lysozyme encoded by a compatible plasmid, pLysS, will bind to any residual T7 RNA polymerase made in the absence of induction and inactivate it. Also, the presence of *lac* operators between the T7 promoter and the cloned gene will further reduce transcription of the cloned gene in the absence of the inducer IPTG. Reprinted with permission from the *Novagen Catalog*, p. 24, Novagen, Madison, Wis., 1995.

not all changes in the DNA sequence will be equally represented (see chapters 13 and 14).

SITE-SPECIFIC MUTAGENESIS WITH SYNTHETIC OLIGONUCLEOTIDE PRIMERS

DNA synthesis in vitro offers a way to make specific sequence changes in a DNA clone. These methods involve oligononucleotides, which serve as primers for the synthesis of new DNA, with the cloned DNA as a template. If the primer has a sequence slightly different from that of the DNA being mutagenized, the newly synthesized DNA will have the sequence of the primer rather than that of the original DNA template.

The original method for this type of site-specific mutagenesis is based on making the double-stranded replicative (RF) form of a single-stranded phage DNA such as M13mp18 (see chapter 7). As shown in Figure 15.19, the DNA to be mutagenized is first cloned in the single-stranded phage vector. An oligonucleotide com-

plementary to the region to be mutagenized, except for the change to be made, is synthesized and then hybridized to the single-stranded DNA. DNA polymerase uses the oligonucleotide as a primer to synthesize the complementary, or minus, strand of the M13 DNA, including the clone. The RF DNA is then ligated to give the covalently closed RF form (see chapter 7). When this DNA is transfected into cells, some of the progeny phage will have the sequence of the oligonucleotide primer rather than the original sequence in the clone. This method of site-specific mutagenesis can be used only to make minor changes, such as single-base-pair changes, in the DNA sequence. With larger changes, the primer will no longer hybridize to the clone to be mutagenized.

The major difficulty with this method lies in finding the mutated M13 clones among those with the original wild-type sequence. At most, only half of the M13 plaques could be due to M13 with the mutated clone, since only one of the two complementary strands of each

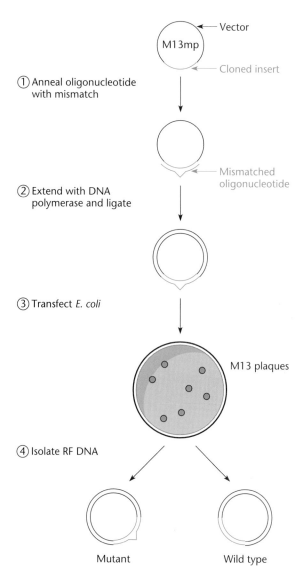

① Anneal oligonucleotide with mismatch

② Extend with DNA polymerase and ligate

③ Transfect *E. coli*

④ Isolate RF DNA

Figure 15.19 Site-specific mutagenesis with M13 and a mismatched oligonucleotide primer. See the text for more details.

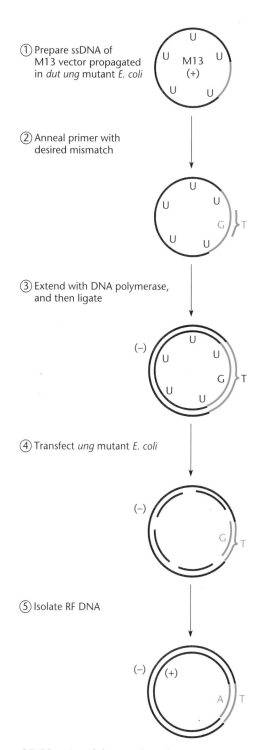

① Prepare ssDNA of M13 vector propagated in *dut ung* mutant *E. coli*

② Anneal primer with desired mismatch

③ Extend with DNA polymerase, and then ligate

④ Transfect *ung* mutant *E. coli*

⑤ Isolate RF DNA

Figure 15.20 Use of the uracil-*N*-glycosylase to eliminate the wild-type sequence after site-specific mutagenesis. See the text for details. ssDNA, single-stranded DNA; DNA Pol, DNA polymerase.

RF has the mutated sequence. The numbers are usually much lower than this, however. Even under the best of conditions, not all the M13 DNAs will have formed RFs, either because the replication is not complete or because the primer will not have hybridized to all the M13 templates. In the absence of any selection for phage with the mutant clone, we are left with the dismal prospect of screening the phage in many plaques by DNA sequencing or by hybridization with the mutated oligonucleotide to find phage clones with the mutation.

Consequently, several ways have been devised to increase the frequency of mutated M13 clones, including the one illustrated in Figure 15.20. In this method, the thymines in the M13 phage cloning vector are replaced with uracils by propagating the phage in a dUTPase- and uracil-*N*-glycosylase-deficient (Dut⁻ Ung⁻) host, in which many of the thymines in the DNA will be replaced with uracil (see chapter 3). The DNA from these phage is purified and used as the plus-strand template to make the minus strand with the mutated oligonucleotide primer as before. The double-stranded RF molecules are then transfected into Ung⁺ cells. The uracil-*N*-glycosylase in the Ung⁺ bacteria will recognize the uracils in the template plus strand and remove them, breaking the template strand. Most of the DNA strands that survive will be the minus strands, which contain the mutated sequence, so that the phage of most of the plaques will contain the mutated clone.

Another method for eliminating phage with the wild-type sequence is to use two primers and a phage or plasmid cloning vector that has a restriction site in the region of complementarity to one of the primers, usually in an antibiotic resistance gene (Figure 15.21). The primer complementary to the region of the restriction site differs by one nucleotide in the restriction site, so that the mutated sequence will not be cut by the restriction endonuclease. Both primers are used simultaneously to synthesize the new strand, as shown in Figure 15.21. If this DNA is used to transfect cells and the double-stranded RF phage DNA is isolated and cut with the restriction endonuclease, only the RFs that are descended from the newly synthesized strand and that therefore have the mutated sequence will not be cut by the restriction endonuclease and will survive.

RANDOM MUTAGENESIS WITH "SPIKED" PRIMERS

In the example shown in Figure 15.19, site-specific mutagenesis was used to change one predetermined base pair in the DNA sequence. This method can also be adapted to allow random mutagenesis of a DNA region so that every possible base pair change will be represented in the population of molecules. Instead of a well-defined primer, a mixture of "spiked," or contaminated, oligonucleotide primers is used to mutagenize the DNA. These oligonucleotide primers are synthesized with the deliberate intention of making mistakes. The nucleotide added at each step of the synthesis is deliberately contaminated with a low concentration of the other three nucleotides; the concentration of the contaminating nucleotides is adjusted to make one mistake, on average, in each of the oligonucleotides. When these misfits are used as primers for the synthesis of complementary strands, as above, the DNA synthesized will have a random collection of changes in the region being mutagenized. This method has the advantage that it can be used to make all of the possible base pair changes in a region without preferentially mutagenizing hot spots like chemical mutagens do.

Once the desired change has been made in the cloned gene, it can be recloned from M13 into an expression vector to determine the effect of the change on the activity of the protein product of the gene.

Polymerase Chain Reaction

The **polymerase chain reaction (PCR)** is one of the most useful methods in molecular genetics and molecular biology. This method makes it possible to selectively amplify regions of DNA out of much longer DNAs. The method is so sensitive that it can detect and amplify sequences from just a few molecules of DNA and has applications in many areas, including physical mapping of DNA, gene cloning, mutagenesis, and DNA sequencing.

The principles behind PCR are outlined in Figure 15.22. The method takes advantage of the properties of DNA polymerases—that they require a primer and a template and polymerize DNA from the 3' end of the primer to make a complementary copy of the template

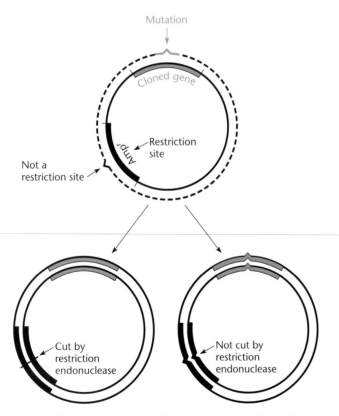

Figure 15.21 Use of two primers to eliminate the wild-type sequence after site-specific mutagenesis. See the text for details.

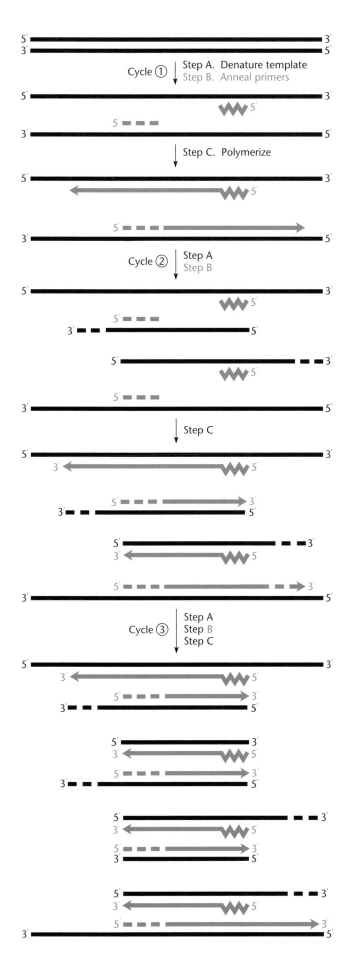

Figure 15.22 The steps in a PCR. In the first cycle, the template is denatured by heating. Primers are added; they hybridize to the separated strands for the synthesis of the complementary strand. The strands of the DNA are separated by the next heating cycle, and the process is repeated. The DNA polymerase will survive the heating steps because it is from a thermophilic bacterium.

(see chapter 1). PCR uses two primers, one that has the same sequence as the coding strand upstream of the region to be amplified and another that has the same sequence as the noncoding strand downstream of the region to be amplified. The DNA is denatured to separate the strands and hybridized to the primers. One primer will prime the synthesis of DNA over the region to be amplified, and the other primer will then hybridize to this newly synthesized strand and prime replication back over it to make a double-stranded DNA. The strands of this double-stranded DNA are then separated by heating, the primers are reannealed to the single-stranded DNAs, and the process is repeated until the sequences in the DNA between the regions to which the two primers are complementary has been amplified many times. By using PCR, micrograms of a DNA region can be synthesized starting from only a few molecules containing the DNA region.

The first attempts at such techniques were performed with the DNA polymerase of *E. coli*. However, since the *E. coli* DNA polymerase is inactivated by the temperatures required to separate the DNA strands, new DNA polymerase had to be added at each step, an impracticality. The insight that made PCR so useful was to do the reaction with the DNA polymerase of a thermophilic bacterium, which could grow at extremely high temperatures. The enzymes of thermophilic bacteria are generally much more resistant to heating than are the enzymes of mesophilic bacteria like *E. coli* and so will survive the high temperatures required to separate the strands of the DNA at each step. The most commonly used DNA polymerase, the *Taq* polymerase, is from the extreme thermophile *Thermus aquaticus*.

Uses of PCR

PCR MUTAGENESIS
PCR can be used either to make specific changes in a DNA sequence or to randomly mutagenize a region of DNA. Making specific changes by PCR is similar to the other means of site-specific mutagenesis. A complementary primer is made but with the desired change in the sequence. When the polymerase uses the primer to amplify the region, the specfic change will be made in the sequence.

PCR can be used to make random changes in a sequence because the *Taq* polymerase makes many mistakes, particularly in the presence of manganese ions. In fact, the mistake level during normal amplification by

Taq polymerase is so high that clones made from PCR fragments should usually be sequenced to be certain that no unwanted mutations have been introduced. The *Taq* polymerase makes so many mistakes because it lacks an editing function (see chapter 1). The DNA polymerases from some other thermophilic bacteria have editing functions and so make many fewer mistakes. Unfortunately, the other DNA polymerases that have been tested function somewhat less well for amplification than the *Taq* polymerase does. They are nevertheless becoming popular for some applications.

Introducing Restriction Sites

PCR is also useful for adding sequences to the ends of the amplified fragment. Although the primers used for PCR amplification must be complementary to the sequence being amplified at the 3′ end, they need not be complementary at their 5′ end. Therefore, they can include, for example, sites for specific restriction endonucleases, which could be used to subclone a fragment with defined end points.

Figure 15.23 illustrates how we can use PCR amplification to introduce restriction sites for this purpose. As shown, just the ORF of a gene is being amplified from a larger clone, and *Bam*HI sites (GGATC/CTAGG) are being introduced both upstream and downstream of the gene. First, two oligonucleotide primers are made. One has the same sequence as the coding strand of the DNA to be amplifed, including the initator of the gene,

ATG. This primer also has extra sequences, including the *Bam*HI restriction site at its 5′ end. The other oligonucleotide primer has the sequence of the template strand, including the terminator codon, TAG; therefore, it will hybridize to the other end of the gene. It also has extra sequences at its 5′ end, including the sequence of the *Bam*HI restriction site. When these two primers are used to PCR amplify the DNA, one oligonucleotide will serve as a primer for the first round of replication, making a DNA strand that continues past the region to which the other primer is complementary. In the next round, this new strand is used as a template with the other oligonucleotide as a primer. This fragment will now include *Bam*HI restriction site sequences at both ends, when there were none in the original. Amplification will continue, and after many cycles of heating and cooling, millions of copies of this fragment will be made that are all the same length and all have *Bam*HI recognition sites at both ends. The amplified fragment can be purified by gel electrophoresis, cut with *Bam*HI, and cloned into a compatible restriction site in an expression vector. Moreover, the primers can be designed so that when the amplified fragment is cloned into the expression vector, the AUG codon for the gene will be approximately 5 to 7 bases downstream of the Shine-Dalgarno sequence in the cloning vector, or the ORF of the gene will be in frame with the upstream or downstream sequences encoding a tag (see above).

Figure 15.23 Use of PCR to add convenient restriction sites to the ends of an amplified fragment. The primers contain sequences at their 5′ ends that are not complementary to the gene to be cloned but, rather, contain the sequence of the cleavage site for *Bam*HI (underlined). The amplified fragment will contain a *Bam*HI cleavage site at both ends. See the text for details.

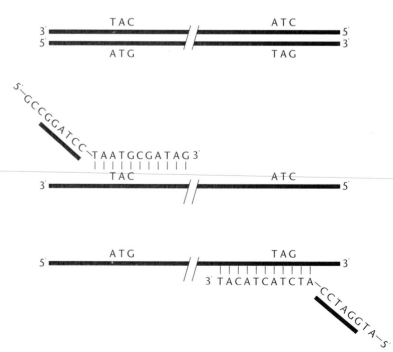

SUMMARY

1. Type II restriction endonucleases recognize defined sequences in DNA and cut at a defined position in or near the recognition sites, which has made them very useful in physical mapping of DNA.

2. The physical map of a DNA shows the location of restriction sites in the DNA. For some types of bacteria and viruses, the physical map has been correlated with the genetic map, showing the position of genes relative to the restriction sites.

3. Most type II restriction endonucleases cut at the same position in both strands of symmetric sequences. If the cuts are not immediately opposite each other, single-stranded sticky ends will form that can hybridize to any other sticky end cut with the same or a compatible enzyme. This property has made these enzymes very useful for DNA cloning and for DNA manipulations in vitro.

4. Other type II restriction endonucleases cut in symmetric sequences but cut the two strands exactly opposite each other, leaving blunt ends. These enzymes are also very useful because blunt ends are, in a sense, all compatible with each other and so can be ligated together. It is often useful to create blunt ends from the sticky ends left by other enzymes by filling in with DNA polymerase or removing the single-stranded ends with single-strand-specific nucleases.

5. Cloned DNA consists of multiple copies of a sequence descended from a single molecule. If a piece of DNA is inserted into a cloning vector, the DNA will replicate along with the vector, making millions of clones of the original DNA.

6. A DNA library is a collection of clones that, among themselves, contain all the DNA sequences of the organism.

7. Clones containing a particular gene can be identified by screening libraries by means of such methods as plate hybridization, marker rescue, and complementation. The method used depends on the situation.

8. There are a number of methods available to locate a gene on a large clone, including transposon mutagenesis, subcloning, and making deletions in vitro.

9. Expression vectors can be used to synthesize the product of a cloned gene. These vectors have promoters and translation initiation sites that can be used to express foreign genes.

10. PCR amplifies defined regions of DNA. Micrograms of DNA can be made from a few molecules by this method. PCR can be used for mutagenesis and for introduction of restriction sites for cloning. PCR uses the DNA polymerase from thermophilic bacteria.

QUESTIONS FOR THOUGHT

1. Why do you suppose some bacteria encode eight-hitter restriction endonucleases in their chromosome? How would you test your hypothesis?

2. Many phages and plasmids encode antirestriction proteins that help them neutralize host restriction systems. These proteins often bind to the restriction enzymes and inhibit them. List some other ways you think phages and plasmids could avoid host restriction.

3. A current question in paleontology is whether dinosaurs were warm-blooded reptiles, unlike the poikilothermic reptiles alive today. If you could get some dinosaur DNA, how would you use PCR and DNA cloning in expression vectors to determine if dinosaurs were warm-blooded?

PROBLEMS

1. Which of the following sequences is most likely to be the recognition site for a type II restriction endonuclease: GAATCG, GATATC, AAATTT, or ACGGCA? The sequence of only one of the two strands is shown.

2. What percentage of the hybrid sites made by ligating Sau3A-cut DNA to BamHI-cut DNA can be cut by BamHI?

3. One way to create blunt ends out of sticky ends with 5′ overhangs is to fill in the sticky ends with the Klenow fragment of DNA polymerase. This method also offers a way of removing restriction sites from DNAs. If a DNA is cut with a restriction endonuclease that leaves 5′ overhangs and the ends are filled in and religated, the site will usually be lost. For

which of the following restriction endonucleases is the site not lost after such a treatment: BamHI (G*GATCC), TaqI (T*CGA), or BssHII (G*CGCGC)? The asterisk shows where the sequence is cut.

4. You have isolated a nonrevertible hemA mutant of E. coli that requires δ-aminolevulinic acid to make hemes required for growth on succinate. You wish to use your mutant to clone the hemA gene. You do a partial digest of E. coli DNA with Sau3A, in which the average-sized piece is about 2 kbp, and clone the pieces into plasmid pBR322 cut with BamHI and treated with phosphatase. The ligation mix is then used to transform the hemA mutant, selecting the ampicillin resistance

gene on the plasmid. The colonies are then tested for growth on plates containing δ-aminolevulinic acid. How many Amp^r transformants should you have to test to have a reasonable chance of finding a colony that contains bacteria no longer requiring δ-aminolevulinic acid for growth on succinate? There is about 4,500 kbp of DNA in the *E. coli* genome.

5. One of the transformant colonies you test contains a few bacteria that no longer require δ-aminolevulinic acid, but most of the bacteria in the colony still require it. Do you think the clone has all of the *hem*A gene on it? Why or why not? Is the HemA$^+$ phenotype in these few transformants due to recombination or complementation?

6. You have cloned the gene for a protease from *Serratia marcescens* that you suspect is a zinc-dependent metallopro-

tease because it contains two histidines and a glutamate that might coordinate with the zinc. Outline how you would change one of the histidines and see if it is essential for the activity of the protease.

7. What are some of the reasons why a gene might be excluded from a DNA library?

8. You have isolated Thy$^-$ mutants of *E. coli* that presumably have mutations in the *thy*A gene. Outline how you would use PCR to amplify the mutant *thy*A genes from the chromosome to determine the sequence change in the mutations. The wild-type sequence of the *thy*A gene of *E. coli* is known.

SUGGESTED READING

Cohen, S. N., A. C. Y. Chang, H. W. Boyer, and R. B. Helling. 1973. Construction of biologically functional bacterial plasmids *in vitro*. *Proc. Natl. Acad. Sci. USA* **70**:3240–3244.

Danna, K. J., and D. Nathans. 1971. Specific cleavage of simian virus 40 DNA by restriction endonuclease of *Haemophilus influenzae*. *Proc. Natl. Acad. Sci. USA* **68**:2913–2917.

Jackson, D. A., R. H. Symons, and P. Berg. 1972. Biochemical method for inserting new genetic information into DNA of simian virus 40: circular SV40 DNA molecules containing lambda phage genes and the galactose operon of *Escherichia coli*. *Proc. Natl. Acad. Sci. USA* **69**:2904–2909.

Kohara, Y., K. Akiyama, and K. Isono. 1987. The physical map of the whole *E. coli* chromosome: application of a new strategy for rapid analysis and sorting of a large genomic library. *Cell* **50**:495–508.

Linn, S., and W. Arber. 1968. Host specificity of DNA produced by *Escherichia coli* X: *in vitro* restriction of phage fd replicative form. *Proc. Natl. Acad. Sci. USA* **59**:1300–1306.

Wilson, G. G. 1991. Organization of restriction-modification systems. *Nucleic Acids Res.* **19**:2539–2566.

CLONING MANUALS

Ausubel, F. M., M. R. Brent, R. E. Kingston, D. D. Moore, J. G. Seidman, J. A. Smith, and K. Struhl. 1987. *Current Protocols in Molecular Biology*. Greene Publishing Associates and Wiley Interscience, New York.

Sambrook, J., E. F. Fritsch, and T. Maniatis. 1989. *Molecular Cloning: a Laboratory Manual*, 2nd ed. Cold Spring Harbor Laboratory Press, Cold Spring Harbor, N.Y.

Chapter 16

Molecular Genetic Analysis and Biotechnology

T HE LAST CHAPTER reviewed some methods that have been developed to clone genes and to manipulate DNA in the test tube. These methods included site-specific mutagenesis in which certain regions or even individual base pairs of a cloned DNA can be changed in predetermined ways. These methods are based on the use of bacteria, which also serve as the source of most of the enzymes involved. With the advent of techniques for reintroducing DNA into living cells, it has often become possible to determine the effects that predetermined changes in DNA sequence have on the living organism. This general approach, called **molecular genetic analysis,** has revolutionized our understanding of how living cells and organisms develop and function at the molecular level. It has also allowed the spectacular developments in biotechnology that we read about every day.

Throughout the book, we have used examples based on the molecular genetic approach. This was necessary because much of what is known about bacterial genes and their regulation has come from this approach. However, we have not yet discussed these methods in detail or how the data are actually obtained and interpreted. In this chapter, we go into more detail about the methods of molecular genetics as applied to bacteria and also give some current examples of how these methods are being applied in modern biotechnology.

Deletion Mapping of Protein Functional Domains

Many proteins have more than one function that is carried on the same polypeptide chain. These functions are often performed by separate, identifiable regions of the protein called **functional domains**. Mutants can sometimes be isolated with mutations that specifically inactivate one of these functions without affecting the others. Mapping these mutations within the gene may help localize the functional domains of a protein. We already applied this approach to finding the domains of the LacI protein, the repressor of the *lac* operon, discussed in chapter 14. However, the classical genetic approach used in this analysis was laborious because each deletion had to be selected and its endpoints had to be mapped within the *lacI* gene. By making it easier to isolate deletions with known endpoints, molecular genetics has greatly facilitated the mapping of functional domains. The deletions can be isolated in the cloned gene in the test tube, and their endpoints can be determined by measuring the size of what remains of the gene. The deleted clones can then be crossed with mutations in the chromosome to determine if the mutation lies outside the deleted region.

Mapping the Thymidylate Synthetase Enzyme

The thymidylate synthetase enzyme, which makes dTMP for DNA synthesis in *Escherichia coli* (see Figure 1.5), provides a good example of a protein with multiple functions. In catalyzing the synthesis of dTMP, the enzyme must bind the substrate dUMP and the methyl donor tetrahydrofolate. The enzyme is also inhibited by 5-fluoro-dUMP, an analog of dTMP. Therefore, we know that the enzyme has at least three functions, which may reside in separable functional domains.

The *thyA* gene of *E. coli* is particularly amenable to molecular genetic analysis because a positive selection (see chapter 14) exists for *thyA* mutants, which can multiply in the presence of the dihydrofolate reductase inhibitor trimethoprim provided that thymine is added to the medium. Wild-type *thyA*⁺ *E. coli* cannot multiply under these conditions. This positive selection makes it easy to isolate large numbers of *thyA* missense mutations, some of which may specifically inactivate one of the specific functions of the thymidylate synthetase.

Mutations affecting the various functions of the thymidylate synthetase were mapped by being crossed with deletions extending known distances into the *thyA* gene (see Belfort and Pedersen-Lane, Suggested Reading). However, rather than selecting mutations in vivo, these investigators made the deletions in a clone of the *thyA* gene in vitro. A plasmid vector containing a clone of the *thyA* gene was cut with a restriction endonuclease and digested with *Bal* 31 exonuclease (see chapter 15 for a description of the methods). The endpoints of the deletions in the *thyA* gene were then determined by measuring the size of the remaining clone in each deletion (Figure 16.1).

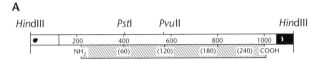

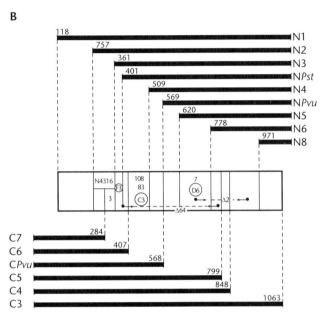

Figure 16.1 (A) Map of the *thyA* gene of *E. coli*. (B) Mutations in the chromosome were mapped by being crossed with deletions extending various distances into the cloned *thyA* gene of *E. coli*. Solid bars show the regions deleted in each of the constructs. N is the amino terminus of the gene, and C is the carboxy terminus. Redrawn from M. Belfort and J. Pedersen-Lane, *J. Bacteriol.* **160:**371–378, 1984.

Once they had constructed the deleted clones, these investigators introduced the deletions of the *thyA* gene into bacterial cells with one of the chromosomal *thyA* missense mutations by using an M13 phage cloning vector. They introduced deleted clones in two ways. One was to reclone the deleted DNA into an M13 phage cloning vector (see chapter 7). Phage containing each of the deleted clones were used individually to infect cells with each of the chromosomal *thyA* missense mutations to be mapped. Phage M13 does not kill the host; therefore, in general, infected cells can form colonies. However, most of these infected cells will be unable to multiply to form a colony on minimal medium without thymine because they are Thy⁻ since the deleted copy of the *thyA* gene in the M13 vector and the mutated *thyA* gene in the chromosome are both inactive. However, if the missense mutation in the *thyA* gene in the chromosome lies outside the deleted region in the M13 clone, a few rare Thy⁺ recombinants will appear that can multiply to form colonies under these conditions. Thus, these investigators could map a large number of *thyA* mutations to small regions of the *thyA* gene. The positions of the mutations in the *thyA* gene could then be correlated with the function of the thymidylate synthetase that they specifically inactivate.

Regulation of Plasmid Replication

Molecular genetics has been particularly useful in mapping plasmid mutations and in studies of the control of plasmid replication and copy number. Plasmids are particularly amenable to molecular genetic analysis because they are relatively small and can replicate autonomously. Two types of plasmids, ColE1 and iteron plasmids, which use very different mechanisms of replication control, provide an excellent illustration of how molecular genetic techniques have been used to study copy number control.

Replication Control of ColE1-Derived Plasmids

To reiterate our discussion in chapter 4, the ColE1-type plasmids have a high copy number and use two plasmid-encoded RNAs, RNA I and RNA II, to regulate their replication. RNA II is the primer for initiation of replication; it must be processed before it can function. RNA I inhibits replication by binding to RNA II and preventing its processing. The more copies of the plasmid the cell contains, the more RNA I is made and the more plasmid replication is inhibited. Thus, the plasmid replicates until its concentration reaches its copy number. Then the high concentration of RNA I inhibits further replication of the plasmid.

The RNA I of a ColE1 plasmid is also responsible for determining the incompatibility (Inc) group of the plasmid. Members of a given Inc group cannot stably coexist in a culture (see chapter 4). All ColE1-derived plasmids are members of the same Inc group because they synthesize the same RNA I and will inhibit each other's replication.

Genetic studies of plasmid replication control are complicated because mutations that inactivate the plasmid orgin of replication or allow its regulation to run uncontrolled cannot be isolated since they will either make production of the plasmid impossible or kill the host cell, respectively. Therefore, a molecular genetic approach was necessary to study the copy number control and incompatibility determinants of ColE1-type plasmids (see Lacatena and Cesareni, Suggested Reading). To isolate mutations in the ColE1 plasmid origin of replication, these investigators constructed a **phasmid** (part phage and part plasmid) with two *ori* regions, one from ColE1 and the other from λ phage (Figure 16.2),

by inserting the ColE1 plasmid into a λ cloning vector. The phasmid also contained *attB* and *attP* sites on either side of the plasmid. Thus, in cells containing an excess of the λ integrase enzyme, recombination between the *attB* and *attP* sites promoted by the integrase would excise the plasmid. Once out of the phasmid, the mutant plasmids could be studied to determine whether they could regulate replication.

With two origins of replication, the phasmid can replicate and form plaques under conditions that inactivate either one of its origins of replication, as long as the other origin remains functional. The plasmid origin of replication will be inactive if the cells contain another ColE1-derived plasmid, and the λ phage origin of replication will be inactive if the phasmid is plated on a λ lysogen. In the first case, the RNA I from the resident ColE1 plasmid will inhibit replication from the ColE1 *ori* site of the phasmid, but the phasmid will still be able to replicate from the phage origin (Figure 16.2). In the latter case, the repressor made by the λ prophage prevents the synthesis of the λ O and P proteins as well as the transcription of the λ *ori* region, which are all required for replication from the λ origin (see chapter 7). Under these conditions, the phasmid can replicate from the plasmid origin of replication. As a consequence, the phasmid can replicate and form plaques on cells that either are λ lysogens or contain a ColE1-derived plasmid (Figure 16.2, steps 1 and 2). However, if the cells harbor both a λ prophage and a ColE1-derived plasmid, plaques will not form, because the phasmid will be unable to replicate from either origin of replication.

These investigators reasoned that it might be possible for the phasmid to replicate on a λ lysogen carrying a ColE1-type plasmid if the ColE1 replication origin were mutated so that its replication was no longer inhibited by the resident plasmid. Accordingly, they plated millions of mutagenized phasmids on a λ lysogen carrying a ColE1 plasmid (Figure 16.2, step 3). A few plaques formed, presumably from phasmids with mutations in either the plasmid or the λ origin of replication (Figure 16.2, step 4). Phasmids picked from these plaques were then used to infect cells with excess λ integrase (Figure 16.2, step 5) in order to excise the plasmid, as described above, and determine whether its control of replication had been altered. Two types of mutations in the ColE1 origin were observed and are discussed below.

MUTATIONS THAT PREVENT INTERACTION OF RNA I AND RNA II

No transformants appear when cells are transformed with plasmids with this type of mutation. Presumably, these mutations cause the transformants to be killed,

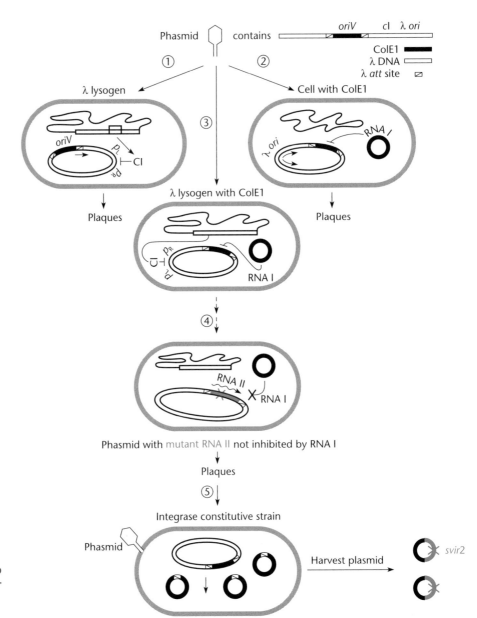

Figure 16.2 A phasmid method for isolating copy number mutants of the ColE1 plasmid. The phasmid has two origins of replication and so can replicate in either a λ lysogen or cells harboring a ColE1-derived plasmid.

apparently by uncontrolled, runaway replication of the plasmid. Presumably, the RNA I is no longer able to interact with RNA II to inhibit primer formation, and the plasmid continues to replicate until the cells are killed. Note that such a mutation will not prevent plaque formation by the phasmid, since the phasmid replication does not need to be controlled for a plaque to form.

MUTATIONS THAT CHANGE THE Inc GROUP
The other type of mutation is more interesting. As shown in Figure 16.3A, the plasmid in these mutant phasmids can be excised and used to transform bacteria, where they will be stably maintained even in the pres-

ence of a wild-type ColE1-derived plasmid or another mutated plasmid excised from a different mutant phasmid. However, the mutant phasmids will *not* form plaques on cells containing their *own* excised plasmid.

We can deduce what is happening by examining the model for copy number control of ColE1 plasmids shown in Figure 4.4. According to the model, RNA I must bind to RNA II through complementary base pairing to inhibit plasmid replication. Because the two RNAs are made from opposite strands of the DNA in the same region, a mutant RNA I will be completely complementary to an RNA II made from the same plasmid. However, the mutant RNA I will not be completely

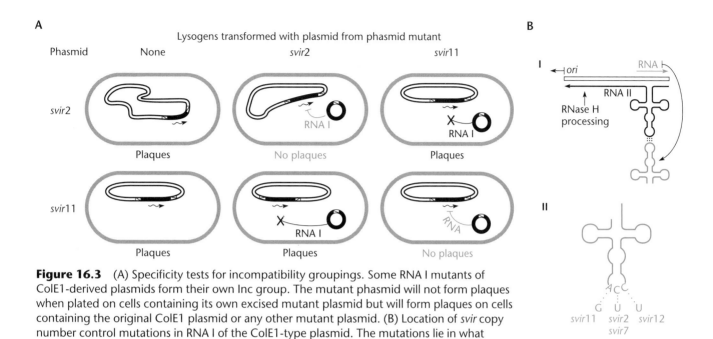

Figure 16.3 (A) Specificity tests for incompatibility groupings. Some RNA I mutants of ColE1-derived plasmids form their own Inc group. The mutant phasmid will not form plaques when plated on cells containing its own excised mutant plasmid but will form plaques on cells containing the original ColE1 plasmid or any other mutant plasmid. (B) Location of *svir* copy number control mutations in RNA I of the ColE1-type plasmid. The mutations lie in what would be the anticodon region of a tRNA-like structure. Redrawn with permission from R. M. Lacatena and G. Cesareni, *Nature* (London) **294**:623–626, 1981.

complementary to an RNA II made by a different plasmid, and so it cannot prevent that RNA II from being processed. Hence, the mutant RNA I will not regulate the replication of another ColE1 plasmid, but it will regulate its own replication. Apparently, the origin of replication of the mutant plasmid now belongs to a different Inc group from other ColE1-derived plasmids. Indeed, it now defines its own Inc group! In any case, the fact that changing RNA I and RNA II in complementary ways can create new incompatibility groups offers a dramatic confirmation of the model for ColE1 regulation presented in chapter 4.

The positions of some of the changes that alter the incompatibility group are also intriguing. Figure 16.3B shows some of these changes. RNA I can be drawn as a cloverleaf structure reminiscent of a tRNA, and the changes occur in what would correspond to the anticodon loop. However, it is not clear whether this cloverleaf structure ever forms and what relationship, if any, this structure (if it does form) has to the regulation. In any case, the RNA I nucleotides changed in each of the mutant phasmids must be important for the pairing of RNA I with RNA II.

Replication Control of Iteron Plasmids

The replication of the iteron plasmids is regulated very differently from replication of the ColE1-type plasmids (see chapter 4). Iteron plasmids use a protein, often

called RepA, rather than an RNA like RNA I to regulate their copy number. This protein binds to a number of repeated sequences, called iterons, close to their origin of replication. The presence of iteron sequences is the distinguishing feature of these plasmids.

EFFECT OF INCREASING THE RepA CONCENTRATION

Recent experimentation, mostly on the iteron plasmids P1, R6K, and RK2, led to the coupling model for how the replication of the iteron plasmids is regulated (see Figure 4.7 and McEachern et al., Suggested Reading). Some of the experiments that suggested the coupling model were designed to determine the effect of varying the concentration of RepA protein on the copy number of the plasmid. The argument was that if the copy number is regulated exclusively through the concentration of the RepA protein, increasing the concentration of RepA should increase the plasmid copy number. In the experiments, the RepA protein was expressed from a clone of the *repA* gene in another plasmid of a different Inc group (Figure 16.4A). Thus, the investigators could evaluate the effects of RepA concentration on plasmid copy number without the complication of the amount of RepA also depending on the number of plasmid copies. The experiments showed that increasing the concentration of the RepA protein even hundreds of times (by expressing the *repA* gene from a very strong

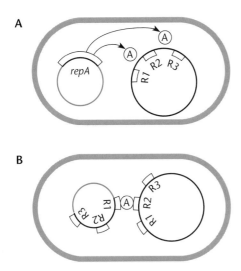

Figure 16.4 Molecular genetic analysis of the regulation of iteron plasmids. (A) The RepA protein is expressed from a clone of the *repA* gene in an unrelated plasmid cloning vector. (B) Additional iteron sequences in an unrelated plasmid can lower the copy number of an iteron plasmid. R1, R2, and R3 are iteron sequences.

promoter on the other plasmid) led to only a modest increase in the copy number.

To explain this observation, the investigators first proposed that the RepA protein exerts both positive and negative effects on replication. At low concentrations, the RepA protein might act as an activator of replication, but at high concentrations it might act as an inhibitor, preventing further replication. In support of this idea, mutants that had mutations in the *repA* gene and whose copy number was higher were isolated. It was proposed that these mutations, called copy-up mutations, inactivated the negative function of the RepA protein on replication without affecting its positive role in activating replication.

EFFECT OF INTRODUCING MORE ITERON SEQUENCES

Additional evidence concerning the regulation of iteron plasmids could not be easily explained by this initial, simple model. This evidence indicated that replication of the plasmid was sensitive not only to the concentration of RepA but also to the concentration of the plasmid itself or, more precisely, to the concentration of iteron sequences in the cell. When another plasmid of a different Inc group, containing a few copies of the iteron sequence but no *repA* gene, was introduced into the same cell with the original plasmid (Figure 16.4B), the copy number of the original plasmid decreased. Somehow, the presence of extra copies of the iteron sequence in the cell reduces the plasmid copy number, even when the additional sequences are in an unrelated plasmid that does not depend on RepA for its replication.

One possible explanation for this result is that extra copies of the iteron will **titrate**, or in a way soak up, the RepA protein, so that not enough is available to initiate replication. If so, further overproduction of the RepA protein until more is available than can be titrated by the extra iterons should overcome the negative effect of the extra iteron sequences. However, even if RepA protein is grossly overproduced, the effect of the extra iteron sequences on plasmid copy number remains.

COUPLING MODEL FOR REGULATION OF ITERON PLASMIDS

The coupling model for the regulation of iteron plamids discussed in chapter 4 provides an explanation for the puzzling observations described above. According to this model, RepA can bind to the iteron sequences of more than one plasmid at the same time, effectively coupling the plasmids, or handcuffing them together. Higher levels of RepA protein increase the amount of coupling between plasmids, inhibiting their replication. Moreover, the presence of iteron sequences on an unrelated plasmid will cause coupling between the two plasmids, lowering the copy numbers of both.

The coupling model also explains the behavior of the copy-up mutations in the *repA* gene. According to this model, these copy-up mutations change the RepA protein so that it binds more weakly to the iteron sequences. Then the model predicts that higher than normal concentrations of these mutant RepA proteins would be required to overcome the effect of additional iteron plasmids in the cell. This prediction was confirmed. Increasing the concentration of copy-up RepA protein does overcome the effect of extra iterons, unlike increasing the concentration of the wild-type RepA protein, which has no effect.

More direct support for the coupling model in the replication control of iteron plasmids has come from electron microscopic pictures of purified iteron plasmids mixed with the purified RepA protein for that plasmid. In these pictures, two plasmid molecules can often be seen linked to each other by RepA protein.

Molecular Genetics and Transposition

The study of transposons and the development of molecular genetics have enjoyed a mutually advantageous relationship. Many techniques in molecular genetics depend upon transposons, and much of what we know about transposons has come from molecular genetic studies.

The usefulness of transposons in molecular genetics derives from their ability to hop randomly into DNA and introduce new selectable genes and restriction sites for genetic and physical mapping. They can also be used to detect genes subject to particular types of regulation and for in vivo cloning. In this section, we discuss some of the applications of transposons in molecular genetics.

Genetic Requirements for Transposition

Like plasmids, transposons are relatively small and so are particularly amenable to molecular genetic analysis. One application of molecular genetics is the determination of the genetic requirements for transposition of a transposon.

IDENTIFYING REGIONS ENCODING TRANSPOSITION FUNCTIONS

The first analysis of the genetic requirements for transposition used transposon Tn3, which transposes by a replicative mechanism (see Gill et al., Suggested Reading, and Figure 8.8 for a picture of Tn3). To determine the genetic requirements for transposition of Tn3, a plasmid containing the transposon was cut randomly with DNase and then DNA linkers containing the recognition sequence for a restriction endonuclease were ligated into the plasmid (see chapter 15 for a description of the methods). The presence of the restriction site on the insertion makes the site of the mutation easy to map physically (see chapter 15). These insertions also disrupt the DNA sequences and, if inserted into a translated open reading frame (ORF), usually cause frameshifts or create in-frame nonsense codons.

Insertion mutations scattered in various places around the transposon were tested for their effects on transposition by the mating-out assay (see chapter 8). As illustrated in Figure 16.5, cells containing both a small, nonmobilizable plasmid carrying the mutant Tn3 and a larger self-transmissible plasmid were mixed with recipient cells, and tranconjugants resistant to ampicillin were selected. Because the smaller plasmid was not mobilizable, ampicillin-resistant transconjugants could be produced only by donor cells in which the transposon had hopped from the smaller plasmid into the larger, self-transmissible one. When no ampicillin-resistant transconjugants were observed, the mutation in the transposon must have prevented transposition into the self-transmissible plasmid. When larger than normal numbers of ampicillin-resistant transconjugants were observed, the mutation must have increased the frequency of transposition.

The effect of a mutation depends on where it is in the transposon. As shown in Figure 16.6, mutations in the inverted repeat (IR) sequences and mutations that disrupted the *tnpA* ORF can totally prevent transposition. In contrast, mutations that disrupt the *tnpR* ORF result in higher than normal rates of transposition and the formation of cointegrates, in which the self-transmissible and nonmobilizable plasmids are joined and are transferred together into the recipient strain. Mutations in the short sequence called *res* (for *res*olution sequence) also gave rise to cointegrates, but unlike *tnpR* mutations, they result in normal rates of transposition.

COMPLEMENTATION TESTS WITH TRANSPOSON MUTATIONS

Complementation tests were used to determine which mutations in transposon Tn3 disrupt *trans*-acting functions and which disrupt *cis*-acting sequences or sites. The complementation tests used the mating-out assay illustrated in Figure 16.5, except that the cell in which the transposition was to occur also contained another Tn3-related transposon inserted into its chromosome (Figure 16.7). If the mutation in the plasmid transposon inactivates a *trans*-acting function, this transposon should be able to transpose because the chromosomal transposon can furnish any *trans*-acting functions that the mutant transposon cannot make for itself. However, if the mutation in the plasmid transposon inactivates a *cis*-acting site, it will not be able to transpose properly, even in the presence of the chromosomal transposon, since mutations that inactivate *cis*-acting sites cannot be complemented. The Tn3-like transposon used to provide complementing functions was one that could transpose but lacked a functional ampicillin resistance gene. Thus, the transposition of only the original mutant transposition would be indicated by the appearance of ampicillin-resistant transconjugants after the mating.

The complementation tests revealed that mutations that inactivate both *tnpA* and *tnpR* genes could be complemented to give normal transposition. However, neither mutations in the IR sequences at the ends of the transposon nor those in the sequence called *res* could be complemented. The investigators concluded that *tnpA* and *tnpR* encode *trans*-acting proteins while IR and *res* are *cis*-acting sites on the DNA.

These genetic data prompted formulation of the model presented in chapter 8 for replicative transposition of Tn3-like transposons. Briefly, mutations in the *tnpA* gene prevent transposition because the *tnpA* gene encodes the transposase TnpA, which promotes transposition. Moreover, *tnpA* mutations can be complemented because the transposase can be furnished in *trans* by another transposon. Mutations in the inverted repeats at the ends of the transposon also prevent trans-

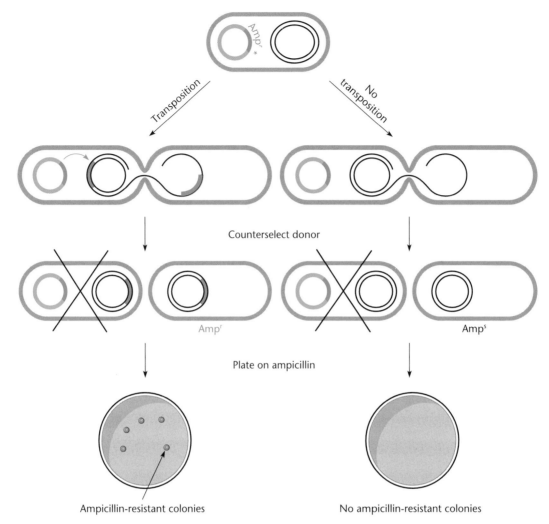

Figure 16.5 Molecular genetic analysis of transposition of the replicative transposon Tn*3*. The transposon is in blue. The asterisk marks the position of the mutation. Amp^r, ampicillin resistance. See the text for details.

position, because these are the sites at which the TnpA transposase acts to promote transposition. However, IR mutations cannot be complemented, because they affect a *cis*-acting site.

The behavior of mutations in *tnpR* and *res* was more difficult to explain. To reiterate, *tnpR* mutations not only cause higher than normal rates of transposition but also cause the formation of cointegrates. They can be complemented, resulting in normal rates of transposition and no cointegrates. Mutations in *res* also cause cointegrates to form but do not affect the frequency of transposition and cannot be complemented.

To explain these results, the investigators proposed that *tnpR* encodes a protein with two functions. First, the TnpR protein acts as a repressor (like the LacI repressor), which represses the transcription of the *tnpA*

gene for the transposase. Therefore, *tnpR* mutations cause higher rates of transposition by allowing more TnpA synthesis. Second, the TnpR protein acts as a recombinase that resolves cointegrates by promoting site-specific recombination between the *res* sequences in the two copies of the transposon in the cointegrate (see Figure 8.10). This function explains why both *tnpR* and *res* mutations cause the accumulation of cointegrates. Either type will prevent the site-specific recombination that resolves the cointegrates. It also explains why *tnpR* mutations but not *res* mutations can be complemented to give normal transposition; the former encodes a diffusible gene product, and the latter is a site. The TnpR protein can be provided by another transposon in the cell, but the *res* mutations affect a *cis*-acting site and so cannot be complemented.

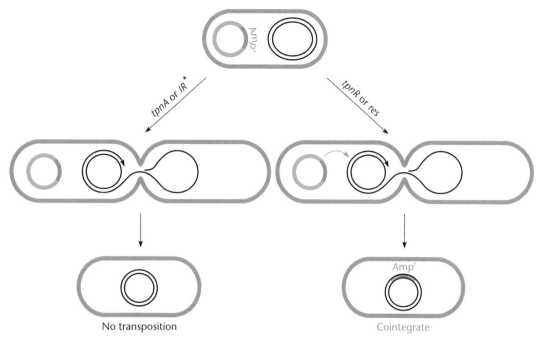

Figure 16.6   Effects of mutations in different genes required for transposition of Tn*3*. In the left-hand pathway, a *tpnA* or *IR* mutation prevents transposition, and so no Amp^r transconjugants form. In the right-hand pathway, transposition by a *tnpR* or *res* mutant leads to the formation of Amp^r transconjugants that contain the mobilizable and self-transmissible plasmids fused to each other in a cointegrate.

Transposon Mutagenesis

One of the most important uses of transposons is for transposon mutagenesis. A gene that has been marked with a transposon is relatively easy to map by genetic crosses or by physical mapping with restriction endonucleases. Furthermore, genes marked with a transposon are also relatively easy to clone by using plate hybridizations or by selecting for selectable genes carried on the transposon.

Not all types of transposons are equally useful for mutagenesis. A transposon used in this technique should have the following properties:

1. it should transpose at a fairly high frequency;
2. it should not be selective in its target sequence; that is, it should hop almost randomly into target sequences;
3. it should carry an easily selectable gene, such as one for resistance to an antibiotic;
4. it should have a broad host range for transposition, if it is to be used in a variety of bacteria.

Transposon Tn*5* is ideal for random mutagenesis of gram-negative bacteria because it embodies all of these features. Not only does Tn*5* transpose with a fairly high frequency, but also it has almost no target specificity

and will transpose in essentially any gram-negative bacterium. It also carries a kanamycin resistance gene that is expressed in most gram-negative bacteria. Phage Mu is another transposon-like element that can hop in many types of gram-negative bacteria. So far, equally useful transposons have not been identified in gram-positive bacteria, although Tn*917* is a transposon that hops in some gram-positive bacteria.

Transposon mutagenesis also requires that the transposon be introduced into cells in a suicide vector, which is a piece of DNA that cannot replicate in the cells being mutagenized (see chapter 8). If the suicide vector contains the origin of transfer (*oriT*) of a promiscuous plasmid, such as RK2 or another IncP plasmid, it can be introduced into a wide range of gram-negative bacteria. This feature is essential to studies of uncharacterized bacteria, as we discuss later in the section on genetics of all bacteria.

TRANSPOSON MUTAGENESIS OF PLASMIDS

Transposons are particularly useful for the mutagenesis of plasmids because the relatively small size of plasmids makes physical mapping easier. Also, it is often fairly easy to isolate large numbers of transposon insertions in a plasmid.

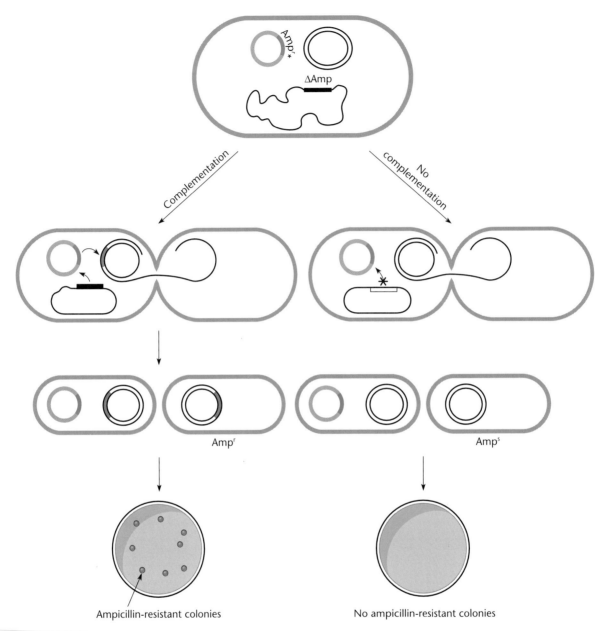

Figure 16.7 Complementation tests of transposition-defective Tn3 mutations. The mutant Tn3 transposon being complemented is in blue, and the asterisk indicates the mutation. See the text for details.

Figure 16.8 illustrates the steps in the selection of plasmids with transposon insertions in *E. coli*. A suicide vector containing the transposon is introduced into cells harboring the plasmid. Cells in which the transposon has hopped into cellular DNA, either the plasmid or the chromosome, are then selected by plating on medium containing the antibiotic to which a transposon gene confers resistance, in this case kanamycin. Only the cells in which the transposon has hopped to another DNA will become resistant to the antibiotic, since the suicide vector cannot replicate and will be lost. In most of the kanamycin-resistant bacteria, the transposon will have hopped into the chromosome rather than into the plasmid, simply because the chromosome is the larger target. Therefore, most of the plasmids will be normal. However, the plasmids with transposon insertions can be isolated by pooling the colonies due to kanamycin-resistant bacteria and preparing the plasmids from them by one of the procedures outlined in chapter 4. This mixture of plasmids is then used to transform another

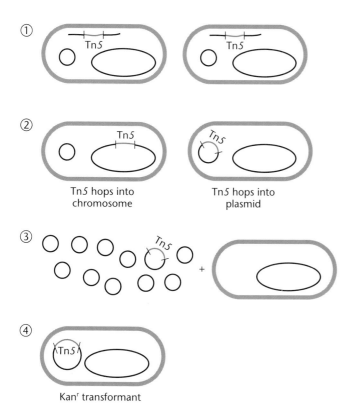

Figure 16.8 Random transposon mutagenesis of a plasmid. In step 1, the transposon Tn*5* is introduced into cells on a suicide vector. In step 2, the culture is incubated, allowing the Tn*5* time to hop, either into the chromosome (large circle) or into a plasmid (small circle). Plating on kanamycin-containing medium will result in the selection of cells in which a transposition has occurred. In step 3, plasmid DNA is prepared from Kan^r cells and used to transform a Kan^s recipient. In step 4, selection for Kan^r will allow the identification of cells that have acquired a Tn*5*-carrying plasmid.

strain of *E. coli*, and kanamycin-resistant transformants are selected. All the kanamycin-resistant transformants should contain the plasmid with a transposon inserted somewhere in it, since linear chromosomal DNA cannot transform wild-type *E. coli* (see the section on gene replacements, below). In a few simple steps, plasmids with transposon insertion mutations have been isolated. The method can be used to isolate many different plasmids, with the transposon inserted at a different position in each plasmid.

Physical Mapping of the Site of Transposon Insertions in the Plasmid

Once plasmids with transposon insertion mutations have been isolated, we can map the sites of the insertions relative to the positions of known restriction sites on the plasmid. This mapping depends on the fact that

the size and number of DNA fragments obtained with a restriction endonuclease are dependent on the number and location of the sites for that restriction endonuclease in the DNA being cut. The insertion of the transposon into the plasmid will introduce new restriction sites and change the size of some of the fragments.

As an example, consider the small plasmid pAT153 and the transposon Tn*5* (Figure 16.9). Use of a small plasmid like pAT153 makes intrepetation of the physical mapping data easier, but the same general methods are applicable to much larger plasmids.

To locate the site of insertion, we need to use two different restriction endonucleases, both of which cut in the transposon and the original plasmid. The restriction endonucleases *Pst*I and *Hin*dIII fulfill this requirement (Figure 16.9). Plasmid pAT153 is 3,600 bp (3.6 kb) long and has only one *Hin*dIII site and one *Pst*I site. The transposon is 5.6 kb long and has two *Hin*dIII and four *Pst*I sites. Cutting the original pAT153 plasmid without the transposon insertion with either *Hin*dIII or *Pst*I should give one band of 3.6 kb. However, the plasmid with the transposon inserted is much larger (3.6 kb + 5.6 kb = 9.2 kb); therefore, cutting this larger molecule with *Hin*dIII should yield three bands, and cutting it with *Pst*I should yield five bands, because of the *Hin*dIII and *Pst*I sites in the transposon. With each restriction endonuclease, the fragments should add up to 9.2 kb. However, the sizes of some of the fragments will vary, and these sizes will tell us where the transposon has inserted into the plasmid.

Restriction fragments that come from inside the transposon, called the **internal fragments**, will be the same size no matter where the transposon has inserted. *Hin*dIII leaves one internal fragment of 3.3 kb. The three internal fragments left by *Pst*I are 1.08, 0.92, and 2.40 kb long. The sizes of the remaining two fragments will vary depending on where the transposon has inserted. These fragments are called the **junction fragments** because they include both plasmid and transposon DNA as well as the junction between them. It is the size of the junction fragments that concern us.

The representative data in Figure 16.10 illustrate how the site of insertion of the Tn*5* transposon can be determined for a particular insertion mutation. To obtain Figure 16.10A, the plasmid with the transposon insertion was digested separately with *Hin*dIII and *Pst*I and the fragments were applied to an agarose gel, along with marker DNA fragments of known size. During electrophoresis, smaller fragments move faster than larger fragments, and the rate of migration decreases exponentially with size (see chapter 15). Therefore, on semilog paper, a plot of distance versus size for the marker DNAs should give a standard curve approxi-

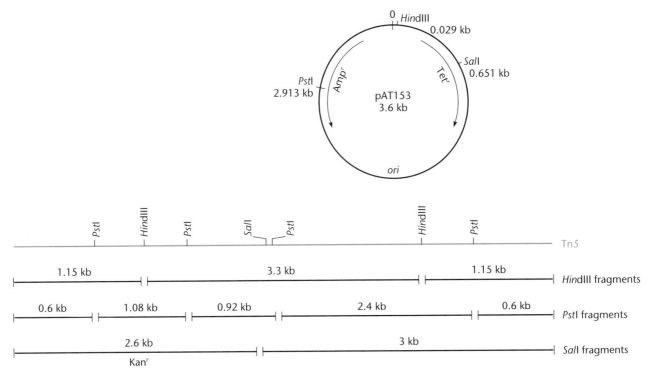

Figure 16.9 Restriction maps of plasmid pAT153 and transposon Tn5. The positions of the *Pst*I, *Sal*I, and *Hin*dIII sites are shown.

mating a straight line, as shown in Figure 16.10B. From this standard curve, we can estimate the size of the unknown fragments by plotting the distance they travel and reading over to obtain their size. Thus, according to Figure 16.10B, the three *Hin*dIII fragments are approximately 4.60, 3.45, and 1.40 kb, adding up to 9.45 kb, which is close enough to the expected 9.2 kb. We estimate the five *Pst*I fragments to be 3.40, 2.20, 1.60, 1.00, and 0.87 kb, which add up to 9.07 kb, close enough again to 9.2 kb. Therefore, all the restriction fragments seem to be present and accounted for.

Our next step is to identify the two junction fragments in each digestion, since the sizes of these fragments reveal where the transposon inserted into the plasmid. By first identifying the internal fragments, by a process of elimination we will have identified the junction fragments. After *Hin*dIII digestion, the internal fragment of 3.30 kb is probably the *Hin*dIII fragment that we have estimated to be 3.45 kb. The junction fragments with *Hin*dIII are therefore the 4.60- and 1.40-kb fragments. Similarly, the internal *Pst*I fragments that should be 1.08, 0.92, and 2.40 kb are probably the fragments that we have estimated to be 1.00, 0.87, and 2.20 kb, leaving the 3.40- and 1.60-kb fragments to be the *Pst*I junction fragments.

The next step is examine the junction fragments to determine how far the insertion site for the transposon is from one of the restriction sites. Let us start with the *Hin*dIII site on the plasmid. The smaller of the two *Hin*dIII junction fragments is 1.40 kb, which must be the distance from the *Hin*dIII site in the original plasmid to the nearest *Hin*dIII site in the transposon. Since 1.15 kb of the junction fragment is taken up by transposon DNA, the actual site of insertion of the transposon is 1.4 − 1.15 = 0.25 kb from the plasmid *Hin*dIII site. However, from the *Hin*dIII data alone, the transposon could be inserted either 0.25 kb clockwise or 0.25 kb counterclockwise of the *Hin*dIII site on the plasmid as the plasmid map is drawn in Figure 16.9. To determine the side of the *Hin*dIII site on which the transposon is inserted requires an analysis of the *Pst*I data, since the size of the *Pst*I junction fragments will depend on which side of the *Hin*dIII site the transposon inserted. From the size of the smallest *Pst*I junction fragment, which we estimated to be 1.60 kb, the transposon must be inserted 1.60 − 0.60 = 1.00 kb from the plasmid *Pst*I site, since 0.60 kb of this junction fragment is taken up with transposon DNA. The only way the transposon could be inserted both approximately 0.25 kb from the *Hin*dIII site and 1.00 kb from the *Pst*I site is to be in-

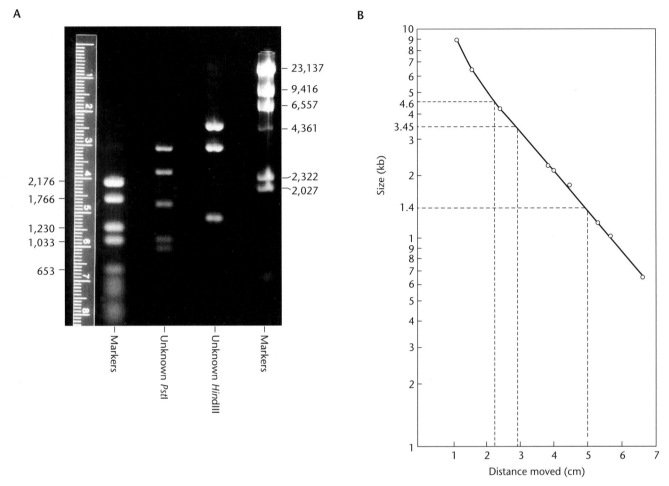

Figure 16.10 (A) Agarose gel electrophoresis of a Tn*5*-mutagenized pAT153 plasmid cut with *Hin*dIII and *Pst*I. The sizes of the marker fragments are also shown. The DNA fragments on the gel were stained with ethidium bromide and photographed under UV illumination. (B) Standard curve of the size of the known fragments plotted against the distance moved in the gel shown in panel A. The sizes of the unknown fragments can be estimated from the standard curve derived from the marker fragments by plotting the distance that the unknown fragments moved on the gel (dotted lines). Only the positions of the *Hin*dIII fragments are shown.

serted approximately 0.25 kb clockwise of the *Hin*dIII site, as shown in Figure 16.11.

Using Transposon Insertions To Identify and Map Genes on a Plasmid

Insertion of a transposon into a gene will almost invariably inactivate the gene. If the site of insertion of the transposon is then mapped as above, the site of the gene on the plasmid can be determined. This general procedure can be used to locate either genes on the plasmid itself or genes in a larger DNA fragment cloned in the plasmid.

To illustrate, we shall use the identification and mapping of *tra* genes on plasmid pKM101 (see Winans and

Walker, Suggested Reading, and Figure 5.1). As discussed in chapter 5, the products of the *tra* genes are required for plasmid transfer during conjugation. Every self-transmissible plasmid must encode its own Tra functions, and a large percentage of these plasmids is devoted to encoding such functions.

1. Isolating tra *mutant plasmids.* The first step the investigators used to identify the *tra* genes of plasmid pKM101 was to mutagenize the plasmid with the Tn*5* transposon to obtain a large number of transferless (Tra⁻) mutants. Of the two methods used to isolate plasmids with Tn*5* insertions, the first was the same as that

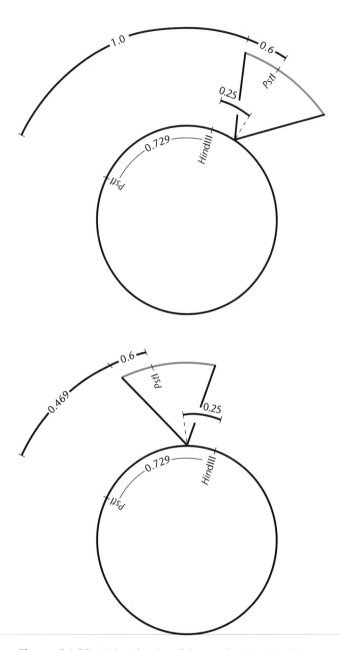

Figure 16.11 Using the size of the smallest junction fragments from Figure 16.10 to determine the site of insertion of the transposon in the plasmid. The size of the smallest *Hind*III junction fragment indicates that the transposon must have inserted 0.25 kb from the *Hind*III site in the plasmid. The smallest *Pst*I junction fragment indicates that the transposon must be positioned clockwise of the plasmid *Hind*III site.

tant transformants, each of which contained the plasmid with a transposon insertion in a different region of the plasmid, were isolated. The second method began in the same way, in that the λ suicide vector carrying the transposon was used to infect cells containing the plasmid. Now, however, after being infected, the cells were incubated only long enough to allow the transposon time to hop. The cells were then mixed with recipient bacteria, and kanamycin-resistant transconjugants were isolated. In general, only cells in which the transposon had hopped into the plasmid would be able to transfer kanamycin resistance to the recipient, since usually only the plasmid will be transferred. Even if the transposon had happened to hop into a *tra* gene in the plasmid, the plasmid would still transfer, because after such a short incubation, the cells would still contain some of the Tra functions, owing to phenotypic lag (see chapter 3). Using conjugation rather than transformation to isolate plasmids with transposon insertions makes it easier to isolate many mutant plasmids, because conjugation is often more efficient than transformation.

Once they had isolated several transposon insertions in the plasmid, the investigators needed to determine which of the transposon insertions had inactivated a *tra* gene. If the transposon had hopped into, and therefore inactivated, a *tra* gene, the plasmid would no longer be able to transfer itself. Therefore, each of the mutagenized plasmids was tested individually for its ability to transfer.

To streamline the process of screening the mutants, the investigators devised the replica-plating test for transfer illustrated in Figure 16.12. In the example, a number of plasmid-containing bacteria were patched onto a plate so that each patch would have bacteria containing a different mutagenized plasmid. After incubation to allow multiplication of the bacteria in the patches, this plate was replicated onto another plate containing rifampin and rifampin-resistant recipient bacteria. After incubation to allow any potential transconjugants to form and multiply, the second plate was replicated onto a third plate containing media with both kanamycin and rifampin. If the Tn5 insertion in the plasmid did not inactivate a *tra* gene, the plasmid will have transferred into some of the recipient bacteria on the second plate, so that colonies will appear on the third plate owing to the multiplication of transconjugants that are both kanamycin- and rifampin-resistant. However, if the transposon had inserted into a *tra* gene, no kanamycin- and rifampin-resistant transconjugants should appear on the third plate. In this way, several mutant plasmids were identified that had a transposon insertion in one of their *tra* genes. The site of the trans-

illustrated in Figure 16.8. Cells containing plasmid pKM101 were infected with a λ suicide vector containing the transposon, and plasmids with transposon insertion mutations were isolated by preparing the plasmids and using them for transformation. Kanamycin-resis-

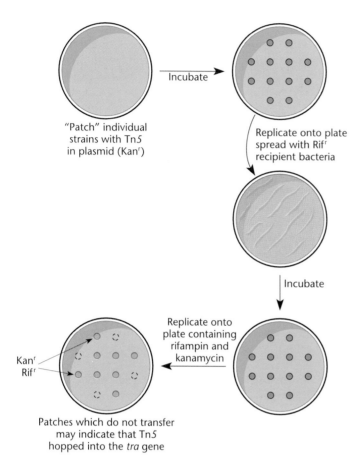

"Patch" individual
strains with Tn5
in plasmid (Kanʳ)

Incubate

Replicate onto plate
spread with Rifʳ
recipient bacteria

Incubate

Replicate onto
plate containing
rifampin and
kanamycin

Kanʳ
Rifʳ

Patches which do not transfer
may indicate that Tn5
hopped into the *tra* gene

Figure 16.12 Replica plating to find *tra* plasmid mutants resulting from insertion of a transposon. Cells containing the plasmid with the transposon inserted in different places are patched onto a plate without antibiotics, and the plates are incubated to allow the formation of colonies. The following steps are as indicated in the illustration. Failure of a colony to develop on the third plate after incubation indicates that the transposon is in a *tra* gene of the plasmid, so that no trans-conjugants had formed on the second plate.

poson insertions in these *tra* plasmids could then be physically mapped by methods such as that described above.

2. Complementation tests to determine the number of tra *genes.* Mapping the site of insertion of the transposon in a *tra* mutant plasmid reveals where a *tra* gene is located on the plasmid but does not, by itself, reveal the number of *tra* genes. Only complementation tests can determine the number of genes that have been mutated. For these tests, plasmids containing two different *tra* mutations must be introduced into the same cell. If the mutations inactivate different genes, each mutant plasmid will furnish the Tra function that the other one

lacks, and one or both plasmids will be able to transfer. If, however, the two mutations inactivate the same *tra* gene, neither plasmid will be able to make the product of that gene and neither plasmid will be able to transfer. To save time, these complementation tests need be done only between *tra* mutants with insertion mutations close to each other on the plasmid. Transposon insertion mutations that are farther apart, or have transposon insertions between them that do not inactivate a *tra* gene, are almost certainly not in the same *tra* gene, and so they do not need to be tested by complementation.

Complementation tests between different *tra* mutations in the same plasmid are complicated by the fact that any two mutant plasmids will have been derived from the same pKM101 plasmid and so are in the same Inc group. They cannot therefore stably coexist (see chapter 4). Figure 16.13 shows one way to overcome this difficulty. The region of plasmid pKM101 containing one of the *tra* mutations (*tra₁* in Figure 16.13) was inserted into a plasmid cloning vector from a different Inc group. The plasmid containing the clone was then used to transform cells carrying the pKM101 plasmid with the other *tra* mutation (*tra₂*). Since the plasmids carrying the two *tra* mutations are members of different Inc groups, they will stably coexist, and if the two *tra* mutations are in different genes, they will complement each other and the mutant PKM101 plasmid will transfer into recipient cells. However, the cloned *tra₁* region must include the entire gene, and the gene must be expressed on the cloning vector plasmid.

The other way to avoid the problem of incompatability is by "transient heterozygosis." Even if plasmids are in the same Inc group, they can coexist transiently, allowing time for complementation. Therefore, if one mutant plasmid is used to transform cells containing the other mutant plasmid and the cells are immediately mixed with a recipient, any kanamycin-resistant transconjugants will be evidence of complementation between the insertion mutations in the two plasmids.

By using these procedures, the investigators estimated the number of *tra* genes on pKM101. However, this number may be an underestimation of the total number of *tra* genes because it is possible that not all the *tra* genes have transposon insertions. Moreover, transposon insertion mutations are sometimes polar (see chapter 2 for an explanation of polarity), so that two mutations in different genes can behave as though they were in a single gene in complementation tests. In spite of these difficulties, the investigators identified most of the *tra* genes of plasmid pKM101 and, by mapping the site of insertions that inactivated each of the *tra* genes, obtained the map of pKM101 shown in Figure 5.1.

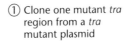
① Clone one mutant *tra*
region from a *tra*
mutant plasmid

② Transform into a cell
with a different
tra mutant plasmid

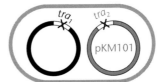

Figure 16.13 Testing for complementation
between two *tra* mutations. The region
containing one of the *tra* mutations is cloned
into a cloning vector from a different Inc group.
If the two mutations complement each other,
the plasmid can transfer, as determined by
methods such as that outlined in Figure 16.12.
The cloning vector is shown in black.

③ Test for transfer
to a recipient. If
transfer occurs,
the mutations
complement

Recipient

TRANSPOSON MUTAGENESIS OF THE BACTERIAL CHROMOSOME

The same methods used to mutagenize a plasmid with a transposon can also be used to mutagenize the chromosome. A gene with a transposon insertion is much easier to map or clone than a gene with another type of mutation, making this a popular method for mutagenesis of chromosomal genes. The only caveat is that transposon insertions will usually inactivate a gene, a lethal event in a haploid bacterium if the gene is essential for growth. Therefore, this method can generally be used only to mutate genes that are nonessential or essential under only some conditions. However, one way in which essential genes could be mapped is to isolate transposon insertion mutations that are not in the gene itself but close to it in the DNA. If the insertion is close enough, it might be used to map or clone the gene (see chapter 14 and 15). It is also important to remember that insertion of some transposons may increase expression of genes at the insertion site (see chapter 8).

Physical Mapping of Chromosomal Transposon Insertions

The physical mapping of transposon insertion mutations in the chromosome is more difficult than mapping of insertions in plasmids, because the chromosome is so large. Usually, eight-hitter restriction nucleases must be used instead of six-hitter enzymes, so that there are larger but fewer fragments. However, the fragments obtained with eight-hitter restriction endonucleases are usually too large to resolve on normal agarose gels, and

techniques such as pulsed-field gel electrophoresis (PFGE) (see chapter 4) must be used to separate them.

Even if the fragments can be separated, fragments containing an inserted transposon are often difficult to identify because the few thousand base pairs that a transposon adds will usually not make a significant difference in the size of such large fragments. However, if the transposon has a restriction site for the eight-hitter enzyme, any fragment containing the transposon will have a new site for the eight-hitter enzyme, making it easier to identify. Fortuitously, transposon Tn5 has a site for the eight-hitter *Not*I. Wherever transposon Tn5 has inserted in the chromosome, a new *Not*I site is introduced, which can be used in the physical mapping of the transposon insertion. Once the eight-hitter fragment containing the transposon has been identified, the site of the insertion of the transposon in this somewhat smaller piece can be further localized by using six-hitters and more standard techniques.

Cloning Genes Mutated with a Transposon Insertion

Genes that have been mutated by transposon insertion are usually relatively easy to clone by cloning the easily identified antibiotic resistance gene in the transposon. Since some antibiotic genes, for example the kanamycin resistance gene in Tn5, are expressed in many types of bacteria, this method could even be used to clone genes from one bacterium in another. This is particularly desirable because most cloning vectors and recombinant DNA techniques have been designed for *E. coli*. To clone a gene mutated by a transposon from a bacterium distantly related to *E. coli,* the DNA from the mutagenized strain is cut with a restriction endonuclease that does not cut in the transposon and ligated into an *E. coli* plasmid cloning vector cut with the same or a compatible enzyme. The ligation mixture is then transformed into *E. coli,* and the transformed cells are spread on a plate containing the antibiotic to which the transposon confers resistance. Only cells containing the mutated, cloned gene will multiply to form a colony.

Transposons engineered to make cloning of transposon insertions more efficient contain an *E. coli* plasmid origin of replication so that the DNA containing the transposon need only cyclize to replicate autonomously in *E. coli.* The use of such a transposon for cloning transposon insertions is illustrated in Figure 16.14. In the example, the transposon carrying a plasmid origin of replication has inserted into the gene to be cloned. The chromosomal DNA is isolated from the mutant cells and cut with a restriction endonuclease that does not cut in the transposon, *Eco*RI in the example. When the cut DNA is religated, the fragment containing the transposon will become a circular replicon with the plasmid origin of replication. If the ligation mixture is used to transform *E. coli* and the ampicillin resistance gene on the transposon is selected for, the chromosomal

DNA surrounding the transposon will have been cloned. Since the restriction endonuclease cuts outside the transposon, any clones of the transposon cut from the chromosome will also include sequences from the gene of interest into which the transposon had inserted.

Once the gene containing the transposon insertion has been cloned, it can be used in several ways. We may want to directly sequence the gene with primers complementary to the ends of the transposon. Alternatively, we might need a clone of the wild-type gene without the transposon insertion; in this case, we could use the clone with the transposon insertion as a probe to identify the wild-type gene by screening a library of wild-type DNA in a plate hybridization (see chapter 15). This method allows the cloning of genes about which nothing is known except the phenotype of mutations that inactivate the gene, and it can be easily adapted to clone genes from any bacterium in which the transposon can hop to create the original chromosomal mutation. Once mutants are obtained with the transposon inserted in the gene of interest, the remainder of the cloning is performed in *E. coli.* We shall give examples involving such methods in the section on genetics of all bacteria, below.

USING TRANSPOSON MUTAGENESIS TO MAKE RANDOM GENE FUSIONS

The ability of some transposons to hop randomly into DNA has made them very useful for making **random gene fusions** to reporter genes (see chapter 2). Fusing a gene to a reporter gene can make regulation of the gene much easier to study or can be used to identify genes subject to a certain type of regulation or those localized to certain cellular compartments. We discussed the use of gene fusions to study protein transport and other cellular functions in previous chapters, but we go into more detail about these methods here.

As discussed in chapter 2, gene fusions can be either transcriptional or translational. In a transcriptional fusion, one gene is fused to the promoter for another gene so that the two genes are transcribed into mRNA together. In a translational fusion, the ORFs for the two genes are fused to each other in the same reading frame so that translation, initiated at the translational initiation region (TIR) for one protein, will continue into the ORF for the other protein, making a fusion protein.

Transposons engineered to make random gene fusions include Tn*3HoHo1,* Tn*5lac,* and Tn*5lux.* These transposons carry a reporter gene at one end that either has its own TIR, if the transposon is to make transcriptional fusions, or lacks a TIR, if the transposon is to make translational fusions. When the transposon hops into the chromosome, it will fuse its reporter gene to whatever gene it hops into.

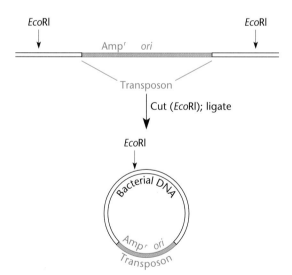

Figure 16.14 Cloning genes mutated by insertion of a transposon. A transposon used for mutagenesis of a chromosome contains a plasmid origin of replication (*ori*), and the chromosome is cut with the restriction endonuclease *Eco*RI and religated. If the ligation mix is used to transform *E. coli,* the resulting plasmid in the Amp^r transformants will contain the sequences that flanked the transposon insertion in the chromosome. Chromosomal sequences are shown in black, and transposon sequences are shown in blue.

Mud(Amp^r, *lac*)

The prototype transposon for making random gene fusions is the Mud(Amp^r, *lac*) transposon (see Casadaban and Cohen, Suggested Reading), which will serve to illustrate the basic principles involved. The Mud(Amp^r, *lac*) transposon is derived from phage Mu, which will transpose with almost no target specificity (see chapter 8, Box 8.1). Phage Mu also has a quite broad host range and will productively infect and transpose in many gram-negative bacteria, including *Erwinia carotovora*, *Citrobacter freundii*, and *Escherichia coli*.

Figure 16.15 shows the essential features of random gene fusions created by the Mud(Amp^r, *lac*) transposon. Most of the phage Mu DNA is removed, except the ends and the transposase, which is the product of Mu genes *A* and *B*. The transposase and the ends of phage Mu are sufficient for transposition of the phage DNA into the chromosome after infection. Close to one of its ends, the Mud(Amp^r, *lac*) transposon carries the *lacZ* gene of *E. coli* as its reporter gene. The *lacZ* gene has no promoter of its own and so will not normally be transcribed. However, if the transposon has hopped into a target DNA in such a way that the *lacZ* gene is positioned in the correct orientation downstream of a promoter, the *lacZ* gene will be turned on and will express β-galactosidase, which is easily detected by colorimetric assays with dyes such as X-Gal or ONPG. The Mud(Amp^r, *lac*) transposon also has an ampicillin resistance gene that can be used to select only those cells that have the transposon inserted somewhere in their DNA.

The procedure for using the Mud(Amp^r, *lac*) transposon to make random gene fusions is illustrated in Figure 16.16. The first step is to prepare a lysate by inducing a Mu prophage in cells that also contain the Mud(Amp^r, *lac*) transposon. The cell must also contain the normal phage Mu to furnish all the proteins needed to make a phage, including the head and tail proteins, which the Mud(Amp^r, *lac*) transposon cannot make. Since Mu cannot distinguish its own DNA from the Mud(Amp^r, *lac*) transposon DNA, which has the same ends, the

replicating phage Mu will sometimes package the Mud(Amp^r, *lac*) transposon instead of its own DNA. Therefore, when the cells lyse, some of the phage particles that are released will contain the Mud(Amp^r, *lac*) transposon. This phage lysate is then used to infect cells at a low multiplicity of infection (MOI). Any ampicillin-resistant transductants, selected on ampicillin plates, will have the Mud(Amp^r, *lac*) transposon inserted somewhere in their chromosome. If the Mud(Amp^r, *lac*) transposon has inserted downstream of an active promoter, the *lacZ* gene on the transposon will be transcribed from the promoter and the cells will make β-galactosidase.

The fact that the Mud(Amp^r, *lac*) transposon expresses *lacZ* only if the gene into which it hops is transcribed has been used to identify genes that are turned on only under certain conditions. One example is the identification of the *din* (damage-inducible) genes of *E. coli*, which are turned on only after DNA damage (see Kenyon and Walker, Suggested Reading). To use Mud(Amp^r, *lac*) to identify *din* genes, investigators first used the method described above to isolate ampicillin-resistant (Amp^r) transductants with Mud(Amp^r, *lac*) inserted somewhere in their genome. They then needed to identify transductants with the transposon inserted in the correct orientation into a gene that is induced when the DNA is damaged. To accomplish this, they replicated the plates containing the Amp^r transductants onto two other sets of plates, one containing only X-Gal and the other containing X-Gal plus a DNA-damaging agent such as mitomycin. The colonies that were blue on the set of plates containing the DNA-damaging agent but not on the plates with X-Gal alone were known to have the transposon inserted in a *din* gene in the right orientation. By mapping the ampicillin resistance gene on the transposon, the investigators could map a number of *E. coli din* genes that are turned on after DNA damage.

In Vivo Cloning

Transposons can also be used for in vivo cloning. Like other cloning procedures, in vivo cloning requires a library of recombinant DNA. However, these libraries do not need to be made in vitro with restriction endonucleases and DNA ligase as outlined in chapter 15. Instead, in vivo cloning relies on genetic methods and lets bacteriophages and transposons do most of the work of making the library.

Most in vivo cloning procedures are also based on phage Mu and the fact that Mu replicates by replicative transposition without resolving the cointegrates. After Mu replicates, copies of the phage genome are inserted all over the bacterial chromosome (see chapter 8, Box

Figure 16.15 Structure of random gene fusions created by the original Mud(Amp^r, *lac*) transposon. The essential transposon elements are the ends of Mu, the reporter gene *lacZ*, and a gene for ampicillin resistance (Amp^r). The *lacZ* gene lacks its own promoter and so will be transcribed only from a promoter outside the transposon (*p*). The sequences outside the transposon are shown as dashed lines, the Mu sequences of the transposon in blue.

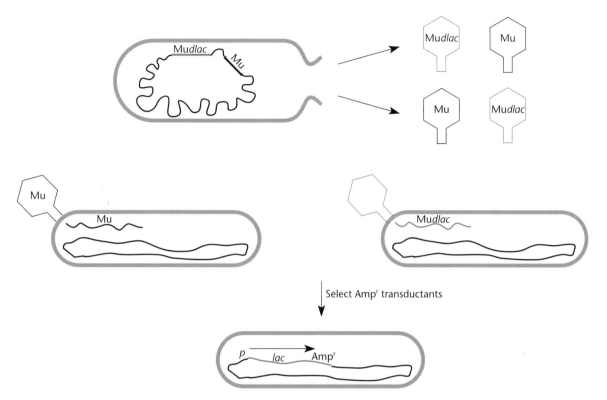

Figure 16.16 Isolating random gene fusions with Mud(Amp^r, *lac*). The Mu prophage is induced in a cell containing the Mud(Amp^r, *lac*) transposon (in blue), which will be packaged into some of the phage. All Amp^r transductants have Mud(Amp^r, *lac*) somewhere in their chromosome, and in cells that form blue colonies on X-Gal plates, the Mud(Amp^r, *lac*) transposon has hopped downstream of a promoter in such a way that the *lacZ* gene is transcribed.

8.1). During the normal infection cycle, these phage genomes are then packaged into phage heads by using specific *pac* sites at the ends of the phage DNA.

CONSTRUCTING MINI-Mu ELEMENTS

Because the phage Mu packaging system recognizes the ends of phage DNA, it can, in principle, recognize the outside ends of two phage genomes lying close to each other in the chromosome and package both phage Mu DNAs plus any chromsomal DNA between them. This does not normally happen, because the phage Mu head is large enough to hold only a single Mu genome. Shortened versions of Mu, called **mini-Mu** elements, have been constructed to allow room to package chromosomal DNA (see Groisman and Casadaban, Suggested Reading). These mini-Mu elements lack all of the functions of Mu except the ends required for transposition and packaging. In addition, to make them useful for in vivo cloning, the mini-Mu elements have a plasmid origin of replication and an antibiotic resistance gene cloned into them, whose significance of which will become apparent later.

Figure 16.17 illustrates how mini-Mu elements may be used for in vivo cloning. A mini-Mu element is first introduced into a cell by transformation on a plasmid and then induced to replicate (transpose) by infection of the cells with wild-type Mu, which contributes the transposase functions. Both the wild-type Mu and the mini-Mu elements will then transpose around the chromosome, making many copies of themselves. Later in the infection, while the normal phage DNAs are being cut out and packaged, some pairs of mini-Mu elements, which happen to be separated by a phage Mu length of DNA, will also be cut out and packaged along with the chromosomal DNA between them. When the cells lyse, the released phage will be a mixture of phage containing normal Mu DNA and others containing chromosomal DNA packaged between copies of the mini-Mu element.

The plasmid origin of replication and the selectable gene on the mini-Mu now come into play. When new cells are infected with the phage lysate at a low MOI, some of the cells will be infected by phage containing the mini-Mu elements and chromosomal DNA. The

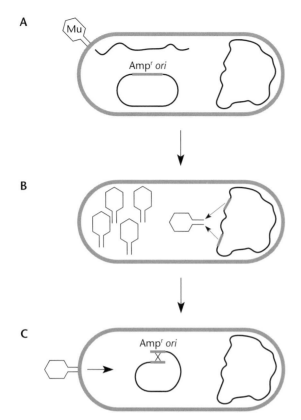

Figure 16.17 In vivo cloning with mini-Mu. In the first cell, a mini-Mu (in blue) on a plasmid is induced to replicate by infection with a helper phage Mu, allowing the mini-Mu to hop randomly around the chromosome as it replicates. In the second cell, pairs of mini-Mu elements, along with chromsomal DNA lying between them, are packaged into some phage Mu heads. The sites of cleavage for packaging the two mini-Mu elements are slightly outside the ends of the mini-Mu elements (arrows). Upon infection of the third cell, the two mini-Mu elements can recombine to form a circular DNA that can replicate because of the origin of replication on the mini-Mu elements. The ampicillin resistance gene also carried by the mini-Mu allows selection of transductants.

DNA injected into these cells can then cyclize by recombination between the identical mini-Mu sequences at the ends to form a circular DNA that can replicate from the plasmid origin of replication. Because the mini-Mu element also contains the gene for ampicillin resistance, transductants can be selected by plating on ampicillin-containing plates. If enough phage in the lysate carried chromosomal DNA, every region of the bacterial DNA will be represented in the ampicillin-resistant transductants, constituting a library of the bacterial DNA. Once a library has been obtained, a clone containing the desired gene can be found in the library through complementation or hybridization techniques, such as those discussed in chapter 15. This general method has been used in a variety of gram-negative bacteria, including *E. coli*, *Klebsiella* spp., and *Erwinia* strains (for some examples, see Van Gijsegem and Toussaint, and Chen and Maloy, Suggested Reading)

For a reference that lists and discusses many useful transposons for *E. coli* as well as other bacteria, see Berg and Berg, Suggested Reading.

Gene Replacement and Reverse Genetics

Gene replacement allows us to make predetermined changes in a DNA sequence and then introduce the altered DNA into a living cell to see what effect these changes have. An organism into which DNA has been introduced from outside is called a **transgenic organism**. The introduced DNA could either replace the original, normal DNA or be integrated somewhere else in the cellular DNA. If the introduced DNA with the gene mutated in vitro replaces the normal gene in the chromosome, the process is called **gene replacement**.

Gene replacement occurs through homologous recombination between the cloned DNA and homologous sequences in the chromosome. Therefore, identical sequences must exist in the cloned DNA and the region in the chromosome that the cloned DNA is to replace. The entire sequences need not be identical, however, but only the sequences on either side of the region to be inserted (the **flanking sequences**). Because identical flanking sequences generally occur only at the site of the cloned region, the recombination between flanking sequences in the clone and in the chromosome will usually occur only where the clone resided in the chromosome.

Bacteria and some lower eukaryotes are ideal for gene replacements because the recombination that results in insertion of the cloned DNA is restricted to regions of extensive homology in the flanking sequences. This is not the case for all organisms. In higher organisms, recombination often results in insertion of the clone at sites other than its original place in the chromosome. This phenomenon is poorly understood but has complicated gene replacements in these organisms. Gene replacements have been successful in a few higher organisms such as mice and chickens. Nevertheless, they are much easier in bacteria.

In this section, we discuss gene replacement methods in detail, showing some actual procedures for how bacteria with gene replacements can be selected.

Selecting Gene Replacements in Bacteria

In gene replacement, the DNA is introduced into the cell in a suicide vector. Because it was introduced in a suicide vector, the DNA will be lost unless it recombines with the chromosome. Note that this procedure is similar to transposon mutagenesis but depends on recombination between identical sequences rather than transposition.

The recombination events that lead to gene replacement are relatively rare, and only a few cells in a population ever have their normal gene replaced by the mutated copy. However, gene replacements are relatively easy to select if the mutation that inactivates the cloned gene is an insertion of a selectable gene such as a gene for antibiotic resistance. Genes for resistance to various antibiotics are commercially available and are sometimes called **antibiotic resistance gene cassettes**. Cloning an antibiotic resistance cassette into the gene on the clone will both disrupt the gene in the clone and almost certainly inactivate the gene product; however, it will also introduce a selectable marker, which can be used later to select cells in which the replacement has occurred.

In the example shown in Figure 16.18, a kanamycin resistance cassette has been introduced into a cloned gene in a suicide vector that carries a gene conferring ampicillin resistance. The cells are then plated on medium containing kanamycin. Because the suicide vector cannot replicate in the cells, the cells can become kanamycin resistant only if the plasmid DNA integrates into the chromosome by recombination between the cloned DNA and the corresponding endogenous gene in the chromosome.

Two types of recombination products are possible, depending on whether one or two crossovers occur between the cloned gene and the chromosomal gene. If a single crossover occurs, the plasmid will integrate to form the structure shown at the top of Figure 16.18, in which two copies of the cloned gene bracket the cloning vector. Only one of these genes is disrupted by the kanamycin resistance cassette, and the other is normal. However, a second crossover between the duplicated copies of the cloned region will remove the cloning vector and leave only one copy of the cloned region, as shown at the bottom of Figure 16.18. Depending on where the second crossover occurs, only the disrupted copy of the gene containing the kanamycin resistance cassette will remain in the chromosome, replacing the normal, wild-type sequence, or only the normal copy will remain, returning the cell to normal. In either case, the excised DNA will be lost from the cells since it has the replication defect of the original suicide vector.

In most of the kanamycin-resistant cells, a single crossover will have occurred, for the simple reason that two crossovers are less frequent than one crossover. However, for gene replacement, we want the cells in which two crossovers have occurred and the disrupted

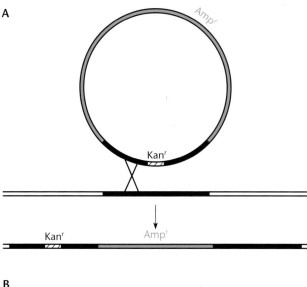

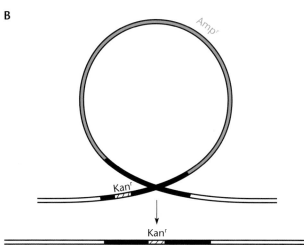

Figure 16.18 Gene replacement. With a single crossover, the cloning vector will integrate into the chromosome and the cells will become ampicillin resistant (Amp^r) and kanamycin resistant (Kan^r). With a second crossover, the cells lose the cloning vector. Depending on where this second crossover occurs, the cells may be left with only the cloned sequence with the kanamycin resistance cassette and be resistant to kanamycin alone. See the text for details. Plasmid cloning vector sequences are in blue. The cloned region of the chromosome is in black.

gene has replaced the normal one. Fortunately, cells in which only one crossover has occurred can be easily distinguished from those with two crossovers if the suicide vector also contains an antibiotic resistance gene, in the example ampicillin resistance. The kanamycin-resistant cells with only one crossover will still have the cloning vector sequences and so will be ampicillin resistant, while those in which the second crossover has occurred will have lost the cloning vector sequences and will be ampicillin sensitive. Therefore, by screening for kanamycin-resistant cells that are ampicillin sensitive, we can select bacteria in which a double crossover has resulted in a chromosome with the mutated gene. To be certain that the inactivated gene has replaced the endogenous gene in these cells, we may need to perform a Southern blot hybridization with the cloned gene as a probe (see chapter 15). No band corresponding to the intact gene in the cellular DNA should hybridize to the probe, only a band corresponding to the disrupted gene. Once a strain in which the disrupted gene has replaced the endogenous gene has been isolated, any mutant phenotypes exhibited by the strain can probably be attributed to the loss of the function of the product of the gene that was cloned, although the inserted resistance cassette may cause a polar effect on expression of a downstream gene in an operon.

SELECTING GENE REPLACEMENTS BY USING THE LETHAL EFFECT OF A SECOND REPLICATION ORIGIN

Most methods for gene replacement require that a selectable gene cassette be introduced into the cloned gene. Otherwise, it is difficult to select those few cells in which a second crossover has replaced the chromosomal gene with the mutant cloned copy. However, sometimes we may want to introduce another type of mutation into the chromosome.

An alternative gene replacement method (see Hamilton et al., Suggested Reading) is based on the fact that *E. coli* cells are killed if their chromosome contains two active replication origins. The mutant gene is first cloned into a derivative of the plasmid vector pSC101 that confers tetracycline resistance (Tet^r) and has a temperature-sensitive mutation in its *repA* gene [*repA*(Ts)]. Because of the *repA*(Ts) mutation, the pSC101 *ori* region is not active at higher temperatures (around 42°C), so that at these temperatures the plasmid will be lost from the cells. Only at lower temperatures (around 30°C) will the plasmid origin be active.

Figure 16.19 presents the details of this method. First, the plasmid with the *repA*(Ts) mutation and the cloned mutant gene is transformed into *E. coli,* and transformants are selected on tetracycline plates at the high temperature of 42°C. Since the plasmid cannot replicate at this temperature, the only cells that become Tet^r are those in which the plasmid has integrated into the chromosome by a single crossover between the cloned mutant gene and the normal gene in the chromosome. These cells can be selected by plating on tetracycline. The chromosome in these transformants will have two potential origins of replication, its own, *oriC,* and the

origin on the integrated plasmid, but the cells survive because the plasmid *ori* is inactive owing to the high temperature. However, when these Tet^r cells are shifted to 30°C, at which the pSC101 origin is active, the only cells that survive are those in which a second crossover has occurred, excising the plasmid from the chromosome, as shown in Figure 16.19, and leaving only the chromosomal origin of replication. Therefore, to select the few cells in which a second crossover has excised the plasmid, the cells need only be plated at 30°C. As in the previous example, some of these cells will have the normal gene restored in the chromosome and some will now have the mutant copy instead of the normal gene, depending on where the second crossover occurred.

This method has an advantage over the previously described method in that the excised plasmid remains in the cells because it is a replicon at 30°C. This feature allows us to determine the cells in which the normal gene has been replaced with the mutant copy by ascertaining whether the corresponding plasmids contain the normal or mutant gene. Hence, plasmids are purified from some bacteria taken from each colony, and the inserted DNA is sequenced. The bacteria whose plasmids contain the normal sequence presumably have the mutant sequence in their chromosome, but this should still be confirmed by some method, for example by direct PCR sequencing of the chromosome (see chapter 15).

Genetics of All Bacteria

Bacteria play a central role in the ecology of the Earth, being responsible, for example, for carbon and nitrogen cycling, the degradation of organic materials, the generation of oxygen, the reduction of metals, and the generation and removal of methane and CO_2. Because bacteria are the most physiologically diverse organisms on Earth, they offer many untapped resources for biotechnology. Bacteria are already used in many technological applications including the manufacture of products such as vitamins and the cleanup of toxic waste.

Molecular genetics plays an important role in modern biotechnology. Adapting a physiological function of a given bacterium to make it more useful often involves cloning and studying the genes involved in the function. Recently developed methods allow us to perform genetic experiments with bacteria about which little to nothing is known. These methods include conjugation with promiscuous plasmids, transposon mutagenesis, the isolation of broad-host-range prime factors, and gene fusion techniques. In this section, we discuss some techniques of bacterial molecular genetics that can potentially be applied to all types of bacteria and devote

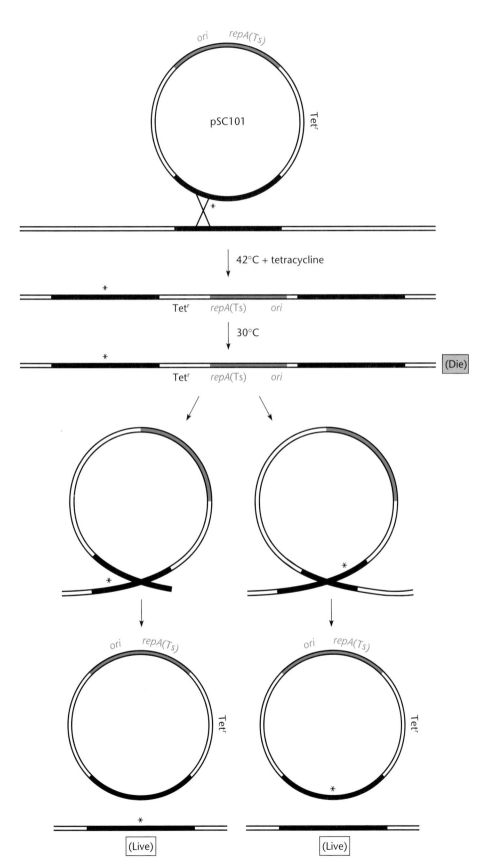

Figure 16.19 A method of selecting for the second crossover that removes the cloning vector and replaces the normal sequence with the sequence mutated in vitro. See the text for details. The asterisk shows the position of the mutation; *ori* denotes the plasmid origin of replication.

the remainder of the chapter to actual applications of bacteria in biotechnology.

Cloning Genes of Uncharacterized Bacteria

To study a gene in detail, we first have to clone it. In chapter 15, we described ways to identify clones containing a gene of interest, such as by hybridization to a heterologous probe derived from a similar gene from a closely related organism or to a complementary probe derived from the amino acid sequence of part of the purified protein product of the gene. These methods can usually be used only if the protein product of the gene can be detected and hence purified or if the gene is known to be related to another gene already cloned from a different organism. However, the molecular function of the cloned gene is often completely unknown, and we must rely on transposon mutagenesis and/or gene expression and complementation in the host of origin to identify it. Here we review these methods.

DNA Libraries

As discussed in chapter 15, DNA libraries must contain enough different clones to be representative, so that each region of the bacterial DNA is represented by at least one clone. We can establish libraries for bacteria that have not been well characterized by cloning their genes in *E. coli* as before but using a broad-host-range cloning vector that is capable of being mobilized into and replicating in the original bacterium. For gram-negative bacteria, cloning vectors derived from IncP or IncQ plasmids usually fulfill these criteria (see chapter 4), and many plasmids of gram-positive bacteria also have broad host ranges.

To facilitate the cloning process and obtain enough clones to be representive of the DNA of the unknown bacterium, some broad-host-range cloning vectors have been made into cosmids that contain the packaging, or *cos*, site of phage λ. The cloned DNA can thus be packaged into phage λ heads in vitro (see chapter 7). After ligation of the bacterial DNA to a cosmid, the ligation mix can be introduced into *E. coli* by infection, a process that is more efficient than transformation, making it easier to get enough clones to have a representative library. The library can then be mobilized into the original bacterium by using the transfer functions of a promiscuous plasmid.

Cloning Transposon Insertions

As discussed earlier in this chapter and in chapter 15, cloning genes inactivated by a transposon insertion can be relatively easy if the transposon carries a selectable

gene. Therefore, adapting transposon mutagenesis to bacteria that are not well characterized greatly facilitates attempts to clone their genes.

As mentioned earlier, transposons can be introduced into various cells by means of suicide vectors engineered to have the *oriT* sites of promiscuous plasmids such as RP4 and other IncP plasmids. Once they are in the cell, many transposons, such as Tn5 or phage Mu, can hop because they are broad host range for transposition. Figure 16.20 illustrates a popular method for transposon mutagenesis of gram-negative bacteria other than *E. coli* (see Simon et al., Suggested Reading). In addition to the broad host range of Tn5 and the promiscuity of RP4, this method takes advantage of the narrow host range of ColE1-derived plasmids, which will replicate only in *E. coli* and a few other closely related species.

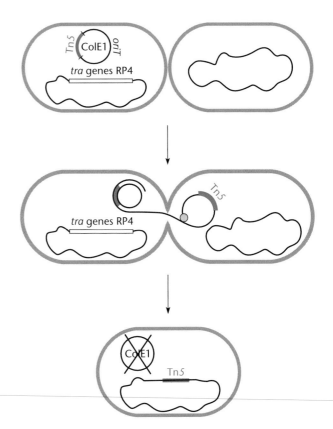

Figure 16.20 A standard protocol for transposon mutagenesis of gram-negative bacteria. A suicide ColE1-derived plasmid containing the *oriT* sequence of promiscuous plasmid RP4 and transposon Tn5 is mobilized into the bacterium by the products of the RP4 transfer genes, which are inserted in the chromosome. The transposon hops into the chromosome of the recipient cell, and the ColE1 plasmid is lost because it cannot replicate. The Tn5 transposon is shown in blue.

In the procedure, a small ColE1-derived plasmid containing Tn5 and the *oriT* of RP4 is transformed into an *E. coli* strain that has the transfer (*tra*) genes of RP4 integrated into the chromosome. Because the ColE1 plasmid contains the RP4 *oriT,* it can be mobilized into other bacteria by the transfer functions encoded by these genes (see chapter 5). It is preferable to have the *tra* genes in the chromosome rather than in another plasmid to prevent their loss from the cells and to avoid the complication of having another plasmid be transferred during mating. Thus, when these *E. coli* cells are mixed with the gram-negative bacterial cells to be mutagenized, the ColE1 plasmid carries the transposon into the recipient cells, where ColE1 is likely to be a suicide vector because of its narrow host range. In that case, the recipient bacteria can become kanamycin resistant only if the transposon hops from the small plasmid into the chromosome or another replicon in the cell, such as an indigenous plasmid. Should the transposon happen to hop into the gene of interest, it will inactivate it. Therefore, to find the desired mutants, we screen the kanamycin-resistant transconjugants for the desired phenotype, for example the inability to grow on a chlorinated hydrocarbon.

Once mutants with transposon insertion mutations of the gene in question have been isolated, the mutant DNA can be isolated and the gene containing the transposon insertion can be cloned in *E. coli,* as described above. However, the cloned gene will be interrupted by the transposon. The regions flanking the transposon in the chromosome can be sequenced directly by using primers homologous to the ends of the transposon. These flanking sequences will be the sequences of the gene of interest. We can obtain the wild-type form of the gene by using the interrupted gene as a probe in plate hybridizations with a library made from the DNA of the wild-type bacterium.

Cloning transposon insertions has many advantages, the major one being that the gene with the transposon insertion can be cloned in *E. coli,* the bacterium for which most cloning procedures are designed. This makes it unnecessary to find cloning vectors that will function in other bacteria.

Complementation in the Host of Origin

Interesting and potentially useful functions often cannot be mutagenized efficiently with a transposon, or for other reasons it is not always possible to isolate genes with transposon insertions. In such cases, we may have to identify a cloned gene by its ability to complement a mutation in the chromosome. The major advantage of this method is that a cloned gene will almost always be expressed in its host of origin. This method can also be

used to identify the cloned gene, even if the molecular basis for the mutation being complemented is not known.

CLONING THE METHANOL-UTILIZING GENES OF *METHYLOBACTERIUM ORGANOPHILUM*

We shall illustrate cloning by complementation in the host of origin by using a specific example, the cloning of genes required for methanol utilization (*meOH* genes) in the gram-negative bacterium *Methylobacterium organophilum* (see Allen and Hanson, Suggested Reading). When this work was begun, almost nothing was known about this bacterium except that it could multiply on methanol as the sole carbon and energy source. For unknown reasons, this strain cannot be mutagenized efficiently with Tn5; therefore, cloning a transposon insertion was not possible.

To clone the genes required for methanol utilization, the investigators first isolated MeOH⁻ mutants of *M. organophilum* that could no longer use methanol as the sole carbon and energy source but could use other carbon-containing compounds. The bacteria were mutagenized with chemical mutagens, the mutagenized bacteria were spread on a plate containing another carbon source, and the colonies that arose were then replicated onto plates containing methanol as the sole carbon source (see chapter 14 for a description of methods). The colonies that did not appear on the methanol plate were MeOH⁻ mutants. The investigators next introduced a library of clones of wild-type *M. organophilum* DNA into the mutant cells and identified cells in which the mutation in the chromosome was complemented by the gene on the cloning vector, so that the mutant could form a colony on plates containing methanol.

The cloning vector used in this screening assay had to be one that could be introduced into *M. organophilum*. Accordingly, a broad-host-range cosmid was constructed carrying antibiotic resistance genes for tetracycline and kanamycin, with a unique *Bgl*II restriction site in the kanamycin resistance gene that could be used for insertional inactivation. The cosmid also had the *oriT* site of plasmid RK2 (also called RP4), so that it could be mobilized by RK2. The cosmid was then used to make a library of *M. organophilum* DNA in *E. coli* by partially digesting the *M. organophilum* DNA with *Sau*3A and ligating it into the cosmid, which had been cut with the compatible restriction endonuclease *Bgl*II. The ligation mix was then packaged into λ, the λ was used to infect *E. coli,* and tetracycline-resistant (Tet^r) transductants were selected. These transductants were then screened for kanamycin sensitivity (Kan^s) to determine which ones contained the cosmid with an insert of *M.*

organophilum DNA and which contained only a recyclized cosmid. Of the Kan^s transductants, approximately 1,000 were saved as the library. Because their cosmid held about 29 kb of DNA, the investigators reasoned that 1,000 clones would cover the entire *M. organophilum* genome, assuming that this species had approximately as much DNA as most bacteria.

To determine which of the cosmid clones in the library could complement *meOH* mutations in the chromosome, each was mobilized separately into each of the individual MeOH⁻ mutants, and the transconjugants were screened to identify those able to use methanol as the sole carbon and energy source. The investigators streamlined the screening by using the triparental cross illustrated in Figure 16.21 (see chapter 5 for the principle behind a triparental cross). First, the *E. coli* cells containing each of the 1,000 or so cosmid clones were patched separately on plates. Using a grid, the investigators could patch about 50 different cosmid-containing strains on the same plate; therefore, they needed about 20 plates to patch all of them. On 20 other plates, they spread two strains of bacteria. One was an *E. coli* strain containing a self-transmissible plasmid derived from RK2. The other was *M. organophilum* with one of the *meOH* mutations. Each one of the first group of plates containing potential clones of *meOH* genes was replicated onto a plate from the second group, which was in turn incubated. During the incubation, RK2 was transferred from the *E. coli* strain that harbored it into the *E. coli* strain containing the cosmid clone. Then the cosmid containing the clone was, in turn, mobilized into the *meOH* mutant of *M. organophilum* on the same plate. When bacterial colonies appeared, this second plate was replicated onto a third plate containing methanol as the sole carbon source. Any growth on the third plate may result from an *M. organophilum* strain having received a complementing clone from the corresponding patch of *E. coli* on the first plate, since *E. coli* cells cannot use methanol as a carbon source even when they have all the genes for methanol utilization from *M. organophilum*. As suspected, most clones could not complement the *meOH* mutation, and most of the cells remained MeOH⁻ and unable to form a colony on the methanol plates. However, a few colonies did appear at the position of some of the patches on the first plate. By aligning the first and last plates, the investigators identifed *E. coli* on the first plate that contained the complementing clone. The cosmid containing a partial clone of a gene for a MeOH-utilizing enzyme was then isolated from the *E. coli* strain by standard techniques.

Once the cosmid clones that complement *meOH* mutations were identified, the clones were mutagenized

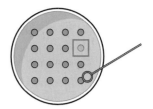

Plate 1   Patch Tet^r cosmid clones in *E. coli*

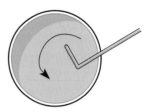

Plate 2   Spread MeOH⁻ mutant of *Methylobacterium* and *E. coli* with RK2

Replicate plate 1 onto plate 2.
Then plate 2 onto plate 3

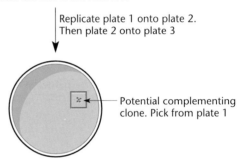

Potential complementing clone. Pick from plate 1

Plate 3   Methanol as sole carbon source

Figure 16.21   Triparental matings to isolate a cosmid clone of a methanol utilization gene of *Methylobacterium organophilum*. Plate 1 contains patched *E. coli* containing individual clones of *M. organophilum* DNA in a broad-host-range cosmid containing the *oriT* sequence of a promiscuous plasmid. The plate is incubated overnight until the colonies develop. The next day, this plate is replicated onto plate 2, on which has been spread an MeOH⁻ *M. organophilum* mutant, which is unable to use methanol, and an *E. coli* strain containing a plasmid with the transfer functions of RK2. Plate 2 is also incubated overnight and then replicated onto plate 3, which contains only methanol as a possible carbon and energy source. After incubation, the presence of any bacterial colonies on the third plate at the position of a patch on the original plate is an indication that the cosmid clone on the first plate is complementing the *meOH* mutation (shown in the blue box).

with a transposon in *E. coli*. The mutagenized clones were then mobilized into the *meOH* mutant, and the investigators looked for loss of complementation. Transposon insertions that inactivate the responsible *meOH* gene could be mapped by methods discussed earlier in the chapter. In this way, the region of the cosmid clone responsible for the complementation could be identified and the gene for methanol utilization could be localized. The gene was then subcloned and sequenced, and, *voila,* a methanol-utilizing gene of *M. organophilum* had been identified and characterized.

Manufacturing Bioproducts

Bacteria make a enormous number of different compounds, many of which have important uses. Some of these compounds are polymers and proteins that are too complicated to make chemically. This is a very promising area of biotechnology, and useful new compounds are continually being discovered. Unfortunately, we have space for only a few examples of how molecular genetics contributes to the synthesis of such bioproducts.

BIODEGRADABLE PLASTICS

Bacteria are being used for the synthesis of biodegradable plastics from polymers of hydroxyacids, of which polyhydroxybutyrate is the best known. Bacteria synthesize and store hydroxyacids for much the same reason plants store starch and animals store fat, i.e., to use as carbon and energy sources during times of deprivation. In bacteria such as *Alcaligenes eutrophus,* as much as 80% of the cell (by dry weight) can be polyhydroxybutyrate under some growth conditions. Plastics made from polyhydroxybutryate are already in commercial production under the name Biopol, and their major advantages are that they are biodegradable (by the same bacteria that make them) and heat stable.

Biotechnology may help in the manufacture of biodegradable plastics by allowing us to change the chemical composition of these products. Currently, polyhydroxybutyrate is made by growing *A. eutrophus* cells in glucose without a nitrogen source. Under these conditions, they accumulate large amounts of polyhydroxybutyrate, which can then be extracted and used to make plastics. However, the plastics made from polymerized 4-hydroxybutyrate alone are too brittle for many commercial applications. Less brittle plastics are made by replacing some of the hydroxybutyrates in the polymer with 3-hydroxyvalerate or 3-hydroxybutyrate. At present, this can be accomplished only by growing the bacteria in glucose plus a mixture of organic acids. However, with the right combination of mutations, bacteria that make polyhydroxybutyrate substituted to various degrees with other hydroxy acids can be obtained. This might obviate the need for adding organic acids and might make the process more dependable.

Also, molecular genetics might help reduce the cost of making polyhydroxybutyrate. If the *A. eutrophus* genes for polyhydroxybutyrate synthesis are cloned and ex-

pressed in *E. coli,* the *E. coli* cells will accumulate larger than normal amounts of polyhydroxybutyrate (see Slater et al., Suggested Reading). A number of easily lysed mutants of *E. coli* have been isolated, and one of these may facilitate the extraction of polyhydroxybutyrate.

Another approach to lowering the cost of making biodegradable plastics is to attempt to clone polyhydroxybutyrate-synthesizing genes of bacteria into plants such as potatoes and corn, where the compounds would be easier to produce in large amounts. The use of transgenic plants is another application of biotechnology that involves bacterial molecular genetics (see below).

XANTHAN GUMS

The plant pathogen *Xanthomonas campestris* produces a group of mixed polysaccharides called xanthan gums. The bacterium makes this gummy material to plug up plant wounds, and it is important in pathogenesis. Xanthan gum already has numerous uses in industry, and approximately 10,000 tons is produced from these bacteria every year.

The xanthan gum polymer is like cellulose in that it is composed of a backbone of glucose held together by $1,4\beta$-linkages but with repeating side chains composed of mannose and glucuronic acid extending from every other glucose in the chain. The polymer is synthesized much like fatty acids: a unit chain of five glucose molecules is altered by addition of the side chains, and then these units are added to the growing chain by using carrier molecules.

Molecular genetics might also be useful for altering the properties of xanthan gum (see Vanderslice et al., Suggested Reading). Mutations in the genes for the transferases and other enzymes involved in the pathway can lead to the assembly of smaller subunits and drastic changes in the properties of the gum. These altered gums may be appropriate for other commercial applications, which could be forthcoming.

INSECT TOXINS OF *BACILLUS THURINGIENSIS*

Bacterial molecular genetics has enabled the synthesis of natural insecticides in quantities large enough for commercial use. The soil bacteria *Bacillus thuringiensis* and, to some extent, *Bacillus sphaericus* make specfic proteins with insecticide activity (see Hofte and Whiteley, Suggested Reading). These proteins, sometimes called Bt toxins, accumulate as crystals when the bacteria sporulate and will kill susceptible insects that ingest the spores. The protoxin is converted to a toxin by proteases in the insect midgut and acts by destroying the epithelial cells. After killing the insect, the bacterium can germinate and multiply on the dead insect tissue.

Different *B. thuringiensis* strains produce toxins with different specificities. Most are active against Lepidoptera, such as moths, but others are active against Diptera, such as flies. The bacterial spores commercially produced for use as insecticides are spread on crops to kill agricultural pests or added to water to kill mosquito larvae. The specifity of these toxins offers the advantage that they, in general, kill only the type of insect for which they are designed. They are also not toxic to humans and other animals and are rapidly biodegradable.

Usually, the genes to make Bt toxins are carried on plasmids that are self-transmissible and so can be transferred to other species. This is where molecular genetics aids in the manufacture of these toxins. Plasmids carrying toxins of different specificities can be conjugated into the same bacterium so that it can synthesize toxins that act against more than one type of insect. The genes for some of these toxins have also been cloned, and genetic manipulation may result in increased yields and higher potency of the insecticides. Molecular genetics can also be used to facilitate the expression of the Bt toxin genes in transgenic plants (see below).

Agrobacterium tumefaciens and the Genetic Engineering of Plants

In chapter 5, we mentioned that some plasmids can transfer themselves between kingdoms and gave an example of the Ti plasmid of the soil bacterium *Agrobacterium tumefaciens*. This plasmid can transfer a part of itself, called the T-DNA (Figure 16.22), into many plants, where it integrates into the chromosome of the plant cell. To date, this is the only well-documented example of a natural gene transfer from one biological kingdom (eubacteria) to another (plants), although the relatedness of some bacterial and eukaryotic genes suggests that they may have been transferred relatively recently.

Integration of the T-DNA of the Ti plasmid of *A. tumefaciens* into a plant chromosome causes a tumor called a crown gall to form on the plant. The tumor results from the expression of genes on the T-DNA for the plant growth hormones auxin and cytokinin, which cause the plant cells to proliferate, giving rise to the tumor. Other genes of the T-DNA encode enzymes to make small compounds called opines, which can be used by the infecting bacteria as a carbon, nitrogen, and energy source. Each type of Ti plasmid encodes its own opine and the genes to degrade that particular opine. As a consequence, usually only those bacteria containing the same Ti plasmid as the one that caused the tumor can use the opines given off by that plant tumor, which precludes competition from other bacteria.

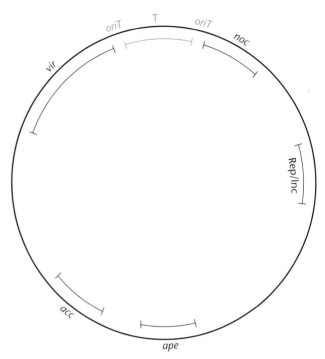

Figure 16.22 Structure of a wild-type Ti plasmid of *Agrobacterium tumefaciens*. The *noc* region encodes enzymes for the degration of the opine nopaline. Rep/Inc is the replication origin. *ape* encodes a phage exclusion system. *acc* encodes enzymes involved in degrading the opine agropine. *vir* denotes the genes involved in plasmid transfer into the plant. T is the region transferred into the plant.

One of the most remarkable features of the Ti plasmids is that they contain some genes that can be expressed only in bacteria and other genes that can be expressed only in plants. As discussed in chapter 2, bacteria and eukaryotes require very different promoters and TIRs for their genes. The genes in the T-DNA, which are expressed in the plant, have eukaryote-type promoters and TIRs, whereas the genes in the remainder of the Ti plasmid have bacterium-type promoters and TIRs.

The mechanism of transfer of the T-DNA portion of the Ti plasmid to plant cells is remarkably similar to that of conjugative transfer of a plasmid from one bacterium to another (see Zambryski, Suggested Reading, and chapter 5). The Ti plasmid encodes Tra functions (called Vir functions in the Ti plasmid [Figure 16.22]) that promote the transfer. Some of these Vir functions sense the presence of phenolic compounds given off by a wounded plant cell. Other Vir functions then cut the Ti plasmid at two *oriT*-like sites bracketing the T-DNA region (Figure 16.22). A single strand of the T-DNA is then displaced and transferred to the plant cell, much like the transfer of a single plasmid strand during bacterium-to-bacterium transfer. Once it is in the plant, however, special functions are required to integrate it into the plant cell DNA. It is thought that two of the Vir proteins of the plasmid accompany the T-DNA into the plant cell and may help it enter the cell nucleus, so that it can integrate into the plant genome.

Not only are the mechanism of transfer from bacterium to bacterium and the mechanism of transfer from bacterium to plant similar, but also the *oriT* sequences and some of the *vir* genes of the Ti plasmid are highly homologous to the *oriT* and *tra* genes of the IncP plasmids (see Waters et al., Suggested Reading). In fact, the sites are similar enough that the plasmid RSF1010, which can be mobilized by IncP plasmids into other bacteria, can also be mobilized by the Ti plasmid into plant cells. Once in the plant cell, however, the RSF1010 plasmid is lost because it cannot replicate and lacks the means to be integrated into the plant chromosome.

MAKING TRANSGENIC PLANTS

It was recognized early that the Ti plasmid of *A. tumefaciens* may offer a way to make transgenic plants with foreign genes inserted into their genomes. Any gene cloned into the T-DNA region of the Ti plasmid should be integrated into the plant genome when the T-DNA is transferred. If the gene is a plant gene or a gene engineered to be expressed from a plant-type promoter and TIR in the T-DNA, it should be expressed in the plant cells.

One difficulty in making transgenic plants or animals is that the gene must be introduced into the germ line of the plant or animal to be passed on to future generations. Usually, the gene must be introduced into an ovum, or egg cell, because only these cells will develop into a mature organism and produce gametes that carry the introduced gene. However, it is difficult to inject DNA into single cells. Yet some plants can be regenerated from individual somatic cells. If a gene can be introduced into the somatic cells and the whole plant regenerated from these cells, all of the cells of the plant will have the introduced gene in their chromosome, and the gene can then be passed on to subsequent generations, leading to a strain of transgenic plants.

Engineering the Ti Plasmid

The original Ti plasmid was not ideal for use in the creation of transgenic plants, and so it was adapted to this purpose. For convenience, the changes were made in *E. coli*, using cloned pieces of the Ti plasmid DNA, and the altered DNAs were then transferred to Ti plasmids in *A. tumefaciens* by gene replacement techniques such as those discussed earlier. The first change inactivated the

T-DNA genes for plant tumor formation so that the transgenic plants would not develop tumors. Then a gene selectable in plants was cloned into the T-DNA. Early experiments used the kanamycin resistance gene from the Tn5 transposon because kanamycin and its relatives will kill plant cells, and the kanamycin resistance gene will make them resistant if it is expressed. However, to be expressed in the plant, this or any other gene must be cloned into the T-DNA in such a way that it will be expressed from a plant promoter and TIR.

Introducing the Engineered T-DNA into the Plant Cell Chromosome

Figure 16.23 diagrams a general procedure for making transgenic plants (see Horsch et al., Suggested Reading). In our example, the T-DNA contains the Kan[r] gene.

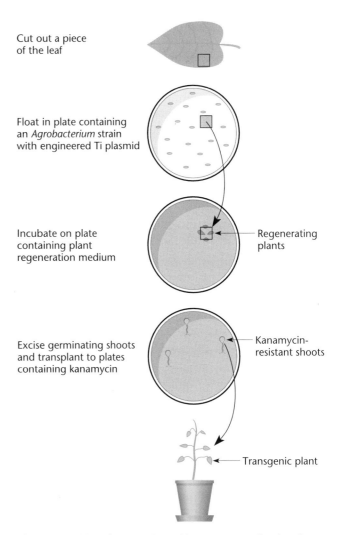

Cut out a piece of the leaf

Float in plate containing an *Agrobacterium* strain with engineered Ti plasmid

Incubate on plate containing plant regeneration medium — Regenerating plants

Excise germinating shoots and transplant to plates containing kanamycin — Kanamycin-resistant shoots

— Transgenic plant

Figure 16.23 The steps in making a transgenic plant by using an engineered Ti plasmid. See the text for details.

This basic method has worked well with any plant that can be infected by *A. tumefaciens* and that can be regenerated from a few cells of leaf tissue.

In the first step, a piece of leaf is cut out and floated in a bath of the *A. tumefaciens* bacteria. During this time, the T-DNA will be transferred into the chromosome of some of the plant cells at the edge of the leaf piece. The plant tissue is then allowed to regenerate for a few days on a disc on a plate containing a special regeneration medium. The plant cells into which the T-DNA has integrated can be identified by their resistance to kanamycin. Hence, when this disc is transferred to another plate containing the same medium but with kanamycin, only cells that are kanamycin resistant because they contain the T-DNA should multiply further to form shoots. Some of the shoots can be transplanted to allow the formation of roots and eventually of transgenic plants. For proof of gene transfer, the plants can be shown by Southern hybridizations to harbor the engineered T-DNA and to give off the opine(s) whose synthesis is encoded by the T-DNA.

At first, the T-DNA will have integrated into only one of the chromosomes of a homologous pair in the diploid plant. The other chromosome in the pair will lack the integrated T-DNA, and so the plants will be heterozygous. To make a homozygous transgenic plant, we have to cross the heterozygous plants and identify the homozygous progeny.

Efforts to introduce a number of useful genes into plants by using the Ti plasmid and *A. tumefaciens* are under way. Examples include genes for hot and cold resistance; genes for polyhydroxybutyrate synthesis, as we have mentioned; and genes for disease resistance.

Introducing *Bacillus thuringiensis* Insect Toxin Genes into Plants

One very promising application of *A. tumefaciens* gene transfer is the introduction of the genes for Bt toxins directly into crop plants. The transgenic plants then synthesize the particular Bt toxin made by the gene that was introduced, making the plant naturally resistant to the insect that is susceptible to that toxin. This approach to pest control may offer many advantages over spraying with synthetic insecticides. As mentioned, the Bt toxins are not toxic to humans and are very specific, so that humans and most beneficial insects would not be harmed. Most important, however, is that the plant itself is making the toxin, so that the toxin need not be added to the environment. Consequently, no insect that was not guilty of eating the crop plant would be harmed.

One obstacle to expressing *B. thuringiensis* genes in plants is that these genes are very rich in AT (adenosine and thymine) base pairs, and so their codon usage is very different from that of plants, which are only about 50% AT. Also, the strings of thymines in the bacterial DNA are often mistaken for plant polyadenylation signals, and transcription is terminated. To avoid these problems, a *B. thuringiensis* insect toxin gene has been altered to introduce different cognate codons richer in G's and C's for many of the amino acids, so that the gene will look more like a plant gene (see Perlak et al., Suggested Reading).

Making New Antibiotics

Another promising application of bacterial molecular genetics is in making new antibiotics and other bioactive compounds. Soil bacteria, notably those of the genus *Streptomyces*, make thousands of different antibiotics and other bioactive compounds, presumably to kill competing soil microorganisms. Most of these bioactive compounds are made when the bacteria cease growth and begin to sporulate, and many are large, complicated molecules whose synthesis requires the products of dozens of genes. Some of the bioactive compounds made by soil bacteria have found useful applications in medicine and agriculture, including in antibiotic therapy against bacteria, fungi, and nematodes; as antitumor drugs; and as immunosuppressants. In chapters 1 and 2, we described examples of these compounds, including antibiotics made by soil bacteria that block bacterial gene expression and DNA replication.

A major effort in the pharmaceutical industry in the past few decades has been to search soil samples from around the world to find bacterial strains that make new, useful antibiotics. However, of the thousands of antibiotics discovered in this way, most are not useful, either because they are too toxic or because they do not enter the affected regions of the host. Further searching seems only to reveal repeats of the same antibiotics. Another problem is that many bacteria and other infectious agents often develop resistance to the known antibiotics. Chemical approaches may alter some antibiotics, enhancing their properties, but these compounds are often too complicated to allow easy chemical manipulation of their structure. However, bacterial molecular genetics offers a means of creating antibiotics with different properties.

MAKING NEW POLYKETIDE ANTIBIOTICS

The polyketide antibiotics are particularly amenable to alteration by molecular genetic techniques. They are a large, diverse class of molecules, many of which are well-known compounds, including the antibiotics tetracycline and erythromycin, the fungal aflatoxins, the antiparasitic drug avermectin, the immunosuppressive drug FK-506, and the antitumor drug adriamycin.

The backbone molecule of the large, complicated polyketide antibiotics is made through condensation reactions of hydroxyacids, much like fatty acids are synthesized. The hydroxyacid building blocks, which usually are acetate, propionate, or butyrate, are activated by coenzyme A and then polymerized. After synthesis, the backbone molecule is altered by oxidation and reduction reactions to give the final molecule. How the correct subunits are chosen to be polymerized and how the polymerization of the subunits is limited to polymers of the correct length to give the backbone molecule are not completely understood.

The pathway of the synthesis of the polyketide antibiotic actinorhodin made by *Streptomyces coelicolor* is illustrated in Figure 16.24. The synthesis of this antibiotic requires at least eight different enzymes encoded by almost 20 different ORFs. As with the other polyketide antibiotics, the first step in actinorhodin synthesis is the polymerization of the acetyl and malonyl subunits by a process analogous to fatty acid synthesis. Then this backbone molecule is progressively altered by other enzymes to give the final antibiotic.

Mutationally Altering Polyketide Antibiotics

In some cases, mutations that inactivate the various enzymes of the pathway can give rise to intermediates that are themselves antibiotics. For example, a mutation that inactivates the *actVB* gene of the actinorhodin pathway will cause the accumulation of an intermediate, kalafungin. This is also an antibiotic and is made naturally by a different species, *Streptomyces tanashiensis*.

Hybrid Antibiotics

Mixing and matching genes for enzymes from different antibiotic synthesis pathways, taken from different *Streptomyces* species, offers another way to alter existing antibiotics (see McDaniel et al., Suggested Reading). If some of the genes for the synthesis of different antibiotics are put together in the same cell, the enzymes from one pathway will sometimes act on intermediates from the other pathway to make a novel antibiotic. An example is the synthesis of the novel antibiotic mederrhodin A (see Hopwood et al., Suggested Reading) (Figure 16.25). Introducing a clone of the *S. coelicolor* gene *actVA* for actinorhodin synthesis into a different *Streptomyces* species, strain AM-7161, which makes the related antibiotic medermycin, leads to the synthesis of

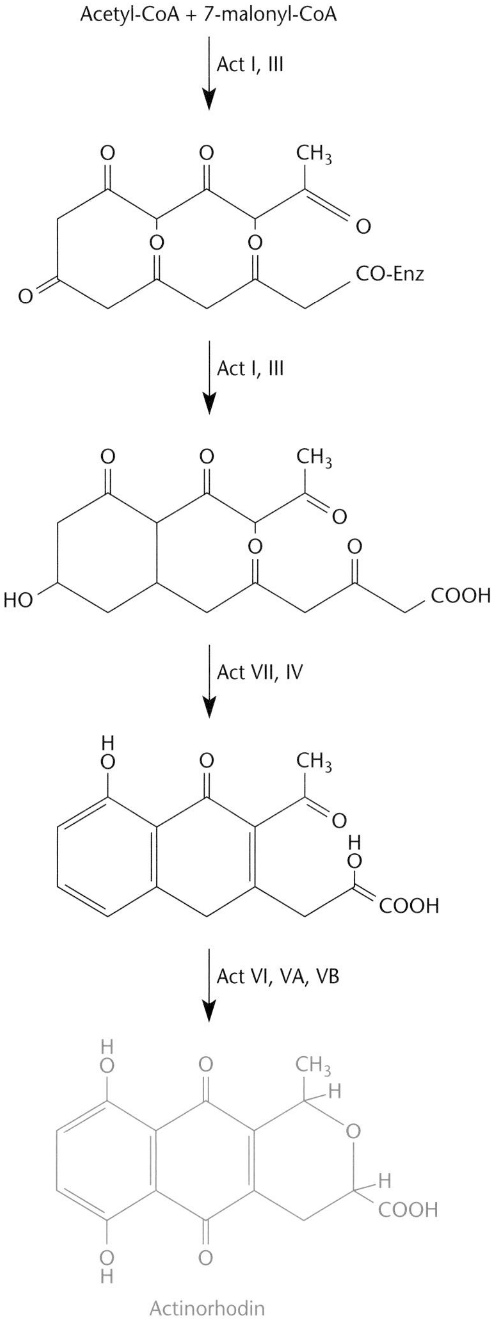

Figure 16.24 The steps in the synthesis of the polyketide antibiotic actinorhodin. CoA, coenzyme A. Adapted with permission from L. Katz and S. Donadio, *Annu. Rev. Microbiol.* 47:875–912, 1993.

mederrhodin A, which differs by a hydroxyl group at position C-8 of medermycin. A major problem with this general approach, however, is that the bacterium is now making a novel antibiotic to which it might not be resistant, causing its own death!

Actinorhodin

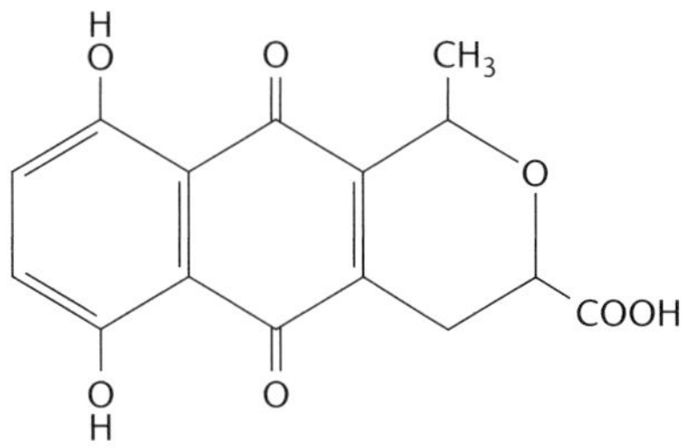

Medermycin

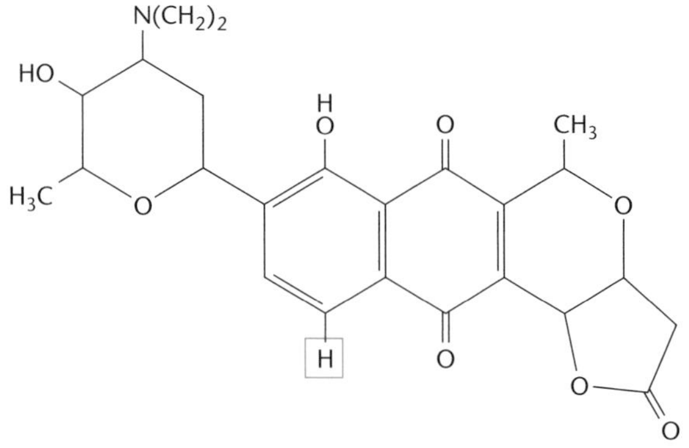

Mederrhodin

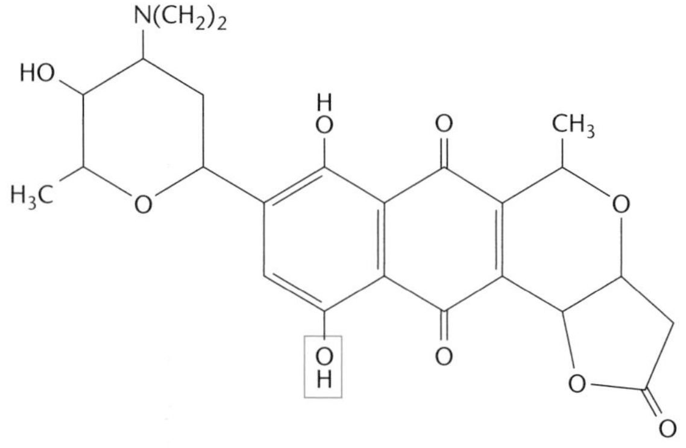

Figure 16.25 Comparison of the structures of some polyketide antibiotics. Introducing one of the genes for medermycin synthesis into *Streptomyces coelicolor* will cause the synthesis of an entirely new antibiotic, mederrhodin. The hydroxyl group that distinguishes medermycin and mederrhodin is boxed. Adapted with permission from L. Katz and S. Donadio, *Annu. Rev. Microbiol.* **47:**875–912, 1993.

Horizons for Molecular Genetics Applications

These are only a few selected examples of how the molecular genetics of bacteria can be used to understand and manipulate the natural world for our benefit. Because they are the simplest organisms on Earth, studies of the basic physiological mechanisms of bacteria have provided the foundation for understanding the cell biology and development of much more complex multicellular organisms like ourselves. At a very basic level, all organisms are remarkably similar, and when a process is understood in bacteria, it is quite often found to function in a similar way in all other organisms. Bacteria are also the largest untapped natural resource on Earth and will continue to provide us with surprises and technologies that we cannot now even imagine. Molecular genetics has been, and will continue to be, a major contributor to these developments, and it stands as one of the major accomplishments of humankind.

SUMMARY

1. Molecular genetics combines techniques for manipulating DNA in vitro with classical genetics to analyze cellular and organismal functions.

2. Marker rescue can be used to map mutations in the chromosome. Marker rescue occurs through recombination between a cloned piece of DNA and the corresponding gene in the chromosome, yielding wild-type recombinants and rescuing the cell from the effects of the mutation.

3. In transposon mutagenesis, a gene is disrupted by insertion of a transposon, which can introduce a selectable marker and additional restriction sites at the site of insertion that can be used for genetic and physical mapping of the gene.

4. Especially engineered transposons carrying reporter genes can be used to make random gene fusions. Insertion of the transposon into a gene can lead to expression of the reporter gene on the transposon from the promoter or TIR of the disrupted gene, depending on whether the fusion is transcriptional or translational.

5. Genes that have been mutated by insertion of a transposon are often easy to clone in *E. coli* if an antibiotic resistance gene is on the transposon. Some transposons have been engineered to contain a plasmid origin of replication so that the restriction fragment containing the transposon

(continued)

SUMMARY (continued)

does not need to be cloned in another cloning vector but only cyclized after ligation to form a replicon in *E. coli*.

6. In reverse genetics, a cloned gene is mutated in vitro and then used to replace the normal gene in the chromosome. Any changes in phenotype are attributed to the mutation introduced in vitro.

7. Molecular genetic techniques are being adapted to bacteria other than the standard laboratory strains. Many of these techniques utilize promiscuous, broad-host-range plasmids and transposons that will transpose in many hosts.

8. Bacteria make useful biopolymers such as biodegradable plastics and gums. The techniques of molecular genetics can be used to alter the composition of these polymers, thereby broadening their applications.

9. The techniques of molecular genetics can help make natural insect toxins from *Bacillus thuringiensis* and alter the properties of the toxins.

10. The techniques of molecular genetics are also being used to make transgenic plants. The bacterium *Agrobacterium tumefaciens* contains a plasmid called the Ti plasmid that can transfer part of itself, the T-DNA, into plant cells, where it integrates into the plant chromosomal DNA. The T-DNA can be engineered to include other genes, such as genes for resistance to disease and insects, that can be introduced into plants.

11. The techniques of molecular genetics are also being used to make novel antibiotics. This includes altering antibiotics by changing the genes in the pathways for their synthesis and making hybrid antibiotics, whereby genes for the enzymes from one pathway for synthesizing an antibiotic in one species can often act on intermediates from a different pathway from a different species.

PROBLEMS

1. List the advantages and disadvantages of transposon mutagenesis over chemical mutagenesis to obtain mutations.

2. Outline how you would use transposon mutagenesis to mutagenize the plasmid pBR322 in *E. coli* with the transposon Tn5. Determine what size of junction fragments would be obtained with *Pst*I and *Hin*dIII if the transposon hopped into the 1-kb position clockwise of 0 on the plasmid. See Chapter 4 for a map of pBR322. The plasmid is 4.36 kb, and the *Hin*dIII sites are at 0.029 and 3.673 kb, respectively. See Figure 16.9 for a map of Tn5.

3. The ampicillin resistance gene of plasmid RK2 is unregulated. The more copies of the gene a bacterium has, the more gene product will be made. In this case, the resistance of the cell to ampicillin will be higher the more of these genes it has. Use this fact to devise a method to isolate mutants of RK2 that have a higher than normal copy number. Determine whether your mutants have mutations in the *trfA* gene.

4. How would you determine whether a new transposon you have discovered integrates randomly into DNA?

5. In the example shown in Figures 16.9 to 16.11, digestion with the restriction endonuclease *Sal*I gives two fragments of 2.971 and 6.229 kb. In what orientation has the transposon inserted? Draw a picture of the transposon inserted in the plasmid showing the position of the kanamycin resistance gene.

6. The cryptic DNA element e14 resides at 25 min in the *E. coli* DNA and has an invertible sequence. How would you determine if its invertase can invert the invertible sequences of Mu, P1, and *Salmonella typhimurium*?

7. You have isolated a strain of *Pseudomonas putida* that can grow on the herbicide 2,4-dichlorophenoxyacetic acid (2,4-D) as the sole carbon and energy source. Outline how you would clone the genes for 2,4-D utilization (**a**) by transposon mutagenesis and (**b**) by complementation in the original host.

SUGGESTED READING

Allen, L. N., and R. S. Hanson. 1985. Construction of broad-host-range cosmid cloning vectors: identification of genes necessary for growth of *Methylobacterium organophilum* on methanol. *J. Bacteriol.* **161:**955–962.

Belfort, M., and J. Pedersen-Lane. 1984. Genetic system for analyzing *Escherichia coli* thymidylate synthase. *J. Bacteriol.* **160:**371–378.

Berg, C. M., and D. E. Berg. 1996. Transposable element tools for microbial genetics, p. 2588–2612. *In* F. C. Neidhardt, R. Curtiss III, J. L. Ingraham, E. C. C. Lin, K. B. Low, B. Magasanik, W. S. Reznikoff, M. Riley, M. Schaechter, and

H. E. Umbarger (ed.), *Escherichia coli and Salmonella: Cellular and Molecular Biology,* 2nd ed. ASM Press, Washington, D.C.

Casadaban, M. J., and S. N. Cohen. 1979. Lactose genes fused to exogenous promoters in one step using a Mu-*lac* bacteriophage: in vivo probe for transcriptional control sequences. *Proc. Natl. Acad. Sci. USA* **76:**4530–4533.

Chen, L.-M., and S. Maloy. 1991. Regulation of the proline utilization in enteric bacteria: cloning and characterization of the *Klebsiella put* control region. *J. Bacteriol.* **173:**783–790.

Gill, R., F. Heffron, G. Dougan, and S. Falkow. 1978. Analysis of sequences transposed by complementation of two classes of transposition-deficient mutants of Tn3. *J. Bacteriol.* **136**:742–756.

Groisman, E. O., and M. J. Casadaban. 1986. Mini-Mu bacteriophage with plasmid replicons for in vivo cloning and *lac* gene fusing. *J. Bacteriol.* **168**:357–364.

Hamilton, C. M., M. Aldea, B. K. Washburn, P. Babitzke, and S. R. Kushner. 1989. A new method for generating deletions and gene replacements in *Escherichia coli*. *J. Bacteriol.* **171**:4617–4622.

Hofte, H., and H. R. Whiteley. 1989. Insecticidal crystal proteins of *Bacillus thuringiensis*. *Microbiol. Rev.* **53**:242–255.

Holloway, B. W. 1993. Genetics for all bacteria. *Annu. Rev. Microbiol.* **47**:659–684.

Hopwood, D. A., F. Malpartida, H. M. Kieser, H. Ikeda, J. Duncan, I. Fujii, B. A. M. Rudd, H. G. Floss, and S. Omura. 1985. Production of "hybrid" antibiotics by genetic engineering. *Nature* (London) **314**:642–644.

Horsch, R. B., J. E. Fry, N. L. Hoffmann, D. Eichholtz, S. G. Rogers, and R. T. Fraley. 1985. A simple and general method for transferring genes into plants. *Science* **227**:1229–1231.

Katz, L., and S. Donadio. 1993. Polyketide synthesis. *Annu. Rev. Microbiol.* **47**:875–912.

Kenyon, C. J., and G. C. Walker. 1980. DNA-damaging agents stimulate gene expression at specific loci in *Escherichia coli*. *Proc. Natl. Acad. Sci. USA* **77**:2819–2823.

Lacatena, R. M., and G. Cesareni. 1981. Base pairing of RNA 1 with its complementary sequence in the primer precursor inhibits ColE1 replication. *Nature* (London) **294**:623–626.

McDaniel, R., S. S. Ebert-Khosla, D. A. Hopwood, and C. Khosla. 1995. Rational design of aromatic polyketide natural products by recombinant assembly of enzymatic subunits. *Nature* (London) **375**:549–554.

McEachern, M. J., M. A. Bott, P. A. Tooker, and D. R. Helinski. 1989. Negative control of plasmid R6K replication: possible role of intermolecular coupling of replication origins. *Proc. Natl. Acad. Sci. USA* **86**:7942–7946.

Perlak, F. J., R. L. Fuchs, D. A. Dean, S. L. McPherson, and D. A. Fischhoff. 1991. Modification of the coding sequence enhances plant expression of insect control protein genes. *Proc. Natl. Acad. Sci. USA* **88**:3324–3328.

Simon, R., U. Priefer, and A. Puhler. 1983. A broad host range mobilization system for in vivo genetic engineering: transposon mutagenesis in gram negative bacteria. *Bio/Technology* **1**:784–790.

Slater, S., T. Gallaher, and D. Dennis. 1992. Production of poly-(3-hydroxybutyrate-Co-3-hydroxyvalerate) in a recombinant *Escherichia coli* strain. *Appl. Environ. Microbiol.* **58**:1089–1094.

Vanderslice, R. W., D. H. Doherty, M. A. Capage, M. R. Betlach, R. A. Hassler, N. M. Henderson, J. Ryan-Graniero, and M. Tecklenburg. 1993. Genetic engineering of poysaccharide structure in *Xanthomonas campestris*, p. 145–156. *In* V. Crescenzi, I. C. M. Dea, and S. S. Stivola (ed.), *Recent Developments in Industrial Polysaccharides*. Gorden and Breach Science, New York.

Van Gijsegem, F., and A. Toussaint. 1983. In vivo cloning of *Erwinia carotovora* genes involved in catabolism of hexuronates. *J. Bacteriol.* **154**:1227–1235.

Waters, V. L., K. H. Hirata, W. Pansegrau, E. Lanka, and D. G. Guiney. 1991. Sequence identity in the nick regions of IncP plasmid transfer origins and T-DNA borders of *Agrobacterium* Ti plasmids. *Proc. Natl. Acad. Sci. USA* **88**:1456–1460.

Winans, S. C., and G. C. Walker. 1985. Conjugal transfer system of the Inc N plasmid, pKM101. *J. Bacteriol.* **161**:402–410.

Zambryski, P. 1992. Chronicles from the *Agrobacterium*-plant cell DNA transfer story. *Annu. Rev. Plant Physiol. Mol. Biol.* **43**:465–490.

Answers to Questions for Thought and Problems

Chapter 1

Questions for Thought

1. The two strands of bacterial DNA probably aren't replicated in the 3'-to-5' direction simultaneously because replicating a DNA as long as the chromosome in this manner would leave single-stranded regions that were so long that they would be too unstable or susceptible to nucleases.

2. DNA molecules may be very long because if cells contained many short pieces of DNA, each one would have to be segregated individually into the daughter cells.

3. DNA may be the hereditary material instead of RNA because double-stranded DNA has a slightly different structure than double-stranded RNA. The B-form structure of DNA may have advantages for replication, etc. The use of DNA instead of RNA may allow the primer, which is made of RNA, to be more easily identified and removed from a DNA molecule by the editing functions. The editing functions do not operate when the 5' end is synthesized. By removing and resynthesizing any RNA regions, using upstream DNA as primer, mistakes can be minimized.

4. A temperature shift should cause the rate of DNA synthesis to drop but not too abruptly. Each cell would complete the rounds of replication that were under way at the time of the shift but would not begin another round. The synthesis rate would drop exponentially. If cells are growing rapidly, the drop would be less steep because a number of rounds of DNA replication would be under way in each molecule and each would have to complete the cycle.

5. The gyrase of *Streptomyces sphaeroides* might be naturally resistant to novobiocin. You could purify the gyrase and test its ability to introduce supercoils into DNA in vitro in the presence of novobiocin.

6. How chromosome replication and cell division are coordinated in bacteria like *Escherichia coli* is not well understood. One hypothesis is that a protein required for cell division might be encoded by a gene located close to the termination region and this gene is transcribed into RNA only when it replicates. Cell division could then begin only when the termination region has replicated. You could move the terminus of replication somewhere else in the chromosome and see if this affects the timing of cell division.

7. There is no obvious answer why the terminus of chromosome replication seldom carries essential genes.

Problems

1. 5′GGATTA3′

2. 5′GGAddT3′

   5′GGATddT3′

   5′GGATTACGGddT3′

   5′GGATTACGGTAAGGddT3′

3. *I* = 25 min, *C* = 40 min, *D* = 20 min

4. *I* = 90 min, *C* = 40 min, *D* = 20 min

5. The *topA* mutant lacks topoisomerase I, which removes negative supercoils, so that there should be more negative supercoils in the DNA of the mutant.

Chapter 2
Questions for Thought

1. Some people think RNA came first because the peptidyl-transferase that links amino acids together to make protein is the 23S rRNA, and some other enzymes are also RNA. However, the question remains open to speculation.

2. The genetic code may be universal because once the code was established, too many components—aminoacyl-tRNA synthetases, etc.—were involved in translating the code to change all of them.

3. Why eukaryotes do not have polycistronic mRNAs is open to speculation.

4. The genetic code of mitochondrial genes differs from the chromosomal code of eukaryotes because mitochondria were once bacteria with their own simple translation apparatus, ribosomes, etc. This translation apparatus has remained independent of that for the chromosomal genes.

5. Selenocysteine may be a relic of what was once a useful process in an earlier organism from which all other organisms evolved.

6. The translation apparatus is very highly conserved evolutionarily, so that an antibiotic that inhibits the translation apparatus of one type of bacteria is apt to inhibit the translation apparatus of all bacteria.

Problems

1. 5′CUAACUGAUGUGAUGUCAACGUCCUACU-CUAGCGUAGUCUAA3′

2. Translation should begin in the second triplet GUG because this is followed by a long ORF, although this hypothesis would have to be tested.

3. The answer is (a). Both sequences have a string of A's, but only in (a) is this preceded by an inverted repeat that could form a hairpin loop in the mRNA.

Chapter 3
Questions for Thought

1. Why the genetic maps of *Salmonella* and *Escherichia* spp. are similar is unknown. Perhaps there is an optimal way to arrange genes on a chromosome, with genes that are expressed at high levels closest to the origin of replication and transcribed in the same direction in which they are replicated. Inversions would then be selected against.

2. Duplication mutations would allow the number of genes of an organism to increase. The duplicated genes could then evolve so that their products could perform novel functions. Sometimes organisms with a duplication of a particular region may have a selective advantage in a particular environment, and so the duplication would be preserved.

3. The cells of higher organisms may be more finely tuned because runthrough proteins resulting from translation through the ends of genes may create more problems for eukaryotic cells, which, being more complicated, may be less tolerant of aberrant proteins.

4. It is not known how directed, or adaptive, mutations might occur. In their purest form, adaptive mutations would require that the cell somehow sense that a mutation would be desirable and change the DNA sequence accordingly.

Problems

1. Arginine auxotrophy would have a higher mutation rate because many genes encode enzymes to make arginine and any mutation that inactivates the product of one of these genes would make the cell Arg⁻. Rifampin resistance, however, can be caused by only a few mutations in the gene for the β subunit of RNA polymerase because very few amino acids can be changed and have the RNA polymerase no longer bind rifampin but still be active for transcription.

2. Approximately 8×10^{-10}

3. 5.5×10^{-8}

4. a. *arg-1* is probably a leaky missense or another type of base pair change mutation since it seems to retain some activity of the gene product and it reverts.

b. *arg-2* is probably a deletion mutation since it is not leaky and does not revert.

c. *arg-1* could be a frameshift mutation close to the carboxy end of the gene, so that the gene product is not totally inactivated. The mutant with *arg-2* could be a double mutant with two missense mutations in *arg* genes, so both of them would seldom revert.

5. You could make a double mutant with a *dam* mutation and an *arg* mutation like *arg-1*. Then you could compare the rever-

sion frequency to Arg⁺ of this double mutant with that of the single *arg* mutant without the *dam* mutation.

Chapter 4
Questions for Thought

1. If essential genes were carried on a plasmid, cells that were cured of the plasmid would die. By having nonessential genes on plasmids, the chromosome can be smaller so that the cells can multiply faster yet the species can adapt to a wide range of environments because of selection for those cells with a particular plasmid. You would not expect genes encoding enzymes of the tricarboxylic acid cycle such as isocitrate dehydrogenase or genes for proteins involved in macromolecular synthesis such as RNA polymerase to be carried on plasmids. Genes involved in using unusual carbon sources such as the herbicide 2,4-D or genes required for resistance to antibiotics such as ampicillin might be expected to be on plasmids.

2. Why some but not all plasmids have a broad host range is unknown. A broad-host-range plasmid can parasitize more species of bacteria. A narrow-host-range plasmid can develop a better commensal relationship with its unique host.

3. Perhaps the plasmid binds to a unique site on the membrane of the cell. When the cell is ready to divide, there are two copies of this site, each of which can bind a copy of the plasmid. To ensure that the number of copies of the plasmid never exceeds the number of such membrane sites, perhaps the plasmid can replicate only once and then only when it is bound to one of these sites.

4. If the genes required for replication of the plasmid are not all closely linked to the *ori* site, you could find them by isolating temperature-sensitive mutants of the plasmid that cannot replicate at high temperature and then look for pieces of plasmid DNA that can help the mutant plasmid replicate at the high temperature when introduced into the cell in a cloning vector.

5. You could determine which of the replication genes of the host *E. coli* (*dnaA*, *dnaC*, etc.) are required for replication of a given plasmid by introducing the plasmid into cells with temperature-sensitive mutations in each of the genes and then determining whether the plasmid can replicate at the high nonpermissive temperature for the mutant.

Problems

1. You introduce your plasmid into cells with the other plasmid and grow the cells in the absence of kanamycin. After a few generations, you plate the cells and test the colonies for kanamycin resistance. If more of the cells are kanamycin sensitive than would be the case if your plasmid were replicating in the same type of cell without the other plasmid, your plasmid is probably an IncQ plasmid.

2. 1/2,048

3. You take the colonies due to the transformants on ampicillin plates and test them on tetracycline plates to determine whether they are tetracycline resistant. The plasmids in the cells that are ampicillin resistant but tetracycline sensitive probably have an insert in the *Bam*HI site, since the *Bam*HI site of pBR322 is in the Tetʳ gene.

Chapter 5
Questions for Thought

1. If plasmids of the same Inc group all have the same Tra functions, the exclusion functions will prevent a plasmid from transferring into a cell that already contains a plasmid of the same Inc group, where one or the other would be lost.

2. If the *tra* genes and the *oriT* site on which they act are close to each other, recombination will seldom occur between the *oriT* site and the genes for the Tra functions. Separating the Tra functions from the *oriT* site on which they act will render them nonfunctional.

3. Perhaps plasmids with certain *mob* sites can be transferred only by certain corresponding Tra functions because only these functions will allow the mobilizable plasmid to recognize the signal that the cell has made contact with a potential recipient cell and be cut.

4. Self-transmissible plasmids encode their own primases, so they can transfer themselves into a host cell with an incompatible primase and still synthesize their complementary strand.

5. The absence of an outer membrane in gram-positive bacteria may make a pilus unnecessary. The question is open to speculation.

6. The answer is not known. Perhaps the role of the pilus in helping transmit DNA into the cell can be easily subverted by the phage to transfer its own DNA into the cell.

7. Plasmids are generally either self-transmissible or mobilizable because if they were neither, they would not be able to move to cells that did not already contain them. By being promiscuous, plasmids can expand their host range.

Problems

1. The recipient strain is the one that becomes recombinant and retains most of the characteristics of the original strain. If the transfer is due to a prime factor, the apparent recombinants will become donors of the same genes.

2. Determining which of the *tra* genes of a self-transmissible plasmid encodes the pilin protein is not easy. For example, phage-resistant mutants will not necessarily have a mutation in the pilin gene. They could also have a mutation in a gene whose product is required to assemble the pilus on the cell surface. You could purify the pili and make antibodies to them. Then *tra* mutants in the pilin gene won't make an antigen that will react with the antibody. Similarly, to determine which *tra* mutants don't make the DNase that nicks the DNA at the *ori* region or the helicase that separates the DNA strands, you may have to develop assays for these enzymes in crude extracts and determine which *tra* mutants don't make the enzyme in your assay.

3. You can show that only one strand of donor DNA enters a recipient cell by using a recipient that has a temperature-sensitive mutation in its primase gene. The plasmid DNA should remain single stranded after transfer into such a strain, provided, of course, that the plasmid can't make its own primase. Single-stranded DNA is more sensitive to some types of DNases and behaves differently from double-stranded DNA during gel electrophoresis.

4. If the tetracycline resistance gene is in a plasmid, the tetracycline-resistant recipient cells should have acquired a plasmid. If it is in a conjugative transposon, no transferred plasmid will be in evidence.

Chapter 6
Questions for Thought

1. The chapter lists some possible reasons for the development of competence in bacteria. At this time we do not know the correct answer.

2. To discover whether the competence genes of *Bacillus subtilis* are turned on by UV irradiation and other types of DNA damage, you could make a gene fusion with a reporter gene to one of the competence factor-encoding genes and see if the reporter gene is induced following UV irradiation.

3. To determine whether antigenic variation in *Neisseria gonorrhoeae* results from transformation between bacteria or recombination within the same bacterium, you could introduce a selectable gene for antibiotic resistance into one of the antigen genes and see if it is transferred naturally under conditions where antigenic variation occurs.

Problems

1. To determine whether a given bacterium is naturally competent, you would isolate an auxotrophic mutant, such as a Met⁻ mutant, and mix it with DNA extracted from the wild-type bacterium. The mixture would then be plated on medium without methionine. The appearance of colonies due to Met⁺ recombinants would be evidence of transformation.

2. To isolate mutants defective in transformation, you would take your Met⁻ mutant, mutagenize it, and repeat the test above on individual isolates. Any mutants that do not give Met⁺ recombinants when mixed with the wild-type DNA might be mutants with a second mutation in a competence gene.

3. To discover whether a naturally transformable bacterium can take up DNA of only its own species or any DNA, you could make radioactive DNA and mix it with your competent bacteria. Any DNA taken up by the cells would become resistant to added DNase, and the radioactivity would be retained with the cells on filters. Try this experiment with radioactive DNA from the same species as well as from different species.

4. If the bacterium can take up DNA of only the same species, it must depend upon uptake sequences from that species. The

experiment should be done as in problem 3 but with only a piece of DNA instead of the entire molecule.

5. If the DNA of a phage successfully transfects competent *E. coli*, plaques will appear when the transfected cells are plated with bacteria sensitive to the phage.

Chapter 7
Questions for Thought

1. If phages made the proteins of the phage particle at the same time as they made DNA, the DNA would be prematurely packaged into phage heads, leaving no DNA to replicate.

2. Phages that make their own RNA polymerase can shut off host transcription by inactivating their host RNA polymerase without inactivating their own RNA polymerase. However, phages that use the host RNA polymerase can take advantage of the ability of the host molecule to interact with other host proteins, allowing more complex regulation.

3. Perhaps the λ prophage uses different promoters to transcribe the *cI* repressor gene immediately after infection and in the lysogenic state because this may allow the repressor gene to be transcribed from a strong, unregulated promoter immediately after infection but then be transcribed from a weaker, regulated promoter in the lysogenic state.

4. By making two proteins, the cell can use the Int protein to promote recombination for both integration and excision. Then the smaller Xis protein need only recognize the hybrid *att* sites at the ends of the prophage.

5. By duplicating host functions, the phage can extend its host range because the host functions need not be compatible with sites such as promoters and *ori* regions on the phage DNA. However, a phage that uses host functions can have a smaller genome with fewer genes.

6. Toxins and other proteins that confer pathogenicity may often be encoded by phages for the same reason that only conditionally essential genes are carried on plasmids (see chapter 4). Many of the same bacteria can also live outside the host, and by having the virulence genes carried on prophages, not all the bacteria of the species need carry them and so the bacteria can have a smaller genome and replicate more rapidly.

7. Why some types of prophage can be induced only if another phage of the same type infects the lysogenic cell containing them is unknown. Perhaps there is some way of inducing them that hasn't been tried.

Problems

1. The regulatory gene is probably gene *M*, because mutations in this gene can prevent the synthesis of many different gene products. The other genes probably encode products required for the assembly of tails and heads.

2. Amber mutations introduce a nonsense UAG codon into the coding sequence of an mRNA, stopping translation and

leading to synthesis of a shortened gene product. The *ori* sequence does not encode a protein, and so an amber mutation could not be isolated from it.

3. If the clear mutant has a *vir* mutation, it will form plaques on a λ lysogen.

4. A specialized transducing phage carrying the *bio* operon of *E. coli* would be isolated in the same way as the λd*gal* phage in the text, except that a Bio⁻ mutant would be infected and plated on medium without biotin. The Bio⁺ bacteria would be isolated, and the phage would be induced. It might be necessary to add a wild-type helper phage before induction, since *bio* substitutions extend into the *int* and *xis* genes. The mutant should form plaques, because no replication genes should be substituted. A λ phage with *vir* mutations in the o_1 sites of o_L and o_R should form plaques on a λ lysogen because the repressor cannot bind to the mutant operators.

5. Both Int and Xis are required to integrate phage λ transducing particles next to an existing prophage because the recombination occurs between two hybrid *attP-attB* sites.

6. To determine whether the *Staphylococcus aureus* toxin is encoded by a prophage, you could take the strain of *S. aureus* and try to induce a phage from it by UV irradiation, etc. The cells could be filtered out, and the cell-free medium could be plated on a closely related *S. aureus* strain to see if plaques form. You could then use the phage in the plaques to try to isolate lysogens of the related *S. aureus* strain to determine whether the lysogens produced the toxin, as evidenced by the acquired ability to kill human cells in culture.

Chapter 8

Questions for Thought

1. Perhaps replicative transposons do not occur in multiple copies around a genome because their resolution functions cause deletions between repeated copies of the transposon, resulting in the death of cells with more than one copy. Also, a poorly understood phenomenon called target immunity inhibits transposition of a transposon into a DNA that already contains the same transposon.

2. The transposon Tn3 and its relatives may have spread throughout the bacterial kingdom on promiscuous plasmids.

3. Transposons sometimes, but not always, carry antibiotic resistance or other traits of benefit to the host. They may also help the host move genes around, as in the construction of plasmids carrying multiple drug resistance.

4. The origin of the genes within integrons is not known. Perhaps they come from plasmids or other transposons.

5. It is possible that invertible sequences rarely invert because very little of the invertase enzyme is made or because the enzyme works very inefficiently on the sites at the ends of the invertible DNA sequence.

Problems

1. You would integrate a λd*gal* at the normal λ attachment site close to the *gal* operon with one of the *gal* mutations by selecting for Gal⁺ transductants. Sometimes the *gal* genes in the chromosome would recombine with the *gal* genes on the integrated λ and the cell would become Gal⁻ by gene conversion (homogenoting). When the λd*gal* are induced, their DNA should be longer and the phage denser owing to the inserted DNA.

2. The colonies should not be sectored because only one strand, either the *lac* or *lac⁺* strand, of the original heteroduplex transposon will have been inserted.

3. By methods such as those outlined in chapter 15, you could clone the pigment gene and use it as a probe in Southern blot analyses to see whether the physical map of the DNA around the gene changes when it is in the pigmented as opposed to the nonpigmented form.

4. You could make a plasmid that has two copies of the Mu phage by cloning a piece of DNA containing Mu into a plasmid cloning vector also containing Mu. You could then see if the two repeated Mu elements could resolve themselves in the absence of the host recombination functions.

Chapter 9

Questions for Thought

1. Recombination might help speed up evolution by allowing new combinations of genes to be tried or may help in repair of DNA damage.

2. The reason that RecBCD recombination is so complicated is unknown. The advantages of *chi* sites or of having a recombinase enzyme enter the DNA only at double-stranded ends are not clear.

3. The different pathways of recombination may function under different conditions for recombination between short and long DNAs or between circular and linear DNAs, etc.

4. The RecF pathway may be preferred under conditions different from the ones normally used in the laboratory to measure recombination. The SbcB and SbcC functions may interfere only under the conditions normally used in the laboratory.

5. By encoding their own recombination functions, phages can increase their rate of recombination. Also, some phages use the recombination functions for replication and, by encoding their own functions, can inhibit the host recombination functions to prevent them from interfering with phage replication, as in the case of the RecBCD function and λ rolling-circle replication.

Problems

1. To determine if recombinants in an Hfr cross have a *recA* mutation, you could take the individual recombinants and make a streak with them on a plate. Then half of each streak

could be covered with glass (glass is opaque to UV) and the plate irradiated before incubation. If the recombinant is RecA⁻, it will grow only in the part of the streak that was covered by the plate.

2. To determine what, if any, other genes participate in the *recG* pathway, start with a *ruvABC* mutant and isolate mutants with additional mutations that make them deficient in recombination. See if any have mutations in genes other than the *recG* gene.

3. The recombination promoted by homing double-stranded nucleases to insert an intron occurs in the same manner as in Figure 9.4, except that the double-stranded break that initiates the recombination occurs at the site in the target DNA into which the intron will home. The invading DNA will then pair with the homologous flanking DNA on one side of the transposon in the donor DNA and replicate over the transposon until it meets the other 5' end.

4. In a RecB⁺C⁺D⁻ host, compare recombination between the same two markers in λ DNAs, one with a *chi* mutation and another without. If there is no difference, *chi* sites only stimulate recombination because of the RecD function.

Chapter 10
Questions for Thought

1. The answer is unknown. Perhaps the alternative to methylation of the bases is some other type of covalent modification, such as methylation of the phosphates, or a protein or RNA may remain bound to the newly synthesized DNA strand for a period after it has been synthesized.

2. Different repair pathways may work better depending upon where the damage occurs or the extent of the damage. For example, if damage is so extensive that lesions occur almost opposite each other in the two strands of the DNA, it might be easier to repair the lesions with excision repair than with recombination repair.

3. The SOS mutagenesis pathway might exist to allow the cells to survive damage other than that due to UV irradiation, or it might be more effective under culture conditions different from those used in the laboratory.

Problems

1. Assuming that you can grow the organism in the laboratory (you might have to grow it under high pressure), you could irradiate it in the dark and then divide the culture in half and expose half of the cells to visible light before diluting and measuring the surviving bacteria. If more of the cells survive after they have been exposed to visible light, the bacterium has a photoreactivation system.

2. The procedure is explained in the text. Briefly, to show that the mismatch repair system preferentially repairs the unmethylated base, you could make heteroduplex DNA of λ phage. One strand should be unmethylated and heavier than

the other because the λ from which this strand was derived were propagated on Dam⁻ *E. coli* cells grown in heavy isotopes. The two λ used to make the heteroduplex DNA should also have mutations in different genes, so that there will be mismatches at these positions. After the heteroduplex λ DNA is transfected into cells, test the progeny phage to determine which genotype prevails, the genotype of the phage from which the unmethylated DNA was prepared, or the genotype of the phage with methylated DNA.

3. To determine whether the photoreactivating system is mutagenic, perform an experiment similar to that in problem 1 but with *E. coli*. Measure the frequency of mutations (such as reversion of a *his* mutation) among the survivors of UV irradiation in the dark as opposed to those that have been exposed to visible light after UV irradiation. More cells should survive if they are exposed to visible light, but a higher *frequency* of these survivors should be His⁺ revertants if photoreactivation is mutagenic. If photoreactivation is not mutagenic, a lower frequency should be His⁺ revertants because the photoreactivation system will have removed some of the potentially mutagenic lesions. It may be better to do this experiment with a *umuCD* mutant to lower the background mutations due to SOS mutagenesis.

4. To find out whether the nucleotide excision repair system can repair damage due to aflatoxin B, treat wild-type *E. coli* cells and a *uvrA*, *uvrB*, or *uvrC* mutant with aflatoxin B. Dilute and plate. Compare the survivor frequencies of the *uvr* mutant and the wild type.

5. Express *umuC* and *umuD* from a clone that has a constitutive operator mutation so that the cloned genes are not repressed by LexA. Also, be sure that part of the *umuD* gene has been deleted so that UmuD', rather than the complete UmuD, will be synthesized. This clone can be put into isogenic RecA⁻ and RecA⁺ strains of *E. coli* that have a *his* mutation. After UV irradiation, the frequency of His⁺ revertants among the surviving bacteria can be compared for each strain. If the RecA⁺ strain shows a higher frequency of His⁺ revertants, the RecA protein may have a role in UV mutagenesis other than cleaving LexA and UmuD.

6. You could make a transcriptional fusion of the *recN* gene to a reporter gene such as *lacZ* and then determine whether more of the reporter gene product is synthesized after UV irradiation, as it should be if *recN* is an SOS gene. A strain that also has a *lexA*(Ind⁻) mutation should not show this induction if *recN* is an SOS gene.

Chapter 11
Questions for Thought

1. Why operons are regulated both positively and negatively is not clear. However, there may be different advantages to the two types of regulation. For example, negative regulation might require more regulatory protein but might allow more complete repression, while it might be easier to achieve inter-

mediate levels of expression with positive regulation. There may be less interaction between negative regulatory systems than between positive regulatory systems, in which a regulatory protein might inadvertently turn on another operon. Also, constitutive mutants will be rarer with positive regulation.

2. The genes for regulatory proteins may be autoregulated to save energy. If they are autoregulated, only the amount of regulatory protein needed will be synthesized. Also, in the case of positive regulators, more can be made after induction to further increase the expression of the operons under their control.

3. Regulation by attenuation of transcription of amino acid biosynthetic operons offers the advantage that the ability of the cell to translate codons for that amino acid can be exploited to regulate the operon. The regulatory system can be designed so that transcription will continue into the structural genes of the operon if ribosomes stall in the leader region at codons for the amino acid because not enough of the amino acid is available. A major disadvantage of this type of regulation is that it is wasteful. A short RNA is always made from the operon, even if it is not needed.

Problems

1. To isolate a *lacI*s mutant, take advantage of the fact that *lacI*s mutations, while rare, are dominant Lac⁻ mutations. Mutagenize a *galE* mutant that contains an F' factor with the *lac* operon, and plate it on Pgal medium containing another carbon source such as maltose. The survivors that form colonies are good candidates for *lacI*s mutants because inactivation of the *lacZ* genes in both the chromosome and the F' factor requires two independent mutations, which should be even rarer. The mutants can be further tested by mating the F' factor from them into other strains whose chromosome contains a wild-type *lac* operon. If the F' factor makes the other strain Lac⁻, the F' factor must contain a *lacI*s mutation.

2. AraC must be in the P1 state, since it represses the *ara* operon.

3. To determine whether *phoA* is negatively or positively regulated, you could first isolate constitutive mutants to determine how frequent they are. Since PhoA turns XP blue, you could mutagenize cells and isolate mutants that form blue colonies on XP-containing medium even in the presence of excess phosphate in the medium. If *phoA* is negatively regulated, constitutive mutants should be much more frequent than if it is positively regulated. Also, at least some of these constitutive mutants should have null mutations, deletions, etc., that inactivate the regulatory gene.

4. Plate wild-type *E. coli* cells in the presence of low concentrations of 5-methyltryptophan in the absence of tryptophan. Only constitutive mutants can multiply to form colonies under these conditions, because 5-methyltryptophan is a corepressor of the *trp* operon but cannot be used for protein synthesis, so that the nonmutant wild-type *E. coli* cells will starve for tryptophan. To isolate mutants defective in feedback inhibition of

tryptophan synthesis, plate a constitutive mutant in the presence of higher concentrations of 5-methyltryptophan in the absence of tryptophan. Even constitutive mutants cannot multiply to form colonies under these conditions, because the first enzyme of tryptophan synthesis will be feedback inhibited by the 5-methyltryptophan. Only mutants that are defective in feedback inhibition will form colonies.

Chapter 12
Questions for Thought

1. Perhaps it is important to synthesize more of the proteins involved in synthesizing new proteins so that the rate of protein synthesis will increase after heat shock, allowing more rapid replacement of the proteins irreversibly denatured as a result of the shock.

2. Perhaps *Salmonella* species, which are normal inhabitants of the vertebrate intestine, are usually in an environment where amino acids are in plentiful supply but NH_3 is limiting. *Klebsiella* species may usually be free-living, where NH_3 is present but amino acids are not.

3. Perhaps the genes for corresponding sensor and response regulator genes are similar to allow cross talk between regulatory pathways. If the genes are similar, a signal from one pathway can be passed to the other pathway, allowing coordinate regulation in response to the same external stimulus. However, there is no good evidence for the importance of cross talk. Another possible explanation is that the genes had a common ancestor in evolution and still retain many of the same properties.

4. The enzymes responsible for ppGpp synthesis during amino acid starvation and during growth rate control may be different because the enzymes involved in stringent control and growth rate regulation must be in communication with different cellular constituents. The RelA protein works in association with the ribosome, where it can sense amino acid starvation, while the enzyme involved in synthesizing ppGpp during growth rate regulation might have to sense the level of catabolites. SpoT might be involved both in synthesizing and degrading ppGpp if the equilibrium of the reaction is somehow shifted. All enzymes function by lowering the activation energy and so in a sense catalyze both the forward and backward reaction. However, because the equilibrium favors the forward reaction, this is the reaction that predominates. If the equilibrium were shifted, perhaps by sequestering the ppGpp as it is made, SpoT could synthesize ppGpp rather than degrade it.

Problems

1. Determine if the mutant can use other carbon sources such as lactose and galactose. If it has a *cya* or *crp* mutation, it should not be able to induce other catabolite-sensitive operons and so will not be able to grow on these other carbon sources.

2. **a.** Gln⁻ (glutamine requiring); Ntr constitutive (express Ntr operons even in presence of NH_3)

b. Ntr⁻ (can't express Ntr operons even at low NH₃ concentrations). Make intermediate levels of glutamine synthetase independent of the presence or absence of NH₃.

c. Gln⁻, Ntr⁻

d. Gln⁻, Ntr⁻

e. Ntr constitutive

f. Temperature sensitive (won't grow at high temperature)

g. Constitutive expression of diphtheria toxin and other virulence determinants, even in Fe²⁺

h. No ppGpp; grows very slowly

3. As described in chapter 15, you could clone the gene for the toxin and then perform Southern blot analysis to show that the gene is carried on a large region of DNA that is not common to all the members of the species. Using the methods in chapter 7, you could also try to induce a phage from the cells and show that production of the toxin requires lysogeny by the phage.

4. If the *rpoH* gene for σ³² is transcriptionally autoregulated, the same amount of RNA should be made from the gene when it exists in two or more copies as is made when it exists in only one copy. Introduce a clone of *rpoH* in a multicopy plasmid into cells, and measure the amount of RNA made on the gene by DNA-RNA hybridization. If more RNA is made from *rpoH* under these conditions, the gene is not transcriptionally autoregulated.

5. To show which ribosomal protein is the translational repressor, introduce an in-frame deletion into the *rplJ* gene and determine if the synthesis of L12 increases. Similarly, introduce a mutation (any inactivating mutation will do) into *rplL* and determine if L10 synthesis increases.

Chapter 13

Problems

1. 0.01 or 1%

2. At the extreme end of the *rIIA* gene, the only portion not deleted by *r1241*

3. Mutations *r736* and *r866* will not necessarily complement. *r736* might be at the extreme end of the *rIIA* gene very close to *r26* in the adjacent end of the *rIIB* gene, while *r866* might be at the other end of *rIIA*, much farther from *r736* but not complemented by it, since they are in the same gene.

4. Cross the mutants with each other on *E. coli* B, and plate the progeny on *E. coli* B. Pick plaques, and spot them on *E. coli* K-12λ to identify mutants that cannot grow on *E. coli* K12λ. Then test these mutants individually by crossing them with both *r26* and *r686*. *rII* mutants that do not give *r⁺* recombinants when crossed with *r26* and *r686* are double mutants with both the *r26* and *r686* mutations.

5. The order will be *r686-r736-r26* because with this order two crossovers are required to make a wild-type *r⁺* recombinant. With the other order, a single crossover between *r26* and *r736* would be sufficient to make an *r⁺* recombinant, and we already know from problem 3 above that the recombination frequency between *r26* and *r736* is 0.1%, much higher than 0.005%.

6. The order is A-Q-M because with this order most of the *am⁺* recombinants with a crossover between *amA* and *amQ* would have the Ts mutation in gene M. A second crossover between Q and M would be required to give the wild-type recombinant.

7. *r291* could be a frameshift or a nonsense mutation.

8. The viruses are diluted and plated with indicator cells under the given set of conditions. From the number of plaques, the concentration of viruses that actually infect cells under these conditions can be determined. The ratio of this concentration to the actual concentration of the viruses determined by counting under the electron microscope is the effective multiplicity of infection.

9. T1 will have a linear genetic map but with some genes repeated at the ends. ABCDEFG - - - - XYZABC.

10. To show that a *nutL* mutation is *cis* acting, infect a P2 lysogen simultaneously with λ that has a *nutL* mutation and another λ with a *red gam* mutation. If the phage can multiply, the *nutL* mutation cannot be complemented to make the Red and Gam proteins from the same DNA.

Chapter 14

Problems

1. Bacteria offer the advantages for genetic analysis that they are haploid, so that recessive mutations show their phenotypes without the need to make strains homozygous for the mutations; they multiply in colonies asexually, so that it is easy to get purebred strains; millions of bacteria can be analyzed on single petri plates, which enables the selection of very rare mutants and recombinant types; and many bacteria have convenient genetic systems that make it possible to do crosses. The disadvantages are that haploidy makes complementation tests difficult. The small size of bacteria also makes visual phenotypes difficult to detect, and we must rely on the observation of colonies containing bacteria many generations removed from the original bacterium that formed the colony.

2. Mutants must be isolated from different cultures to obtain independent mutations because the mutants in the same culture will often be descendants of the same original mutant and thus will have inherited the same set of mutations.

3. The mutations in those cultures with only a few mutant bacteria are more likely to be independent. In cultures with an exceptionally large number of mutants, a mutant must have arisen early in the growth of the culture, and most of the mutants will be descendants or siblings of this mutant.

4. a. Make a plate containing medium with a concentration of coumermycin just sufficient to kill the wild-type bacterium. Spread millions of wild-type bacteria on the plate. Only coumermycin-resistant mutants should multiply to form a colony.

b. Make a plate containing minimal medium with all of the growth requirements of the *trp* mutant except tryptophan. Spread millions of the mutant bacteria on the plate. Only bacteria in which the *trp* mutation has reverted or been suppressed will multiply to form a colony. To distinguish true revertants from suppressed mutations, transduce or otherwise transfer the region of the original *trp* mutation from one of these apparent revertants into another strain. If no *trp* recombinants appear, the mutation had probably reverted rather than been suppressed.

c. Plate millions of the bacteria with the Ts mutation at the nonpermissive temperature for the mutant. Only bacteria in which the Ts mutation has reverted or been suppressed will multiply to form a colony. To distinguish those that have been suppressed, transduce or otherwise transfer the *dnaA* gene from some of the apparent revertant strains separately into another strain. If temperature-sensitive recombinants appear, the Ts mutation in *dnaA* in that strain had probably been suppressed rather than reverted.

d. Plate the *araD* mutant on plates containing arabinose plus another carbon source such as lactose. Any colonies that arise are due to mutant bacteria that are no longer sensitive to arabinose. In those that are still Ara⁻, as evidenced by the fact that they cannot multiply to form colonies with arabinose as the sole carbon source, the mutation was probably suppressed rather than reverted. The suppressor mutation could be localized to the *araA* gene by complementation tests.

e. Polarity of the mutation in *hisC* will make the strain appear *hisB* as well, so it will not be complemented by mutations in either *hisB* or *hisC*. Therefore, a partially diploid bacterium with a prime factor containing the *his* region with the polar *hisC* mutation and a *hisB* mutation in the chromosome will be His⁻ and will be unable to grow on media lacking histidine. If large numbers of these partially diploid bacteria are plated on media lacking histidine, the few colonies that appear may be due to mutants with a *suA* mutation, which relieves polarity and so allows the expression of the *hisB* gene from the prime factor, complementing the *hisB* mutation in the chromosome. To prove this, you could show that some of the apparent His⁺ revertants have suppressing mutations, as in problems 4c and 4d above, and that the suppressing mutations map to the *suA* gene, using methods described in the text in the section on mapping suppressor mutations.

5. Raise the temperature to the nonpermissive temperature for the mutants you want to enrich for, add ampicillin for a period of time, and then remove the ampicillin and lower the temperature.

6. Very close to the *thyA* gene

7. A little past 44 min

8. The cotransduction frequency of the *argH* and *metB* markers is 37%; that of the *argH* and Rifʳ markers is 20%. The order is *metB1–argH5–rif-8*. The order *argH5–metB1–rif-8* is consistent with some of the data, but the *argH* marker is probably in the middle, since most of the Rifʳ transductants are not Met⁺.

9. Close to *argG*

10. Nonsense suppressors are dominant because the suppressor tRNA will suppress nonsense codons even in the presence of the normal tRNA.

11. If the mutant *trp* operon has a *cis*-acting operator mutation, it will express the *trp* structural genes constitutively only from the DNA with the mutation. Therefore, if you introduce a *trpA* mutation into the chromosome with the constitutive mutation, it will no longer constitutively express the tryptophan synthetase product of the wild-type *trpA* gene on the prime factor in the same cell. Also, a *trpR⁻ᵈ* mutation will map in *trpR*, some distance from the *trp* operon, unlike *trp* operator mutations that map in the *trp* operon.

12. To determine which of the regions of the EnvZ protein are in the periplasm and which are in the cytoplasm, make translational fusions of *phoA* to various regions of the *envZ* gene. See if bacteria containing the fusion form blue colonies on XP plates. If PhoA is fused to a region of EnvZ in the periplasm, the colonies will be blue. If it is fused to a region in the cytoplasm, the colonies will be colorless.

13. A signal sequence mutation in *malE* will make the cells Mal⁻ and unable to transport maltose to use as a carbon source. A suppressor of the signal sequence mutation will make them Mal⁺. Spread millions of the Mal⁻ signal sequence mutants on medium containing maltose as the sole carbon source. The colonies that arise will be due to Mal⁺ bacteria that either have reverted the signal sequence mutation or have a second-site mutation that suppresses the effect of the signal sequence mutation. As in problems 4c and 4d, if the mutation is suppressed, when the *malE* region of the apparent revertants is introduced into another strain by transduction or other means, the other strains will become Mal⁻.

Chapter 15

Questions for Thought

1. The question of why bacteria encode eight-hitters in their chromosome has no obvious answer. The role of these enzymes could not be to restrict infecting phages, since most phage DNAs will not contain even a single site for an eight-hitter. One possibility is that the eight-hitter is actually encoded by a plasmid integrated into the chromosome and is part of a plasmid addiction system.

2. Some phages have modified bases in their DNA that make them immune to restriction endonucleases. Another way to avoid restriction systems might be to inhibit the SAMase, the enzyme that makes *S*-adenosylmethionine, since some restric-

Glossary

Activator. A protein that positively regulates transcription of an operon by interacting with RNA polymerase at the promoter and allowing RNA polymerase to begin transcribing.

Activator site. Sequence in DNA upstream of the promoter to which the activator protein binds.

Adaptive response. Activation of transcription of the genes of the Ada regulon, which is involved in the repair of some types of alkylation damage to DNA.

Adenine (A). One of the two purine (two-ringed) bases in DNA and RNA.

Alkylating agent. A chemical that reacts with DNA and thereby forms a carbon bond to one of the atoms in DNA.

Allele. One of the forms of a gene, e.g., the gene with a particular mutation. Can refer to the wild-type or mutant form.

Allele-specific suppressor. A second-site mutation that restores the wild-type phenotype but only in strains with a particular type of mutation in a gene.

Amber codon. The codon UAG.

Amber mutation. A mutation that causes the codon UAG to appear in frame in the coding region of an mRNA.

Amino group. The NH_2 chemical group.

Amino terminus. *See* N terminus.

Antibiotic. A chemical that kills cells or inhibits their growth.

Antibiotic resistance gene cassette. A fragment of DNA, usually bracketed by restriction sites, that contains a gene whose product confers resistance to an antibiotic.

Anticodon. The three-nucleotide sequence in a tRNA that pairs with the codon in mRNA by complementary base pairing.

Antiparallel. Referring to the fact that moving in one direction along a double-stranded DNA or RNA, the phosphates in one strand are attached 3' to 5' to the sugars while the phosphates in the other strand are attached 5' to 3'.

Antisense RNA. An RNA that contains a sequence complementary to a sequence in an mRNA.

Antitermination. A process in which changes in the RNA polymerase allow it to transcribe through transcription termination signals in DNA.

AP endonuclease. A DNA-cutting enzyme that cuts next to a deoxynucleotide that has lost its base, i.e., an apurinic or apyrimidinic site.

Aporepressor. A protein that can be converted into a repressor if a small molecule called the corepressor is bound to it.

Archaea. A separate kingdom of prokaryotic single-celled organisms that share some of the features of both eukaryotes and prokaryotes and usually inhabit extreme environments.

Assimilatory reduction. Addition of electrons to nitrogen-containing compounds to convert them into NH_3 for incorporation into cellular constituents.

Attenuation. Regulation of an operon by premature termination of transcription.

Autocleavage. Process by which a protein cuts itself.

Autokinase. A protein able to transfer a PO_4 group from ATP to itself.

Autophosphorylation. Process by which a protein transfers a PO_4 group to itself, independent of the source of the PO_4 group.

Autoregulation. Process through which a gene product controls the level of its own synthesis.

Auxotrophic mutant. A mutant that cannot make or use a growth substance that the normal or wild-type organism can make or use.

Backbone. The chain of phosphates linked to deoxyribose sugars that holds the DNA chain together.

Bacterial lawn. The layer of bacteria on an agar plate that forms when many bacteria are plated and the bacterial colonies grow together.

Bacteriophage. A virus that infects bacteria.

Base. Carbon-, nitrogen-, and hydrogen-containing chemical compounds with structures composed of one or two rings that are constituents of the DNA or RNA molecule.

Base analog. A chemical that resembles one of the bases and so is mistakenly incorporated into DNA or RNA during synthesis.

Base pair. Each set of opposing bases in the two strands of double-stranded DNA or RNA that are held together by hydrogen bonds and thereby help hold the two strands together. Also used as a unit of length.

Base pair change. A mutation in which one type of base pair in DNA (e.g., an AT pair) is changed into a different base pair (e.g., a GC pair).

Binding. Process by which molecules are physically joined to each other by noncovalent bonds.

Bioremediation. The removal of toxic chemicals from the environment by microorganisms.

Biosynthesis. Synthesis of chemical compounds by living organisms.

Biosynthetic operon. An operon composed of genes whose products are involved in synthesizing compounds rather than degrading them.

Blot. The process of transferring DNA, RNA, or protein from a gel or agar plate to a filter. Also refers to the filter that contains the DNA, RNA, or protein.

Blunt end. A double-stranded DNA end in which the 3' and 5' termini are flush with each other, that is, with no overhanging single strands.

Branch migration. The process by which the site at which two double-stranded DNAs are held together by crossed-over strands moves, changing the regions of the two DNAs that are paired in heteroduplexes.

Broad host range. The ability of a phage, plasmid, or other DNA element to enter and/or replicate in a wide variety of bacterial species.

CAP binding site. The sequence on DNA to which the catabolite activator or CAP protein binds.

CAP regulon. All of the operons that are regulated, either positively or negatively, by the CAP protein.

Capsid. The protein and/or membrane coat that surrounds the genomic nucleic acid (DNA or RNA) of a virus.

Carboxy group. The chemical group COOH of amino acids.

Carboxy terminus. *See* C terminus.

Catabolic operon. An operon composed of genes whose products degrade organic compounds.

Catabolism. The degradation of an organic compound, such as a sugar, to make smaller molecules with the concomitant production of energy.

Catabolite. A small molecule produced by the degradation of larger organic compounds such as sugars.

Catabolite activator protein (CAP). The DNA and cAMP binding protein that regulates catabolite-sensitive operons in enteric bacteria by binding to their promoter regions.

Catabolite repression. The reduced expression of some operons in the presence of high cellular levels of catabolites.

Catabolite-sensitive operons. Operons whose expression is regulated by the cellular levels of catabolites.

CCC. *See* Circular and covalently closed.

Cell division. The splitting of a cell into two daughter cells.

Cell division cycle. The events occurring between the time a cell is created by division of its mother cell and the time it divides.

Cell generation. The making of a new cell by the growth and division of its mother cell.

Central Dogma. The tenet that protein is translated from RNA that was transcribed from DNA.

Chaperone. A protein that binds to other proteins and helps them fold correctly or prevents them from folding prematurely.

chi **mutation.** A mutation that causes the sequence of a *chi* site to apppear in the DNA.

chi **site.** The sequence 5'GCTGGTGG3' in DNA. Stimulates nearby recombination by the RecBCD nuclease in *E. coli*.

Chromosome. In a bacterial cell, the DNA molecule that contains most of the genes required for cellular growth and maintenance.

CI repressor. The phage λ-encoded protein that binds to the phage operator sequences close to the p_R and p_L promoters and prevents transcription of most of the genes of the phage.

Circular and covalently closed (CCC). Referring to a circular double-stranded DNA with no breaks or discontinuities in either of its strands.

cis **acting.** Referring to an element that affects only the DNA molecule in which it occurs and not other DNA molecules in the same cell.

cis-**acting site.** A functional region on a DNA molecule that does not encode a gene product and so affects only the DNA molecule in which it resides (e.g., an origin of replication).

Classical genetics. The study of genetic phenomena by using only intact living organisms.

Clonal. Referring to a situation in which all the descendants of an organism or replicating DNA molecule remain together, as in colonies on an agar plate.

Clone. A collection of DNA molecules or organisms that are all identical to each other because they result from replication or multiplication of the same original DNA or organism.

Cloning vector. An autonomously replicating DNA, usually a phage or plasmid, into which can be introduced other DNA molecules that are not capable of replicating themselves.

Coding strand. The strand of DNA in a gene that has the same sequence as the mRNA transcribed from the gene.

Codon. A 3-base sequence in mRNA that stipulates one of the amino acids.

Cognate aminoacyl-tRNA synthetase. The enzyme that attaches a particular amino acid onto a tRNA.

Cointegrate. A DNA molecule with another DNA molecule inserted into it.

Cold-sensitive mutant. A mutant that cannot live and/or multiply in the lower temperature ranges at which the normal or wild-type organism can live and/or multiply.

Colony. A small lump or pile made up of millions of multiplying cells on an agar plate.

Colony papillation. Process leading to sectors or sections in a colony that appear different from the remainder of the colony.

Colony purification. Isolation of individual bacteria on an agar plate so that all the cells in a colony that forms after incubation will be descendants of the same bacterium.

Compatible restriction endonucleases. Restriction endonucleases that leave the same overhangs after cutting a DNA molecule. The resulting ends can pair, allowing the molecules to be ligated to each other.

Competence pheromones. Small peptides given off by bacterial cells. Required to induce competence in neighboring cells.

Competent. The state of cells capable of taking up DNA.

Complementary base pair. A pair of nucleotides that can be held together by hydrogen bonds between their bases, i.e., dGMP and dCMP or dAMP and dTMP.

Complementation. The restoration of the wild-type phenotype when two DNAs containing different mutations that cause the same mutant phenotype are in the cell together. Usually means the two mutations affect different genes.

Complementation group. A set of mutations of which none will complement any of the others.

Composite transposon. A transposon made up of two insertion (IS) elements plus any DNA between them.

Concatemer. Two or more almost identical DNA molecules linked tail to head.

Conditional lethal mutation. A mutation that inactivates an essential cellular component, but only under a certain set of circumstances, for example, a temperature-sensitive mutation that inactivates RNA polymerase only at relatively high temperatures.

Conjugation. The transfer of DNA from one bacterial cell to another by the transfer functions of a self-transmissible DNA element such as a plasmid.

Conjugative transposon. A transposon that encodes functions that allow it to transfer itself into other bacteria.

Consensus sequence. A nucleotide sequence in DNA or RNA, or an amino acid sequence in protein, in which each position in the sequence has the nucleotide or amino acid that has been found most often at that position in molecules with the same function.

Conservative. Referring to a reaction involving double-stranded DNA in which the molecule retains both of its original strands.

Constitutive mutant. A mutant in which the genes of an operon are transcribed whether or not the inducer of the operon is present.

Context. The sequence of nucleotides in DNA or RNA surrounding a particular sequence.

Cooperative binding. Process in which the binding of one protein molecule to a DNA site greatly enhances the binding of another protein molecule of the same type to an adjacent site.

Coprotease. A protein that binds to another protein and thereby activates the second protein's protease activity.

Copy. A molecule of a particular type identical to another in the same cell.

Copy number. The number of copies of a plasmid per cell immediately after cell division. Also the ratio of the number of plasmids of a particular type in the cell to the number of copies of the chromosome.

Core polymerase. The part of the DNA or RNA polymerase that actually performs the polymerization reaction and functions independently of accessory and regulatory proteins that cycle on and off the protein.

Corepressor. A small molecule that binds to an aporepressor and converts it into a repressor.

cos **site.** The sequence of deoxynucleotides at the ends of λ DNA in the phage head. A staggered cut in this sequence at the time the phage DNA is packaged gives rise to complementary or cohesive ends that can base pair with each other.

Cosmid. A plasmid that carries the sequence of a *cos* site so that it can be packaged into λ phage heads.

Cotranscribed. Referring to two or more contiguous genes transcribed by a single RNA polymerase molecule from a single promoter.

Cotransducible. Referring to two genetic markers that are close enough together on the DNA that they can be carried in the same phage head during transduction.

Cotransduction. Occurs when tranductants that were selected for being recombinant for one marker in DNA are also recombinant for a second marker.

Cotransduction frequency. The percentage of transductants selected for one genetic marker from the donor that have also received another genetic marker from the donor. A measure of how far apart the markers are on DNA.

Cotransformable. As in cotransducible, but the regions of two genetic markers are close enough together to be carried on the same piece of DNA during transformation.

Cotransformation frequency. As in cotransduction frequency, except that the percentage is of transformants. A measure of how far apart the markers are on DNA.

Counterselection of donor. Selection of transconjugants under conditions in which the donor bacterium cannot multiply to form colonies.

Coupling model. A model for the regulation of replication of iteron plasmids in which two or more plasmids are joined by binding to the same Rep protein through their iteron sequences. *See* Handcuffing model.

Covalent bond. Bonds that hold two atoms together by a sharing of their electron orbits.

Covalently closed circular. *See* Circular and covalently closed.

Cross. Any means of exchange of DNA between two organisms.

Crossing. Allowing the DNAs of two strains of an organism to enter the same cell so they can recombine with each other.

Crossover. The site of the breaking and rejoining of two DNA molecules during homologous recombination.

C-terminal amino acid. The amino acid on one end of a polypeptide chain that has a free carboxyl group unattached to the amino group of another amino acid.

C terminus. The end of a polypeptide chain with the free carboxyl (COOH) group.

Cured. Referring to a cell that has lost a DNA element such as a plasmid.

Cut and paste. A mechanism of transposition in which the entire transposon is excised from one place in the DNA and inserted into another.

Cyclic AMP (cAMP). Adenosine monophosphate with the phosphate attached to both the 3′ and 5′ carbons of the ribose sugar.

Cyclobutane ring. A ring structure of four carbons held together by single bonds. Present in some types of pyrimidine dimers in DNA.

Cytoplasmic domain. The regions of a polypeptide of a transmembrane protein that are in the interior or cytoplasm of the cell.

Cytosine (C). One of the pyrimidine (one-ringed) bases in DNA and RNA.

Daughter cell. One of the cells arising from division of another cell.

Daughter DNA. One of the two DNAs arising from replication of another DNA.

Deaminating agent. A chemical that reacts with DNA, causing the removal of amino (NH_2) groups from the bases in DNA.

Deamination. The process of removing amino (NH_2) groups from the bases in DNA.

Defective prophage. A DNA element in the bacterial chromosome that presumably was once capable of being induced to form phages but has lost genes essential for lytic development.

Degenerate probe. A chemically synthesized oligonucleotide that is made to be complementary to a certain protein-coding sequence in DNA or RNA but in which the third base in some codons has been randomized to include all the codons that could encode each amino acid.

Degradative operon. Like a catabolic operon, an operon whose genes encode enzymes required for the breakdown of molecules into smaller molcules with the concomitant release of energy.

Deletion mapping. Procedure in which mutants that have mutations to be mapped are crossed with mutants that have deletion mutations with known endpoints. Wild-type recombinants will appear only if the unknown mutation lies outside the deleted region.

Deletion mutation. A mutation in which a number of base pairs have been removed from the DNA.

Deoxyadenosine. An adenine base attached to a deoxyribose sugar.

Deoxyadenosine methylase (Dam methylase). An enzyme that attaches a CH_3 (methyl) group to the adenine base in DNA. The enzyme from *E. coli* methylates the A in the sequence GATC.

Deoxycytidine. A cytosine base attached to a deoxyribose sugar.

Deoxyguanosine. A guanine base attached to a deoxyribose sugar.

Deoxynucleoside. A base (A, G, T, or C) attached to a deoxyribose sugar.

tion factor IF2 and responds to the initiator codons AUG and GUG and, more rarely, to other codons in a TIR.

Four-hitter. A type II DNA restriction endonuclease that recognizes and cuts at a 4-bp sequence in DNA.

Frameshift mutation. Any mutation that adds or removes one or more (but not a multiple of 3) base pairs from DNA, whether or not it occurs in the coding region for a protein.

Functional domain. The region of a polypeptide chain that performs a particular function in the protein.

Gel electrophoresis. Procedure for separating proteins, DNA, or other macromolecules that involves the application first of the macromolecules to a gel made of agarose, acrylamide, or some other gelatinous material and then of an electric field, forcing the electrically charged macromolecules to move toward one or the other electrode. How fast the macromolecules move will depend upon their size and their charge.

Gene. A region on DNA encoding a particular polypeptide chain or functional RNA such as an rRNA or a tRNA.

Gene conversion. Nonreciprocal apparent recombination associated with mismatch repair on heteroduplexes that are formed between two DNA molecules during recombination. The name comes from genetic experiments with fungi in which the alleles of the two parents were not always present in equal numbers in an ascus, as though an allele of one parent had been "converted" into the allele of the other parent.

Gene dosage experiment. An experiment in which the number of copies of a gene in a cell is increased to determine what effect this has on the amount of gene product that is synthesized.

Gene replacement. A molecular genetic technique in which a cloned gene is altered in the test tube and then reintroduced into the organism, selecting for organisms in which the altered gene has replaced the corresponding normal gene in the organism.

Generalized transduction. The transfer, via phage transduction, of essentially any region of the DNA from one bacterium to another.

Generation time. The time it takes for the number of cells in an exponentially growing culture to double.

Genetic code. The assignment of each mRNA nucleotide triplet to an amino acid.

Genetic linkage map. An ordering of the genes of an organism solely on the basis of recombination frequencies between mutations in the genes in genetic crosses.

Genetic marker. A difference in sequence of the DNAs of two strains of an organism that causes the two strains to exhibit different phenotypes that can be used for genetic mapping.

Genetic recombination. The joining of genetic markers or DNA sequences in new combinations.

Genetics. The science of studying organisms on the basis of their genetic material.

Genome. The nucleic acid (DNA or RNA) of an organism or virus that includes all the information necessary to make a new organism or virus.

Genotype. The sequence of nucleotides in the DNA of an organism, usually discussed in terms of the alleles of its genes.

Global regulatory mechanism. A regulatory mechanism that affects many genes scattered around the genome.

Glucose effect. The regulation of genes involved in carbon source utilization based on whether glucose is present in the medium.

Glutamate dehyrogenase. An enzyme that adds ammonia directly to α-ketoglutarate to make glutamate. Responsible for assimilation of nitrogen in high ammonia concentrations.

Glutamate synthase. An enzyme that transfers amino groups from glutamine to α-ketoglutarate to make glutamate.

Glutamine synthetase (GS). An enzyme that adds ammonia to glutamate to make glutamine. Responsible for the assmilation of nitrogen in low ammonia concentrations.

Gradient of transfer. In a conjugational cross, the decrease in the transfer of chromosomal markers the farther they are from the origin of transfer of an integrated plasmid.

Gram-negative bacteria. Bacteria characterized by an outer membrane and a thin peptidoglycan cell wall that stains poorly with a stain invented by the Danish physician Hans Christian Gram in the 19th century.

Gram-positive bacteria. Bacteria characterized by having no outer membrane and a thick peptidoglycan layer that stains well with the Gram stain.

Growth rate regulation of ribosomal synthesis. The regulation of ribosomal synthesis that ensures that cells growing more slowly have fewer ribosomes.

Guanine (G). One of the two purine (two-ringed) bases in DNA and RNA.

Guanosine. The base guanine with a ribose sugar attached. A nucleoside.

Guanosine pentaphosphate (pppGpp). The nucleoside guanosine with two phosphates attached to the 3' carbon and three phosphates attached to the 5' carbon of the ribose sugar.

Guanosine tetraphosphate (ppGpp). The nucleoside guanosine with two phosphates attached to each of the 3' and 5' carbons of the ribose sugar.

Gyrase. A type II topoisomerase capable of introducing negative supercoils two at a time into DNA with the concomitant cleavage of ATP. Apparently unique to bacteria.

Hairpin. A secondary structure formed in RNA or single-stranded DNA when one region of the polynucleotide chain folds back on itself and pairs by complementary base pairing with another region a few nucleotides away.

Handcuffing model. A model for the regulation of replication of iteron plasmids in which two plasmid molecules are held together by binding to the same Rep protein through their iteron sequences. *See* Coupling model.

Haploid. The state of a cell that has only one copy of each of its chromosomal genes. *See* Diploid.

Haploid segregant. A haploid cell or organism derived from multiplication of a partially or fully diploid or polyploid cell.

Headful packaging. A mechanism of encapsulation of DNA in a virus head in which the concatemeric DNA is cut after uptake of a length of DNA sufficient to fill the head.

Heat shock protein (Hsp). One of a group of proteins whose rate of synthesis markedly increases after an abrupt increase in temperature or certain other stresses on the cell.

Heat shock regulon. The group of *E. coli* genes under the control of σ^{32}, the heat shock sigma.

Heat shock response. The cellular changes that occur in the cell after an abrupt rise in temperature.

Helix-destabilizing protein. A protein that preferentially binds to single-stranded DNA and so can help keep the two complementary strands of DNA separated during replication, etc.

Helper phage. A wild-type phage that furnishes gene products that a deleted form of the phage cannot make, thereby allowing the deleted form to multiply and form phage.

Hemimethylated. When only one strand of DNA is methylated at a sequence with twofold symmetry, such as when only one of the two A's in the sequence GATC/CTAG is methylated.

Heterodimer. A protein made of two polypeptide chains that are encoded by different genes. *See* Homodimer.

Heteroduplex. A double-stranded DNA region in which the two strands come from different DNA molecules.

Heteroimmune. Referring to related lysogenic phages that carry different immunity regions and therefore cannot repress each other's transcription.

Heterologous probe. A DNA or RNA hybridization probe taken from the same gene or region of a different organism. It is usually not completely complementary to the sequence being probed. *See* Hybridization probe.

Heteromultimer. A protein made of several polypeptide chains that are encoded by different genes. *See* Heterodimer.

Hfr strain. A bacterial strain that contains a self-transmissible plasmid integrated into its chromosome and which thus can transfer its chromosome by conjugation.

HFT lysate. The lysate of lysogenic phage containing a significant percentage of transducing phage with bacterial DNA substituted for phage DNA.

High multiplicity of infection (MOI). Referring to a virus infection in which the number of viruses greatly exceeds the number of cells being infected.

High negative interference. A phenomenon in which a crossover in one region of the DNA greatly increases the probability of an apparent second crossover close by.

Holliday junction. An intermediate in homologous recombination in which one strand from each of two DNAs crosses over and is joined to the corresponding strand on the opposite DNA.

Holliday model. A model for homologous recombination developed by Robin Holliday, stating that one strand of each DNA is cut at exactly the same place and crosses over to be joined to the corresponding strand on the other DNA. The resulting Holliday junction can then isomerize and be cut in the crossed strands to recombine the flanking DNA sequences.

Homeologous recombination. Homologous recombination in which the deoxynucleotide sequences of two participating regions are somewhat different from each other, usually because they are in different regions of the DNA or because the DNAs come from different species. *See* Ectopic recombination.

Homodimer. A protein made up of two polypeptide chains that are identical to each other, usually because they are encoded by the same gene. *See* Homomultimer.

Homologous recombination. A type of recombination that depends on the two DNAs having identical or at least very similar sequences in the regions being recombined because complementary base pairing between strands of the two DNAs must occur as an intermediate state in the recombination process.

Homomultimer. A protein made up of more than one identical polypeptide, usually encoded by the same gene. *See* Homodimer.

Host range. All of the types of host cells in which a DNA element, plasmid, phage, etc., can multiply.

Hot spot. A position in DNA that is particularly prone to mutagenesis by a particular mutagen.

Hybridization probe. A DNA or RNA that can be used to detect other DNAs and RNAs because it shares a complementary sequence with the DNA or RNA being sought and so will hybridize to it by base pairing.

Hypoxanthine. A purine base derived from the deamination of adenine.

IF2. *See* Initiation factor 2.

Incompatibility. The interference of plasmids with one another's replication and/or partitioning.

Incompatibility (Inc) group. A set of plasmids that interfere with each other's replication and/or partitioning and so cannot be stably maintained together in the same culture.

Induced mutations. Mutations that are caused by irradiating cells or treating cells or DNA deliberately with a mutagen such as a chemical.

Inducer. A small molecule that can increase the transcription of an operon.

Inducible. Referring to an operon capable of having its transcription increased by an inducer.

Initiation codon. The 3-base sequence in an mRNA that specifies the first amino acid to be inserted in the synthesis of a polypepetide chain; in prokaryotes, the 3-base sequence (usually AUG or GUG) within a translation initiation region for which formylmethionine is inserted to begin translation. In eukaryotes, the AUG closest to the 5' end of the mRNA is usually the initiation codon and methionine is inserted to begin translation.

Initiation factor 2 (IF2). The specialized EF-Tu-like protein that binds formylmethionyl-tRNA ($tRNA_f^{Met}$) and brings it into the ribosome in response to an initiation codon that is part of a translation initiation region.

Insertion element. *See* Insertion sequence element.

Insertion mutation. A change in a DNA sequence due to the incorporation of another DNA sequence such as a transposon or antibiotic resistance cassette.

Insertion sequence element (IS element). A small transposon in bacteria that carries only genes for the enzymes needed to promote its own transposition.

Integrase. A type of site-specific recombinase that promotes recombination between two defined sequences in DNA, causing the integration of a circular DNA into another DNA, such as the integration of a phage DNA into the chromosome.

Intergenic. In different genes.

Intergenic suppressor. A suppressor mutation that is in a gene different from that containing the mutation it suppresses. Also called extragenic suppressor.

Internal fragments. Fragments, created by cutting DNA containing a transposon or other DNA element with a restriction endonuclease, that come from entirely within the DNA element.

Interstrand cross-links. Covalent chemical bonds between the two complementary strands of DNA in a double-stranded DNA.

Intragenic. In the same gene.

Intragenic complementation. Complementation between two mutations in the same gene. Rare and allele specific. Usually occurs only if the protein product of the gene is a homomultimer.

Intragenic suppressor. A suppressor mutation that occurs in the same gene as the mutation it is suppressing.

Inversion junctions. The points where the recombination event occurred that inverted a sequence.

Inversion mutation. A change in DNA sequence as a result of flipping a region within a longer DNA so that it lies in reverse orientation. Usually due to homologous recombination between inverted repeats in the same DNA molecule.

Inverted repeat. Two nearby sequences in DNA that are the same or almost the same when read 5' to 3' on the opposite strands.

Invertible sequence. A sequence in DNA that inverts often owing to the action of a site-specific recombinase protein that promotes recombination between inverted repeats at the ends of the sequences.

In vitro mutagen. A mutagen that reacts only with purified DNA or with viruses. Cannot be used to mutagenize intact cells.

In vitro packaging. The incorporation of DNA or RNA into virus heads in the test tube.

In vivo mutagen. A mutagen that will enter and mutagenize the DNA of intact cells.

IS element. *See* Insertion sequence element.

Isomerization. The changing of the spatial conformation of a molecule without breaking any bonds.

Iteron sequences. Short DNA sequences, often repeated many times in the origin region of some types of plasmids, that play a role in the regulation of replication of the plasmid.

Junction fragments. Fragments, created by cutting a DNA containing a transposon or other DNA element, that contain sequences from one of the ends of the DNA element as well as flanking sequences from the DNA into which the element has inserted.

Kinase. An enzyme that transfers a phosphate group from ATP to another molecule.

Lagging strand. During DNA replication, the newly synthesized strand that must be synthesized in the direction opposite the overall movement of the replication fork.

Late gene. A gene that is expressed only relatively late in the course of a developmental process, e.g., a late gene of a phage.

Lawn. *See* Bacterial lawn.

Leader region. The region 5' of the coding region of the first structural gene on the mRNA for an operon.

Leader sequence. An RNA sequence within a leader region.

Leading strand. During DNA replication, the newly synthesized strand that is made in the same direction as the overall direction of movement of the replication fork.

Leaky mutation. A mutation in a gene that does not completely inactivate the product of the gene, hence leaving some residual activity.

Lesion. Any change in a DNA molecule as a result of chemical alteration of a base, sugar, or phosphate.

Linked. A genetic term referring to the fact that two markers are close enough together that they are usually not separated by recombination.

Locus. A region in the genome of an organism.

Low multiplicity of infection (MOI). Referring to a virus infection in which the number of cells almost equals or exceeds the number of viruses, so that most cells remain uninfected or are infected by at most one or very few viruses.

Lyse. To break open cells and release their cytoplasm into the medium.

Lysogen. A strain of bacterium that harbors a prophage.

Lysogenic conversion. A property of a bacterial cell caused by the presence of a particular prophage.

Lysogenic cycle. The series of events that follow infection by a bacteriophage and culminate in the formation of a stable prophage.

Lytic cycle. The series of events that follow infection by a bacteriophage or induction of a prophage and culminate in lysis of the bacterium and the release of new phage into the medium.

Macromolecule. A large molecule such as DNA, RNA, or protein.

Major groove. In the DNA double helix, the larger of the two grooves between the two strands of DNA wrapped around each other.

Male bacterium. A bacterium harboring a self-transmissible plasmid.

Male-specific phage. A phage that infects only cells carrying a particular self-transmissible plasmid. The plasmid produces the sex pilus used by the phage as its adsorbtion site.

Maltodextrins. Short chains of glucose molecules held together by α1-4 linkages. Breakdown products of starch.

Maltotriose. A chain of three glucose molecules held together by α1-4 linkages.

Map expansion. A phenomenon that occurs in genetic linkage experiments in which two markers appear to be farther apart than they are because of hyperactive apparent recombination.

Map unit. A distance between genetic markers corresponding to a recombination frequency of 1% between the markers.

Marker effect. A difference in the apparent genetic linkage between the site of a mutation and other markers depending on the type of mutation at the site.

Marker rescue. The acquisition of a genetic marker by the genome of an organism or virus through recombination with a cloned DNA fragment containing the marker.

Maxam-Gilbert sequencing. A method for DNA sequencing that depends on the ability of certain chemicals to react with and cleave DNA at particular bases.

Membrane protein. A protein that at least partially resides in, or is tightly bound to, one of the cellular membranes.

Membrane topology. The distribution of the various regions of a membrane protein between the membrane and the two surfaces of the membrane.

Merodiploid. Referring to a bacterial cell that is mostly haploid but is diploid for some regions of the genome. *See* Partial diploid.

Messenger RNA (mRNA). An RNA transcript that includes the coding sequences for at least one polypeptide.

Methionine aminopeptidase. An enzyme that removes the N-terminal methionine from newly synthesized polypeptides.

Methyl-directed mismatch repair system. A repair system in enteric bacteria that recognizes mismatches in newly replicated DNA and specifically removes and resynthesizes the new strand, which is distinguishable from the old strand because it is not methylated at nearby GATC sequences.

Methyltransferase. In DNA repair, an enzyme that removes a CH_3 (methyl) or CH_3CH_2 (ethyl) group from a base in DNA by attaching the group to itself.

Mini-Mu. A shortened version of phage Mu DNA in which most of the phage DNA has been deleted except the inverted repeat ends and the transposase genes, leav-ing it unable to replicate or be packaged into a phage head without the assistance of a helper wild-type phage mu. Other DNAs, such as genes for antibiotic resistance and a plasmid origin of replication, can be inserted between copies of the mini-Mu.

Minor groove. In double-stranded DNA, the smaller of the two gaps between the two strands of DNA wrapped around each other in a helix.

Minus (–) strand. In a virus with a single-stranded nucleic acid genome (DNA or RNA), the strand that is complementary to the strand in the virus head.

–10 sequence. In a bacterial σ^{70}-type promoter, a short sequence that lies about 10 bp upstream of the transcription start site. The canonical or consensus sequence is TATAAT/ATATTA.

–35 sequence. In a bacterial σ^{70}-type promoter, a short sequence that lies about 35 bp upstream of the transcription start site. The canonical or consensus sequence is TTGACA/AACTGT.

Mismatch. When the normal bases in DNA are not paired properly, e.g., an A opposite a C.

Mismatch repair system. Any pathway for removing mismatches in DNA and replacing them with the correctly paired bases.

Missense mutation. A base pair change mutation in a region of DNA encoding a polypeptide that changes an amino acid in the polypeptide.

***mob* region.** A region in DNA carrying an origin of transfer (*oriT* sequence) and genes whose products allow the plasmid to be mobilized by self-transmissible plasmids.

Mobilizable. A plasmid or other DNA element that cannot transfer itself into other bacteria but can be transferred by other self-transmissible elements.

Molecular genetic analysis. Any study of cellular or organismal functions that involves manipulations of DNA in the test tube.

Molecular genetic techniques. Methods for manipulating DNA in the test tube and reintroducing the DNA into cells.

Mother cell. A cell that divides or differentiates to give rise to a new cell or spore.

mRNA. *See* Messenger RNA.

Multimeric protein. A protein that consists of more than one polypeptide chain.

Multiple cloning site. A region of a cloning vector that contains the sequences cut by many different type II restriction endonucleases. Also called polyclonal site.

Multiplicity of infection (MOI). The ratio of viruses to cells in an infection.

Mutagen. A chemical or type of irradiation that causes mutations by damaging DNA.

Mutagenic chemicals. Chemicals that cause mutations by damaging DNA.

Mutagenic repair. A pathway for repairing damage to DNA that sometimes changes the sequence of deoxynucleotides as a consequence.

Mutant. An organism that differs from the normal or wild type as a result of a change in the sequence (mutation) of its DNA.

Mutant allele. The mutated gene of a mutant organism that makes it different from the wild type.

Mutant enrichment. A procedure for increasing the frequency of a particular type of mutant in a culture.

Mutant phenotype. A characteristic that makes a mutant organism different from the wild type.

Mutation. Any heritable change in the sequence of deoxynucleotides in DNA.

Mutation rate. The probability of occurrence of a mutation causing a particular phenotype each time a newborn cell grows and divides.

N-terminal amino acid. The amino acid on the end of a polypeptide chain whose amino (NH_2) group is not attached to another amino acid in the chain through a peptide bond.

N terminus. The end of a polypeptide chain with the free amino (NH_2) group.

Narrow host range. If the range of hosts in which a DNA element can replicate includes only a few closely related types of cells.

Natural competence. The ability of some types of bacteria to take up DNA at a certain stage in their growth cycle without chemical or other treatments.

Naturally transformable bacteria. Bacteria that have a growth stage during which they are naturally competent.

Negative regulation. A type of regulation in which a protein or RNA molecule, in its active form, inhibits a process such as the transcription of an operon or translation of a mRNA.

Negative selection. The process of detecting a mutant on the basis of the inability of the mutant to multiply under a certain set of conditions in which the normal or wild-type organism can multiply.

Negatively supercoiled. Referring to two strands of a DNA molecule that are wrapped around each other less than about once every 10.5 bp.

Nicked DNA. Double-stranded DNA in which one strand contains a broken phosphate-deoxyribose bond in the phosphodiester backbone.

Noncomposite transposon. A transposon in which the transposase genes and the inverted-repeat ends are included in the minimum transposable element and are not part of autonomous IS elements.

Noncovalent change. Any change in a molecule that does not involve the making or breaking of a covalent bond in the molecule.

Nonhomologous recombination. The breaking and rejoining of two DNAs into new combinations which does not necessarily depend on the two DNAs having similar sequences in the region of recombination.

Nonpermissive temperature. A temperature at which the wild-type organism or virus but not the mutant organism or virus can multiply.

Nonselective. Referring to conditions or media in which both the mutant and wild-type strains of an organism or virus can multiply.

Nonsense codons. Usually the codons UAG, UGA, and UAA. These codons do not stipulate an amino acid in most types of organisms but, rather, trigger the termination of translation.

Nonsense mutation. In a region of DNA encoding a protein, a base pair change mutation that causes one of the nonsense codons to be encountered in frame when the mRNA is translated.

Nonsense suppressor. A suppressor mutation that allows an amino acid to be inserted at some frequency for one or more of the nonsense codons during the translation of mRNAs.

Nonsense suppressor tRNA. A mutation in the gene for a tRNA that allows the tRNA to pair with one or more of the nonsense codons in mRNA during translation and therefore causes an amino acid to be inserted for the nonsense codon. This type of mutation usually changes the anticodon on the tRNA.

Ntr (nitrogen regulation) system. A global regulatory system that regulates a number of operons in response to the nitrogen sources available.

Nucleoid. A compact, highly folded structure formed by the chromosomal DNA in the bacterial cell and in which the DNA appears as a number of independent supercoiled loops held together by a core.

Nucleotide excision repair. A system for the repair of DNA damage resulting in a relatively large distortion in the DNA helix. A cut is made on either side of the damage on the same strand, and the damaged strand is removed and resynthesized.

Ochre codon. The nonsense codon UAA.

Ochre mutation. In a region of DNA encoding a polypeptide, a base pair change mutation that causes the codon UAA to appear in frame in the mRNA for the polypeptide.

Ochre suppressor. A mutation that causes an amino acid to be inserted at some frequency wherever the codon UAA is encountered during mRNA translation.

Okazaki fragments. The short pieces of DNA that are initially synthesized during replication of the lagging strand at the replication fork.

Oligopeptide. A short polypeptide only a few amino acids long.

Opal codon. The codon UGA. Also called umber codon.

Open reading frame (ORF). A sequence on DNA, read three nucleotides at a time, that is unbroken by any nonsense codons.

Operator. Usually a sequence on DNA to which a repressor protein binds. More generally, any sequence in DNA or RNA to which a negative regulator binds.

Operon. A region on DNA encompassing genes that are transcribed into the same mRNA, as well as any adjacent *cis*-acting regulatory sequences.

Operon model. The model proposed by Jacob and Monod for the regulation of the *lac* operon in which transcription of the structural genes of the operon is prevented by the LacI repressor binding to the operator region and thereby preventing access of the RNA polymerase to the promoter. In the presence of lactose, the inducer binds to LacI and changes its conformation so that it can no longer bind to the operator, and as a result, the structural genes are transcribed.

oriC. The site in the bacterial chromosome at which initiation of a round of replication occurs.

Origin of replication. The site on a DNA at which replication initiates, including all of the surrounding *cis*-acting sequences required for initiation.

pac **site.** The sequence in phage DNA at which packaging of the phage DNA into heads begins.

Packaging sites. *See pac* site.

Papilla. A section of a bacterial colony with an appearance different from that of most of the colony.

par **function.** A site or a gene that encodes a product required for the partitioning of a plasmid.

Parent. An organism participating in a genetic cross.

Parental types. Progeny of a genetic cross that are genetically identical to one or the other of the parents.

Partial digest. A restriction endonclease digestion of DNA in which not all the available sites are cut, either because the amount of endonuclease enzyme is limiting or because the time of incubation is too short.

Partial diploid. A bacterium that has two copies of part of its genome, usually because a plasmid or prophage in the bacterium contains some bacterial DNA. Also called merodiploid.

Partitioning. An active process by which at least one copy of a replicon (plasmid, chromosome, etc.) is distributed into each daughter cell at the time of cell division.

PCR. *See* Polymerase chain reaction.

Peptide bond. A covalent bond between the amino (NH_2) group of one amino acid and the carboxyl (COOH) group of another.

Peptide deformylase. An enzyme that removes the formyl group from the amino-terminal formylmethionine of newly synthesized polypeptides.

Peptidyl tRNA hydrolase. An enzyme that removes polypeptides from tRNA. It is not associated with the ribosome, and so it may be a scavenger enzyme that is involved in the recycling of polypeptides that are prematurely released during translation.

Periplasm. The space between the inner and outer membranes in gram-negative bacteria.

Periplasmic domain. A region of a polypeptide that is located in the periplasm of the cell.

Periplasmic protein. A protein that is located in the periplasm.

Permissive temperature. A temperature at which both a temperature-sensitive mutant (or cold-sensitive mutant) and the wild type can multiply.

Phage. *See* Bacteriophage.

Phage genome. The nucleic acid (DNA or RNA) that is packaged into the phage head.

Phase variation. The reversible change of one or more of the cell surface antigens of a bacterium at a high frequency.

Phasmid. A hybrid DNA element containing both plasmid and phage sequences.

Phenotype. Any identifiable characteristic of a cell or organism that can be altered by mutation.

Phenotypic lag. The delay between the time a mutation occurs in the DNA and the time the resulting change in the phenotype of the organism becomes apparent.

Phosphate. The chemical group PO_4.

Photolyase. An enzyme that uses the energy of visible light to split pyrimidine butane dimers in DNA, restoring the original pyrimidines.

Photoreactivation. The process by which cells exposed to visible light after DNA damage achieve greater survival rates than cells kept in the dark.

Physical map. A map of DNA showing the actual distance in deoxynucleotides between sites.

Pilin. A protein in pili. *See* Pilus.

Pilus. A protrusion or filament attached to the surface of a bacterial cell. *See* Sex pilus.

Plaque. The clear spots in a bacterial lawn caused by phage killing and lysing the bacteria as the bacterial lawn is forming.

Plaque purification. The isolation of a pure strain of a phage by plating to obtain individual plaques.

Plasmid. Any DNA molecule in cells that replicates independently of the chromosome and regulates its own replication so that the number of copies of the DNA molecule remains relatively constant.

Plasmid incompatibility. *See* Incompatibility.

Pleiotropic mutation. A mutation that causes many phenotypic changes in the cell.

Plus (+) strand. In a virus with a single-stranded genome (DNA or RNA), the strand that is packaged in the phage head.

Polarity. A condition in which a mutation that causes a nonsense codon in one gene reduces the transcription of a downstream gene that is cotranscribed into the same mRNA.

Polycistronic mRNA. An mRNA that contains more than one translational initiation region so that more than one polypeptide can be translated from the mRNA.

Polyclonal site. *See* Multiple cloning site.

Polymerase chain reaction (PCR). A technique that uses the DNA polymerase from a thermophilic bacterium and two primers to make many copies of a given region of DNA occurring between sequences complementary to the primers.

Polymerization. A reaction in which small molecules are joined together in a chain to make a longer molecule.

Polymorphism. A difference in DNA sequence between otherwise closely related strains.

Polypeptide. A long chain of amino acids held together by peptide bonds. The product of a single gene.

Porin. A protein that forms channels in the outer membrane of gram-negative bacteria.

Positive regulation. A type of regulation in which the gene is expressed only if the active form of a regulatory protein (or RNA) is present.

Positive selection. Conditions under which only the mutant being sought and not the normal or wild type can multiply.

Positively supercoiled. Referring to a DNA in which the two strands of the double helix are wrapped around each other more than once every 10.5 bp.

Postreplication repair. *See* Recombination repair.

Posttranscriptional regulation. Any regulation in the expression of a gene that occurs after the mRNA has been synthesized from the gene.

Precise excision. The removal of a transposon or other foreign DNA element from a DNA in such a way that the original DNA sequence is restored.

Precursor. The smaller molecules that are polymerized to form a polymer.

Primary structure. The sequence of nucleotides in an RNA or of amino acids in a polypeptide.

Primase. An enzyme that synthesizes short RNAs to prime the synthesis of DNA chains.

Prime factor. A self-transmissible plasmid carrying a region of the bacterial chromosome.

Primer. A single-stranded DNA or RNA that can hybridize to a single-stranded template DNA and provide a free 3' hydroxyl end to which DNA polymerase can add deoxynucleotides to synthesize a chain of DNA complementary to the template DNA.

Primosome. A complex of proteins involved in making primers for the initiation of synthesis of DNA strands.

Probe. A short oligonucleotide (DNA or RNA) that is complementary to a sequence being sought and so will hybridize to the sequence and allow it to be identified from among many other sequences.

Prokaryotes. Organisms whose cells do not contain a nuclear membrane and visible nucleus or many of the other organelles characteristic of the cells of higher organisms. They include the eubacteria and archaea.

Promiscuous plasmid. A self-transmissible plasmid that can transfer itself into many types of bacteria, some of which are only distantly related to each other.

Promoter. A region on DNA to which RNA polymerase binds to initiate transcription.

Prophage. The state of phage DNA in a lysogen in which the phage DNA is integrated into the chromosome of the bacterium or replicates as a plasmid.

Protein disulfide isomerase. An enzyme that catalyzes the oxidation of the sulfhydryl groups of cysteines in polypeptides, cross-linking the cysteines to each other.

Protein export. The transport of proteins through the cellular membranes to the outside medium.

Protein secretion. The transport of proteins into or through cellular membranes.

Pseudoknot. An RNA tertiary structure with interlocking loops held together by regions of hyrogen bonding between the bases.

Purine. A base in DNA and RNA with two ring structures.

Pyrimidine. A base in DNA and RNA with only one ring.

Pyrimidine dimer. A type of DNA damage in which two adjacent pyrimidines are covalently joined by chemical bonds.

Quaternary structure. The complete three-dimensional structure of a protein including all the polypeptide chains making up the protein and how they are wrapped around each other.

Random gene fusions. A technique in which transposon mutagenesis is used to fuse reporter genes to different regions in the chromosome. A transposon containing a reporter gene hops randomly into the chromosome, resulting in various transposon insertion mutants that have the reporter gene on the transposon fused either transcriptionally or translationally to different genes or to different regions within each gene.

Random-mutation hypothesis. A hypothesis explaining the adaptation of organisms to their environment. It states that mutations occur randomly, free of influence from their consequences, but that mutant organisms preferentially survive and reproduce themselves if the mutations inadvertently confer advantages under the circumstances.

Reading frame. Any sequence of nucleotides in RNA or DNA read three at a time in succession, as during translation of an mRNA.

Rec⁻ (recombination-deficient) mutant. A mutant in which DNA shows a reduced capacity for recombination.

Recessive. A mutation or other genetic marker that does not exert its phenotype in an organism that is diploid for the region because it also contains the corresponding region from the wild-type organism.

Recessive mutation. Referring to complementation tests, a mutation that does not exhibit its phenotype in the presence of a wild-type copy of the same region of DNA.

Recipient. In a genetic cross between two bacteria, the bacterium that receives DNA from another bacterium.

Recipient allele. The sequence of a gene or allele as it occurs in the recipient bacterium.

Recipient strain. Bacteria with the genotype of those that were used as recipients in a genetic cross.

Reciprocal cross. A genetic cross in which the alleles of the donor and recipient strain are reversed relative to an earlier cross. An example would be a transduction in which the phage were grown on the strain that had the alleles of what was previously the recipient strain and used to transduce a strain with the alleles of what was previously the donor strain.

Recombinant DNA. A DNA molecule derived from ligation of the sequences of two different DNAs to each other in a test tube.

Recombinant type. In a genetic cross, progeny that are genetically unlike either parent in the cross because they have DNA sequences that are the result of recombination between the parental DNAs.

Recombinase. An enzyme that specifically recognizes two sequences in DNA and breaks and rejoins the strands to cause a crossover within the sequences.

Recombination. The breakage and rejoining of DNA into new combinations.

Recombination frequency. In a genetic cross, the number of progeny that are recombinant types for the two parental markers divided by the total number of progeny of the cross.

Recombination repair. Another name for postreplication repair. A mechanism by which the cell tolerates DNA damage by using recombination to transfer an undamaged strand to the other daughter DNA containing a damaged strand, so that the damaged strand will be opposite a good complementary strand.

Redundancy. *See* Terminal redundancy.

Regulation of gene expression. Control of the rate of synthesis of the active product of a gene, so that the gene product can be synthesized at different rates, depending, for example, on the developmental stage of the organism or the state in which the organism finds itself.

Regulatory cascade. A strategy for regulating the expression of genes during developmental processes in which the products of genes expressed during one stage of development regulate the expression of genes for the next stage of development.

Regulatory gene. A gene whose product regulates the expression of other genes.

Regulon. The set of operons that are all regulated by the product of the same regulatory gene.

Relaxed control. Occurs when the synthesis of rRNA and other stable RNAs continues even after protein synthesis is blocked by starvation for an amino acid.

Relaxed plasmid. A plasmid that has a high copy number, so that its replication need not be tightly controlled.

Relaxed strain. A strain of bacteria that exhibits relaxed control of rRNA and other stable RNA synthesis. Does not synthesize ppGpp in response to amino acid starvation.

Release factors. Proteins that are required for the termination of polypeptide synthesis when the ribosome encounters a specific in-frame nonsense codon in the mRNA. Also required for the release of the newly synthesized polypepetide from the ribosome.

Replica plating. A technique in which bacteria grown on one plate are transferred to a fuzzy cloth and then are transferred from the fuzzy cloth onto another plate so that the bacteria on the first plate will be transferred to the corresponding position on the second plate.

Replication fork. The region in a replicating double-stranded DNA molecule where the two strands are separating to allow synthesis of the complementary strands.

Replicative bypass. Occurs when the replication fork moves past damage to the DNA that interferes with proper base pairing, either by skipping over the damage and leaving a gap in the newly synthesized strand or by inserting deoxynucleotides at random opposite the damaged bases.

Replicative form (RF). The double-stranded DNA or RNA that forms by synthesis of the complementary minus strand after infection by a virus that has a single-stranded genome.

Replicative transposition. A type of transposition in which single-stranded nicks are made at each end of the transposon and a staggered double-strand break is made in the target DNA. The free 3' ends at the ends of the transposon are ligated to the free 5' ends of the target DNA and the free 3' ends of the target DNA are used as primers to synthesize over the transposon, giving rise to a cointegrate.

Replicon. A DNA molecule capable of autonomous replication because it contains an origin of replication that functions in the cell.

Reporter gene. A gene whose product is stable and easy to assay and so is convenient for detecting and quantifying the expression of genes to which it is fused.

Repressor. A protein that, in its active state, binds to operator sequences close to the promoter for an operon and thereby prevents transcription of the operon. More generally, a protein or RNA that negatively regulates transcription or translation.

Resolution of a cointegrate. Separation of the two DNAs joined in a cointegrate by recombination between repeated sequences in the two copies of the transposon, leaving each DNA with one copy of the transposon.

Resolvase. A type of site-specific recombinase that breaks and rejoins DNA in *res* sequences in the two copies of the transposon in a cointegrate, thereby resolving the cointegrate into separate DNAs, each with one copy of the transposon.

Response regulator. A protein that is part of a two component system and that upon receiving a signal from another protein, the sensor protein, performs a regulatory function, e.g., activates transcription of operons.

Restriction fragment. A piece of DNA obtained by cutting a longer DNA with a restriction endonuclease.

Restriction fragment length polymorphism (RFLP). A difference in the size of restriction fragments obtained by cutting DNA from two different strains with the same restriction endonuclease.

Restriction modification system. A complex of proteins that, among them, recognize specific sequences in DNA, methylate a base in the sequence, and cut in or near the sequence if it is not methylated.

Retroregulation. Occurs when the amount of an RNA in the cell is determined by an event that occurs at the 3′ end of an RNA rather than at the 5′ end.

Retrotransposon. A type of transposon that hops to a new location in DNA by first making an RNA copy of itself, then making a DNA copy of this RNA with a reverse transcriptase, and then inserting its DNA copy into the target DNA.

Reverse genetics. The process by which the function of the product of a gene is determined by first cloning the gene and then mutating the gene in the test tube before reintroducing it into a cell to see what effect the mutation has on the organism.

Reversion. The restoration of a mutated sequence in DNA to the wild-type sequence.

Reversion rate. The probability that a mutated sequence in DNA will change back to the wild-type sequence each time the organism multiplies.

Revert. *See* Reversion.

Revertant. An organism in which the mutated sequence in its DNA has been restored to the wild-type sequence.

RF. *See* Replicative form.

RFLP. *See* Restriction fragment length polymorphism.

Ribonucleoside triphosphate. A base (usually A, U, G, or C) attached to a ribose sugar with three phosphate groups attached in tandem to the 5′ carbon of the sugar.

Ribonucleotide reductase. An enzyme that catalyzes the reduction of nucleoside diphosphates to deoxynucleoside diphosphates by removing the hydroxy group at the 2′ carbon of the ribonucleoside diphosphate and replacing it with a hydrogen.

Riboprobe. A hybridization probe made of RNA rather than DNA.

Ribosomal proteins. The proteins that, in addition to the rRNAs, make up the structure of the ribosome.

Ribosomal RNA (rRNA). Any one of the three RNAs (16S, 23S, and 5S in bacteria) that make up the structure of the ribosome.

Ribosome. The cellular organelle, made up of about 50 different proteins and 3 different RNAs, that is the site of protein synthesis.

Ribosome binding site. *See* Translational initiation region.

Ribosome release factor. *See* Release factors.

RNA modification. Any covalent change to RNA that does not involve the breaking and joining of phosphate-phosphate or phosphate-ribose bonds in the backbone of the RNA.

RNA polymerase. An enzyme that polymerizes ribonucleoside triphosphates to make RNA chains by using a DNA or RNA template.

RNA processing. Covalent changes to RNA that involve the breaking and joining of phosphate-phosphate or phosphate-ribose bonds in the backbone of the RNA.

Rolling-circle replication. A type of replication of circular DNAs in which a single-stranded nick is made in one strand of the DNA and the 3′ hydroxyl end is used as a primer to replicate around the circle, displacing the old strand.

Round of replication. Cycle of replication of a circular DNA in which one copy of the DNA is made.

Sanger dideoxy sequencing. A method for DNA sequencing in which the chain-terminating property of the dideoxynucleotides is used.

Satellite virus. A naturally occurring virus that depends on another virus for its multiplication.

Screening. The process of testing a large number of organisms for a particular mutant type.

***sec* gene.** One of the genes whose products are required for transport of proteins through the inner membrane.

Secondary structure. A structure of a polynucleotide or polypeptide chain that results from noncovalent pairing between nucleotides or amino acids in the chain.

Secretory protein. A protein that enters the membrane.

Selected marker. A difference in DNA sequence between two strains in a bacterial or phage cross that is used to select recombinants. The cross is plated so that only recombinants that have received the donor sequence can multiply.

Selection. Procedure in which bacteria or viruses are grown under conditions in which only the wild type or the desired mutant or recombinant can multiply, allowing the isolation of even very rare mutants and recombinants.

Selectional genetics. Genetic analysis in which selection of mutants or recombinants is used.

Selective conditions. Conditions under which only the wild type or the desired mutant can multiply.

Selective media. Media that have been designed to allow multiplication of only the desired mutant or wild type. Such media either lack one or more nutrients or contain a substance that is toxic.

Selective plate. An agar plate made with selective media.

Self-transmissible. Referring to a plasmid or other DNA element that encodes all the gene products needed to transfer itself to other bacteria through conjugation.

Semiconservative replication. A type of DNA replication in which the daughter DNAs are composed of one old strand and one newly synthesized strand.

Sensitive cell. A type of cell that can serve as the host for a particular type of virus.

Sensor. The protein in a two-component system that detects changes in the environment and communicates this information to the response regulator, usually in the form of a phosphoryl group.

Sequestration. The entry of the origin of replication (*oriC*) of the bacterial chromosome into an incompletely understood dormant state after a round of chromosome replication has initiated.

Serial dilution. A procedure in which an aliquot of a solution is first diluted into one vessel and then an aliquot of the solution in this vessel is diluted into a second vessel, and so forth. The total dilution is the product of each of the individual dilutions.

7,8-Dihydro-8-oxoguanine. *See* 8-OxoG.

Sex pilus. A rod-like structure that forms on the surface of bacterial cells containing a self-transmissible plasmid and that facilitates transfer of the plasmid or other DNAs into another bacterium.

Shine-Dalgarno sequence. A short sequence, usually about 10 nucleotides upstream of the initiation codon in a bacterial translation initiation region, that is complementary to a sequence in the 3' end of the 16S rRNA. It helps position the ribosome for initiation of translation. Named after the persons who first noticed it.

Shuttle vector. A plasmid cloning vector that contains two origins of replication that function in different types of cells so that the plasmid can replicate in both types of cells.

Siblings. Two cells or viruses that arose from the multiplication of the same cell or virus.

Signal sequence. A short sequence, composed of mostly hydrophobic amino acids, that is located at the N terminus of some secreted proteins and that is removed as the protein passes into or through the membrane.

Signal transduction pathway. A set of proteins that pass a signal, usually a phosphoryl group, from one to the other in reponse to a change in the cellular conditions.

Silent mutation. A change in a DNA sequence of a gene encoding a protein that does not change the amino acid sequence of the protein, usually because it changes the last base in a codon, thereby changing it to another codon but one that encodes the same amino acid.

Single mutant. A mutant organism that has only one of the two or more mutations being studied.

Site-specfic mutagenesis. One of many methods for mutagenizing DNA in such a way that the change is localized to a predetermined base pair or small region in the DNA.

Site-specific recombinase. An enzyme that catalyzes the breaking and rejoining of DNA molecules but only at certain sequences in the DNA.

Site-specific recombination. Recombination that occurs only between defined sequences in DNA. Usually performed by recombinases.

6-4 lesion. A type of damage to DNA in which the carbon at the 6 position of a pyrimidine is covalently bound to the carbon at the 4 position of an adjacent pyrimidine.

Six-hitter. A type II restriction endonuclease that recognizes and cuts at a specific 6-bp sequence in DNA.

SOS gene. A gene that is a member of the LexA regulon, so that its transcription is normally repressed by LexA repressor.

SOS mutagenesis. Mutagenesis that occurs at an increased rate in cells in which the SOS genes are induced.

SOS response. Induction of transcription of the SOS genes in response to DNA damage.

Southern blot hybridization. A procedure for transferring DNA from an agarose gel to a filter for hybridization. Named after the person who developed the procedure.

Specialized transduction. A type of transduction caused by lysogenic phages in which only DNA sequences close to the attachment site of the prophage in the chromosome are transduced. It results from mistakes made during the excision of the prophage that substitute some chromosomal DNA sequences for some of the phage DNA sequences that are normally packaged in the phage head.

Spontaneous mutations. Mutations that occur in organisms without deliberate attempts to induce them by irradiation or chemical treatment.

Start point of transcription. The deoxynucleotide in DNA with which the first ribonucleotide pairs to begin making an RNA.

Starve. To deprive an organism of an essential nutrient that it cannot make for itself.

Sticky end. The short single-stranded DNA that sticks out from the end of the DNA molecule after it has been cut with a type II restriction endonuclease that makes a staggered break in the DNA.

Strain. A subdivision of species. A group of organisms that are identical to each other but differ genetically from other organisms of the same species.

Strand passage. A reaction performed by topoisomerases in which one or two strands of a DNA are cut and the ends of the cut DNA are held by the enzyme to prevent rotation while other strands of the same or different DNAs are passed through the cuts.

Stringent control. Occurs when the synthesis of rRNA and other stable RNAs in the cell ceases when the cells are starved for an amino acid.

Stringent plasmid. A plasmid that exists in only one or very few copies per cell, so that its replication must be very tightly controlled.

Structural gene. One of the genes in an operon for a pathway that encodes one of the enzymes of the pathway.

Subclone. A smaller clone obtained by cutting a clone with restriction endonucleases and cloning one of the pieces.

Sugar. A simple carbohydrate with the general formula $(CH_2O)_n$; as found in nature, n is 3 to 9.

Suicide vector. A DNA element, usually plasmid or phage DNA, that cannot replicate in the cells into which it is being introduced.

Supercoiled. Referring to DNA in which the two strands of the double helix are wrapped around each other either more or less often than predicted from the Watson-Crick structure of DNA.

Superinfection. The infection of cells by a virus when the same cells are already infected by the same type of virus.

Suppression. The alleviation of the effects of a mutation by a second mutation elsewhere in the DNA.

Suppressive effect. Occurs if a clone, when introduced into a cell in high copy number, alleviates the effect of a mutation indirectly, rather than complementing the mutation.

Suppressor mutation. A mutation elsewhere in the DNA that alleviates the effects of another mutation.

Symmetric sequence. A sequence of deoxynucleotides in double-stranded DNA that reads the same in the 5'-to-3' direction on both strands.

Synapse. In recombination, a structure in which two DNAs are held together by pairing between their strands.

Synchronize. To treat a culture of cells so that they will all be at approximately the same stage in their cell cycle at the same time.

Tag. A sequence of amino acids added to a protein so that the protein can be purified more easily.

Tag vector. A cloning vector designed so that an amino acid sequence that is easy to purify will be added to a protein if the gene for the protein is cloned into the vector in frame with the translation initiation region on the vector and with no intervening nonsense codons.

Tandem duplication. A type of mutation that causes a sequence in DNA to be followed immediately by the same sequence.

Target DNA. The DNA into which a transposon hops.

Tautomer. A form of a molecule in which the electrons are distributed differently among the atoms.

Temperature-sensitive (Ts) mutant. A mutant that cannot grow in the higher-temperature ranges in which the wild type can multiply.

Template strand. The strand of DNA in a region from which RNA is synthesized that has the complementary sequence of the RNA and so serves as the template for RNA synthesis.

Terminal redundancy. A term used to describe a DNA molecule that has direct repeats at both ends; that is, the sequences at both ends are the same in the direct orientation.

Termination of replication. The process by which the replication apparatus leaves the DNA when DNA replication is completed and the daughter DNAs are separated.

Termination of transcription. The process by which the RNA polymerase leaves the DNA and the RNA chain is released at a transcription termination site in the DNA.

Termination of translation. The process by which the ribosome leaves the mRNA and the polypeptide is released when a nonsense codon in the mRNA is encountered in frame.

Tertiary structure. The three-dimensional structure of a polypeptide or polynucleotide.

Theta replication. A type of replication of circular DNA in which the replication apparatus initiates at an origin of replication and proceeds in one or both directions around the circle with leading and lagging strands of replication. The molecule in an intermediate state of replication resembles the Greek letter theta (θ).

Three-factor cross. A type of genetic cross used to order three closely linked mutations and in which one parent has two of the mutations and the other parent has the third mutation.

3' end. The terminus of a polynucleotide chain (DNA or RNA) ending in the nucleotide that is not joined at the 3' carbon of its ribose to the phosphate of another nucleotide.

3' exonuclease. An enzyme that degrades a polynucleotide from its 3' end by removing nucleotides one or more at a time.

3' hydroxyl end. In a polynucleotide, a 3' end that has a hydroxyl group on the 3' carbon of the ribose sugar of the last nucleotide without a phosphate group attached.

3' overhang. An unpaired single strand extending from the strand with a free 3' end in a double-stranded DNA.

3' untranslated region. In an mRNA, the sequences downstream or toward the 3' end of the last sequence that encodes a protein.

Thymine (T). One of the pyrimidine (one-ringed) bases.

TIRs. *See* Translational initiation region.

Titration. A process of increasing the concentration of one of two types of molecules that bind to each other until all of the other type of molecule is bound.

Topoisomerase. An enzyme that can alter the topology of a DNA molecule by cutting one or both strands of DNA, passing other DNA strands through the cuts while holding the cut ends so that they are not free to rotate, and then resealing the cuts.

Topology of DNA. The relationship of the strands of DNA to each other in space.

Tra functions. Gene products encoded by the *tra* genes of self-transmissible DNA elements that allow the plasmids to transfer themselves into other bacteria.

trans-acting function. A gene product that can act on DNAs in the cell other than the one from which it was made.

trans-acting mutation. A mutation that affects a gene product that leaves the DNA from which it is made and so can be complemented.

Transconjugant. A recipient cell that has received DNA from another cell by conjugation.

Transcribe. To make an RNA that is a complementary copy of a strand of DNA.

Transcribed strand. In a region of a double-stranded DNA that is transcribed into RNA, the strand of DNA that is used as a template and so is complementary to the RNA.

Transcript. An RNA made from a region of DNA.

Transcription bubble. A region that follows the RNA polymerase during transcription of DNA in which the two strands of the DNA are separated and the newly synthesized RNA forms a short RNA-DNA duplex with the transcribed strand of DNA.

Transcriptional activator. A protein that is required for transcription of an operon. The protein makes contact with the RNA polymerase and allows the RNA polymerase to initiate transcription from the promoter of the operon.

Transcriptional autoregulation. The process by which a protein regulates the transcription of its own gene.

Transcriptional fusion. The introduction of a gene downstream of the promoter for another gene or genes so that it will be transcribed from the promoter for the other gene(s) but will be translated from its own translational initiation region.

Transcriptional regulation. When the amount of product of a gene that is synthesized under certain conditions is determined by how much mRNA is made from the gene.

Transducing phage. A type of phage that sometimes packages bacterial DNA during infection and introduces it into other bacteria upon infection of the other bacteria.

Transductant. A bacterium that has received DNA from another bacterium by transduction.

Transduction. A process in which DNA other than phage DNA is introduced into a bacterium by infection by a phage containing the DNA.

Transfection. Initiation of a virus infection by introducing virus DNA into a cell by transformation rather than by infection.

Transfer RNA (tRNA). The small stable RNAs in cells to which specific amino acids are attached by aminoacyl tRNA synthetases. The tRNA with the amino acid attached enters the ribosome and base pairs through its anticodon sequence with a 3-nucleotide codon sequence in the mRNA to insert the correct amino acid in the growing polypeptide chain.

Transformant. A cell that has received DNA by transformation.

Transformasomes. Globular structures that appear on the surfaces of some types of bacteria into which DNA first enter during natural transformation of the bacteria.

Transformation. The introduction of DNA into cells by mixing the DNA and the cells.

Transformylase. The enzyme that transfers a formyl (CHO) group to the amino group of methionine to make formylmethionine.

Transgenic organism. An organism that has inherited foreign DNA sequences that have been experimentally introduced into its ancestors. The introduced DNA sequences are passed down from generation to generation because they are inserted into a stably inherited DNA, such as a chromosome.

Transition mutation. A type of base pair change mutation in which the purine base has been changed into the other purine base and the pyrimidine base has been changed into the other pyrimidine base (e.g., AT to GC or GC to AT).

Translated region. A region of an mRNA that encodes a protein.

Translation elongation factor G (EF-G). The protein required to move the peptidyl tRNA from the A site to the P site on the ribosome, with the concomitant cleavage of GTP to GDP after the peptide bond has formed.

Translation elongation factor Tu (EF-Tu). The protein that binds to aminoacyl tRNA and accompanies it onto the ribosome. It then cycles off the ribosome with the concomitant cleavage of GTP to GDP.

Translational coupling. A gene arrangement in which the translation of one protein coding sequence on a polycistronic mRNA is required for the translation of the second, downstream coding sequence.

Translational fusion. The fusion of parts of the coding regions of two genes so that translation initiated at the translational initiation region for one polypeptide on the mRNA will continue into the coding region for the second polypeptide in the correct reading frame for the second polypeptide. A polypeptide containing amino acid sequences from the two genes that were joined to each other will be synthesized.

Translational initiation region (TIR). The initiation codon, the Shine-Dalgarno sequence, and any other surrounding sequences in mRNA that are recognized by the ribosome as a place to begin translation. Also called a ribosome-binding site (RBS).

Translational regulation. The variation, under different conditions, in the amount of synthesis of a polypeptide due to variation in the rate at which the polypeptide is translated from the mRNA.

Translationally autoregulated. Referring to a protein that can affect the rate of translation of its own coding sequence on its mRNA. Usually in such cases the protein binds to its own translational initiation region or that of an upstream gene to which it is translationally coupled; hence, the protein represses its own translation.

Translocation. During translation, the movement of the tRNA with the polypeptide attached from the A site to the P site on the ribosome after the peptide bond has formed.

Transmembrane domain. The region in a polypeptide between a region that is exposed to one surface of a membrane and a region that is exposed to the other surface. This region must traverse and be embedded in the membrane. Usually, transmembrane domains have a stretch of at least 20, mostly hydrophobic, amino acids that is long enough to extend from one face of a bilipid membrane to the other.

Transmembrane protein. A membrane protein that has surfaces exposed at both sides of the membrane.

Transposase. An enzyme encoded by a transposon that cuts the target DNA and the DNA at both ends of the transposon and joins the cut ends to each other during transposition.

Transposition. The act of a transposon moving from one place in DNA to another.

Transposon. A DNA sequence that can move from one place in DNA to a different place with the help of transposase enzymes. Should be distinguished from homing DNA elements that usually move only into the same sequence in another DNA and depend upon homologous recombination.

Transposon mutagenesis. Technique in which a transposon is used to make random insertion mutations in DNA. The transposon is usually introduced into the cell in a suicide vector, and so it must transpose into another DNA in the cell to become established.

Transversion. A type of base pair change mutation in which the purine in the base pair is changed into the pyrimidine and vice versa, e.g., GC to TA or GC to CG.

Triparental mating. A conjugational mating for introducing mobilizable plasmids into cells, in which three strains of bacteria are mixed. One strain contains a self-transmissible plasmid, and the second strain contains the mobilizable plasmid, which is then mobilized into the third strain.

Triple-stranded helix. A type of DNA structure promoted by RecA protein as an intermediate in homologous recombination, in which three strands of DNA, each of which is complementary to another strand in the triplex, are wrapped around each other in a helix, which is extended relative to the helix of the Watson-Crick structure of DNA.

tRNA$_f^{Met}$. The tRNA to which formylmethionine is added and that pairs with the initiator codon in a translational initiation region to initiate translation of a polypeptide in bacteria.

Two-component regulatory system. A pair of proteins, one of which, the sensor, undergoes a change in response to a change in the environment and communicates this change, usually in the form of a phosphate, to another protein, the response regulator, which then causes the appropriate cellular response. Different two-component systems are often highly homologous to each other.

Umber codon. The codon UGA. Also known as the opal codon.

Unselected marker. A difference between the DNA sequences of two bacteria or phages involved in a genetic cross that can be used for genetic mapping. Is a difference other than the difference used to isolate recombinants. Mapping information can be obtained by testing recombinants that have been selected for being recombinant for one marker, the selected marker, to determine if they have the sequence of the donor or the recipient for another marker, an unselected marker.

Upstream. From a given point, sequences that lie in the 5′ direction on RNA or in the 5′ direction on the coding strand of a DNA region from which an RNA is made.

Upstream activator sequence (UAS). A sequence in DNA upstream of a promoter that increases transcription from the promoter. Removal of a UAS from the DNA reduces transcription from the promoter.

Uptake sequence. A short sequence in DNA that allows DNA containing the sequence to be taken up by some types of bacteria during natural transformation.

Uracil (U). One of the pyrimidine (one-ringed) bases.

Uracil-N-glycosylase. An enzyme that removes the uracil base from DNA by cleaving the bond between the base and the deoxyribose sugar.

UvrABC endonuclease (UvrABC exinuclease). A complex of three proteins that cuts on both sides of any DNA lesion causing a significant distortion of the helix, as a first step in excision repair of the damage.

Very short patch (VSP) repair. A type of repair in enteric bacteria that removes the mismatched T in the sequence CT(A/T)GG/GG(T/A)CC and replaces it with a

C. This prevents mutagenesis at this site due to deamination of the C, which is methylated in enteric bacteria, to a T, which will not be recognized by the uracil-N-glycosylase. A very short stretch of the DNA strand around the mismatched T is removed and resynthesized during the repair, hence the name.

VSP repair.   *See* Very short patch repair.

Watson-Crick structure of DNA.   The double helical structure of DNA first proposed by James Watson and Francis Crick. The two strands of the DNA are antiparallel and held together by hydrogen bonding between the bases.

Weigle mutagenesis (W-mutagenesis).   Another name for SOS mutagenesis. Named after Jean Weigle, who first observed it.

Weigle reactivation (W-reactivation).   Applies to the increased ability of phages to survive UV irradiation damage to their DNA if the cells they infect have been previously exposed to UV irradiation. Named after Jean Weigle, who first observed it.

Wild type.   The organism as it was first isolated from nature. In a genetic experiment, the strain from which mutants were derived. The normal type.

Wild-type allele.   The form of a gene as it exists in the wild-type organism.

Wild-type phenotype.   The particular outward trait characteristic of the wild type that is different in the mutant.

W-mutagenesis.   *See* Weigle mutagenesis.

Wobble.   The property of the genetic code that codons for the same amino acid often differ only in the last (third) nucleotide. Reflects the fact that the base of the first nucleotide in the anticodon of a tRNA can often pair with more than one base in the third nucleotide of a codon in the mRNA.

Xanthine.   A purine base that results from deamination of guanine.

Zero frame.   In the coding region of a gene, the sequence of nucleotides, taken three at a time, in which the polypeptide encoded by the gene is translated.

Index